# *Ecosystem*
# Management
# in the Boreal Forest

PRESSES DE L'UNIVERSITÉ DU QUÉBEC
Le Delta I, 2875, boulevard Laurier, bureau 450
Québec (Québec) G1V 2M2
Telephone : (418) 657-4399 • Fax : (418) 657-2096
Email : puq@puq.ca • Website : www.puq.ca

Diffusion / Distribution :

**CANADA and other countries**

PROLOGUE INC.
1650, boulevard Lionel-Bertrand
Boisbriand (Québec) J7H 1N7
Telephone : (450) 434-0306 / 1 800 363-2864

**FRANCE**
AFPU-DIFFUSION
SODIS

**BELGIQUE**
PATRIMOINE SPRL
168, rue du Noyer
1030 Bruxelles
Belgique

**SUISSE**
SERVIDIS SA
Chemin des Chalets
1279 Chavannes-de-Bogis
Suisse

# Ecosystem Management in the Boreal Forest

*Edited by*

Sylvie Gauthier

Marie-Andrée Vaillancourt

Alain Leduc

Louis De Grandpré

Daniel Kneeshaw

Hubert Morin

Pierre Drapeau

Yves Bergeron

*Preface by*

James Fyles

**2009**

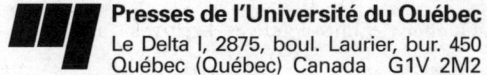

**Presses de l'Université du Québec**
Le Delta I, 2875, boul. Laurier, bur. 450
Québec (Québec) Canada   G1V 2M2

*Bibliothèque et Archives nationales du Québec*
*and Library and Archives Canada cataloguing in publication*

Main entry under title:

  Ecosystem management in the boreal forest

  Translation of: Aménagement écosystémique en forêt boréale.

  Includes bibliographical references and index.

  ISBN 978-2-7605-2381-4

  1. Taigas - Management - Environmental aspects - Canada. 2. Sustainable forestry - Canada.
3. Logging - Environmental aspects - Canada. 4. Forest conservation - Canada. I. Gauthier, Sylvie, 1961- .

  SD567.A4313 2009          634.9'20971          C2009-940697-7

We are grateful for the financial assistance received from the Government
of Canada under the Book Publishing Industry Development Program (BPIDP).


Publication of this book was made possible through the financial support
of Société de développement des entreprises culturelles (SODEC).

Layout: Infoscan Collette-Québec

Cover – Design: Richard Hodgson
         Shots: 1 – Philippe Duval
                2 – Antoine Nappi
                3 – Claude Bouchard
                4 – Marie-Noëlle Caron

```
        ┌───┐
        │ 1 │
    ┌───┼───┤
    │ 3 │ 2 │
    ├───┴───┘
    │ 4 │
    └───┘
```

**1** 2 3 4 5 6 7 8 9 PUQ 2009 9 8 7 6 5 4 3 2 **1**

Legal deposit – 2nd quarter 2009
Bibliothèque et Archives nationales du Québec / Bibliothèque et Archives Canada
Printed in Canada

# Preface

*Over the past two decades, the boreal forest has changed, in the minds of Canadians, from an untouched, remote land of trees and rivers, to a source of paper and wood products and an ecosystem threatened by industrial development. This changing awareness has brought with it a change in expectations for what the boreal forest can provide to Canadian society: construction materials and paper products, certainly, but also biological diversity, fresh water, recreation, and spiritual and cultural values. Society is asking more of its forests and of its forest managers.*

*Changing expectations have challenged forest researchers and managers to move beyond the long-held views of sustained-yield management for wood products. The concepts of uniformly applied even-age management, and "normal" forest with equal representation of forest age-classes up to harvest age, were well adapted to landscapes in which wood was the only value product of the forests, but could not effectively meet multiple objectives. Social pressures in the 1990s, particularly on the west coast of North America, focused on biodiversity values associated with old-growth forests and drove the search for management systems that could provide those values on the landscape while allowing exploitation of timber values to continue. This discussion was brought into public view in Quebec by the release in 1999 of the film L'Erreur boréale, which questioned many of the long-held assumptions of forest management. The recommendations of the Coulombe Commission promoted the concept of ecosystem management (aménagement écosystémique) as a way to better meet the many demands that society is placing on the forest.*

*From its inception in 1995, the Sustainable Forest Management Network Centre of Excellence has been concerned with creating the knowledge required to develop a "new forestry" for the Canadian boreal forest. The Network's early research recognized that disturbance by agents such as fire, insects, and wind was a process associated with all boreal forests, and sought to provide an understanding of disturbance mechanisms and resulting patterns as a basis for the development of forest management systems that would be more effective in sustaining biodiversity. The results of this early research led to further projects to design harvesting and silviculture techniques that would create or maintain, in managed forests, the critical ecological features that support biodiversity in unmanaged forest landscapes.*

*This book draws together the results and experience from several research projects that have been conducted since 2000 and, in particular, a major team effort led by Sylvie Gauthier, initiated in 2003. These projects aimed to lay a solid foundation for ecosystem management, first by characterizing natural forest at the landscape and stand levels, and second, by considering the ways that operational forestry could mimic or "emulate" the features of the natural forest. The underlying hypothesis is that populations of forest organisms will be sustained best by providing ecological conditions that are similar to those under which the organisms evolved and to which they are adapted. Since forest management for wood products cannot create purely "natural" conditions, the functional hypothesis which is being tested in different ways across the country is that forest management can recreate enough of the key ecological features of the natural forest to sustain the diversity of organisms and ecosystems.*

*The first challenge facing researchers and managers developing ecosystem management is to come to a common understanding. The suite of chapters in this book provide a broad statement of the many dimensions of the concept. The first part provides focused discussion of what ecosystem management is and why it is needed. The concept is challenging because it requires both a landscape-level and a stand-level perspectives, with many different features to be considered at each level, and regional differences in each feature. Translating these concepts, with the level of detail required, into a simplified framework applicable to operational forestry is a further challenge and the long-term goal of the research represented in this book.*

*That disturbance regimes vary considerably between forest regions is well known in general, but specific knowledge of disturbance patterns and processes has been lacking in many regions. This lack of knowledge has been a constraint on the development of ecosystem management. Part 2 provides a wealth of detail on the disturbance regimes of different regions of Québec and Manitoba, and considers the implications of each regime from the perspective of ecosystem management. The emerging view is of disturbance regimes that involve several agents, each influencing the forest at different scales, and in some cases interacting with each other. This is a good foundational concept on which to build approaches to forest management.*

*Strategies for the implementation of ecosystem management must be adapted to the nature of the forest and the values that management aims to sustain. The last part of the book explores silvicultural systems designed with reference to the disturbance regimes and features of the forests to which they are applied. Together these chapters show the range of possibilities from stand-level interventions including partial harvesting, treatment of dead wood, regeneration, and soil protection, to practices in planning and harvesting that create landscape patterns similar to those created by natural disturbance.*

*Several chapters report the results of experiments that have tested the effects of ecosystem management on indicator organisms. The emerging conclusion is that, as might be expected in a complex forest system, species respond to different disturbances in different ways. In many cases the ecosystem management approach supports a biodiversity that is closer to the natural forest than traditional harvesting and silvicultural methods. Hence, the results are a promising indication that adoption of ecosystem approaches to forest management will increase the possibility of sustaining biodiversity in "working" landscapes.*

*The forests and societies of Canada are changing and with them the opportunities that Canadians will see in the forest and demands that we will place on forested landscapes. If we are to continue to derive the wide range of benefits from the forest that we have become accustomed to, we will need to adopt forest management systems that support the new realities. The process will not be easy and will require a broad discussion of advantages and disadvantages, benefits and costs. The results of the research described in this volume provide a solid background of knowledge to inform this debate.*

James W. Fyles
Scientific Director, Sustainable Forest Management Network Centre of Excellence
Professor and Tomlinson Chair in Forest Ecology
Department of Natural Resource Sciences
McGill University

# Acknowledgments

This book is the result of a large collaborative effort that was possible because of the great dedication of many people. We are thus grateful to everyone who participated in the preparation of this book.

The initial idea of a book dedicated to ecosystem management was brought up by Hubert Morin during a meeting held on the North Shore region in Québec at the end of the summer of 2005. We were summarizing the work that had been done in the context of the research project initiated following a Sustainable Forest Management Network (SFMN) grant involving 10 researchers and 70 students. This idea would not have come to fruition without the contribution of all those who participated in the definition of this project as well as in its realization. Although we cannot name each participant, they are all gratefully acknowledged.

Dominique Boucher, who coordinated this SFMN project for the first four years, has all our gratitude. Ahn Thu Pham, who participated in the first stages of this book, is also thanked.

The scientific quality of the text would not have been as high without the much appreciated contribution of several reviewers who agreed to take some of their time to comment on and improve this book's chapters: André Arseneault (BC Forest Service), Marilou Beaudet (UQAM), Michel Campagna (MRNF), Elizabeth Campbell (BC Ministry of Forests and Range), Michel Chabot (MRNF), Han Chen (Lakehead University), Barry Cooke (NRCan – CFS), Mathieu Côté (Consortium en foresterie Gaspésie-Les-Îles), Benoit Courbeau (Cemagref), Rhéaume Courtois (MRNF), Louis Dumas (Tembec), Jacques Duval (MRNF), Elston Dzus (Alberta Pacific), Michelle Garneau (UQAM), Pierre Grondin (MRNF),

Michel Huot (MRNF), Robert Jobidon (MRNF), Gordon Kayahara (OMNR), Pierre LaRue (MRNF), Marc Leblanc (MRNF), Jean-Martin Lussier (NRCan – CFS), Christian Messier (UQAM), Alison Munson (Université Laval), Jean Noël (MRNF), Étienne Vézina (Domtar), Michel Villeneuve (Bureau du Forestier en chef), and Mike Wotton (NRCan – CFS). We are extremely grateful to all of them.

Yan Boucher, from the Direction de la recherche forestière of the Ministère des Ressources naturelles et de la Faune du Québec (MRNF), accepted enthusiastically to read the whole manuscript and also contributed to improving it. We sincerely thank him.

A book that contains 20 chapters written by more than 60 authors could not have been produced without the involvement of each contributor. We are thankful to them, and appreciate their collaboration and their professionalism.

Some of the work presented in this book and the financial help that promoted this collaborative work came from the SFMN. In addition, the book production was made easier by the important logistical and financial support provided by the Canadian Forest Service and the Centre d'Étude de la Forêt. Funds from the fifth North American Forest Workshop (NAFEW), Université du Québec à Montréal, Université du Québec à Chicoutimi, and Université du Québec en Abitibi-Témiscamingue also helped with the coordination and the editing of this book.

We also acknowledge Héloïse Le Goff who coordinated the translation of the book.

Finally, we acknowledge the contribution of Marie-Noëlle Germain and Céline Fournier from Les Presses de l'Université du Québec.

Sylvie Gauthier
Marie-Andrée Vaillancourt
Alain Leduc
Louis De Grandpré
Daniel Kneeshaw
Hubert Morin
Pierre Drapeau
Yves Bergeron

# Table of Contents

# Acronym List

| | |
|---|---|
| AAC | Annual allowable cut |
| ASP | Adapted silvicultural practices |
| BEC | Biogeoclimatic ecosystem community |
| BF | Balsam fir |
| BS | Black spruce |
| CAMC | *Coupe adaptée maintenant le couvert* (adapted cuts maintaining canopy cover) |
| CCFM | Canadian Council of Forest Ministers |
| CLAAG | Careful logging around advance growth |
| CPHRS | *Coupe avec protection de la haute régénération et des sols* (cuts with protection of tall regeneration and soils) |
| CPPTM | *Coupe avec protection des petites tiges marchandes* (cut protecting small merchantable stems) |
| CPRS | *Coupe avec protection de la régénération et des sols* (careful logging protecting advance regeneration and soils) |
| DBH | Diameter at breast height |
| DC | Drought Code |
| DMPF | Duck Mountain Provincial Forest |
| DMPP | Duck Mountain Provincial Park |
| ESSF | Engelmann Spruce Subalpine Fir zone |

| | |
|---|---|
| FML | Forest Management License |
| FMU | Forest Management Unit |
| FTC | Forest tent caterpillar |
| FVS | Forest Vegetation Simulator |
| FWI | Fire-Weather Index |
| HARP | Harvest with regeneration protection |
| ICH | Interior Cedar-Hemlock zone |
| IRM | Integrated resources management |
| JP | Jack pine |
| LDF | Lake Duparquet Forest |
| LIA | Little Ice Age |
| LP | Louisiana Pacific |
| MH | Machine hour |
| MRNF | *Ministère des Ressources naturelles et de la faune* (Ministry of Natural Resources and Wildlife) |
| NDM | Natural disturbance management |
| NRV | Natural range of variability |
| OGMA | Old-growth management area |
| OSB | Oriented strand board |
| PW | PATCHWORK |
| RP | Red pine |
| SBS | Sub-boreal spruce zone |
| SBW | Spruce budworm |
| SI | Site index |
| SFM | Sustainable forest management |
| SLAM | Spatial landscape assessment models |
| SSI | Successional Stage Index |
| TA | Trembling aspen |
| TRF | Tall residual forest |
| UQAM | Université du Québec à Montréal |
| UQAT | Université du Québec en Abitibi-Témiscamingue |
| YB | Yellow birch |
| WB | White birch |
| WP | White pine |
| WS | White spruce |

# Figures

# Tables

# Introduction

## Ecological Issues Related
## to Forest Management*

*Jean-Pierre Jetté, Marie-Andrée Vaillancourt, Alain Leduc,
and Sylvie Gauthier*

\* We thank Pamela Cheers, Benoît Arseneault and Isabelle Lamarre from Natural Resources Canada
for editing the text. We also aknowledge the financial support of Natural Resources Canada,
the Sustainable Forest Management Network, and the Center of Forest Research. The photos
on this page were graciously provided by Marie-Ève Sigouin, Virginie-Arielle Angers and Michel
Robert (Canadian Wildlife Service).

# 1. CURRENT ISSUES CONCERNING CANADIAN FORESTRY

At the beginning of the 21st century, Canadian forestry has to face several social and economic issues to meet various societal needs (see box 1). Ecological concerns have been added to these issues by forest scientists who have been studying boreal forest ecosystems for decades. By observing forest landscape changes following the intensification and extension of forestry activities, they have been able to identify some ecological issues that must be addressed in the near future. Obviously, all these issues, whether economic, social or ecological, are interrelated. Although the main objective of this book is to understand these ecological issues,• we have to keep in mind that ecosystem management generates a shift in the way we conceive and manage forest ecosystems that allows us to address simultaneously several types of issues by considering the forest in a holistic fashion.

• Social and economic issues will be indirectly addressed in several chapters of the book.

## Box 1
## Examples of Social and Economic Issues in Forestry in Canada

◆ Native ancestral rights

◆ Increasing timber supply costs and maintenance of forest industry competitiveness in terms of international markets

◆ Increasing world trade competition and protectionism

◆ Increasing demand for certified forest products

◆ Increasing economic activity in sectors other than forestry (outfitters, ecotourism, non-timber products, etc.)

◆ Use of the land by multiple users

Ecological issues can be defined as problems – real or apprehended – that could affect the long-term viability of forest ecosystems. There is an agreement on the fact that maintaining viable ecosystems is the best guarantee that we have to ensure the durability of forest goods and services (including timber supply) and preserve all the potential it could offer in the future. Furthermore, preserving biodiversity and ecological processes is essential to ensuring forest ecosystem resilience following environmental changes, and this is especially true in the context of imminent climate change (IPCC 2007).

To identify ecological issues, managed landscapes have to be compared with natural forest landscapes in order to determine the main differences. It is worth stressing that this idea is based on the following premise: preserving natural forest landscape attributes is the best guarantee we have to maintain biodiversity (Seymour and Hunter 1999). Although the detailed reasoning of this statement will be discussed in the first chapters of the book, we wish to present briefly the main ecological issues to which ecosystem management can provide solutions.

## 2.  ECOLOGICAL ISSUES AND APPREHENDED EFFECTS

Divergence between managed and natural landscapes exists because the nature and frequency of disturbances generated by forest practices are different from those of natural disturbances. We will see later in this book how natural disturbance cycles are longer than planned forest revolutions and how their effects are complex and diverse compared with forest management systems involving mainly low-retention silvicultural treatments (i.e., clearcutting, careful logging preserving advance regeneration). Consequently, important changes in key attributes like stand vertical and horizontal structures or forest composition and configuration can be apprehended. This raises concerns because such attributes are essential for the maintenance of biodiversity and ecological processes. A brief overview of the main ecological issues concerning boreal forest ecosystems and examples of related apprehended effects will be presented in this section.

## 2.1.  Age Structure at the Landscape Level

Forest management aimed at normalizing boreal forest landscapes is truncating forest-stand age-class distributions. Mature and old forests are harvested and management strategies are not designed with the intent to maintain them. Consequently, the old-growth forest proportion will inevitably be lower than that of the natural landscape, even in regions where fire frequency is high. Mature and old forests are characterized by particular habitat attributes on which several species depend. These species could be threatened by the rarefaction of these habitats at the landscape level.

Fennoscandian countries have historically intensively managed a large part of their territory. Their forest landscapes were highly transformed and old-growth forests have drastically diminished (Östlund et al. 1997), which resulted in threats to many old-growth-forest-dwelling species (Berg et al. 1994). In contrast to Fennoscandia, old-growth forests in the Canadian eastern boreal forest are far from the levels reached by northern European countries. However, the pace of anthropic changes and the disappearance of the last large intact forest landscapes in some portions of the Canadian boreal forest (Lee 2007) are signs of an eventual major land transformation that could have impacts on forest health.

> ### *Apprehended Effect*
> Rarefaction of mature and old-growth stands to be replaced by more regenerating stands.

## 2.2.  Vegetation Composition

Harvesting techniques and their effects on vegetation dynamics as well as cutting cycle could modify forest vegetation composition. Some cover types or species could become under- or over-represented at the landscape scale compared with what was observed in natural forests. The various cover types offer distinct habitat attributes (e.g., food, shelter, nest structures) that are used by different species. The rarefaction or overabundance of specific cover types or species at the landscape level could have consequences for the abundance and distribution of these species (animals or vegetation) as well as for ecological processes (e.g., nutrient cycle).

### Apprehended Effects

Increase in shade-intolerant species to the detriment of shade-tolerant species.

In specific sectors, balsam fir increases to the detriment of black spruce.

Forest composition homogenization at the stand and landscape levels.

Rarefaction of old-growth associated tree species (e.g., white spruce, northern white-cedar).

In specific sectors, invasion of clearcutting areas by ericaceous shrubs.

## 2.3.　Stand Internal Structure

The use of even-aged management practices and intermediate treatments such as thinning generates a higher proportion of regular stands. The prevalence of younger stands showing a simpler vertical and horizontal structure could lead to habitat loss for species depending on specific structure attributes found in irregular stands (e.g., dead wood, lateral obstruction, large-diameter trees).

### Apprehended Effects

Rarefaction of complex-structured stands (uneven-aged, two-storied, etc.) to be replaced by simple-structured stands.

Decreasing wildlife tree availability (e.g., large-diameter trees, standing and down dead wood).

## 2.4.　Spatial Configuration at the Stand and Landscape Levels

At the stand level, managed areas contain very few standing trees compared with naturally disturbed stands that comprise various biological legacies. In managed forest landscapes, old forest stands are distributed in patches of forest isolated in a matrix dominated by young stands, which contrast with naturally disturbed landscapes. These configuration changes could have an impact on connectivity of old forest patches and affect forest-dwelling wildlife movements (for lichens, mosses, fungi, insects, mammals, etc.). Additionally, large old-growth forest landscapes are found increasingly far from residual forests in harvested areas because of the receding cutting limit. This could affect interior forest species living in residual forests by limiting individual dispersion and genetic exchanges, for instance.

***Apprehended Effects***

Rarefaction or fragmentation of large old-growth forest landscapes.

Creation of vast regenerating areas (cutblock agglomeration) larger than those generated by natural disturbances.

Decreasing availability of biological legacies (e.g., green patches, snags) in managed areas compared with natural disturbance patches.

Increasing edge effect and interior habitat loss caused by linear forest retention (e.g. riparian buffer strips and cutblock separators).

Lower residual forest habitat connectivity.

## 2.5. Forest Soil Productivity

Soil structure and associated processes are modified by natural disturbances such as fire that affects the soil's physical (e.g., exposing mineral soil) and chemical attributes (releasing nutrients). Forest management practices retrieve crowns and boles that would naturally decompose and fertilize forest soil. This could modify soil processes and have a long-term impact on forest productivity, although up to now there is no strong evidence to show this. In other respects, forest operations, in particular for regions and sites at risk, could increase natural phenomena such as paludification (i.e., the accumulation of organic matter), stand canopy opening, and lichen woodland formation.

***Apprehended Effects***

Depletion of fertility for certain forest soils.

Increase in natural phenomena such as paludification and stand canopy opening.

## 2.6. Recently Disturbed Forests

The increasing use of salvage logging operations following major natural disturbances (fires, severe insect outbreaks) raises concerns with regard to their impacts on forest ecosystems. Natural disturbances create particular habitat characteristics (e.g., sudden increase in deadwood availability) that are crucial for various species. Some species even depend on such habitat attributes. Moreover, knowledge of ecological functions of these disturbances and their structural legacies is still fragmentary.

***Apprehended Effect***

Rarefaction of ecological attributes specific to naturally disturbed stands.

To address the concerns listed above and determine if the apprehensions are well founded and require changes in current forest management practices, we need relevant knowledge of natural disturbance regimes and associated forest dynamics. Then it will be possible to measure the differences between natural and managed landscapes regarding forest attributes highlighted by these ecological issues. This will thereafter allow us to define management objectives aimed at reducing these differences. These three steps briefly illustrate an ecosystem management framework. The EM rationale will be explained in depth in the first chapters of the book.

## 3.　BACKGROUND ON THE BOOK

Several scientists who are specialists on forest and natural disturbance ecology in different provinces have worked together for five years in the context of a vast research project financed by the Sustainable Forest Management Network (SFMN). To meet the project objectives, researchers shared their works to increase knowledge with respect to natural disturbance regimes across several Canadian regions. They attempted to define and implement an ecosystem management framework in collaboration with industrial partners and government agencies. The idea to write this book originates from the researchers' desire to synthesize their results for readers who are interested in understanding the ecosystem management concept and its rationale in order to facilitate its large-scale implementation.

The research conducted during this project was mainly done on boreal forest ecosystems. This book will therefore be oriented towards the boreal zone which covers approximately 757 million hectares in Canada, corresponding to more than 50% of the country (CCFM 2005). In Canada, the boreal zone runs from east to west from Newfoundland to the Yukon Territory. In the eastern portion, it covers approximately latitudes 48° to 58° N but extends to 67° N in the Yukon (figure 1). Although ecosystem-management-related concepts can be applied to a wide range of forest ecosystems, we will mainly use examples from research projects conducted in boreal forests.

This book presents an overview of the natural disturbance and forest-dynamic knowledge acquired previously and during this SFMN project that supports forest ecosystem management implementation. The book is divided into three parts:

### Forest Ecosystem Management: An Approach Based on Natural Disturbances

The first part deals with the basics of ecosystem management. The first chapter covers the socio-historical context that led to this new approach and explains the ecological principles supporting it. Chapter 2 discusses natural disturbances and how their characteristics can guide ecosystem management implementation. Chapter 3 addresses fire regimes. It shows differences between those and forest management regimes based on even-aged management systems traditionally used in boreal forests and suggests solutions that can minimize these differences.

**Figure 1**
**Canadian boreal forest zone**

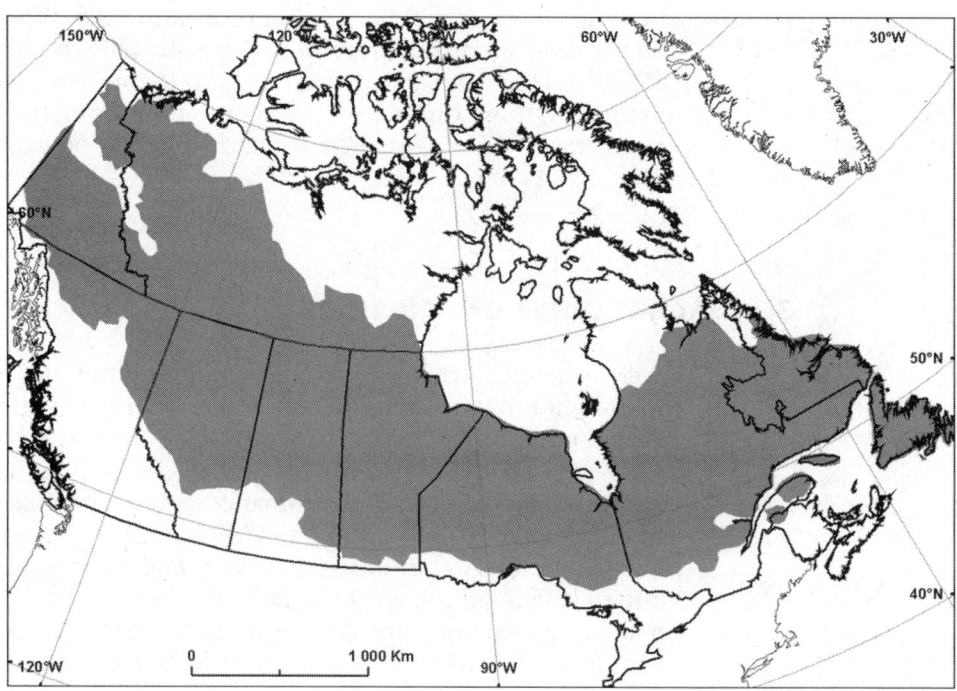

Based on Rowe 1972. Map produced by J. Morissette, CFS.

## Spatio-Temporal Variations of Disturbance Regimes

This part looks at disturbance regimes from two angles. First, five chapters review the knowledge of the major disturbance regimes, i.e. those related to the main disturbance agents in the eastern and central Canadian boreal forest. Three chapters discuss features that influence or result from fire regimes, such as the role of climate in regulating fire activity (chapter 4), the impacts of climate change and adaptation strategies that will have to be implemented (chapter 5) and the forest-landscape spatial distribution resulting from fire activity (chapter 6). Chapters 7 and 8 synthesize current knowledge available for two important defoliator insects of central and eastern boreal forest (i.e., spruce budworm and forest tent caterpillar, respectively).

Secondly, four chapters present a current-knowledge synthesis of disturbance regimes for different regions of the boreal forest, from Gaspésie (chapter 9) and the North Shore region (chapter 10) in eastern Québec – two regions characterized by low fire frequency – as well as northwestern Abitibi (chapter 11) and Manitoba (chapter 12) – characterized by higher fire frequencies that have strongly influenced landscape structure. Each of these chapters will discuss forest dynamic specificities and ecological issues related to their specific context and their implications for ecosystem management.

## Ecosystem Management Implementation

This part introduces ecosystem management implementation in three parts. First, chapter 13 illustrates how different silvicultural treatments can be put together to develop management systems aimed at maintaining natural stand dynamics and addressing the main ecological issues. Chapter 14 presents an ecological monitoring approach – an essential component of ecosystem management – that makes it possible to validate the achievements of ecosystem management objectives and the preservation of biodiversity in managed areas. Chapter 15 presents a silvicultural and ecological evaluation of different types of partial cuts implemented in the context of an experimental network. This evaluation includes biodiversity monitoring using different indicator species.

Secondly, two chapters discuss the development and use of modelling tools that are essential features in EM implementation. Chapter 16 presents new modelling tools developed to predict tree growth in uneven-aged management systems – where conditions are more complex than in traditional even-aged systems. Chapter 17 describes a scenario comparison process used in the development of a 20-year management plan. The main objective is to assess the capacity of different management scenarios to meet various management objectives while using spatial modelling tools.

Finally, three chapters present experimental ecosystem management projects implemented in different forest regions of Canada, i.e., the mixedwood boreal forest (chapter 18) and black spruce forest (chapter 19) of Abitibi, and interior wet forest of British Columbia (chapter 20). Initial objectives, development stages, and preliminary results at the silvicultural and ecological levels are presented when available.

## Perspectives

In the concluding chapter, an overview of the main ecological issues identified in the introduction section is presented and will emphasize the fact that boreal forests were considerably rejuvenated over the past decades and that ecosystem management must be implemented within a short time frame. This approach can provide short-term results, and can also offer mid- and long-term possibilities. The needs and developments required are discussed, including the necessity to establish large protected territories in order to increase our knowledge on ecosystem functions' resilience when facing natural and anthropic disturbances.

## REFERENCES

Berg, A., Ehnström, B., Gustafsson, L., Hallingbäck, T., Jonsell, M., and Weslien, J. 1994. Threatened plant, animal, and fungus species in Swedish forests: distribution and habitat associations. Conserv. Biol. **8**: 718–731.

Canadian Council of Forest Ministers (CCFM). 2005. National Forestry Database. [Online] <pndf.ccfm.org> (accessed November 10, 2008).

International Panel on Climate Change (IPCC). 2007. Climate change 2007: impacts, adaptation and vulnerability. Working group II: Contribution to the Intergovernmental panel on climate change. Fourth assessment report. Summary for policymakers. Geneva, Switzerland. [Online] <www.ipcc-wg2.org> (accessed November 8, 2007).

Lee, P. 2007. 1990–2006 anthropogenic changes within the Pascagama site in Québec's boreal forest: summary of results. Global Forest Watch Canada, Edmonton, Alta.

Östlund, L., Zackrisson, O., and Axelsson, A.-L. 1997. The history and transformation of a Scandinavian boreal forest landscape since the 19th century. Can. J. For. Res. 27: 1198–1206.

Rowe, J.S. 1972. Forest regions of Canada. Canadian Forestry Service, Ottawa, Ontario, Publication No. 1300.

Seymour, R.S. and Hunter, M.L., Jr. 1999. Principles of ecological forestry. *In* Maintaining biodiversity in forest ecosystems. *Edited by* M.L. Hunter, Jr., Cambridge University Press, Cambridge, UK, pp. 22–61.

# Part 1

**Forest Ecosystem Management**
An Approach Inspired
by Natural Disturbances

# Chapter 1

# Forest Ecosystem Management*
## Origins and Foundations

*Sylvie Gauthier, Marie-Andrée Vaillancourt, Daniel Kneeshaw, Pierre Drapeau, Louis De Grandpré, Yves Claveau, and David Paré*

\* We thank Michel Chabot, Christian Messier and Alain Leduc for their comments on the early drafts of the manuscript. Many discussions with a large number of people involved in and dedicated to forestry provided food for thought for our reflections, and we thank them warmly. Lastly, we thank the Sustainable Forest Management Network, Natural Resources Canada, and the Centre for Forest Research (CFR-CEF) for their financial support. The photos on this page were graciously provided by Marie-Andrée Vaillancourt, Marie-Ève Sigouin, Héloïse Le Goff, and Danielle Charron.

# 1. INTRODUCTION

In the last few years, there has been a growing will across Canada to use an ecosystem-based management approach in forest management (Perera et al. 2004). In addition, forest ecosystem management has become the focus of discussions among researchers interested in how a forest functions and an increasingly important topic for governments, industry, cooperative associations, and other organizations concerned with the management of forest resources. This approach, which is based on an ecosystem model, has emerged in response to concerns about the changes occurring in managed forests. The trend toward an ecosystem focus is also based on societal recognition that biodiversity maintenance, the integral conservation of certain habitats and the aesthetics of forest landscapes are all values that should be preserved and must be taken into account in forest management.

Forest ecosystem management (FEM) is not a new concept in the forest sector. It is the result of a lengthy process of reflection and knowledge acquisition on ecosystem functioning, which has led to the following conclusion: If we want to maintain forest ecosystem functions and processes over the long term, we must manage forests holistically. Conventional timber harvesting as it has been practiced in recent decades – specifically, the extensive use of even-aged (or low-retention) harvesting practices such as clearcutting –raises concerns and questions about the preservation of biodiversity and ecological processes, which are a guarantee of ecosystem resilience and sustained forest productivity.

The primary purpose of this chapter is to define FEM. First, we will briefly review the evolution of forest management in eastern Canada and the development of public awareness of the ecological issues associated with forestry, in order to understand the context for FEM. We will then describe the ecological foundations of this management approach. Lastly, we will examine the elements involved in an FEM approach and the reasons for harnessing knowledge of natural disturbance regimes in implementing this approach.

## 1.1. Overview of the History of Forestry in Eastern Canada

Forests have figured prominently in Canada's economic, social, and environmental landscape for several centuries. In this section, we will briefly characterize the different periods in forestry, from the early colonization of eastern North America onward (see Burton et al. 2003 for more details). Readers should note that this description concerns Canadian forests in general and not just the boreal forest, which gradually began to be exploited as southern Canadian timber supplies dwindled and new markets such as pulp and paper emerged.

From pre-settlement until the late 18th century, Aboriginal peoples and the early European settlers used the forest mainly for subsistence activities. Due to the low population density, clearing the land for farming and cutting timber for houses and firewood did not threaten the sustainability of timber resources, and so there were no incentives to establish laws and regulations pertaining to forests. Although during the French regime (1534–1760), the intendants (district administrators) had demanded that the "noble species" – large, long-lived forest tree species such as oaks (*Quercus* spp.) and pines (*Pinus* spp.) – be protected in order to ensure supplies of squared timber for France's shipbuilding needs, these species were still relatively unexploited.

Large-scale commercial harvesting of forests in Canada did not begin until the early 19th century (Drushka 2003). The timber trade intensified during Napoleon's naval blockade of Great Britain (1806–1812), when England could no longer obtain supplies from the Baltic and had to turn to its colonies to meet its shipbuilding needs. The lack of regulations on timber harvesting led, in the 1800s, to a scarcity of eastern white pine (*Pinus strobus* L.) and a decrease in the size of trees available in southeastern Canada in general (Drushka 2003; Hébert 2006). The timber trade with the United States increased and lumber gradually became Canada's main export to its southern neighbour (Gaudreau 1999). Around the same time, the first forest reserves were created in Québec and Ontario, and their main role was to protect forest land from other uses to ensure timber supplies for the industry (Bouthillier 1998; Hébert 2006). Forestry regulations were still non-existent. The logging companies cut what they needed on Crown land without any government interference.

During the 19th century, forestry entered a so-called regulatory or administrative period, when the need to regulate arose in order to preserve timber supplies and safeguard the stability of the forest industry. The year 1849 saw the first piece of forestry legislation enacted in Canada (*Act for the Sale and Betterment of Timber upon Public Lands, Crown Timber Act*), which allowed the government to clarify its appropriation of forest land (Crown land) and encourage logging companies to plan their harvests on the forest tenures allotted to them. With the advent of Confederation in 1867, jurisdiction over forestry was transferred to the provinces and most of them adopted forestry policies establishing the first forest regimes, which constitute the basis of current policies. Forest inventories and allowable cut calculations were developed and forest fire protection became a major concern. At this time, pulp and paper production was quickly outpacing lumber production as the volume of available timber decreased and the U.S. demand for newsprint increased (Bouthillier 1998; Gaudreau 1999). As a result, spruce (*Picea* spp.) became the new "noble species" because of its suitability for newsprint production. It was probably around this time that harvesting of the boreal forest really began. Since the pulp and paper industry did not require trees of large diameter, logging became increasingly intensive and spread to more and more parts of the country.

Facing an imminent decline in timber capital that first became apparent at the turn of the 20th century, forestry entered the era of sustained yield management. Sustained yield forest management, which is based on European forestry practices, results in the agriculturalization of the forest. Under this approach, harvesting is bolstered by silvicultural treatments (plantations, thinning, and site preparation), and yields (in timber volume) are calculated to ensure a stable long-term supply of timber. Between 1937 and 1949, most of the provinces modified their forestry regimes to incorporate this management approach (Natural Resources Canada 1998; Drushka 2003). In Québec, although the notion of sustained yields was incorporated in forestry practices at this time, it did not make its way into the law until 1986 (Bouthillier 1998). This management approach placed emphasis on "normalizing" the boreal forest by eliminating overmature stands, considered less productive, and by harvesting mature stands but also ensuring that annual harvests did not exceed what the forest produced. In other words, sustained yield management aims to harvest the interest but preserve the forest capital.

In the second half of the 20th century, with the allocation of new forest areas made accessible by the construction of logging roads, conflicts between uses multiplied, along with the number of users demanding recreational and aboriginal rights in forest management units. As a consequence, in forestry planning, multiple uses had to be taken into account on land previously used exclusively for producing timber. In Québec, among other things, this led to the creation of controlled harvesting zones (*zones d'exploitation controlée*, ZEC), public land on which the planning and management of hunting and fishing activities were delegated to game and wildlife conservation organizations. Measures were also taken by most provincial governments, although they differed from province to province, to protect certain types of habitats (e.g., retention patches) for game species and to protect water quality (e.g., riparian buffer strips). Timber production remains a priority in public forests, however, and very little attention has been paid to other values of forest ecosystems, including values that do not involve the harvesting of popular animal or plant species, such as the preservation of biodiversity and the maintenance of ecological processes.

The expansion of forestry activities and particularly silvicultural strategies that substantially favour the extensive use of even-aged (or low-retention) cutting such as clearcutting and careful logging, when applied to all Crown land in the boreal forest, raises serious questions about the capacity of forest ecosystems to recover after such changes to the landscape. In recent years, in the face of growing doubts over the ability of timber-production-centred forest management to preserve ecological integrity,• and therefore maintain the productivity of forest land, a new paradigm for forestry has emerged, the objective of which is more comprehensive and integrated forest management. This new approach gives greater emphasis to forest values other than timber production, and is based on our knowledge of forest ecology.

• Ecological integrity: maintenance of a system's wholeness, including the presence of all appropriate elements and occurrence of all processes at appropriate rates (p. 39 *in* Kohm and Franklin 1997).

## 1.2.  Role of Ecological Knowledge

In parallel with the changes occurring in forestry, in the field of ecology, certain land management and conservation issues came to the fore beginning in the 1930s. According to a 1932 report on conservation areas by the Ecological Society of America (ESA), in order to achieve species conservation objectives, it is important to keep in mind that habitats are dynamic and need to be protected (Shelford 1933). Consequently, to maintain species for the long term, a wide range of habitats must be preserved. The American naturalist Aldo Leopold (1949) was one of the first researchers to caution against the loss of forest habitats and the consequences for native flora and fauna, favouring an ethic of land use that simply enlarges the boundaries of the community to include soils, waters, plants, and animals, or collectively: the land."

For years, conservation biologists have observed that protected areas and ecological reserves established to preserve biodiversity are often not large enough to successfully support all species, particularly large mammals (Leopold 1949; Grumbine 1992; Fischer et al. 2006); the areas protected are too small in number, too isolated, too static or too small. Consequently, some populations' gene pools cannot be maintained solely in a network of reserves since the fragmentation around these reserves reduces the genetic exchanges that formerly occurred between different populations. This genetic isolation may threaten the survival of some animal populations, particularly after major environmental fluctuations. This failure of protected areas is illustrated by the example of grizzly bears in North Cascades National Park in the U.S. Pacific Northwest, where a territory larger than 35,000 km² is not large enough on its own to maintain the current bear population (Grumbine 1992; Romain-Bondi et al. 2004). Given the inability of such protected areas to ensure integral conservation, it is essential to determine how biodiversity – whether at the genetic, species or ecosystem level – can be protected within the matrix of a managed forest around protected areas (Harris 1985; Lindenmayer and Franklin 2002).

Since the early 1990s, the importance of the link between biodiversity and proper ecosystem functioning has been increasingly recognized (Johnson et al. 1996; Chapin et al. 1997; Schwartz et al. 2000). The worldwide biodiversity crisis, caused by habitat loss from urbanization and land use, is a source of deep concern, given the potential environmental, social, and economic consequences (Ehrlich and Wilson 1991). Commercial logging is a part of this global crisis and both the public and the scientific community are encouraging forestry stakeholders to change their methods of managing timber resources in order to stem the loss of biodiversity. In particular, the controversy over the Spotted Owl (*Strix occidentalis caurina*), a threatened species associated with mature forests in the U.S. Northwest and sensitive to forest fragmentation (see Grumbine 1992), has strongly contributed to the rise of alternative approaches in forestry (Galindo-Leal and Bunnell 1995).

The Rio Earth Summit in 1992, from which the notion of sustainable development emerged, greatly increased public awareness of the importance of considering environmental values in all spheres of economic activity. In the wake of the summit, the governments of most industrialized countries moved to integrate this principle into their forestry policies and regulations. In 1995, the Canadian Council of Forest Ministers (CCFM) adopted six criteria for sustainable forest development (CCFM 1995), four of which involve ecological knowledge:

1. Conservation of biological diversity;

2. Maintenance and enhancement of forest ecosystem condition and productivity;

3. Conservation of soil and water resources;

4. Forest ecosystem contributions to global ecological cycles;

5. Multiple benefits of forests to society;

6. Accepting society's responsibility for sustainable development.

Currently, most of the provinces have integrated the concept of sustainable management into their forest policies, although full implementation of the underlying principles, which requires the integration of acquired scientific knowledge and the consideration of the non-economic values of forest ecosystems, has yet to be fully achieved.

This paradigm shift in forestry is therefore part of an international movement that places great importance on integrating an understanding of ecosystem functioning in the implementation of "sound" natural resource development and management practices (WCED 1987; Christensen et al. 1996). Scientific understanding of the structure and functions of forest ecosystems has advanced significantly in the last two decades, particularly for the boreal forest (Engelmark et al. 1993; Bergeron et al. 1998; Engelmark et al. 2000; Kuuluvainen 2002a; Macdonald 2004; Bergeron et al. 2006). Consequently, integrating ecological knowledge into the development of new forest management practices should become increasingly feasible.

## 2. ECOLOGICAL BASES FOR A PARADIGM SHIFT IN FORESTRY

In order to better understand the importance of preserving some characteristics of forest ecosystems, we will first review how natural disturbances act as agents of change in forest ecosystems at various levels. Ecosystem responses to disturbances are characterized by two qualities: resistance and resilience. We will examine in turn how these two characteristics are ensured by (1) the diversity found in biological systems, whether at the gene, species, community or landscape level; (2) the interactions observed between these different levels of biological organization; and (3) the biological legacies that result from disturbances. Special attention will be paid to illustrating these elements through studies conducted under similar conditions to those found in the boreal forests of eastern Canada.

### 2.1. Ecosystems

First, it would be useful to define the concept of the ecosystem. According to Whittaker (1962), an ecosystem is a "functional system that includes an assemblage of interacting organisms (plants, animals, and decomposers) and their environment, which acts on them and on which they act." The concept of the ecosystem therefore incorporates interactions, structures and processes linking biotic and physical elements (figure 1.1A). Ecosystems are made up of several hierarchical levels (figure 1.1B). A species consists of individuals with a similar genetic background, while a community is an assemblage of different species living and interacting with one another. The result is a variety of habitats with linkages at the landscape level. Ecosystems are anything but static, since the processes occurring in them trigger changes operating at different levels and over different time frames (figure 1.1A). Certain processes found in a forest environment (e.g., growth and mortality) occur at the individual tree level, while others affect individuals of the same species or different species (e.g., competition and symbioses). Yet other processes occur at the landscape level, including

**Figure 1.1**
## Diagram of (A) the concept of an ecosystem showing certain key elements in a simplified way and (B) different levels of organization of elements making up the ecosystem

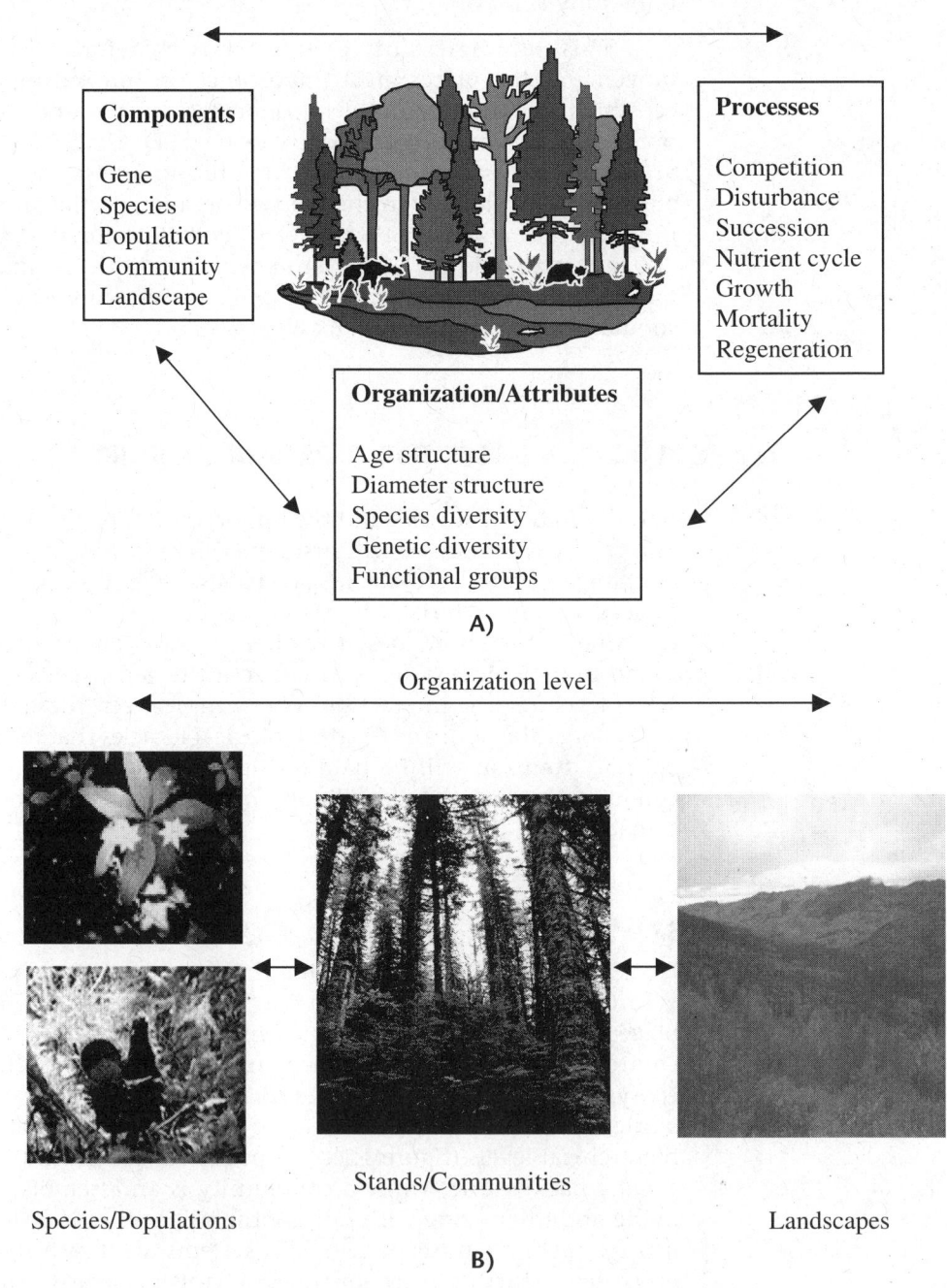

**Components**

Gene
Species
Population
Community
Landscape

**Processes**

Competition
Disturbance
Succession
Nutrient cycle
Growth
Mortality
Regeneration

**Organization/Attributes**

Age structure
Diameter structure
Species diversity
Genetic diversity
Functional groups

A)

Organization level

Species/Populations

Stands/Communities

Landscapes

B)

disturbances such as fire and insect outbreaks. These diverse processes generate forest patterns and attributes that are also expressed at various levels, such as the horizontal and vertical structure of a stand, the association between soil fauna and a particular type of soil, or the age structure of stands at the landscape level. An ecosystem therefore has no explicitly defined spatial limits and is constantly evolving in time (Kimmins 1987). In addition, an ecosystem is an open entity that interacts with other ecosystems.

## 2.2. Disturbances

It has been known for over three decades that disturbances are important agents of change that occur inevitably in ecosystems (Pickett and White 1985; Frelich 2002), including the boreal forest (Heinselman 1973; Holling 1973; Kuuluvainen 1994). In a given region, different types of disturbances of biotic (e.g., insect outbreaks, diseases) and abiotic (e.g. fire, wind, ice storms) origin can operate independently or interdependently at different temporal and spatial scales. These disturbances are themselves influenced by climate, characteristics of the physical environment (topography, type of surficial deposit, etc.), and biotic factors. By occurring at different times and in different places, disturbances can induce changes in forest composition, reinitiate succession, and create a diversified stand mosaic (see Vaillancourt et al., chapter 2).

The fact that forest ecosystems have been subject to natural disturbances for thousands of years suggests that some aspects of their functioning are key to their capacity to adapt to environmental fluctuations. What mechanisms allow these ecosystems to recover from disturbances and maintain their integrity? This is a particularly important question in forestry since our actions affecting the forest also constitute disturbances – disturbances that we hope will not affect the ecosystem's continued productivity or its ability to adapt to unexpected environmental fluctuations.

## 2.3. Ecosystem Resistance and Resilience

Resistance and resilience are two key notions related to the ability of ecosystems to recover from disturbances. Resistance is the ability of a system to withstand small perturbations and keep them from increasing in magnitude (Perry and Amaranthus 1997). For example, trees' chemical defences against insects and the presence of natural enemies that help to keep insect populations in check are two mechanisms that enable stands to resist spruce budworm (*Choristoneura fumiferana Clem.*) attacks. When resistance mechanisms are weakened, the system's capacity to remain in equilibrium also diminishes. In extensive monospecific coniferous stands in the boreal mixedwood forest, Cappuccino et al. (1998) noted that natural enemies of the spruce budworm were present at lower population levels and that this decreased the stands' resistance. Such stands had more severe outbreaks than those in landscapes with a more complex mosaic in terms of forest composition (presence of mixedwood stands, pure softwood stands, and pure hardwood stands within the mosaic). At the stand level, decreased balsam fir defoliation (Su et al. 1996) and mortality (MacLean 1980; Bergeron et al. 1995) were observed when hardwoods were also present in the softwood-dominated forests. Studies on the forest tent caterpillar (*Malacosoma disstria* Hbn.) have shown that forest fragmentation affects stand resistance by affecting the efficiency with which natural enemies parasitize their host (Roland and Taylor 1997), thus contributing to an increase in the duration of outbreaks in fragmented landscapes (Roland 1993; Cooke and Roland 2000).

Ecological resilience is a property of ecosystems that relates to their ability to reorganize after a disturbance. Resilient ecosystems can reorganize on their own while undergoing change or disturbances (Holling 1973; Gunderson and Holling 2002). This notion also relates to how quickly the ecosystem can return to original conditions after a disturbance (Begon et al. 1996; Perry and Amaranthus 1997). An example of ecosystem resilience is the regeneration of a jack pine (*Pinus banksiana* Lamb.), poplar (*Populus* spp.) or black spruce (*Picea mariana* [Mill.] B.S.P.) forest after fire (Johnson 1992; Greene and Johnson 1999). Even if the condition of a forest at a specific site 100 years after a disturbance is not an exact replica of a 100-year-old stand on the same site before the disturbance, the main stand characteristics (e.g., stand structure, dominant species and form, type of habitats created) and processes (e.g., water and nutrient cycles, energy flow) would be essentially the same.

Resistance and resilience are therefore two related properties that allow ecosystems to dynamically maintain their functions and structure, thus ensuring long-term reorganization after a disturbance (Drever et al. 2006). These properties can be influenced by characteristics and mechanisms operating at both the landscape and stand level or even between different ecosystems. For example, the return of some species after a fire may depend on the presence of refugia for these species in the landscape as well as on the presence of survivors within the disturbed site itself.

## 2.3.1.  Key Ecosystem Elements Conferring Resistance and Resilience

Another goal of forest management is to conserve forest ecosystems' capacity to recover fully from environmental changes so that they are able to continue to provide forest goods and services that have commercial, aesthetic, recreational or other values. In other words, the aim is to maintain ecosystems that are resistant and resilient in the face of environmental variations such as those induced by climate change. Two key attributes contribute to these ecosystems' resilience and resistance: biodiversity and biological legacies.

### Biodiversity

Biodiversity is a crucial element that allows forest ecosystems to withstand stress, fluctuations, and disturbances. It should be emphasized that biodiversity refers not only to the number of species inhabiting a particular type of habitat but also to the relations and functions provided by diversity at the different organizational levels of the ecosystem (genes, species, communities, and landscapes; Blondel 1995). Current knowledge suggests that the ability of ecosystems to reorganize after disturbance and stress is linked to the range of responses of the organisms making up these ecosystems. Similar ecological functions can be shared by more than one species (redundancy) (McCann 2000) in an ecosystem, and different species will be favoured depending on the fluctuations in environmental conditions. As a result, biodiversity provides ecosystems with flexibility in their ability to respond to disturbances. A certain degree of redundancy provided by diversity would also confer a greater ability to respond to natural disturbances and stresses as well as to new disturbance agents. For example, there are a number of natural enemies of spruce budworm whose abundance may vary with fluctuations in environmental conditions, providing systems with flexibility in the face of such fluctuations. Given new threats such as global warming, it is prudent to maintain as much flexibility as possible in potential responses by promoting the maintenance of all biodiversity (see box 1.1).

### Box 1.1
## Hypotheses on the link between biodiversity and ecological processes*

1) Redundancy hypothesis: most species are ecologically redundant (their ecological functions can be carried out by another species) and only some species play a critical role in the system as a whole (Walker 1992; Lawton and Brown 1993).

2) Rivet hypothesis: all species play a key role in the ecosystem. All species depend on one another just as the parts of an airplane all play a specific role and together hold the plane together (Ehrlich and Ehrlich 1981).

3) Idiosyncratic response hypothesis: biodiversity and ecological functions are linked. However, interactions are too complex to predict the direction and magnitude of the system's response should some species become extinct (Lawton 1994).

Regardless of which hypothesis turns out to be true, it can be confidently stated that, given our incomplete knowledge of the roles of all organisms, the precautionary principle calls on us to conserve the largest possible amount of biodiversity. The expected changes in climate will certainly modify the conditions to which species are adapted, making it difficult to predict their responses. Species with roles that appear unimportant now may play a crucial role under extreme conditions (Schulze and Mooney 1993). For the time being, to ensure that forest ecosystems maintain the ability to adapt to changes, it is essential to conserve as many organisms as possible and to maintain interconnected ecosystems characterized by relatively natural conditions, as this will help to maintain ecosystem resilience.

* See Vitousek and Hoper 1993; Lawton 1994; and Perry and Amaranthus 1997 for more details.

The importance of genetic diversity has long been recognized in forestry. Studies on geographic variation (provenance tests) have been carried out for a number of species, allowing rules for seed source movement (seed zones) to be formulated that facilitate adaptation to the recipient environment (Beaulieu et al. 1996; Carter 1996; Matyas 1996; Beaulieu and Rainville 2005). It is also well known that, in certain cases, harvesting and management activities may reduce genetic diversity, which may hinder a species' ability to adapt to changing environmental conditions. For example, studies on eastern white pine have shown a reduction in genetic diversity in fragmented populations on the St. Lawrence Plain compared with more continuous populations in the Ottawa Valley (Beaulieu and Simon 1994). Although, in these cases, the genetic markers used to measure genetic diversity were not markers for adaptive traits, Beaulieu et al. (1996) observed in an analysis of a provenance/progeny test that the trees in St. Lawrence populations were shorter than those in Ottawa Valley populations at the age of ten years.

At the stand level, site quality together with species diversity probably has an impact on primary productivity, nutrient retention, and disease resistance (Perry and Amaranthus 1997). For example, Paré and Bergeron (1995) and

Bergeron and Harvey (1997) demonstrated that, in the balsam fir domain, mixed stands are the most productive in terms of standing biomass. In addition, in these mixed stands, softwood mortality from spruce budworm outbreaks is low compared with that observed in pure softwood stands. De Grandpré and Bergeron (1997) have shown that herbaceous plant communities are more diversified in this type of forest and, after cutting, this diversity provides greater resistance to so-called invasive species.

At the landscape level, the diversity of forest types may help to limit or control some disturbances. For example, the presence of hardwood stands may slow the spread of fire (Hély et al. 2000; Cumming 2001; Kafka et al. 2001). Similarly, natural enemies of the spruce budworm, such as certain parasitoids, are probably more abundant in hardwood stands and mixedwood stands, thus conferring greater resistance to forest mosaics that contain a wide range of stand types (i.e., mixed and hardwood stands) (Perry and Amaranthus 1997; Cappuccino et al. 1998). The same is true for avian predators of spruce budworm, which are more abundant and occur in greater species diversity in mixed stands (Hobson and Bayne 2000) than in pure hardwood or softwood stands; the same phenomenon holds true at the landscape level in forest mosaics composed of a combination of mixedwood, hardwood, and softwood stands (Drapeau et al. 2000). In addition, Crawford and Jennings (1989) observed that, during a spruce budworm outbreak, insectivorous birds can significantly curb the growth of budworm populations, although this effect may vary depending on the species composition of avian predator communities. In addition, they found that mixed, multi-layered stands provide better habitat for avian predators than younger and denser forests with little shrub cover. Conversely, several studies on continental forest bird trends show population increases during budworm outbreaks (Hagan et al. 1992; Patten and Burger 1998). Although there is no consensus in these studies on the existence of top-down control (in which predation controls community organization) or bottom-up control (prey availability controls community organization), there is agreement in the scientific community that demographic parameters in these two taxonomic groups are mutually influential (Bolgiano 2004).

### Biological Legacies

After a disturbance, the affected forest still retains some characteristics inherited from the pre-disturbance forest. Living trees, standing dead trees, downed trees, seeds, or other propagules in the soil and disturbed duff layers are still present in the ecosystem. These components make up what are called the biological legacies (*sensu* Franklin et al. 2000; Lindenmayer and Franklin 2002) of the predecessor forest, and they have an effect on the successional pathway that a forest follows after a disturbance.

Biological legacies occur at different scales. For example, in a major fire, areas burned to varying degrees will provide a major portion of legacies (Bergeron et al. 2002; Perron et al., chapter 6). Patches of trees preserved in the burn also serve as refugia for species that are not fire adapted, also providing legacies (e.g. balsam fir [*Abies balsamea* (L.) Mill.] and white spruce [*Picea glauca* (Moench) Voss]) that will eventually colonize regenerating burns (Galipeau et al. 1997; Kafka et al. 2001). When identifying biological legacies, it is equally important to consider the type and scale of the disturbance (see Vaillancourt et al., chapter 2). It has been observed that, unlike forests where a burn has occurred, in mature forests affected by lower-intensity or smaller-scale disturbances (outbreaks, gaps or senescence), standing and downed dead trees, distributed within a predominantly forested matrix, constitute biological legacies.

The litter and organic matter in the soil (duff) also play an important role in ecosystem resistance and resilience. For example, in a forest fire, the fire reduces the thickness of the duff layer, creating seedbeds conducive to the establishment of trees and other plants that normally appear after fire (Johnson 1992; Greene et al. 2006). Duff reduction is spatially heterogeneous and depends on combustion conditions and the initial depth of the organic matter layer (Miyanishi and Johnson 2002; Ryan 2002). This layer of organic matter also affects ecosystem productivity by limiting erosion (Ryan 2002). Similarly, at the stand level, the uprooting of trees exposes the mineral soil in areas where organic matter has accumulated. The spatial heterogeneity in mineral soil exposure and the nearby presence of duff are particularly important in the establishment and growth of some tree species (Gastaldello et al. 2007).

In the boreal forest, dead trees are a key component in natural forest dynamics (Sturtevant et al. 1997; Hély et al. 2000; Kuuluvainen 2002b; Drapeau et al. 2005). Between the initial stage of mortality (gradual or sudden) and the time they disappear completely, standing or fallen dead trees serve as a substrate for numerous organisms that are involved in the decomposition of deadwood (Sanders 1970; Harmon et al. 1986; Bull et al. 1997; Martikainen et al. 1998; Hammond et al. 2001; Grove 2002; DellaSala et al. 2004). The presence of dead wood in forest stands is equally important in maintaining processes such as drainage and the control of soil erosion as well as the nutrient and carbon cycles (Harmon et al. 1986; Ågren and Bosatta 1996; Hyvönen and Ågren 2001). In many countries, the reduced availability of dead wood in managed forests is currently considered one of the main causes of the loss of biodiversity (Siitonen 2001; Grove 2002). In Fenno-scandinavian forests, one fourth of threatened species are associated with dead trees (Virkkala and Toivonen 1999).

Entities that survive a disturbance in different forms (e.g., living trees, seeds, and root and stump sprouts) and allow the system to quickly reinitiate succession are another example of biological legacies. The seeds in the serotinous cones of jack pine and the semi-serotinous cones of black spruce, along with the roots of poplars that quickly resprout after fire, allow burned sites to be recolonized and the forest reconstituted fairly quickly (Greene and Johnson 1999; Greene et al. 2006). Observations suggest that when salvage logging is done without taking such biological legacies into account, the system's resilience is short-circuited in part by hindering any natural regeneration that might otherwise become established (Purdon et al. 2004; Donato et al. 2006; Greene et al. 2006). Biological legacies are therefore crucial in forests affected by major disturbances such as fire since they ensure rapid recolonization of the site.

Although we still do not understand the full importance of biological legacies in relation to ecosystem resilience and resistance, it can be stated that such legacies play a determining ecological role in the structure and functions of ecosystems that are regenerating after a disturbance. Biological legacies must therefore be taken into account in implementing forest management approaches emulating natural ecosystem functioning (Franklin et al. 2002). In an ecosystem management approach, maintaining these elements is essential to ensure that forest ecosystems have the flexibility needed to withstand future environmental fluctuations such as disturbances and climate change.

### Maintaining Processes Inherent in Disturbances

The physical and chemical action of natural disturbances sometimes results in processes that cannot be maintained solely by ensuring biodiversity and biological legacies in managed areas. For example, the action of fire on soil results in processes that are crucial for maintaining soil fertility in regions where an accumulation of organic matter reduces forest productivity (see Simard et al., chapter 11). The combustion of organic matter increases soil pH and the availability of several nutrients (MacLean et al. 1983; Certini 2005). By exposing the mineral soil and releasing nutrients, fire contributes to ensuring the productivity of peaty forest soils. Processes inherent in disturbances, such as the biological, chemical, and physical effects of fire, are difficult to reproduce with traditional silvicultural treatments. In FEM, the goal is to ensure that these processes are maintained at both the stand and landscape levels in order to preserve the ability of disturbed forests to maintain their productivity.

## 3. FOREST ECOSYSTEM MANAGEMENT (FEM)

## 3.1. Main Elements of FEM

The main aim of the conceptual framework of FEM is to strike a balance between the harvesting of timber and the maintenance of ecosystem structure and functions by managing the forest as a whole (Kimmins 2004) to ensure its integrity and sustainability. Although it is difficult to formulate a definitive definition of ecosystem management, we propose the following one:

> A management approach that aims to maintain healthy and resilient forest ecosystems by focusing on a reduction of differences between natural and managed landscapes to ensure long-term maintenance of ecosystem functions and thereby retain the social and economic benefits they provide to society.

Despite the wide range of definitions proposed in the literature, a certain consensus has been reached regarding several essential elements that are part of the ecosystem management approach (Grumbine 1994; Galindo-Leal and Bunnell 1995; Christensen et al. 1996; Kimmins 2004). Most of the elements encountered in forest ecosystem management are defined in box 1.2.

Since the principal objective of FEM is to maintain the ecological integrity of ecosystems, it focuses on their resistance and resilience. To achieve this, an expanded spatial and temporal frame of reference is required to take account of ecosystem dynamics. While the main concern of conventional forest management is maximizing timber productivity and harvests, FEM views the forest in its broader sense, planning not only what is to be extracted but also what should be maintained in the landscape. This involves managing the forest at multiple levels and taking account of the scale of influence on different organisms.

Box 1.2
## Common elements involved in FEM approaches

1) **Sustainability of all components of forest ecosystems and maintenance of ecological integrity**

Preserve biodiversity at the gene, species, and population level, along with the ecological processes that take place in the ecosystem, with a long-term perspective in mind.

2) **Multiple spatial and temporal scales**

Since forest ecosystems are much larger than stands and even conventional management units, ecosystem management must take the connection between the various spatial scales into account. Since changes to ecosystems occur over long periods of time, management must be planned with a longer time frame in mind.

3) **Establishment of ecological targets based on the best available knowledge**

Based on current scientific knowledge (local and regional), ecological targets must be set based on the desired results. Targets must also be measurable to make it possible to verify the success of operations themselves and the maintenance of key attributes.

4) **Comparison of managed and natural ecosystems**

This step involves determining major differences, particularly with regard to key attributes, between natural and managed ecosystems and, based on these findings, evaluating various silvicultural strategies for mitigating these differences.

5) **Implementation plan involving joint action by agencies involved**

To integrate all forest ecosystem components into a single management framework, all organizations working in the territory concerned must cooperate (timber resource, traplines, mines, etc.). In addition, in this step, ecological targets can be modified to take account of other social or economic objectives in the area.

6) **Monitoring**

To attain the previously defined ecological targets, the impact of forestry practices on various ecosystem components must be assessed using quantitative measurements so that adjustments can be made if targets are not met. This also involves ensuring that the hoped-for effects on organisms are also achieved.

7) **Adaptability**

Owing to the constant evolution of scientific knowledge, an ecosystem-based management framework must be flexible and lend itself to adjustment when new findings related to forest ecosystem dynamics are obtained by the scientific community. For example, it may be realized that some attributes that were targeted for protection are not as important as was previously thought, while others may be more crucial.

To achieve this goal, establishing maintenance targets – based on the best available ecological knowledge – is just as important as establishing harvesting targets. For example, one common objective of conventional boreal forest management is to "normalize" age classes and completely harvest stands when they are mature, whereas under FEM, the aim is to ensure that each type of stand, including overmature ones, is adequately represented at the landscape level. This makes it possible to maintain the necessary proportions of mature forest and forests with different structures and compositions in the landscape. However, the FEM approach does not require that every stand be restored to the state it was in before the management interventions were carried out. In other words, stands at various stages of succession may be found at various locations within the landscape, but their respective proportions must remain within the natural limits determined by disturbance regimes (the so-called shifting mosaic steady state, *sensu* Bormann and Likens [1979]).

Under the ecosystem-based approach, management targets (e.g., composition and structure) are established based on knowledge of the variability of conditions in a natural forest environment, instead of the fixed prescriptive framework used for planning timber harvests under the conventional approach. FEM more closely resembles a management-by-objectives approach adapted to regional characteristics of forest functioning rather than a prescriptive framework in which fixed rules are applied to the entire territory. Comparison of natural ecosystems and managed forests can provide a basis for identifying key processes or attributes that are at the greatest risk under current practices and that call for adjustments when developing silvicultural systems or management strategies.

Management targets set under an ecosystem-based approach are used to shed light on available resource development options for ensuring that managed ecosystems remain within the limits of natural variability (see Vaillancourt et al., chapter 2). Such options can obviously be adjusted under a forest land management and governance approach that incorporates participatory democracy (Gauthier 2005). Although social and economic considerations may indeed eventually influence management options, initially the emphasis must be on setting ecologically based management targets in order to make wise decisions. Therefore, ecological targets are evaluated and modified in order to achieve a plan that is unique to the territory in question during the land management planning stage.

To ensure that objectives set under an FEM system are met, monitoring is crucial for assessing the success or failure of the implementation of the management system. A monitoring program must be established not only to ensure that ecological targets are met but also to measure target organism responses to the system established. Similarly, to enable proposed changes to be carried out based on the results obtained, the FEM system must be flexible enough to allow refinements and the development and incorporation of new scientific knowledge.

## 3.2.   FEM and Other Forest Management Strategies

To understand how FEM fits into a forestry context, it should be compared with other forest management concepts developed in recent years. Ecosystem management indeed fits in with the notion of sustainable forest development, which has been defined by the Canadian Council of Forest Ministers (CCFM) as management that aims to "maintain and enhance the long-term health of our forest ecosystems for the benefit of all living things [...] while providing environmental, economic, social, and cultural opportunities for the benefit of present and future generations" (CCFM 1992). In FEM, the environment, as it is the support for resource production, is the foundation for the economic and social components of sustainable development. Consequently, no compromises can be made in terms of the sustainability of the resource and therefore the health of the ecosystem.

FEM offers a novel natural resources management framework by facilitating the formulation of environmental issues and targets that have to be sustained or achieved within harvesting systems. Thus, FEM is not a resource management process like integrated resources management (IRM) for instance. Instead, FEM takes place prior to the resource management process for which it is not a substitute; FEM aims at maintaining the whole ecosystem's integrity. It however provides new guidance for territorial planning by emphasizing on the maintenance of processes that are responsible for forest resilience, so as to guarantee the conservation of goods and services provided by the forest ecosystem in contrast to IRM, which primarily aims at developing specific forest resources.

Lastly, FEM also includes zoning approach (e.g., TRIAD), under which different types of land use are assigned to managed areas. Under this approach, portions of the forest are set aside for integral conservation or for intensive timber production on appropriate sites. On the remaining land, an extensive FEM approach is used to maintain ecological integrity, to meet economic needs through the harvesting of certain resources and to satisfy social values (recreation, aesthetics, traditional activities, etc.).

## 3.3. The Precautionary Principle

Although substantial progress has been made in boreal forest research in the last decade (Engelmark et al. 1993; Bergeron et al. 1998; Engelmark et al. 2000; Kuuluvainen 2002a; Macdonald 2004; Bergeron et al. 2006), understanding of how boreal ecosystems function is still in its infancy. Therefore, when we are managing the boreal forest, obviously we do not understand or fully grasp all the implications of our actions on the ecosystem and its ability to recover from disturbances or stress. Furthermore, our knowledge of the maintenance of biodiversity and related processes is still very piecemeal, which is another reason for exercising caution. Based on the precautionary principle, we must try to maintain all diversity since we do not yet fully understand the impact on ecosystem functioning of a loss of one or more elements (see box 1.1). In addition, it is important to realize that the harmful consequences of a loss of biodiversity (whether at the genetic, species or ecosystem level) or biological legacies may not be noticeable in the short term, as it may become obvious only once a new environmental stress occurs, under new climatic conditions, for example.

## 3.4. Fine and Coarse Filters

Because of our fragmentary understanding of ecosystems, one precautionary strategy is the hierarchical application of fine and coarse filters in forest management (*sensu* Hunter 1990). A coarse filter approach involves maintaining a variety of habitats representative of natural forests along with some of their key characteristics (table 1.1). The aim is to conserve most of the biodiversity. For rare species or species with special habitat requirements, a more targeted fine filter approach should also be implemented. FEM focus on natural disturbances fits into the fine and coarse filter approaches since the conditions created by appropriate silvicultural strategies aim at maintainig the key characteristics associated with natural disturbances. As table 1.1 shows, many of the key attributes of habitats that are to be maintained are linked to the disturbance regime in the region concerned (see Vaillancourt et al., chapter 2).

## 3.5. Ecosystem Management Based on Natural Disturbance Dynamics

In implementing FEM, inspiration can be drawn from regional disturbance regimes, which are responsible, along with climatic, biogeographical, and physical characteristics, for the biodiversity observed at a regional scale. Forest management based on an understanding of natural disturbances rests on two premises: (1) disturbances occur frequently enough to have an impact on biodiversity (from the genetic to the landscape level; Harper 1977) and (2) ecosystems (organisms and processes) are resilient in the face of these disturbances (Attiwill 1994; Angelstam 1998).

**Table 1.1**

**Links between the disturbance regime and forest elements associated with key attributes favouring forest ecosystem resilience and resistance in response to disturbances and environmental stress**

| | Links with disturbance characteristics | |
|---|---|---|
| **Forest elements** | **Stand level** | **Landscape level** |
| Age structure of stands and proportion of old-growth forests | | Cycle or mean interval between disturbances |
| Forest composition | Time since last disturbance<br>Type and severity of disturbances | Frequency or mean interval between disturbances |
| Stand structure | Size and severity of disturbances | Average, and variation in, size of disturbances<br>Spatial distribution of disturbances |
| Residual patches and stands in disturbed areas | Type and severity of disturbances | Type and severity of disturbances |
| Living or dead residual trees | Type and severity of disturbances | Type and severity of disturbances |
| Soil organic matter | Type and severity of disturbances | Type and severity of disturbances |

In the boreal forest, natural disturbances occur frequently and are an integral part of forest ecosystems, although they were long perceived as external factors. Fires, for example, have been around for so long and occurred so frequently that species like jack pine have been able to develop adaptations such as serotinous cones (Gauthier et al. 1993, 1996). Disturbances operate at several levels, from a pathogen attacking a single tree to the forest fire that kills several thousand trees. These disturbances generate a variety of forest attributes (table 1.1) that are important, if not essential, in maintaining biodiversity and the processes making the system resilient to disturbances.

### 3.5.1.  Key Attributes for Preserving Ecosystem Resistance and Resilience

Key attributes that are important in maintaining ecosystem capacity to withstand stress and disturbances were identified based on findings concerning the role of biodiversity and biological legacies in forest ecosystems (table 1.1). All of these attributes are closely linked to the characteristics of disturbance regimes found in the regions studied. They can be divided into four main groups: forest composition, forest structure, dead wood or woody debris (standing or downed), and organic matter in the soil (duff). All these attributes, which occur at several different levels, are subject to a great deal of manipulation in managed forests.

#### Forest Composition

Several habitat characteristics are influenced by the species of trees making up a stand. Forest composition is therefore a particularly important attribute, since it may be a determining factor in stand productivity (Légaré et al. 2005) and the diversity of understory communities (De Grandpré et al. 1993, 2003). It is also in part responsible for wildlife habitat characteristics, at both the stand and landscape levels (Man and Lieffers 1999; Drapeau et al. 2000). At the stand level, forest composition is determined not only by the condition of the physical environment (soils, drainage, climate, etc.) but also by the time elapsed since the last disturbance (Johnson 1992; Bergeron 2000; Le Goff and Sirois 2004). At the landscape level, the variety of stand types and successional stages represented also depends on the physical characteristics of the environment as well as the average interval between major disturbances and the range of intervals possible. In planning harvesting, it is important to set targets for stand species composition and age structure at the landscape level; they should be based mainly on the fire regime in the region (see Gauthier et al., chapter 3).

### Forest Structure

Forest structure, at the stand and landscape levels, is recognized to be one of the forest characteristics that create habitat diversity benefitting a wide variety of organisms (Drapeau et al. 2000; Kashian and Barnes 2000; Payer and Harrison 2000). At the landscape level, the aim is to preserve structural patterns found in a natural disturbance regime (Landres et al. 1999), such as a percentage of old-growth forests similar to that found under a particular fire cycle (see Gauthier et al., chapter 3). At the stand level, the key attributes of natural forests (e.g. wide range of diameter classes, and presence of snags and coarse woody debris) must be maintained (Franklin 1993; Spies 1998). These structural elements vary from region to region depending on the disturbance regime and the physical environment. Forest structure (horizontal or vertical) is another attribute that is altered by management activities in order to create regular, even-aged stands. The result is a homogenization of forest structure at both the stand and landscape levels.

### Dead Wood

Standing and downed dead wood have also proven to be key attributes of forest ecosystems. Dead wood is important for maintaining biodiversity in forests and ecosystem capacity to recover from disturbance (Bull et al. 1997; Watt and Caceres 1999; Drapeau et al. 2002), and is considerably affected by forestry operations (Drapeau et al. 2002; Bütler et al. 2004; Gjerde et al. 2005; Vaillancourt et al. 2008). In addition, woody debris provides a source of food and shelter for a large variety of birds and insects (Imbeau and Desrochers 2002; Nappi et al. 2003; Saint-Germain et al. 2004; Koivula et al. 2006). It influences stand dynamics by favouring the establishment and initial survival of many plant and tree species (Harmon et al. 1986; Simard et al. 2003). The salvage logging of burns (Nappi et al. 2004; Donato et al. 2006) and the harvesting of so-called healthy dry wood and of overmature stands greatly influence the availability of woody debris at the stand and landscape levels.

### Soil Organic Matter

The depth of the soil organic layer and its distribution are important attributes at the stand level. Disturbances like fire reduce the thickness of this layer, thus creating seedbeds favourable for the establishment of certain species (Johnson 1992; Greene et al. 2006). Cutting and site preparation methods can also affect the quantity and distribution of organic matter in the soil, expose the mineral soil, and thus favour the regeneration of many species. Rootlet access to the organic layer seems to be a key factor in the survival and growth of regeneration (Gastaldello et al. 2007). In the boreal forest, Thiffault et al. (2007) observed that post-harvest soil conditions and nutrients for trees were closer to those found after fire when conventional, stem-only harvesting was used (in other words when the organic matter in the foliage and branches was left in the cutover), than when whole-tree harvesting was used. Under other circumstances, low-intensity disturbance to the soil (for example, during low-severity burns or winter cutting) can accelerate the paludification process in forests susceptible to this phenomenon and accelerate the transition to unproductive forested peatlands (Simard et al. 2007; see Simard et al., chapter 11). Severe fires, site preparation or prescribed burns may control or reverse this paludification process (Lavoie et al. 2005).

### Large-Scale and Long-Term Planning

Finally, it should be noted that, in planning forest management activities, special attention must be paid to maintaining the distribution of key attributes in **space** and **time**. Several examples discussed above show that what occurs at the stand level has repercussions at the landscape level. With regard to the temporal aspect of forest management, for example, in terms of species composition, some species need more time to recover after major disturbances such as crown fires. This is particularly true for eastern white-cedar (*Thuja occidentalis* L.), which does not become re-established quickly on severely disturbed sites. Other attributes associated with old-growth forests such as large-diameter dead wood or epiphytic lichens also require long intervals after major disturbances to become re-established. When short-rotation management is implemented, some of these species or processes may be eliminated from the ecosystem. A better definition of the attributes linked to natural disturbance (table 1.1) and the development of indicators (see box 1.3) will in turn provide a better understanding of the functions linked to these attributes and facilitate their maintenance when required.

---

**Box 1.3**
## Monitoring and adaptability – example of a soil-linked indicator

This is an example of how empirical validation and new scientific knowledge can be used to adjust indicators so that they perform better, but are not necessarily more complex.

Paré et al. (2002) developed an indicator for site sensitivity in terms of impacts of whole-tree harvesting on soil fertility. This indicator was assessed by taking account of the difference between the quantity of nutrients lost during harvesting and the soil's capacity to provide these nutrients. The quantity of nutrients lost during harvesting was found to be largely dependent on the type of species and its growth rate. Subsequently, this indicator was subject to empirical validation under four different situations (Thiffault et al. 2006) revealing that site sensitivity to whole-tree harvesting was unaffected or only slightly affected by the type of stand present before harvest and that a simpler indicator, which only took into account the soil's capacity to provide nutrients, performed better in this regard. One of the cases studied was jack pine stands on sandy soil. Harvesting these stands resulted in little nutrient loss but they were established on very poor soil in the first place. Empirical validation showed that these stands were at risk for nutrient loss and that taking only the richness of the soil into account provided a better indicator of this risk (Thiffault et al. 2006).

This example illustrates how monitoring can support continual improvement of management strategies. While the initial indicator required the measurement of three variables, empirical validation in the field allowed the indicator to be simplified. This not only required the measurement of fewer variables, but resulted in better indicator performance.

# 4. CONCLUSION

The implementation of FEM systems based on natural disturbances in public forests in eastern and central Canada will require profound changes in forestry practices and strategies. To be able to reproduce key forest habitat conditions similar to those found under natural disturbance regimes, it is necessary to have a sound scientific knowledge base, to use longer-term planning horizons, and, in managed areas, to pay equal attention to resources to be retained and those to be extracted, at the stand and landscape levels.

In managing large areas like those making up the boreal forest, our duty is to adopt practices that more closely resemble those occurring in nature. Obviously, this task is quite a complex one and demands a degree of reflection. Managing the huge area that is the boreal forest cannot be done mechanically and unthinkingly without detrimentally affecting biodiversity and therefore the long-term productivity of forest ecosystems.

Throughout this book, we will see how the forest ecosystem management (FEM) approach, which draws inspiration from natural disturbances regimes, can help us halt the depletion of mature and overmature forests through the development of management approach and strategies that attempt to recreate the conditions found in these forests. In some regions, however, we are aware that it can be difficult to implement such practices since the forests have already been extensively modified. In this case, FEM can be used to help establish restoration objectives for second-growth forests, in order to recreate key attributes and conditions similar to those found in the past, so that ecosystem capacity to cope with future environmental stresses and fluctuations is maintained.

# REFERENCES

Ågren, G.I. and Bosatta, E. 1996. Theoretical ecosystem ecology: understanding element cycles. Cambridge University Press, Cambridge, UK.

Angelstam, P.K. 1998. Maintaining and restoring biodiversity in European boreal forests by developing natural disturbance regimes. J. Veg. Sci. **9**: 593–602.

Attiwill, P.M. 1994. The disturbance of forest ecosystems: the ecological basis for conservative management. For. Ecol. Manag. **63**: 247–300.

Beaulieu, J., Plourde, A., Daoust, G., and Lamontagne, L. 1996. Genetic variation in juvenile growth of *Pinus strobus* in replicated Quebec provenance-progeny tests. For. Genet. **3**: 103–112.

Beaulieu, J. and Rainville, A. 2005. Adaptation to climate change: genetic variation is both a short- and a long-term solution. For. Chron. **81**: 705–709.

Beaulieu, J. and Simon, J.-P. 1994. Genetic structure and variability in *Pinus strobus* in Quebec. Can. J. For. Res. **24**: 1726–1733.

Begon, M., Harper, J.L., and Townsend, C.R. 1996. Ecology: individuals, populations and communities. 3rd edition. Blackwell Scientific Publications, Oxford, UK.

Bergeron, Y. 2000. Species and stand dynamics in the mixed woods of Quebec's southern boreal forest. Ecology, **81**: 1500–1516.

Bergeron, Y., Engelmark, O., Harvey, B., Morin, H., and Sirois, L. (Eds.) 1998. Key issues in disturbance dynamics in boreal forests. J. Veg. Sci. **9**: 463–610.

Bergeron, Y. and Harvey, B. 1997. Basing silviculture on natural ecosystem dynamics: an approach to the southern boreal mixedwoods of Québec. For. Ecol. Manag. **92**: 235–242.

Bergeron, Y., Leduc, A., Harvey, B., and Gauthier, S. 2002. Natural fire regime: a guide for sustainable forest management in the Canadian boreal forest. Silva Fenn. **36**: 81–95.

Bergeron, Y., Leduc, A., Morin, H., and Joyal, C. 1995. Balsam fir mortality following the last spruce budworm outbreak in northwestern Quebec. Can. J. For. Res. **25**: 1375–1384.

Bergeron, Y., Macdonald, E., Engelmark, O., Kuuluvainen, T., and Shorohova, E. (Eds.) 2006. Disturbance dynamics in boreal forest. Écoscience, **13**: iii.

Blondel, J. 1995. Biogéographie. Approche écologique et évolutive. Masson, Paris, France.

Bolgiano, N.C. 2004. Cause and effect: changes in boreal bird irruptions in eastern North America relative to the 1970s spruce budworm infestation. Am. Birds **58**: 26–33.

Bormann, F.H. and Likens, G.E. 1979. Patterns and process in a forested ecosystem. Springer-Verlag, New York, USA.

Bouthillier, L. 1998. Brève histoire du régime forestier québécois. Faculté de foresterie et de géomatique, Université Laval, Que.

Bull, E.L., Parks, C.G., and Torgersen, T.R. 1997. Trees and logs important to wildlife in the Interior Columbia River Basin. Gen. Tech. Rep. PNW-GTR-391. U.S. Department of Agriculture, Pacific Northwest Research Station, Forest Service, Portland, Ore., USA.

Burton, P.J., Messier, C., Weetman, G.F., Prepas, E.E., Adamowicz, W.L., and Titler, R. 2003. The current state of boreal forestry and the drive for change. *In* Towards sustainable management of the boreal forest. *Edited by* P.J. Burton, C. Messier, D.W. Smith, and W.L. Adamowicz. NRC Research Press, Ottawa, Ont., pp. 1–40.

Bütler, R., Angelstam, P., Ekelund, P., and Shkaepfer, R. 2004. Deadwood threshold values for the three-toed woodpecker presence in boreal sub-alpine forest. Biol. Conserv. **119**: 305–319.

Canadian Council of Forest Ministers (CCFM). 1992. Sustainable forests: a Canadian commitment. Canadian Council of Forest Ministers, Natural Resources Canada, Ottawa, Ont.

Canadian Council of Forest Ministers (CCFM). 1995. Defining sustainable forest management: a Canadian approach to criteria and Indicators. Canadian Council of Forest Ministers, Natural Resources Canada, Ottawa, Ont.

Cappuccino, N., Lavertu, D., Bergeron, Y., and Régnière, J. 1998. Spruce budworm impact, abundance and parasitism rate in a patchy landscape. Oecologia, **114**: 236–242.

Carter, K.K. 1996. Provenance tests as indicators of growth response to climate change in 10 north temperate tree species. Can. J. For. Res. **26**: 1089–1095.

Certini, G. 2005. Effects of fire on properties of forest soils: a review. Oecologia, **143**: 1–10.

Chapin, F.S., Walker, B.H., Hobbs, R.J., Hooper, D.U., Lawton, J.H., Sala, O.E., and Tilman, D. 1997. Biotic control over the functioning of ecosystems. Science, **277**: 500–503.

Christensen, N.L., Bartuska, A.M., Brown, J.H., Carpenter, S., D'Antonio, C., Francis, R., Franklin, J.F., MacMahon, J.A., Noss, R.F., Parsons, D.J., Peterson, C.H., Turner, M.G., and Woodmansee, R.G. 1996. The report of the Ecological Society of America committee on the scientific basis for ecosystem management. Ecol. Appl. **6**: 665–691.

Cooke, B.J. and Roland, J. 2000. Spatial analysis of large-scale patterns of forest tent caterpillar outbreaks. Écoscience, **7**: 410–422.

Crawford, H.S. and Jennings, D.T. 1989. Predation by birds on spruce budworm *Choristoneura fumiferana*: functional, numerical, and total responses. Ecology, **70**: 152–163.

Cumming, S.G. 2001. Forest type and wildfire in the Alberta boreal mixedwood: what do fires burn? Ecol. Appl. **11**: 97–110.

De Grandpré, L. and Bergeron, Y. 1997. Diversity and stability of understory communities following disturbance in the southern boreal forest. J. Ecol. **85**: 777–784.

De Grandpré, L., Bergeron, Y., Nguyen, T., Boudreault, C., and Grondin, P. 2003. Composition and dynamics of the understory vegetation in the boreal forests of Quebec. *In* The herbaceous layer of forests of eastern North America. *Edited by* F.S. Gilliam and M.R. Roberts. Oxford University Press, UK, pp. 238–261.

De Grandpré, L., Gagnon, D., and Bergeron, Y. 1993. Changes in the understory of the Canadian southern boreal forest after fire. J. Veg. Sci. **4**: 803–810.

DellaSala, D.A., Williams, J.E., Williams, C.D., and Franklin, J.F. 2004. Beyond smoke and mirrors: a synthesis of fire policy and science. Conserv. Biol. **18**: 976–986.

Donato, D.C., Fontaine, J.B., Campbell, J.L., Robinson, W.D., Kauffman, J.B., and Law, B.E. 2006. Post-wildfire logging hinders regeneration and increases fire risk. Science, **311**: 352–354.

Drapeau, P., Leduc, A., Giroux, J.-F., Savard, J.-P.L., Bergeron, Y., and Vickery, W.L. 2000. Landscape-scale disturbances and changes in bird communities of boreal mixed-wood forests. Ecol. Monogr. **70**: 423–444.

Drapeau, P., Nappi, A., Giroux, J.-F., Leduc, A., and Savard, J.-P.L. 2002. Distribution patterns of birds associated with coarse woody debris in natural and managed eastern boreal forests. *In* Ecology and management of deadwood in western forests. *Edited by* B. Laudenslayer, W.F. Laudenslayer, Jr., P.J. Shea, B.E. Valentine, C.P. Weatherspoon, and T.E. Lisle. USDA Forest Service General Technical Report PSW-GTR 181. USDA Forest Service Pacific Southwest Research Station, Albany, N.Y., USA, pp. 193–205.

Drapeau, P., Nappi, A., Saint-Germain, M., and Angers, V.-A. 2005. Les régimes naturels de perturbations, l'aménagement forestier et le bois mort dans la forêt boréale québécoise. *In* Bois mort et à cavités, une clé pour des forêts vivantes. *Edited by* D. Vallauri, J. André, B. Dodelin, R. Eynard-Machet, and D. Rambaud, WWF/Tec & Doc, Paris, France, pp. 45–55.

Drever, C.R., Peterson, G., Messier, C., Bergeron, Y., and Flannigan, M. 2006. Can forest management based on natural disturbances maintain ecological resilience? Can. J. For. Res. **36**: 2285–2299.

Drushka, K. 2003. Canada's forest: A history. McGill-Queen's University Press, Montréal, Que.

Ehrlich, P.R. and Ehrlich, A.H. 1981. Extinction: The causes and consequences of disappearance of species. Random House, New York, USA.

Ehrlich, P.R. and Wilson, E.O. 1991. Biodiversity studies: Science and policy. Science, **253**: 758–762.

Engelmark, O., Bradshaw, R., and Bergeron, Y. (Eds.) 1993. Disturbance dynamics in boreal forest. J. Veg. Sci. **4**: 729–832.

Engelmark, O., Gauthier, S., and van der Maarel, E. (Eds.) 2000. Disturbance dynamics in boreal and temperate forests: introduction. J. Veg. Sci. **11**: 779–780.

Fischer, J., Lindenmayer, D.B., and Manning, A.D. 2006. Biodiversity, ecosystem function, and resilience: ten guiding principles for commodity production landscapes. Front. Ecol. Environ. **4**: 80–86.

Franklin, J.F. 1993. Preserving biodiversity: species, ecosystems, or landscape? Ecol. Appl. **3**: 202–205.

Franklin, J.F., Lindenmayer, D.B., MacMahon, J.A., McKee, A., Magnusson, J., Perry, D.A., Waide, R., and Foster, D.R. 2000. Threads of continuity: ecosystem disturbances, biological legacies and ecosystem recovery. Conserv. Biol. Pract. **1**: 8–16.

Franklin, J.F., Spies, T.A., Van Pelt, R., Carey, A.B., Thornburgh, D.A., Berg, D.R., Lindenmayer, D.B., Harmon, M.E., Keeton, W.S., Shaw, D.C., Bible, K., and Chen, J. 2002. Disturbances and structural development of natural forest ecosystems with silvicultural implications, using Douglas-fir forests as an example. For. Ecol. Manag. **155**: 399–423.

Frelich, L.E. 2002. Forest dynamics and disturbance regimes – studies from temperate evergreen – deciduous forests. Cambridge University Press, Cambridge, UK.

Galindo-Leal, C. and Bunnell, F.L. 1995. Ecosystem management: implications and opportunities of a new paradigm. For. Chron. **71**: 601–606.

Galipeau, C., Kneeshaw, D., and Bergeron, Y. 1997. White spruce and balsam fir colonization of a site in the southeastern boreal forest as observed 68 years after fire. Can. J. For. Res. **27**: 139–147.

Gastaldello, P., Ruel, J.-C., and Paré, D. 2007. Microvariations in yellow birch (*Betula alleghaniensis*) growth conditions after patch scarification. For. Ecol. Manag. **238**: 244–248.

Gaudreau, G. 1999. Les récoltes des forêts publiques au Québec et en Ontario: 1840–1900. McGill-Queen's University Press, Montréal, Que.

Gauthier, M. 2005. Gestion intégrée de l'environnement en milieu urbain : vers un renouvellement des pratiques planificatrices? Organisations et territoires (Numéro spécial : Aménagement des territoires), **14**: 59–67.

Gauthier, S., Bergeron, Y., and Simon, J.-P. 1996. Effects of fire regime on the serotiny level of jack pine. J. Ecol. **84**: 539–548.

Gauthier, S., Gagnon, J., and Bergeron, Y. 1993. Population age structure of *Pinus banksiana* at the southern edge of the Canadian boreal forest. J. Veg. Sci. **4**: 783–790.

Gjerde, I., Saetersdal, M., and Nilsen, T. 2005. Abundance of two threatened woodpecker species in relation to the proportion of spruce plantations in native pine forests of western Norway. Biodivers. Conserv. **14**: 377–393.

Greene, D.F., Gauthier, S., Noël, J., Rousseau, M., and Bergeron, Y. 2006. A field experiment to determine the effect of post-fire salvage on seedbeds and tree regeneration. Front. Ecol. Environ. **4**: 69–74.

Greene, D.F. and Johnson, E.A. 1999. Modelling recruitment of *Populus tremuloides*, *Pinus banksiana*, and *Picea mariana* following fire in the mixedwood boreal forest. Can. J. For. Res. **29**: 462–473.

Grove, S.J. 2002. Saproxylic insect ecology and the sustainable management of forests. Annu. Rev. Ecol. Syst. **33**: 1–23.

Grumbine, R.E. 1992. Ghost bears: exploring the biodiversity crisis. Island Press, Washington, D.C., USA.

Grumbine, R.E. 1994. What is ecosystem management? Conserv. Biol. **8**: 27–38.

Gunderson, L. and Holling, C.S. (Eds.). 2002. Panarchy: understanding transformations in human and natural systems. Island Press, Washington, D.C., USA.

Hagan, J.M. III, Lloyd-Evans, T.L., Atwood, J.L., and Wood, D.S. 1992. Long-term changes in migratory landbirds in the northeastern United States: evidence from migration capture data. *In* Ecology and conservation of neotropical migrant landbirds. *Edited by* J.M. Hagan III and D.W. Johnston. Smithsonian Institution Press, Washington, D.C., USA, pp. 115–130.

Hammond, H.E., Langor, D.W., and Spence, J.R. 2001. Early colonization of *Populus* wood by saproxylic beetles (*Coleoptera*). Can. J. For. Res. **31**: 1175–1183.

Harmon, M.E., Franklin, J.F., Swanson, F.J., Sollins, P., Gregory, S.V., Lattin, J.D., Anderson, N.H., Cline, S.P., Aumen, N.G., Sedell, J.R., Lienkaemper, G.W., Cromack, K., Jr, and Cummins, K.W. 1986. Ecology of coarse woody debris in temperate ecosystems. Adv. Ecol. Res. **15**: 133–302.

Harper, J.L. 1977. Population biology of plants. Academic Press, New York, USA.

Harris, L. 1985. The fragmented forest. University of Chicago Press, Chicago, Ill., USA.

Hébert, Y. 2006. Une histoire de l'écologie au Québec: les regards sur la nature des origines à nos jours. Les Éditions du GID, Québec, Que.

Heinselman, M.L. 1973. Fire in the virgin forests of the Boundary Waters Canoe Area, Minnesota. Quat. Res. **3**: 329–382.

Hély, C., Bergeron, Y., and Flannigan, M.D. 2000. Effects of stand composition on fire hazard in mixed-wood Canadian boreal forest. J. Veg. Sci. **11**: 813–824.

Hobson, K.A. and Bayne, E.M. 2000. Breeding bird communities in boreal forest of western Canada: consequences of "unmixing" the mixedwoods. Condor, **102**: 759–769.

Holling, C.S. 1973. Resilience and stability of ecological systems. Annu. Rev. Ecol. Syst. **4**: 1–23.

Hunter, M.L., Jr. 1990. Wildlife, forests, and forestry: principles of managing forests for biological diversity. Prentice-Hall, Englewood Cliffs, N.J., USA.

Hyvönen, R. and Ågren, G.I. 2001. Decomposer invasion rate, decomposer growth rate, and substrate chemical quality: how they influence soil organic matter turnover. Can. J. For. Res. **31**: 1594–1601.

Imbeau, L. and Desrochers, A. 2002. Foraging ecology and use of drumming trees by Three-toed Woodpeckers. J. Wild. Manag. **66**: 222–231.

Johnson, E.A. 1992. Fire and vegetation dynamics: studies from the North American boreal forest. Cambridge University Press, Cambridge, UK.

Johnson, K.H., Vogt, K.A., Clark, H.J., Schmidt, O.J., and Vogt, D.J. 1996. Biodiversity and the productivity and stability of ecosystems. Trends Ecol. Evol. **11**: 372–377.

Kafka, V., Gauthier, S., and Bergeron, Y. 2001. Fire impacts and crowning in the boreal forest: study of a large wildfire in western Quebec. Int. J. Wildland Fire, **10**: 119–127.

Kashian, D.M. and Barnes, B.V. 2000. Landscape influence on the spatial and temporal distribution of the Kirtland's warbler at the Bald Hill burn, northern Lower Michigan, U.S.A. Can. J. For. Res. **30**: 1895–1904.

Kimmins, J.P. 1987. Forest Ecology. Macmillan Publishing Company, New York, USA.

Kimmins, J.P. 2004. Emulating natural forest disturbances: what does this mean? *In* Emulating natural forest landscape disturbances: concepts and applications. *Edited by* A.H. Perera, L.J. Buse, and M.G. Weber. Columbia University Press, New York, USA, pp. 8–28.

Kohm, K.A. and Franklin, J.F. 1997. Creating a forestry for the 21st century: the science of ecosystem management. Island Press, Washington, D.C., USA.

Koivula, M., Cobb, T., Déchêne, A.D., Jacobs, J., and Spence, J.R. 2006. Responses of two *Sericoda* Kirby, 1837 (*Coleoptera*: *Carabidae*) species to forest harvesting, wildfire and burn severity. Entomol. Fenn. **17**: 315–324.

Kuuluvainen, T. 1994. Gap disturbance, ground microtopography, and the regeneration dynamics of boreal coniferous forests in Finland: a review. Ann. Zool. Fenn. **31**: 35–51.

Kuuluvainen, T. (Ed.) 2002a. Disturbance dynamics in boreal forests: Defining the ecological basis of restoration and management of biodiversity. Silva Fenn. **36**: 5–11.

Kuuluvainen, T. 2002b. Natural variability of forests as a reference for restoring and managing biological diversity in boreal Fennoscandia. Silva Fenn. **36**: 97–125.

Landres, P.B., Morgan, P., and Swanson, F.J. 1999. Overview of the use of natural variability concepts in managing ecological systems. Ecol. Appl. **9**: 1179–1188.

Lavoie, M., Paré, D., Fenton, N., Groot, A., and Taylor, K. 2005. Paludification and management of forested peatlands in Canada: a literature review. Environ. Rev. **13**: 21–50.

Lawton, J.H. 1994. What do species do in ecosystems? Oikos, **71**: 367–374.

Lawton, J.H. and Brown, V.K. 1993. Redundancy in ecosystems. *In* Biodiversity and ecosystem function. *Edited by* E.D. Schulze and H.A. Mooney. Springer-Verlag, New York, USA, pp. 255–269.

Légaré, S., Paré, D., and Bergeron, Y. 2005. Influence of aspen on forest floor properties in black spruce-dominated stands. Plant Soil, **275**: 207–220.

Le Goff, H. and Sirois, L. 2004. Black spruce and jack pine dynamics simulated under varying fire cycles in northern boreal forest of Quebec, Canada. Can. J. For. Res. **34**: 2399–2409.

Leopold, A. 1949. A Sand County Almanac. Ballantine Books, New York, USA.

Lindenmayer, D.B. and Franklin, J.F. 2002. Conserving forest biodiversity: a comprehensive multiscaled approach. Island Press, Washington, D.C., USA.

Macdonald, E. (Ed.) 2004. Foreword, 4th International Workshop on Disturbance Dynamics in Boreal Forests: Disturbance Processes and their Ecological Effects special issue. Can. J. For. Res. **34**: v–vi.

MacLean, D.A. 1980. Vulnerability of fir-spruce stands during uncontrolled spruce budworm outbreaks: a review and discussion. For. Chron. **56**: 213–221.

MacLean, D.A., Woodley, S.J., Weber, M.G., and Wein, R.W. 1983. Fire and nutrient cycling. *In* The role of fire in northern circumpolar ecosystems. *Edited by* R.W. Wein and D.A. MacLean. John Wiley and Sons, New York, USA, pp. 111–132.

Man, R. and Lieffers, V.J. 1999. Are mixtures of aspen and white spruce more productive than single species stands? For. Chron. **75**: 505–513.

Martikainen, P., Kaila, L., and Haila, Y. 1998. Threatened beetles in White-backed Woodpecker habitats. Conserv. Biol. **12**: 293–301.

Matyas, C. 1996. Climate adaptation of trees: rediscovering provenance tests. Euphytica, **92**: 45–54.

McCann, K.S. 2000. The diversity-stability debate. Nature, **405**: 228–233.

Miyanishi, K. and Johnson, E.A. 2002. Process and patterns of duff consumption in the mixedwood boreal forest. Can. J. For. Res. **32**: 1285–1295.

Nappi, A., Drapeau, P., Giroux, J.-F., and Savard, J.-P.L. 2003. Snag use by foraging Black-backed Woodpeckers (*Picoides arcticus*) in a recently burned eastern boreal forest. The Auk, **120**: 505–511.

Nappi, A., Drapeau, P., and Savard, J.-P.L. 2004. Salvage logging after wildfire in the boreal forest: is it becoming a hot issue for wildlife? For. Chron. **80**: 67–74.

Natural Resources Canada. 1998. State of Canada's forests. Ottawa, Ont.

Paré, D. and Bergeron, Y. 1995. Aboveground biomass accumulation along a 230-year chronosequence in the southern portion of the Canadian boreal forest. J. Ecol. **83**: 1001–1008.

Paré, D., Rochon, P., and Brais, S. 2002. Assessing the geochemical balance of managed boreal forests. Ecol. Indicat. **1**: 293–311.

Patten, M.A. and Burger, J.C. 1998. Spruce budworm outbreaks and the incidence of vagrancy in eastern North American Wood Warblers. Can. J. For. Res. **76**: 433–439.

Payer, D.C. and Harrison, D.J. 2000. Structural differences between forests regenerating following spruce budworm defoliation and clear-cut harvesting: implications for marten. Can. J. For. Res. **30**: 1965–1972.

Perera, A.H., Buse, L.J., and Weber, M.G. 2004. Emulating natural forest landscape disturbances: concept and applications. Columbia University Press, New York, USA.

Perry, D.A. and Amaranthus, M.P. 1997. Disturbance, recovery and stability. *In* Creating a forestry for the 21st century: the science of ecosystem management. *Edited by* K.A. Kohm and J.F. Franklin. Island Press, Washington, D.C., USA, pp. 31–56.

Pickett, S.T.A. and White, P.S. (Eds.). 1985. The ecology of natural disturbance and patch dynamics. Academic Press, Orlando, Fla., USA.

Purdon, M., Brais, S., and Bergeron, Y. 2004. Initial response of understorey vegetation to fire severity and salvage-logging in the southern boreal forest of Québec. Appl. Veg. Sci. **7**: 49–60.

Roland, J. 1993. Large-scale forest fragmentation increases the duration of tent caterpillar outbreak. Oecologia, **93**: 25–30.

Roland, J. and Taylor, P.D. 1997. Insect parasitoid species respond to forest structure at different spatial scales. Nature, **386**: 710–713.

Romain-Bondi, K.A., Wielgus, R.B., Waits, L., Kasworm, W.F., Austin, M., and Wakkinen, W. 2004. Density and population size estimates for North Cascade grizzly bears using DNA hair-sampling techniques. Biol. Conserv. **117**: 417–428.

Ryan, K.C. 2002. Dynamic interactions between forest structure and fire behavior in boreal ecosystems. Silva Fenn. **36**: 13–39.

Saint-Germain, M., Drapeau, P., and Hébert, C. 2004. Xylophagous insect species composition and patterns of substratum use on fire-killed black spruce in central Quebec. Can. J. For. Res. **34**: 677–685.

Sanders, C.J. 1970. The distribution of carpenter ant colonies in the spruce-fir forests of northwestern Ontario. Ecology, **51**: 865–873.

Schulze, E.D. and Mooney, H.A. (Eds.) 1993. Biodiversity and ecosystem function. Springer-Verlag, New York, USA.

Schwartz, M.W., Brigham, C.A., Hoeksema, J.D., Lyons, K.G., Mills, M.H., and van Mantgem, P.J. 2000. Linking biodiversity to ecosystem function: implications for conservation biology. Oecologia, **122**: 297–305.

Shelford, V.E. 1933. The preservation of natural biotic communities. Ecology, **14**: 240–245.

Siitonen, J. 2001. Forest management, coarse woody debris and saproxylic organisms: Fennoscandian boreal forests as an example. Ecol. Bull. **49**: 11–41.

Simard, M., Lecomte, N., Bergeron, Y., Paré, D., and Bernier, P.-Y. 2007. Forest productivity decline caused by successional paludification of boreal forests. Ecol. Appl. **17**: 1619–1637.

Simard, M.-J., Bergeron, Y., and Sirois, L. 2003. Substrate and litterfall effects on conifer seedling survivorship in southern boreal stands of Canada. Can. J. For. Res. **33**: 672–681.

Spies, T.A. 1998. Forest structure: a key to the ecosystem. Northwest Sci. **72**: 34–39.

Sturtevant, B.R., Bissonnette, J.A., Long, J.N., and Roberts, D.W. 1997. Coarse woody debris as a function of age, stand structure, and disturbance in boreal Newfoundland. Ecol. Appl. **7**: 702–712.

Su, Q., MacLean, D.A., and Needham, T.D. 1996. The influence of hardwood content on balsam fir defoliation by spruce budworm. Can. J. For. Res. **26**: 1620–1628.

Thiffault, É., Bélanger, N., Paré, D., and Munson, A.D. 2007. How do forest harvesting methods compare with wildfire? A case study of soil chemistry and tree nutrition in the boreal forest. Can. J. For. Res. **37**: 1658–1688.

Thiffault, É., Paré, D., Bélanger, N., Munson, A., and Marquis, F. 2006. Harvesting intensity at clear-felling in the boreal forest: impact on soil and foliar nutrient status. Soil Sci. Soc. Am. J. **70**: 691–701.

Vaillancourt, M.-A., Drapeau, P., Gauthier, S., and Robert, M. 2008. Availability of standing trees for large cavity nesting birds in the eastern boreal forest of Québec, Canada. For. Ecol. Manag. **255**: 2272–2285.

Virkkala, R. and Toivonen, H. 1999. Maintaining biological diversity in Finnish forests. The Finnish Environment 278, Finnish Environment Institute, Helsinki, Finland.

Vitousek, P.M. and Hoper, D.U. 1993. Biological diversity and terrestrial ecosystem biogeochemistry. *In* Biodiversity and ecosystem function. *Edited by* E.D. Schulze and H.A. Mooney. Springer-Verlag, New York, USA, pp. 3–14.

Walker, B.H. 1992. Biodiversity and ecological redundancy. Biol. Conserv. **6**: 18–23.

Watt, W.R. and Caceres, M.C. 1999. Managing for snags in the boreal forests of northeastern Ontario. Ontario Ministry of Natural Resources, Toronto, NEST Technical Note TN–016.

Whittaker, R.H. 1962. Classification of natural community. Bot. Rev. **28**: 1–29.

World Commission on Environment and Development (WCED). 1987. Our common future. Oxford University Press, Oxford, UK.

# Chapter 2

# How Can Natural Disturbances Be a Guide for Forest Ecosystem Management?*

*Marie-Andrée Vaillancourt, Louis De Grandpré, Sylvie Gauthier, Alain Leduc, Daniel Kneeshaw, Yves Claveau, and Yves Bergeron*

* The authors acknowledge the financial support of the Sustainable Forest Management Network, Natural Resources Canada, and the Centre for Forest Research (CFR-CEF) for the writing and translation of this chapter. We thank Pamela Cheers and Isabelle Lamarre from Natural Resources Canada for the linguistic revision of the manuscript. The photos on this page were graciously provided by Pierre Drapeau and Jacques Morissette.

# 1. INTRODUCTION

Natural disturbances are ecological processes inherent to forest ecosystems (White and Pickett 1985; Attiwill 1994), including boreal forests (MacLean 1980; Morin 1994; Johnson et al. 1998; Kneeshaw 2001; McCarthy 2001; Wooster and Zhang 2004), where they are important drivers of forest dynamics and shape the diversity of habitats at the landscape level (see Gauthier et al., chapter 1). Ecosystem management based on natural disturbances is not intended to emulate, in every way, the effects and processes of natural disturbances. Instead, it aims at using knowledge of disturbance regimes to define the current issues resulting from forestry practices. The main goal of adopting an ecosystem management framework is to minimize the differences between natural forest conditions and those generated by our forestry practices in order to limit their impacts on biodiversity and ecosystem productivity. Therefore, implementation of ecosystem management must rely on a good understanding of the main disturbance regimes' effects on forest ecosystems at various spatial and temporal scales.

Effects of disturbance on forest characteristics (such as composition or structure) result from three main disturbance features: return interval (and other temporal descriptors), size (and other configuration descriptors), and severity (their intrinsic nature). These three types of descriptors define the range of variability generated by disturbance regimes (figure 2.1). There is now a consensus on the fact that a forest management regime reduces the variability of these

---

Box 2.1
## Conceptual model illustrating primary features describing disturbance regimes

A conceptual model illustrating primary features describing disturbance regimes is presented to help us better visualize how to meet the objective of reducing the disparity between managed forest and natural disturbance regimes.

As an example, we can represent the natural variability of a fire disturbance regime using three axes (return interval, severity, and size; figure 2.1A). In the boreal forest, considerable amplitude may exist for each of these axes, which also vary across regions. Since fire occurrence is almost random, one site can burn several times within a short time period whereas another site can be spared for several hundred years. Likewise, a burn area can cover less than 1 hectare up to several thousand square kilometres. Finally, while some surface fires can affect only understory vegetation, crown fires can cause widespread mortality among the canopy and consume the humus layers down to the mineral soil. The combination of these characteristics (return interval between fires, fire severity, fire size), as well as others not illustrated here but presented later (table 2.1), constitute a fire regime and is unique to a given forest region. This disturbance regime, and its interaction with environmental features (such as thermal, water, and nutrient regimes), is responsible for most of the habitat diversity within a region. It determines the coarse filter* on which maintenance of biodiversity should rely.

In contrast to natural variability, we can easily represent theoretical variability created by an intensive management regime including, for example, the widespread use of plantations and stand-tending treatments. In this context, harvest rotations, harvest block size, and severity constitute a management regime with a narrow variability compared with that of the natural disturbance regime (figure 2.1B). This narrow range can impose constraints on biodiversity and forest productivity.

---

* This concept is defined in chapter 1 (p. 29).

Although the main objective of forest ecosystem management is to respect the variability inherent to natural disturbance regimes, in practical terms it often becomes a social and economic compromise within the limits of historical variability (see last section of this chapter; Bergeron et al. 2004a). This management target is generally located between the great variability generated by natural disturbances and the homogeneity generated by forest management regimes aimed primarily at sustaining fibre yield (figure 2.1C).

### Figure 2.1
## Conceptual model describing disturbance regime variability in natural ecosystems (A) and managed ecosystems (B and C)

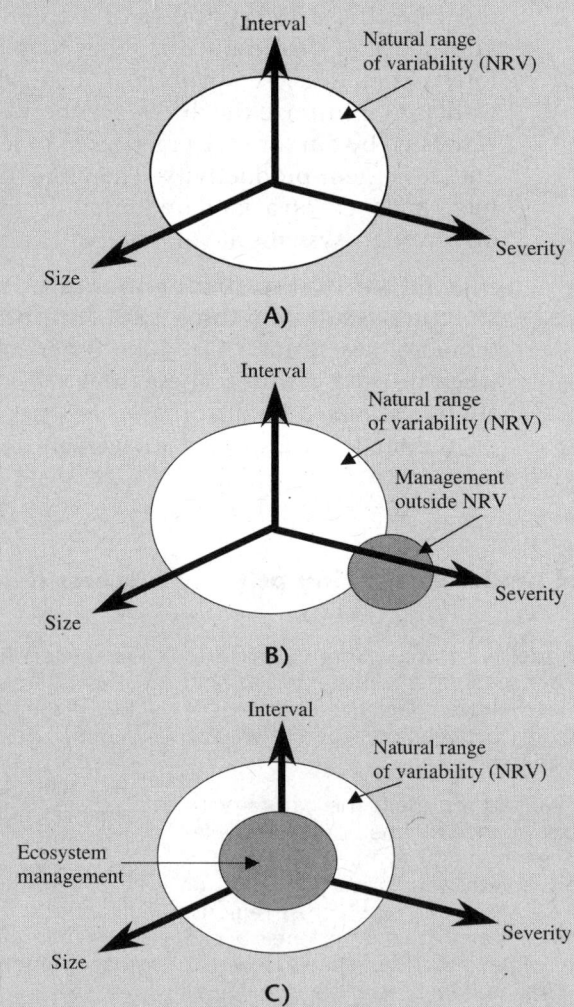

In figure A, there is a great variation in return interval between successive disturbances, in disturbance sizes as well as disturbance severity. In figure B, the forest management regime reproduces little variability regarding disturbance interval, size and severity. In figure C, the management regime creates conditions similar to those of the natural regime and maintains variability, but without reaching the extremes. Adapted from Bergeron et al. (2007).

features, which creates a homogenization of the forest conditions (McCrae et al. 2001; Hauessler and Kneeshaw 2003). This homogenization results from forest management strategies using regular harvesting rates and intensity and systematic harvesting of mature and old-growth forest stands. By using knowledge of natural disturbance regime variability, it is possible to compare forest conditions generated by natural and anthropic disturbance regimes. Appropriate management targets must then be established based on the differences observed at the stand and landscape levels (see box 2.1).

Knowledge of natural disturbance regimes of the eastern Canadian boreal forest is currently developing quickly. The second part of this book presents an ecological knowledge synthesis with regard to the major disturbance regimes (fire and insects) as well as regional examples of disturbance regimes and their implications for ecosystem management implementation. First of all, it is crucial to define a disturbance regime and explain how its characteristics can be used to identify ecological issues and set forest management objectives. In this chapter, we will define disturbance regime characteristics and illustrate how they influence forest conditions using examples from the effects of two disturbance agents of the boreal forest. Finally, we will clarify the historic and regional aspects of natural variability.

## 2. NATURAL DISTURBANCE REGIME CHARACTERISTICS

Natural disturbances can be defined as relatively distinct events that affect the structure of an ecosystem, a community, or a population, and modify resources, substrate availability, or the physical environment (White and Pickett 1985). Some natural disturbances such as intense crown fires are often described as "catastrophic" at the human scale. However, on an evolutionary scale, a catastrophe is a very rare event that does not exert selection pressure on the organisms living within the affected ecosystem (Harper 1977). The majority of disturbance events affecting boreal forests have occurred frequently for thousands of years and living organisms are well adapted to them. Thus, such disturbances cannot be defined as catastrophic.

In the boreal forest there is a great variety of disturbance agents having impacts at different levels. Among these, fire is probably the most documented one. In fact, a large number of studies have been conducted on fire regimes across central and eastern Canada (Bergeron et al. 2001, 2004a, 2006; Kafka et al. 2001; Lauzon et al. 2007; Tardif 2004). However, fire is not predominant across all the North American boreal forest and other disturbance agents can also play an important role in forest dynamics (Bergeron et al. 1998; Engelmark et al. 2000; McCarthy 2001). Indeed, in Canada insect outbreaks affect areas larger than those affected by fire and forest management combined (Kneeshaw 2001; CCFM 2006). The main insect outbreak regimes having a significant impact on timber supply in the eastern and central Canadian boreal forests are those of the spruce budworm (*Choristoneura fumiferana* Clem.; see Morin et al., chapter 7), forest tent caterpillar (*Malacosoma disstria* Hbn.; see Sutton et al., chapter 8), and hemlock looper (*Lambdina fiscellaria* Guen.). At a more local scale, windthrow, ice storms, pathogens, or senescence are disturbance agents that can also drive forest dynamics. Some of these agents affect the forest at a particular development stage. For example, senescence can occur in a synchronized manner at the stand level when a post-fire tree cohort becomes older than the tree species' mean longevity (Kneeshaw and Gauthier 2003). This senescence can be accelerated by the occurrence of various diseases, such as heart rot or root rot, which are common pathogens affecting balsam fir trees (*Abies balsamea* [L.] Mill.).

A disturbance regime consists of a combination of all the characteristics generated by one or several disturbance agents acting within a given land area (Frelich 2002). Thus, this regime can be described by several descriptors (spatial, temporal, or others) that are specific to a disturbance agent or to a region under study. An exhaustive list of the descriptors commonly used is presented in table 2.1. In the following sections, we will illustrate some of these descriptors, focusing on those that could be useful within an ecosystem management framework. Although each descriptor is important for the global comprehension of disturbance regimes, they are not all relevant in the development of forest management strategies (see section 3).

## 2.1.  Temporal Descriptors

Several descriptors of disturbance regime designate temporal properties such as the return interval, periodicity, cycle, annual proportion affected, and seasonality (defined in table 2.1). Among these, cycle is a descriptor commonly used to describe fire regimes and corresponds to the time needed to burn an area equivalent to the area under study (Johnson and Gutsell 1994). The inverse of the cycle (annual burned area proportion) corresponds to the fire frequency or annual burn rate. The fire interval is the time elapsed between two successive fires at the same location. Although it is often stated in the literature that fire cycle corresponds to the mean return interval between two successive fires measured at several sites, Reed (2006) showed that this statement involves a simplification of fire size distribution that rarely occurs in reality. Throughout the Canadian boreal forest, fire cycle can vary from 50 years to more than 500 years (Haeussler and Kneeshaw 2003). At the stand level, return intervals as short as 10 years (Van Wagner 1983) and as long as 1,000 years (Cyr et al. 2005) have been observed.

Temporal descriptors used to characterize insect outbreaks include return interval, as well as periodicity because of the cyclic nature of these phenomena. Like fires, outbreak return interval is the time elapsed between two successive outbreaks at the same location whereas periodicity corresponds to the return of outbreaks in a larger time frame without taking into account their location, severity, and dispersion. At the landscape level, spruce budworm outbreak periodicity is approximately 35 years (Royama 1984; Boulanger and Arseneault 2004) although several forest sites are not affected during each outbreak. However, spruce budworm outbreak periodicity seems to be characteristic only of the last two centuries (see Morin et al., chapter 7). At the stand level, the interval between two outbreaks can vary from 30 to 120 years, depending on forest composition, stand age, and random factors (Morin 1994; MacLean 2004).

Table 2.1

**Definition of the principal descriptors related to natural disturbance regimes**

| Descriptors | Definition |
|---|---|
| **Disturbance event** | |
| Size | Size of a discrete disturbed area. This area is created by a single disturbance agent during a single disturbance event. |
| Duration | The period of time (minutes to years) from the beginning to the end of a single disturbance event. |
| Intensity | "Physical force of the event per area per time" (e.g., heat for fire, windspeed for hurricane) (White and Pickett 1985). |
| Severity | "Impact [of a single disturbance event] on the organism, community, or ecosystem" (e.g., mortality) (White and Pickett 1985). |
| **Disturbance regime** | |
| Occurrence | Number of disturbance events that occurred during a given time frame, in a given territory, regardless of their size. |
| Distribution | "Spatial distribution including relationship to geographic, topographic, environmental, and community gradients" (White and Pickett 1985). |
| Size distribution | Combination of all the disturbance event sizes for a given disturbance agent on a given territory. |
| Predictability | "A scaled inverse function of variance in the return interval" (White and Pickett 1985). |
| Interval | Time elapsed between two successive events caused by a disturbance agent within a specific site. |
| Periodicity | Time distribution for a recurring disturbance agent at regular intervals. |
| Cycle | Time required to affect a land area equivalent to the study area by a single disturbance agent. |
| Burn rate frequency | Mean proportion of area affected annually. |
| Seasonality | The time distribution of events according to the season. |
| Specificity | The selective nature of a disturbance agent towards one or several types of habitat (or towards a species). |

## 2.2.  Spatial or Configuration Descriptors

Disturbances such as fires or windthrow can be described by their size (forest area affected), their shape, and the spatial distribution of the various disturbance events. Size distribution of all disturbance events can also characterize a disturbance regime for a given land area. For example, fire size distribution of a region (from less than 1 hectare to several thousand square kilometres) provides information on the prevalence of large events compared with that of small ones, as well as the proportion of area affected by small or large fires. In general, fire size distribution follows a negative exponential function (Boychuk et al. 1997; Li et al. 1999; Wimberly et al. 2000), i.e. it is greatly extended towards large fires. Thus, although the majority of fires are small, they only contributed to a small fraction of the total area burned (Johnson et al. 1998; Stocks et al. 2002; Bergeron et al. 2004b). For example, even if 90% of all fires cover less than 50 ha in the western boreal forest of Québec, these fires contribute to less than 1% of burned areas during the last 40 years (Lefort et al. 2004). Furthermore, 37% of the burned areas resulted from fires affecting more than 15,000 ha (Kneeshaw et al. 2000). On the other hand, it is more difficult to characterize spatial and configuration features for insect outbreaks because they create extended and complex networks of defoliated areas at diverse intensity levels through time that can occasionally cover very large areas. However, the gap size distribution resulting from an outbreak has a negative exponential distribution like the one for fires (Kneeshaw and Bergeron 1998; de Römer et al. 2007).

Characteristics related to the size of disturbances provide information on the spatial configuration of the landscape such as the prevalence and maximum size of regeneration areas. For example, simulation work conducted by Belleau et al. (2007) suggests that the proportion of regeneration areas increases with the fire cycle length. In summary, Belleau et al. (2007) showed that for a given fire cycle, a fire regime dominated by large fires (which are less numerous) will increase the distance between regeneration areas compared with a fire regime dominated by small fires. Finally, the spatial distribution of all disturbance events is a descriptor that provides information on the spatial configuration of the forest landscape and on the availability of large forest tracts (see Perron et al., chapter 6).

## 2.3.  Intrinsic Descriptors: Specificity and Severity

Specificity describes the selective nature of a disturbance agent in terms of a particular habitat or species. For example, mature balsam fir stands are more vulnerable to being severely defoliated during a spruce budworm outbreak (MacLean 1980). Likewise, some tree species, such as balsam fir, are also more vulnerable to windthrow in boreal forest stands (Ruel 2000).

The last descriptor illustrated here is severity. Severity can be measured by direct impacts on living organisms or on the soil during a disturbance event. Tree mortality within a forest stand or the depth of consumed organic matter following a fire are two examples of severity indicators (Gauthier 2002). Disturbance regimes are characterized by a variable range of disturbance event severity and the severity gradient can be used to characterize the forest ecosystem response after a disturbance event. Frelich and Reich (1998) have defined four categories of disturbance severity, from low to very high (box 2.2). Fire is generally recognized as the most severe disturbance in the boreal forest as it frequently exposes mineral soil and causes widespread mortality within the canopy over large areas, which reinitiates stand succession with a return to the initial stage of a stand (i.e., first cohort, *sensu* Bergeron et al. 1999). However, fire severity is variable and some fires can have a lower impact on the soil or on the canopy, which does not result in a response commonly observed following a severe disturbance. Furthermore, the impact of fire on the soil and on the canopy is not always correlated. For example, a spring fire can have a large impact on the canopy, killing virtually all the trees, while it has a low impact on frozen soil.

When fire frequency is low within a given forest landscape, other disturbances become major factors in the disturbance regime. Insect outbreaks, windthrow, pathogens, and tree senescence can generate significant impacts but rarely cause total mortality of stands over large areas. These disturbances do not

---

### Box 2.2
## Severity categories as defined by Frelich and Reich (1998)

**Low severity**: disturbance kills only a few scattered trees, either in the overstory or understory, leaving an essentially intact forest with a few gaps (e.g., individual mortality, senescence).

**Moderate severity**: disturbance kills most of the overstory or most of the understory, leaving either a canopy layer or seedling/seedbank layer intact (e.g., surface fire that affects soil and regeneration but not overstory).

**High severity**: disturbance kills most of both the overstory and understory, leaving only a few scattered survivors and portions of the seedbank (e.g., intense crown fire).

**Very high severity**: disturbance removes virtually all individuals, including the seedbank and greatly affects the environment (e.g., glaciation, landslide, and volcanic eruption).

---

generally affect soil as fire does, and seldom reinitiate secondary stand succession. As for fire, there is a wide severity gradient caused by these disturbances, especially for insect outbreaks, that can have a severe impact on the canopy but no impact on the soil. These disturbances are often called moderate severity disturbances or canopy gap disturbances (see Kneeshaw et al., chapter 9) depending on their severity and the gap size created within the forest matrix.

Severity varies among disturbances but also within the same disturbance event. For example, a great variation of severity levels can be observed within a single burn. Living trees are dispersed in the burned area and scattered patches of intact or lightly burned forests are observed (Van Wagner 1983; Kafka et al. 2001; Schmiegelow et al. 2006). Likewise, during a spruce budworm outbreak, the resulting forest matrix shows patches with high mortality rates and patches of partially affected stands as well as intact forest areas (D'Aoust et al. 2004). Factors responsible for this severity variability are numerous. In the case of fires, it can be explained by variables such as weather, forest composition (Cumming 2001), and topography (Turner and Romme 1994; Cyr et al. 2007). During spruce budworm outbreaks, severity variability within an affected area can be in part explained by species composition, as well as by forest age because of the specificity of this disturbance agent (MacLean 1980; Bergeron et al. 1995; Su et al. 1996; MacKinnon and MacLean 2004).

## 3. DISTURBANCE REGIME AND FOREST ECOSYSTEM MANAGEMENT

Every descriptor defined in table 2.1 and in the previous section is important for the general understanding of disturbance regimes. However, they are not all relevant in the development of ecosystem management strategies based on natural disturbances as only a few of them can address the ecological issues identified in the Introduction. One way of selecting the descriptors that are more appropriate to use when defining forest management objectives is to target those that have a strong impact on key forest attributes (i.e., forest composition, forest structure, coarse woody debris, and soil organic matter; see Gauthier et al., chapter 1). It is worth noting that the disturbance regime descriptors that are not retained at a given time are not necessarily useless. A better understanding may show that some descriptors have an important, as yet underestimated, effect on forest dynamics. In this context, it is necessary to work in an adaptive forest management framework to adjust management objectives as knowledge develops.

Based on the main effects of two major disturbance agents in the forest – i.e. crown fires and spruce budworm outbreaks – we suggest that the most relevant descriptors for the development of ecosystem management strategies in regions affected by these agents are: the return interval, the severity and the size and spacing of disturbances, as well as the specificity with respect to spruce budworm outbreaks. We illustrate in this section how the variation of these descriptors has an impact on key forest attributes at different spatial scales (from stand to landscape).

## 3.1. Effects of Disturbance Descriptors on Key Forest Ecosystem Attributes

### 3.1.1. Return Interval and Annual Burn Rate

At the stand level, the return interval influences stand composition and structure, as well as coarse woody debris availability and soil organic matter. A long fire return interval can result in the transition from an even-aged stand to an uneven-aged stand dominated by shade-tolerant species (Bergeron and Dubuc 1988; De Grandpré et al. 2000; Gauthier et al. 2000). This structural transition is often associated with an increasing coarse woody debris volume (Harper et al. 2005), changes in understory plant (De Grandpré et al. 1993), bryophyte, and lichen communities (Boudreault et al. 2002), as well as changes in stand productivity (see Simard et al., chapter 11). Finally, the regional context can also influence succession within a stand. For example, in a landscape with a long fire cycle, shade-tolerant species prevail and there is a high probability of finding these species in forest patches that escaped fire (Kafka et al. 2001). They can thus reinvade a disturbed site earlier during succession than in a landscape characterized by a short fire cycle.

At the landscape level, the variation of burn rates determines age structure and influence landscape composition. For example, young forest stands prevail in a region where the burn rate is high (short fire cycle) compared with one characterized by a long fire cycle that is dominated by mature and old-growth stands (Bergeron et al. 1999). Likewise, shade-intolerant species (e.g. trembling aspen [*Populus tremuloides* Michx.] and white birch [*Betula papyrifera* Marsh.]) and fire-adapted species (e.g., jack pine [*Pinus banksiana* Lamb.]) are dominant in a forest mosaic characterized by a high burn rate (short cycle) whereas shade-tolerant species (e.g., balsam fir) dominate in a region where burn rate is low (Bergeron and Dansereau 1993; De Grandpré et al. 2003). Annual burn rate will help determine the proportion of a region to be managed under different management systems when implementing an ecosystem management framework based on natural disturbances (see Gauthier et al., chapter 3). The portion to be managed with even-aged silviculture should be reduced in landscapes characterized by low burn rates because the natural proportions of even-aged stands are historically low.

With regard to spruce budworm outbreaks, the interval between two outbreaks at a same site affects the structure and dead wood recruitment at the stand and landscape levels. However, since the cyclic interval seems to be specific to the last two centuries and is closely linked to disturbance severity and specificity (see the following section), it is currently difficult to integrate this characteristic into forest management practices. However, Côté and Bélanger (1991) found that there can be a lack of regeneration in second-growth balsam fir stands. Harvest rotations shorter than outbreak intervals could potentially reduce natural stocking in balsam fir stands. The influence of outbreak return interval seems to be less obvious than that of fire, but future knowledge could lead to management targets based on this characteristic.

## 3.1.2.  Spatial Configuration: Size and Spacing of Disturbance

Disturbance size has an impact on forest composition, structure, and coarse woody debris availability. In the case of fires, mean fire size influences the size of regeneration areas (thus, age structure and stand composition) and the spacing between them. Considering two regions of the same size and annual burn rate, the one with a lower mean fire size will have smaller regeneration areas that are closer to each other (the number of fires will be higher). On the other hand, in the region with a higher mean fire size, fires will be less frequent and farther from each other (Bergeron et al. 2004b; Belleau et al. 2007). In that case, large areas of mature forest are generated within the landscape.

At the regional level, the spatial distribution of disturbances shapes the landscape structure. In an ecosystem management framework, annual burn rate helps to determine the landscape age structure, but we must know the size distribution of disturbance events and how they are spatially distributed. As seen earlier, fire frequency is a guide for even-aged management proportion, whereas mean fire size is a guide for determining the size variability of regeneration areas that we aim to recreate within a forest management regime and their spacing across the landscape.

Fire cycle can interact with spruce budworm outbreak severity. For example, a forest dynamic driven by frequent fires maintains a composition dominated by fire-adapted species (Blais 1983; Anderson et al. 1987). On the other hand, shade-tolerant species like balsam fir – the most vulnerable species to spruce budworm – prevail in regions characterized by a long fire cycle. Thus, the size and spatial distribution of regeneration areas created by spruce budworm outbreaks are mainly determined by the severity and specificity attributes (pre-disturbance composition) of this type of disturbance. A light outbreak that generated small openings is less likely to lead to a massive recruitment of shade-intolerant species as a severe one would. On the other hand, small openings contribute to increasing horizontal structure complexity, whereas large openings tend to homogenize and simplify stand structure. Availability of coarse woody debris also increases with gap size. By using silvicultural treatments like careful logging that protects regeneration, we maintain and increase balsam fir in the understory of forest stands, which creates conditions similar to those following a spruce budworm outbreak, rather than those following fires (Baskerville 1975).

Other ecosystem management aspects with regard to spruce budworm outbreaks are found in the second part of the book (Morin et al., chapter 7; Kneeshaw et al., chapter 9; De Grandpré et al., chapter 10).

### 3.1.3.  Severity

Severity affects the majority of forest ecosystem key attributes. Fire severity in the overstory will determine the proportion of live and dead trees within the disturbed area. In the case of large fires, the variability in severity within an event results in patches of forest that escaped fire, which adds to the structural complexity of disturbed forest stands (Kafka et al. 2001). Variability in canopy mortality and duff consumption affects stand regeneration (Arseneault 2001; Jayen et al. 2006). For example, intense fire that consumes most of the duff layer influences forest composition and post-fire stand structure (Johnson 1992; Ryan 2002).

Live and dead trees (snags) are the two main residual structures following a disturbance. In contrast to fire, few residual structures remain following clearcutting. Comparing fire and clearcuts is thus essential for establishing retention targets for live and dead trees within even-aged management systems (Sougavinski and Doyon 2002; Serrouya and D'Eon 2004) as well as for salvage logging activities. Furthermore, fire severity at the soil level is not reproduced by careful logging practices. In the context of ecosystem management, soil interventions – such as scarification and prescribed burning – must be considered to maintain fire ecological processes (see Simard et al., chapter 11).

For spruce budworm outbreaks, severity and specificity (because of the selective nature of this insect towards host species) will influence forest composition and structure, as well as coarse woody debris availability. Because specific stands show higher vulnerability (e.g. mature balsam fir stands), pre-outbreak forest composition is a good predictor of the severity level and the impacts on key attributes during the outbreak. A severe outbreak generates an increased amount of coarse woody debris and promotes shade-intolerant species establishment both at the stand and landscape levels (Kneeshaw and Bergeron 1998; MacLean 2004). In the case of a moderate outbreak (i.e., in a forest mosaic where balsam fir stands are few or immature), a complex forest mosaic is created, with patches of forest showing high mortality distributed according to balsam fir stand distribution (MacLean 1980; D'Aoust et al. 2004). Following a light outbreak, the forest mosaic generally consists of stands with complex structures and composition patterns because of the gaps created throughout the landscape.

Regions characterized by long fire cycles and for which forest dynamics are driven mostly by insect outbreaks must develop forest management systems integrating treatments that maintain residual structures typical of those found in post-disturbance forest stands. For example, one can consider reserves of green patches composed of non-host species and reserves of residual trees distributed throughout managed areas, which emulates a higher severity level. Proportion and distribution of residual structures can vary according to stand age and composition (depending on vulnerability). Such practices could contribute to minimizing forest vulnerability during future insect outbreaks and reducing outbreak phases in comparison with those observed during the last century (see Morin et al., chapter 7).

## 4.   NATURAL DISTURBANCE REGIME VARIABILITY: REGIONAL AND HISTORICAL PERSPECTIVES

The main objective in establishing the natural variability of a disturbance regime is to understand how disturbances vary across sites and regions and how the underlying processes have shaped the ecosystem in the past, how they will today and in the future (Landres et al. 1999). To delimit this natural variability, two fundamental elements need to be taken into account: the time period and the geographic range considered.

### 4.1.   Historical Variability

Disturbance regimes vary greatly through time. This historical variability can be an advantage in an ecosystem management context, especially when current conditions are intermediate compared with the total range of variability (see Gauthier et al., chapter 3, where this situation is illustrated with regard to fire frequency). However, knowledge on historical variability must be available and relevant in the forest management context. To achieve that, the description and quantification of this temporal variability should be done on an appropriate time scale. To define the adequate time period – i.e. the period when current ecosystems were well adapted – we need to determine a period when species were similar, large-scale climatic changes did not occur, and human influence was negligible.

Determining historical variability requires a great amount of information on current and past disturbance regimes. Current conditions can be quantified using inventories and archive data (on fires, insect outbreaks; aerial photographs, data bases, etc.) whereas historical variability requires complex methods such as dendrochronology (see Girardin et al., chapter 4) or paleoecology (see Morin et al., chapter 7). Once the necessary information is gathered, it is possible to quantify characteristics such as fire frequency or outbreak severity during the past and to compare them with current disturbance regime characteristics.

When the time comes to establish management targets using historical variability, several elements must be considered. The range of historical variability is inevitably made up of extremes (very high severity, very large area affected, etc.). Since these events are particularly extreme and rare, it is not relevant to consider them as maximum and minimum ranges when establishing natural variability in the context of sustainable management (Cyr et al. 2009). First, extreme disturbance events will happen in the future even though there are active suppression programs and, second, human interventions of such magnitude would not be acceptable from a social perspective (see box 21.1).

Knowledge on historical variability consists of a starting point for ecosystem management because it allows the establishment of the range within which management objectives will be determined. After considering the social, ecological, and aesthetic values we want to maintain, a subset of this variability will ensure that management practices respect past conditions, i.e. those under which forest has developed.

## 4.2. Regional Variability

In addition to temporal variations, disturbance regime, for a given disturbance agent and within a similar ecosystem, will show significant regional variations. This complexity is explained by several physical (e.g., climate and topography) and biological variables (e.g., soil development, forest composition and structure). Because of these variations, natural disturbance regime characterization must be done for a region where climate is relatively homogenous. This arbitrarily defined area must be large enough to contain two to three times the largest disturbance documented (Johnson 1992). A given region will therefore be characterized by a unique disturbance regime that consists of the combined action of several disturbance agents.

In central and eastern Canada, a general pattern of decreasing burn rate is observed from Manitoba to eastern Québec (see Gauthier et al., chapter 3). Although this is a rough trend and significant intra-regional variations can be observed (see De Grandpré et al., chapter 10), this fire risk gradient involves significant variation in the prevalence of disturbance agents like insects, pathogens, windthrow, and senescence. In the following chapter, we will see how fire frequency can be a basis for ecosystem management framework implementation (Gauthier et al., chapter 3). In addition, regional examples involving different disturbance regimes, including some where fire frequency is low, are presented in part 2 (see Kneeshaw et al., chapter 9; De Grandpré et al., chapter 10), and regional implications of ecosystem management strategies suggested for these regions are discussed.

## 5. CONCLUSION

Boreal forests are complex systems because of their numerous components and their interactions. Consequently, simple and unique solutions cannot be applied universally without affecting essential functions that contribute to maintaining productivity and biodiversity. Natural disturbances are inherent processes that modify forest ecosystems and the responses observed following different levels of alteration provide reliable information. Thus, natural disturbances can be used as guides to determine the forest conditions we wish to recreate with forest management practices in order to respect ecosystem resilience.

Natural disturbance regimes are variable in time and space. Using homogenous silvicultural treatments over large forest landscapes – no matter which kind of treatment (clearcuts, partial cuts, plantations, etc.) is used – will result in an increasing risk of altering ecosystem functions. Diversifying forestry practices and basing management systems on disturbance regime and natural variability (spatial and temporal) knowledge is the way that has been widely adopted to maintain a viable forestry industry over the long term while preserving other ecological and social values.

# REFERENCES

Anderson, L., Carlson, C.E., and Wakimoto, R.H. 1987. Forest fire frequency and western spruce budworm outbreaks in western Montana. For. Ecol. Manag. **22**: 215–260.

Arseneault, D. 2001. Impact of fire behavior on postfire forest development in a homogeneous boreal landscape. Can. J. For. Res. **31**: 1367–1374.

Attiwill, P.M. 1994. The disturbance of forest ecosystems: the ecological basis for conservative management. For. Ecol. Manag. **63**: 247–300.

Baskerville, G.L. 1975. Spruce budworm: super silviculturist. For. Chron. **51**: 138–140.

Belleau, A., Bergeron, Y., Leduc, A., Gauthier, S., and Fall, A. 2007. Using spatially explicit simulations to explore size distribution and spacing of regenerating areas produced by wildfires: recommendations for designing harvest agglomerations for the Canadian boreal forest. For. Chron. **83**: 72–83.

Bergeron, Y., Cyr, D., Drever, C.R., Flannigan, M., Gauthier, S., Kneeshaw, D., Lauzon, È., Leduc, A., Le Goff, H., Lesieur, D., and Logan, K. 2006. Past, current and future fire frequencies in Quebec's commercial forests: implications for the cumulative effects of harvesting and fire on age-class structure and natural disturbance-based management. Can. J. For. Res. **36**: 2737–2744.

Bergeron, Y. and Dansereau, P.-R. 1993. Predicting the composition of Canadian southern boreal forest in different fire cycles. J. Veg. Sci. **4**: 827–832.

Bergeron, Y., Drapeau, P., Gauthier, S., and Lecomte, N. 2007. Using knowledge of natural disturbances to support sustainable forest management in the northern Clay Belt. For. Chron. **83**: 326–337.

Bergeron, Y. and Dubuc, M. 1988. Succession in the southern part of the Canadian boreal forest. Plant Ecol. **79**: 51–63.

Bergeron, Y., Engelmark, O., Harvey, B., Morin, H., and Sirois, L. (Eds.) 1998. Key issues in disturbance dynamics in boreal forests. J. Veg. Sci. **9**: 463–610.

Bergeron, Y., Flannigan, M., Gauthier, S., Leduc, A., and Lefort, P. 2004a. Past, current and future fire frequency in the Canadian boreal forest: implications for sustainable forest management. Ambio, **33**: 356–360.

Bergeron, Y., Gauthier, S., Flannigan, M., and Kafka, V. 2004b. Fire regimes at the transition between mixedwood and coniferous boreal forest in Northwestern Quebec. Ecology, **85**: 1916–1932.

Bergeron, Y., Gauthier, S., Kafka, V., Lefort, P., and Lesieur, D. 2001. Natural fire frequency for the eastern Canadian boreal forest: consequences for sustainable forestry. Can. J. For. Res. **31**: 384–391.

Bergeron, Y., Harvey, B.D., and Gauthier, S. 1999. Forest management guidelines based on natural disturbance dynamics: stand and forest level considerations. For. Chron. **75**: 49–54.

Bergeron, Y., Leduc, A., Morin, H., and Joyal, C. 1995. Balsam fir mortality following the last spruce budworm outbreak in northwestern Quebec. Can. J. For. Res. **25**: 1375–1384.

Blais, J.R. 1983. Trends in the frequency, extent, and severity of spruce budworm outbreaks in eastern Canada. Can. J. For. Res. **13**: 539–547.

Boudreault, C., Bergeron, Y., Gauthier, S., and Drapeau, P. 2002. Bryophytes and lichen communities in mature to old-growth stands in eastern boreal forests of Canada. Can. J. For. Res. **32**: 1080–1093.

Boulanger, Y. and Arseneault, D. 2004. Spruce budworm outbreaks in eastern Québec over the last 450 years. Can. J. For. Res. **34**: 1035–1043.

Boychuk, D., Perera, A.H., Ter-Mikaelian, M.T., Martell, D.L., and Li, C. 1997. Modelling the effect of spatial scale and correlated fire disturbances on forest age distribution. Ecol. Model. **95**: 145–164.

Canadian Council of Forest Ministers (CCFM). 2006. Compendium of Canadian Forestry Statistics. Canadian Council of Forest Ministers, Ottawa, Ont. [Online] <nfdp.ccfm.org/compendium/index_e.php> (accessed November 12, 2007).

Côté, S. and Bélanger, L. 1991. Variations de la régénération préétablie dans les sapinières boréales en fonction de leurs caractéristiques écologiques. Can. J. For. Res. **21**: 1779–1795.

Cumming, S.G. 2001. Forest type and wildfire in the Alberta boreal mixedwood: what do fires burn? Ecol. Appl. **11**: 97–110.

Cyr, D., Bergeron, Y., Gauthier, S., and Larouche, A.C. 2005. Are the old-growth forests of the Clay Belt part of a fire-regulated mosaic? Can. J. For. Res. **65**: 65–73.

Cyr, D., Gauthier, S., and Bergeron, Y. 2007. Scale-dependent influence of topography on fire frequency in a coniferous boreal forest of Eastern Canada. Landsc. Ecol. **22**: 1325–1339.

Cyr, D., Gauthier, S., Bergeron, Y., and Carcaillet, R. 2009. Forest management is driving the eastern North American boreal forest outside its natural range of variability. Front. Ecol. Environ. 7: doi10.1890/080088.

D'Aoust, V., Kneeshaw, D.D., and Bergeron, Y. 2004. Characterization of canopy openness before and after a spruce budworm outbreak in the southern boreal forest. Can. J. For. Res. **34**: 339–352.

De Grandpré, L., Bergeron, Y., Nguyen. T., Boudreault, C., and Grondin, P. 2003. Composition and dynamics of the understory vegetation in the boreal forests of Quebec. *In* The herbaceous layer of forests of eastern North America. *Edited by* F.S. Gilliam and M.R. Roberts. Oxford University Press, Oxford, UK, pp. 238–261.

De Grandpré, L., Gagnon, D., and Bergeron,Y. 1993. Changes in the understory of the Canadian southern boreal forest after fire. J. Veg. Sci. **4**: 803–810.

De Grandpré, L., Morissette, J., and Gauthier, S. 2000. Long-term post-fire changes in the northern boreal forest of Quebec. J. Veg. Sci. **11**: 791–800.

de Römer, A.H., Kneeshaw, D.D., and Bergeron, Y. 2007. Small gap dynamics in the southern boreal forest of Eastern Canada: do canopy gaps influence stand development? J. Veg. Sci. **18**: 815–826.

Engelmark, O., Gauthier, S., and van der Maarel, E. (Eds.) 2000. Disturbance dynamics in boreal and temperate forests: Introduction. J. Veg. Sci. **11**: 779–780.

Frelich, L.E. 2002. Forest dynamics and disturbance regimes: studies from temperate evergreen-deciduous forests. Cambridge University Press, Cambridge, UK.

Frelich, L.E. and Reich, P.B. 1998. Disturbance severity and threshold responses in the boreal forest. Conserv. Ecol. [Online] **2**: 7. <www.consecol.org/vol2/iss2/art7> (accessed September 11, 2008).

Gauthier, S. 2002. Le régime de feu au Québec. *In* Proceedings of the seminar *L'Aménagement forestier et le feu.* April 9–11, 2002, Chicoutimi, Que. *Edited by* the Canadian Forest Service, the Société pour la protection contre les feux and the Sustainable Forest Management Network. Ministère des Ressources naturelles, Québec, Que., pp. 21–26.

Gauthier, S., De Grandpré, L., and Bergeron, Y. 2000. Differences in forest composition in two boreal forest ecoregions of Quebec. J. Veg. Sci. **11**: 781–790.

Harper, J.L. 1977. Population and biology of plants. Academic Press, New York, USA.

Harper, K., Bergeron, Y., Drapeau, P., Gauthier, S., and De Grandpré, L. 2005. Structural development following fire in black spruce boreal forest. For. Ecol. Manag. **206**: 293–306.

Haeussler, S. and Kneeshaw, D.D. 2003. Comparing forest management to natural processes. *In* Towards sustainable management of the boreal forest. *Edited by* P.J. Burton, C. Messier, D.W. Smith, and W.L. Adamovicz. NRC Research Press, Ottawa, Ont., pp. 307–368.

Jayen, K., Leduc, A., and Bergeron, Y. 2006. Effect of fire severity on regeneration success in the boreal forest of northwest Québec, Canada. Écoscience, **13**: 143–151.

Johnson, E.A. 1992. Fire and vegetation dynamics: studies from the North American boreal forest. Cambridge University Press, Cambridge, UK.

Johnson, E.A. and Gutsell, S.L. 1994. Fire frequency models, methods and interpretations. Adv. Ecol. Res. **25**: 239–287.

Johnson, E.A., Miyanishi, K., and Weir, J.M.H. 1998. Wildfires in the western Canadian boreal forest: landscape patterns and ecosystem management. J. Veg. Sci. **9**: 603–610.

Kafka, V., Gauthier, S., and Bergeron, Y. 2001. Fire impacts and crowning in the boreal forest: study of a large wildfire in western Quebec. Int. J. Wildland Fire, **10**: 119–127.

Kneeshaw, D.D. 2001. Are non-fire disturbances important to boreal forest dynamics? *In* Recent research develop-ments in ecology. *Edited by* S.G. Pandalarai. Transworld Research Press, pp. 43–58.

Kneeshaw, D. and Bergeron, Y. 1998. Canopy gap characteristics and tree replacement in the southeastern boreal forest. Ecology, **79**: 783–794.

Kneeshaw, D. and Gauthier, S. 2003. Old growth in the boreal forest: a dynamic perspective at the stand and landscape level. Environ. Rev. **11**: S99-S114.

Kneeshaw, D., Messier, C., Leduc, A., Drapeau, P., Carignan, R., Paré, D., Ricard, J.-P., Gauthier, S., Doucet, R., and Greene, D. 2000. Towards an ecological forestry: A proposal for indicators of SFM inspired by natural disturbances. Sustainable Forest Management Network. Edmonton, Alta.

Landres, P.B., Morgan, P., and Swanson, F.J. 1999. Overview of the use of natural variability concepts in managing ecological systems. Ecol. Appl. **9**: 1179–1188.

Lauzon, È., Kneeshaw, D.D., and Bergeron, Y. 2007. Reconstruction of fire history (1680–2003) in Gaspesian mixedwood boreal forests of eastern Canada. For. Ecol. Manag. **244**: 41–49.

Lefort, P., Leduc, A., Gauthier, S., and Bergeron, Y. 2004. Recent fire regime (1945–1998) in the boreal forest of western Quebec. Écoscience, **11**: 433–445.

Li, C., Corns, I.G.W., and Yang, R. 1999. Fire frequency and size distribution under natural conditions: a new hypothesis. Landsc. Ecol. **14**: 533–543.

MacKinnon, W.E. and MacLean, D.A. 2004. Effect of surrounding forest and site conditions on growth reduction of balsam fir and spruce caused by spruce budworm defoliation. Can. J. For. Res. **34**: 2351–2362.

MacLean, D.A. 1980. Vulnerability of fir-spruce stands during uncontrolled spruce budworm outbreaks: a review and discussion. For. Chron. **56**: 213–221.

MacLean, D.A. 2004. Predicting forest insect disturbance regimes for use in emulating natural disturbances. *In* Emulating natural forest landscape disturbances: concepts and applications. *Edited by* A.H. Perera, L.J. Buse, and M.G. Weber. Columbia University Press, New York, USA, pp. 69–82.

McCarthy, J. 2001. Gap dynamics of forest trees: A review with particular attention to boreal forests. Environ. Rev. **9**: 1–59.

McCrae, D.J., Duchesne, L.C., Freedman, B., Lynham, T.J., and Woodley, S. 2001. Comparisons between wildfire and forest harvesting and their implications in forest management. Environ. Rev. **9**: 223–260.

Morin, H. 1994. Dynamics of balsam fir forests in relation to spruce budworm outbreaks in the boreal zone of Quebec. Can. J. For. Res. **24**: 730–741.

Reed, W.J. 2006. A note on fire frequency concepts and definitions. Can. J. For. Res. **36**: 1884–1888.

Royama, T. 1984. Population dynamics of the spruce budworm *Choristoneura fumiferana.* Ecol. Monogr. **54**: 429–462.

Ruel, J.-C. 2000. Factors influencing windthrow in balsam fir forests: from landscape studies to individual tree studies. For. Ecol. Manag. **135**: 169–178.

Ryan, K.C. 2002. Dynamic interactions between forest structure and fire behavior in boreal ecosystems. Silva Fenn. **36**: 13–39.

Schmiegelow, F.K.A., Stepnisky, D.P., Stambaugh, C.A., and Koivula, M. 2006. Reconciling salvage logging of boreal forests with a natural-disturbance management model. Conserv. Biol. **20**: 971–983.

Serrouya, R. and D'Eon, R. 2004. Variable retention forest harvesting: research synthesis and implementation guidelines. Sustainable Forest Management Network, University of Alberta, Edmonton, Alta. [Online] <www.sfmnetwork.ca/docs/e/SR_200405serrouyarvari_en.pdf> (accessed August 26, 2008).

Sougavinski, S. and Doyon, F. 2002. Variable retention: research findings, trial implementation and operational issues. Sustainable Forest Management Network, University of Alberta, Edmonton, Alta. [Online] < www.sfmnetwork.ca/docs/e/Variable%20Retention-final%20english.pdf> (accessed August 26, 2008).

Stocks, B.J., Mason, J.A., Todd, J.B., Bosch, E.M., Wotton, B.M., Amiro, B.D., Flannigan, M.D., Hirsch, K.G., Logan, K.A., Martell, D.L., and Skinner, W.R. 2002. Large forest fires in Canada, 1959–1997. J. Geophys. Res. **107**: 8149.

Su, Q., Needham, T.D., and MacLean, D.A. 1996. The influence of hardwood content on balsam fir defoliation by spruce budworm. Can. J. For. Res. **36**: 1620–1628.

Tardif, J. 2004. Fire history in the Duck Mountain Provincial Forest. Western Manitoba Sustainable Forest Management Network. Project Report 2003/2004 Series. University of Alberta, Edmonton, Alta. [Online] <www.sfmnetwork.ca/docs/e/PR_200304tardifjfire6fire.pdf> (accessed November 29, 2007).

Turner, M.G. and Romme, W.H. 1994. Landscape dynamics in crown fire ecosystems. Landsc. Ecol. **9**: 59–77.

Van Wagner, C.E. 1983. Fire behaviour in northern conifer forests and shrublands. *In* The role of fire in northern circumpolar ecosystems. *Edited by* R.W. Wein and D.A. MacLean. John Wiley and Sons, New York, USA, pp. 65–80.

Wimberly, M.C., Spies, T.A., Long C.J., and Whitlock, C. 2000. Simulating historical variability in the amount of old forest in the Oregon Coast Range. Conserv. Biol. **14**: 167–180.

White, P.S. and Pickett, S.T.A. 1985. Natural disturbance and patch dynamics: an introduction. *In* The ecology of natural disturbance and patch dynamics. *Edited by* S.T.A. Pickett and P.S. White. Academic Press, New York, USA, pp. 3–13.

Wooster, M.J. and Zhang, Y.H. 2004. Boreal forest fires burn less intensely in Russia than in North America. Geophys. Res. Lett. **31**: 1–3.

# Chapter 3

# Fire Frequency and Forest Management Based on Natural Disturbances*

*Sylvie Gauthier, Alain Leduc, Yves Bergeron, and Héloïse Le Goff*

\* We thank Anh Thu Pham for writing parts of this chapter as well as Marie-Andrée Vaillancourt for its revision and translation. Comments from Jean-Pierre Jetté, Alison Munson, and Louis De Grandpré have greatly improved this chapter. We also want to acknowledge the work of several of our students (D. Cyr, R. Drever, V. Kafka, È. Lauzon, P. Lefort, and D. Lesieur) as well as the financial contribution of the Sustainable Forest Management Network. Lastly, we thank Pamela Cheers, Benoît Arsenault, and Isabelle Lamarre of Natural Resources Canada for editing the manuscript. The photos on this page were graciously provided by Michel Robert (Canadian Wildlife Service) and Danielle Charron.

# 1. INTRODUCTION

In the first chapter of this book, several questions were raised with regard to the impact of current forest management practices. The quasi-exclusive use of clearcutting combined with a short rotation period result in more or less drastic changes in age-class distribution of forest landscapes. These changes are especially important in natural forest landscapes where fire cycles are longer than operational harvest rotations (Bergeron et al. 2001, 2006). Because forest harvesting targets mature and old-growth stands, logging activities rarefy these development stages while creating a loss or a fragmentation of large interior forest areas at the landscape level. Important changes are also anticipated at the stand level. A younger forest cover implies a decline of key attributes essential to biodiversity conservation, including the loss of snags, down woody debris, and large diameter trees (also known as wildlife trees). The important differences between managed and natural landscapes raise questions as to the potential impact of forest management on biodiversity, which in turn leads to the implementation of mitigation, protection or even restoration measures for landscapes that have been deeply modified by harvesting.

Fires play a predominant role in eastern Canadian boreal forest dynamics because of their severity and spatial distribution over large forest areas. Until recently, the idea that boreal forests were almost exclusively composed of post-fire even-aged stands was commonly held (Johnson 1992). This idea has often been used to justify the general use of clearcutting. Indeed, in the Canadian boreal forest clearcutting and careful logging practices[*] are perceived as regeneration mechanisms that are similar to fire (Natural Resources Canada 2004). However, two conditions must be met in order to replace a fire regime by an even-aged management system: 1) wildfires must be entirely suppressed to maintain a constant harvesting level through time; and 2) the patterns and processes created by fires must be reproduced using management strategies (landscape level) and silvicultural practices (stand level).

[*] The term "careful logging" is used to designate all types of cut that aim at protecting established regeneration, for example, the CPRS (cut with protection of regeneration and soil) in Québec and the CLAAG (careful logging around advanced growth) in Ontario.

Forest landscapes managed using even-aged silvicultural systems, which are widely used as fire substitutes, are different from those historically driven by fire. In this chapter, we explain why harvesting systems that are mostly based on even-aged practices are neither possible nor desirable to manage eastern Canadian boreal forests in a sustainable manner. First, we compare the age structure of natural landscapes controlled by wildfires to the age-class distribution generated by an even-aged system in order to highlight the main differences. Second, we identify potential solutions with respect to the loss of mature and old-growth forests as well as the loss of large forested landscapes. For these solutions to work, natural disturbance knowledge must be integrated into the development phase of the different management strategies. This aspect will be discussed more deeply in the following chapters.

## 2.   FOREST MANAGEMENT PRACTICES ARE CONSIDERABLY ALTERING AGE-CLASS DISTRIBUTION

The conditions created by even-aged management can be similar to those resulting from natural disturbances such as fire and insect outbreaks when the harvesting rate is similar to the natural disturbance cycle. However, the age-class distribution generated by forest management is not equivalent to that observed in natural ecosystems. In fact, the objective of even-aged management is to standardize the age structure in order to produce a constant supply of mature stands through time. In theory, a forest area managed under a 100-year rotation cycle will not include stands older than 100 years and will have an average stand age of 50 years. However, a forest area characterized by a fire cycle of 100 years will have a very different age-class distribution. This difference can be explained by the quasi-random nature of fire events: over a 100-yr period, some forest stands of a landscape may escape from forest fire and become older, while others can be affected by fire several times. If we consider that the probability to burn is independent of stand age, which is what boreal forest studies generally report (van Wagner 1978; Johnson 1992), more than 35% of all forested areas should include stands that are older than 100 years. In other words, when comparing a fire cycle with an equivalent forest rotation length, forest harvesting will cause the loss of almost all stands that are older than 100 years whereas fire will maintain a large part of these stands in the landscape.

While older stands disappear, in most cases, regeneration areas that result from logging activities add up to the ones already created by fire. Indeed, it is now generally recognized that complete fire suppression has some limits. Fire suppression is not very efficient in large forested landscapes with limited access and when fires are ignited during very dry seasons, i.e. when fire risk is high (Leduc 2002; Le Goff et al. 2005). Large fires will inevitably affect northern forest areas in the future, thus contributing to forest rejuvenation. Additionally, salvage logging practices are limited in those areas because of restricted road access and the presence of wood-boring insects that affect timber quality. Besides, as mature and old-growth stands decrease in size and numbers because of harvesting, fire will affect younger stands and will make salvage logging non-profitable (see Le Goff et al., chapter 5). In short, harvesting pressures associated with both fire and logging result in important changes in age-class distribution (loss of older age-class and strong increase in the abundance and areas of young age classes) compared with what is observed in natural forest landscapes.

## 3.   PAST AND CURRENT FIRE REGIMES IN EASTERN CANADIAN BOREAL FOREST

The difference observed between forest landscapes managed under an even-aged system and those managed under a natural fire regime is fundamental. It implies either the loss of old-growth forests that are essential to biodiversity conservation, or a reduction in allowable cuts when the forest rotation period is extended (Burton et al. 1999). This apprehension of old forests decline is well justified by the fact that many boreal forest regions comprise large proportions of forest stands that have escaped fire for longer time periods than the usual rotation period and also because there is a variety of regional contexts (Gauthier et al. 2001; Lefort et al. 2004). To provide a global idea of fire regime variations across eastern Canada, we compiled the historical fire regime reconstructions conducted over vast forest regions to obtain annual burn rates (i.e., mean annual proportion of burned area) (figure 3.1). Different periods were evaluated: the historical period, which covers at least 200 years (except for northwest Ontario), and the recent period that generally covers years 1940 to 2003 (see table 3.1 for details). With regard to the historical period, we estimated annual burn rates using the mean time since fire that was derived from the age-class distribution observed

### Figure 3.1
## Study site locations in different bioclimatic regions of eastern and central Canada

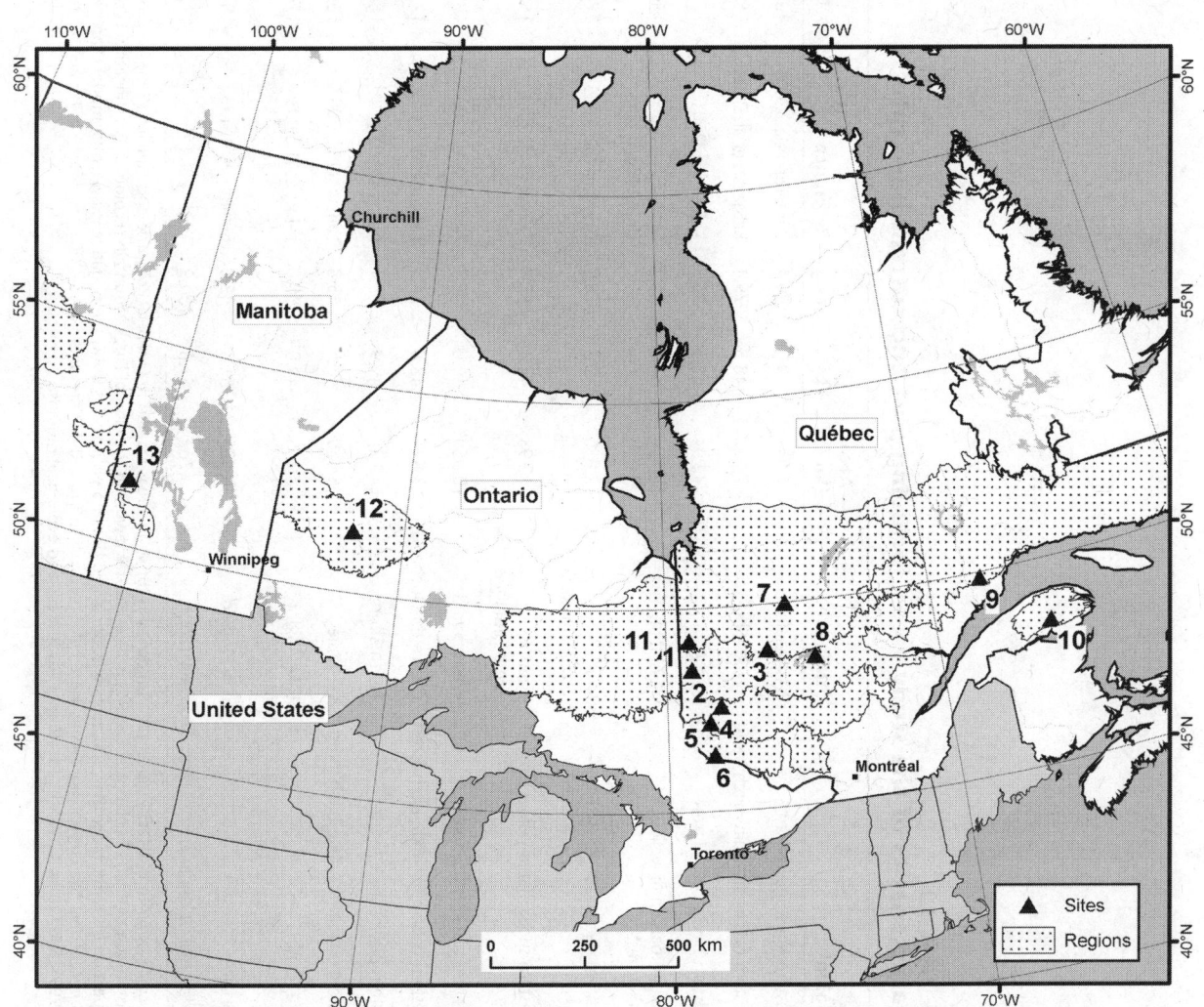

Region numbers correspond to those indicated in table 3.1. Map produced by J. Morissette, CFS.

## Table 3.1
## Characteristics of the study areas, including past and current burn rates

| Site | Region[1] | Study area | Area (km²) | Lat. | Long. | Reference | Period | Past rate | % forest > 100 yrs[2] | Current rate[3] |
|---|---|---|---|---|---|---|---|---|---|---|
| 1 | Western black spruce forest | NW Abitibi | 7,942 | 49.2 | 79.1 | Bergeron et al. 2004 | 1680–1997 | 0.528 | 60 | 0.239 |
| 2 | Western balsam fir–white birch forest | SW Abitibi | 7,777 | 48.5 | 79.1 | Bergeron et al. 2004 | 1530–1996 | 0.604 | 55 | 0.258 |
| 3 | Western black spruce forest | East Abitibi | 3,294 | 48.9 | 76.3 | Kafka et al. 2001 | 1770–1995 | 0.708 | 49 | 0.239 |
| 4 | Western balsam fir–yellow birch forest | SE Abitibi | 13,156 | 47.6 | 78.1 | Lesieur et al. Unpublished data | 1800–2004 | 0.389 | 68 | 0.048 |
| 5 | Western balsam fir–yellow birch forest | North Témiscamingue | 2,850 | 47.2 | 78.5 | Grenier et al. 2005 | 1740–2003 | 0.454 | 63 | 0.048 |
| 6 | Western sugar maple–yellow birch forest | South Témiscamingue | 1,793 | 46.4 | 78.4 | Drever et al. 2006 | 1580–2004 | 0.319 | 73 | 0.036 |
| 7 | Western black spruce forest | Waswanipi | 10,950 | 50.0 | 75.5 | Le Goff et al. 2007 | 1720–2003 | 0.781 | 46 | 0.239 |
| 8 | Western balsam fir–white birch forest | Centre-du-Québec | 3,844 | 48.6 | 74.5 | Lesieur et al. 2002 | 1720–1998 | 0.665 | 51 | 0.258 |
| 9 | Eastern black spruce forest | North Shore | 14,135 | 49.9 | 68.1 | Cyr et al. 2007 | 1640–2003 | 0.356 | 70 | 0.155 |
| 10 | Eastern balsam fir–white birch forest | Gaspésie | 6,480 | 48.5 | 65.8 | Lauzon et al. 2007 | 1680–2003 | 0.621 | 54 | 0.205 |
| 11 | Lake Abitibi | Lake Abitibi Model Forest | 8,245 | 49.0 | 80.0 | Lefort et al. 2003 | 1740–1998 | 0.580 | 56 | 0.051 |
| 12 | Cat Lake | NW Ontario | 30,625 | 51.5 | 92.0 | Suffling et al. 1982 | ~1870–1974 | 1.920 | 15 | 0.691 |
| 13 | Mid-boreal Uplands | Duck Mountain | 3,760 | 51.6 | 100.9 | Tardif 2004 | 1723–2002 | 0.787 | 45 | 0.625 |

1. Regions correspond to MRNFQ's bioclimatic subdomains for study areas located in Québec (Robitaille and Saucier 1998), to Ecological Land Classification's ecoregions in Ontario, and to an ecoregion of Canada's Ecological Land Classification Framework in the case of Duck Mountain (Ecological Stratification Working Group 1996).
2. The proportion of stands older than 100 years was calculated based on burn rates using the negative exponential model (Van Wagner 1978).
3. For sites in Québec, the current burn rates were calculated based on MRNFQ fire data (1940–2003); for sites 11, 12, and 13 rates were calculated based on the Large Fire Database (1959–1999). See Figure 3.1 for the location of study areas and ecological regions.

in these study sites. According to Gauthier et al. (2002) and Bergeron et al. (2004), mean time since fire is equivalent to the global burn rate (i.e., for the time frame analyzed and the whole region studied) and takes into account temporal variations induced by climate variations. Thus, mean time since fire includes part of the variations in burn rate caused by climate change and is more stable than the fire cycle. As for the current annual burn rates, they were calculated using record data from Québec for this province's ecological regions whereas the Large Fire Database (Stocks et al. 2002) was used for Ontario and Manitoba's ecozones.

The data presented in table 3.1 show that all regions covered by the compiled studies currently have lower annual burn rates (based on the last 40 to 60 years) compared with past burn rates (measured for the last 200 to 400 years). The majority of the 13 study areas surveyed have a proportion of old stands (≥100 years) higher than 50%. In fact, when past burn rates are low, the old-growth forest proportion is high. On the North Shore (9) and in South Témiscamingue (6), for instance, this proportion is higher than or equal to 70% whereas in other regions it exceeds 50%, except for northwestern Ontario (12) and Mid-boreal Uplands in Manitoba (13) where the proportion of old-growth forest is 15 and 45%, respectively. It is therefore obvious that the decline or disappearance of old-growth forests in these regions could pose a threat to biodiversity conservation.

## 4. CHANGES IN THE FIRE REGIME PROVIDE OPPORTUNITIES... BUT IT IS NOT ENOUGH

The lower annual burn rates currently observed in several boreal regions across eastern Canada compared with past rates (table 3.1) seem to result from climate change (see Girardin et al., chapter 4) and also in part from improved fire control strategies. Additionally, the forest landscape fragmentation resulting from agriculture and road network development has increased accessibility and created fire breaks. The recent reduction in fire frequency makes it possible to use even-aged forest management to reproduce part of the age structure associated with forests that were shaped by shorter fire intervals in the past. This situation provides some flexibility because it becomes possible to "replace" the proportion that would have burned in the past by forest areas created using even-aged management systems. However, it is worth noting that although the differences observed in burn rates justify the use of even-aged management, its exclusive use is not recommended. In fact, for the majority of the compiled studies, past burn rates are less than 1%. Such a rate (1%), which is equivalent to a rotation period of 100 years, implies that, for instance, on a 300,000-ha management unit, an average of 3,000 ha would burn annually. The observed rates, which are lower than 1%, indicate that pre-industrial landscapes were characterized by an important proportion of forests older than the rotation length (70 to 100 years) commonly used in commercial forests. This proportion is even higher if we consider that fires affect forest stands in a random fashion in contrast to forest management that systematically targets mature stands. Considering an annual burn rate of 1%, more than a third of the forest stands would be over 100 years old in a territory driven by wildfires whereas no forest stands would be this old in an entirely managed area.

Figure 3.2 shows the proportion of the study area available for even-aged management annually, taking into consideration past and current burn rates in various regions across Québec, Ontario, and Manitoba. This proportion takes into account the fact that even if the current burn rates are lower than the historical ones, fire still affects some areas every year. If we take the northwestern Abitibi region (1) as an example, the past burn rate was 0.528 (illustrated by the entire bar). This rate has now dropped to 0.238, which represents the mean amount of forest that currently burns annually. The proportion of this study region available for even-aged management aiming at emulating fire would be 0.289 (illustrated by the gray portion of the bar). For all regions represented in figure 3.2 (except for northwestern Ontario [12]), if we take into account the current burn rates the annual even-aged management objective should be modified from the current 1 to 1.4% rate to one that varies between 0.20 and 0.54%.

**Figure 3.2**
**Annual land area proportion available for even-aged harvesting (%)**

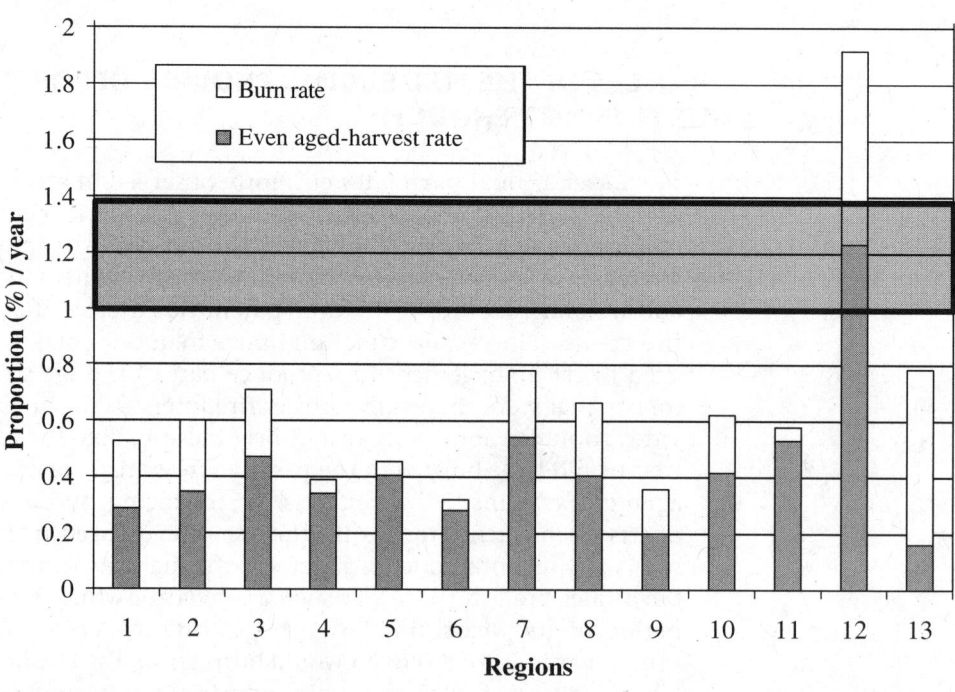

Each complete bar represents the historical proportion of forest burned annually for each region. The white portion corresponds to the current burn rate (1940–2003). The gray portion corresponds to the proportion of the area available for even-aged harvesting. The halftone rectangle represents the current even-aged harvest rate. Region numbers correspond to those indicated in table 3.1.

## 4.1. Constraints Related to the Partial Substitution of Fire by Forest Management

The possibility to replace the historical proportion of regions affected by fire with areas managed according to clearcutting practices and the impact of such a change on old-growth forest conservation must be evaluated jointly. To this end, we developed a simple model based on the comparison between past and current burn rates that takes into account the constraints related to biodiversity conservation and those related to wood production (yield constraint) (figure 3.3A). The $x$ axis of the model corresponds to the current annual burn rate. It also represents a yield constraint as fire competes with logging more intensively when the average burn rate increases (Armstrong 1999). The $y$ axis corresponds to the past burn rate. Low values indicate that a higher proportion of old-growth forest has historically been observed, thus resulting in a "biodiversity constraint" related to old-growth forests conservation. Indeed, it becomes more difficult to reproduce past age-class distributions in those areas through the sole use of even-aged management strategies, and it is therefore more difficult to conserve their associated indigenous species assemblages and stand structures.

The diagonal line shown in figure 3.3A (line 1:1) indicates cases where past and current annual burn rates are equal. The area below this line shows situations where the current burn rate is higher than the past one, which decreases the potential for even-aged management substitution. At the opposite, the area over the diagonal corresponds to situations where the current burn rate is lower than the past one, which makes it possible to replace fire with even-aged management. The dotted lines that cut the two axes at their middle correspond to an arbitrary annual harvest rate of 1%, which is roughly what is used in boreal forests. These lines separate the graphic in four quadrants representing different constraint levels:

- Northwestern Ontario is located in the zone with low constraints (upper left; table 3.1) because the past burn rate of 2% is higher than the current rate, possibly because of climate change, land-use changes, and more efficient fire suppression. In such a situation, even-aged management practices would make it possible to maintain the historical forest age structure while having few biodiversity constraints.

- The western balsam fir bioclimatic subzone of Québec is located in the lower left quadrant because the past burn rate is lower than 1% and has been decreasing (table 3.1). This situation represents a risk for biodiversity conservation should clearcutting be widely used because the past and current burn rates are lower than the desired 1%.

- The lower right zone represents the least desirable scenario because a historical burn rate lower than the current one means that the pre-industrial landscape was mainly composed of mature and old-growth forests but that climate change, land-use changes, and other factors have resulted in a higher current burn rate. This situation generates a lot of pressure on biodiversity. Additionally, logging is competing with fire in this zone, which is accelerating changes in the forest age structure.

**Simple graphic model based on the comparison between past and current burn rates illustrating biodiversity and wood production constraints (A), and comparing past and current burn rates for compiled study areas in the same model (B)**

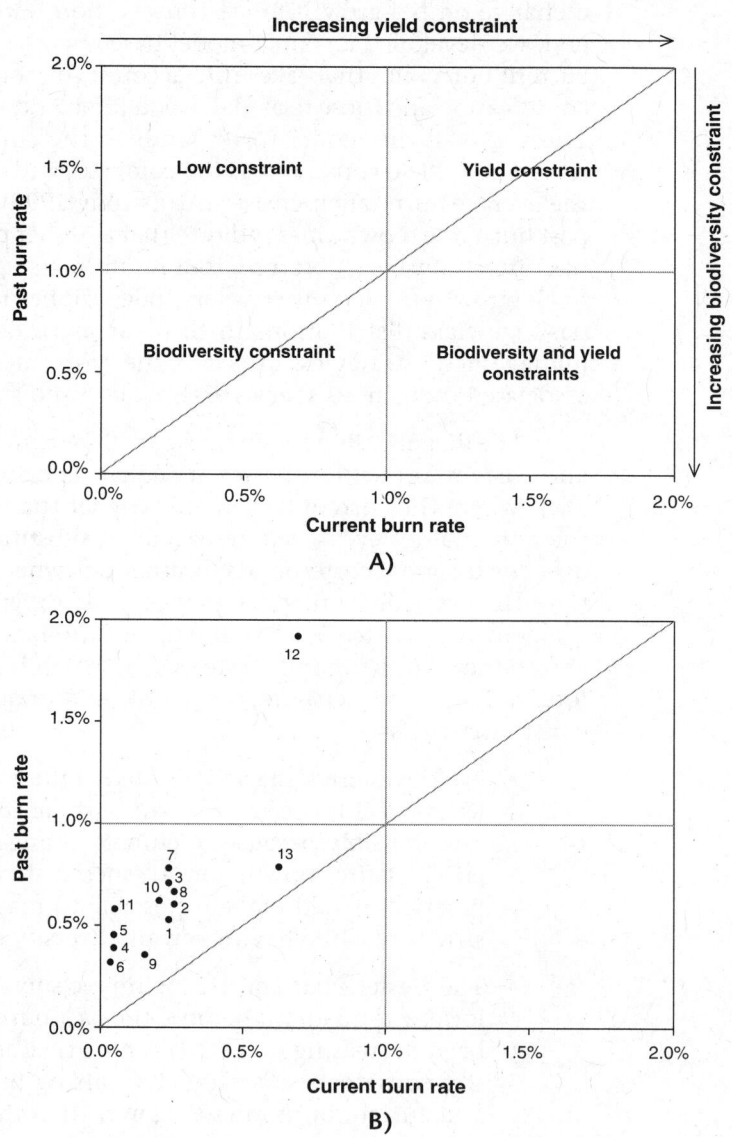

A)

B)

A) Over the 1:1 line, the current burn rate is lower than the past one; it is therefore possible to substitute fire by logging. As we go further above the 1:1 line, biodiversity constraints are replaced with production constraints. B) The majority of the regions studied have a biodiversity constraint. Region numbers correspond to those indicated in table 3.1.

- Finally, the upper right zone corresponds to a scenario of high yield constraints because the past and current burn rates are higher than the desired harvest rate (1 to 1.4%) and even-aged management is competing with fire.

All the study areas analyzed, except for the one studied by Suffling et al. (1982) in northwestern Ontario, are located in the lower left quadrant (figure 3.3B), i.e. in a biodiversity constraint scenario. For all these regions, it would not be possible to maintain the historical proportion of old-growth forests using a management system based exclusively on even-aged regeneration practices and short rotations. In those areas where forest management is limited by objectives related to the maintenance of old and irregular forests characteristics, it is essential to develop approaches that allow timber harvesting while also maintaining the structural characteristics associated with mature and old-growth forest at the stand and landscape levels.

## 5. PROPOSED SOLUTIONS: COHORT MANAGEMENT USING ADAPTED SILVICULTURAL PRACTICES

Since almost all eastern Canadian boreal regions fall in the lower left quadrant of the model presented in the previous section (figure 3.3B), the ecosystem management approach, as defined in chapter 1, prompts the adoption of a lower even-aged harvest rate. This must be done jointly with the establishment of old-growth forest conservation targets (including the use of longer forest rotations) and the development of silvicultural treatments that allow timber harvesting in mature and old-growth forests while also maintaining their associated key attributes. The data presented in the previous section makes it possible to fix targets in terms of the amount of old-growth forests (100 years old and more) that should be maintained in the region (table 3.1). This implies that young forests that have structural and compositional attributes that are similar to those of old-growth forests can also contribute to reach this objective. The proportion of old-growth forests exceeds 45% in all the regions presented here except for northwestern Ontario, where the past burn rate is considerably high compared with the other study areas.

In regions where the age structure has not yet been too altered there exist several means of minimizing the differences between a managed forest landscape and one that is under a natural fire regime. In western Canada, the suggested approach was to increase the forest rotation length (Burton et al. 1999; DeLong 2007), thus allowing stands to become older. However, because boreal species of eastern Canadian forests have a shorter longevity than those of western Canada, the general use of this approach is not desirable from an economical standpoint. Silvicultural practices aiming at maintaining old-growth stands structure characteristics in managed stands could promote biodiversity conservation while barely modifying allowable cut estimates. The Three-Cohort Model initially proposed by Bergeron and Harvey (1997) and Bergeron et al. (1999) was developed along those lines (see figure 3.4). Briefly, this concept is based on the use of various adapted silvicultural treatments aimed at maintaining the forest

Figure 3.4
## Schematic representation of the natural dynamics and associated silvicultural treatments according to the Three-Cohort Model developed for the western boreal forest of Québec (Coniferous and Mixedwood)

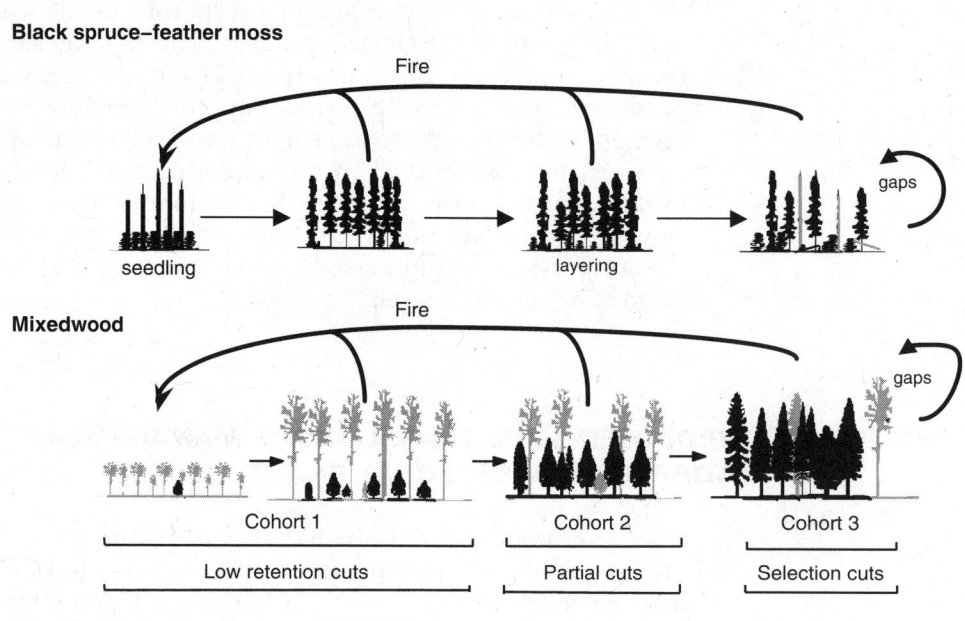

Modified from Bergeron et al. (2002).

composition and structure associated with the different development stage characteristics of the forest of the studied region. In this model, cohorts consist of forest stands at different development stages following fire. Stands of the first cohort are composed of pioneer species that colonize stands after fire and that quickly reach the canopy. Stands of the second cohort still harbour pioneer trees in the canopy, but these are being gradually replaced by other individuals that reach the canopy as first cohort trees begin to die. Finally, the third-cohort stage is reached when there are no more surviving post-fire individuals within the forest canopy.

Figure 3.4 illustrates the natural dynamics observed in the Black spruce and Balsam fir–white birch bioclimatic domains.* To obtain detailed scenarios, interested readers can consult Harvey et al. (2002) and Gauthier et al. (2004). The natural dynamics of black spruce forests can be summarized as follows:

* These forest domains are part of the Québec hierarchical ecological classification as defined by Saucier et al. 1998.

- Provided that fires are severe, especially with regard to their effects on the soil organic matter layer, and that post-fire regeneration is dominated by black spruce, the first cohort is characterized by a dense and even-aged black spruce stand stemming from seeds.

- In the absence of fire, black spruce trees reach maturity and begin dying. They are gradually replaced by a forest that is more open and that still comprises a portion of the post-fire trees. The regeneration mainly originates from layering. This stage corresponds to second-cohort stands.

- Still in the absence of fire, these stands will evolve towards an uneven-aged structure mainly sustained by layering. The stand is characterized by a complex vertical structure and the accumulation of coarse woody debris. This stage corresponds to third-cohort stands.

With respect to the Balsam fir–white birch domain, forest development is characterized by a replacement of species:

- First-cohort stands generally show an even-aged structure and are composed of shade-intolerant deciduous species with an understory regeneration composed of shade-tolerant conifer species.

- When pioneer trees reach maturity, the trees in the canopy are gradually replaced by conifers, which results in mixedwood transition stands. This stage corresponds to second-cohort stands.

- Finally, still in the absence of fire, conifer and shade-tolerant species will prevail in the forest cover, forming stands characterized by a complex vertical structure and important amounts of dead wood. This stage corresponds to third-cohort stands.

Note that in this model, cohorts are seen as groups of trees and species that will successively dominate the canopy rather than groups of stems that chronologically establish themselves in the understory.

In boreal mixedwood forests, an overabundance of first-cohort stands generated by even-aged forest management will result in an excess of deciduous species in the forest cover (Grondin et al. 2003; Boucher et al. 2006). To overcome this problem, various silvicultural treatments could be used to maintain stands that have a structure and composition similar to first-, second-, and third-cohort stages (figure 3.4). For example, using clearcutting or careful logging practices followed by seedling or planting would make it possible to maintain first-cohort stands that are generally composed of shade-intolerant species (e.g., aspen, white birch, jack pine). Another part of the forest unit could be managed using adapted silvicultural treatments that emulate the natural development (succession) of old-growth stands (second-cohort stands). The remaining part could be managed using selection logging or other treatments aimed at maintaining a continuous forest cover to simulate old-growth forest dynamics. Moreover, it is worth noting that some silvicultural treatments should be used to create a transition from first-cohort stands to second- or third-cohort stands (stand-tending treatments) whereas others can be used to maintain a stand in the same cohort (see also Bouchard, chapter 13; Harvey et al., chapter 18; Harvey et al. 2002).

Cohort proportions to be maintained in the landscape can be established as a function of past and current annual burn rates, as explained in the previous sections. The proportions of first cohorts to be maintained is easily determined because it is directly derived from burn rates (figure 3.2). Second- and third-cohort proportions are however more difficult to establish because the rate of transition between cohorts 1 and 2 and between cohorts 2 and 3 can vary depending on the region, stand composition (and understory regeneration), and site. Furthermore, natural disturbances other than fire (e.g. insect outbreaks, windthrow) can considerably affect forest landscapes in some regions. This will have important effects on second- and third-cohort stand characteristics. Under such regimes, specific adjustments with regard to silvicultural treatments may be necessary (see Kneeshaw et al., chapter 9 for an example).

Lastly, it is worth stressing that the Three-Cohort Model deals with a long-term strategic planning that makes it possible to determine the landscape proportion that will be managed under even-aged and uneven-aged systems with respect to the fire regime and other disturbances affecting the area. Various operational tactics can be adopted in order to meet the general objectives. The simplified version of cohort management presented here will be described in more detail and adapted to regional situations in several subsequent chapters (Simard et al., chapter 11; Bouchard, chapter 13; Fenton et al., chapter 15; Harvey et al., chapter 18; Belleau and Légaré, chapter 19).

## 5.1. Even-Aged Harvesting Rate vs. Burn Rate: Minimizing the Differences

In the previous section it was suggested to develop silvicultural practices that could maintain the composition and structure characteristics typical of second- and third-cohort stands. Under an even-aged management system (first cohort), it is also important to ensure that the regenerating forests maintain their quality. Both at the stand and landscape levels, even-aged treatments (such as clearcutting, careful logging, regular shelterwood cutting) and fire have different ecological effects (see McRae et al. 2001). At the landscape level, the spatial configuration of cutblocks does not reproduce the natural spatial patterns created by fire (see Perron et al., chapter 6). Such management practices can lead to landscape fragmentation and to the loss of mature and old-growth forest landscapes (DeLong and Tanner 1996). Forest management solutions that address these concerns are suggested in chapter 13 (Bouchard).

At the stand level, even-aged management can lead to the rarefaction of essential structural features for biodiversity (e.g. large-diameter trees, snags, winter shelter, and escape structures) (Franklin et al. 2000). Various retention strategies can however be used to maintain these structures within managed areas (see Drapeau et al., chapter 14). Additionally, the absence of soil disturbance in those stands during winter logging poorly reproduces fire processes (e.g. organic-layer reduction, mineral soil exposition) and can contribute to reducing productivity and have a negative impact on fire-adapted species (McRae et al. 2001; Haeussler et al. 2002). This phenomenon is particularly preoccupying for the Claybelt region in northwestern Québec and northeastern Ontario where organic matter accumulates rapidly following fire (Lecomte et al. 2006). This process cannot be reversed by logging and can lead to a decrease in forest productivity (Fenton et al. 2005; Lavoie et al. 2005; see Simard et al., chapter 11). Thus, fire and logging effects on ecological processes must be carefully assessed in order to define guidelines with respect to low-retention harvesting in the context of sustainable forest management (OMNR 2001; Bergeron et al. 2002).

## 6. CONCLUSION

Recent studies comparing past burn rates show that an important proportion of eastern Canadian pre-industrial forests were composed of mature and old-growth forests that current forest practices tend to rarefy. The fact that current annual burn rates are lower than in the past provides some flexibility by allowing the use of even-aged management practices over large areas, although this approach alone will not make it possible to maintain pre-industrial forest characteristics. The Three-Cohort Model is proposed to overcome this difficulty in order to maintain the key characteristics associated with mature and old-growth forests while also allowing timber harvesting. It is now essential and urgent to implement strategies that take into consideration regional specificities in the eastern and central Canadian boreal forests by adopting adapted silvicultural practices for territories that are mainly composed of pristine forests. Adopting this approach will certainly be less costly than having recourse to restoration strategies such as in the case of Fennoscandia (Kuuluvainen 2002). Moreover, the suggested approach remains appropriate for areas that have already been managed and could be implemented with the use of restoration strategies, especially to reintroduce mature and old-growth forests' key attributes.

## REFERENCES

Armstrong, G.W. 1999. A stochastic characterisation of the natural disturbance regime of the boreal mixedwood forest with implications for sustainable forest management. Can. J. For. Res. 29: 424–433.

Bergeron, Y., Cyr, D., Drever, C.R., Flannigan, M., Gauthier, S., Kneeshaw, D., Lauzon, È., Leduc, A., Le Goff, H., Lesieur, D., and Logan, K. 2006. Past, current and future fire frequency in Quebec's commercial forests: implications for ecosystem management. Can. J. For. Res. 36: 2737–2744.

Bergeron, Y., Gauthier, S., Flannigan, M., and Kafka, V. 2004. Fire regimes at the transition between mixedwood and coniferous boreal forest in Northwestern Quebec. Ecology, 85: 1916–1932.

Bergeron, Y., Gauthier, S., Kafka, V., Lefort, P., and Lesieur, D. 2001. Natural fire frequency for the eastern Canadian boreal forest: consequences for sustainable forestry. Can. J. For. Res. 31: 384–391.

Bergeron, Y. and Harvey, B. 1997. Basing silviculture on natural ecosystem dynamics: an approach applied to the southern boreal mixedwood forest of Quebec. For. Ecol. Manag. 92: 235–242.

Bergeron, Y., Harvey, B., Leduc, A., and Gauthier, S. 1999. Forest management guidelines based on natural disturbance dynamics: stand- and forest-level considerations. For. Chron. 75: 49–54.

Bergeron, Y., Leduc, A., Harvey, B.D., and Gauthier, S. 2002. Natural fire regime: a guide for sustainable management of the Canadian boreal forest. Silva Fenn. 36: 81–95.

Boucher, Y., Arseneault, D., and Sirois, L. 2006. Logging-induced change (1930–2002) of a preindustrial landscape at the northern range limit of northern hardwoods, eastern Canada. Can. J. For. Res. 36: 505–517.

Burton, P.J., Kneeshaw, D.D., and Coates, K.D. 1999. Managing forest harvesting to maintain old growth in boreal and subboreal forests. For. Chron. 75: 623–631.

Cyr, D., Gauthier, S., and Bergeron, Y. 2007. Scale-dependent influence of topography on fire frequency in a coniferous boreal forest of Eastern Canada. Landsc. Ecol. 22: 1325–1339.

DeLong, S.C. 2007. Implementation of natural disturbance-based management in northern British Columbia. For. Chron. 83: 338–346.

DeLong, S.C. and Tanner, D. 1996. Managing the pattern of forest harvest: lessons from wildfire. Biodivers. Conserv. 5: 1191–1205.

Drever, R., Messier, C., Bergeron, Y., and Doyon, F. 2006. Fire and canopy species composition in the Great Lakes–St. Lawrence forest of Témiscamingue, Québec. For. Ecol. Manag. 231: 27–37.

Ecological Stratification Working Group. 1996. A national ecological framework for Canada. Agriculture and Agri-Food Canada, Research Centre for Land and Biological Resources Research, and Environment Canada, State of the Environment Directorate, Ecozone Analysis Branch, Ottawa, Ont./Hull, Que.

Fenton, N., Lecomte, N., Légaré, S., and Bergeron, Y. 2005. Paludification in black spruce (*Picea mariana*) forests of eastern Canada: potential factors and management implications. For. Ecol. Manag. **213**: 151–159.

Franklin, J.F., Lindenmayer, D.B., MacMahon, J.A., McKee, A., Magnusson, J., Perry, D.A., Waide, R., and Foster, D.R., 2000. Threads of continuity: ecosystem disturbances, biological legacies and ecosystem recovery. Conserv. Biol. Pract. **1**: 8–16.

Gauthier, S., Leduc, A., Harvey, B., Bergeron, Y., and Drapeau, P. 2001. Les perturbations naturelles et la diversité écosystémique. Nat. Can. **125**: 10–17.

Gauthier, S., Lefort P., Bergeron Y., and Drapeau, P. 2002. Time since fire map, age-class distribution and forest dynamics in the Lake Abitibi Model Forest. Information Report, Laurentian Forestry Center, Québec Region, Canadian Forest Service, No. LAU-X-125E.

Gauthier, S., Nguyen, T.-X., Bergeron, Y., Leduc, A., Drapeau, P., and Grondin, P. 2004. Developing forest management strategies based on fire regimes in Northwestern Quebec, Canada. *In* Emulating natural forest landscape disturbances: concepts and applications. *Edited by* A.H. Perera, L.J. Buse, and M.G. Weber. Columbia University Press, New York, USA, pp. 219–229.

Grenier, D.J., Bergeron, Y., Kneeshaw, D., and Gauthier, S. 2005. Fire frequency for the transitional mixedwood forest of Timiskaming, Quebec, Canada. Can. J. For. Res. **35**: 656–666.

Grondin, P., Bélanger, L., Roy, V., Noël, J., and Hotte, D. 2003. Envahissement des parterres de coupe par les feuillus de lumière (enfeuillement). *In* Les enjeux de biodiversité relatifs à la composition forestière. *Edited by* P. Grondin and A. Cimon. Ministère des Ressources naturelles, de la Faune et des Parcs, Direction de la recherche forestière et Direction de l'environnement forestier. Québec, Que., pp. 131–174.

Haeussler, S., Bedford, L., Leduc, A., Bergeron, Y., and Kranabetter, J.M. 2002. Silvicultural disturbance severity and plant communities of the southern Canadian boreal forest. Silva Fenn. **36**: 307–327.

Harvey, B.D., Leduc, A., Gauthier, S., and Bergeron, Y. 2002. Stand-landscape integration in natural disturbance-based management of the southern boreal forest. For. Ecol. Manag. **155**: 369–385.

Johnson, E.A. 1992. Fire and vegetation dynamics: Studies from the North American boreal forests. Cambridge study in ecology, Cambridge University Press, Cambridge, UK.

Kafka, V., Gauthier, S., and Bergeron, Y. 2001. Fire impacts and crowning in the boreal forest: study of a large wildfire in western Quebec. Int. J. Wildland Fire, **10**: 119–127.

Kuuluvainen, T. 2002. Natural variability of forests as a reference for restoring and managing biological diversity in boreal Fennoscandia. Silva Fenn. **36**: 97–125.

Lauzon, È., Kneeshaw, D.D., and Bergeron, Y. 2007. Reconstruction of fire history (1680–2003) in Gaspesian mixedwood boreal forests of eastern Canada. For. Ecol. Manag. **244**: 41–49.

Lavoie, M., Paré, D., Fenton, N., Groot, A., and Taylor, K. 2005. Paludification and management of forested peatlands in Canada: a literature review. Environ. Rev. **13**: 21–50.

Lecomte, N., Simard, M., and Bergeron, Y. 2006. Effects of fire severity and initial tree composition on stand structural development in the coniferous boreal forest of northwestern Québec, Canada. Écoscience, **13**: 152–163.

Leduc, A. 2002. Effet de la suppression des incendies forestiers sur les régimes des feux. *In* Proceedings of the seminar *L'Aménagement forestier et le feu*, April 9–11, 2002, Chicoutimi, Que. *Edited by* the Canadian Forest Service, the Société de protection des forêts contre le feu and the Sustainable Forest Management Network. Ministère des Ressources naturelles, Québec, Que., pp. 85–90.

Lefort, P., Gauthier, S., and Bergeron, Y. 2003. The influence of fire weather and land use on the fire activity of the Lake Abitibi area, Eastern Canada. For. Sci. **49**: 509–521.

Lefort, P., Leduc, A., Gauthier, S., and Bergeron, Y. 2004. Recent fire regime (1945–1998) in the boreal forest of western Quebec. Écoscience, **11**: 433–445.

Le Goff, H., Flannigan, M.D., Bergeron, Y., and Girardin, M.P. 2007. Historical fire regime shifts related to climate teleconnections in the Waswanipi area, central Quebec, Canada. Int. J. Wildland Fire, **16**: 607–618.

Le Goff, H., Leduc, A., Bergeron, Y., and Flannigan, M. 2005. The adaptive capacity of forest management to changing fire regimes in the boreal forest of Quebec. For. Chron. **81**: 582–592.

Lesieur, D., Gauthier, S., and Bergeron, Y. 2002. Fire frequency and vegetation dynamics for the south-central boreal forest of Quebec, Canada. Can. J. For. Res. **32**: 1996–2009.

McRae, D.J., Duchesne, L.C., Freedman, B., Lynham, T.J., and Woodley, S. 2001. Comparisons between wildfire and forest harvesting and their implications in forest management. Environ. Rev. **9**: 223–260.

Natural Resources Canada. 2004. The state of Canada's forests 2003–2004. Ottawa, Ont.

Ontario Ministry of Natural Resources (OMNR). 2001. Forest management guide for natural disturbance pattern emulation, Version 3.1. Queen's Printer for Ontario, Toronto, Ont.

Robitaille, A. and Saucier, J.-P. 1998. Paysages régionaux du Québec méridional. Les publications du Québec, Sainte-Foy, Que.

Stocks, B.J., Mason, J.A., Todd, J.B., Bosch, E.M., Wotton, B.M., Amiro, B.D., Flannigan, M.D, Hirsch, K.G., Logan, K.A., Martell, D.L., and Skinner, W.R. 2002. Large forest fires in Canada, 1959–1997. J. Geophys. **108**: D1: FFR5, 1–12.

Suffling, R., Smith, B., and Dal Molin, J. 1982. Estimating past forest age distributions and disturbance rates in North-Western Ontario: a demographic approach. J. Environ. Manag. **14**: 45–56.

Tardif, J. 2004. Fire history in the Duck Mountain Provincial Forest. Western Manitoba Sustainable Forest Management Network. Project Report 2003/2004 Series. University of Alberta, Edmonton, Atla. [Online] <www.sfmnetwork.ca/docs/e/PR_200304tardifjfire6fire.pdf> (accessed November 29, 2007).

Van Wagner, C.E. 1978. Age-class distribution of the forest fire cycle. Can. J. For. Res. **8**: 220–227.

# Part 2

## Spatio-Temporal Variations of Disturbance Regimes

Marie-Andrée Vaillancourt,
Louis De Grandpré,
and Sylvie Gauthier

Over the past few years, the understanding of natural dynamics with regard to major disturbance regimes affecting the boreal forest – especially fire and insect outbreaks – has greatly increased. Impacts and landscape structures created by these disturbance agents can guide the development of alternative forest management strategies aimed at minimizing the differences observed between natural and managed landscapes. The first chapters of this book have illustrated how natural disturbances can serve as the basis of forest ecosystem management development and implementation. This can be done by using knowledge of main characteristics of disturbance regimes (e.g., return interval, severity, and size) and their impacts on forest conditions. Thus, establishing forest ecosystem management targets that are ecologically relevant requires a good knowledge of the conditions created by major disturbance regimes affecting the ecosystem and also of their regional and historical variability (chapters 2 and 3).

## MAJOR DISTURBANCE REGIMES IN BOREAL FOREST

The first chapters of this second part focus on the variability of three important disturbance agents in the boreal forest: fire, spruce budworm (*Choristoneura fumiferana* Clem.), and forest tent caterpillar (*Malacosoma disstria* Hübner). Disturbance regime variability is caused by numerous factors. In the case of fire regimes, climate plays an important role at both the regional and temporal levels (chapter 4). In the context of ecosystem management, knowledge of the proximal link between climate and the fire frequency leads to the possibility of predicting future variability under climate change and adapting forest management practices in order to cope with the changes to come (chapter 5). Characteristics that are linked to fire regime variability for a given region will influence spatial distribution of regenerating areas and residual forests that escaped fire (chapter 6). Thus, by comparing spatial distribution of regenerating areas under a fire regime and a forest management regime, it is possible to define ecosystem management objectives that aim to reduce these differences and thereby minimize potential threats on biodiversity.

Insect outbreak regimes in eastern Canada generally have a specific impact that is less severe than fire but can affect very large tracts of forests. Two chapters in this section will present a portrait of current knowledge with regard to the characteristics and the regional and historical variability of spruce budworm regimes (chapter 7) and forest tent caterpillar regimes (chapter 8), two major defoliator insects in the eastern boreal forest of Canada that affect respectively softwood and hardwood species.

## DISTURBANCE REGIMES AND FOREST DYNAMICS: REGIONAL EXAMPLES

The second half of this part of the book will present regional examples from various localities of eastern and central Canada to illustrate the role of disturbance regimes in shaping forest dynamics and regional specificities. Their disturbance regimes vary in terms of disturbance agent preponderance and variability, which inevitably generate specific forest dynamics characteristics, which involves adopting different management strategies and targets in the context of ecosystem management.

The general pattern of fire cycle lenghtening from west to east in eastern and central Canada (see chapter 3) allows us to illustrate the resulting forest dynamics. In Gaspésie and the North Shore, two regions of eastern Québec where fires are not frequent, we observe a prevalence of "smaller-scale" disturbances such as spruce budworm, windthrow, pathogens, and senescence (chapters 9 and 10). The Clay Belt region of Québec, which lies in the boreal forest in northeastern Ontario and northwestern Québec, is characterized by a specific soil problem. Paludification, which consist of an accumulation of organic matter in forest stands through time, is a common process in this region and is part of a reduction in forest productivity (chapter 11). Finally, the Duck Mountain region in Manitoba was characterized by a short fire cycle in the past that has considerably lengthened during the last century. This has generated changes in forest dynamics and has consequences for the timber supply, especially for companies that use mainly hardwood species (chapter 12).

Throughout this part, avenues for the development of ecosystem management strategies that meet the various challenges caused by specific forest dynamics are presented.

# Chapter 4

# Climate, Weather, and Forest Fires*

*Martin P. Girardin, Mike D. Flannigan, Jacques C. Tardif, and Yves Bergeron*

\* We thank the Sustainable Forest Management Network, the Fonds québécois de la recherche sur la nature et les technologies (FQRNT), and the Canadian Forest Service for funding the work done by Mr. Girardin. We also thank Pamela Cheers and Isabelle Lamarre of Natural Resources Canada for editing the manuscript. The photos on this page were graciously provided by Brian Stocks (Canadian Forest Service) and Danielle Charron.

# 1. INTRODUCTION

The past decades has seen an increasing interest in forest management based on historical or natural disturbance dynamics. The rationale is that management that favours landscape compositions and stand structures similar to those found historically should also maintain biodiversity and essential ecological functions (see Gauthier et al., chapter 1). For instance, in fire-dominated landscapes, a substitution of fire by even-aged forest management could occur without elevating the overall frequency of disturbance or affecting forest ecosystem functioning. This approach is feasible only if current and future fire activities are sufficiently low compared with pre-industrial fire activity (see Gauthier et al., chapter 3; Le Goff et al., chapter 5). In the advent of greater fire activity, other adaptation options exist to assist forest management to cope with new constraints and their consequences (see Le Goff et al., chapter 5). Reconstructing past, understanding current, and forecasting future fire-conducive climate and fire activities is thus of central importance for effective implementation of ecosystem management. The development and implementation of strategies appropriate for ecosystem management require a thorough understanding of the relationship between climate and disturbances, their features (past and present) and a better anticipation of their future development, particularly with respect to forest fires.

Wildfire is a primary natural process organizing the physical and biological attributes of the forest, shaping landscape diversity and influencing biogeochemical cycles (Weber and Flannigan 1997; Bourgeau-Chavez et al. 2000). The mosaics of different vegetation types are to a large extent an expression of their respective fire regimes and many boreal tree species show adaptation to fire. Fire activity responds rapidly to changes in weather and climate in comparison to vegetation – the rate and magnitude of fire-regime-induced changes to the boreal forest landscape can greatly exceed anything expected from climate change alone (Weber and Flannigan 1997).

Weather consists of short-term (minutes to days) variations in the atmosphere. Usually, weather is thought of in terms of temperature, humidity, precipitation, cloudiness, visibility, and wind (American Meteorological Society 2000). Weather influences daily fire characteristics because of its impact on fuel moisture and the effects of precipitation (particularly its frequency), relative humidity, air temperature, wind speed, and lightning (Flannigan and Harrington 1988; Flannigan and Van Wagner 1991; Harrington et al. 1991; Johnson 1992; Agee 1997; Weber and Flannigan 1997; Bergeron et al. 2001). Weather is particularly important as a control on the occurrence of forest fire; fires spread rapidly when the fuels are dry and the weather conditions are warm, dry, and windy. Despite the increasing importance of human activity as a source of fire ignition over the last few decades (Stocks et al. 2003), dry forest fuels and wind remain critical factors influencing the occurrence of large stand-replacing fires (e.g. Johnson et al. 1990; Masters 1990; Johnson 1992; Westerling et al. 2006). Nearly 81% of all burned area in Canada is caused by lightning-ignited forest fires.

As distinguished from weather, climate is typically characterized in terms of suitable averages of the climate system over periods of a month or more, taking into consideration the variability in time of these averaged quantities (American Meteorological Society 2000). Climate is not constant, but rather consists of warming and cooling cycles at intervals of several decades. As climate varies, the corresponding weather variables can vary in magnitude and direction. As such, climate change can be defined as any systematic change in the long-term statistics (average or variability) of climate elements (such as temperature, pressure, and winds) sustained over several decades or longer (American Meteorological Society 2000). Climate change may be due to 1) natural external forcing, such as changes in solar output or slow changes in the Earth's orbital elements, 2) natural internal processes of the climate system, or 3) anthropogenic forcing (American Meteorological Society 2000; IPCC 2001; Meehl et al. 2003; Scafetta and West 2006). For instance, the net progressive warming of the Northern Hemisphere after the 1850s, the cooler-than-average temperatures from the late 1930s to mid-1960s and the steady temperature increase since 1970 are indications of the existence of a dynamic climate system (Mann et al. 1998; IPCC 2001; Cook et al. 2004; Esper et al. 2004; Moberg et al. 2005; Smith and Reynolds 2005).

With a dynamic climate and the strong linkage between climate, weather, and forest fires, variations in historical observations of fire activity due to changes in the climate are expected (Flannigan and Harrington 1988; Johnson 1992; Swetnam 1993; Carcaillet et al. 2001; Gillett et al. 2004; Flannigan et al. 2005; Girardin et al. 2006a; Girardin 2007). In the past four decades or so, significant progress has been made in characterizing the fire-weather and fire-conducive climate variability in Canada. These advances include the development of a Fire-Weather Index system (FWI), which allows monitoring fire weather across Canada on a daily basis (Van Wagner 1987). Additionally, the gathering of fire statistics has led to a number of fire studies relating numerous fires to climate variability over many years (e.g., Flannigan and Harrington 1988; Skinner et al. 1999, 2002, 2006; Flannigan et al. 2005; Girardin 2007). These fire studies have spatial domain ranges from hundreds of square kilometres to continental scale and a temporal range of 10 years to about 100 years. The temporal scale in fire studies is often limited by the availability of fire statistics and meteorological data. In this matter, much work has been done on the long-term estimation of past fire-conducive climate variability and fire activity variability from tree rings (e.g., Girardin 2007; Girardin et al. 2004a, 2006a, 2006b), time-since-fire maps (e.g., Van Wagner 1978; Johnson and Larsen 1991; Larsen 1997; Weir et al. 2000; Bergeron et al. 2001, 2004; Tardif 2004), proxy climate records (Westerling and Swetnam 2003; Girardin et al. 2006c), and charcoal in lake sediments (e.g. Gajewski et al. 1993; Larsen and MacDonald 1998; Carcaillet et al. 2001). These studies were conducted to increase our understanding of the link fluctuations in climate and fire activity over centuries at local, regional, and national scales (Campbell and Flannigan 2000).

The objective of this chapter is to highlight the connection between climate, weather, and forest fires. We have used examples primarily from eastern boreal Canada to illustrate our points. The chapter is divided into sections that describe the methodological approaches for monitoring fire-conducive weather, including ground and upper-atmospheric features. The coupling of fire-weather conditions to global ocean temperatures is briefly presented. We also try to identify key knowledge gaps in the fire and weather/climate relationship, with particular emphasis on the limited temporal coverage of fire statistics and climate recordings. The chapter closes with a discussion about the contribution of tree-ring-based drought and fire activity reconstructions and time-since-last-fire maps to fill these gaps.

## 2.  MONITORING FIRE WEATHER

Wildfire occurrence and spread follow the day-to-day variations in weather and are often the result of complex interactions between precipitation, temperature, humidity, solar radiation, ignition agents, and wind (Van Wagner 1987; Flannigan and Harrington 1988). Also, these features depend largely on the horizontal and vertical state of the atmosphere. The strength, location, and movement of surface anticyclones (high-pressure systems) and cyclones (low-pressure systems) and associated warm/cold fronts are functions of the three-dimensional atmosphere. Thus surface weather conditions at the fire site are greatly influenced by upper air features. Additionally, many large forest fires are not constrained to the ground surface in that these fires may have convection columns that extend many kilometres into the atmosphere (a convection column is the thermally produced, ascending column of gases, smoke, and debris produced by a fire). The interaction between the convection column and the vertical structure of the atmosphere can have a significant impact on fire behaviour and fire growth (Jenkins et al. 2001).

Although this chapter has a strong focus on weather and climate, it is important to emphasize that several other factors can influence fire activity, notably ignition sources (human or lightning), fuel characteristics and vegetation (Hély et al. 2000; Krawchuk et al. 2006), land use (Westerling et al. 2006), topography, and other physiographic features (Turner and Romme 1994; Cyr et al. 2007). Discussion of these factors is beyond the scope of this chapter.

### 2.1.  Surface Weather

In the 1920s, J.G. Wright began research on fire danger rating systems by tracking day-to-day susceptibility of the forest to fire (Van Wagner 1987). In the following decades, four different fire danger systems were developed for various regions of Canada, with each version based on field research on the fuel types of local importance. During the late 1960s there was an increasing demand by forest fire control agencies for the development of a new fire danger rating index. The result was called the Canadian Forest Fire Weather Index (FWI) system (see box 4.1). This system retained a solid link with previous systems by building on their best features and adding new components where necessary (Van Wagner 1987). Today, the FWI system is used daily across Canada by fire management agencies to monitor forest fire danger and expansion of the monitoring system has begun in other countries (Fire Ecology Research Group 2005). Forest fire researchers from Canada, Russia, and Germany have recently developed methodologies for electronically gathering daily weather data and producing daily fire weather and fire behaviour potential maps for large portions of northern Europe and northern Asia.

The remainder of this chapter will often refer to the Drought Code (DC) component, which is the "seasonal" drought component of the FWI system. The DC was specifically developed to serve as an index of the water stored in the soil, on average about 20 cm deep, and to warn when lower layers of deep duff may be drier than the upper ones. At this soil depth, drought is a determining factor for forest fire severity because it allows deep burn and smouldering. Therefore, via its influence on fire severity, it becomes an important controlling factor of post-fire ecosystem structure and function through a direct impact on underground plant roots, reproductive tissues, and soil seed banks (Weber and Flannigan 1997).

---

### Box 4.1
## Description of fire weather indices and the Fire Weather Index (FWI) system

The FWI system consists of six components that account for the effects of fuel moisture and wind on fire behaviour (see figure 4.1). The first three components are the fuel moisture codes:

◆ Fine Fuel Moisture Code represents the numeric ratings of the moisture content of litter and other fine fuels in a forest stand, in a layer of dry weight ~0.25 kg m$^{-2}$ (time constant about 2/3 days). It is an indicator of sustained flaming ignition and fire spread.

◆ Duff Moisture Code represents the average moisture content of loosely compacted, decomposing organic layers of moderate depth weighing ~5 kg m$^{-2}$ when dry (time constant about 12 days). It relates to the probability of lightning ignition and fuel consumption.

◆ Drought Code (DC) represents the average moisture content of deep, compact organic layers (about 10 to 25 cm from surface) weighing ~25 kg m$^{-2}$ when dry (time constant about 52 days). It relates to the consumption of heavier fuels and the effort required to extinguish a fire.

The remaining three components are fire behaviour indices and are computed from the preceding indices and wind velocity. Their values rise as the fire danger increases:

◆ Initial Spread Index combines the effects of wind and the Fine Fuel Moisture Code. It is a numerical rating of fire spread.

◆ Build-Up Index combines the Duff Moisture Code and the DC, and measures the fuel available for combustion.

◆ Fire Weather Index combines the Initial Spread Index and the Buildup Index. It is a numeric rating of frontal fire intensity.

Many drought indices rely on the same principle and basic mathematical functions, but substantially vary in terms of water-holding capacity, drying rates, and weather input. This is the case in the FWI components relative to other numerical drought indicators used in hydrology and agriculture, such as the Crop Moisture Index (Palmer 1968) and the Palmer Drought Severity Index (Palmer 1965). The FWI system, by being particularly sensitive to daily weather, is more realistic for determining the fire danger because it is the distribution of rainfall events, rather than the total amount, that is important when monitoring fire weather conditions (Flannigan and Harrington 1988).

### Figure 4.1
## Components of the Fire Weather Index (FWI) system

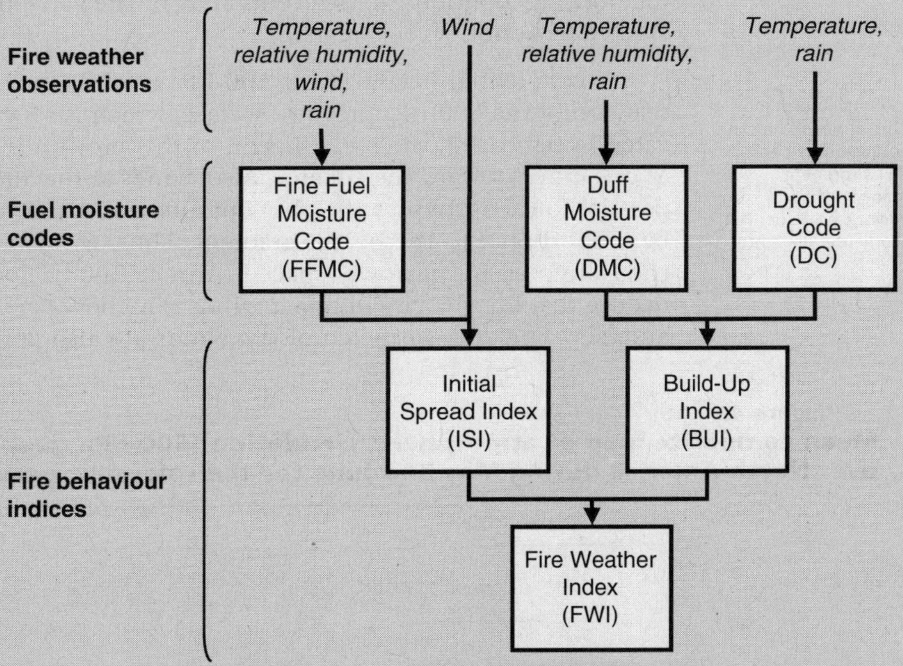

**Fire weather observations**

*Temperature, relative humidity, wind, rain* — *Wind* — *Temperature, relative humidity, rain* — *Temperature, rain*

**Fuel moisture codes**

Fine Fuel Moisture Code (FFMC) — Duff Moisture Code (DMC) — Drought Code (DC)

**Fire behaviour indices**

Initial Spread Index (ISI) — Build-Up Index (BUI)

Fire Weather Index (FWI)

Calculation of the components is based on consecutive daily observations of temperature, relative humidity, wind speed, and 24-hour rainfall. The six standard components provide numeric ratings of relative potential for wildland fire. Modified from Van Wagner (1987).

## 2.2. Fire and the Upper Air Level

Atmospheric circulation plays a key role in creating drought conditions that are conducive to fire and can also act as a determining factor for fire ignition. Area burned relates to upper air features in two ways: atmospheric circulation at 500 hectopascals (hPa; also often expressed in millibars [mb]) and the vertical structure of the atmosphere.

### 2.2.1. Atmospheric Circulation

Large forest fires in boreal Canada are associated with prolonged blocking high-pressure systems in the upper atmosphere over or upstream from the affected regions (Skinner et al. 1999, 2002). These systems are defined by the American Meteorological Society (2000) as "anomalous" circulation patterns that typically remain nearly stationary or move slowly westward and persist for a week or more. The relationship to atmospheric circulation is most often expressed with 500-hPa geopotential height composite and correlation charts (figure 4.2). Composite charts are often used in synoptic climatology to reveal wind-flow patterns during signature years (for instance, during years of high fire activity, figure 4.3). Correlation charts are produced to represent the typical relationship between temporal distributions in fire events and climate variability for the entire period studied (covering several years).

Geopotential height at the 500-hPa level approximates the height above sea level of the 500-hPa pressure surface, which is roughly at 5.5 km.[*] As indicated in figure 4.2, the mean height of this pressure surface tends to increase as one moves toward the equator. Also, winds at the 500-hPa level tend to flow parallel to the isohypses (height contours) and their velocity (speed) tends to be proportional to the height gradient. Thus, wind velocity is greatest where the isohypses are narrowly spaced (Ahrens 2003). Looking at figure 4.2, we deduce that the velocity of upper winds is higher over the North Pacific Ocean and New England. In meteorology, winds are also defined according to their

• Additional information on the atmosphere can be obtained by consulting the Environment Canada Web site: <www.qc.ec.gc.ca/Meteo/ Documentation_e.htm>.

### Figure 4.2
## Mean composite map of atmospheric circulation (500-hPa geopotential height) over North America during May and June for the reference period 1968–1996

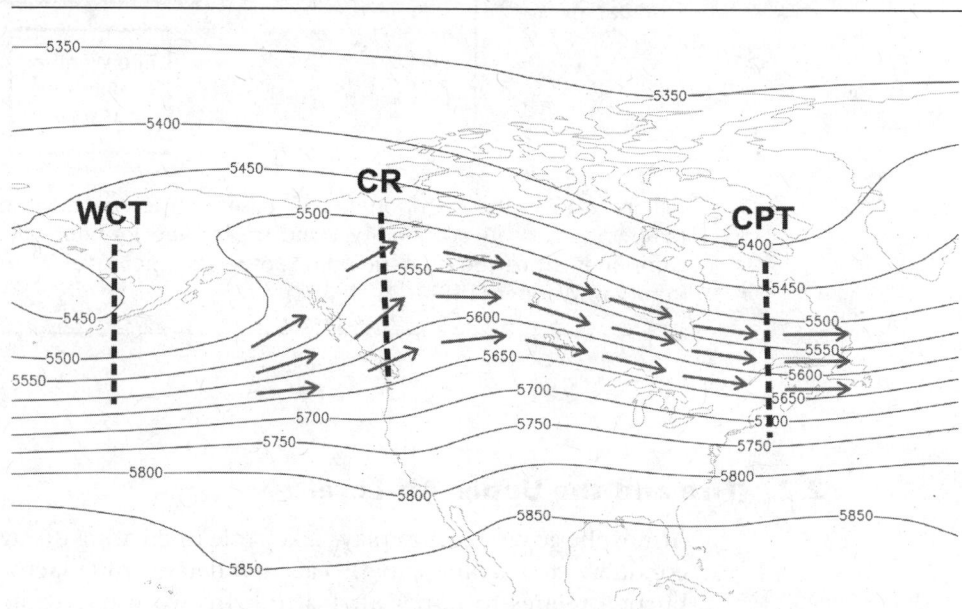

⟶ Direction of winds in the upper atmosphere

The thin lines are 50-m contour intervals; the 500-hPa pressure surface is thus at a higher altitude closer to the equator. Note the long wave pattern from west to east, and the direction of major westerlies indicated with arrows. Thick vertical lines indicate the location of the main features affecting the North American climate. From west to east these are the West Coast Trough (WCT), the Continental Ridge (CR), and the Canadian Polar Trough (CPT). This map was obtained from the Climate Explorer of the Royal Netherlands Meteorological Institute (KNMI) (<climexp.knmi.nl>) using the NCEP/ NCAR reanalysis data (Kalnay et al. 1996). Daily composite maps in figure 4.3 should be compared with this climatology.

flow. The winds are called meridional when the isohypses form a strong wave-like pattern, and zonal when the isohypses are nearly parallel to the lines of latitude (Ahrens 2003).

The term "long-wave" is used with respect to atmospheric circulation to denote northward or southward displacements in the major belt of westerlies characterized by large wavelength and significant amplitude, also known as planetary or Rossby waves. Typically, 3 to 5 long waves can be found encircling the Northern Hemisphere at any given time. Within these waves, an elongated area of positive height anomalies is known as a ridge and an elongated area of negative height anomalies is a trough (figure 4.2). The mean features of the upper atmospheric circulation over Canada include the presence of troughs located over the North Pacific (West Coast Trough, WCT) and northeastern Canada (Canadian Polar Trough, CPT) and a ridge across the northeastern Pacific to the West Coast (Continental Ridge, CR) (figure 4.2). It is during strong meridionality (i.e., increased wave-like pattern) that obstruction of the normal west-to-east movement of migratory storms occurs. The ridges cause air subsidence in the upper atmosphere resulting in typically sunny, warm days that create dry fuel conditions that can extend over several hundreds of kilometres (Newark 1975; Johnson and Wowchuk 1993; Bessie and Johnson 1995; Skinner et al. 1999, 2002). The breakdown of these upper ridges is often accompanied by convective activity leading to numerous lightning strikes, as much shorter waves or a cold front move along the west side of the ridge. Additionally, as the ridge breaks down, strong and gusty surface winds are common.

The example shown in figures 4.3 and 4.4 illustrates the relationship between upper-level ridging and weather at the surface conducive to wildfire. It is the daily changes in weather conditions that led to the 2005 fire season in Québec. This fire season ranked about fourth in importance since 1922 and the largest since that of 1941 (reference period: 1922 to 2005; MRNFQ 2006). After high indices of fire weather severity in northern Québec and several days (May 23 to 28) under blocking ridges, more than 38,000 thunderbolts struck part of Québec from May 30 to June 6, igniting 141 fires (MRNFQ 2006). This period of intense lightning activity occurred after the breakdown of the ridge (May 29 and 30; figure 4.3) with the passage of a short wave in the upper atmosphere. The sectors most affected by the fires (figure 4.4) were between Matagami and the Manicouagan Reservoir (northeast corner of the map), more specifically north of Lac Saint-Jean. The critical period of fire activity went beyond June 6 and continued until June 19. The 141 fires burned 345,679 hectares of forest (MRNFQ 2006).

## 2.2.2.  Vertical Structure of the Atmosphere

Wildfires may be either convection-column driven or wind driven (Nelson and Adkins 1988; Nelson 1993). Fires for which the convection column is well developed are more sensitive to the atmospheric decrease in temperature with height (lapse rate) because these columns extend further up into the atmosphere. Conversely, wind-driven fires are influenced by strong winds at the surface and typically have a rapid change of speed and/or direction in any direction, which is called vertical wind shear. Typically there is a significant increase in wind speed with height above the ground as friction reduces the wind speed near the Earth's surface. Under such conditions, the convection column is sheared and as such does not play a significant role in fire spread and behaviour.

## Figure 4.3
## Daily 500-hPa composite maps

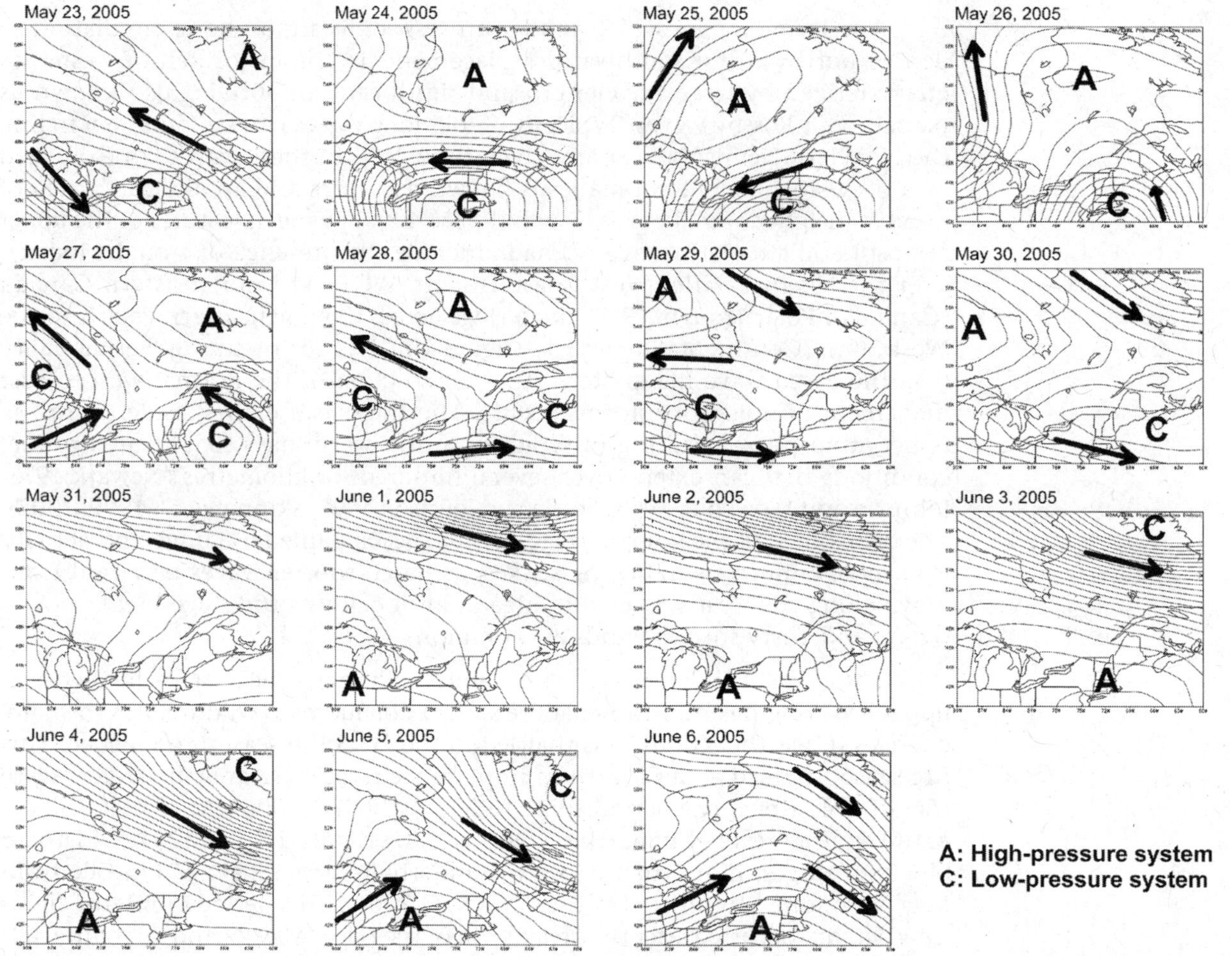

The maps show the movement of weather patterns over eastern Ontario and Québec from May 23 to June 6, 2005. Direction of wind vectors of magnitude greater than 12 m/s is indicated with arrows. Compare these maps with the average climatology in figure 4.2. Weather conditions at the surface are analyzed in figure 4.4. Images were provided by the NOAA-CIRES Climate Diagnostics Center, Boulder, Colorado (<www.cdc.noaa.gov>) from the NCEP/NCAR reanalysis data (Kalnay et al. 1996).

**A: High-pressure system**
**C: Low-pressure system**

• Fire whirls are spinning vortex column of ascending hot air rising from a fire and having the structure and behaviour of a tornado, carrying aloft smoke, debris, and flame over several meters high.

Additionally, there is a dynamic interaction between the fire and the atmosphere that can lead to some erratic fire behaviour (Jenkins et al. 2001) such as fire whirls, horizontal roll vortices, spotting (by transport of sparks or embers) and the development of pyrocumulus clouds including thunderstorms generated by the fire (Stocks and Flannigan 1987; Jenkins et al. 2001). In the absence of strong winds, dry and unstable air promotes a well-developed convection column, which may produce spotting and fire whirls.• When wind speeds are strong near the Earth's surface, the instability created by convection and turbulence due to friction allows these high winds to be mixed with those at the surface. This mixed layer promotes fire spread and horizontal roll vortices (Haines et al. 1983).

Figure 4.4
**Maps of hot spots and daily severity of Fire Weather Index (FWI)**

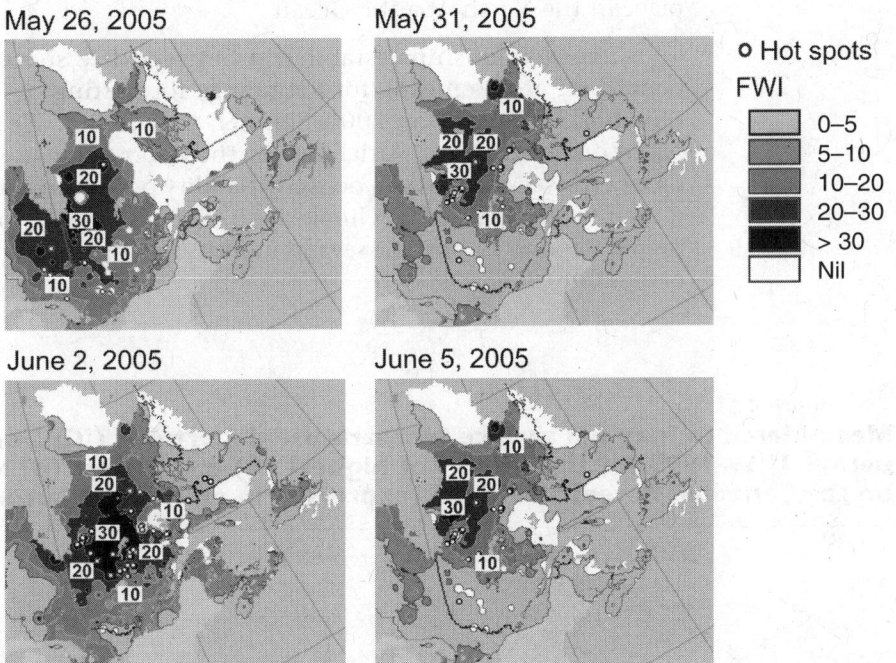

The FWI intervals of 10, 20, and 30 units are given (a high index level is an indication of conditions conducive to fire). A hot spot is a pixel of a satellite image with an infrared intensity typical of the vegetation being burned. The maps were obtained from Natural Resources Canada Web site (<fire.cfs.nrcan.gc.ca>).

## 2.2.3.   Influence of the Oceans

It is now becoming clearer that the oceans exert a significant influence on fire activity in Canada through their effect on large-scale atmospheric circulation. The most pronounced relationship occurs between the Pacific Ocean sea surface temperatures (El Niño, La Niña, and the Pacific Decadal Oscillation) and fire weather and climate conditions across Canada (Skinner et al. 2006). The relatively high temporal variability of weather and climate parameters over land masses and the associated area burned are related to the more slowly varying ocean. Atmospheric response to sea surface temperature anomalies in the equatorial Pacific determines ocean conditions over the remainder of the world's oceans and ultimately affects the large-scale atmospheric circulation over land (Schneider et al. 2002; Kumar and Hoerling 2003; Wu and Liu 2003; Yang and Zhang 2003; Lau et al. 2004; Shabbar and Skinner 2004). However, effects from this interaction are still uncertain. By means of fast atmospheric transport and/or slow overturning circulation in the upper ocean, extratropical-tropical linkages could also act as conveyors for transporting extratropical climatic anomalies (poleward of ~30°) to the tropics (Wu et al. 2007). The example shown in figure 4.5 demonstrates that years of high fire activity on the Boreal Shield• are synchronous

• The Boreal Shield is a broad U-shaped zone that extends from northern Saskatchewan east to Newfoundland, passing north of Lake Winnipeg, the Great Lakes, and the St. Lawrence River.

to ocean surface temperatures warmer along the West Coast and cooler in the centre of the Pacific Ocean. The opposite situation is found during years of low fire activity. A similar pattern of reversal of ocean surface temperatures takes place in the North Atlantic Ocean.

The relationships established between fire signature years and ocean conditions are promising for long-range forecasting of fire-conducive weather (Flannigan and Wotton 2001; Skinner et al. 2006). Weather prediction limits based on the intrinsic variability in the atmosphere are thought to be on the order of 2 weeks. However, ocean variability occurs on a much slower time scale; it mainly affects seasonal climate, and, as such, implies that seasonal fire danger prediction may be possible several months in advance(Uppenbrink 1997; Skinner et al. 2006).

**Figure 4.5**

**Mean March to May sea surface temperature departures (°C) from the reference period 1959–1999 for the 6 years of highest (A) and lowest (B) area burned on the Canadian Boreal Shield (C) (expressed in millions of burned hectares)**

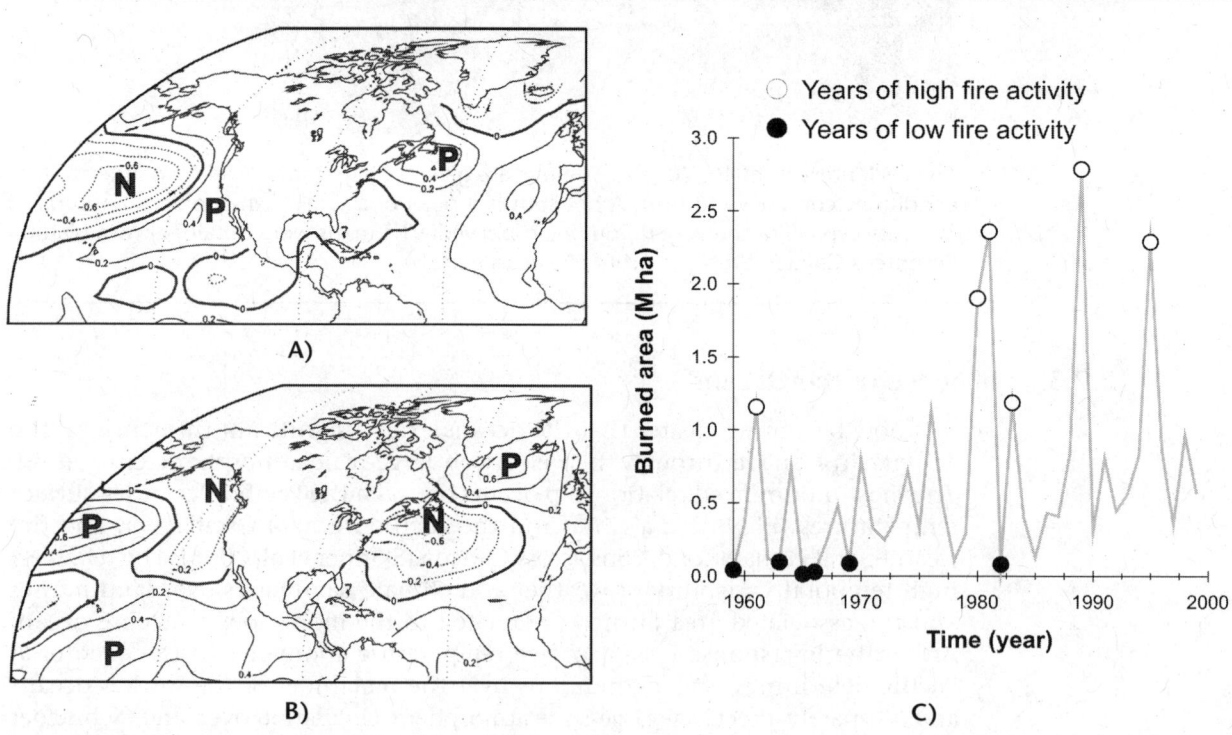

Based on this analysis, high area burned on the Boreal Shield is, on average, positively related to warm sea surface temperatures (P: positive departures) along the coast and cooler ones (N: negative departures) in the interior North Pacific, and vice versa for low fire years. Note also the changing pattern of sea surface temperatures over the North Atlantic basin. Images were created using the Royal Netherlands Meteorological Institute (KNMI) Climate Explorer and the Kaplan sea surface temperature data (<climexp.knmi.nl>). Adapted from Skinner et al. (2006); area burned data from Stocks et al. (2003).

## 3. FIRE AND CLIMATE

### 3.1. The Recent Period (1920 to Present)

Area burned in Canada has been steadily increasing since 1970 and it was suggested that the trend reflects a detectable influence of human-induced climate warming (Gillett et al. 2004). Trends in area burned are consistent across the boreal region, although western regions experienced greater increases in larger fire years compared with eastern regions (Kasischke and Turetsky 2006). Much of the area-burned trend is due to increases in the frequency of large fire years.

There remain nonetheless uncertainties about the atmospheric factors responsible for the increase in area burned. Indeed, area burned in Canada varies more with temperature on long-term time scales (decades to centuries) than it does on interannual time scales (Gillett et al. 2004). Hence, in terms of statistical confidence, the current fire records (1920 to present) are too short to verify if large-scale temperature variations are indeed linked to variations in area burned. Fire records of area burned have also been criticized because of changing fire reporting and suppression practices over the past century (Van Wagner 1988; Bourgeau-Chavez et al. 2000; Podur et al. 2002). It is thus inherently difficult to relate trends in climate, increased weather variability, or some other factor, to the area-burned trend.

In addition, while the progress in characterizing the spatial and temporal variability of fire and fire-conducive weather and climate in Canada is significant (e.g., Skinner et al. 1999, 2002, 2006; Amiro et al. 2004; Girardin et al. 2004b), the information is based upon observational data and limited to periods not exceeding a few decades. However, for some Canadian regions, fire statistic recording began during the second half of the 20th century, i.e. well after the advent of the ecological processes that shaped the land as we see it today. The boreal forest structure rises from climatic conditions and ecological processes spreading over periods of 30 years and going beyond 500 years. This is the case in northwestern Québec boreal forests, where forest inventories show that over 80% of forest stands originated from forest fires that took place prior to 1920 and over 47% prior to 1850 (Bergeron et al. 2004). Furthermore, the period of largest area burned, the 1910s–1920s (approximately 22% of total forested area), has no equivalent in the present day in terms of extent (see Bergeron et al. 2004), and its climatic cause is confounded by the increase in ignition potential that followed intensive European colonization (Lefort et al. 2003). While over short time periods we understand fire events very well, our current knowledge of fire regimes might not reflect the true amplitude of fire disturbance. Thus, our understanding of current and recent fire regimes provides only a partial portrait of how they are influenced by climate and of their magnitude. We have to go back further in the past to gain a better understanding of the relationship between fire and climate.

## 3.2.  Fire and Climate of the Past: Contributions from Tree Rings

Tree rings can serve as biological substitutes for past fire-conducive climate and thereby provide information that enhances the limited temporal coverage of observational records. Trees growing in temperate regions produce annual radial increments, where changes in ring width from one year to the next reflect changes in precipitation, temperature, and drought, as well as other factors (Fritts 2001). Trees can also sense climate variability that promotes fire activity (Girardin 2007; Girardin et al. 2006a). This may be the case for variations in the amplitude of the major belt of westerlies (figures 4.2 and 4.3) and their effect on severity of fire weather indices (figure 4.4; Girardin and Tardif 2005).

Because of mortality (including that caused by wildfires), most of the trees sampled on the Boreal Shield for dendroclimatic purposes are less than 160 years old (Contributors to the International Tree-Ring Data Bank 2004; Girardin et al. 2006b). However, trees from species like red pine (*Pinus resinosa* Ait.) and jack pine (*Pinus banksiana* Lamb.) can survive fires over periods of several centuries (see examples in figure 4.6). Tree species well adapted to wetlands, such as black spruce (*Picea mariana* [Mill.] B.S.P.) and white-cedar (*Thuja occidentalis* L.), can also reach honourable ages (sometimes over 300 years; Girardin et al. 2006b). The use of dead trees can also help in extending tree-ring records further back in the past.

From patterns of year-to-year changes in ring width as documented across well-replicated networks of tree-ring data, one can infer fire-conducive climate variability (known as dendroclimatic reconstruction) and extend records back to times during which there was no weather and fire reporting. Tree-ring substitutes for past fire-conducive climate variability can provide valuable means for understanding past changes in fire activity (Bergeron and Archambault 1993; Larsen 1996; Westerling and Swetnam 2003) and establishing future fire regimes under the projected warming of the Northern Hemisphere.

Recent achievements in dendroclimatology include the development of five reconstructions of the July DC and one reconstruction of mean July to August temperature going back to the early 1700s (figure 4.7; Girardin et al. 2006b). Note that the DC is a "seasonal" component of the Fire-Weather Index (FWI) system. Moisture losses in the DC are the result of daily evaporation and transpiration, while daily precipitation accounts for moisture gains. Each year, the period of melting snow is simulated and it is assumed that the deep layers of organic matter are completely recharged with water. Evaporation and transpiration losses are first estimated as maximum potential evapotranspiration based on temperature and seasonal day length. Second, this maximum potential evapotranspiration value is scaled by the available soil moisture to reflect the fact that as soil moisture content is reduced, evaporation becomes increasingly difficult (Turner 1972). In the process of water recharge, 2.80 mm of total precipitation are withdrawn per 24-hour period to accommodate canopy and surface fuel interceptions. The scale of the DC is cumulative, with each unit representing a decrease of 0.254 mm of water available in the soil. The maximum holding capacity of the soil is 100 mm for an organic layer whose dry weight is 25 kg $\cdot$ m$^{-2}$, which equates to approximately 400% of water per unit dry mass. There are no absolute guidelines as to the meaning of the DC values but generally speaking, values below 200 are considered low and 300 may be moderate in most parts of

Figure 4.6
**Cross-section of a jack pine tree from eastern Ontario dating back to 1763 (A) and ring width measurements along increment cores collected from three live red pine trees (*Pinus resinosa* Ait.) on a neighbouring site (B)**

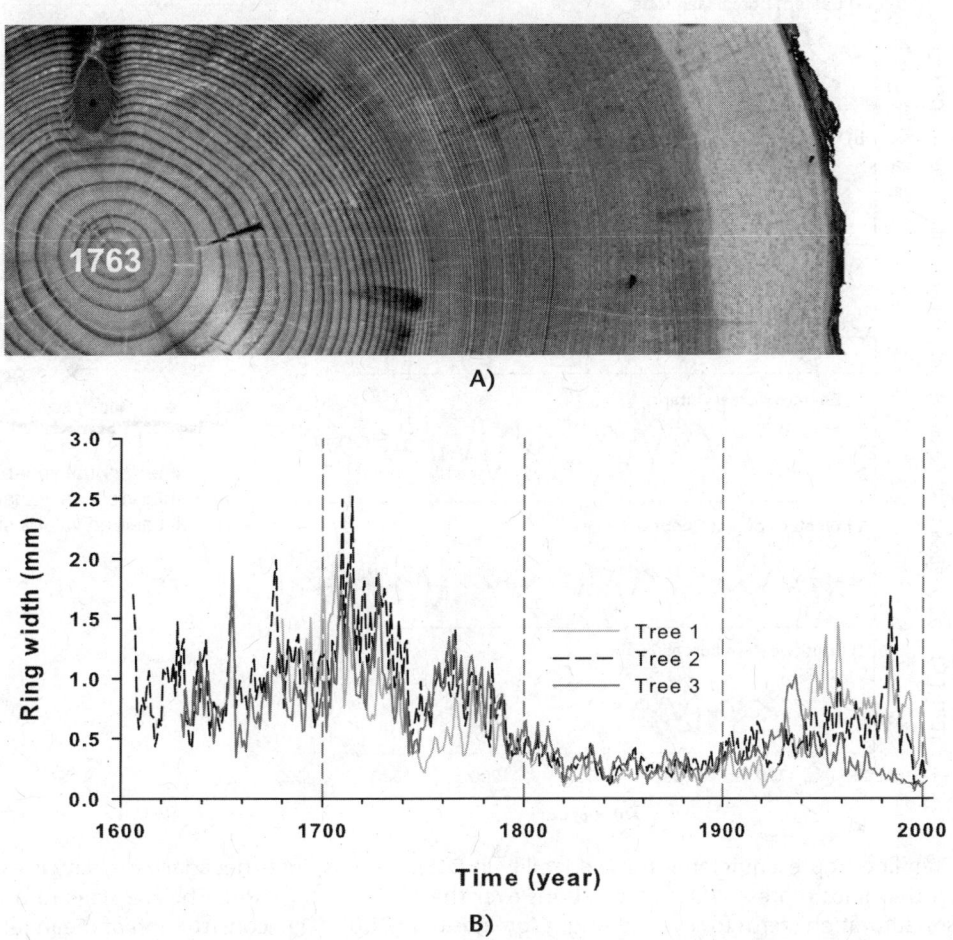

A)

B)

The longest measurement series dates from 1605 to 2002. A section of a tree trunk is made up of rings, each of which corresponds to the growth of the tree during a year. The inter-annual variability in the thickness of the rings reflects a climate favourable or not to the growth of that tree, and therefore provides information on temperatures and precipitation that took place during the years the rings were formed. Note that this variability should not be confused with the long-term trend linked to the age of the tree, or to changes in forest cover. The study of tree rings makes it possible to reconstruct particular climatic parameters, such as temperature, precipitation, and drought indices.

Canada. A DC rating of 300 or more indicates that fire will involve burning of deep sub-surface and heavy fuels. Because of its cumulative scale, the July DC is a rough estimate of water content of deep organic layers for a period ranging from about May to July. In Canada, 78% of the area burned from 1959 to 1998 occurred in June and July, while 8% occurred during May and 13% during August

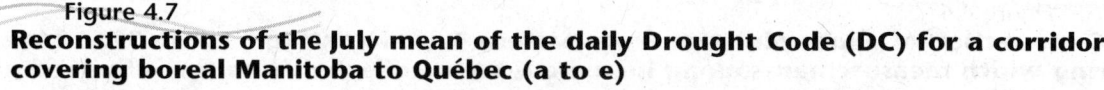

Figure 4.7
**Reconstructions of the July mean of the daily Drought Code (DC) for a corridor covering boreal Manitoba to Québec (a to e)**

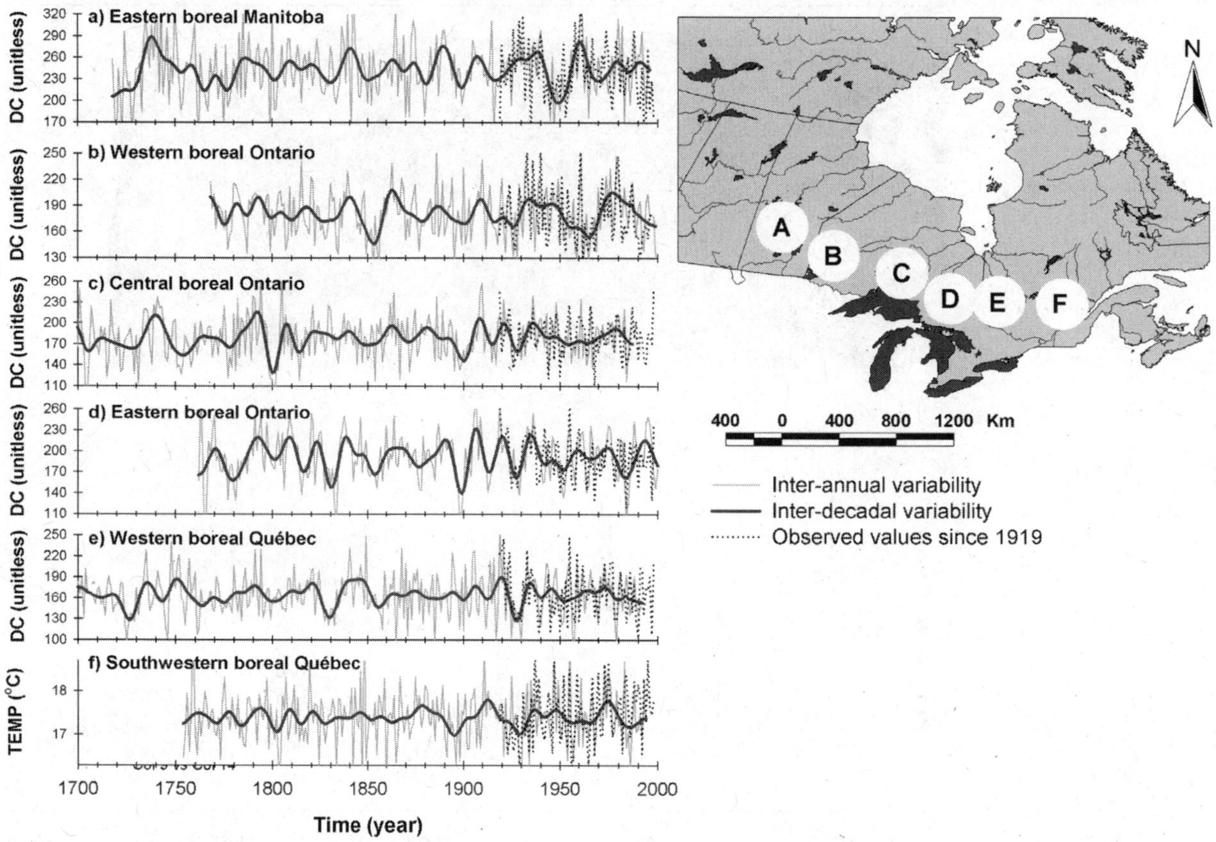

Thin lines represent inter-annual variability in DC estimates; inter-decadal variability is expressed using thick lines. Dotted lines represent DC observations over the 1919–1998 period. The DC scale (a unitless index) ranges from soil saturation (zero) to severe drought (greater than 300). (f) Reconstruction of mean July to August temperatures for southwestern boreal Québec (TEMP, expressed in °C).

(Stocks et al. 2003). Therefore, the season covered by the reconstructions may be a good proxy for fire weather conditions at the time of greatest area burned. More details on the DC can be obtained by consulting Van Wagner (1987) and Girardin et al. (2004b).

The six reconstructions presented in figure 4.7 were developed from a network of 120 well-replicated site tree-ring width chronologies distributed mainly on the Boreal Shield and extending back to at least AD 1866. Site tree-ring width chronologies are defined as averages of annual ring width measurements for one to several cores per tree and for several trees growing on similar ecological sites. The development of these tree-ring records represents the achievement of over a decade of research, sampling efforts, and laboratory work by a number of researchers. In total, over 4,400 ring width measurement series from 13 species were gathered. Most sampled trees belonged to the genus *Pinus* (pines;

45% of all sites). The genus *Picea* (spruces; 27% of all sites) and *Thuja occidentalis* L. (white cedar; 15% of all sites) also contributed to a large proportion of the total network (Girardin et al. 2006b).

Indications from these tree-ring-based reconstructions (figure 4.7) suggest high variability in the year-to-year and decade-to-decade drought severity across the boreal forest of Manitoba, Ontario, and Québec. Historically, multi-year droughts have always covered vast land areas. Notable drought episodes include the 1730s and 1790s that covered much of the boreal corridor (table 4.1). But what is intriguing is the absence of prolonged drought in the western part of boreal Québec since the 1850s (with the exception of the 1910s–1920s), as opposed to eastern Manitoba and western Ontario (table 4.1; figure 4.7). Based on the synoptic characteristics of recent droughts, the change in variability may translate into a response to an increasing frequency of upper-level ridging (troughing) over western (eastern) Canada since ca. 1850 (Girardin et al. 2006b). Specifically, this suggests an amplification of the long wave illustrated in figure 4.2. Increasing cyclonic activity (deeper troughing) in eastern Canada and incursion of moist air masses may have favoured climate conditions that are less suitable to fires.

The change in drought variability across the corridor is also well correlated with past changes in North Pacific sea surface temperatures (figure 4.8; see also Girardin et al. 2004a, 2006b; Skinner et al. 2006). Biological substitutes for Pacific sea surface temperatures and western coast air temperatures did suggest striking changes in variability around 1850 (D'Arrigo et al. 2001; Evans et al. 2002; Finney et al. 2000; Wilson and Luckman 2003), which is similar in timing to documented changes in fire activity across eastern boreal Canada (see Gauthier et al., chapter 3; Bergeron et al. 2001, 2004). The timing leads us to believe that these two phenomena are embedded in a context of climate change at a relatively global spatial scale (Girardin et al. 2006b).

Table 4.1
## Prolonged drought episodes for a corridor covering boreal Manitoba–Québec

| Eastern boreal Manitoba | Western boreal Ontario | Central boreal Ontario | Eastern boreal Ontario | Western boreal Québec | Southwestern boreal Québec |
|---|---|---|---|---|---|
| 1735 to 1743 | N/A | 1736 to 1744 | N/A | 1734 to 1738 | N/A |
| | N/A | | N/A | 1748 to 1754 | N/A |
| | 1791 to 1795 | 1787 to 1795 | 1790 to 1795 | 1789 to 1792 | 1791 to 1795 |
| | | 1806 to 1809 | 1807 to 1811 | | |
| | | | | 1819 to 1822 | |
| 1838 to 1843 | 1838 to 1842 | | 1837 to 1840 | 1837 to 1849 | |
| | 1860 to 1867 | | | | |
| | | | | | 1876 to 1882 |
| 1887 to 1892 | | | 1889 to 1892 | | |
| | 1908 to 1911 | 1907 to 1911 | 1905 to 1909 | | 1909 to 1916 |
| | | 1920 to 1923 | 1919 to 1922 | 1917 to 1922 | |
| 1936 to 1940 | 1932 to 1937 | 1934 to 1938 | 1933 to 1937 | | |
| 1958 to 1963 | | | | | |
| | 1973 to 1983 | | | | 1971 to 1978 |
| | | 1991 to 1995 | 1992 to 1995 | | |

This table was inferred from the six tree-ring-based drought reconstructions (see figure 4.7). From Girardin et al. (2006c). N/A: not available.

**Figure 4.8**

**Normalized eastern boreal Manitoba minus normalized western boreal Québec smoothed Drought Code reconstructions (figure 4.7a minus figure 4.7e) versus the leading principal component of North Pacific sea surface temperatures (SST, in standardized departures) north of 20°N**

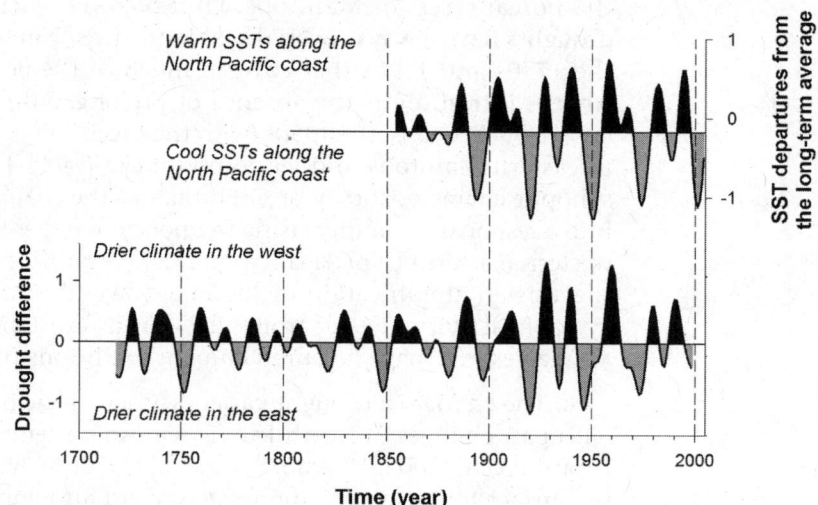

Note the increasing contrast in drought severity between the two regions since 1850 and the striking similarity with the North Pacific SST over their common period. Data from Girardin et al. (2006a).

## 3.3.   Reconstruction of Past Fire Activity

The relative effect of climate changes over the past two centuries with regard to annual fire activity on the Boreal Shield has recently been translated by rescaling the tree-ring-based Canadian Drought Code reconstructions to the mean and variance of the fire activity on the Boreal Shield (Girardin et al. 2006c). This "statistical reconstruction of the fire activity" indicated striking changes in variability in the course of the past 200 years, namely, extensive fire activity was estimated for 1789–1796, 1820–1823, 1837–1841, 1862–1866, 1906–1912, 1919–1922, 1933–1938, and 1974–1977 (figure 4.9). The most notable changes were seen with the occurrence rates of extreme fire years, with the prevalence of a period of low occurrence of extreme fire years from the 1850s to the early 1900s. The second half of the 20th century was also marked by low occurrence of extreme fire years.

These variations were validated with a stand age distribution derived from a regional time-since-last-fire map from the western boreal Québec region (figure 4.9). Indeed, the stand age distribution shows the recruitment of several forest stands during the decades 1810 to 1840 and 1910 to 1920. For these periods, dendroclimatic estimates suggest several successive years of high fire activity. Interestingly, the strong coherence between the fire activity estimates of the Boreal Shield and the stand age distribution derived from the time-since-last-fire

Figure 4.9
**A) Statistical reconstruction of the area burned on the Boreal Shield.**
**B) Occurrence rate of extreme-area-burned events (n = 24) on the Boreal Shield with 90% confidence bands; high values (1800–1850 and 1910–1940) are indicative of periods during which many years of extreme wildfire risk have occurred.**
**C) Stand age distribution for an area of 15,000 km² located at the transition zone between the mixedwood and coniferous boreal forests of southwestern Québec (vertical bars) in 10-year age classes (expressed in % of total study area)**

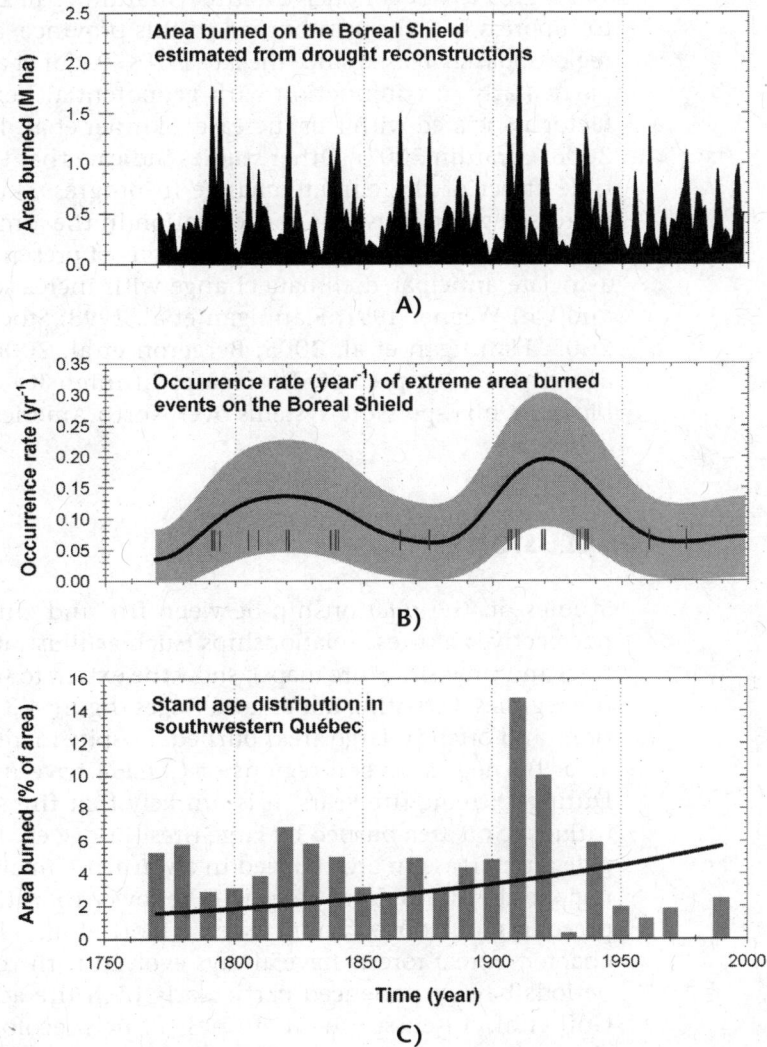

The negative exponential curve shows the theoretical stand age distribution expected under a constant fire regime. Values of area burned below the theoretical curve (notably during the late 20th century) are indicative of low fire activity, and vice versa. Modified from Girardin et al. (2006b).

map suggested that the 1910s–1920s fires in the Abitibi Plains occurred at a time when the climate was very conducive to fire. Although human activities may have contributed to the high fire activity by acting as an ignition source, atmospheric conditions (persistent ridging) and drought have certainly been decisive factors.

In contrast, the period 1950–2000 saw rather low fire activity compared with historical levels, as illustrated by the stand age distribution (figure 4.9c). Further dendroclimatological work done in Ontario highlighted the period 1940–1969 as not conducive to fire (Girardin et al. 2006a). Despite greater efforts to suppress fire, the area burned in this province, as in several other Canadian regions, has increased since the late 1970s (Podur et al. 2002). Climatic variability, particularly in conjunction with geopotential heights, seems to be the main factor associated with this increase (Skinner et al. 1999; Kasischke and Turetsky 2006; Girardin 2007). Other studies indicate that this trend would be a perceptible effect of the climate change in progress, known to be associated with increased greenhouse gas concentration in the atmosphere (Gillett et al. 2004). Indeed, on an ongoing basis, simulations of present and future rates of burning associate anticipated climate change with increases in fire activity (Flannigan and Van Wagner 1991; Flannigan et al. 1998; Stocks et al. 1998; Wotton et al. 2003; Flannigan et al. 2005; Bergeron et al. 2006). These predictions are in agreement with predicted increases in intensity, frequency, and duration of blocking high-pressure systems over North America (Meehl and Tebaldi 2004).

## 4.   CONCLUSION

Studies on the relationship between fire and climate, and on the historical perspectives of these relationships (such as illustrated with the growth rings of trees and time-since-fire maps), show the extent to which climate affects regional fire regimes. Persisting upper-level ridges (figure 4.3) result in extreme fire conditions and often in large areas burned. Despite increased fire suppression efforts, areas burned in several regions of Canada have increased steadily since 1970. During extreme fire years, it is unlikely that fire suppression has had a major influence on area burned by large fires (Bridge et al. 2005). Although we anticipate an increase in area burned in a warmer climate, it is possible that for some regions the future burn rates will be lower than those seen during the period preceding our observations (see Le Goff et al., chapter 5). As shown in this chapter, boreal forests have always evolved with the presence of fire and some periods have experienced particularly high fire activity. The next chapter (Le Goff et al.) discusses the usefulness of paleoecological data illustrated in this chapter with the objective of developing adaptation strategies to fire in a context of climate change. The authors place particular emphasis on ecosystem management and solutions that exist to help forest managers to cope with new climate constraints and their consequences.

# REFERENCES

Agee, J.K. 1997. The severe weather wildfire: too hot to handle? Northwest Sci. **71**: 153–156.

Ahrens, C.D. 2003. Meteorology today: an introduction to weather, climate, and the environment, 7th edition. Brooks/Cole – Thomson Learning, Pacific Grove, Calif., USA.

American Meteorological Society (AMS). 2000. Glossary of meteorology, 2nd edition. Allen Press, Boston, USA.

Amiro, B.D., Logan, K.A., Wotton, B.M., Flannigan, M.D., Todd, J.B., Stocks, B.J., and Martell, D.L. 2004. Fire weather index system components for large fires in the Canadian boreal forest. Int. J. Wildland Fire, **13**: 391–400.

Bergeron, Y. and Archambault, S. 1993. Decreasing frequency of forest fires in the southern boreal zone of Québec and its relation to global warming since the end of the "Little Ice Age." Holocene, **3**: 255–259.

Bergeron, Y., Cyr, D., Drever, C.R., Flannigan, M., Gauthier, S., Kneeshaw, D., Lauzon, È., Leduc, A., Le Goff, H., Lesieur, D., and Logan, K. 2006. Past, current, and future fire frequencies in Quebec's commercial forests: implications for the cumulative effects of harvesting and fire on age-class structure and natural disturbance-based management. Can. J. For. Res. **36**: 2737–2744.

Bergeron, Y., Gauthier, S., Flannigan, M., and Kafka, V. 2004. Fire regimes at the transition between mixed-wood and coniferous boreal forest in northwestern Quebec. Ecology, **85**: 1916–1932.

Bergeron, Y., Gauthier, S., Kafka, V., Lefort, P., and Lesieur, D. 2001. Natural fire frequency for the eastern Canadian boreal forest: consequences for sustainable forestry. Can. J. For. Res. **31**: 384–391.

Bessie, W.C. and Johnson, E.A. 1995. The relative importance of fuels and weather on fire behavior in subalpine forests. Ecology, **76**: 747–762.

Bourgeau-Chavez, L.L., Alexander, M.E., Stocks, B.J., and Kasischke, E.S. 2000. Distribution of forest ecosystems and the role of fire in the North American boreal region. *In* Fire, climate change and carbon cycling in North American boreal forests. *Edited by* E. Kasischke and B.J. Stocks, Springer-Verlag, p. 111–131.

Bridge, S.R.J., Miyanishi, K., and Johnson, E.A. 2005. A critical evaluation of fire suppression effects in the boreal forest of Ontario. For. Sci. **51**: 41–50.

Campbell, I.D. and Flannigan, M.D. 2000. Long-term perspectives on fire-climate-vegetation relationships in the North American boreal forest. *In* Fire, climate change and carbon cycling in North American boreal forests. *Edited by* E. Kasischke and B.J. Stocks, Springer-Verlag, pp. 151–172.

Carcaillet, C., Bergeron, Y., Richard, P.J., Fréchette, B., Gauthier, S., and Prairie, Y.T. 2001. Change of fire frequency in the eastern Canadian boreal forests during the Holocene: does vegetation composition or climate trigger the fire regime? J. Ecol. **89**: 930–946.

Contributors to the International Tree-Ring Data Bank. 2004. IGBP PAGES/World Data Center for Paleoclimatology, NOAA/NCDC Paleoclimatology Program, Boulder, Colo., USA.

Cook, E.R., Esper, J., and D'Arrigo, R. 2004. Extra-tropical Northern Hemisphere land temperature variability over the past 1000 years. Quat. Sci. Rev. **23**: 2063–2074.

Cyr, D., Gauthier, S., and Bergeron, Y. 2007. Scale-dependent determinants of heterogeneity in fire frequency in a coniferous boreal forest of eastern Canada. Landsc. Ecol. **22**: 1325–1339.

D'Arrigo, R., Villalba, R., and Wiles, G. 2001. Tree-ring estimates of Pacific decadal climate variability. Clim. Dyn. **18**: 219–224.

Esper, J., Frank, D.C. and Wilson, R.J.S. 2004. Climate reconstructions: low-frequency ambition and high-frequency ratification. Eos Trans. AGU **85**: 113–120.

Evans, M.N., Kaplan, A., and Cane, M.A. 2002. Pacific sea surface temperature field reconstruction from coral δ18O data using reduced space objective analysis. Paleoceanography, **17**: 1007.

Finney, B.P., Gregory-Eaves, I., Sweetman, J., Douglas, M.S.V., and Smol, J.P. 2000. Impacts of climatic change and fishing on Pacific salmon abundance over the past 300 years. Science, **290**: 795–799.

Fire Ecology Research Group. 2005. Experimental climate prediction center: fire weather index forecast. The Global Fire Monitoring Center (GFMC), Freiburg University, Freiburg, Germany.

Flannigan, M.D., Bergeron, Y., Englemark, O., and Wotton, B.M. 1998. Future wildfire in circumboreal forests in relation to global warming. J. Veg. Sci. **9**: 469–476.

Flannigan, M.D. and Harrington, J.B. 1988. A study of the relation of meteorological variables to monthly provincial area burned by wildfire in Canada (1953–80). J. Appl. Meteorol. **27**: 441–452.

Flannigan, M.D., Logan, K.A., Amiro, B.D., Skinner, W.R., and Stocks, B.J. 2005. Future area burned in Canada. Clim. Change, **72**: 1–16.

Flannigan, M.D. and Van Wagner, C.E. 1991. Climate change and wildfire in Canada. Can. J. For. Res. **21**: 66–72.

Flannigan, M.D. and Wotton, B.M. 2001. Climate, weather, and area burned. *In* Forest fires: behavior and ecological effects. *Edited by* E.A. Johnson and K. Miyanishi, Academic Press, New York, USA, pp. 351–373.

Fritts, H.C. 2001. Tree rings and climate. Blackburn Press, Caldwell, N.J., USA.

Gajewski, K., Payette, S., and Ritchie, J.C. 1993. Holocene vegetation history at the boreal-forest–shrub-tundra transition in north-western Québec. J. Ecol. **81**: 433–443.

Gillett, N.P., Weaver, A.J., Zwiers, F.W., and Flannigan, M.D. 2004. Detecting the effect of climate change on Canadian forest fires. Geophys. Res. Lett. **31**: L18211.

Girardin, M.P. 2007. Interannual to decadal changes in area burned in Canada from 1781 to 1982 and the relationship to Northern Hemisphere land temperatures. Global Ecol. Biogeogr. **16**: 557–566

Girardin, M.P., Bergeron, Y., Tardif, J.C., Gauthier, S., Flannigan, M.D., and Mudelsee, M. 2006c. A 229-year dendroclimatic-inferred record of forest fire activity for the Boreal Shield of Canada. Int. J. Wildland Fire, **15**: 375–388.

Girardin, M.P. and Tardif, J. 2005. Sensitivity of tree growth to the atmospheric vertical profile in the Boreal Plains of Manitoba, Canada. Can. J. For. Res. **35**: 48–64.

Girardin, M.P., Tardif, J., and Flannigan, M.D. 2006a. Temporal variability in area burned for the province of Ontario, Canada during the past 200 years inferred from tree rings. J. Geophys. Res. **111**: D17108.

Girardin, M.P., Tardif, J., Flannigan, M.D., and Bergeron, Y. 2004a. Multicentury reconstruction of the Canadian Drought Code from eastern Canada and its relationship with paleoclimatic indices of atmospheric circulation. Clim. Dyn. **23**: 99–115.

Girardin, M.P., Tardif, J.C., Flannigan, M.D., and Bergeron, Y. 2006b. Synoptic-scale atmospheric circulation and boreal Canada summer drought variability of the past three centuries. J. Clim. **19**: 1922–1947.

Girardin, M.P., Tardif, J., Flannigan, M.D., Wotton, B.M., and Bergeron, Y. 2004b. Trends and periodicities in the Canadian Drought Code and their relationships with atmospheric circulation for the southern Canadian boreal forest. Can. J. For. Res. **34**: 103–119.

Haines, D.A., Main, W.A., Frost, J.S., and Simard, A.J. 1983. Fire-danger rating and wildfire occurrence in the northeastern United States. For. Sci. **29**: 679–696.

Harrington, J., Kimmins, J., Lavender, D., Zoltai, S., and Payette, S. 1991. The effect of climate change on forest ecology in Canada. *In* Proceedings of the 10th World Forestry Congress, September 16–17, 1991, Paris. École nationale du génie rural, des eaux et des forêts, Nancy, France. Revue Forestière Française (special issue), **2**: 49–58.

Hély, C., Bergeron, Y., and Flannigan, M.D. 2000. Effects of stand composition on fire hazard in mixed-wood Canadian boreal forest. J. Veg. Sci. **11**: 813–824.

Intergovernmental Panel on Climate Change (IPCC). 2001. Climate Change 2001: the scientific basis. Cambridge University Press, Cambridge, UK.

Jenkins, M.A., Clark, T., and Coen, J. 2001. Coupling atmospheric and fire models. *In* Forest fires: behavior and ecological effects. *Edited by* E.A. Johnson and K. Miyanishi, Academic Press, New York, USA, pp. 258–303.

Johnson, E.A. 1992. Fire and vegetation dynamics: studies from the North American boreal forest. Cambridge University Press, Cambridge, UK.

Johnson, E.A., Fryer, G.I., and Heathcott, M.J. 1990. The influence of man and climate on frequency of fire in the Interior Wet Belt Forest, British Columbia. J. Ecol. **78**: 403–412.

Johnson, E.A. and Larsen, C.P.S. 1991. Climatically induced change in fire frequency in the southern Canadian Rockies. Ecology, **72**: 194–201.

Johnson, E.A. and Wowchuk, D.R. 1993. Wildfires in the southern Canadian Rocky Mountains and their relationship to mid-tropospheric anomalies. Can. J. For. Res. **23**: 1213–1222.

Kalnay, E., Kanamitsu, M., Kistler, R., Collins, W., Deaven, D., Gandin, L., Iredell, M., Saha, S., White, G., Woollen, J., Zhu, Y., Chelliah, M., Ebisuzaki, W., Higgins, W., Janowiak, J., Mo, K.C., Ropelewski, C., Wang, J., Leetmaa, A., Reynolds, R., Jenne, R., and Joseph, D. 1996. The NCEP/NCAR 40-year reanalysis project. Bull. Am. Meteorol. Soc. **77**: 437–471.

Kasischke, E.S. and Turetsky, M.R. 2006. Recent changes in the fire regime across the North American boreal region: spatial and temporal patterns of burning across Canada and Alaska. Geophys. Res. Lett. **33**: L09703.

Krawchuk, M.A., Cumming, S.G., Flannigan, M.D., and Wein, R.W. 2006. Biotic and abiotic regulation of lightning fire initiation in the mixedwood boreal forest. Ecology, **87**: 458–468.

Kumar, A. and Hoerling, M.P. 2003. The nature and causes for the delayed atmospheric response to El Niño. J. Clim. **16**: 1391–1403.

Larsen, C.P.S. 1996. Fire and climate dynamics in the boreal forest of northern Alberta, Canada, from AD 1850 to 1989. Holocene, **6**: 449–456.

Larsen, C.P.S. 1997. Spatial and temporal variations in boreal forest fire frequency in northern Alberta. J. Biogeogr. **24**: 663–673.

Larsen, C.P.S. and MacDonald, G.M. 1998. Fire and vegetation dynamics in a jack pine and black spruce forest reconstructed using fossil pollen and charcoal. J. Ecol. **86**: 815–828.

Lau, K.M., Lee, J.Y., Kim, K.M., and Kang, I.S. 2004. The North Pacific as a regulator of summertime climate over Eurasia and North America. J. Clim. **17**: 819–833.

Lefort, P., Gauthier, S., and Bergeron, Y. 2003. The influence of fire weather and land use on the fire activity of the Lake Abitibi area, eastern Canada. For. Sci. **49**: 509–521.

Mann, M.E., Bradley, R.S., and Hughes, M.K. 1998. Global-scale temperature patterns and climate forcing over the past six centuries. Nature, **392**: 779–787.

Masters, A.M. 1990. Changes in forest fire frequency in Kootenay National Park, Canadian Rockies. Can. J. Bot. **68**: 1763–1767.

Meehl, G.A. and Tebaldi, C. 2004. More intense, more frequent, and longer lasting heat waves in the 21st century. Science, **305**: 994–997.

Meehl, G.A., Washington, W.M., Wigley, T.M.L., Arblaster, J.M., and Dai, A. 2003. Solar and greenhouse gas forcing and climate response in the twentieth century. J. Clim. **16**: 426–444.

Ministère des Ressources naturelles et de la Faune du Québec (MRNFQ). 2006. Insectes, maladies et feux dans les forêts québécoises en 2005. Gouvernement du Québec, Québec, Que.

Moberg, A., Sonechkin, D.M., Holmgren, K., Datsenko, N.M., Karlen, W., and Lauritzen, S.E. 2005. Highly variable Northern Hemisphere temperatures reconstructed from low- and high-resolution proxy data. Nature, **433**: 613–617.

Nelson, R.M. Jr. 1993. Byram derivation of the energy criterion for forest and wildland fires. Int. J. Wildland Fire, **3**: 131–138.

Nelson, R.M. Jr. and Adkins, C.W. 1988. A dimensionless correlation for the spread of wind-driven fires. Can. J. For. Res. **18**: 391–397.

Newark, M.J. 1975. The relationship between forest fire occurrence and 500 mb longwave ridging. Atmosphere, **13**: 26–33.

Palmer, W.C. 1965. Meteorological drought. Research Paper No. 45, U.S. Department of Commerce Weather Bureau, Washington, D.C., USA.

Palmer, W.C. 1968. Keeping track of crop moisture conditions, nationwide: the new crop moisture index. Weatherwise, **21**: 156–161.

Podur, J., Martell, D.L., and Knight, K. 2002. Statistical quality control analysis of forest fire activity in Canada. Can. J. For. Res. **32**: 195–205.

Scafetta, N. and West, B.J. 2006. Phenomenological solar signature in 400 years of reconstructed Northern Hemisphere temperature record. Geophys. Res. Lett. **33**: L17718.

Schneider, N., Miller, A.J., and Pierce, D.W. 2002. Anatomy of North Pacific decadal variability. J. Clim. **15**: 586–605.

Shabbar, A. and Skinner, W. 2004. Summer drought patterns in Canada and the relationship to global sea surface temperatures. J. Clim. **17**: 2866–2880.

Skinner, W.R., Flannigan, M.D., Stocks, B.J., Martell, D.L., Wotton, B.M., Todd, J.B., Mason, J.A., Logan, K.A., and Bosch, E.M. 2002. A 500 hPa synoptic wildland fire climatology for large Canadian forest fires, 1959–1996. Theor. Appl. Clim. **71**: 157–169.

Skinner, W.R., Shabbar, A., Flannigan, M.D., and Logan K. 2006. Large forest fires in Canada and the relationship to global sea surface temperatures. J. Geophys. Res. **111**: D14106.

Skinner, W.R., Stocks, B.J., Martell, D.L., Bonsal, B., and Shabbar, A. 1999. The association between circulation anomalies in the mid-troposphere and area burned by wildland fire in Canada. Theor. Appl. Clim. **63**: 89–105.

Smith, T.M. and Reynolds, R.W. 2005. A global merged land-air-sea surface temperature reconstruction based on historical observations (1880–1997). J. Clim. **18**: 2021–2036.

Stocks, B.J. and Flannigan, M.D. 1987. Analysis of the behaviour and associated weather for a 1986 Northwestern Ontario wildfire: Red Lake. *In* Proceedings of the ninth conference on fire and forest meteorology (April 21), San Diego, USA, pp. 94–100.

Stocks, B.J., Fosberg, M.A., Lynham, T.J., Mearns, L., Wotton, B.M., Yang, Q., Jin, J.-Z., Lawrence, K., Hartley, G.R., Mason, J.A., and McKenney, D.W. 1998. Climate change and forest fire potential in Russian and Canadian boreal forests. Clim. Change, **38**:1–13.

Stocks, B.J., Mason, J.A., Todd, J.B., Bosch, E.M., Wotton, B.M., Amiro, B.D., Flannigan, M.D., Hirsch, K.G., Logan, K.A., Martell, D.L., and Skinner, W.R. 2003. Large forest fires in Canada, 1959–1997. J. Geophys. Res. **108**: 8149, 10.1029/2001JD000484.

Swetnam, T.W. 1993. Fire history and climate change in giant sequoia groves. Science, **262**: 885–889.

Tardif, J. 2004. Fire history in the Duck Mountain Provincial Forest, western Manitoba Sustainable Forest Management Network. Project Reports 2003/2004 series. University of Alberta, Edmonton, Alta. [Online] <www.sfmnetwork.ca/docs/e/PR_200304tardifjfire6fire.pdf> (accessed November 29, 2007).

Turner, J.A. 1972. The drought code component of the Canadian Forest Fire Behaviour System. Environment Canada, Canadian Forest Service Publication 1316, Ottawa, Ont.

Turner, M.G. and Romme, W.H. 1994. Landscape dynamics in crown fire ecosystems. Landsc. Ecol. **9**: 59–77.

Uppenbrink, J. 1997. Climate: nota bene: seasonal climate prediction. Science, **277**: 1952.

Van Wagner, C.E. 1978. Age-class distribution and the forest fire cycle. Can. J. For. Res. **8**: 220–227.

Van Wagner, C.E. 1987. Development and structure of the Canadian Forest Fire Weather Index System. Forestry Technical Report 35, Canadian Forestry Service, Ottawa, Ont.

Van Wagner, C.E. 1988. The historical pattern of annual burned area in Canada. For. Chron. **64**: 182–185.

Weber, M.G. and Flannigan, M.D. 1997. Canadian boreal forest ecosystem structure and function in a changing climate: impact on fire regimes. Environ. Rev. **5**: 145–166.

Weir, J.M.H., Johnson, E.A., and Miyanishi, K. 2000. Fire frequency and the spatial age mosaic of the mixedwood boreal forest in western Canada. Ecol. Appl. **10**: 1162–1177.

Westerling, A.L., Hidalgo, H.G., Cayan, D.R., and Swetnam, T.W. 2006. Warming and earlier spring increase western U.S. forest wildfire activity. Science, **313**: 940–943.

Westerling A.L. and Swetnam, T.W. 2003. Interannual to decadal drought and wildfire in the western United States. Eos Trans. AGU **84**: 545–560.

Wilson, R.J.S. and Luckman, B.H. 2003. Dendroclimatic reconstruction of maximum summer temperatures from upper treeline sites in Interior British Columbia, Canada. Holocene, **13**: 851–861.

Wotton, B.M., Martell, D.L., and Logan, K.A. 2003. Climate change and people-caused forest fire occurrence in Ontario. Clim. Change, **60**: 275–295.

Wu, L. and Liu, Z. 2003. Decadal variability in the North Pacific: the eastern North Pacific mode. J. Clim. **16**: 3111–3131.

Wu, L., Liu, Z., Li, C., and Sun, Y. 2007. Extratropical control of recent tropical Pacific decadal climate variability: a relay teleconnection. Clim. Dyn. **28**: 99–112.

Yang, H.J. and Zhang, Q. 2003. On the decadal and inter-decadal variability in the Pacific Ocean. Adv. Atmos. Sci. **20**: 173–184.

## Chapter 5

# Management Solutions to Face Climate Change*
## The Example of Forest Fires

*Héloïse Le Goff, Mike D. Flannigan, Yves Bergeron, Alain Leduc, Sylvie Gauthier, and Kim Logan*

* We thank Julie Fortin (MNRFQ) for providing data on Québec fires and Alan Cantin (CFS, Great Lakes Forestry Centre), who supplied the Fire Weather Index components for Duck Mountain, northwestern Ontario, and the Lake Abitibi Model Forest. We also thank Jacques Morissette (CFS, Laurentian Forestry Centre), who produced figure 5.2. Pierre Bernier (CFS, Laurentian Forestry Centre), Michel Campagna (MRNFQ), Michelle Garneau (UQAM), and Karelle Jayen (UQAM) improved this chapter with their comments and suggestions. The photos on this page were graciously provided by the SOPFEU, Sophie Périgon, and Danielle Charron.

# 1. INTRODUCTION

The forestry sector hesitated for a long time over integrating the problem of climate change into its policies and planning. Given the uncertainty about predictions of climate change effects at the forest management unit level, such hesitation is understandable (Burton 1998; Johnston et al. 2006). However, forest industries already face operational difficulties resulting from climate change, such as a shorter season for use of winter roads and longer forest fire seasons. Moreover, the planning horizon for forest management involves a time scale at which the effects of climate change are predictable (Spittlehouse 1997). The forestry sector must better integrate future climate change into its planning to avoid the economic cost that will result if it does not introduce measures to adapt to climate change (Johnston et al. 2006).

While a great deal of research has been devoted to the study of the potential impact of climate change on ecosystems, the implementation of concrete solutions to deal with such impacts has too often been neglected (Hulme 2005). As a result, the consequences of climate change for the management of forestry resources have been seen unclearly and unavoidably harmful. However, management tools already exist that will help us to deal with climatic conditions and their expected impacts on managed forests. In spontaneously adapting to certain climatic limitations in the past, forest managers have already begun to build a capacity to adapt to future climate change. Routine management practices such as fire suppression, salvage logging, and artificial regeneration can reduce the vulnerability of forest management to fires that are mainly determined by climate and its variations.

This chapter considers the impacts of climate change on the boreal forest as well as the measures that could be implemented to adapt to new climate conditions based on the example of the changes these new conditions will impose on fire regimes. Fire is one of the main natural disturbances in the boreal forest. Climate and weather conditions are determining factors that regulate fire activity (Girardin et al., chapter 4). Numerous paleoecology studies, as well as analysis of recent fire records, show that past climate variability has modified regional fire regimes (Girardin et al., chapter 4), and simulations of future climate changes indicate that they will continue to affect fire activity. In the first part, we describe how regional fire regimes will be affected by climate change; then we look at concrete solutions that will enable us to deal with either an increase or a decrease of fire activity due to climate change. While this study deals with the example of forest fires to evaluate one aspect of forest management vulnerability to climate change, a true risk analysis would require consideration of all the components that affect the vulnerability of forests and forest management to the effects of climate change.

Adaptation is a concept that is broader than and complementary to the mitigation of the climate change impacts (Le Goff et al. 2005). While mitigation consists in reducing concentrations of greenhouse gases by reducing sources and developing sinks, the concept of adaptation applies to systems that must deal with the impacts of climate change in a shorter time frame. Adaptation involves understanding and reducing the negative impacts of climate change, but also identifying and taking advantage of the benefits of the new climate conditions (McCarthy et al. 2001; Burton et al. 2002). Whether speaking of adaptation or mitigation, the concept of a long-term adaptive strategy is a basic principle for dealing with climate change. Such a strategy implies an iterative assessment framework that allows for adjusting strategies and practices on the basis of developments in our knowledge and results achieved by the measures implemented (figure 5.1).

The possibilities for adaptation depend on the vulnerability of a system to climatic conditions. Vulnerability (i.e. the degree to which a system is incapable of coping with the adverse effects of climate change, including variability and extreme conditions; McCarthy et al. 2001) depends on the nature and magnitude of the climate change, the sensitivity of the system to such change, and the ability of the system to adjust to the change (its adaptive capacity). It is recognized that the adaptive capacity of a system depends not only on available scientific and technical knowledge, but also on the social, economic, and political aspects of implementing planned adaptation strategies (Yohe and Tol 2002). However, in this chapter, the focus is on scientific and technical aspects of adaptation.

### Figure 5.1
## Conceptual framework illustrating adaptation to climate change

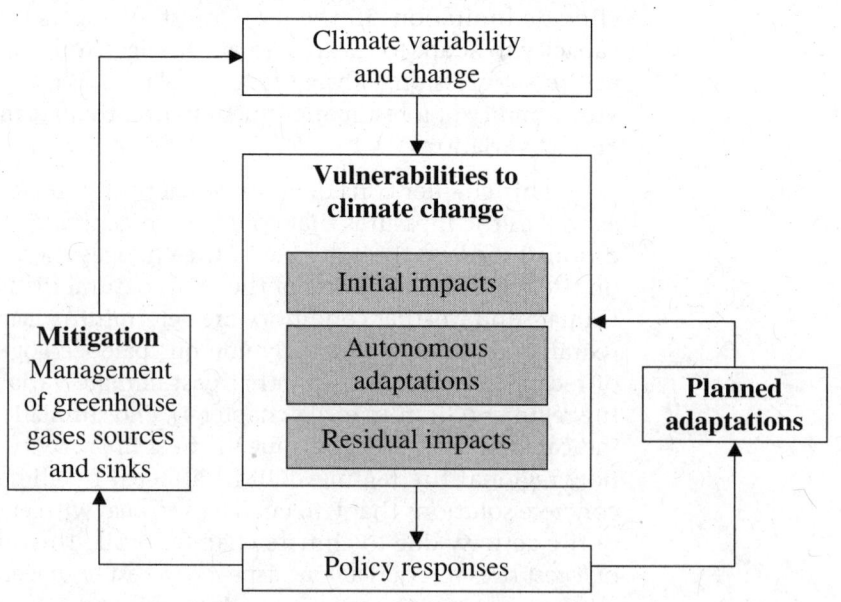

The adaptive capacity of a system facing climate variations and changes depends on initial impacts (determined by climate change magnitude and by the sensitivity of the system to these changes) and spontaneous adaptations of the system facing these changes (natural resilience of the system to the variation in environmental conditions). Political answers will thus concern the residual impacts. Adaptation is a political response complementary to mitigation measures. It aims at reducing the system's vulnerability to climate change. Modified from IPCC (2001).

## 2. CLIMATE CHANGE AND ITS IMPACTS

The changing seasons and their varying lengths set the pace for the growth and dynamics of our forests. Climate change disrupts the length and intensity of the seasons and this is especially clear in winter. In the northern hemisphere, winters have grown shorter and milder, with a reduced period of snow cover (about 10% shorter than in the late 1960s) and two weeks less of ice cover on lakes than was seen in the early 20th century (Consortium Ouranos 2004). Average global temperature increased by 0.74°C (0.56–0.92°C) between 1906 and 2005 (IPCC 2007). Eleven of the past 12 years (1995–2006) are ranked among the 12 warmest years recorded since 1850. In addition, there has been an increase in total precipitation since 1900 in eastern North America (IPCC 2007).

According to climate simulation models, temperatures will continue to rise and the increase will be greater in northern regions than in more southern regions (IPCC 2007; Plummer et al. 2006). Boreal forests are therefore particularly exposed to the impacts of climate change. For boreal forests in central and eastern Canada (from Manitoba to the Atlantic provinces), recent simulations using the Canadian Regional Climate Model predict an increase in average temperature of about 3.5°C in winter, 2°C in spring and fall, and 2.5°C in summer, as well as an increase of about 5% in precipitation in all seasons (Plummer et al. 2006). Johnson et al. (1999) reported that the fire season in Saskatchewan (Prince Albert National Park) would start earlier in the spring, compared with 1945. According to simulations carried out with atmospheric general circulation models, this trend toward a longer fire season could continue under the influence of the climate conditions predicted for Canada. Thus, for a $2\times CO_2$ scenario (corresponding to the period 2040–2060), the length of the fire season could increase by 16 to 17% (about 25 days) in central and eastern Canada, starting earlier in the spring and extending further into the fall (Wotton and Flannigan 1993; Stocks et al. 1998).

Climate change affects not only average conditions but also variability and climate extremes, such as the frequency of droughts or heavy rains (IPCC 2007). While species are generally well adapted to average climate conditions, they are especially sensitive and vulnerable to extreme conditions (Burton 1998; Parker et al. 2000). Forest fire activity is mainly affected by extreme weather conditions. In Canada, only 3% of forest fires account for 97% of the area burned over the past 30 years (Stocks et al. 2003). These very large fires occur most frequently during periods of extreme drought (Flannigan and Harrington 1988).

The influence of climate on forests is very complex and involves a variety of mechanisms (see box 5.1):

- direct mechanisms, such as the influence of temperature or season length on growth, reproduction success and species migration;
- indirect mechanisms, through changes in natural disturbance regimes (fires, insect epidemics, windfall, and disease).

The effects of changes in disturbance regimes on the structure and composition of forest stands are much more immediate than the direct effects on the distribution, extinction or migration of species (Weber and Flannigan 1997). That is why fire regimes are considered to be agents of change in the boreal forest.

Box 5.1

## Examples of direct or indirect impacts of climate change on vegetation

The influence of climate on forests is complex, involving direct mechanisms, such as the influence of temperature or season length on growth, reproduction success, and species migration, and indirect mechanisms, through changes in the natural disturbance regime (fire, insect outbreaks, windfall, and disease). The effects of changes in disturbance regimes on the structure and composition of stands are much more immediate than the direct effects on the distribution, extinction or migration of species (Weber and Flannigan 1997).

### Examples of direct impacts

The influence of warmer temperatures on tree growth and on the net primary productivity of forests is complex, because it depends on many factors. Longer growing seasons and a smaller number of summer frost periods in the boreal forest have a particularly beneficial effect on net growth of trees in cooler regions by prolonging the period of photosynthetic activity. Earlier warming may also encourage early bud burst (Raulier and Bernier 2000). However, the tree populations subjected to weather conditions warmer than those to which they are adapted may not be able to obtain maximum benefit from these more temperate conditions (Beaulieu and Rainville 2005). The benefits of climate change on forest productivity also depend on the degree of interference with natural disturbances, such as forest fires and insect outbreaks, which are also modified by climate change (Johnston et al. 2006). Changes in temperature and precipitation will also modify other parameters of forest dynamics, such as reproductive success, distribution and abundance of species, species assemblages, and species migration.

The current forest management system is based on a constant long-term yield, even though the calculation of allowable cut does not take into account the consequences of future climate change for forest productivity. Yield tables used to calculate allowable cut reflect the spatial variations resulting from historical climate but do not reflect future climate conditions that regenerating stands will have to face. That is why researchers are now developing models such as 3PG (Landsberg and Waring 1997), Standleap (Raulier et al. 2003), and TRIPLEX (Peng et al. 2002), which will make it possible to calculate the effects of future climate change on forest productivity.

### Examples of indirect impacts

It is generally accepted that climate change will have an impact on disturbance regimes and that this indirect impact of climate change on vegetation will be very considerable (IPCC 2007). It is believed that the incidence and the size of areas affected by insect defoliation and disease could increase under warmer and drier weather conditions (Volney and Hirsch 2005; Candau and Fleming 2005). However, the effect of climate change on these two types of disturbance is complex, since variations in weather conditions can change the development cycle of pathogens and insects (see Carroll et al. 2006 on the pine beetle), the resistance of host species or the distribution range of the pathogen or the host species (Johnston et al. 2006).

Weather events such as drought, ice storms, cycles of frost and thaw, and wind storms will also change in magnitude and frequency under the influence of climate change. These events have their own impact on vegetation, but they may also interact with other natural disturbances such as insect outbreaks and fire. Drought, for example, could make some individuals weaker and more vulnerable to disease.

Climate change leads to a longer growing season, which generally also implies a longer fire season (Wotton and Flannigan 1993). However, season length is not the only parameter of the fire regime that will be affected by climate change. Flannigan et al. (2005) expect an increase in annual area burned all across Canada. Price and Rind (1994) report that the number of lightning fires could rise as a result of increased lightning activity. However, other studies suggest that climate conditions favourable to fire could remain at the same level or even decrease in eastern Canada (Flannigan et al. 1998, 2001). All these studies agree that the effects of climate change on fires will vary from one region to another, because while a fire regime is mainly governed by climate, which varies from region to region, it also depends on more local characteristics such as fuel (structure and composition of forest stands), topography, and hydrography (presence of a fire break).

## 3. HOW WILL CLIMATE CHANGE AFFECT FUTURE FOREST FIRE ACTIVITY IN EASTERN CANADA?

We have used the model presented by Gauthier et al. in chapter 3 to compare past, current, and future annual burn rates in different regions of central and eastern Canada (figures 5.2 and 5.3A; see box 5.2 for notes on methodology). The response of fire regimes to future climate change varies from one region to another in terms of magnitude (table 5.1, figures 5.3B and 5.4), but for most of the

### Figure 5.2
**Location of study areas for which past, current, and future annual burn rates are compared**

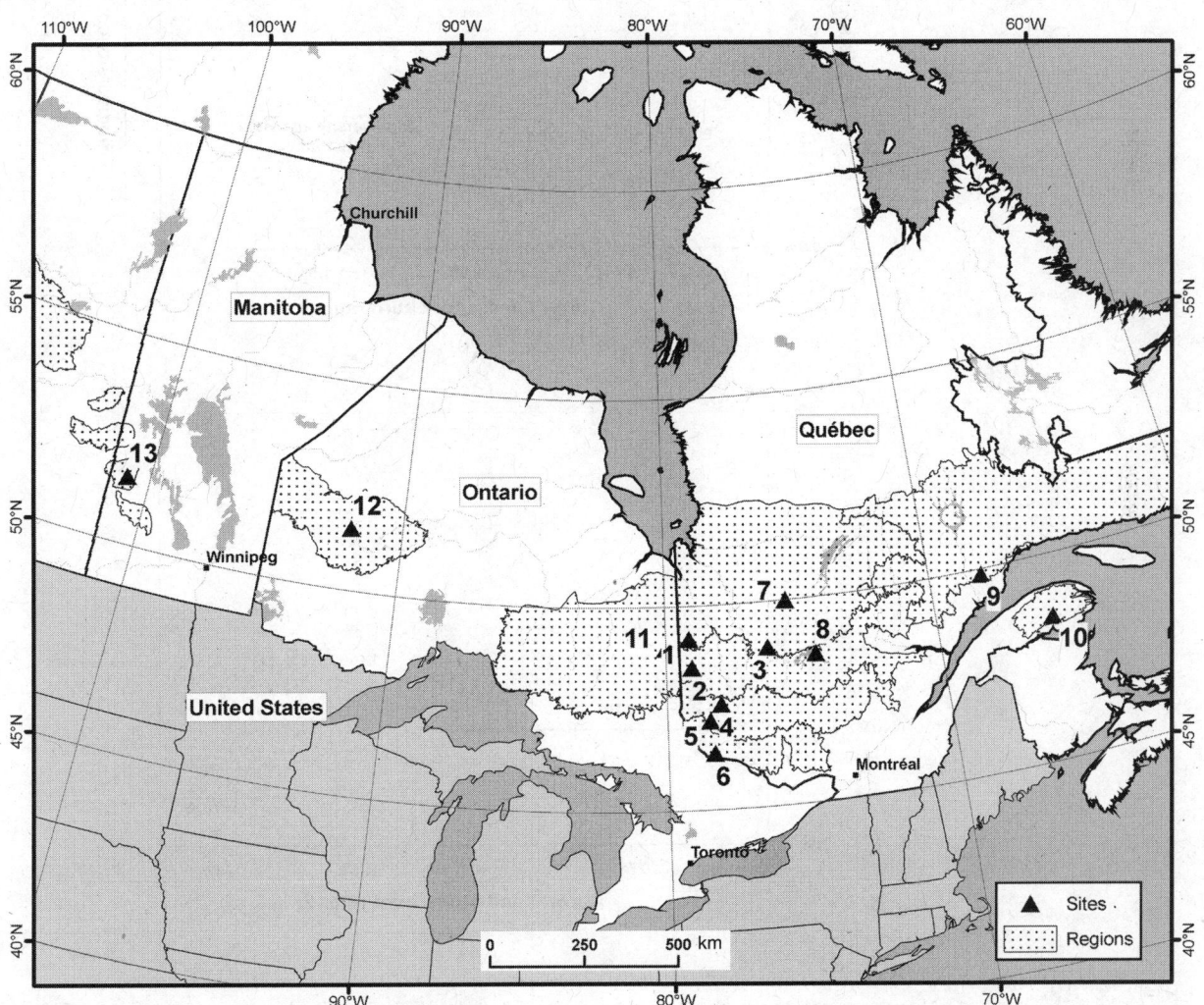

See table 5.1 for the names of numbered study areas and for ecological regions in which they are located (map produced by J. Morissette, CFS).

**Figure 5.3**
**A) Basic graphic model for the comparison of past, current, and future annual burn rates illustrating the constraints relative to biodiversity conservation and sustaining fibre yield. B) Current rates (black), and future 2×CO₂ (gray) and 3×CO₂ (white) rates as a function of past burn rates**

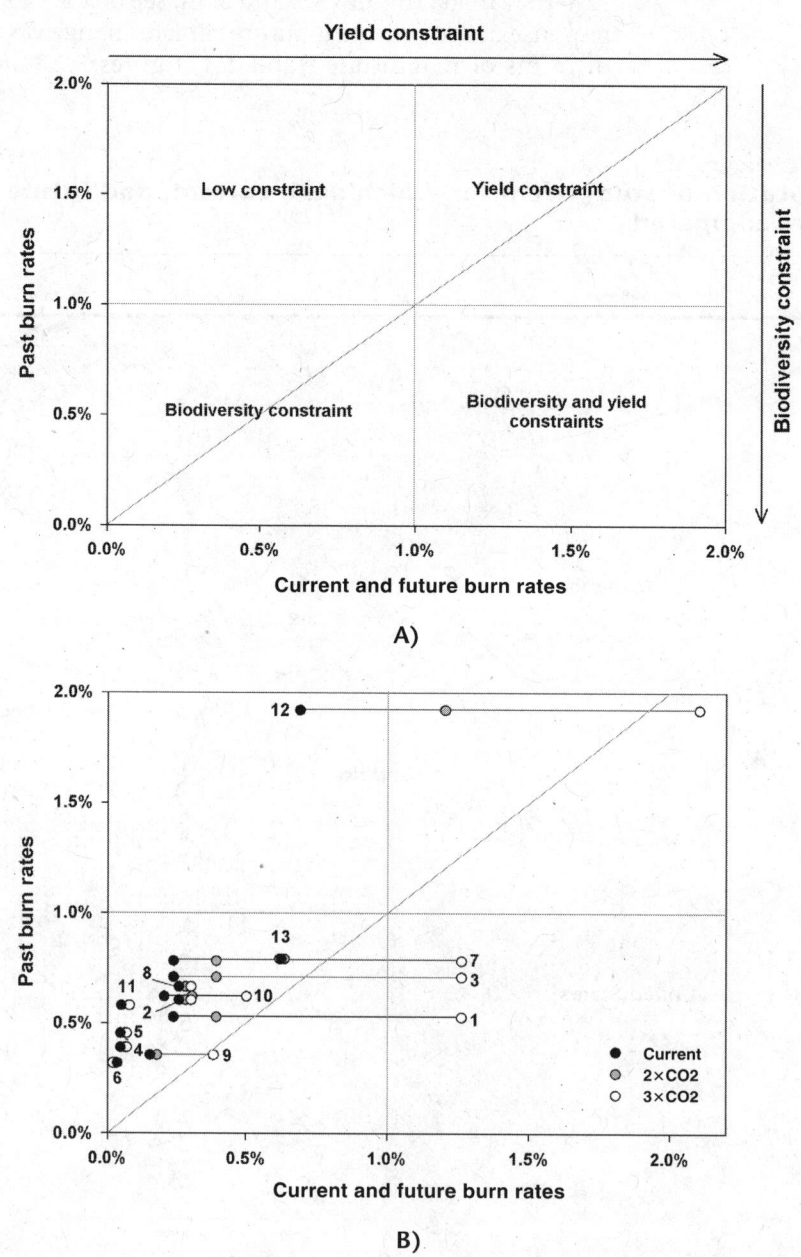

A)

B)

Numbers correspond to sites presented in table 5.1. For site locations, see figure 5.2.

Box 5.2
## Comparing annual burn rates: past, present, and future

The disturbance rate is the proportion of a territory disturbed by fire or harvesting. We have calculated the annual burn rate and the proportion of forests older than 100 years old for different future reference periods in order to compare them with current and historical values:

1. The historical relation between monthly burned areas and the Fire Weather Index (FWI) components (Van Wagner 1987; Girardin et al., chapter 4) was calculated with weather data (temperature, precipitation, relative humidity, and wind speed) from the nearest stations, using stepwise linear regressions for each ecological unit (see below).

2. The values of the FWI components were calculated from weather variables (temperature, precipitation, relative humidity, and wind speed) calculated using the Canadian Atmospheric General Circulation Model for two future reference periods: $2\times CO_2$ (doubling of the present $CO_2$ atmospheric concentration, corresponding to the period 2040 to 2060) and $3\times CO_2$ (during the period from 2080 to 2100). This weather model includes the forcing of both greenhouse gases and sulphate aerosols, which contribute to a 1% increase in $CO_2$

per year. The FWI components were then integrated into the regional regression models previously established to determine future burn rates.

3. The relationships between current and future FWI values ($2\times CO_2$ or $3\times CO_2$) were applied to current burn rates to calculate projected future rates.

Current burn rates were calculated for the period 1940–2003 using fire data from the Ministère des Ressources naturelles for sites in Québec, and from the Canadian Forest Service's Large Fire Database (fires > 200 ha, 1959–1999) for the three other sites. Current burn rates and regression equations were calculated at the following scales:

◆ bioclimatic sub-domains for Québec (Saucier et al. 1998),

◆ ecoregions of the Ontario Land Classification System (Van Sleeuwen 2006) for areas in the Lake Abitibi Model Forest and for northwestern Ontario,

◆ the Mid-Boreal Uplands ecoregion of Canada's Ecological Land Classification for the Duck Mountain site (Ecological Stratification Working Group 1996).

Figure 5.4
## Proportion of study area occupied by forest stands older than 100 years for the past, current, and future ($2\times CO_2$ and $3\times CO_2$) periods

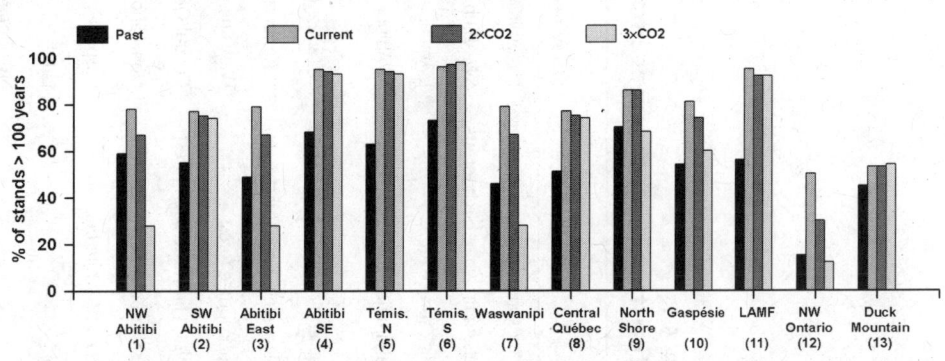

These proportions were calculated as a function of burn rates (table 5.1) using the negative exponential formula (Van Wagner 1978). The proportion of stands older than 100 years decreases as burn rate increases. For site location, see figure 5.2.

Table 5.1

# Characteristics of study areas including past, current, and future burn rates

| Site | Region[1] | Study area | Area (km²) | Lat. | Long. | Reference | Period | Past | Current[3] | 2×CO₂ | 3×CO₂ |
|---|---|---|---|---|---|---|---|---|---|---|---|
| 1 | Western black spruce forest | NW Abitibi | 7,942 | 49.2 | 79.1 | Bergeron et al. 2004 | 1680–1997 | 0.528 | 0.239 | 0.394 | 1.262 |
| 2 | Western balsam fir–white birch forest | SW Abitibi | 7,777 | 48.5 | 79.1 | Bergeron et al. 2004 | 1530–1996 | 0.604 | 0.258 | 0.282 | 0.302 |
| 3 | Western black spruce forest | E Abitibi | 3,294 | 48.9 | 76.3 | Kafka et al. 2001 | 1770–1995 | 0.708 | 0.239 | 0.394 | 1.262 |
| 4 | Western balsam fir–yellow birch forest | SE Abitibi | 13,156 | 47.6 | 78.1 | Lesieur et al. Unpublished data | 1800–2004 | 0.389 | 0.048 | 0.064 | 0.072 |
| 5 | Western balsam fir–yellow birch forest | N Témiscamingue | 2,850 | 47.2 | 78.5 | Grenier et al. 2005 | 1740–2003 | 0.454 | 0.048 | 0.064 | 0.072 |
| 6 | Sugar maple–yellow birch forest | S Témiscamingue | 1,793 | 46.4 | 78.4 | Drever et al. 2006 | 1580–2004 | 0.319 | 0.036 | 0.030 | 0.020 |
| 7 | Western black spruce forest | Waswanipi | 10,950 | 50.0 | 75.5 | Le Goff et al. 2007 | 1720–2003 | 0.781 | 0.239 | 0.394 | 1.262 |
| 8 | Western balsam fir–white birch forest | Central Québec | 3,844 | 48.6 | 74.5 | Lesieur et al. 2002 | 1720–1998 | 0.665 | 0.258 | 0.282 | 0.302 |
| 9 | Eastern black spruce forest | North Shore | 14,135 | 49.9 | 68.1 | Cyr et al. 2007 | 1640–2003 | 0.356 | 0.155 | 0.179 | 0.384 |
| 10 | Eastern balsam fir–white birch forest | Gaspésie | 6,480 | 48.5 | 65.8 | Lauzon et al. 2007 | 1680–2003 | 0.621 | 0.205 | 0.306 | 0.503 |
| 11 | Lake Abitibi | LAMF | 8,245 | 49.0 | 80.0 | Lefort et al. 2003 | 1740–1998 | 0.580 | 0.051 | 0.079 | 0.081 |
| 12 | NW Ontario | Cat Lake | 30,625 | 51.5 | 92.0 | Suffling et al. 1982 | ~1870–1974 | 1.920 | 0.691 | 1.205 | 2.106 |
| 13 | Mid-Boreal Uplands | Duck Mountain | 3,760 | 51.6 | 100.9 | Tardif 2004 | 1723–2002 | 0.787 | 0.625 | 0.637 | 0.618 |

1. Regions correspond to MRNFQ bio-climatic subdomains for the study areas in Québec, (Saucier et al. 1998); ecoregions of the Ecological Land Classification in the case of Ontario; and an ecoregion of Canada's Ecological Land Classification in the case of Duck Mountain (Ecological Stratification Working Group 1996).
2. Proportions of forest stands more than 100 years old have been calculated as a function of burn rate using the negative exponential model (Van Wagner 1978).
3. For sites in Québec, current burn rates have been calculated from MRNFQ fire data (1940–2003); for sites 11, 12, and 13, rates have been calculated from the Large Fire Database (1959–1999). See figure 5.1 for locations of the study areas and ecological regions.

territories considered, future burn rates for a $2\times CO_2$ period are higher than current burn rates (figure 5.3B). However, they remain lower than past rates (figure 5.3B). Four major types of response to climate change can be distinguished:

1. For the western black spruce–feather moss forest and for northwestern Ontario (regions 1, 3, 7, and 12), a $3\times CO_2$ period would lead to annual burn rates higher than past rates. However, it must be noted that the western black spruce forest is one of the largest ecological regions used in calculating current annual burn rates (figure 5.2), and there are therefore great spatial variations in fire activity. In addition, this region is intersected by the boundary of the intensive fire protection zone. The northern part of the region includes forests where fire activity is much greater than in the south (MRNQ 2000). Thus, it is difficult to interpret results for the whole of this region, since the great spatial variability may have led to an overestimation of burn rates for certain parts of the region, and underestimation for other parts. Moreover, in places where future rates are likely to be very high, there could potentially be major impediments to maintaining biodiversity and sustained yield, since fire activity would be competing with harvesting for mature stands.

2. Where burn rates remain lower than past rates (points located above diagonal 1:1 line in figure 5.3B), the historical impact of fire on forest management is reduced, in part, by the effects of climate change (regions 2, 4, 5, 8, and 11). This situation could represent an attractive opportunity for implementing forest strategies and practices that will maintain the historical age-class distribution at the landscape scale (Gauthier et al., chapter 3). However, even if the risk of loss of harvest volume does not arise from fire, other disturbances could dominate the forest dynamics and present new risks of volume loss.

3. In the case of Témiscamingue South, the model suggests that the future burn rate could be equal to or lower than the past burn rate. This result is at first surprising since even though regional fire regimes do not respond with the same magnitude, one would expect them to show the same trend in response to a global climatic stimulus. Drever et al. (in press) foresee a slight increase in future burn rates using regressions based on a larger area. Many factors could contribute to this result: accessibility of fire suppression methods, a history of earlier colonization, and a greater deciduous component than in other regions considered. These factors all contribute to older and more efficient fire suppression than in the other regions studied. The small amount of fire activity in the historical period limits our ability to estimate future burn rates.

4. Finally, we find areas with an intermediate situation, such as the North Shore, the Gaspésie, and Duck Mountain, where the future burn rate could be equivalent to past rates (figure 5.3B). In that situation, it would be difficult to adapt the harvest system to make up for decreased fire disturbance, since the proportion of forest stands more than 100 years old for $3\times CO_2$ period remains essentially the same as for the historical period (figure 5.3). Fire suppression and salvage logging would have to be based on an integrated strategy designed to avoid adding the effects of fire to those of forest management.

These results, arising as they do from climate simulations, must be interpreted with caution. Above all, they give a qualitative idea of the effects of climate change (e.g., direction, order of magnitude; Flato et al. 2000). Simulations of future values of the Fire Weather Index (Van Wagner 1987; Girardin et al., chapter 4) for different climate change scenarios carried out to date anticipate maintenance of current values or a reduction in the frequency of fires depending on the regions considered (Flannigan et al. 1998, 2001). In contrast, regression methods that use burned areas as predictive variables (as presented in this chapter) suggest that annual burned area will increase in all of Canada's ecozones (Flannigan et al. 2005). Simulation results suggest that the effects of climate change on fire activity are complex and that they vary from one region to another. By themselves, warmer temperatures would lead to an increase in fire activity. However, at the local level, changes in precipitation regimes modulate the response of fire activity to an increase in temperature. Indeed, increases in total precipitation could compensate for and even exceed the effect of warmer temperatures, thereby leading to a fire risk that is equivalent to or lower than the current risk in some areas (Flannigan et al. 1998; Bergeron et al. 2006).

While the models allow us to anticipate future trends in fire activity on a large scale, a great deal of uncertainty remains at the management unit level. Thus, it is unlikely that there would be a uniform decrease in fire frequency in black spruce forests in Québec. Western areas might react differently from those in the east, because of topographic and climatic differences. For that reason, at the forest management unit level, one must be prepared for either an increase or decrease in forest fire activity, and responses for both situations must be developed. A first step would probably involve adapting forest management to current burn rates. Despite the protection system, fires will likely continue to affect large areas, whether dry seasons are more or less frequent.

## 4. ADAPTATION OPTIONS TO FACE CHANGING FIRE ACTIVITY

Fires represent a risk of loss of merchantable volume that is difficult to predict locally because of their random occurrence and variability from year to year. The risk should nevertheless be considered when calculating annual allowable cut because of the weather dependence of fires and their impact on timber supply. Including fire risk into the calculations would also make it possible to regulate timber supply (Armstrong et al. 1999; Fall et al. 2004). Fire represents a risk that increases the vulnerability of managed forests to climate change and variability.

### 4.1. Re-examining Current Practices

Adaptation starts with a review of routine practices, followed by development of new strategies such as Fire-Smart Management or a Triad approach, which have been developed to improve our ability to react more effectively to changes in a fire regime (see also box 5.3, p. 119).

### 4.1.1.  Fire Suppression

In studies on the capacity of the forestry sector to adapt to climate change, increased fire suppression is often recommended as a means of reducing losses in allowable cut resulting from increased forest fire activity (Dale et al. 2001; Spittlehouse and Stewart 2003). Our fire suppression system was developed at a time when fire activity was lower than it is today, and fire activity has been continually increasing (1960–1970, see Girardin et al., chapter 4). Across Canada, fire frequency has been increasing for the past 30 years (Skinner et al. 1999, 2002; Gillett et al. 2004), which contributes to increase overflowing situations for fire protection agencies. The initial attack is considered to be the best practice to reduce the burned area (Hirsch et al. 1998; Lemaire 2002). However, McAlpine and Hirsch (1999) suggest that even with massive investment in forest fire suppression, 3 to 4% of forest fires would still burn out of control. Such fires would leave behind large burned areas, because they occur under extreme weather conditions (Larsen 1997; Drolet 2002; Gauthier et al. 2005). As a result, in years when the Fire Weather Index is high, burned areas remain large in spite of suppression efforts (McAlpine and Hirsh 1999; Leduc et al. 2002). Our current ability to suppress forest fires already faces economic, logistical, and ecological limits (Hirsch et al. 2001; Wheaton 2001; Chapin et al. 2003). These limits suggest the need for a greater rationalization of suppression efforts rather than a simple increase of these efforts everywhere, because fire regimes vary from one region to another (Lefort et al. 2004; Stocks et al. 1998; Parker et al. 2000; Wheaton 2001). In 2005, the Canadian Council of Forest Ministers formulated a national strategy for forest fire management that recognized the logistical limits of suppression and the ecological role of fire (CCFM 2005). This document also established criteria for rationalizing fire suppression efforts. While fire control is one available adaptation tool, adopting a strategy for total exclusion of fire is not only ecologically undesirable, but also economically unrealistic. Strategies must therefore be developed that give priority to interventions in zones of high economic, social, and ecological value while options have to be envisioned to take advantage of fire occurrence in other places.

Finally, recognizing that forest fires will always occur, even with our suppression efforts, it is frequently being suggested that fire risk be included in the calculation of annual allowable cut. When this is done, annual timber supply is stabilized even if it is reduced (Armstrong et al. 1999; Fall et al. 2004).

### 4.1.2.  Salvage Logging

In evaluations of the forest sector's adaptation to climate change, general and systematic salvage logging operations are often recommended as a step to adapting to anticipated increases in fire activity (Wheaton 2001; Spittlehouse and Stewart 2003; Johnston et al. 2006). However, it must be emphasized that in the wake of a fire, the salvage rate depends directly on the age structure of the forest: the higher the proportion of mature and over-mature stands, the higher the proportion of burned forest suitable for salvage will be. Traditional forest management, with its overall effect of rejuvenating the forest through harvesting, reduces the number of stands suitable for salvage logging (Leduc et al. 2004), since many burned trees will not have reached commercial size at the time of the fire. Despite a persistent belief that fire particularly affects mature forests, fire in the boreal forest acts independently of stand age (Johnson 1992). Thus it must be recognized that a large portion of stands that have been burned will not be suitable for salvage logging.

As in other provinces that practice salvage logging, when Québec's commercial forests are affected by fire, the Ministère des Ressources naturelles et de la Faune (MRNF) evaluates the volumes to be harvested and develops special salvage plans for the areas affected by the fire. Priority is given to the most accessible locations. In addition, holders of Timber Supply and Forest Management Agreements must comply with the special plan. Failing implies that the timber volume allocated under the contract will be reduced by an amount equal to the volume that would have been salvaged under the special plan (MRNFQ 2006). The compulsory nature of these special plans led to an increase in volumes of timber salvaged after fires from 3 m³/ha in 1991 to 10 m³/ha in 1996 (Purdon et al. 2002). In Québec, between 1995 and 1998, 13 to 40% of burned commercial forest areas over 1,000 ha were subject to salvage logging operations. During the same period, the intensity of salvage logging in burned areas increased by more than 50% (Nappi et al. 2004). Since then, the salvage rate has varied in line with the annual area burned. For 2005–2006, which was marked by many large fires, the figure was more than 6 million cubic metres, the equivalent of about 30% of total harvest volume in Québec during that period.

Salvage logging must take place soon after a fire, since the quality of timber is reduced once insects start to colonize it. Emergency salvage plans for burned areas affect operating costs by disrupting established forestry plans (Martell 2002). For example, they involve construction of forest roads that were not planned and the use of specific equipment to avoid contaminating traditionally managed forest products (Patry 2002). These plans, which are often developed in great haste after a fire, would be more effective if they were part of a strategy established well before fire occurs (Lindenmayer et al. 2004). Efforts are now underway in Québec, Ontario, Saskatchewan, and Alberta to improve planning for salvage logging and to carry out such logging on the basis of ecosystem management criteria (Schmiegelow et al. 2006).

Several studies emphasize the ecological limits of salvage logging as it is now practised (Lindenmayer et al. 2004; Brais et al. 2006; Donato et al. 2006; Greene et al. 2006; Lindenmayer 2006). Salvage logging can act as an additional disturbance to soil and vegetation, at a particularly critical stage since the operation takes place during the first stages of post-fire regeneration (Johnston et al. 2006). While species and ecosystems are generally well adapted to the recurrence periods of forest fires with which they have evolved, a fire followed by a harvesting operation could exceed the limits of their resilience, resulting, for example, in regeneration accidents. Salvage logging can also act as an additional disturbance to soils that have been made particularly fragile by a reduction in the thickness of organic matter as a result of combustion. Forest stands that have been subjected to salvage logging present a different array of available nutrients from stands that were burned but not harvested (Brais et al. 2000; Purdon et al. 2004). Finally, as currently practised, salvage logging reduces the availability of post-fire habitat and also limits the availability of prey for insect-eating birds

(Morissette et al. 2002). In an ecosystem approach to forest management, the negative impacts of salvage logging would have to be mitigated in order to set management objectives that are not limited to regeneration.

In some cases, salvage logging produces ecological benefits in addition to its economic benefits. For example, if a burned area has not been naturally regenerated because the fire interval is too short, or the fire was not sufficiently severe, salvage logging followed by reforestation will limit the loss of forest cover. This aspect is especially important at the limit of the commercial forest (Jasinski and Payette 2005). In short, salvage logging is a potential tool for adapting to climate change but it does have related logistic, economic, and ecological limitations.

### 4.1.3.   Improving Regeneration

Some stands regenerate well after fire, especially when the fire has been severe enough to release seeds and create suitable germination beds (by burning sufficient organic matter on the ground). In such cases, artificial regeneration may be unnecessary and the machinery used for it could hinder natural regeneration (Donato et al. 2006; Greene et al. 2006). In contrast, other stands affected by less severe fires will regenerate poorly and require artificial regeneration (Jayen et al. 2006). Regeneration accidents generally result from interaction between the regeneration potential of the species at the time of the fire (stand age) and the characteristics of the fire. The fire interval and the severity determine both the regeneration potential of the species at the time of the fire (Le Goff and Sirois 2004), and the quality and quantity of germination beds (Jayen et al. 2006). Weather conditions after a fire will influence germination and seedling establishment (Thomas and Wein 1985; Lavoie and Sirois 1998). Since the regeneration potential of species at the time of the fire is an essential factor, it should be noted that the younger the forest, the more susceptible it will be to regeneration accidents, a fact that necessarily demands an increased plantation effort (Leduc et al. 2004). In areas where fire activity increases because of climate change, the interval between two consecutive fires in the same location could become shorter, which would increase the probability of regeneration accidents.

If regeneration accidents increase under the influence of climate change, the need for artificial regeneration will be more frequent. Genetically improved seedlings could be used to restock stands after harvest, using better performing individuals (Grossnickle and Sutton 1999). Recent advances in technology also make it possible to consider varietal forestry in order to achieve much better productivity than what can be achieved through improved first generation seed trees (Park 2002). In addition, advances in ground preparation methods (Hébert et al. 2006) and proper characterization of sites allow for better selection of species for replanting that are more suited to conditions at the regeneration site. For example, in some regions, planters work with a mixed basket of plants that allows them to select pine seedlings for sites where mineral soil is uncovered and on well-drained small hills, and spruce seedlings for rich microsites where the litter layer is intermediate and the soil remains moist during the summer (microsite planting, Lieffers et al. 2003). Finally, in zones of frequent fire occurrence, the use of species such as jack pine (*Pinus banksiana* Lamb.), which is able to regenerate effectively after fire at a very young age, should be encouraged.

## 4.1.4.   Prescribed Burning

Prescribed burning is a tool that can serve different management and conservation objectives:

- Reducing the amount of available fuel, which reduces the risk of fire. For example, in British Columbia, in 1988, two prescribed burns were used on 37,000 ha of near-urban forest (Forest Practices Board 2006).

- Improving regeneration conditions of many pioneer species (Weber and Taylor 1992), by making nutrients available through elimination of competing vegetation (Tellier et al. 1995) and creating suitable germination beds (Nguyen-Xuan et al. 2000).

- Restoring or maintaining a pyrophilous species or community (Johnston and Elliott 1996; Whittle et al. 1997; Quenneville and Thériault 2002).

- Controlling invasive exotic species (DiTomaso et al. 2006).

- Maintaining wildlife habitat (Weber and Taylor 1992). The British Columbia Ministry of the Environment uses prescribed burns for this purpose on some 200,000 ha each year (Forest Practices Board 2006).

In Québec, prescribed burning is only used to achieve conservation objectives. For the past 15 years, Parks Canada has used prescribed burning as part of its program to restore stands of eastern white pine (*Pinus strobus* L.) in La Mauricie National Park (Quenneville and Thériault 2002). There has been a great deal of resistance to the use of fire as a management tool. This resistance is chiefly due to a heritage of European forest management and a policy of systematic exclusion of fires (Blanchet 2003). This mistrust of fire was strongly reinforced by a great increase in forest fire activity across Canada in the years from 1980 to 1990. However, the International Crown Fire Modelling experiment has considerably advanced our understanding of fire behaviour (Stocks et al. 2004).

The eastern boreal forest is subject to a crown fire regime that has little dependence on ground fuel accumulation (Hély et al. 2000). For this reason, it is generally recognized that fire risk does not depend on the age of the stand (Johnson 1992). Since fuel accumulation does not increase the risk of severe fire as a function of the time elapsed since the previous fire, prescribed burning has little attraction for reducing fuel loads in the boreal forest of western Québec. However, Lavoie et al. (2005) suggest that it could be used in northern Abitibi, alone or in combination with drainage activities, to reduce the accumulation of organic matter where forest litter is moderately thick, and this would contribute to improved timber productivity (see Simard et al., chapter 11).

In the case of reduced fire frequency, prescribed burning could be considered a means of maintaining the ecological benefits of fire. It could also find many applications in ecosystem forest management, particularly in promoting biodiversity and the regeneration of certain species. However, more exploration of such applications is needed in Québec. The development and implementation of such measures will require a policy on information and education of the public and forestry managers to overcome existing resistance to prescribed burns.

## 4.2. Developing New Adaptation Options

### 4.2.1. Ecosystem Management Based on Natural Disturbances

Adaptation to climate change and ecosystem management share common principles in that both are founded on an understanding of the natural, integrated dynamics of ecosystems. They require an adaptive management framework in order to progress and improve in line with our understanding and with the effectiveness of the measures implemented to achieve objectives. Within a forest ecosystem management context, it is useful to examine the potential effects of climate change and assess the resulting vulnerability (see box 5.3).

Even if forest ecosystems are well adapted to their natural disturbance regime, harvesting often represents a pressure that is added to the effect of natural disturbances. The combination may well exceed the resilience of the forest ecosystem. To maintain the organisms and functions of these ecosystems, it is important to keep total disturbance rate (harvesting plus natural disturbances) within its historical limits of variability. A sustainable annual allowable cut could correspond to the difference between the average disturbed area with and without fire suppression (Armstrong et al. 1999). This type of calculation is one of the solutions for integrating fire risk a priori into the estimation of annual allowable cut. Meanwhile, in forests where the current annual burn rate is lower than the past burn rate, the Three-Cohort approach, presented by Bergeron et al. (1999)

---

**Box 5.3**
## Is managed boreal forest more vulnerable to climate change than natural forest?

A managed boreal forest is thought to be less vulnerable to climate change because it is generally younger (and therefore considered more vigorous) and suitable for implementing planned adaptation strategies, such as salvage logging or selection of species better adapted to current and future climate conditions (Leduc et al. 2004; see section 4). Moreover, it is often argued that planned adaptation strategies can only be applied within managed forests (Spittlehouse 1997). However other adaptation measures, such as fire suppression and insect outbreak control, can be carried out over large forest areas even when imminent harvesting is not planned.

Natural and managed forests differ in terms of susceptibility to the impacts of climate change because of differences in age structure and composition. The harvesting pressure that forest management exercises on the natural forest mosaic produces a forest in which the proportion of pre-mature and regenerating stands is greater than in a natural forest dominated by fire (see Gauthier et al., chapter 3). Such rejuvenation also implies a large proportion of early succession species. In the case of careful logging techniques such as the Cutting with Regeneration and Soil Protection (CPRS), as practised in black spruce stands (*Picea mariana* [Mill.] B.S.P.) with an understory of balsam fir (*Abies balsamea* [L.] Mill.), the proportion of balsam fir in the canopy may be higher than in similar stands under natural succession (Leduc et al. 2004). This situation may increase stand vulnerability to spruce budworm.

A managed forest does not have the same vulnerabilities to climate change as a natural forest. For example, because of its larger proportion of regenerating stands, a managed forest could be more sensitive to late frost or to successive disturbances that lead to regeneration accident. Because they are younger on average, managed forests are generally less suited to adaptation measures such as salvage logging after a forest fire. However, it is usually easier to suppress forest fires in a managed forest because of greater accessibility and the presence of less flammable hardwood species (Hély et al. 2001). Managed forests are no less vulnerable to climate change than natural forests, but they have different vulnerabilities. As a result, adaptation strategies must target their specific vulnerabilities.

as well as Gauthier et al. (chapter 3), could be implemented to slow rejuvenation of managed forest mosaics and to increase their suitability for salvage logging in the event of a natural disturbance.

## 4.2.2.   The Zoning Approach

In a sustainable forest management context, the spatial distribution of cut blocks represents a compromise between timber values, the cost of road building, and fragmentation of wildlife habitats. The Triad management strategy was proposed as a means of reconciling the objectives of forest management and biodiversity conservation (D'Eon et al. 2004). The Triad strategy distributes protection and intensive management zones within a matrix where forests are managed based on their natural disturbance regimes (Montigny and MacLean 2006). This strategy concentrates intensive management activities within a restricted area that is easier to protect, thereby increasing the volume of timber harvested. This approach also has the advantage of reducing harvesting pressure over a larger part of the region, which can then be assigned to ecosystem conservation or to extensive management based on natural disturbances.

When the Triad approach is applied to a forest management unit, fire protection efforts must be directed primarily toward forest stands of great economic value (e.g. plantations of improved species). Within the framework of this type of forest management, sectors of high economic value must be concentrated within restricted areas having very good accessibility. Such a configuration could greatly enhance the effectiveness of fire suppression measures. The Triad approach could be used to map fire protection priorities within a management unit. By rationalizing suppression efforts in a region, the Triad approach could improve the existing adaptive capacity under changing fire activity. This kind of approach is part of the CCFM's national forest fire strategy (2005).

## 4.2.3.   Fire-Smart Management to Reduce Fire Risk

Fire-Smart forest management uses forestry operations to change the structure and composition of stands in order to reduce their flammability and connectivity (Hirsch et al. 2001). It may involve planting barriers of hardwood stands, which are less flammable than softwood stands (Hély et al. 2000), at strategic locations for future harvesting, or developing a forest road network in order to spread the fire break effect of roads. Fire-Smart forest management has a good potential for adaptation, since it incorporates a number of practices that enable management of fire risk and promote regeneration of burned or harvested stands (Johnston 2001; McKinnon and Kaczanowski 2004). However, application of this approach in eastern Canada would require assessment of the effectiveness of the deciduous component as a firebreak. At the same time, the impacts of changes in the structure and composition of forest stands on biodiversity and on the connectivity of wildlife habitats would have to be assessed to determine the extent to which this management strategy could be integrated into sustainable forest management.

## 5.   RECOMMENDATIONS

### 5.1.   Face Uncertainty with a Proactive Approach

Many studies document an increase in fire frequency across Canada since the 1970s (Skinner et al. 1999, 2002; Gillett et al. 2004; Macias Fauria and Johnson 2006), and this trend could continue in the future (Flannigan et al. 2005, see the section on how climate change will modify future forest fire activity). In this context, increased fire suppression efforts are often recommended, along with systematic salvage logging and practices aimed at improving regeneration. However, the logistical and ecological limits of these adaptation options suggest rationalization and planning of recourse to such practices on the basis of intervention priorities and the need to maintain biodiversity. This kind of approach may require preparation of a detailed map of the risks of lost volumes from different kinds of disturbances. This sort of risk mapping is already being done in New Brunswick to assess, for example, potential losses from epidemics of spruce budworm (*Choristoneura fumiferana* Clem.) (SBW, MacLean et al. 2002). Other vulnerability maps could assess windfall or fire risk. Such maps represent the first step toward an integrated strategy for risk management. They need to be supplemented by other tools, such as guides or action plans for use in emergency situations when there is a fire or an insect outbreak.

### 5.2.   Make a Comprehensive Assessment of Forest Vulnerability

Through its influence on forest conditions, management changes the vulnerability of the forest to climate effects (see box 5.3). This is just as true of vulnerability to fire as it is for other kinds of disturbances, such as windfall, diseases, and insect pests. A proper environmental risk assessment exercise for the forestry sector with respect to climate change would require a thorough combined evaluation of a variety of risk factors. It would involve assessing potential positive and negative effects on various aspects of forest management, such as tree growth and gains or losses of volumes or yields in management units in relation to assessments of future risks of insect epidemics, forest fires or weather anomalies (e.g. late frost, early thaw, or more frequent periods of drought).

A succession of disturbances within a short time span appears to have converted black spruce–feather moss forests into black spruce-lichen woodlands within the boundaries of Québec's commercial forest (Payette et al. 2000). Such waves of disturbances can profoundly affect the structure and composition of a forest. It may involve salvage logging (fire + harvesting) or an outbreak of fire in a regeneration area (harvesting + fire). Added to management practices, climate change threatens to increase the occurrence of successive disturbances, which would change the frequency of regeneration accidents and other types of forest ecosystem conversion. This implies that there is a need to learn to manage a different kind of forest (Spittlehouse 1997). For example, if the proportion of hardwood stands increases, then changes in the susceptibility of forests to insect outbreaks will need to be monitored. A higher proportion of hardwood stands could reduce the risk of a spruce budworm outbreak, while increasing the risk from the forest tent caterpillar outbreak (*Malacosoma disstria* Hbn.). In addition, the large areas affected by an insect outbreak can create a fuel accumulation that would increase the risk of severe large fires in some regions (Fleming et al. 2002). An integrated assessment of the impacts of climate change on forests is needed in relation to the transformations that forests are undergoing, in order to develop effective adaptation strategies and avoid implementing measures that target one particular impact but could aggravate another. This kind of exercise was conducted on a cross-Canada basis (Lemmen and Warren 2004), for Ontario (Colombo et al. 1998), and Québec (Bourque and Simonet 2008). An assessment of the vulnerability of forests and management to climate change, along with integration into forest management practices of the expected effects of such change, are among Québec's climate change priorities (MDDEPQ 2006).

## 5.3.    Choose an Adaptive Framework for Forest Management

Although it is difficult to predict the precise time or place of the next fire or the next insect outbreak, the time frames and spatial horizons of forest operations planning readily lend themselves to assessment and integration of environmental risk. The precautionary principle suggests that adaptation measures be implemented now in order to limit future damage and to better prepare the forest industry for future climate change by increasing its capacity to adapt to new climatic conditions (Spittlehouse and Stewart 2003). In a context where we must act in spite of uncertainty, only an adaptive management framework will enable us to formulate appropriate strategies encouraging rapid adjustments as results are observed and as our understanding improves.

By developing strategies to reduce forest vulnerability to current climate variability, forest managers will be better prepared to respond effectively to the new forest conditions that arise from future climate change (Smit et al. 1999). Finally, it must be accepted that part of the forest's response to climate change cannot be anticipated. This means we must be prepared to periodically revise our expectations about the forest resource (Spittlehouse 2006).

## REFERENCES

Armstrong, G.W., Cumming, S.G., and Adamowicz, W.L. 1999. Timber supply implications of natural disturbance management. For. Chron. 75: 497–504.

Beaulieu, J. and Rainville, A. 2005. Adaptation to climate change: genetic variation is both a short- and a long-term solution. For. Chron. 81: 704–709.

Bergeron, Y., Cyr, D., Drever, C.R., Flannigan, M., Gauthier, S., Kneeshaw, D., Lauzon, È., Leduc, A., Le Goff, H., Lesieur, D., and Logan, K. 2006. Past, current, and future fire frequencies in Quebec's commercial forests: implications for the cumulative effects of harvesting and fire on age-class structure and natural disturbance-based management. Can. J. For. Res. 33: 2737–2744.

Bergeron, Y., Gauthier, S., Flannigan, M., and Kafka, V. 2004. Fire regimes at the transition between mixedwood and coniferous boreal forest in northwestern Quebec. Ecology, 85: 1916–1932.

Bergeron, Y., Harvey, B., Leduc, A., and Gauthier, S. 1999. Forest management guidelines based on natural disturbance dynamics: stand- and forest-level considerations. For. Chron. 75: 49–54.

Blanchet, P. 2003. Feux de forêt: l'histoire d'une guerre. Cantos International Publishing Inc. Montréal, Que.

Bourque, A. and Simonet, G. 2008. Quebec. *In* From impacts to adaptation: Canada in a changing climate 2007. *Edited by* D.S. Lemmen, F.L. Warren, and E. Bush. Government of Canada, Ottawa, Ont.

Brais, S., Paré, D., and Lierman, C. 2006. Tree bole mineralization rates of four species of the Canadian eastern boreal forest: implications for nutrient dynamics following stand-replacing disturbances. Can. J. For. Res. 36: 2331–2340.

Brais, S., Paré, D., and Ouimet, R. 2000. Impacts of wild fire severity and salvage harvesting on the nutrient balance of jack pine and black spruce boreal stands. For. Ecol. Manag. **137**: 231–243.

Burton, I. 1998. Climate adaptation policies for Canada? Policy Options, **May 1998**: 6–10.

Burton, I., Huq, S., Lim, B., Pilifosova, O., and Schipper, E.L. 2002. From impacts assessment to adaptation priorities: the shaping of adaptation policy. Clim. Policy, **2**: 145–159.

Canadian Council of Forest Ministers (CCFM). 2005. Canadian wildland fire: a vision for an innovative and integrated approach to managing the risks. Canadian Council of Forest Ministers, Ottawa, Ont.

Candau, J.N. and Fleming, R.A. 2005. Landscape-scale spatial distribution of spruce budworm defoliation in relation to bioclimatic conditions. Can. J. For. Res. **35**: 2218–2232.

Carroll, A.L., Régnière, J., Logan, J.A., Taylor, S.W., Bentz, B.J., and Powell, J.A. 2006. Impacts of climate change on range expansion by the mountain pine beetle. Mountain Pine Beetle Initiative Working Paper 2006–14. Natural Resources Canada, Canadian Forest Service, Pacific Forestry Centre, Victoria, B.C. [Online] <warehouse.pfc.forestry.ca/pfc/26601.pdf> (accessed October 22, 2007).

Chapin, F.S., III, Rupp, T.S., Starfield, A.M., DeWilde, L., Zavaleta, E.S., Fresco, N., Henkelman, J., and McGuire, A.D. 2003. Planning for resilience: modeling change in human-fire interactions in the Alaskan boreal forest. Front. Ecol. Environ. **1**: 255–261.

Colombo, S.J, Buse, L.J., Cherry, M.L., Graham, C., Greifenhagen, S., McAlpine, R.S., Papadopol, C.S., Parker, W.C., Scarr, T., Ter-Mikaelian, M.T., and Flannigan, M.D. 1998. The impacts of climate change on Ontario's forests. Ontario Forest Research Institute, Forest Research Information Paper, v. 143, no. 50. Sault Ste. Marie, Ont.

Consortium Ouranos. 2004. S'adapter aux changements climatiques. Consortium sur la climatologie régionale et l'adaptation aux changements climatiques. [Online] <www.ouranos.ca> (accessed October 22, 2007).

Cyr, D., Gauthier, S., and Bergeron, Y. 2007. Scale-dependent determinants of heterogeneity in fire frequency in a coniferous boreal forest of eastern Canada. Landsc. Ecol. **22**: 1325–1339.

Dale, V.H., Joyce, L.A., McNulty, S., Neilson, R.P., Ayres, M.P., Flannigan, M.D., Hanson, P.J., Irland, L.C., Lugo, A.E., Peterson, C.J., Simberloff, D., Swanson, F.J., Stocks B.J., and Wotton, B.M. 2001. Climate change and forest disturbances. BioScience, **51**: 723–734.

D'Eon, R.G., Hebert, D., and Viszlai, S.L. 2004. An ecological rationale for sustainable forest management concepts at Riverside Forest Products, southcentral British Columbia. For. Chron. **80**: 341–348.

DiTomaso, J.M., Brooks, M.L., Allen, E.B., Minnich, R., Rice, P.M., and Kyser, G.B. 2006. Control of invasive weeds with prescribed burning. Weed Tech. **20**: 535–548.

Donato, D.C., Fontaine, J.B., Campbell, J.L., Robinson, W.D., Kauffman, J.B., and Law, B.E. 2006. Post-wildfire logging hinders regeneration and increases fire risk. Science, **311**: 352.

Drever, C.R., Bergeron Y., Drever, M.C., Flannigan, M., Logan, T., and Messier, C. in press. Effects of climate on occurrence and size of large fires in a northern hardwood landscape: historical trends, future predictions, and implications for climate change in Témiscamingue, Quebec. Appl. Veg. Sci.

Drever, C.R., Messier, C., Bergeron, Y., and Doyon, F. 2006. Fire and canopy species composition in the Great Lakes-St. Lawrence forest of Témiscamingue, Quebec. For. Ecol. Manag. **231**: 27–37.

Drolet, B. 2002. La protection des forêts contre le feu: bilan et perspectives. *In* the proceedings of the seminar *L'Aménagement forestier et le feu*, April 9–11, 2002, Chicoutimi, Que. *Edited by* the Canadian Forest Service, the Société de protection des forêts contre le feu and the Sustainable Forest Management Network. Ministère des Ressources naturelles du Québec, Québec, Que., pp. 7–17.

Ecological Stratification Working Group 1996. A national ecological framework for Canada. Agriculture and Agri-Food Canada, Research Branch, Centre for Land and Biological Resources Research and Environment Canada, State of Environment Directorate, Ottawa/Hull, Ont.

Fall, A., Fortin, M.-J., Kneeshaw, D.D, Yamasaki, S.H., Messier, C., Bouthillier, L., and Smyth, C. 2004. Consequences of various landscape-scale ecosystem management strategies and fire cycles on age-class structure and harvest in boreal forests. Can. J. For. Res. **34**: 310–322.

Flannigan, M.D., Bergeron, Y., Engelmark, O., and Wotton, B.M. 1998. Future wildfire in circumboreal forests in relation to global warming. J. Veg. Sci. **9**: 469–476.

Flannigan, M., Campbell, I., Wotton, M., Carcaillet, C., Richard, P., and Bergeron, Y. 2001. Future fire in Canada's boreal forest: paleoecology results and general circulation model – regional climate model simulations. Can. J. For. Res. 31: 854–864.

Flannigan, M.D. and Harrington, J.B. 1988. A study of the relation of meteorological variables to monthly provincial area burned by wildfire in Canada (1953–80). J. Appl. Meteorol. **27**: 441–452.

Flannigan, M.D., Logan, K.A., Amiro, B.D., Skinner, W.R., and Stocks, B.J. 2005. Future area burned in Canada. Clim. Change, **72**: 1–16.

Flato, G.M., Boer, G.J., Lee, W.G., McFarlane, N.A., Ramsden, D., Reader, M.C., and Weaver, A.J. 2000. The Canadian Centre for Climate Modelling and Analysis global coupled model and its climate. Clim. Dyn. **16**: 451–467.

Fleming, R.A., Candau, J.-N., and McAlpine, R.S. 2002. Landscape-scale analysis of interactions between insect defoliation and forest fire in central Canada. Clim. Change, **55**: 251–272.

Forest Practices Board. 2006. Forest fuels a burning issue for Interior BC. Forest Practices Board, Victoria, B.C. [Online] <www.fpb.gov.bc.ca/news/releases/2006/06–19.htm> (accessed October 22, 2007).

Gauthier, S., Chabot, M., Drolet, B., Plante, C., Coupal, J., Boivin, C., Juneau, B., Lefebvre, F., Ménard, B., Villeneuve, R., and Gagnon, L. 2005. Groupe de travail sur les objectifs opérationnels de la SOPFEU: rapport d'analyse. Société de protection des forêts contre le feu, Québec, Que.

Gillett, N.P., Weaver, A.J., Zwiers, F.W., and Flannigan, M.D. 2004. Detecting the effect of climate change on Canadian forest fires. Geophys. Res. Lett. **31**: L18211.

Greene, D.F., Gauthier S., Noël, J., Rousseau, M., and Bergeron, Y. 2006. A field experiment to determine the effect of post-fire salvage on seedbeds and tree regeneration. Front. Ecol. Environ. **4**: 69–74.

Grenier, D.J., Bergeron, Y., Kneeshaw, D., and Gauthier, S. 2005. Fire frequency for the transitional mixedwood forest of Timiskaming, Quebec, Canada. Can. J. For. Res. **35**: 656–666.

Grossnickle, S.C. and Sutton, B.C.S. 1999. Applications of biotechnology for forest regeneration. New For. **17**: 213–226.

Hébert, F., Boucher, J.F., Bernier, P.Y., and Lord, D. 2006. Growth response and water relations of 3-year-old planted black spruce and jack pine seedlings in site prepared lichen woodlands. For. Ecol. Manag. **223**: 226–236.

Hély, C., Bergeron, Y., and Flannigan, M.D. 2000. Coarse woody debris in the southeastern Canadian boreal forest: composition and load variations in relation to stand replacement. Can. J. For. Res. **30**: 674–687.

Hély, C., Flannigan, M., Bergeron, Y., and McRae, D. 2001. Role of vegetation and weather on fire behavior in the Canadian mixedwood boreal forest using two fire behavior prediction systems. Can. J. For. Res. **31**: 430–441.

Hirsch, K.G., Corey, P.N., and Martell, D.L. 1998. Using expert judgment to model initial attack fire crew effectiveness. For. Sci. **44**: 539–549.

Hirsch, K., Kafka, V., Tymstra, C., McAlpine, R., Hawkes, B., Stegehuis, H., Quitilio, S., Gauthier, S., and Peck, K. 2001. Fire-smart forest management: A pragmatic approach to sustainable forest management in fire-dominated ecosystems. For. Chron. **77**: 357–363.

Hulme, P.E. 2005. Adapting to climate change: is there scope for ecological management in the face of a global threat? J. Appl. Ecol. **42**: 784–794.

Intergovernmental Panel on Climate Change (IPCC) 2001. Climate change 2001. Impacts, adaptation and vulnerability. Report of Working Group II to IPCC. Summary for Policymakers. Contribution of Working Group II to the Third Assessment Report of the Intergovernmental Panel on Climate Change (IPCC) [Online] <www.ipcc.ch/languageportal/frenchportal.htm#21> (accessed November 14, 2007).

Intergovernmental Panel on Climate Change (IPCC) 2007. Climate change 2007: The physical science basis. Contribution of Working Group I to the Fourth Assessment Report of the Intergovernmental Panel on Climate Change. *Edited by* S. Solomon, D. Qin, M. Manning, Z. Chen, M. Marquis, K.B. Averyt, M. Tignor, and H.L. Miller. Cambridge University Press, Cambridge, UK.

Jasinski, J.P.P. and Payette, S. 2005. The creation of alternative stable states in the southern boreal forest, Quebec, Canada. Ecol. Monogr. **75**: 561–583.

Jayen, K., Leduc, A., and Bergeron, Y. 2006. Effect of fire severity on regeneration success in the boreal forest of northwest Quebec, Canada. Écoscience, **13**: 143–151.

Johnson, E.A. 1992. Fire and vegetation dynamics: Studies from North American boreal forest. Cambridge University Press, Cambridge, UK.

Johnson, E.A., Miyanishi, K., and O'Brien, N. 1999. Long-term reconstruction of the fire season in the mixedwood boreal forest of Western Canada. Can. J. Bot. **77**: 1185–1188.

Johnston, M. 2001. Sensitivity of boreal forest landscapes to climate change. Saskatchewan Research Council, Environmental Branch, SRC Publication No. 11341–7E01. Saskatoon, Sask.

Johnston, M.H. and Elliot, J.A. 1996. Impacts of logging and wildfire on an upland black spruce community in northwestern Ontario. Environ. Monit. Assess. **39**: 283–297.

Johnston, M., Williamson, T., Price, D., Spittlehouse, D., Wellstead, A., Gray, P, Scott, D., Askew, S., and Webber, S. 2006. Adapting forest management to the impacts of climate change in Canada. Final report, BIOCAP Research Integration Program Synthesis Paper, Queen's University, Kingston, Ont. [Online] <www.biocap.ca/rif/report/Johnston_M.pdf> (accessed October 22, 2007).

Kafka, V., Gauthier, S., and Bergeron. Y. 2001. Fire impacts and crowning in the boreal forest: study of a large wildfire in western Quebec. Int. J. Wildland Fire, **10**: 119–127.

Landsberg, J.J. and Waring, R.H. 1997. A generalised model of forest productivity using simplified concepts of radiation-use efficiency, carbon balance and partitioning. For. Ecol. Manag. **95**: 209–228.

Larsen, C.P.S. 1997. Spatial and temporal variations in boreal forest fire frequency in northern Alberta. J. Biogeogr. **24**: 663–673.

Lavoie, M., Paré, D., and Bergeron, Y. 2005. Impact of global change and forest management on carbon sequestration in northern forested peatlands. Environ. Rev. **13**: 199–240.

Lavoie, L. and Sirois, L. 1998. Vegetation changes caused by recent fires in the northern boreal forest of eastern Canada. J. Veg. Sci. **9**: 483–492.

Lauzon, È., Kneeshaw, D., and Bergeron, Y. 2007. Reconstruction of fire history (1680–2003) in Gaspesian mixedwood boreal forests of eastern Canada. For. Ecol. Manag. **244**: 41–49.

Leduc, A. 2002. Effet de la suppression des incendies fores-tiers sur les régimes des feux. *In* the proceedings of the seminar *L'Aménagement forestier et le feu*, April 9–11, 2002, Chicoutimi, Que. *Edited by* the Canadian Forest Service, the Société de protection des forêts contre le feu and the Sustainable Forest Management Network. Ministère des Ressources naturelles du Québec, Québec, Que, pp. 85–90.

Leduc, A., Gauthier, S., Bergeron, Y., and Harvey, B. 2004. La vulnérabilité de la forêt boréale aux changements climatiques: les forêts aménagées sont-elles plus vulné-rables? *In* Effects of climate change on major forest disturbances (fire, insects) and their consequences on biomass production in Canada: synthesis of the current state of knowledge. Workshop held in Quebec City, September 21, 2003. *Edited by* S. Gauthier, D.R. Gray, and C. Li, Natural Resources Canada, Canadian Forest Service, Laurentian Forestry Centre, Québec, Que.

Lefort, P., Gauthier, S., and Bergeron, Y. 2003. The influ-ence of fire weather and land use on the fire activity of the Lake Abitibi area, eastern Canada. For. Sci. **49**: 509–521.

Lefort, P., Leduc, A., Gauthier, S., and Bergeron, Y. 2004. Recent fire regime (1945–1998) in the boreal forest of western Quebec. Écoscience, **11**: 433–445.

Le Goff, H., Flannigan, M.D., Bergeron, Y., and Girardin, M.P. 2007. Historical fire regime shifts related to climate teleconnections in the Waswanipi area, central Quebec, Canada. Int. J. Wildland Fire, **16**: 607–618.

Le Goff, H., Leduc, A., Bergeron, Y., and Flannigan, M. 2005. The adaptive capacity of forest management to changing fire regimes in the boreal forest of Quebec. For. Chron. **81**: 582–592.

Le Goff, H. and Sirois, L. 2004. Black spruce and jack pine dynamics simulated under varying fire cycles in the northern boreal forest of Quebec, Canada. Can. J. For. Res. **34**: 2399–2409.

Lemaire, G. 2002. Lutte directe: portée et limites. *In* the proceedings of the seminar *L'Aménagement forestier et le feu*, April 9–11, 2002, Chicoutimi, Que. *Edited by* the Canadian Forest Service, the Société de protection des forêts contre le feu and the Sustainable Forest Manage-ment Network. Ministère des Ressources naturelles du Québec, Québec, Que., pp. 59–64.

Lemmen, D.S. and Warren, F.J. 2004. Climate change impacts and adaptation: a Canadian perspective. Natural Resources Canada. [Online] <adaptation.nrcan.gc.ca/perspective/index_e.php> (accessed October 22, 2007).

Lesieur, D., Gauthier, S., and Bergeron, Y. 2002. Fire fre-quency and vegetation dynamics for the south-central boreal forest of Quebec, Canada. Can. J. For. Res. **32**: 1996–2009.

Lieffers, V.J., Messier, C., Burton, P.J., Ruel, J.C., and Grover, B.E. 2003. Nature-based silviculture for sustaining a variety of boreal forest values. *In* Towards sustainable management of the boreal forest. *Edited by* P.J. Burton, C. Messier, D.W. Smith, and W.L. Adamovicz. NRC Research Press, Ottawa, Ont., pp. 481–530.

Lindenmayer, D. 2006. Salvage harvesting: past lessons and future issues. For. Chron. **82**: 48–53.

Lindenmayer, D.B., Foster, D.R., Franklin, J.F., Hunter, M.L., Noss, R.F., Schiemegelow, F.A., and Perry, D. 2004. Salvage harvesting policies after disturbance. Science, **303**: 1303.

Macias Fauria, M. and Johnson, E.A. 2006. Large-scale cli-matic patterns control large lightning fire occurrence in Canada and Alaska forest regions. J. Geophys. Res. **111**: G04008.

MacLean, D.A., Beaton, K.P., Porter, K.B., MacKinnon, W.E., and Budd, M.G. 2002. Potential wood supply losses to spruce budworm in New Brunswick estimated using the Spruce Budworm Decision Support System. For. Chron. **78**: 739–750.

Martell, D.L. 2002. Impacts économiques des feux. *In* the proceedings of the seminar *L'Aménagement forestier et le feu*, April 9–11, 2002, Chicoutimi, Que. *Edited by* the Canadian Forest Service, the Société de protection des forêts contre le feu and the Forest Management Net-work. Ministère des Ressources naturelles du Québec, Québec, Que., pp. 59–64.

McAlpine, R.S. and Hirsch, K.G. 1999. An overview of LEOPARDS: the Level of Protection Analysis System. For. Chron. **75**: 615–621.

McCarthy, J.J., Canziani, O.F., Leary, N.A., Dokken, D.J., and White, K.S. 2001. Climate change 2001: impacts, adaptation, and vulnerability. Contribution of Working Group I to the Third Assessment Report of the Inter-governmental Panel on Climate Change. Cambridge University Press, Cambridge, UK.

McKinnon, G. and Kaczanowski, S. 2004. Climate change and forests: making adaptation a reality. Report on a workshop held in Winnipeg, November 18–19, 2003. Winnipeg, Man.

Ministère des Ressources naturelles du Québec (MRNQ). 2000. La limite nordique des forêts attribuables. Ministère des Ressources naturelles du Québec. Québec, Que.

Ministère des Ressources naturelles et de la Faune du Québec (MRNFQ). 2006. Récupération des bois brûlés – Été 2005. Ministère des Ressources naturelles et de la Faune du Québec. [Online] <www.mrn.gouv.qc.ca/forets/fimaq/feu/fimaq-feu-bois.jsp> (accessed November 1, 2007).

Ministère du Développement durable, de l'Environnement et des Parcs du Québec (MDDEPQ). 2006. Le Québec et les changements climatiques, un défi pour l'avenir. Plan d'action 2006–2012. Ministère du Développement durable, de l'Environnement et des Parcs, Québec, Que.

Montigny, M.K. and MacLean, D.A. 2006. Triad forest management: scenario analysis of forest zoning effects on timber and non-timber values in New Brunswick, Canada. For. Chron. **82**: 496–511.

Morissette, J.L., Cobb, T.P., Brigham, R.M., and James, P.C. 2002. The response of boreal forest songbird communities to fire and post-fire harvesting. Can. J. For. Res. **32**: 2169–2183.

Nappi, A., Drapeau, P., and Savard, J.P.L. 2004. Salvage logging after wildfire in the boreal forest: is it becoming a hot issue for wildlife? For. Chron. **80**: 67–74.

Nguyen-Xuan, T., Bergeron, Y., Simard, D., Fyles, J.W., and Paré, D. 2000. The importance of forest floor disturbance in the early regeneration patterns of the boreal forest of western and central Quebec: a wildfire versus logging comparison. Can. J. For. Res. **30**: 1353–1364.

Park, Y.-S. 2002. Implementation of conifer somatic embryogenesis in clonal forestry: technical requirements and deployment considerations. Ann. For. Sci. **59**: 651–656.

Parker, W.C., Colombo, S.J., Cherry, M.L., Greifenhagen, S., Papadopol, C., Flannigan, M.D., McAlpine, R.S., and Scarr, T. 2000. Third millennium forestry: what climate change might mean to forests and forest management in Ontario. For. Chron. **76**: 445–463.

Patry, P. 2002. État des connaissances: récupération des bois après feu. *In* the proceedings of the seminar *L'Aménagement forestier et le feu*, April 9–11, 2002, Chicoutimi, Que. *Edited by* the Canadian Forest Service, the Société de protection des forêts contre le feu and the Sustainable Forest Management Network. Ministère des Ressources naturelles du Québec, Québec, Que., pp. 97–100.

Payette, S., Bhiry, N., Delwaide, A., and Simard, M. 2000. Origin of the lichen woodland at its southern range limit in eastern Canada: the catastrophic impact of insect defoliators and fire on the spruce-moss forest. Can. J. For. Res. **30**: 288–305.

Peng, C., Liu, J., Dang, Q., Zhou, X., and Apps, M. 2002. Developing carbon-based ecological indicators to monitor sustainability of Ontario's forests. Ecol. Indicators, **1**: 235–246.

Plummer, D.A., Caya, D., Frigon, A., Côté, H., Giguère, M., Paquin, D., Biner, S., Harvey, R., and de Elia, R. 2006. Climate and climate change over North America as simulated by the Canadian RCM. J. Clim. **19**: 3112–3132.

Price, C. and Rind, D. 1994. The impact of a $2\times CO_2$ climate on lightning-caused fires. J. Clim. **7**: 1484–1494.

Purdon, M., Brais, S., and Bergeron, Y. 2004. Initial response of understorey vegetation to fire severity and salvage-logging in the southern boreal forest of Québec. Appl. Veg. Sci. **7**: 49–60.

Purdon, M., Noël, J., Nappi, A., Drapeau, P., Harvey, B., Brais, S., Bergeron, Y., Gauthier, S., and Greene, D. 2002. The impact of salvage-logging after wildfire in the boreal forest: lessons from the Abitibi. CRSNG UQAT-UQAM-AFD Industrial Chair. Research Note 4. [Online] <web2.uqat.uquebec.ca/cafd/pdf/fichetech4e. pdf> (accessed October 22, 2007).

Queneville, R. and Thériault, M. 2002. L'utilisation du brûlage dirigé, l'expérience de Parcs Canada. *In* the proceedings of the seminar *L'Aménagement forestier et le feu*,

April 9–11, 2002, Chicoutimi, Que. *Edited by* the Canadian Forest Service, the Société de protection des forêts contre le feu and the Sustainable Forest Management Network. Ministère des Ressources naturelles du Québec du Québec, Québec, Que., pp. 101–106.

Raulier, F. and Bernier, P.Y. 2000. Predicting the date of leaf emergence for sugar maple across its native range. Can. J. For. Res. **30**: 1429–1435.

Raulier, F., Pothier, D., and Bernier P.Y. 2003. Predicting the effect of thinning on growth of dense balsam fir stands using a process-based tree growth model. Can. J. For. Res. **33**: 509–520.

Saucier, J.-P., Bergeron, J.-F., Grondin, P., and Robitaille, A. 1998. The land regions of southern Quebec. 3rd edition. One element in the land classification system developed by the Ministère des Ressources naturelles du Québec. Ministère des Ressources naturelles du Québec, Québec, Que.

Schmiegelow, F.K.A., Stepnisky, D.P., Stambaugh, C.A., and Koivula, M. 2006. Reconciling salvage logging of boreal forests with a natural-disturbance management model. Conserv. Biol. **20**: 971–983.

Skinner, W.R., Flannigan, M.D., Stocks, B.J., Martell, D.L., Wotton, B.M., Todd, J.B., Mason, J.A., Logan, K.A., and Bosch, E.M. 2002. A 500 hPa synoptic wildland fire climatology for large Canadian forest fires, 1959–1996. Theor. Appl. Clim. **71**: 157–169.

Skinner, W.R., Stocks, B.J., Martell, D.L., Bonsal, B., and Shabbar, A. 1999. The association between circulation anomalies in the mid-troposphere and area burned by wildland fire in Canada. Theor. Appl. Clim. **63**: 89–105.

Smit, B., Burton, I., Klein, R.J.T., and Street, R. 1999. The science of adaptation: a framework for assessment. Mitig. Adapt. Strat. Global Change, **4**: 199–213.

Spittlehouse, D.L. 1997. Forest management and climate change. *In* Responding to global climate change in British Columbia and Yukon. *Edited by* E. Taylor and B. Taylor. Environment Canada, Vancouver, B.C., pp. 24-1–24-8.

Spittlehouse, D.L. 2006. Integrating climate change adaptation into forest management. For. Chron. **81**: 691–695.

Spittlehouse, D.L. and Stewart, R.B. 2003. Adaptation to climate change in forest management. BC J. Ecosyst. Manag. [Online] <www.forrex.org/jem/2003/vol4/no1/art1.pdf> (accessed October 22, 2007).

Stocks, B.J., Alexander, M.E., and Lanoville, R.A. 2004. Overview of the International Crown Fire Modelling Experiment (ICFME). Can. J. For. Res. **34**: 1543–1547.

Stocks, B.J., Fosberg, M.A., Lynham, T.J., Mearns, L., Wotton, B.M., Yang, Q., Jin, J.-Z., Lawrence, K., Hartley, G.R., Mason, J.A., and McKenney, D.W. 1998. Climate change and forest fire potential in Russian and Canadian boreal forests. Clim. Change, **38**: 1–13.

Stocks, B.J., Mason, J.A., Todd, J.B., Bosch, E.M., Wotton, B.M., Amiro, B.D., Flannigan, M.D., Hirsch, K.G.,

Logan, K.A., Martell, D.L., and Skinner, W.R. 2003. Large forest fires in Canada, 1959–1997. J. Geophys. Res. **108**: D18149.

Suffling, R., Smith, B., and Dal Molin, J. 1982. Estimating past forest age distributions and disturbance rates in north-western Ontario: a demographic approach. J. Environ. Manag. **14**: 45–56.

Tardif, J. 2004. Fire history in the Duck Mountain Provincial Forest. Western Manitoba Sustainable Forest Management Network. Project Report 2003/2004 Series. University of Alberta, Edmonton, Alta. [Online] <www.sfmnetwork.ca/docs/e/PR_200304tardifjfire-6fire.pdf> (accessed November 29, 2007).

Tellier, R., Duchesne, L.C., McAlpine, R.S., and Ruel, J.-C. 1995. Effets du brûlage dirigé et du scarifiage sur l'établissement des semis et sur leur interaction avec la végétation concurrente. For. Chron. **71**: 621–626.

Thomas, P.A. and Wein, R.W. 1985. Water availability and the comparative emergence of four conifer species. Can. J. Bot. **63**: 1740–1746.

Van Sleeuwen, M. 2006. Natural fire regimes in Ontario. Ontario Ministry of Natural Resources, Queen's Printer for Ontario, Toronto, Ont. [Online] <www.ontarioparks.com/english/pdf/fire_research_2006.pdf> (accessed October 22, 2007).

Van Wagner, C.E. 1978. Age-class distribution and the forest fire cycle. Can. J. For. Res. **8**: 220–227.

Van Wagner, C.E. 1987. Development and structure of the Canadian Forest Fire Weather Index System. Canadian Forestry Service Technical Report 35, Ottawa, ON.

Volney, W.J.A. and Hirsch, K.G. 2005. Disturbing forest disturbances. For. Chron. **81**: 662–668.

Weber, M.G. and Flannigan, M.D. 1997. Canadian boreal forest ecosystem structure and function in a changing climate: impact on fire regimes. Environ. Rev. **5**: 145–166.

Weber, M.G. and Taylor, S.W. 1992. The use of prescribed fire in the management of Canada's forested lands. For. Chron. **68**: 324–334.

Wheaton, E. 2001. Changing fire risk in a changing climate: a literature review and assessment. Saskatchewan Research Council, Saskatoon, Sask., SRC Publ. No. 11341–2E01.

Whittle, C.A., Duchesne, L.C., and Needham, T. 1997. The impact of broadcast burning and fire severity on species composition and abundance of surface vegetation in a jack pine (*Pinus banksiana*) clear-cut. For. Ecol. Manag. **94**: 141–148.

Wotton, B.M. and Flannigan, M.D. 1993. Length of the fire season in a changing climate. For. Chron. **69**: 187–192.

Yohe, G. and Tol, R.S.J. 2002. Indicators for social and economic coping capacity: moving towards a working definition of adaptive capacity. Global Environ. Change, **12**: 25–40.

# Chapter 6

# Spatial Structure of Forest Stands and Remnants under Fire and Timber Harvesting Regimes*

*Nathalie Perron, Louis Bélanger, and Marie-Andrée Vaillancourt*

* We thank an anonymous reviewer for his comments on the manuscript. We are also thankful to Sylvie Gauthier (SCF-CFL), Yves Bergeron (UQAT), and Serge Payette (Laval U.) for their advice and suggestions made during the revision of the doctoral thesis of Nathalie Perron. We also emphasize the contribution of the remote sensing laboratory of UQAC for giving us access to the necessary infrastructure to complete this research. We thank more specifically Raymond Bégin and André Arsenault for their much appreciated contribution and Paul Patry of AbitibiBowater for his full cooperation. To conclude, we underline the valuable contribution of the MNRFQ for its financial support through the Fonds forestier and for allowing use of some of its database. We also acknowledge the financial support of the Consortium de recherche sur la forêt boréale commerciale, the Sustainable Forest Management Network, the Ph.D. support fund and the Centre de recherche en biologie forestière of Laval University, Cascades Inc., and the F.K. Morrow foundation (Richard J. Cullen fellowship). We are also grateful to Michel Saint-Germain for the English translation. The photo on this page was graciously provided by Michel Robert (Canadian Wildlife Service).

# 1. INTRODUCTION

In the past few decades, timber harvesting has emerged as a major source of disturbance in boreal forest ecosystems. This predominance of harvesting over fire as the primary disturbance agent raises questions about the long-term main-tenance of ecological processes (Kimmins 1997). First, fires rarely produce homog-enous post-disturbance conditions. Fire severity is influenced by the physical environment, climatic considerations (drought and wind conditions), and forest structure and composition (Tande 1979; Eberhart and Woodward 1987; DeLong and Tanner 1996; Kay 1997; MRN 2000). Fire usually leaves intact forest patches and partially or totally burned areas (Van Wagner 1983; Turner and Romme 1994; Gauthier et al. 2001; Kafka et al. 2001). As burned area increases, the prob-ability of encountering conditions likely to generate unburned patches (fire breaks such as rivers, wetlands, or strong slopes) also increases (Eberhart 1986). Forest residuals play an important role in post-disturbance maintenance of biodiversity, acting as a source of seeds or propagules allowing the re-colonization of burned areas (Zasada 1971; Boudreault 2001; Kafka et al. 2001) and as a refuge for forest-associated wildlife (Bendell 1974; Ghandi et al. 2001; Imbeau and Desrochers 2002; Nappi et al. 2004).

In opposition to forest fires, harvesting produces more homogenous conditions (Claveau et al. 2007). For example, cutblocks tend to cover about the same area, and harvesting techniques usually yield similar disturbance severity levels in every harvested stand. Also, all stand types are generally harvested on similar timber rotations (MRNFP 2003; Gouvernement du Québec 2007). Although spatial patterns created by fire are generally simple, they are still relatively complex when compared to those created by harvesting (DeLong and Tanner 1996). Disparities are also visible at the stand level. Following fire, surviving trees, snags, and woody debris bring an important contribution to the structural complexity of young natural stands (Franklin et al. 1981; Hansen et al. 1991; Spies et al. 1994). Such structural elements are rarely present in stands regenerat-ing from harvest, and the absence of such biological legacy is one of the major differences between stands of natural and anthropic origins (Hansen et al. 1991; DeLong and Tanner 1996; NCASI 2006).

However, what is the effective magnitude of these disparities at the landscape and disturbance scales? How can current knowledge be used to define an ecosystem-based management strategy? How would the public perceive manage-ment based on the natural disturbance regime? We will attempt to answer these questions using the western spruce-moss forests of Québec as an example. In this chapter we will 1) measure the level of disparity between spatial patterns created by fire and harvesting at the landscape and disturbance scales, 2) establish the basis of an ecosystem-based forest management strategy, and 3) suggest modi-fications to such a strategy to mitigate anticipated negative public perceptions.

## 2.   DESCRIPTION OF THE STUDY AREA

The study area is located in the boreal zone, in the western sub-domain of the spruce-moss forest (Bergeron et al. 1999a). It spans several ecological regions: 6e (Nestaocano River declivity), 5c (Hills of the Haut-Saint-Maurice), and 5d (Hills of Lac-Saint-Jean) to the south, and 6g (Manouane Lake declivity) to the north (Robitaille et Saucier 1998). It covers a total area of 24,000 km² and is essentially devoted to timber harvesting. The territory is mostly depopulated but may be sporadically used by native people, hunters, fishermen, vacationers, etc. Fire cycle estimations[*] made by ecological regions (Gauthier et al. 2001) suggest the study area to be under short- to medium-length fire cycles, i.e. between a 100- and 200-year fire interval. The study area comprises recently burned and harvested areas as well as large, untouched forest tracts (figure 6.1). Black spruce (*Picea mariana* [Mill.] B.S.P.) is the dominant tree species, but jack pine (*Pinus banksiana* Lamb.), balsam fir (*Abies balsamea* [L.] Mill.) and white birch (*Betula papyrifera* Marsh.) also occur (Rowe 1972).

[*] These estimations are based on about 100 years of archived data collected by the Ministère des Ressources naturelles et de la Faune du Québec (MRNF).

## 3.   SPATIAL DISTRIBUTION OF DISTURBANCES

### 3.1.   Landscape-Level Analyses

Exhaustive records of fire and harvesting events from 1973 to 1997 were acquired from the archives of the Ministère des Ressources naturelles et de la Faune du Québec. Data on harvesting varied in terms of precision depending on the year it was compiled. To minimize these disparities in data structure between time periods, cutblocks were merged together in cutblock clusters, which could be defined as zones where cutblocks and residual habitats are concentrated (Perron 2003).

### 3.2.   Landscape-Level Observations

The Landsat imagery dating from 12 August 1996 (figure 6.1) shows a spruce-moss forest dominated by a large, contiguous forest matrix with scattered recent fires and cutblock clusters. The northern harvesting front is also visible. Fires covering areas over 1,000 hectares are scattered over the entire territory (figure 6.2A), while cutblock clusters are located in the southern half of the area (figure 6.2B). Fires appear random in their distribution, while cutblock clusters are, on the contrary, clearly aggregated. Between 1973 and 1997, the nearest-neighbor distance between disturbances over 1,000 ha in area was clearly higher for fires than for cutblock clusters. The distance between fire centroids varied between 18 and 86 km, while it varied between 1 and 14 km for cutblock clusters (figure 6.3A). Median values of distance between centroids are 38 and 3 km for the two disturbance types, respectively. Distance separating a fire border to the next varies between 11 and 55 km, while distance between cutblock clusters varies between 1 and 10 km (figure 6.3B). Median values between borders are 16 and 1 km, respectively.

**Figure 6.1**
## Typical forest mosaic for the western spruce-moss forest of Québec

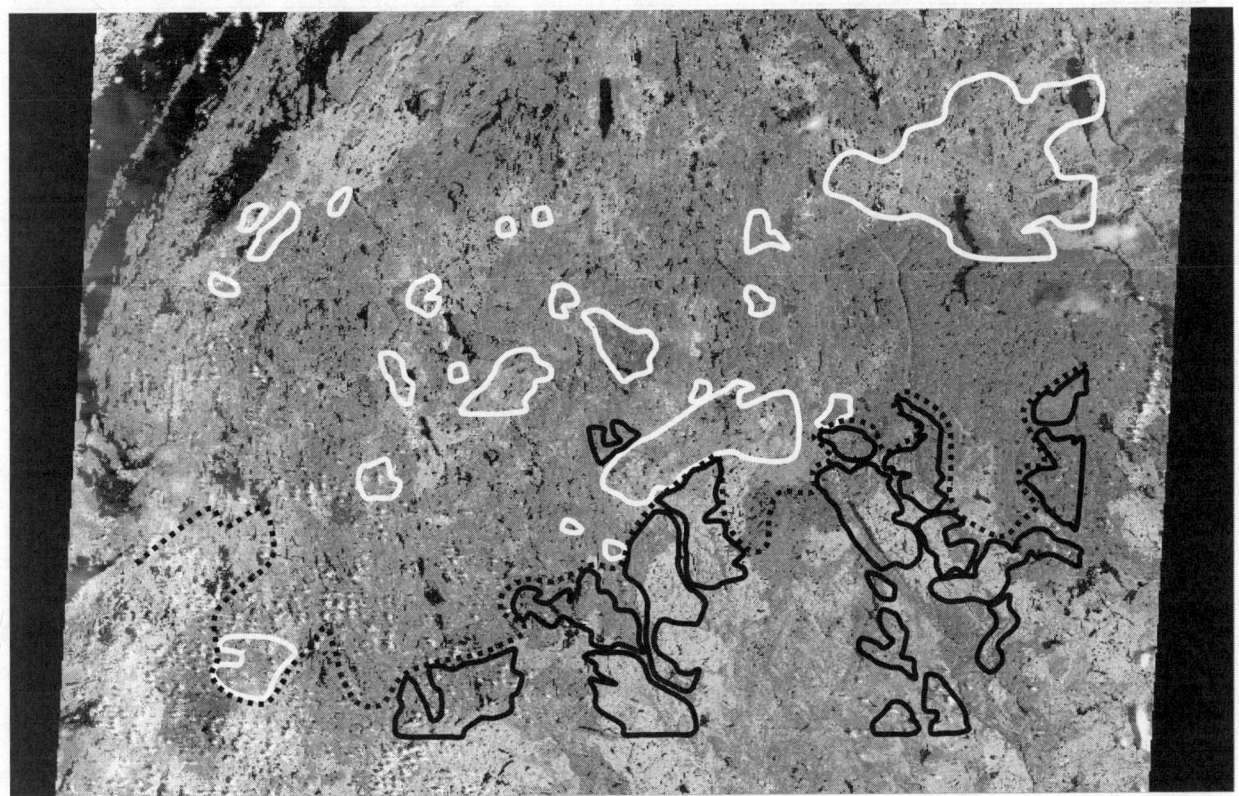

Legend: Land area dominated by large forest tracts with scattered recently burned areas (pale markings) and recently harvested areas (dark markings), showing the northern harvesting front (dotted line). Landsat imagery dating from 12 August 1996 (185 X 185 km).

## 3.3. Alteration of the Forest Mosaic at the Landscape Level Linked with Harvesting

Harvesting has been an important disturbance agent of the western spruce-moss forest landscapes between 1973 and 1997. The juxtaposition of cutblocks in single large areas has led to the creation of cutblock clusters with highly aggregated spatial distribution, in opposition to the random distribution of fire events. Past practices consisting of juxtaposing harvested areas have thus created vast regeneration areas.◆ Through the years, the harvesting front has progressed towards the north, following the prolongation of the road network. Such a practice of "unrolling harvest" brings about a local decline of mature and overmature forest tracts, which can incur important risks for animal and plant species dependent on such habitats (Schmiegelow and Mönkkönen 2002; Hannon 2005; Crites and Dale 1998; NCASI 2006). Comparatively, such boreal, conifer-dominated

◆ Regeneration areas consist of cutblock clusters dating from one or several periods (between 15 and 30 years) of forest management planning which result in large tracts of land area where regeneration does not yet possess structural attributes typical of mature forests.

Figure 6.2.
**Spatial distribution of A) fires and B) cutblock clusters over 1,000 ha between 1973 and 1997**

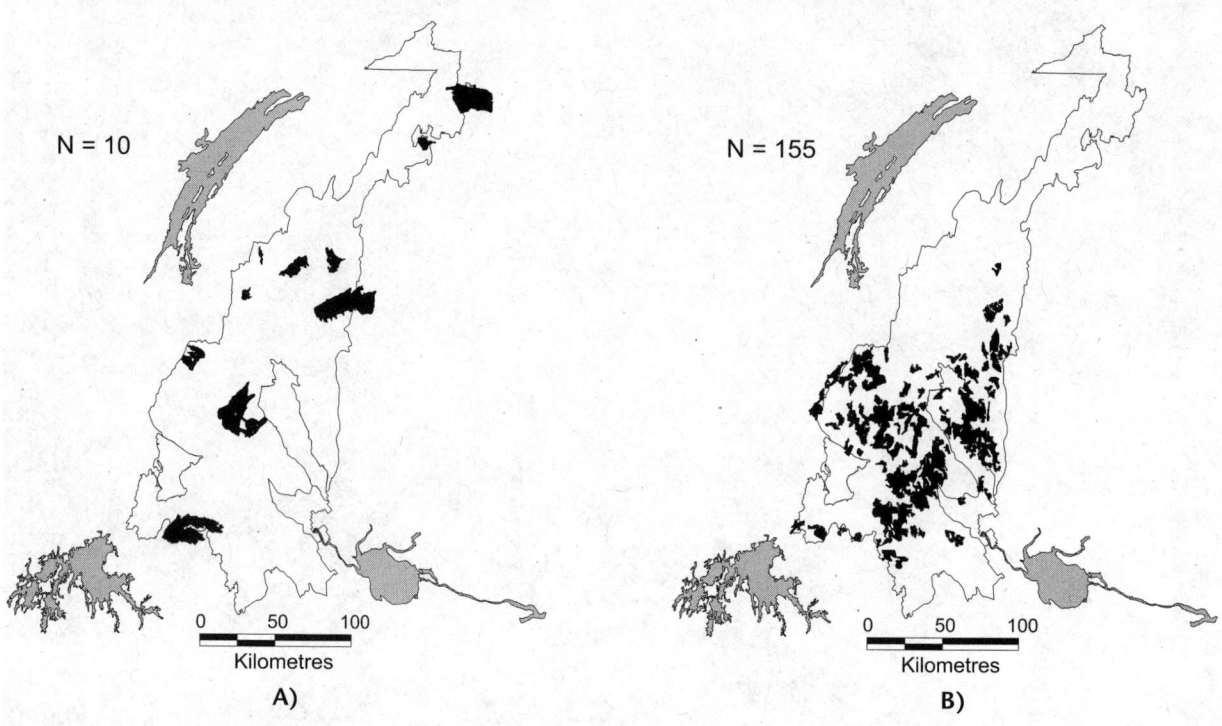

Fires appear to be randomly distributed while cutblock clusters tend to be aggregated.

landscapes driven by natural processes are usually characterized by an important proportion of mature forest tracts even if under the influence of large-scale disturbances (OMNR 1997; see Gauthier et al., chapter 3).

## 4.  SPATIAL CHARACTERISTICS OF FOREST RESIDUALS

### 4.1.  Disturbance-Level Analyses

Landsat satellite imagery taken during the summers of 1984, 1986, 1991, 1993 and 1996 was used to analyze spatial characteristics at the disturbance level. Satellite imagery was corrected and enhanced using methods described by Beaubien (1994), and then classified (Perron 2003). Table 6.1 shows the seven cover classes that were retained. Few classes were retained in order to highlight strongly contrasting structures and thus minimize confusion between classes.

**Figure 6.3**
## Comparison of the distance to the nearest disturbance of >1,000 ha between burns and cutblock clusters from A) their centroid and B) their edge

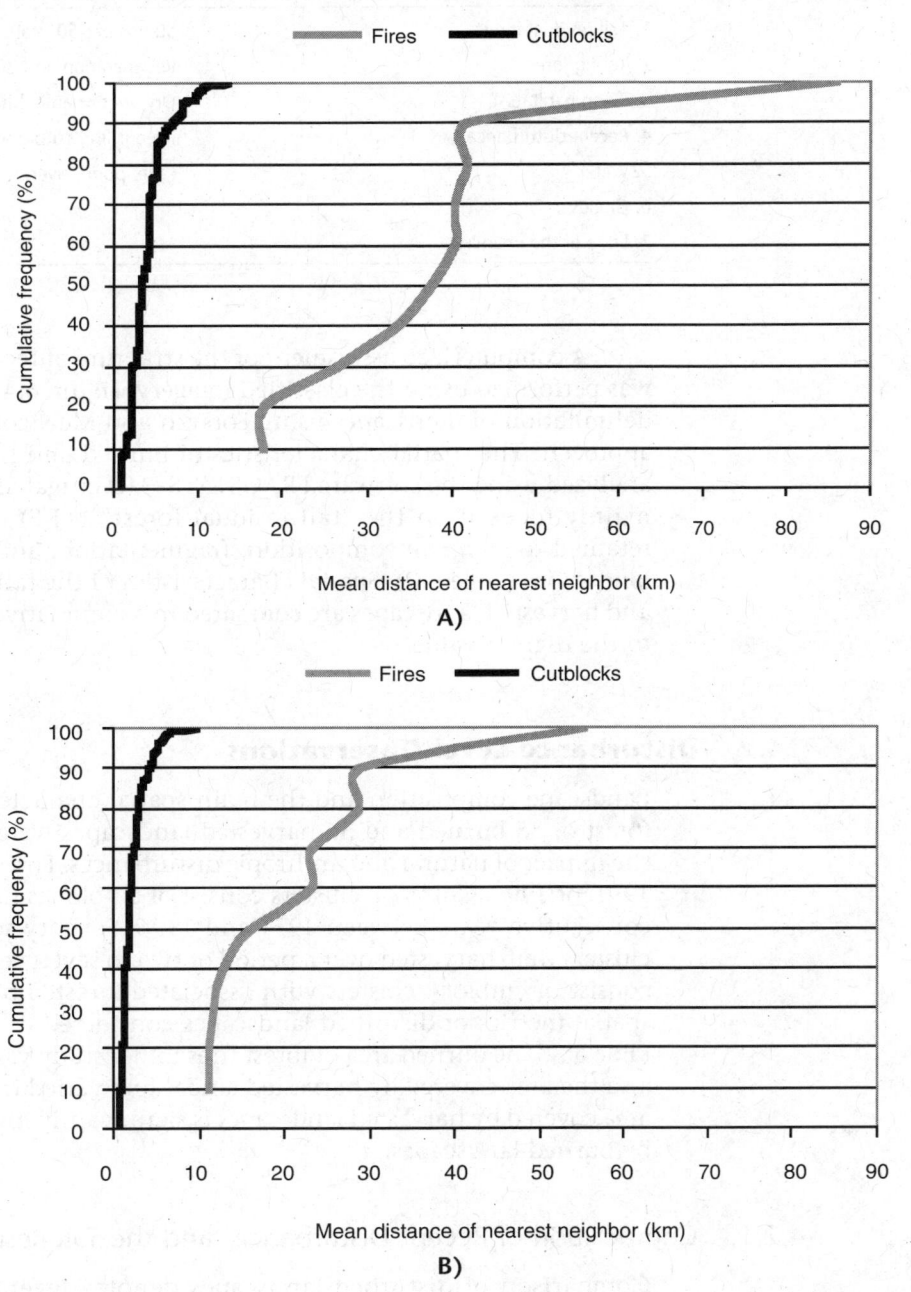

In both cases, the distance between disturbances of >1,000 ha is larger for fires than for cutblock clusters.

**Table 6.1**
**Description of the seven cover classes used**

| Classes | Description |
|---------|-------------|
| 1. Tall residual forest | Stands of >50 years old |
| 2. Young forest | Regeneration and 30-year-old stands |
| 3. Open habitats | Dry, wet open habitats |
| 4. Recent disturbance | Recent fire, cutblock, or road |
| 5. Water | Lake, pond, river |
| 6. Clouds | |
| 7. Edge of the imagery | |

A computerized assessment of the structure of recently disturbed landscape was performed using the classified imagery. Figure 6.4 shows an example of the delimitation of landscapes using Forsyth and McNicol's (1997) methodological approach. The spatial characteristics of burned and harvested landscapes were analyzed using the software FRAGSTATS (McGarigal and Marks 1994). Analyses mainly focused on the "tall residual forest" (TRF) class. Eight metrics were retained to compare composition, fragmentation, and spatial configuration of landscapes (table 6.2). Spatial characteristics of the tall residual forest of burned and harvested landscapes are compared in a cumulative fashion, from the lowest to the highest value.

## 4.2.  Disturbance-Level Observations

Landscape composition and the main spatial characteristics of the tall residual forest of 35 burned and 36 harvested landscapes were compared to investigate the impact of natural and anthropic disturbances. Forest fires took place in 1986, 1991, or 1996. Cutblock clusters consist of cutblocks harvested in one or several consecutive years between 1974 and 1996. Seventy-seven percent of cutblock clusters were harvested over a period of two to seven years. Harvested landscapes consist of cutblock clusters with associated forested borders and residuals. The spatial metrics of disturbed landscapes considered in the analyses are listed in table 6.3. The burned area of forest fires under study varied from 35 to 29,683 ha, and the area covered by harvested landscapes varied from 413 to 38,707 ha. The area covered by harvested landscapes is significantly higher than the one covered by burned landscapes.

### 4.2.1.  Composition of Recent Disturbances and the Tall Residual Forest

Comparison of disturbed landscapes denotes several similarities. Statistical analyses did not show any significant differences between the two types of disturbance in the percent of landscape occupied by tall residual forest or in the largest patch index.

## Figure 6.4
## Detailed view of the delimitation of landscapes recently disturbed by A) fire and B) harvesting

Fire landscape
1996
5,562 ha

Harvest landscape
Cutblocks clustered
from 1994 to 1996
5,598 ha

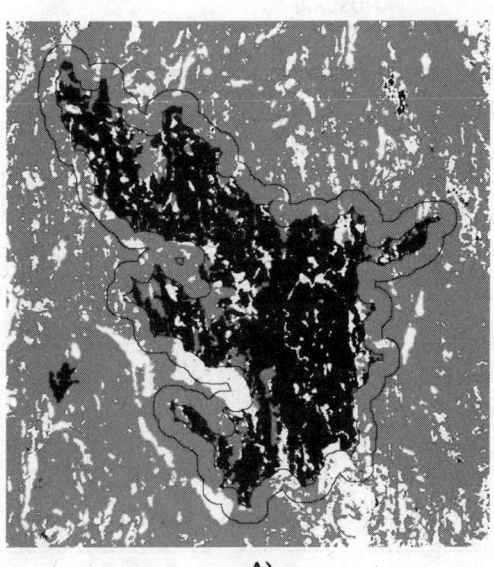
A)

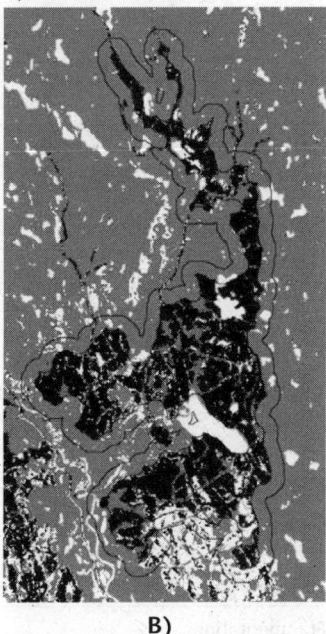
B)

From a Landsat TM imagery classified using Forsyth and McNicol's (1997) method.

## Table 6.2
## Description of the eight indices selected to analyze the composition, fragmentation, and spatial configuration of the tall residual forest of recently disturbed landscapes

| Spatial characteristic of landscape | Metric (units) |
|---|---|
| Composition | 1. Total landscape area (ha) |
| | 2. Percent of landscape |
| | 3. Largest patch index (%) |
| Fragmentation | 4. Patch density (#/100 ha) |
| | 5. Mean patch size (ha) |
| Configuration | |
| Shape | 6. Area-weighted mean shape index[a] |
| Isolation | 7. Mean nearest-neighbor distance (m) |
| Heterogeneity | 8. Interspersion and juxtaposition index (%)[b] |

a.  The shape index compares the perimeter of a landscape unit to that of a square of the same area. It varies from 1.00 to a value which increases as the shape gets more complex.
b.  The interspersion and juxtaposition index reaches its maximum value (100%) when each fragment of a given type is in contact with all other possible types of fragment. It reaches a low value if the number of different contacts remains low.

Table 6.3
## Variability thresholds in the composition of the 35 burned and 36 harvested landscapes analyzed and in the main spatial characteristics of the tall residual forest

| Spatial characteristics | Variability threshold (percentile rank) | | | | | |
| | Fires | | | Cutblocks | | |
| | Min. | 50% | Max. | Min. | 50% | Max. |
| --- | --- | --- | --- | --- | --- | --- |
| **LANDSCAPE** | | | | | | |
| **Composition** | | | | | | |
| Landscape area (ha) * (F < C) | 35 | 375 | 29,683 | 413 | 4,545 | 38,707 |
| Proportion of recent disturbance (%)* (F > C) | 41 | 71 | 89 | 31 | 54 | 72 |
| Proportion of tall residual forest (%) | 7 | 19 | 37 | 9 | 18 | 44 |
| Proportion of young forest (%)* (F < C) | 0 | 1 | 16 | 0 | 9 | 29 |
| Proportion of open habitats (%)* (F < C) | 0 | 2 | 21 | 2 | 9 | 29 |
| **TALL RESIDUAL FOREST** | | | | | | |
| **Composition** | | | | | | |
| Proportion of tall residual forest (%) | | | | | | |
| Total | 7 | 19 | 37 | 9 | 18 | 44 |
| Interior | 0 | 1 | 8 | 0 | 2 | 14 |
| Largest patch index (%) | | | | | | |
| Total | 1 | 4 | 30 | 0 | 2 | 24 |
| Interior | 0 | 1 | 3 | 0 | 1 | 6 |
| **Fragmentation** | | | | | | |
| Patch density (#/100 ha) | | | | | | |
| Total* (F > C) | 8 | 18 | 37 | 5 | 9 | 23 |
| Interior | 0 | 5 | 17 | 0 | 4 | 12 |
| Mean patch size (ha) | | | | | | |
| Total* (F < C) | 0 | 1 | 3 | 1 | 2 | 10 |
| Interior* (F < C) | 0 | 0 | 1 | 0 | 1 | 2 |
| **Configuration** | | | | | | |
| Area-weighted mean shape index | | | | | | |
| Total* (F < C) | 2 | 2 | 7 | 2 | 3 | 6 |
| Interior* (F < C) | 0 | 2 | 3 | 1 | 2 | 4 |
| Mean nearest-neighbor distance (m) | | | | | | |
| Total | 34 | 54 | 85 | 39 | 56 | 82 |
| Interior | 0 | 88 | 700 | 61 | 137 | 649 |
| Interspersion and juxtaposition index (%) | | | | | | |
| Total* (F < C) | 8 | 42 | 75 | 35 | 57 | 80 |
| Interior* (F < C) | 0 | 6 | 74 | 2 | 24 | 76 |

* There is a significant difference between burned and harvested landscapes.
F = fires.
C = cutblocks.

The percent of recent disturbance is shown to be the predominant feature in recently disturbed landscape, ranging from 41 to 89% for fires and from 31 to 72% for harvested landscapes (table 6.3). Recently harvested landscapes are less disturbed than recently burned landscapes, but comprise a larger proportion of young forests and open habitats (table 6.3). **Total**[*] tall residual forest found in burns and harvested areas are a sub-dominant component of the landscape, ranging respectively from 7 to 37% and from 9 to 44%. **Interior**[**] tall residual forest does not exceed 8% of burned landscapes and 14% of harvested landscapes. These percents of total and interior TRF do not differ significantly (figures 6.5A, 6.5B). The largest patch index of the total TRF varies between 1 and 30% in burned landscapes and between 0 to 24% in harvested landscapes. For the interior TRF, this index is less in average and varies little for both burns and harvests, not exceeding 3% for burned landscapes and 6% for harvested landscapes.

> [*] Total tall residual forest comprises all forest tracts left undisturbed.
> [**] Interior tall residual forest comprises only forest fragments isolated within the disturbance.

### 4.2.2.  Fragmentation of the Tall Residual Forest

The tall residual forest usually consists in disturbed landscapes of small isolated patches. Significant differences were seen between burned and harvested landscapes (table 6.3), the residual forest found in burned landscapes being generally more fragmented.

Burned landscapes show a significantly higher patch density of total residual forest, but their mean patch size is significantly lower than in harvested landscapes (figures 6.6A, 6.6B). The maximum patch density of total residual forest is of 37 patches/100 ha for fires and of only 23 patches/100 ha for harvested areas. Median values are respectively of 18 patches/100 ha and 9 patches/100 ha. The maximum mean patch size of total residual forest is of 3 ha for fires and 10 ha for harvested areas. Trends are similar for interior residual forest. The maximum patch density of interior residual forest is of 17 patches/100 ha for fires and of 12 patches/100 ha for harvested areas. Median values are respectively of 5 patches/100 ha for fires and of 4 patches/100 ha for harvested areas. The maximum mean patch size of interior residual forest patches is of 1 ha for fires and 2 ha for harvested areas.

### 4.2.3.  Configuration of the Tall Residual Forest

Significant differences were detected in the spatial configuration of tall residual forest between burned and harvested landscapes in the mean shape, interspersion, and juxtaposition indices, but not in the mean nearest-neighbor distance (table 6.3).

The shape index for total and interior residual forest patches was relatively low, not exceeding 7 and 1 for fires, respectively, and 6 and 2 for harvested areas (figure 6.7). Such results suggest relatively regular[***] shapes for both types of landscape, however a little more irregular for harvested ones, basically as they are narrow bands along roads and rivers.

> [***] With the shape index, regular shape corresponds to a square while irregular shape departs from it.

The mean nearest-neighbor distance of total TRF ranged from 34 to 85 m for fires and from 39 to 82 m for harvested areas. The same metric but for interior TRF ranged from 0 to 700 m for fires and from 61 to 649 m for harvested areas. Patches of residual forest are much closer to each other on the periphery of the

**Figure 6.5**

## Comparison of the proportion of A) total and B) interior tall residual forest left in burned and harvested landscapes

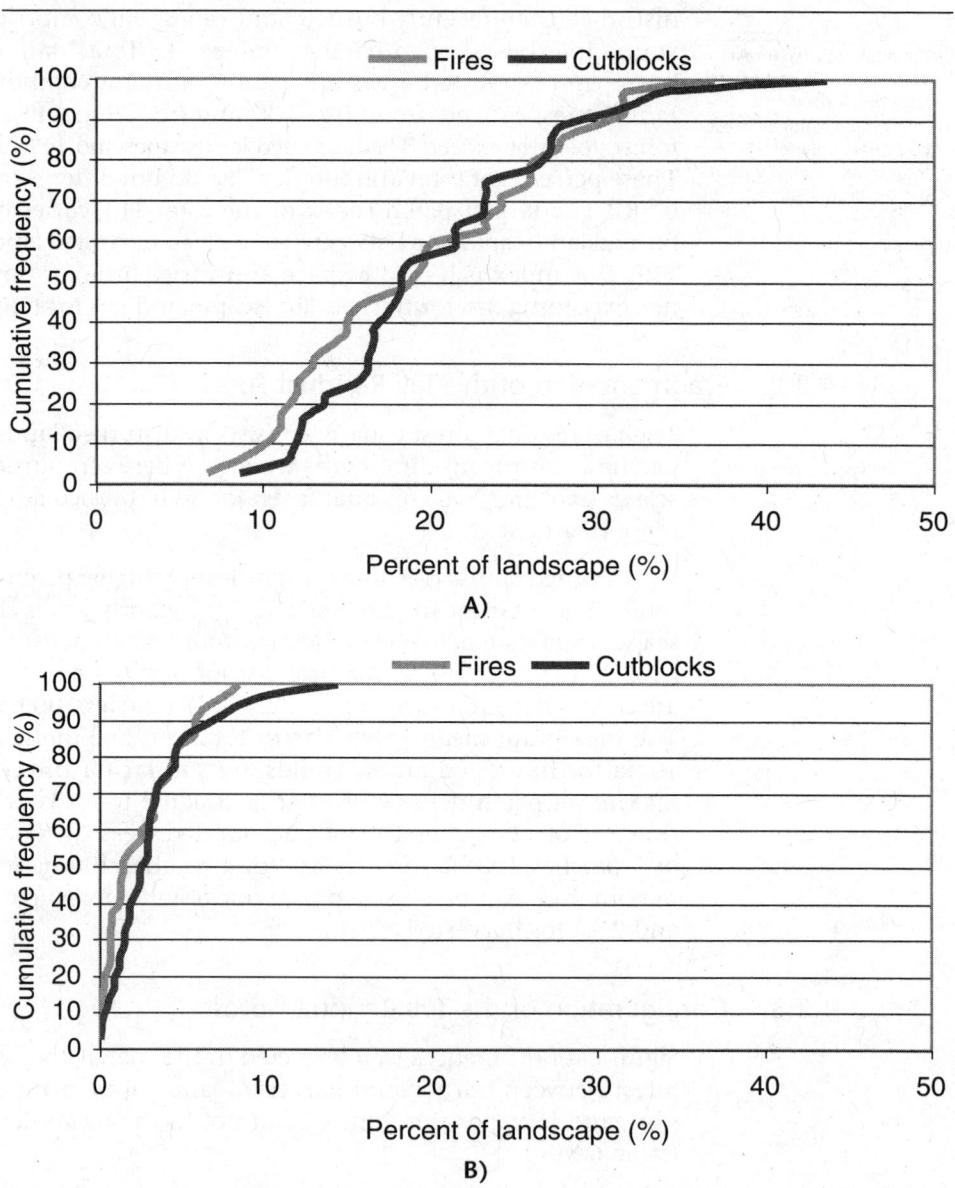

Proportions are similar in the two landscapes.

Figure 6.6
## Comparison of burned and harvested landscapes in terms of the density of A) total and B) interior tall residual forest fragments

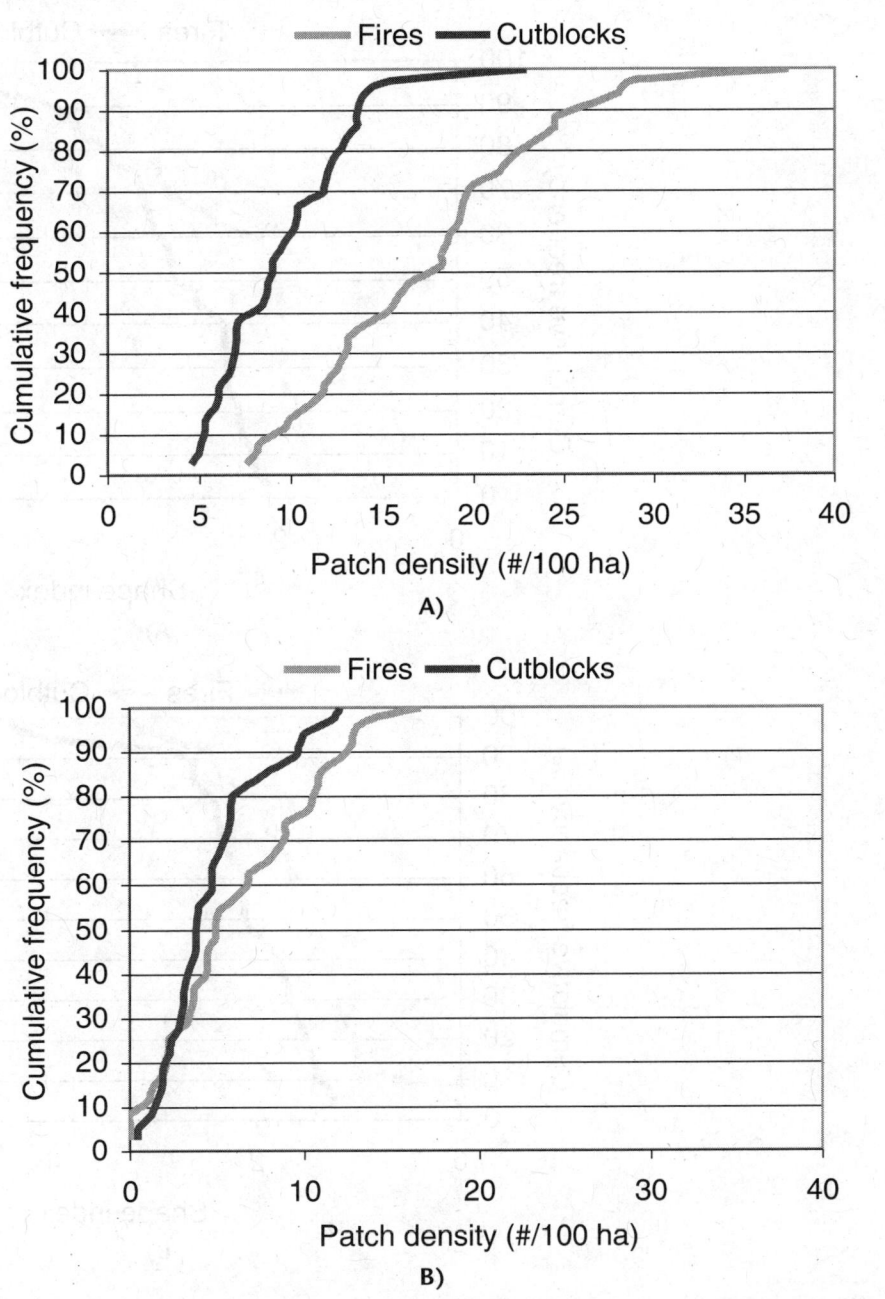

Burned landscapes have significantly more total TRF fragments than harvested landscapes, while numbers of interior TRF patches are similar.

Figure 6.7
**Comparison of burned and harvested landscapes in terms of the shape index of A) total and B) interior tall residual forest fragments**

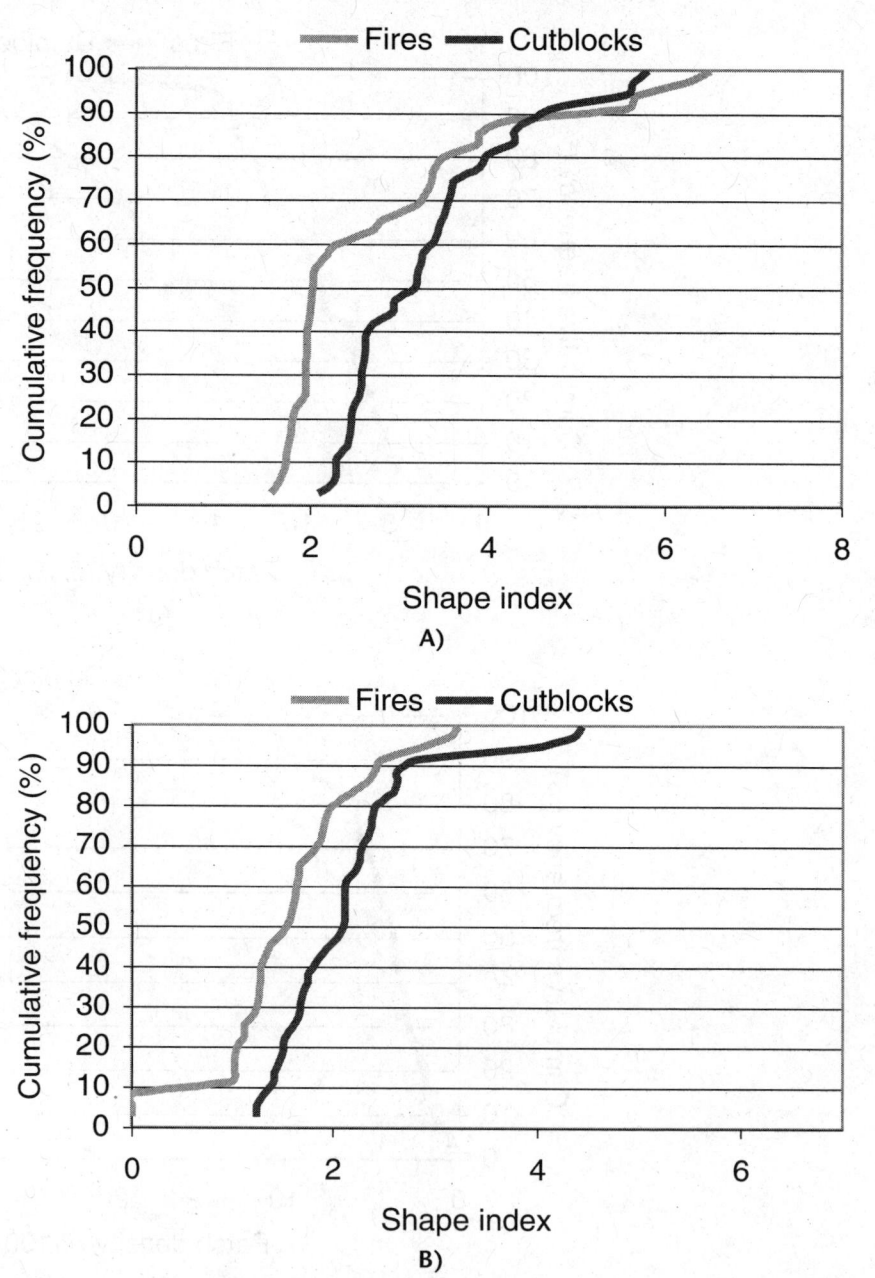

A)

B)

Both landscapes differ significantly, with patches from the harvested landscapes being more linear.

disturbance than on the inside for both types of disturbance. Finally, the interspersion and juxtaposition index for the total TRF varied from 8 to 75% for fires and from 35 to 80% for harvested areas, while for interior TRF it varied from 0 to 74% for fires and from 2 to 76% for harvested areas. These results indicate that residual forest (total and interior) is set in a more heterogeneous environment (i.e., more likely to be surrounded by other cover classes) in harvest areas when compared to fires.

### 4.2.4. Alteration of the Forest Mosaic by Harvesting at the Disturbance Level

Tall residual forest spared by fire constitutes a sub-dominant component of burned landscapes and occurs generally on the periphery of the disturbance. Patches are numerous but are mostly small-sized and irregular-shaped. Although they are relatively distant from each other, TRF patches are not uniformly distributed within the burned area. At the disturbance level, the comparison of burned and harvested landscapes has shown that, despite some similarities, timber harvesting does not fully recreate spatial patterns created by fire. The imprint of recent disturbance is less in harvested landscapes. They comprise a larger proportion of young forest and open areas, but not of tall residual forest. TRF of harvested landscapes is less fragmented and is more irregularly shaped than the one found in burned landscapes.

### 4.2.5. Natural Variability of Residual Forest in Burned Landscapes

The burned landscapes we studied show similarities with those investigated in other Canadian studies (table 6.4). The proportions of residual forest spared by fire in the Western spruce-moss forest falls within the interval of natural variability observed throughout Canada. As for the spatial characteristics of these residual forests, comparisons revealed that they frequently and significantly diverge.

For the western spruce-moss forest, the burned area was not correlated with the proportion of total residual forest found in the landscape. However, it had an influence on the proportion of interior residual forest. These results somewhat differ from those reported in the literature (table 6.5). Eberhart (1986) and DeLong and Tanner (1996) both show an increase in the proportion of total residual forest with increased burned area; however, results from the Ontario Ministry of Natural Resources suggest otherwise (OMNR 1997). These contradictory results are difficult to interpret clearly, given that some of these studies covered time periods in which the mechanization of forest operations was rudimentary; such a factor is likely to have had some influence on residual forest spatial patterns. Differences in methodological approaches and in the range of burned areas found in these studies may also have a role to play in these divergent results.

As shown in some studies conducted in Alberta and Ontario, the size of burned areas has some influence on spatial characteristics of the residual forest. Burned area had a positive influence on the patch density in Québec and Alberta, but not in Ontario where it was negative. The influence was also positive on the mean patch size in Alberta and Québec. Also, in the western spruce-moss forest of Québec, the influence of burned area was negative on the largest-patch index and positive on the shape index; no significant relationships were found for

### Table 6.4
### Comparison of results obtained on burned landscape variability with those from other Canadian studies

| | Eberhart (1986) | DeLong and Tanner (1996) | Gluck and Rempel (1996) | OMNR (1997) | Gauthier et al. (2001) | Perron (2003) |
|---|---|---|---|---|---|---|
| Province | Alberta | British Columbia | Ontario | Ontario | Québec | Québec |
| Ecosystem | Mixed boreal | Boreal | Boreal | Boreal | Boreal | Boreal |
| Data source | Aerial photographs | Aerial photographs | Landsat imagery | Aerial photographs | Aerial survey | Landsat imagery |
| Years | 1970 to 1983 | 1931 to 1950 | 1981 | 1921 to 1950 | 1995 to 1996 | 1991 to 1996 |
| Number of fires | 69 | 9 | 1 | 42 | 16 | 35 |
| **LANDSCAPE** | | | | | | |
| **Composition** | | | | | | |
| **Area (ha)** | | | | | | |
| Minimum | 21 | 159 | – | 54 | 2,567 | 35 |
| Maximum | 17,770 | 2,239 | 18,415 | 52,772 | 66,655 | 29,683 |
| **% of disturbance** | | | | | | |
| Minimum | 82 | – | – | – | 31 | 41 |
| Maximum | 100 | – | 61 | – | 94 | 89 |
| **% of residual forest** | | | | | | |
| Minimum | – | 3 | – | 10 | 1[a] and 5[b] | 7 |
| Maximum | – | 15 | 21 | 50 | 16[a] and 58[b] | 37 |
| **RESIDUAL FOREST** | | | | | | |
| **Fragmentation** | | | | | | |
| **Patch density (#/100 ha)** | | | | | | |
| Minimum | 0 | – | – | 4 | – | 8 |
| Maximum | 2 | – | 8 | 52 | – | 37 |
| **Mean patch size (ha)** | | | | | | |
| Minimum | 1 | – | – | – | – | 0 |
| Maximum | 35 | – | 3 | – | – | 3 |
| **Configuration** | | | | | | |
| **Area-weighted mean shape index** | | | | | | |
| Minimum | – | – | – | – | – | 2 |
| Maximum | – | – | 6 | – | – | 7 |
| **Interspersion and juxtaposition index (%)** | | | | | | |
| Minimum | – | – | – | – | – | 8 |
| Maximum | – | – | 84 | – | – | 75 |

a. Preserved areas.
b. Areas dominated by green treetops.

Table 6.5

## Comparison of results obtained on the influence of burned area on the composition of landscapes and on the spatial characteristics of the residual forest with those from other Canadian studies

| | Eberhart (1986) | DeLong and Tanner (1996) | OMNR (1997) | Perron (2003) | |
|---|---|---|---|---|---|
| Province | Alberta | British Columbia | Ontario | Québec | |
| Ecosystem | Mixed boreal | Boreal | Boreal | Boreal | |
| Data source | Aerial photographs | Aerial photographs | Aerial photographs | Landsat imagery | |
| Years | 1970 to 1983 | 1931 to 1950 | 1921 to 1950 | 1991 to 1996 | |
| Number of fires | 69 | 9 | 42 | 35 | |
| **TOTAL RESIDUAL FOREST** | **INFLUENCE OF BURNED AREA** | | | | |
| **Composition** | | | | Total | Interior |
| % of residual forest | Positive | Positive | Negative | None | Positive |
| Largest patch index | – | – | – | Negative | None |
| **Fragmentation** | | | | | |
| Patch density (#/100 ha) | Positive* | – | Negative | Positive | None |
| Mean patch size (ha) | Positive | – | – | Positive | Positive |
| **Configuration** | | | | | |
| Area-weighted mean shape index | – | – | – | Positive | Positive |
| Mean nearest-neighbor distance (m) | – | – | – | None | None |
| Interspersion and juxtaposition index (%) | – | – | – | None | Positive |

* With the exception of the largest area class (2,001–20,000 ha).

mean nearest-neighbor distance of the same type and for the interspersion and juxtaposition index. In other words, as the burned area gets bigger, patches of residual forest tend to be smaller and more irregularly shaped. Finally, the influence of burned area varied slightly when considering interior TRF spatial characteristics. Only three indices (mean patch size, shape index and mean nearest-neighbor distance of the same type) had significant positive relationships with burned area.

## 5.    STRATEGIES AND PROCEDURES FOR ECOSYSTEM-BASED MANAGEMENT OF THE WESTERN SPRUCE-MOSS FOREST

Although timber harvesting does not alter the forest vocation of a land area, we must keep in mind that harvesting is a significant disturbance. The results we presented so far show that the impact of the forestry practices in use between 1973 and 1997 on spatial metrics of residual forests differed from that of fire. We can thus suggest that timber harvesting has altered the natural dynamics of these landscapes.

The strategies and procedures of ecosystem-based management put forward in this chapter are inspired from patterns produced by forest fires and apply to the western spruce-moss forest of Québec. They are based on a partial substitution of fires by harvesting, and ideally, combined disturbances should not exceed historical burning rates (see Gauthier et al., chapter 3). The dispersed-clustered ecosystem-based harvesting strategy was conceived as an ecologically sound alternative to the traditional harvesting strategies usually applied in the western spruce-moss forest of Québec (Perron 2003). The main divergences observed between natural and anthropic disturbances in terms of spatial metrics of residual forest were highlighted to identify on which metrics to focus to make sure managed landscapes remained within the range of natural variability. The elaborated strategy is based on two scales of perception: that of the landscape and that of the disturbance. The proportion of a landscape to be harvested in an even-aged management approach and the targeted distance between harvested areas should be based on natural disturbance cycles, which are apprehended at the scale of large landscapes (thousands of square kilometres), while at the scale of the disturbance (tens to hundreds of square kilometres), it is the spatial configuration of cutblocks and residual forest which is critical. Guidelines that would contribute to maintaining the spatial characteristics of the Western spruce-moss forest mosaic within their range of natural variability can be drawn from the afore-presented results: 1) clustering cutblocks; 2) distancing cutblock clusters; 3) maintaining tall residual forest within cutblock clusters; and 4) diversifying the configuration of the tall residual forest left behind (table 6.6). Procedures associated with the dispersed-clustered ecosystem-based harvesting strategy (summarized in table 6.7) should be applied to large land areas, e.g. at the forest management unit scale.

## 5.1. Procedures at the Landscape Level: Size and Spacing of Regeneration Areas

The proposed procedures have, as primary goals, to ensure that the management unit remains predominantly forested and that stands aged beyond commerciability are maintained on the land area, as, historically, regeneration areas occupied a sub-dominant position in the forest mosaic of the Western spruce-moss forest. Under a 100- to 200-year fire return interval and an assumed 90-year rotation, between 36 and 59% of the territory should be managed as low-retention harvesting clusters while the remaining should be either temporarily preserved or managed using alternative approaches mimicking less severe disturbances (e.g. insect epidemics, windthrow, natural mortality; see Section 5.3).

For the boreal forest, the main problem has not been clearcutting in itself, but rather how extensively it has been applied (Bélanger 1992). Future practices for the Western spruce-moss forest should favour an ecosystem-based dispersion of agglomerated cutblocks. The maintenance of landscape's natural integrity depends more on the spacing between cutblock clusters than on a reduction of the total harvested area. Large cutblock clusters should remain an option for this part of the boreal forest as they are analogous to disturbances by fire (see Gauthier et al., chapter 3). However, current extent of cutblock clusters exceeds that of the largest fires; the maximal size of such harvested areas should be decreased to better correspond to disturbances by fire. In addition, we would recommend

**Table 6.6**
## Principles of the dispersed-clustered ecosystem-based harvesting strategy as applied for the western spruce-moss forest of Québec

| | |
|---|---|
| **General principle** | Conduct ecosystem-based management based on the fire regime. |
| **Harvest strategy** | Create a managed forest matrix dominated by forest cover but in which cutblock clusters and relatively inaccessible/unaltered areas are interspersed. |
| **Management principles** | |
| 1. *Predominance of forest cover* | Limit regeneration areas originating from timber harvest to a sub-dominant position in the forest matrix. |
| 2. *Dispersion of regeneration areas* | Increase the distance between cutblock clusters to maintain large unaltered forest tracts. |
| 3. *Maintenance of residual forest patches within the regeneration areas* | Maintain a minimal proportion of mature or overmature residual forest within cutblock clusters. |
| 4. *Diversification of the spatial configuration of forest remnants within the regeneration areas* | Vary the size and shape of residual forest fragments left within cutblock clusters. |

**Table 6.7**
## Procedures of the dispersed-clustered ecosystem-based harvesting strategy as applied for the western spruce-moss forest of Québec

| | **Threshold** | |
|---|---|---|
| **Regeneration areas** | | |
| Minimal distance between two areas | 10 to 55 km | |
| Size of regeneration areas | 10–499 ha | 3% |
| | 500–4,999 ha | 6% |
| | 5,000–9,999 ha | 12% |
| | 10,000–39,999 ha | 79% |
| Clustered or dispersed cutblocks | Creation within a single run or span over several years | |
| Type of intervention | Cuts protecting regeneration and soils (CPRS/CPHRS) and cuts with variable retention | |
| Residual forest | 10 to 35%, of which 1 to 8% isolated within the area | |
| Shape | Peninsula (2 to 10 ha) and fragment (<1 ha) | |
| Stand age | Same as the pre-harvest stand | |
| **Forest areas** | | |
| Residual forest | Mainly in large forest tracts (km²) | |
| Silvicultural intervention | Adapted cuts maintaining canopy cover, small-size CPRS | |

minimizing the use of the largest class of cluster size for two reasons. First, it may appear disputable to use catastrophic events (from a human perspective) as a template for management-guiding principles given the impact on soil, flora and fauna of the two disturbances are not the same. Second, extremely large fires are likely to occur again in the future despite efforts to control forest fires.

To create regeneration areas, the foremost methods to be used should be cuts protecting regeneration and soils (CPRS), and cuts protecting tall regeneration and soils (CPHRS).• These approaches should be complemented with strategies of variable retention to maintain some components of tall residual forest. To mimic the variable effect of fire on soils, which can have a determinant influence on the patterns and composition of early regeneration (Johnson 1992; Ryan 2002), specific treatments should be applied to soils (e.g., scarification, prescribed burning) to ensure that harvesting reproduces the entire range of conditions found after fire. Such a combination of approaches should allow at least a partial reproduction of tall vegetation patterns produced by variable severity fires (total mortality, partial mortality and fire skips) at the disturbance scale.

• The CPRS and CPHRS are two careful-logging treatments commonly used in Québec, The CPRS is analogous to the CLAAG (careful logging around advance regeneration) used in Ontario. In contrast with CPRS, CPHRS retains taller but not yet merchantable stems.

Two important aspects should be considered when creating large cutblock clusters: the spacing of clusters and for how long residual forest between clusters should be preserved. Given that regeneration areas created by fires are farther apart than current cutblock clusters, the latter should be more dispersed in the landscape to minimize cumulative effects produced by the juxtaposition of cutblock clusters. It is recommended that borders of regeneration areas should be distanced from each other of at least 10 km. This distance could be as much as 55 km if the regeneration areas are very large (Belleau et al. 2007). Such dispersion of regeneration areas should maintain sufficient tall forest tracts between disturbances. If however the size of cutblock clusters is smaller, then the maximal distance between these areas can be decreased. Between 1973 and 1997, 80% of large fires were distanced from each other by 10 to 30 km. Such thresholds could be seen as satisfactory for the spruce-moss domain studied in this chapter. Finally, we should stress that forest tracts left between harvested areas should not be harvested themselves as long as adjacent regeneration areas have not acquired characteristics typical of mature forests in the landscape. Any premature exploitation of this residual forest could have important negative consequences on plant and wildlife populations depending on mature and overmature forests. The need to maintain irregular forest structures between the cutblock clusters could be coped with by having some of these areas harvested under irregular or uneven-aged harvest systems (selection cutting, adapted cutting practices that maintain canopy cover; see section 5.3; Bouchard, chapter 13).

## 5.2. Procedures at the Disturbance Level: Proportions and Configuration of Residual Forest

According to our results and those of DeLong and Tanner (1996), Gluck and Rempel (1996), OMNR (1997), and Gauthier et al. (2001) (see also Serrouya and D'Eon, 2004 for a review), all regeneration areas should comprise some proportion of residual mature or overmature forest. Although current practices do leave proportions of residual forest similar to those found in recent burns, some aspects of harvesting should be modified in order to create conditions closer to what natural disturbances produce in terms of interior TRF. Residual forest should occupy between 7 and 35% of the regeneration areas, with 1 to 8% isolated within and preserved from further harvesting. Patterns observed following natural disturbance suggest that residual forest should be retained as peninsulas (2–10 ha) or isolated forest patches (<1 ha). The variable severity seen in fires creates some degree of heterogeneity within burned areas (Kafka et al. 2001). Some unburned patches remain and a sometimes important proportion of the area burns only partially. Based on this evidence, we suggest that the minimal threshold of residual forest, based on observed natural variability, should be increased. However, a substantial proportion of the residual forest can be eventually salvaged, as, in natural disturbances, some proportion of partially burned stems eventually die in the following years. A proportion of 3 to 5% of residual forest should however be maintained permanently. The proportion of residual forest plays an important role in the maintenance of animal populations at the landscape level (Imbeau et al. 1999; Drapeau et al. 2003) and also in the natural regeneration of adjacent harvested areas (seedtree patches; Galipeau et al. 1997).

Other than the proportion of residual forest to maintain in cutblock clusters, its configuration and quality (i.e. composition and structure) are key characteristics and affect its effectiveness in maintaining ecological functions within disturbed areas. The configuration of residual forest left unscathed by fire is more variable that the one left by harvesting, which is generally linear, as in the case of cutblock separators and riparian buffers prescribed by the current provincial legislation (Government of Québec 2007). Hence, the spatial configuration of such residual forest should be more similar to patterns created by natural disturbances, i.e. less linear. Because of their linearity and narrowness, these retention elements are subject to edge effect, which usually incurs changes in stand structure. Mascarúa-López et al. (2006) reported a decrease in living stem density and an increase in mortality and windthrow in such linear remnants when compared with interior habitat in spruce-moss forests of Abitibi. These changes in stand structure can influence the potential of these remnants as habitat for wildlife. The width of these linear remnants has been emphasized as being insufficient to maintain some bird species associated with mature forests as well as small mammals (Darveau et al. 1995; Darveau et al. 2001; Hannon et al. 2002; Potvin et al. 2004).

Despite a lack of comparisons in the literature for the spruce-moss domain of residual forests in burned and harvested landscapes in terms of their quality (age, species composition, structural attributes), there are obvious differences in terms of structure between linear remnants left under current legislation and comparable interior forest (Mascarúa-López et al. 2006; Gagné 2006). Species composition and the presence of structural attributes (e.g., large-diameter trees) are key attributes that should be considered in the planning of residual forest in managed landscape. For example, in the balsam fir–white birch domain of the Saguenay region, Vaillancourt et al. (2008) noted that residual forest in managed areas did not reflect the species composition of mature, dominant forest stands, and showed a deficit in large diameter trees, a key structural element for wildlife.

Because they play a capital role in the recolonization of disturbed areas by acting as a seed source and a refuge for flora and wildlife associated with mature forests, residual forest left in harvested areas should carry more variability. For example, riparian buffers could vary in width, and part of the linear separators could be converted to less linear patches (to minimize edge effect) of variable size (based on area classes found in post-fire residual forests). Residual forest should be analogous to those harvested in terms of age, structure and location. All of these measures aim at maintaining in managed landscapes an array of structures and ecosystem processes usually found in pyrogenic landscapes along their development.

## 5.3. Recommended Management Practices to Maintain Large Forest Tracks

The part of the land area which escapes disturbance by fire over long time intervals is instead driven by secondary disturbances such as insect epidemics, windthrow and small-gap dynamics. These stands are of particular value to a large number of plant and animal species which long-term survival could be compromised by a decrease in the proportion of such stands. Silvicultural practices inspired by these secondary disturbances should be applied to this part of the territory (Bergeron et al. 1999b; Bouchard, chapter 13). Adapted cutting patterns intended at maintaining forest cover (e.g., irregular shelterwood cutting) and dispersed small-scale clusters of CPRS could be conducted and adjusted according to local understanding of epidemics and other regional disturbances. Such measures should allow the preservation of the complex structure associated with mature and overmature stands. Also, several large forest tracts (over 100 km$^2$) should be maintained in each forest management unit. To maximize their conservation value, disturbance of such large forest tracts should be kept to a minimum (e.g., by partial cutting, roads). It should be noted that these stands will eventually be converted to regeneration areas when the adjacent landscape matrix will have acquired attributes associated with mature and overmature forests.

## 5.4. Measures Intended to Increase the Social Acceptability of Large Cutblock Clusters

The creation of large regeneration areas is often perceived negatively by the public (Pâquet and Bélanger 1997; Robson et al. 2000). Several options can be considered to increase the social acceptability of timber harvesting in general and, more specifically, that of large cutblock clusters. Visual screening of regeneration areas, combined with information and education of the public, should be prioritized.

To improve the aesthetics of harvested areas, cutblocks with straight lines should be replaced by smoother, more natural shapes which harmonize better with the landscape. To limit visual contrast, a strict exclusion period[*] should be respected, even more so in highly used areas. Wherever local conditions allow it, a combination of CPRS and CPHRS with variable retention strategies should be favoured. In riparian areas, buffers left should be wide enough for the harvested area not to be visible.

[*] The exclusion period is the time necessary for the forest to reach a visually acceptable state (stands >4 m in height) before any other harvesting can be conducted in adjacent areas (Pâquet and Bélanger 1997).

To further minimize the visual impact of large cutblock clusters, they should be cut over several years or several runs. Agglomerated and dispersed cutblock strategies could be combined depending on the level of use of a particular area. Large cutblock clusters should be conducted preferentially in isolated, forestry-committed territories, while avoiding particularly sensitive areas.

Lastly, the volume of information conveyed to the public should be substantially increased. It is necessary to demonstrate by concrete actions that the protection of the environment is taken into account in forest management. It should be explained that ecosystem-based management is a management strategy closer to nature which aims at maintaining the ecological integrity of ecosystems as well as making sure that these activities are sustainable (D'Eon 2007). If informed of the benefits brought about by ecosystem-based management and of the ecological principles on which it rests, chances are that the public will have a much more positive perception of this new approach (Meitner et al. 2005).

## 6.  CONCLUSION

The following observations are based on results presented in this chapter:

1. Timber harvesting has altered western spruce-moss forest landscapes of Québec by creating large harvested areas adjacent to one another, which lead to a local decline of large, mature and overmature forest tracts.

2. To maintain the forest mosaic within its natural range of variability, we must increase the distance between cutblock clusters and continue to maintain tall residual forest in within cutblock clusters as patches of varying size and shape.

3. Adapted silvicultural practices should be increasingly used between cutblock clusters.

The spatial characteristics of landscapes disturbed by fire in the western spruce-moss forest of Québec are similar to those reported in the literature for other regions of the boreal forest. We can thus speculate that these results can be used as a basis to elaborate principles of ecosystem-based management that could be applied to other pyrogenic regions of the boreal forest. However, the implementation of these principles could differ among provinces, being highly dependent on the political will and the governmental orientations embraced concerning ecosystem-based management. Also, the social acceptability of ecosystem-based management still represents an important challenge to tackle. It will depend closely on the scale of disturbances created by harvesting and on the quantity of tall residual forest left.

The measures advocated in this chapter have similarities with management guidelines found in the forest legislation of other Canadian provinces, such as British Columbia (BCMF 1995) and Ontario (OMNR 2001). Also, they can also be found in the National Boreal Standards of the Forest Stewardship Council (FSC) – Canada (2004). Finally, some forestry companies (e.g., Weyerhauser, Alberta Pacific) have integrated these guidelines to their current forestry practices for several years. Others, like Tembec (see Belleau and Légaré, chapter 19) and AbitibiBowater are currently running pilot projects in Québec.

Finally, the precautionary principle should be applied to the western spruce-moss forest of Québec, as the alteration of the landscape at a very large spatial scale would not take more than one rotation, while its restoration would necessitate a much longer time frame (Sachs et al. 1998). Bergeron et al. (1999b) suggest that nature-based practices will probably be less costly on the long term than the restoration of forest landscapes having sustained inadequate interventions.

## REFERENCES

Beaubien, J. 1994. Landsat TM satellite images of forests: from enhancement to classification. Can. J. Rem. Sens. **20**: 17–26.

Bélanger, L. 1992. La forêt mosaïque: une stratégie d'aménagement socialement acceptable pour la forêt boréale du Québec. II – Principes de base pour la sapinière. L'aubelle, **April**: 15–18.

Belleau, A., Bergeron, Y., Leduc, A., Gauthier, S., and Fall, A. 2007. Using spatially explicit simulations to explore size distribution and spacing of regenerating areas produced by wildfires: recommendations for designing harvest agglomerations for the Canadian boreal forest. For. Chron. **83**: 72–83.

Bendell, J.F. 1974. Effects of fire on birds and mammals. *In* Fire and ecosystems. *Edited by* T.T. Kozlowski and C.E. Ahlgren. Academic Press, New York, USA, pp. 73–138.

Bergeron, J.-F., Grondin, P., and Blouin, J. 1999a. Rapport de classification écologique: pessière à mousses de l'ouest. Ministère des Ressources naturelles et de la Faune, Direction des inventaires forestiers, Québec, Que.

Bergeron, Y., Harvey, B., Leduc, A., and Gauthier, S. 1999b. Strategies for forest management based on the dynamics of natural disturbances: considerations on settlement and forest. For. Chron. **75**: 55–61.

British Columbia Ministry of Forests (BCMF). 1995. Forest practices code of British Columbia. Biodiversity guidebook. Forest Service, Victoria, B.C.

Claveau, Y., Kneeshaw, D., and Gauthier, S. 2006. Nos pratiques s'inspirent-elles vraiment des feux? L'Aubelle, **151**: 14–21.

Crites, S. and Dale, M.R.T. 1998. Diversity and abundance of bryophytes, lichens, and fungi in relation to woody substrate and successional stage in aspen mixedwood boreal forests. Can. J. Bot. **76**: 641–651.

Darveau, M., Beauchesne, P., Bélanger, L., Huot, J., and LaRue, P. 1995. Riparian forest strips as a habitat for breeding birds in boreal forest. J. Wild. Manag. **59**: 67–78.

Darveau, M., Labbé, P., Beauchesne, P., Bélanger, L., and Huot, J. 2001. The use of riparian strips by small mammals in a boreal balsam fir forest. For. Ecol. Manag. **143**: 95–104.

DeLong, S.C. and Tanner, C. 1996. Managing the pattern of forest harvest: lessons from wildfire. Biod. Conserv. **5**: 1191–1205.

D'Eon, R. 2007. Ecosystem Management. Sustainable Forest Management Network, Edmonton, Alta. [Online] <www.sfmnetwork.ca/docs/e/E24%20Ecosystem%20 management.pdf> (accessed October 14, 2008).

Drapeau, P., Leduc, A., Bergeron, Y., Gauthier, S., and Savard, J.-P.L. 2003. Bird communities of old spruce-moss forests in the Clay Belt region: problems and solutions in forest management. For. Chron. **79**: 531–540.

Eberhart, K.E. 1986. Distribution and composition of residual vegetation associated with large fires in Alberta. M.Sc. thesis, University of Alberta, Edmonton, Alta.

Eberhart, K.E. and Woodard, P.M. 1987. Distribution of residual vegetation associated with large fires in Alberta. Can. J. For. Res. **17**: 1207–1212.

Forest Stewardship Council (FSC). 2004. National Boreal Forest Management Standard. [Online] <www.fsccanada.org/docs/39146450F65AB88C.pdf> (accessed October 14, 2008).

Forsyth, M.L. and McNicol, J.G. 1997. Governing clearcuts: lessons from fire. *In* Annals of the Ecological Landscape Management Workshop, Canadian Woodland Forum, Canadian Pulp and Paper Association, Fredericton, N.B., pp. 69–73.

Franklin, J.F., Cromack, K., Denison, W., McKee, A., Maser, C., Sedell, J., Swanson, F., and Juday, G. 1981. Ecological characteristics of old-growth Douglas-fir forests. General Technical Report PNW–118. U.S. Department of Agriculture, Forest Service, Pacific Northwest Forest and Range Experiment Station, Portland, Ore., USA.

Gagné, C. 2006. La répartition spatiale des coupes forestières et ses effets sur la distribution et le comportement alimentaire des oiseaux excavateurs. M.Sc. thesis, Université du Québec à Montréal, Montréal, Que.

Galipeau, C., Kneeshaw, D., and Bergeron, Y. 1997. White spruce and balsam fir colonization of a site in the southeastern boreal forest as observed 68 years after fire. Can. J. For. Res. **27**: 139–147.

Gauthier, S., Leduc, A., Harvey, B., Bergeron, Y., and Drapeau, P. 2001. Les perturbations naturelles et la diversité écosystémique. Nat. Can. **125**: 10–17.

Gauthier, S., Lefort, P., Bergeron, Y., and Drapeau, P. 2002. Time since fire map, age-class distribution and forest dynamics in the Lake Abitibi Model Forest. Natural Resources Canada, Canadian Forest Service, Laurentian Forestry Center, Information report LAU-X–125E.

Ghandi, K.J.K., Spence, J.R., Langor, D.W., and Morgantini, L.E. 2001. Fire residuals as habitat reserves for epigaeic beetles (*Coleoptera*: *Carabidae* and *Staphylinidae*). Biol. Conserv. **102**: 131–141.

Gluck, M.J. and Rempel, R.S.. 1996. Structural characteristics of post-wildfire and clearcut landscapes. Env. Mon. Ass. **39**: 435–450.

Hannon, S.J. 2005. Effect of stand vs. landscape level forest structure on species abundance and distribution. Sustainable Forest Management Network, Edmonton, Alta. [Online] www.sfmnetwork.ca/docs/e/SR_200405 hannonseffe_en.pdf (accessed January 7, 2009).

Hannon, S.J., Paszkowski, C.A., Boutin, S., DeGroot, J., Macdonald, S.E., Wheatley, M., and Eaton, B.R. 2002. Abundance and species composition of amphibians, small mammals, and songbirds in riparian forest buffer strips of varying widths in the boreal mixedwood of Alberta. Can. J. For. Res. **32**: 1784–1800.

Hansen, A.J., Spies, T.A., Swanson, F.J., and Ohmann, J.L. 1991. Conserving biodiversity in managed forest. BioScience, **41**: 382–392.

Imbeau, L. and Desrochers, A. 2002. Foraging ecology and use of drumming trees by Three-toed Woodpeckers. J. Wild. Manag. **66**: 222–231.

Imbeau, L., Savard, J.-P., and Gagnon, R. 1999. Comparing bird assemblages in successional black spruce stands originating from fire and logging. Can. J. Zool. **77**: 1850–1860.

Johnson, E.A. 1992. Fire and vegetation dynamics: studies from the North American boreal forest. Cambridge University Press, Cambridge, UK.

Kafka, V., Gauthier, S., and Bergeron, Y. 2001. Fire impacts and crowning in the boreal forest: study of a large wildfire in western Quebec. Int. J. Wildland Fire, **10**: 119–127.

Kay, C.E. 1997. Is aspen doomed? J. For. **95**: 4–11.

Kimmins, H. 1997. Balancing Act. *In* Environmental Issues in Forestry, 2nd Edition, BC Press, Vancouver, B.C.

Mascarúa López, L.E., Harper, K.A., and Drapeau, P. 2006. Edge influence on forest structure in large forest remnants, cutblock separators, and riparian buffers in managed black spruce forests. Écoscience, **13**: 226–233.

McGarigal, K. and Marks, B.J. 1994. FRAGSTATS: a spatial pattern analysis program for quantifying landscape structure. Oregon State University, Corvallis, Ore., USA.

Meitner, M.J., Gandy, R., and D'Eon, R. 2005. Human perceptions of forest fragmentation: implications for natural disturbance management. For. Chron. **81**: 256–264.

Ministère des Ressources naturelles (MRN). 2000. La limite nordique des forêts attribuables. Gouvernement du Québec, Québec, Que.

Ministère des Ressources naturelles, de la Faune et des Parcs (MRNFP). 2003. Le manuel d'aménagement forestier, 4th edition. Gouvernment du Québec, Québec, Que.

Nappi, A., Drapeau, P., and Savard, J.-P.L. 2004. Salvage logging after wildfire in the boreal forest: is it becoming a hot issue for wildlife? For. Chron. **80**: 67–74.

National Council for Air and Stream Improvement, Inc. (NCASI). 2006. Similarities and differences between harvesting- and wildfire-induced disturbances in fire-mediated Canadian landscapes. Technical Bulletin No. 924. Research Triangle Park, N.C., USA.

Ontario Ministry of Natural Resources (OMNR). 1997. Forest management guidelines for the emulation of fire disturbance patterns: analysis results. Unpublished. Toronto, Ont.

Ontario Ministry of Natural Resources (OMNR). 2001. Forest management guide for natural disturbance pattern emulation. Version 3.1. Queen's Printer for Ontario, Toronto, Ont.

Pâquet, J. and Bélanger, L. 1997. Public acceptability thresholds of clearcutting to maintain visual quality of boreal balsam fir landscapes. For. Sci. **43**: 46–55.

Perron, N. 2003. Peut-on et doit-on s'inspirer de la variabilité naturelle des feux pour élaborer une stratégie écosystémique de répartition des coupes à l'échelle du paysage? Le cas de la pessière noire à mousses de l'ouest au Lac-Saint-Jean. Ph.D. thesis, Université Laval, Québec, Que.

Potvin, F. and Bertrand, N. 2004. Leaving forest strips in large clearcut landscapes of boreal forest: a management scenario suitable for wildlife? For. Chron. **80**: 44–53.

Robitaille, A. and Saucier, J.-P. 1998. Paysages régionaux du Québec méridional. Publications du Québec, Que.

Robson, M., Hawley, A., and Robinson, D. 2000. Comparing the social values of forest-dependent, provincial and national publics for socially sustainable forest management. For. Chron. **76**: 615–622.

Rowe, J.S. 1972. Forest regions of Canada. Canadian Forestry Service, Ottawa, Ont., Publication no. 1300.

Ryan, K.C. 2002. Dynamic interactions between forest structure and fire behavior in boreal ecosystems. Silva Fenn. **36**: 13–39.

Sachs, D.L., Sollins, P., and W.B. Cohen. 1998. Detecting landscape changes in the interior of British Columbia from 1975 to 1992 using satellite imagery. Can. J. For. Res. **28**: 23–36.

Serrouya, R. and D'Eon, R. 2005. Variable retention forest harvesting: research synthesis and implementation guidelines. Sustainable Forest Management Network, Edmonton, Alta. [Online] <www.sfmnetwork.ca/docs/e/SR_200405serrouyarvari_en.pdf> (accessed October 15, 2008).

Schmiegelow, F.K.A. and Mönkkönen, M. 2002. Habitat loss and fragmentation in dynamic landscapes: avian perspectives from the boreal forest. Ecol. Appl. **12**: 375–389

Spies, T.A., Ripple, W.J., and Bradshaw, G.A. 1994. Dynamics and pattern of a managed coniferous forest landscape in Oregon. Ecol. Appl. **4**: 555–568.

Tande, G.F. 1979. Fire history and vegetation pattern of coniferous forests in Jasper National Park, Alberta. Can. J. Bot. **57**: 1912–1931.

Turner, M.G. and Romme, W.H. 1994. Landscape dynamics in crown fire ecosystems. Landsc. Ecol. **9**: 59–77.

Vaillancourt, M.-A., Drapeau, P., Gauthier, S., and Robert, M. 2008. Availability of standing trees for large cavity-nesting birds in the eastern boreal forest of Québec, Canada. For. Ecol. Manag. **255**: 2272–2285.

Van Wagner, C.E. 1983. Fire behavior in northern conifer forests. *In* The role of fire in northern circumpolar ecosystems, Scope **18**: 65–80.

Zasada, J.C. 1971. Natural regeneration of interior Alaska forests: seed, seedbed, and vegetative reproduction considerations. *In* Fire in the Northern Environment: a symposium. *Edited by* C.W. Slaughter, R.J. Barney, and G.M. Hansen. USDA Forest Service, Pacific Northwest Forest and Range Experiment Station, Portland, USA, pp. 231–246.

# Chapter 7

## Spruce Budworm Outbreak Regimes in Eastern North America*

*Hubert Morin, Danielle Laprise, Andrée-Anne Simard, and Saida Amouch*

\* The authors thank the Sustainable Forest Management Network, NSERC, the Consortium de recherche sur la forêt boréale commerciale, and the Ministère des Ressources naturelles et de la Faune du Québec for their financial support of this research. Many thanks to Dan Kneeshaw and Marie-Andrée Vaillancourt for the revision and suggestions on an earlier version of the manuscript and to Pierre-Yves Plourde for the drawing of the figures. We are also grateful to Gerardo Reyes for the English translation. The photos on this page were graciously provided by Hubert Morin.

# 1. INTRODUCTION

The spruce budworm (SBW) (*Choristoneura fumiferana* Clem.) is the principal defoliating insect of fir and spruce in the boreal forests of eastern North America. Outbreaks have occurred every 25 to 40 years at the supra-regional scale during the 20th century (Royama 1984; Candau et al. 1998; Jardon et al. 2003; Royama et al. 2005; Campbell 2007). Although SBW is present in all provinces throughout Canada and can sometimes cause severe defoliation to more than half a million hectares, the provinces of eastern Canada are the most affected, especially Québec, Ontario and New Brunswick. The last outbreak (1974–1988) affected more than 55 million hectares of forest, causing the loss of 139 to 238 million cubic metres of fir and spruce in Québec only (Boulet et al. 1996). The area affected reached 32 million hectares in 1975 in Québec, 19 million hectares in 1980 in Ontario, and 3.5 million hectares in 1975 in New Brunswick (Natural Resources Canada 2001).

In eastern Canada, SBW outbreaks are often more important than fire in terms of tree mortality. It used to be believed that the impact of outbreaks was lower in boreal forests dominated by black spruce (*Picea mariana* [Mill.] B.S.P.). However, recent studies have shown that the last four outbreaks were well recorded in the forest structure of the black spruce–feather moss bioclimatic domain. Indeed, almost all natural balsam fir (*Abies balsamea* [L.] Mill.) stands found within the spruce–feather moss zone originated from one of these outbreaks, following a cyclical mechanism of regeneration to which balsam fir is well adapted (MacLean 1984, 1988; Morin 1994; Morin and Laprise 1997; Johnson et al. 2003; Parent et al. 2003; Morin et al. 2007). This cyclical regeneration mechanism is, however, not universal. In areas where the balsam fir seedling bank is less abundant, as well as in more southerly locations, different scenarios have been observed, such as a rapid colonization by non-host species (Kneeshaw and Bergeron 1998, 1999; Bouchard et al. 2006a). Regarding black spruce forests, substantial reductions in black spruce growth have been linked to the last outbreaks (Lussier et al. 2002).

A tri-trophic interaction would exist between the host plant, the insect, and its natural enemies (Royama et al. 2005; Cooke et al. 2007). Balsam fir and white spruce (*Picea glauca* [Moench] Voss.) are the principal host tree species of the SBW, while black and red spruce (*P. rubens* Sarg.) are less affected. In eastern Canada, mature stands of balsam fir are most vulnerable to SBW defoliation (MacLean 1980; MacLean and Ostaff 1989; Bergeron et al. 1995; MacLean and MacKinnon 1997). The proportion of hardwoods within a stand and across the landscape can also affect defoliation levels, with areas having greater hardwood densities being less affected (Bergeron et al. 1995; Su et al. 1996; Cappuccino et al. 1998; MacKinnon and MacLean 2003, 2004; Campbell 2007). A forest mosaic dominated by host trees would favor greater SBW population growth compared with its natural enemies, resulting in a more severe impact across the landscape (Bouchard 2005; Cooke et al. 2007). Conversely, a forest mosaic consisting of alternative host or non-host canopy tree species would result in asynchronous outbreaks having lower impacts and where spread across the landscape is restricted to smaller areas. The entomological cycle caused by the SBW–natural enemy complex would remain the same, but it would be more difficult to detect across the landscape because of the moderate impacts.

The spatio-temporal dynamics of SBW outbreaks is not well understood. The last outbreak (1974–1988) was well documented, but information on the outbreaks occurring in the middle and beginning of the 20th century, in addition to the preceding outbreaks, is still incomplete. Dendrochronological and paleoecological techniques are effective methods for improving our understanding of the spatio-temporal dynamics of SBW outbreaks, allowing us to investigate on historical outbreak periods and, subsequently, to compare outbreak dynamics among centuries. Recent studies indicate that a major shift in the system occurred towards the end of the 19th century. This century was characterized by asynchronous outbreaks at the supra-regional scale that spread very slowly across the landscape, whereas the 20th century outbreaks were more explosive, rapidly affecting large forest areas in a synchronous manner (Blais 1983; Jardon 2002). This shift has also been observed in other systems, and has been attributed to both anthropogenic changes to the forest mosaic (Blais 1954, 1983; Swetnam and Lynch 1993; Williams and Liebhold 2000; Peltonen et al. 2002) and to natural phenomena such as changes in fire frequency (Morin et al. 2007). The forest mosaic would therefore be one of the major factors controlling both population levels of the SBW and the impacts of budworm defoliation (Bouchard 2005; Cooke et al. 2007).

The objectives of this chapter are to: (1) review the knowledge available of the spatio-temporal dynamics of SBW outbreaks in eastern North America from dendrochronological and paleoecological studies, (2) show the impacts of the SBW within the balsam fir–white birch and spruce–feather moss bioclimatic domains, especially on balsam fir, its preferred host, but also on black spruce, and lastly (3) show the implication of these results for forest ecosystem management in the context of climate change.

## 2. TEMPORAL PATTERNS OF SPRUCE BUDWORM OUTBREAKS OVER THE PAST 8,600 YEARS

### 2.1. Dendrochronology

Since Blais' (1954) pioneer work, dendrochronology has been extensively used to detect past outbreaks for a number of defoliating insect species. Eastern and western spruce budworm (Morin and Laprise 1990; Swetnam and Lynch 1993; Krause 1997), forest tent caterpillar (*Malacosoma disstria* Hbn.; see Sutton and Tardif, chapter 8), larch sawfly (*Pristiphora erichsonii* Htg.; Jardon et al. 1994), larch budmoth (*Zeiraphera diniana* Gn. Weber 1997), two-year budworm (*Choristoneura biennis* Free.; Zhang and Alfaro 2002, 2003), and jack pine budworm (*Choristoneura pinus pinus* Free.; Volney 1988) have all been investigated. While data analysis techniques have improved (Swetnam et al. 1985), objects of study have remained the same: the patterns of growth and mortality of the host tree. The maximum age that a host tree can reach thus limits the temporal range of analysis. This is especially important in the case of the SBW since it often kills its host. Cross-dating, i.e. the synchronization of patterns of growth between live and dead trees, has been valuable for extending chronologies from the analysis of old dead trees found in forests or in old buildings (Krause 1997; Boulanger and Arseneault 2004). However, we are currently limited to investigating periods that only go back as far as the 17th century, while data that cover large areas do not extend much further back than the 19th century. Subfossil trees recovered from peatlands surrounded by host trees are now being used to extend analyses to even earlier periods of time (Simard 2003; Simard et al. 2007). One study that covers a 4,800-year time frame showed that outbreaks of the 20th century are easy to distinguish in the growth rings, but that previous outbreaks are more difficult to identify. Specific patterns of growth reductions attributed to the SBW have helped to cross-date the samples to produce two floating chronologies between 4,170 and 4,740 years before present (Simard 2003). Unfortunately, younger samples found between 20- and 60-cm depths exhibited a relatively uniform growth pattern, without distinct growth reductions that would have make it possible to cross-date. This result suggests that, at least for this location, the SBW had less of an impact on spruce growth before the 20th century than during the 20th century. Older chronologies that cover extensive areas are, however, still absent.

Jardon (2002) and Jardon et al. (2003) studied the periodicity, synchronism, and impact of SBW outbreaks occurring over the last two centuries across large areas of Québec province. They showed that outbreaks occurred in a periodic fashion at the supra-regional scale every 25 to 38 years (Jardon et al. 2003). They also showed that the periodicity can change at a local scale, possibly because of the effects of migration, local dynamics of insect populations, or because some sites are able to periodically escape certain infestations. Double waves of outbreaks also occurred. They were similar and generally in synchrony, indicating the presence of a common regional effect known as the Moran effect (Moran 1953; Peltonen et al. 2002). To cover the largest area possible within the context of this chapter, we included an additional 73 white spruce chronologies originating from central Québec to the work of Jardon (2002) (figure 7.1). White spruce is often chosen, when present, for long-term reconstructions because it is a host of the SBW that can generally survive multiple outbreaks, allowing us to use older trees than if we had only used balsam fir trees. The 80 non-host trees used are from four chronologies of eastern white-cedar (*Thuja occidentalis* L.) originating from various regions across Québec.

Conventional dendrochronological techniques were applied (ARSTAN, ITRDB Lib V2.1, Holmes 1999) to standardize the measured series of host and non-host trees, and to subsequently produce a corrected chronology by subtracting non-host from host chronologies (Swetnam et al. 1985; Holmes and Swetnam 1996). The set of regional chronologies shown in figure 7.1 was reorganized to show the percentage of trees affected by SBW across the entire study area (figure 7.2) (see Jardon [2002] for more details).

At the supra-regional scale, figure 7.2 clearly shows the regular periodicity of outbreaks that occurred over the last two centuries. In more southerly areas of Québec, outbreaks occur approximately every 30 years (Jardon 2002; Morin et al. 2007). When examining local or regional chronologies individually, this periodicity is generally observed for the 20th century, but it is more difficult to

**Figure 7.1**
**Locations of host (white spruce) and non-host (eastern white cedar) chronologies of the SBW**

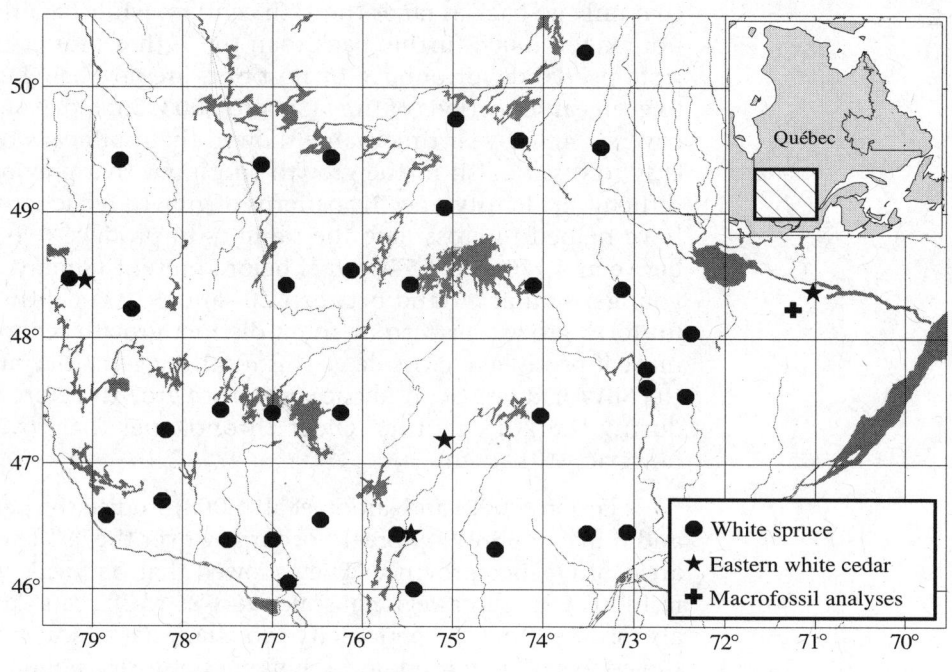

**Figure 7.2**
**Percentage of white spruce trees affected by SBW over the course of their lives in the study area**

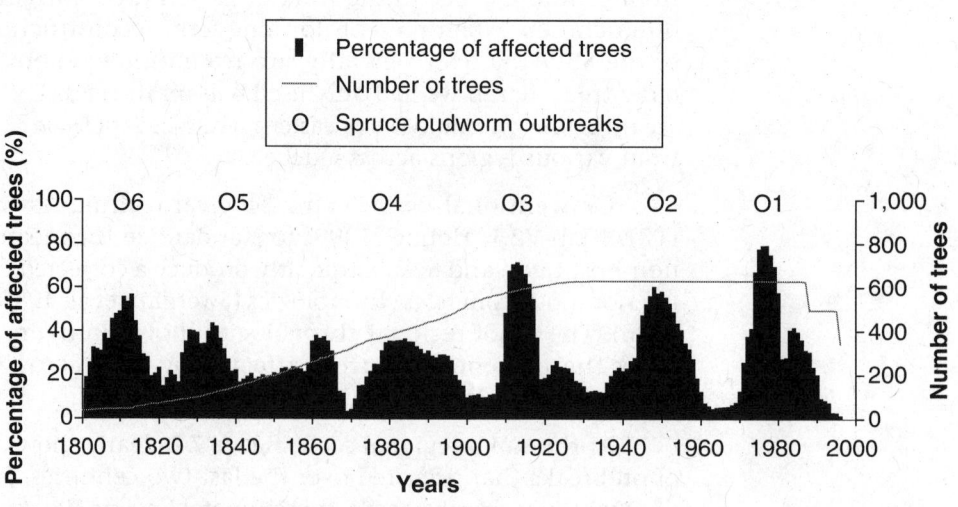

This figure clearly shows six outbreak episodes occurring between 1800 and 2000. From Morin et al. (2007).

observe for the 19th century (Jardon et al. 2003). In fact, in certain locations one or the other of outbreak periods was not registered in growth rings, particularly in the 19th century. Moreover, outbreaks did not necessarily arise in a synchronous fashion throughout the entire study area, as in the Abitibi region for instance (Morin et al. 1993; Campbell 2007). This problem may be linked to the dendrochronological technique itself, since there are fewer trees that could record outbreaks as one goes further back in time (but this point will be addressed later in the text). Using dead trees from old buildings, Krause (1997) and Boulanger and Arsenault (2004) have shown that a periodicity could sometimes be found locally or regionally before the 20th century, but the data cover less and less area as one ventures further back in time.

Outbreaks of the 20th century affected a much larger area relative to those of the 19th century, and the spatial aspects of outbreaks will be covered in more detail in the following section. On the other hand, the number of trees affected at the same time in a given area is indicative of the impact of an outbreak on the area. Our results thus show that outbreaks of the 20th century were generally more severe than those of the 19th century, and that the outbreak in the middle of the 20th century (O2) was less severe than the two others that occurred during the same century (O1 and O3).

## 2.2. Paleoecology

Research by Potelle (1995) and Simard et al. (2002) helped to develop a macrofossil analysis technique that allowed us to assess the presence of the SBW in the sediments. This technique, which uses insect presence indicators, was first tested in fresh forest humus, then in deeper humus layers (Simard et al. 2002). Because the SBW has a rather fragile structure, remains of larval or adult stages that could be used for micro- or macrofossil analyses are not very abundant in the sediments. Head capsules have been used to the detect the presence of several insect species (Davis et al. 1980; Bhiry and Filion 1996) but their numbers are often limited in the case of the SBW, spurring us to add another macrofossil produced in large quantities during outbreaks that can be identified to the species and that is indicative of the SBW presence and abundance: the number of feces (insect frass pellets) (Simard et al. 2002).

Analysis of head capsules and feces from small peat bogs surrounded by host trees has provided an indication of the local abundance of SBW populations over the past 8,000 years in the Saguenay region of Québec (Simard et al. 2006). Obviously, the resolution of this technique is not as precise as denrochronology, which has precision to the nearest year. Macrofossil profiles would rather represent periods of insect activity at the local scale, without necessarily separating outbreak periods. Figure 7.3 is an example of a SBW feces macrofossil profile originating from a small peat bog surrounded by black spruce and balsam fir (Simard et al. 2006). Moreover, the four macrofossil profiles originating from similar small peat bogs all showed the same pattern. First, feces are found along the entire profile, indicating that the SBW was present in the study area since the beginning of peat accumulation. Furthermore, a great abundance of feces is always present in the first few centimetres representing the 20th century. Before this period, several hundreds of years elapsed where feces numbers were very

### Figure 7.3
**Example of a feces macrofossil profile (insect frass pellets) of the SBW originating from a forested peat bog of the Saguenay region**

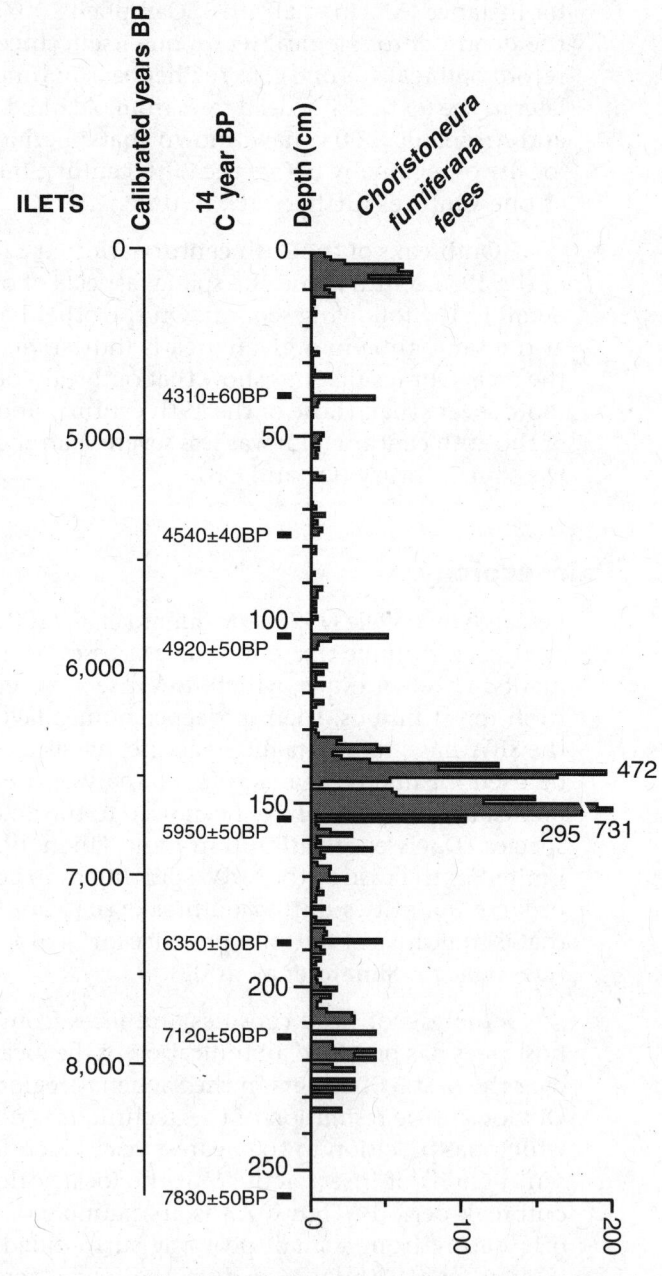

**Number of macrofossils/100 cm³ of peat**

It can be observed that the SBW was present from the beginning of peat accumulation in the study area. Modified from Morin et al. (2007) and Simard et al. (2006).

low, suggesting that local SBW populations were also low. Generally, the profiles indicate that only one to three other periods occurred where feces numbers are as high or are higher than that observed for the 20th century.

Since the problems associated with tree age are not relevant in the case of insect macrofossils found in peat bogs, these results support those obtained with dendrochronology analyses. They do not contradict the hypothesis that outbreaks occurred in a periodic fashion during the Holocene (the last 10,000 years), but the weak temporal resolution and the lack of large spatial coverage prevents the development of a more robust interpretation at this level. The results do suggest, however, that SBW outbreaks had more of an impact during the 20th century than in the previous centuries, and that periods where outbreaks had substantial impacts on the forest during the Holocene were rare. Population levels of the SBW would generally have been at low, endemic levels throughout much of the Holocene, exhibiting only rare episodes of outbreak levels that could have caused very severe local effects.

## 3. SPATIAL PATTERNS OF SBW OUTBREAKS

The area covered by the outbreak chronologies allows us to study their spatial dynamics. For example, two other patterns clearly illustrated in figure 7.2 are the maximum area affected and the time needed to reach this maximum during an outbreak. Note the important difference between outbreaks of the 19th and 20th centuries, and between the outbreak that occurred in the middle of the 20th century (O2) and the other two that occurred during the same century (O1 and O3). Outbreak O2 is believed to have been less severe than O1 and O3, while the O3 outbreak is now recognized as being the most severe outbreak ever documented. The O3 outbreak arose very quickly in each site, as well as across the entire province (Morin and Laprise 1990; Jardon 2002). The number of white spruce trees affected rapidly increased, reaching 80% within four years. This outbreak has killed more trees more rapidly than all other outbreaks ever documented in the province (Boulet et al. 1996).

In contrast with this explosive pattern, the O2 outbreak took 12 years for the insect to affect 60% of trees at the same time during the peak of the outbreak. This difference between outbreak O2 and the other two outbreaks of the 20th century is attributed to the quantity of mature balsam fir stands present across the landscape. Outbreak O3 was so severe that the majority of mature balsam fir (the most vulnerable host species) stands were killed. When the O2 outbreak occurred following the very regular entomological cycle of the SBW (Jardon et al. 2003), mature balsam fir stands were rare. A few survivors of the previous outbreak were present, but the majority of stands were young (less than 30 years of age), and thus less vulnerable, and composed of released advance regeneration able to benefit from canopy openings caused by the mortality of the mature stands (Blais 1983; Morin 1994).

Conversely, outbreak O1, which occurred approximately 30 years later, affected more mature stands and exhibited a spatial pattern similar to outbreak O3, affecting 80% of the trees in the study area within six years. Gray et al.

(2000) precisely documented where and when different patterns of defoliation were produced during this outbreak period using defoliation records in Québec. These models could be used during the next outbreak to predict when defoliation will arise in a given region and will be very useful for the management and protection of vulnerable stands. Elsewhere in Canada, Candau and Fleming (2005) have also used defoliation records from 1967 to 1998 in Ontario to establish the relationship between the spatial distribution of defoliation and bioclimatic conditions. In the northern parts of SBW's range, an elevated frequency of defoliation was associated with cool spring temperatures and dry periods during the month of June, whereas low frequencies were associated with cold winters and generally cool average temperatures. On the other hand, in the southern part of its range, low frequencies of defoliation were associated with low abundance of host species, particularly balsam fir and black and white spruce.

One can presume that the O3 outbreak occurred when there was a great abundance of balsam fir in the canopy. The severity of this outbreak would have trigger the pattern observed during the 20th century in a landscape dominated by balsam fir stands, where an alternation between severe and less severe outbreaks is observed. It is unclear, however, why this pattern was not observed during the 19th century.

Outbreaks of the 19th century displayed a spatial pattern in each stand, as well as across the entire landscape, that was different from that observed during the 20th century. They all exhibited a gradual progression, taking approximately 10 to more than 20 years to reach the maximum number of trees affected. Moreover, this maximum rarely exceeded 40% of the trees affected at the same time. The impact on the forest was gradual, and the outbreak never affected the entire province of Québec at the same time. Note, however, that these results are limited by the dendrochronological technique, since the sampled trees dating back to the 19th century were younger at the time and were thus less vulnerable to infestation. On the other hand, associated researches suggest that this limitation is not the only explanation, indicating that the observed pattern was not simply a sampling artefact. First of all, outbreaks of the 19th century, and the O4 outbreak in particular, were always difficult to identify using dendrochronology (Blais 1965; Morin and Laprise 1990; Krause 1997). This might explain why Blais (1965) had interpreted the O4 outbreak as a period of drought. Morin and Laprise (1990), Krause (1997) and Jardon et al. (2003) later interpreted this period as an outbreak with a more local impact, which never affected vast areas at the same time. Krause (1997) and Boulanger and Arseneault (2004) confirmed this hypothesis using trees originating from old buildings that were already old before the O4 outbreak occurred. Secondly, dendrochronological data were in agreement with paleoecological data that were not affected by the problems of tree age. Indeed, macrofossil analyses could always show the importance of SBW populations during the 20th century. Conversely, except for the 20th century, they also showed that there had not been any substantial increases in fecal pellets of the SBW – that is indicative of an increase in population levels – dating back to at least 5,000 years before present (Simard et al. 2006). These results indicate that a major change in the amplitude of population fluctuations of the SBW occurred at the beginning of the 20th century.

Dendrochronological and macrofossil analyses revealed that, most of the time, SBW populations exhibit cyclical fluctuations that do not reach outbreak levels which rapidly appear in stands and affect vast areas. Local outbreaks would occur here and there but synchronism would not be evident over large areas. As an example of this pattern, the O2 outbreak suggests that the forest composition and structure would have been different. The canopy of mature stands would probably have contained fewer host tree species, especially less balsam firs. This interpretation is in agreement with the tri-trophic interaction model of SBW populations presented by Cooke et al. (2007). In the northern parts of SBW's distribution range, in the black spruce–feather moss bioclimatic domain, black spruce dominates the canopy. Balsam fir is present in the understory, as well as in the canopy of a few rare isolated stands. In a study at the northern limit of the range, Levasseur (2000) observed a general pattern in which the impact of defoliation became more localized and gradual at both stand and landscape scales as one moved northward towards the 53rd parallel. At this latitude, the spatial pattern and impact of the 20th century outbreaks more closely resembled those of the 19th century that occurred in southern locations (Morin et al. 2007). This forest type is maintained by a fire disturbance regime as black spruce, with its semi-serotinous cones, is better adapted to regenerate after fire than balsam fir.

Conversely, the rare periods when the SBW would have reached epidemic levels synchronized over vast areas would be associated with a forest composition and structure in which balsam fir was a major component. Blais (1983) had already observed this change in the outbreak regime. However, available results at that time suggested that there had been an increase in the frequency and severity of outbreaks. According to Blais (1983), the outbreak at the beginning of the 20th century (O3) was less severe than the second outbreak (O2), which in turn was less severe than the most recent outbreak (O1) with respect to the amount of area affected and mortality caused. He mentioned that these results could be explained by an increase in host species, especially balsam fir, in the forests of Québec caused by: 1) forest harvesting practices that favoured the regeneration of balsam fir, 2) forest fire suppression programs that promoted the aging of stands and the increase in the proportion of balsam fir, and 3) spraying programs against SBW outbreaks which allowed for the conservation of mature balsam fir stands. Recent data support another explanation. First, we now know that the O3 outbreak was very severe, probably being the most severe one ever documented in Québec (Boulet et al. 1996), and that the O2 outbreak was less severe. These results are supported by age structure studies of balsam fir stands (Morin 1994) and by numerous chronologies (Morin and Laprise 1990; Krause 1997; Jardon et al. 2003). Furthermore, forest harvesting, fire suppression, and spraying programs against SBW cannot explain the explosive pattern of outbreak O3 that occurred at the beginning of the century. At the time, extensive forest harvesting did occur in the southern portion of the commercial forest only, with very little occurring across the boreal forest region in general. Moreover, fire suppression and spraying programs against SBW outbreaks were at their inception stages and would have had no substantial effects on forest composition and structure (Morin et al. 2007). The impact of man is certainly important on forest composition and structure but effects would only have begun to be felt towards the middle of the century and, undoubtedly, during the last outbreak (O1).

Another factor that had large-scale effects must have been involved in creating an outbreak of the magnitude of the one that occurred at the beginning of the 20th century. If one accepts the hypothesis that an increase in the proportion of balsam fir in the canopy was an impetus, an event that could be responsible for this increase would be a decrease in fire frequency. We proposed that the change in fire frequency that occurred towards the end of the Little Ice Age (around 1850) in eastern North America (Bergeron and Archambault 1993; Bergeron et al. 2001) resulted in the development of forests containing larger proportion of balsam firs during the 19th century, and would have driven the SBW outbreak regime observed during the 20th century (Morin et al. 2007). The climate of the boreal forest during the Little Ice Age would have been under the influence of cold and dry air masses. During the end of this period and as polar front migrated northward, southern parts of the boreal forest would have been subjected to a hot and humid climate which has resulted in a reduction of drought and fires. This period would have been characterized by a reduction in fire ignitions and total area burned, and by the absence of large fires. This reduction in fire frequency would have been responsible for the increase of balsam fir and eastern white cedar during the 20th century. Indeed, the proportion of balsam fir increases in function of time since last fire (Bergeron 1998), as does the mortality due to SBW outbreaks (Bergeron and Leduc 1998).

## 4.   IMPACT OF OUTBREAKS ON THE FOREST

### 4.1.   An Example from Balsam Fir Stands of Boreal Forests North of Saguenay–Lac-Saint-Jean

A good way to determine the impact of disturbances is to study their effects on forest structure, and particularly their effects on the stand age structure. MacLean (1984, 1988) has already described the impact of outbreaks on mature forests stands in the balsam fir–white birch bioclimatic domain. In defoliating and eventually killing mature trees, SBW outbreaks allow balsam fir advance regeneration to accelerate their growth to reach the canopy. Additionally, in boreal mixedwoods, the periodic and synchronous mortality of balsam fir can also favour the emergence of a multi-cohort stand age structure composed of several coniferous and deciduous species (Bouchard 2005; Bouchard et al. 2006b). In the spruce–feather moss zone, this impact is less obvious. We now know that all the natural balsam fir stands studied in the boreal forest region originate from one or another of the last outbreaks (Morin 1994). Indeed, balsam fir stands exhibit an age structure which indicates that the majority of trees established before the period of maximum defoliation (figure 7.4).

The age structures displayed are composed of minimum ages, since it is nearly impossible to precisely date a mature fir tree. Balsam fir can remain in the seedling stage for more than 50 years under the forest cover. Because of a mechanism that gradually buries the stem under accumulating litter, and when the stem is weighed down by snow accumulation, adventitious roots are formed on the stem and allows balsam fir seedlings to remain small for long periods of time (Parent et al. 2000, 2001, 2003). This mechanism permits seedlings to

**Figure 7.4**

**Age structures of three balsam fir stands from the black spruce–feather moss bioclimatic domain and radial growth of dominant trees from stands established following the outbreaks that occurred A) at the middle of the 20th century (O2), B) at the beginning of the 20th century (O3), and C) at the end of the 19th century (O4)**

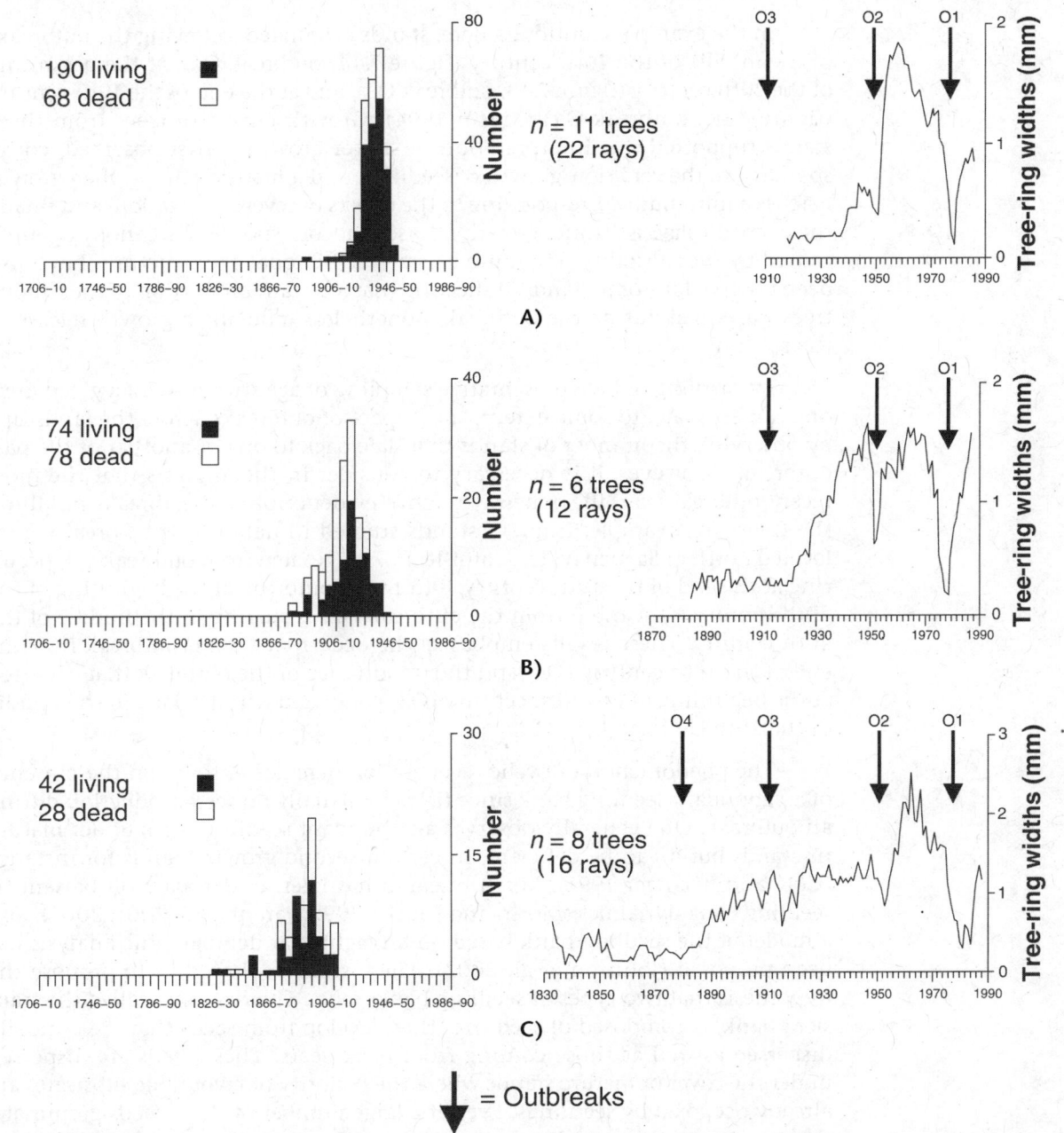

The balsam fir stands exhibit normal-type age structures. Annual growth rings clearly show reductions in growth associated with outbreak periods. Modified from Morin (1994).

maintain an equilibrium between the photosynthetic and non-photosynthetic biomass, to be less demanding on energy resources, and to remain alive for decades to even hundreds of years under the cover of dominant trees (Parent et al. 2005). Yet, these structures indicate that balsam fir stands in the black spruce–feather moss zone established from advance regeneration for which growth have accelerated following the death of mature canopy trees caused by one or another of the SBW outbreaks.

In the example mentioned here, stands originated following the outbreaks of the middle of the 20th century (figure 7.4A, outbreak O2), at the beginning of the 20th century (figure 7.4B, outbreak O3), and at the end of the 19th century (figure 7.4C, outbreak O4) (Morin 1994). Growth curves of trees from these stands supported this interpretation. A slower growth is first observed, corresponding to the very slow growth of seedlings under mature canopy, then growth reaches a minimum corresponding to the effects of severe defoliation, and finally the curve displays a distinct growth release that corresponds to a canopy opening caused by the mortality of mature balsam fir. Bimodal age structures have also been observed in some stands, indicating that a certain proportion of the canopy trees was killed during the outbreak, nonetheless inducing a growth release of the seedlings.

By carrying out a representative sampling of age structures of a given area, one can estimate, to some extent, the impact of outbreaks across the landscape by observing the number of stands that date back to one or another of the past outbreaks. However, it is necessary to consider in the analysis that the most recent outbreaks can in some way remove evidence of past outbreaks in killing the trees. For example, from 17 stands studied in detail in the boreal region located north of Saguenay–Lac-Saint-Jean, 7 originated from outbreak O4, occurring at the end of the 19th century, 8 from the outbreak at the beginning of the 20th century (O3) and 2 from the outbreak O2, occurring in the middle of the 20th century. These results emphasize the occurrence of an outbreak near the end of the 19th century (O4) and the importance of the outbreak that occurred at the beginning of the 20th century (O3) compared with the one in the middle of the 20th century (O2).

The phenomenon of cyclic regeneration depends entirely on the presence of an abundant seedling bank since there is virtually no seed production during an outbreak. One generally observes an abundant seedling bank under mature fir stands but it can be deficient, in certain second-growth stands for instance (Côté and Bélanger 1991). Much research has been undertaken on balsam fir seedling bank dynamics (Morin and Laprise 1997; Parent et al. 2001, 2003), and a model of the seedling bank based on an eight-year demographic analysis has been proposed (Johnson et al. 2003). These studies allowed us to propose the hypothesis that two types of seedling banks exist. The first one, called the transient bank, is composed of seedlings that develop from seeds that are annually dispersed as well as those coming from mast years. These seeds are dispersed under the cover of mature stands where the majority of favourable substrates are already occupied by seedlings. Even if a large number of these seeds germinate, the majority of these seedlings die within a few years since they established on poorer substrates and under a restrictive light environment.

The second type of seedling bank, called the permanent bank, is comprised of the seedlings established under auspicious germination and survival conditions; thus, during periods when favourable light and substrate environments are available. These seedlings would have a greater chance of survival and, once well established, benefit from the characteristics mentioned above and be able to survive under the forest cover for decades.

As a follow-up to a study of balsam fir stands in the black spruce–feather moss zone, a permanent plots monitoring of seedling bank demographics was initiated in 1994 in four stands of different ages (Morin 1994; Duchesneau and Morin 1999). Demographic analysis of seedlings in the permanent plots supports the hypothesis of the presence of two types of seedling banks (figure 7.5). The number of seedlings per square metre originating from good mast years since the implementation of the permanent plots in 1994 has declined exponentially. Conversely, the small number of seedlings that were already present in 1994 were still alive 12 years later, thus displaying a high probability of survival (figure 7.6). Moreover, these seedlings are clearly larger than the seedlings from the cohorts established after 1994.

**Figure 7.5**
**Changes in seedling density for seedlings established before the installation of permanent plots in 1994 north of Lac-Saint-Jean, and the cohorts from the subsequent mast years (1995, 1997, and 2004).**

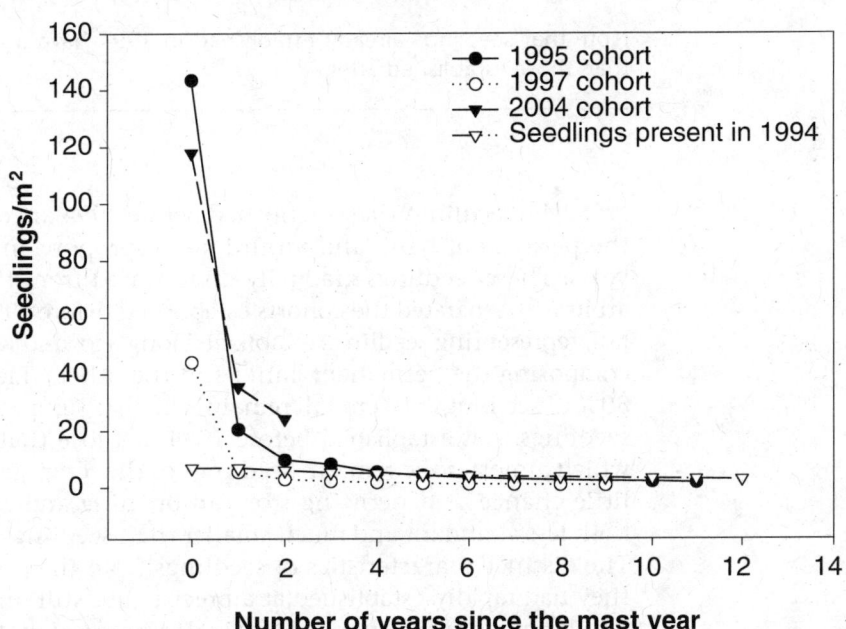

The number of seedlings per square metre establishing from seed mast years decreased exponentially, while the seedlings present in 1994 have maintained their density in the seedling bank.

Figure 7.6
**Probability of survival for seedlings established before 1994 and for the cohorts established from subsequent mast years**

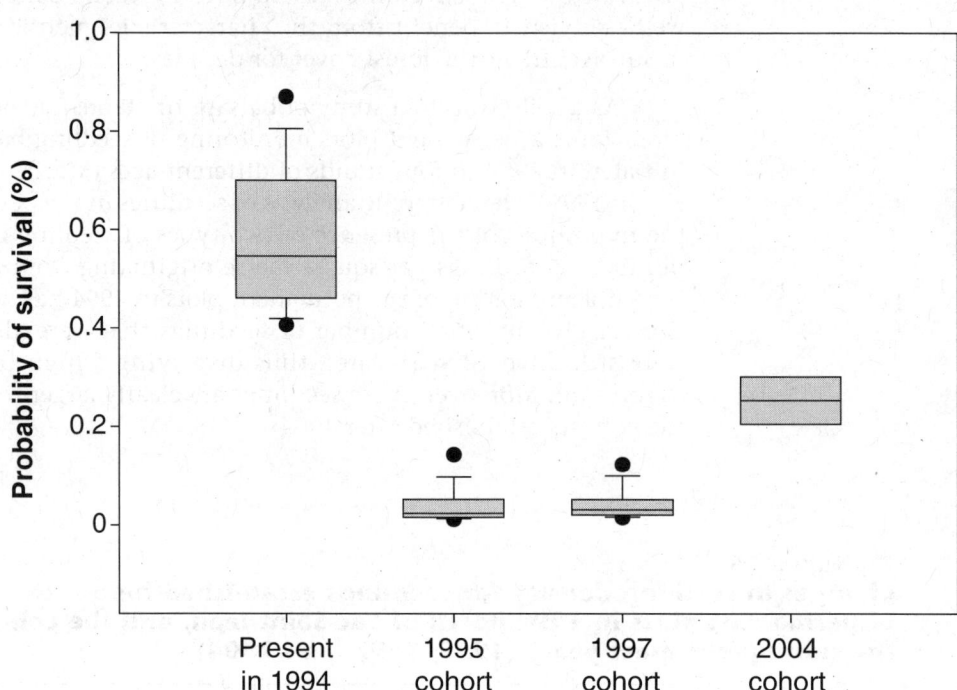

Note that seedlings already established in 1994 have a greater probability of survival than those established after.

The seedlings that established before 1994 are composed of seedlings from the permanent bank and an unknown proportion from previous good mast years. These seedlings gradually disappeared from the cohort. In figure 7.7, we arbitrarily separated the cohorts established before 1994 in two: seedlings >10 cm tall, representing seedlings established long ago, and which have more of a chance composing the permanent bank, and the others. Eleven years later, more than 80% of seedlings >10 cm tall remain, which is significantly higher than the other seedlings that established before 1994 or those that established since 1994, of which almost none remain. Seedlings of the temporary seedling bank thus have little chance of benefitting from an opening and reaching the canopy, being both less abundant and much smaller than seedlings from the permanent bank. The distinct characteristics of seedlings from the permanent bank suggest that they had rapidly established at a precise (but still unknown) time during stand development rather than gradually during good mast years.

These results have direct implications on forest management, in particular on the regeneration of balsam fir populations or mixedwood stands after harvest. Indeed, this indicates that balsam fir seedlings do not establish randomly in space or time as the age structures erroneously suggest, and that conditions of

**Figure 7.7**
**Percent survival of seedlings >10 cm tall**

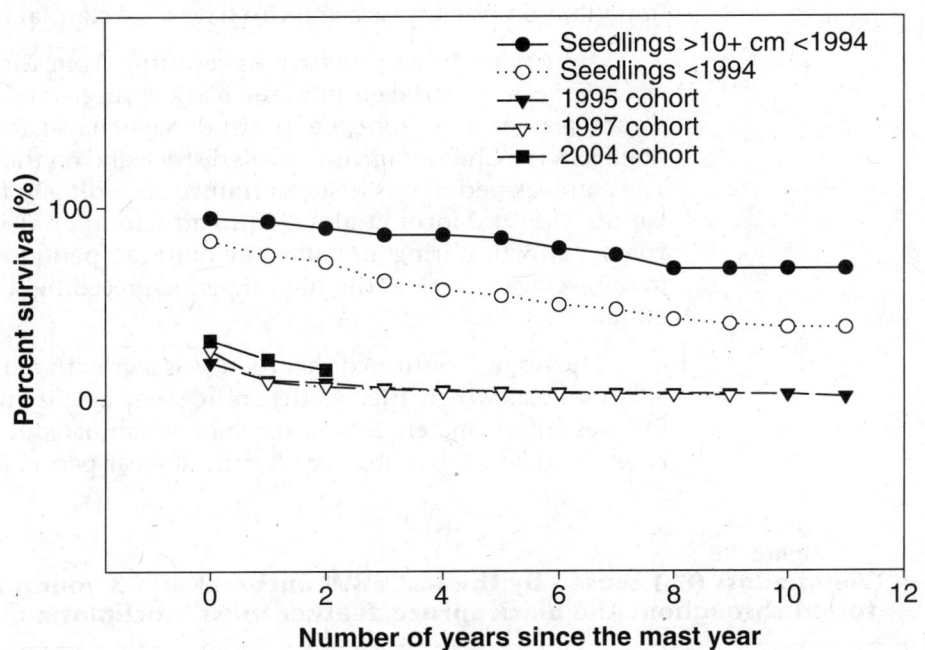

Data were collected at the time of the permanent plots establishment in 1994, for all the seedlings present in 1994 and for the cohorts from subsequent mast years in function of the number of years following the mast year. More than 80% of seedlings >10 cm tall remain present, which is significantly higher than other seedling cohorts established before 1994 and higher than those established since 1994, from which almost none remain.

survival are more restrictive than previously thought. Consequences of forest management strategies that lack in favouring the establishment of a permanent seedling bank under the cover of second-growth forests devoid of regeneration can already be observed (Côté and Bélanger 1991).

## 4.2. Impact in the Black Spruce–Feather Moss Bioclimatic Domain

Black spruce is not the principal host of the SBW. It is generally defoliated by the insect but the mortality in the stands is generally less important, affecting mostly overmature trees in old-growth stands or suppressed saplings under canopy cover (Lussier et al. 2002; De Grandpré et al., chapter 10). The age structure of spruce–feather moss stands is regulated by a fire disturbance dynamics, as many post-fire even-aged stands are observed. Recent studies investigating on the behaviour and feeding requirement of the SBW showed that black spruce is as important as balsam fir or white spruce for egg-laying and the nutritional value of its foliage. The major difference would be linked to the timing of budbreak. Since budbreak happens much later for black spruce compared with its preferred hosts, this leads to an asynchronization with the emergence of the second larval stage in spring, a greater larval mortality level, and consequently a less severe defoliation (Nealis and Régnière 2004). Despite this, defoliation is important on black spruce and can have a substantial impact on tree growth.

A study of growth reductions resulting from the last SBW outbreak has recently been undertaken in mesic black spruce stands found throughout the spruce–feather moss zone, i.e. north of Saguenay–Lac-Saint-Jean, in northern Abitibi, in the Chibougamau-Chapais district, and on the North Shore (figure 7.8). The outbreak period was first determined according to the studies of Morin and Laprise (1990), Morin et al. (1993), and Jardon et al. (2003). The theoretical volume growth during and after the outbreak period was estimated according to the average growth of the 10-year period preceding the outbreak (Lussier et al. 2002).

The results confirmed that there was a growth reduction linked to the last SBW outbreak which affected the entire study area (figure 7.8). On average, this loss was approximately 14% of the volume estimated if growth was maintained at levels equal to that observed for the 10-year period prior to the outbreak for

### Figure 7.8
## Volume loss (%) caused by the last SBW outbreak in 53 young and old stands found throughout the black spruce–feather moss bioclimatic domain in Québec

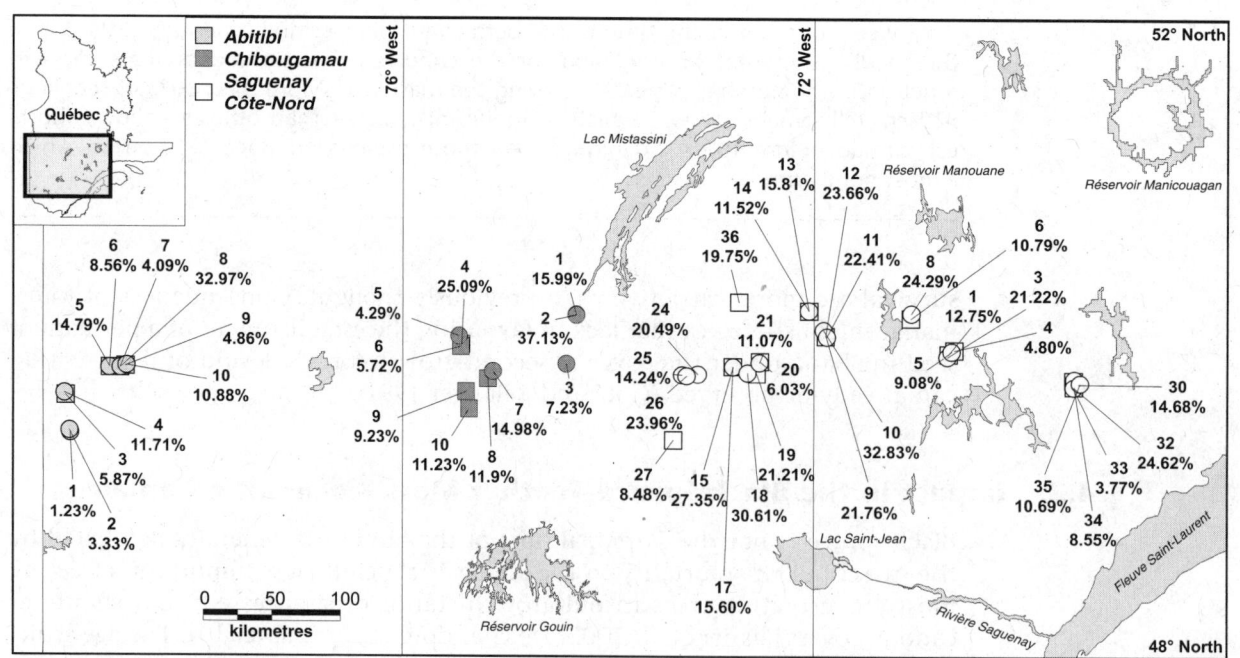

Young (30–85 years (□))
Old (100 years + (○))

Young: 30 to 85 years; old: ≥ 100 years. There is no evident spatial pattern at the regional level examined, although volume losses are greater in old stands.

the dominant stems sampled. Reductions were also not the same throughout the black spruce–feather moss domain, varying between 1 to 34% among dominant stems. Furthermore, there does not seem to be a clear spatial pattern emerging at the regional scale studied. Volume losses are more important in older stands (≥100 years), averaging 17%, in comparison to younger stands (30 to 85 years), which underwent losses averaging 11%.

Several hypotheses can be proposed to explain these geographic variations. Factors such as stand age, site-specific abiotic variables, and the proportion of balsam fir can act at the stand scale. Other factors related to the outbreak dynamics itself, i.e. the local importance of defoliation can also play an important role. It has been suggested that stand age (MacLean 1985) and abiotic variables (Bergeron et al. 1995) explain local variations in balsam fir defoliation levels. Stand age seems to play a role but the impact of abiotic factors has not been clearly shown. The most probable hypothesis to explain defoliation levels in balsam fir stands has been the proportion of balsam fir trees, the preferred host of the SBW, when stands are a mixture of balsam fir and deciduous species (MacLean 1985; Bergeron et al. 1995). However, no study has ever considered stands composed of a mixture of balsam fir and spruce or pure spruce stands in regions where balsam fir is locally present.

In the preceding results (figure 7.8) pure black spruce stands on mesic sites were examined. The proportion of balsam fir in the stand and site-specific abiotic variables were thus not likely to be major contributing factors. These stands contained between 1 and 10% basal area of balsam fir. Yet, it has been shown that in black spruce stands containing an increasing proportion of balsam fir, the impact on black spruce growth increases rapidly when a threshold of 20% fir basal area is reached (Desjardins et al. 2000). The relationship between the proportion of balsam fir and growth reductions was thus not significant in the 53 pure black spruce stands considered.

A study by Amouch (2007) aimed to determine if either the proportion of balsam fir in the landscape or the outbreak dynamics could explain these geographic variations. The study evaluated, in a 3-km radius surrounding each of the 53 study plots, the area occupied by different stand types containing various proportion of balsam fir and hardwood species, as determined from maps of the first decennial forest inventory of Québec carried out between 1970 and 1980. Thus, the majority of stands were inventoried either before the last SBW outbreak or before the SBW could have had a major impact on stand composition. Generally, in combining data for all the study sites, data indicated that the proportion of balsam fir and deciduous stands was very low in many of the sites examined throughout the black spruce–feather moss zone. There was a significant relationship between average stand volume losses (measured from 5 trees) and the proportion occupied by deciduous species, and with the mixture of balsam fir and deciduous species (table 7.1). There also is a significant relationship between each stem volume loss and the proportion of balsam fir, deciduous species, and balsam fir–deciduous species mixture. Lastly, the strongest correlations were observed within the balsam fir–deciduous species group.

Overall, observed correlations were not high. This was expected since the relationship between the proportion of balsam fir in a stand and growth reduction in spruce is not linear (Desjardins et al. 2000). The positive relationship

Table 7.1

**Simple correlations between the proportion of balsam fir and deciduous species found within a 3-km radius and volume loss of spruce stands and stems in the sampled area**

|  | Volume loss | |
|---|---|---|
|  | **Stands** | **Stems** |
| n | 48 | 239 |
| Proportion of fir | 0.2667 | 0.1615* |
| P | 0.0669 | 0.0124 |
| Proportion of broad-leaved deciduous | 0.4139** | 0.2689*** |
| P | 0.0034 | 0.0001 |
| Stand total (fir + broad-leaved deciduous) | 0.3778** | 0.2640*** |
| P | 0.0081 | 0.0001 |

Significant at $p < 0.05$*, $p < 0.01$**, $p < 0.001$***.

between the proportion of deciduous species and the percentage of growth loss in black spruce stems and stands may seem surprising at first, since it is known that the proportion of deciduous species in a balsam fir stand reduces the impact of the SBW (Bergeron et al. 1995; Su et al. 1996; Cappuccino et al. 1998; MacKinnon and MacLean 2003, 2004; Quayle et al. 2003; Campbell 2007). We should stress, however, that the first decennial inventory is rather imprecise at the stand-scale resolution. For example, regarding stands that were classified as deciduous during the first inventory, more than half were classified as "partial outbreak, SS" (>75% balsam fir) or containing a proportion of balsam fir when the mapping of the third decennial inventory (between 1980 and 1990), suggesting that they contained a considerable proportion of balsam fir when the first inventory was conducted.

Amouch (2007) also compared the level of growth reduction in spruce with defoliation intensity and duration using cartography and aerial survey data provided by the Ministère des Ressources naturelles et de la Faune of Québec (MNRF). The analysis was carried out on data collected between 1974 and 1980, years in which an outbreak was recorded in the stands under study. Stands where defoliation was not recorded during the aerial inventories showed significantly lower volume losses (5%) than those wherein outbreak was observed (17%), be it at light, moderate, or severe defoliation levels (figure 7.9A). Furthermore, stands wherein severe defoliation was recorded for three consecutive years or more experienced significantly higher volume losses (24%) compared with those wherein either two years of severe defoliation or one-to-several years of light-to-moderate defoliation occurred (15%), as well as with those not subjected to defoliation (5%) (figure 7.9B). Thus, an outbreak dynamics effect on volume loss was observed. However, since these results are derived from the detailed study of the defoliation surveys of the MRNF, they expose the considerable overall imprecision of using aerial surveys for identifying the damages caused by the SBW on black spruce. The aerial surveys were originally developed to evaluate defoliation levels on balsam fir and white spruce. Flights started in the end of June and continued until mid-July, during the time when damage was most

Figure 7.9
**Mean annual volume loss between 1976 and 1980 according to defoliation level in stands (A), and mean accumulated loss per defoliation class (B)**

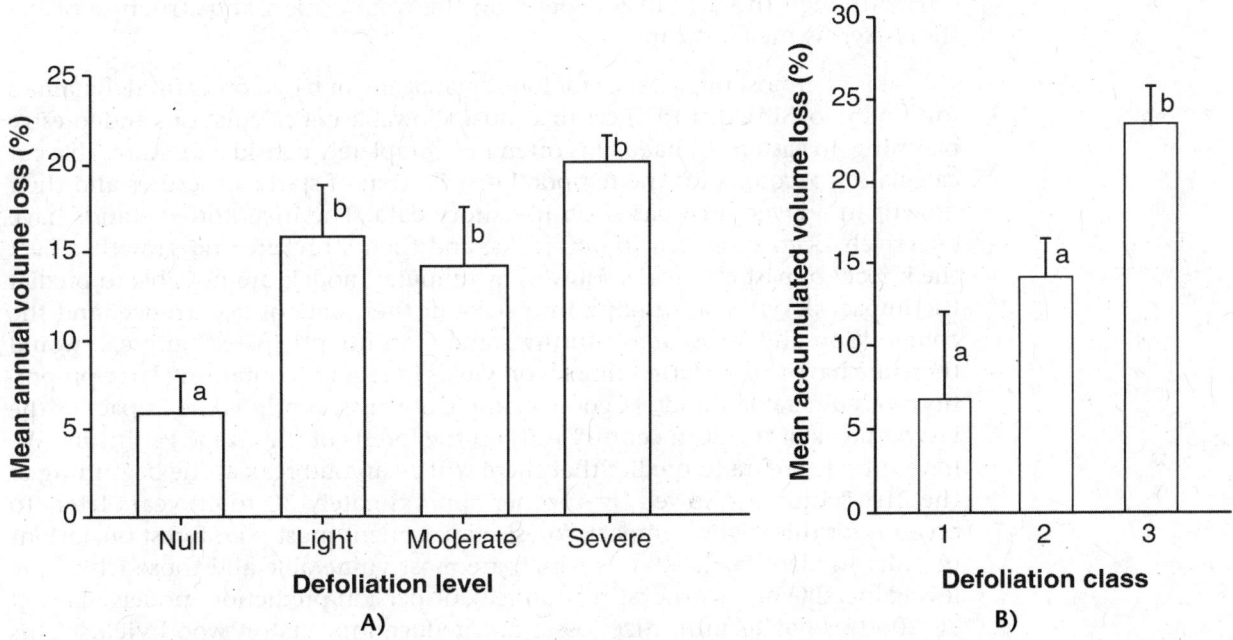

Legend: 1, stands wherein no defoliation was recorded; 2, stands experiencing two years of severe defoliation and/or one to several years of light-to-moderate defoliation; 3, stands subjected to three of more years of severe defoliation. Significant differences are indicated by different letters (Tukey's test).

apparent for these species. When used for evaluating defoliation levels on black spruce, this lack of precision is the result of two principal factors: first, because of the later budburst in black spruce, defoliation begins roughly two weeks after balsam fir; and secondly, defoliation is generally less severe on black spruce.

## 5. MANAGEMENT IMPLICATIONS IN THE CONTEXT OF CLIMATE CHANGE

### 5.1. Impact of Outbreak Cycles

Understanding of the spatial and temporal dynamics of SBW outbreaks is still incomplete. However, interpretations from entomology, dendrochronology, and paleoecology are coming to agreement (Cooke et al. 2007; Morin et al. 2007). While dendrochronological data sometimes lead to different interpretations because of imprecision related to the methodology used to go back in time, a common conclusion emerges from all the researches: outbreaks have recurred in a cyclical manner at the rate of three outbreaks per century during the last two centuries. Thus, one can predict without much incertitude that there will be three SBW outbreaks over the course of the present century. If one accepts the "severe outbreak – moderate outbreak – severe outbreak" cycle, the next outbreak could be less severe than the one observed at the end of the 20th century (O1), although this will also depend on the composition and structure of the stands across the landscape.

It is of utmost importance for forest management based on natural dynamics to account for SBW disturbances in annual allowable cut calculations and forestry planning. In eastern Canada, it is often not completely considered. Allowable cut calculation accounts for the temporal distribution of stand structures and their growth in a given area based on inventory data. The inventoried stands have effectively been subjected to outbreaks, and their structure and growth reflect the impact of past outbreaks. However, computer models are not able to predict the impact of outbreaks taking into account their date of occurrence, and the composition and structure of future stands. An unanticipated outbreak would therefore have a devastating impact on yields in regions containing large proportions of vulnerable stands. A good example of this scenario is the impact of the last outbreak of the 20th century (O1) on the forests of the Gaspé Peninsula. We have sufficient data to predict that there will be an outbreak at the beginning of the 21st century, followed by another approximately 30 to 40 years later. To account for this cyclical impact, forest management must plan forest operations in order to identify the stands which are most vulnerable and those which are less vulnerable by incorporating outbreak dispersion prediction models (Gray et al. 2000), so as to minimize losses and reduce impacts on wood yields. This approach seems paradoxical in a context where one wants to implement natural disturbance-based forest management. Outbreaks are part of the natural dynamics. However, new paleoecological evidence indicates that severe outbreak periods were not common during the course of the Holocene. During most of this era, local outbreaks had less severe effects across a landscape that was probably dominated by less vulnerable stands. Forest management accounting for the vulnerability to SBW is thus an acceptable way of respecting the natural dynamics of SBW outbreaks. Silvicultural approaches adapted to the integrated management of insects are currently under development (MacLean 1996).

Tools such as the decision support system developed by MacLean et al. (2001, 2002) and an updated version which was recently tested in Abitibi (Campbell 2007) will be particularly useful to predict the impacts of an outbreak depending on the stand type, and for the decision of whether or not to protect some targeted stands. This now brings us to discussing the importance of stand structure and composition.

## 5.2. Impact of Stand Structure and Composition

As more data become available, the more they converge towards a common interpretation: in the tri-trophic "plant host – insect – natural enemy" system associated to the SBW, the host plant plays a fundamental role (Bouchard 2005; Cooke et al. 2007). When an abundance of host plants is found in the forest mosaic, the growth of the SBW population is favoured to the detriment of its natural enemies, creating a major disequilibrium in the system. Dendrochronological and paleoecological analyses both suggest that an increase in the proportion of host trees in the canopy leads to an increase in the impact, resulting in outbreaks that quickly affect vast areas. The preferred host species in eastern North America are balsam fir and white spruce. The history of outbreaks demonstrates that an increase in the proportion of balsam fir in the canopy results in explosive outbreaks. Current forestry practices, dominated by cutting practices that protect the regeneration, lead to an increase in the proportion of balsam fir, as it is often abundant under fir-dominated canopies and under spruce canopies. Balsam fir is an important commercial species, but we must be aware that it is also more vulnerable to SBW outbreaks than black spruce. Pothier and Mailly (2006) have determined that the defoliation by the SBW accounts for 6 to 100% of merchantable volume losses caused by mortality in balsam fir forests, depending on outbreak severity. The increase in the proportion of balsam fir in the black spruce zone is of concern. By modifying forest composition in this manner, there are increasing risks of initiating a new outbreak dynamics in areas where it has never been previously observed. Consequently, data from past inventories, from which we calculate wood yields, will not be reliable. Indeed, stands inventoried in the past were subjected to an outbreak regime associated with a different forest composition and structure that had fewer host trees in the canopy.

An increasing number of studies show that the presence of hardwoods within stands and across the landscape reduces impacts of outbreaks on host trees (Bergeron et al. 1995; Su et al. 1996; Cappuccino et al. 1998; MacKinnon and MacLean 2003, 2004; Campbell 2007). Thus, it would be relevant to use adapted silvicultural systems that favour the presence of a proportion of deciduous species in vulnerable stands (Campbell 2007). For example, partial harvesting systems that result in a multi-cohort stand age structure could be used (see Bouchard, chapter 13).

The fact that balsam fir is found under the canopy of several stand types can seem paradoxical, as we suggested earlier in this chapter that this species could have problems regenerating if we did not pay attention to the reconstitution of a permanent seedling bank (i.e. seedlings that have a higher probability of forming the future stand). In the boreal region and particularly in the black spruce–feather moss zone, we are still in the first rotation cycle and therefore harvesting natural forests that were under the influence of natural disturbances and that have an established permanent seedling bank. The problem of establishing a permanent seedling bank particularly arises in second-growth stands (Côté and Bélanger 1991) wherein an elevated density is maintained without adequate germination beds such as coarse woody debris. Regeneration of these stands could be less effective, compromising the future generations of trees in such a context. Natural disturbances such as gaps created during outbreaks produce an abundance of coarse woody debris and other germination beds, which consist of nurseries for seedlings (Parent et al. 2003), and create a heterogenous structure that allows for the survival of a certain number of seedlings. The use of forest management practices favouring the establishment of a seedling bank must thus be considered in second-growth stands. To resolve this situation, ensuring the creation of favourable germination beds by leaving coarse woody debris on the ground or by exposing mineral soil, as well as maintaining a more heterogenous structure that creates variation in light availability at the ground level would be necessary.

Finally, even if it is not the preferred host species of the SBW, black spruce is similarly affected by defoliation. This is never considered in allowable cut calculations, although new data indicate that it should be. Very little is known on this subject, and more precise, in-depth studies are needed. Conclusions from available studies suggest that the impact of an outbreak increases rapidly when the proportion of balsam fir exceeds 15 to 20% in a black spruce stand. Likewise, this impact seems to be influenced by the proportion of balsam fir across the landscape. The lack of reliable aerial surveys to detect defoliation in black spruce stands is disconcerting, and aerial surveys specific to the black spruce–feather moss zone should be planned for in the future.

## 5.3.  In the Context of Climate Change

We proposed that the major change in the outbreak regime occurring at the beginning of the 20th century was caused at least in part by a reduction in fire frequency, which resulted in an increase in host trees, particularly balsam fir, in the forest composition. The SBW outbreak regime appears to be sensitive to changes in the fire regime and, indirectly, to the associated climatic changes (see Gauthier et al., chapter 3). If the current fire frequency is maintained, the outbreak regime observed during the 20th century will recur. The most current simulations of the Canadian Regional Climate Model for the boreal forest of eastern Canada suggests a local rise in mean annual temperature of about 3°C while precipitation levels would increase between 5 and 20% by 2050 (Plummer et al. 2006). Under these circumstances, fire frequency could decrease, which would further increase the dominance of host trees across the forest landscape and subsequently increase the impacts of outbreaks. On the other hand, if the region is subjected to the influence of dry air masses, particularly in spring, an increase in fire frequency, a decrease in the proportion of host trees, and possibly a decrease in outbreak severity would be observed. The regularity of the cycle would remain, but the impact of outbreaks across the landscape would be reduced. This scenario would of course have other devastating consequences on the regeneration and growth of the forest (see Gauthier et al., chapter 3).

These conclusions are valuable in the context where a good portion of the forest remains under the influence of natural disturbances, as it has been the case for the last centuries and the beginning of the 20th century. Because the region is vast and cannot be subjected to complete protection or intensive management, natural disturbances will probably remain the principal factors regulating the forest composition and structure. The effectiveness of fire suppression programs is currently a subject of debate. Proof that these programs have had an effect on the frequency or spread of natural fires, or on forest structure and composition at the landscape scale, has not been convincing (Johnson et al. 2001; Miyanishi and Johnson 2001; Ward et al. 2001). With respect to the protection of forests against insects, and in particular against the SBW, spraying of DDT began in localized areas during the 1950s. However, spraying programs have truly begun in the early 1970s in Québec. These localized sprayings were performed throughout outbreak O1, during which a biological insecticide, Bt (*Bacillus thuringiensis* Berliner), was used. The maximum area sprayed in the province reached 3,960,000 ha in 1973 (SOPFIM, pers. comm.; Armstrong and Cook 1993). There is no evidence that this spraying program had any effect on the outbreak dynamics. The outbreak ended naturally because of an increase in the mortality of the SBW caused by an increase in its natural enemies. In the eventuality of an outbreak occurring in a landscape primarily composed of mature stands wherein host trees such as balsam fir dominate the canopy, one can expect an explosive outbreak having insect populations that increase very rapidly at both the stand and landscape scales in a synchronous fashion, as observed during outbreaks O3 or O1 for example. In this case, spraying programs would not control the outbreak, but could perhaps at most slow down the mortality in some valuable stands. Thus, analogous to the example of "good fire years," when one is not able to contain all the fires that ignite in unison throughout the region because of favourable conditions, one is not able to control the increase in SBW populations in a landscape dominated by mature balsam fir stands. This is why it is important to understand the spatial and temporal dynamics of SBW outbreaks as well as their impacts on balsam fir and black spruce stand dynamics, so that this knowledge can be integrated into our forest management practices.

## REFERENCES

Amouch, S. 2007. Explication de la variation de la perte en volume des pessières noires de la zone boréale après la dernière épidémie de la tordeuse des bourgeons de l'épinette. M.Sc. thesis. Université du Québec à Chicoutimi, Saguenay, Que.

Armstrong, J.A. and Cook, C.A. 1993. Aerial spray applications on Canadian forests: 1945 to 1990. Forest Canada, Information report ST-X–2, Ottawa, Ont.

Bergeron, Y. 1998. Les conséquences des changements climatiques sur la fréquence des feux et la composition forestière au sud-ouest de la forêt boréale québécoise. Géo. Phys. Quat. **52**: 167–173.

Bergeron, Y. and Archambault, S. 1993. Decreasing frequency of forest fires in the southern boreal zone of Quebec and its relation to global warming since the end of the "Little Ice Age." The Holocene **3**: 255–259.

Bergeron, Y., Gauthier, S., Kafka, V., Lefort, P., and Lesieur, D. 2001. Natural fire frequency for the Canadian boreal forest: consequences for sustainable forestry. Can. J. For. Res. **31**: 384–391.

Bergeron, Y. and Leduc, A. 1998. Relationships between change in fire frequency and mortality due to spruce budworm outbreak in the southeastern Canadian boreal forest. J. Veg. Sci. **9**: 493–500.

Bergeron, Y., Leduc, A., Morin, H., and Joyal, C. 1995. Balsam fir mortality following the last spruce budworm outbreak in northwestern Quebec. Can. J. For. Res. **25**: 1375–1384.

Bhiry, N. and Filion, L. 1996. Mid-Holocene hemlock decline in eastern North America linked with phytophagous insect activity. Quat. Res. **45**: 312–320.

Blais, J.R. 1954. The recurrence of spruce budworm infestations in the past century in the Lac Seul area of northwestern Ontario. Ecology, **35**: 62–71.

Blais, J.R. 1965. Spruce budworm outbreaks in the past three centuries in the Laurentides Park, Quebec. For. Sci. **11**: 130–138.

Blais, J.R. 1983. Trends in the frequency, extent, and severity of spruce budworm outbreaks in eastern Canada. Can. J. For. Res. **13**: 539–547.

Bouchard, M. 2005. Dynamique forestière suite aux épidémies de tordeuse des bourgeons de l'épinette dans le nord du Témiscamingue. Ph.D. thesis, Université du Québec à Montréal, Montréal, Que.

Bouchard, M., Kneeshaw, D., and Bergeron, Y. 2006a. Forest dynamics after successive spruce budworm outbreaks in mixedwood forests. Ecology, 87: 2319–2329.

Bouchard, M., Kneeshaw, D., and Bergeron, Y. 2006b. Tree recruitment pulses and long-term species coexistence in mixed forests of western Quebec. Écoscience, **13**: 82–88.

Boulanger, Y. and Arseneault, D. 2004. Spruce budworm outbreaks in eastern Quebec over the last 450 years. Can. J. For. Res. **34**: 1035–1043.

Boulet, B., Chabot, M., Dorais, L., Dupont, A., and Gagnon, R. 1996. Entomologie forestière. *In* Manuel de foresterie. *Edited by* J.A. Bérard and M. Côté. Les Presses de l'Université Laval, Québec, Que., pp. 1008–1043.

Campbell, E. 2007. Patrons temporels et spatiaux de la sévérité des épidémies de la tordeuse des bourgeons de l'épinette en relation aux conditions bioclimatiques de l'est du Canada. Ph.D. thesis, Université du Québec à Montréal, Montréal, Que.

Candau, J.-N. and Fleming, R.A. 2005. Landscape-scale spatial distribution of spruce budworm defoliation in relation to bioclimatic conditions. Can. J. For. Res. **35**: 2218–2232.

Candau, J.-N., Fleming, R.A., and Hopkin, A. 1998. Spatiotemporal patterns of large-scale defoliation caused by the spruce budworm in Ontario since 1941. Can. J. For. Res. **28**: 1733–1741.

Cappuccino, N., Lavertu, D., Bergeron, Y., and Régnière, J. 1998. Spruce budworm impact, abundance and parasitism rate in a patchy landscape. Oecologia, **114**: 236–242.

Cooke, B.J., Nealis, V.G., and Régnière, J. 2007. Insect defoliators as periodic disturbances in northern forest ecosystems. *In* Plant disturbance ecology: the process and the response. *Edited by* E.A. Johnson and K. Miyanishi. Academic Press, Amsterdam, Netherlands, pp. 487–526.

Côté, S. and Bélanger, L. 1991. Variations de la régénération préétablie dans les sapinières boréales en fonction de leurs caractéristiques écologiques. Can. J. For. Res. **21**: 1779–1795.

Davis, R.B., Anderson, R.S., and Hoskins, B.R. 1980. A new parameter for paleoecological reconstruction: head capsules of forest-tree defoliator Microlepidopterans in lake sediment. *In* Abstracts and program of the 6th biennial meeting of the American Quaternary Association, August 18–20, 1980, Institute of Quaternary Studies, University of Maine, Orono, Maine, USA.

Desjardins, O., Morin, H., and Bergeron, Y. 2000. Influence of balsam fir content, at the stand level, on the impact of the last spruce budworm outbreak in the boreal forest of eastern Canada. International Workshop on Disturbance Dynamics in Boreal Forests, August 21–25, 2000, Kuhmo, Finland.

Duchesneau, R. and Morin, H. 1999. Early seedling demography in balsam fir seedling banks. Can. J. For. Res. **29**: 1502–1509.

Gray, D.R., Régnière, J., and Boulet, B. 2000. Analysis and use of historical patterns of spruce budworm defoliation to forecast outbreak patterns in Quebec. For. Ecol. Manag. **127**: 217–231.

Holmes, R.L. 1999. ITRDB, Dendrochronology program library users manual. Laboratory of Tree-Ring Research, University of Arizona, Tucson, Arizona, USA.

Holmes, R.L. and Swetnam, T.W. 1996. Dendroecology program library, Program OUTBREAK users manual. Laboratory of Tree-Ring Research, University of Arizona, Tucson, Arizona, USA.

Jardon, Y. 2002. Analyses temporelles et spatiales des épidémies de la tordeuse des bourgeons de l'épinette au Québec. Ph.D. thesis, Université du Québec à Chicoutimi, Saguenay, Que.

Jardon, Y., Filion, L., and Cloutier, C. 1994. Tree-ring evidence for endemicity of the larch sawfly in North America. Can. J. For. Res. **24**: 742–747.

Jardon, Y., Morin, H., and Dutilleul, P. 2003. Périodicité des épidémies de la tordeuse des bourgeons de l'épinette au cours des deux derniers siècles. Can. J. For. Res. **33**: 1947–1961.

Johnson, E., Miyanishi, K., and Bridge, S.R.J. 2001. Wildfire regime in the boreal forest and the idea of suppression and fuel buildup. Conserv. Biol. **15**: 1554–1557.

Johnson, E., Morin, H., Miyanishi, K., Gagnon, R., and Greene, D.F. 2003. A process approach to understanding disturbance and forest dynamics for sustainable forestry. *In* Towards sustainable management of the boreal forest. *Edited by* P.J. Burton, C. Messier, D.W. Smith, and W.L. Adamowicz. NRC Research Press, Ottawa, Ont., pp. 261–306.

Kneeshaw, D. and Bergeron, Y. 1998. Canopy gap characteristics and tree replacement in the southeastern boreal forest. Ecology, **79**: 783–794.

Kneeshaw, D. and Bergeron, Y. 1999. Spatial and temporal pattern of seedling and sapling recruitment within canopy gaps caused by spruce budworm. Écoscience, **6**: 214–222.

Krause, C. 1997. The use of dendrochronological material from buildings to get information about past spruce budworm outbreaks. Can. J. For. Res. **27**: 69–75.

Levasseur, V. 2000. Analyse dendroécologique de l'impact de la tordeuse des bourgeons de l'épinette (*Choristoneura fumiferana*) suivant un gradient latitudinal en zone boréale au Québec. M.Sc. thesis, Université du Québec à Chicoutimi, Chicoutimi, Que.

Lussier, J.M., Morin, H., and Gagnon, R. 2002. Mortality in black spruce stands of fire or clear-cut origin. Can. J. For. Res. **32**: 539–547.

MacKinnon, W.E. and MacLean, D.A. 2003. The influence of forest and stand conditions on spruce budworm defoliation in New Brunswick, Canada. For. Sci. **49**: 657–667.

MacKinnon, W.E. and MacLean, D.A. 2004. Effects of surrounding forest and site conditions on growth reduction of balsam fir and spruce caused by spruce budworm defoliation. Can. J. For. Res. **34**: 2351–2362.

MacLean, D.A. 1980. Vulnerability of fir-spruce stands during uncontrolled spruce budworm outbreaks: a review and discussion. For. Chron. **56**: 213–221.

MacLean, D.A. 1984. Effects of spruce budworm outbreaks on the productivity and stability of balsam fir forests. For. Chron. **60**: 273–299.

MacLean, D.A. 1985. Effects of spruce budworm outbreaks on forest growth and yield. *In* Recent advances in spruce budworm research. *Edited by* C.J. Sanders, R.W. Stark, E.J. Mullins, and J. Murphy. Proceedings of the CANUSA Spruce Budworm Research Symposium, Bangor, Maine, USA, pp. 148–175.

MacLean, D.A. 1988. Effects of spruce budworm outbreaks on vegetation, structure, and succession of balsam fir forest on Cape Breton Island, Canada. *In* Plant form and vegetation structure: adaptation, plasticity and relation to herbivory. *Edited by* M.J.A. Werger, P.J.M. van der Aart, H.J. During, and J.T.A. Verhoeven. SPB Academic Publishing BV, The Hague, Netherlands, pp. 253–261.

MacLean, D.A. 1996. Silvicultural approaches to integrated insect management: the Green Plan Silvicultural Insect Management Network. For. Chron. **72**: 367–369.

MacLean, D.A., Beaton, K.P., Porter, K.B., MacKinnon, W.E., and Budd, M.G. 2002. Potential wood supply losses to spruce budworm in New Brunswick estimated using the Spruce Budworm Decision Support System. For. Chron. **78**: 739–750.

MacLean, D.A., Erdle, T.A., MacKinnon, W.E., Porter, K.B., Beaton, K.P., Cormier, G., Morehouse, S., and Budd, M. 2001. The Spruce Budworm Decision Support System: forest protection planning to sustain long-term wood supply. Can. J. For. Res. **31**: 1742–1757.

MacLean, D.A. and MacKinnon, W.E. 1997. Effects of stand and site characteristics on susceptibility and vulnerability of balsam fir and spruce budworm in New Brunswick. Can. J. For. Res. **27**: 1859–1871.

MacLean, D.A. and Ostaff, D.P. 1989. Pattern of balsam fir mortality caused by an uncontrolled budworm outbreak. Can. J. For. Res. **19**: 1087–1095.

Miyanishi, K. and Johnson, E.A. 2001. Comment: a re-examination of the effects of fire suppression in the boreal forest. Can. J. For. Res. **31**: 1462–1466.

Moran, P.A.P. 1953. The statistical analysis of the Canadian lynx cycle. Australian J. Zool. **1**: 163–173.

Morin, H. 1994. Dynamics of balsam fir forests in relation to spruce budworm outbreaks in the boreal zone of Quebec. Can. J. For. Res. **24**: 730–741.

Morin, H., Jardon, Y., and Gagnon, R. 2007. Relationships between spruce budworm outbreaks and forest dynamics in Eastern North America. *In* Plant disturbance ecology: The process and the response. *Edited by* E.A Johnson and K. Miyanishi, Academic Press, Amsterdam, Netherlands, pp. 555–564.

Morin, H. and Laprise, D. 1990. Histoire récente des épidémies de la tordeuse des bourgeons de l'épinette au nord du lac St-Jean (Québec): une analyse dendrochronologique. Can. J. For. Res. **20**: 1–8.

Morin, H. and Laprise, D. 1997. Seedling bank dynamics in boreal balsam fir forests. Can. J. For. Res. **27**: 1442–1451.

Morin, H., Laprise, D., and Bergeron, Y. 1993. Chronology of spruce budworm outbreaks near Lake Duparquet, Abitibi region, Quebec. Can. J. For. Res. **23**: 1497–1506.

Natural Resources Canada. 2001. National Forest Inventory. [Online] <nfi.cfs.nrcan.gc.ca/canfi> (accessed November 27, 2007).

Nealis, V.G. and Régnière, J. 2004. Insect-host relationships influencing disturbance by the spruce budworm in a boreal mixedwood forest. Can. J. For. Res. **34**: 1870–1882.

Parent, S. Morin, H., and Messier, C. 2000. Effects of adventitious roots on age determination in balsam fir (*Abies balsamea* (L.) Mill.) regeneration. Can. J. For. Res. **30**: 513–518.

Parent, S., Morin, H., and Messier, C. 2001. Balsam fir (*Abies balsamea* L. (Mill.)) establishment dynamics during a spruce budworm (*Choristineura fumifera* (Clem.)) outbreak in a south-boreal forest: an evaluation of the impact of aging techniques. Can. J. For. Res. **31**: 373–376.

Parent, S., Morin, H., Messier, C., and Simard, M.-J. 2005. Growth, biomass allocation, and adventitious roots of balsam fir seedlings growing in closed-canopy stands. Écoscience, **13**: 89–94.

Parent, S., Simard, M.-J., Morin, H., and Messier, C. 2003. Establishment and dynamics of the balsam fir seedling bank in old forests of northeastern Quebec. Can. J. For. Res. **33**: 597–603.

Peltonen, M., Liebhold, A.M., Bjornstad, O.N., and Williams, D.W. 2002. Spatial synchrony in forest insect outbreaks: Roles of regional stochasticity and dispersal. Ecology, **83**: 3120–3129.

Plummer, D.A., Caya, D., Frigon, A., Côté, H., Giguère, M., Paquin, D., Biner, S., Harvey, R., and De Elia, R. 2006. Climate and climate change over North America as simulated by the Canadian RCM. J. Clim. **19**: 3112–3132.

Potelle, B. 1995. Potentiel de l'analyse des macrorestes pour détecter les épidémies de la tordeuse des bourgeons de l'épinette dans des sols de sapinières boréales. M.Sc. thesis, Université du Québec à Chicoutimi, Chicoutimi, Que.

Pothier, D., and Mailly, D. 2006. Stand-level prediction of balsam fir mortality in relation to spruce budworm defoliation. Can. J. For. Res. **36**: 1631–1640.

Quayle, D., Régnière, J., Cappuccino, N., and Dupont, A. 2003. Forest composition, host population density, and parasitism of spruce budworm *Choristoneura fumiferana* eggs by *Trichogramma minutum*. Ent. Exp. Appl. **107**: 215–227.

Royama, T. 1984. Population dynamics of the spruce budworm *Choristoneura fumiferana*. Ecol. Monogr. **54**: 429–462.

Royama, T., MacKinnon, W.E., Kettela, E.G., Carter, N.E., and Harting, L. 2005. Analysis of spruce budworm outbreak cycles in New Brunswick, Canada, since 1952. Ecology, **86**: 1212–1224.

Simard, S. 2003. La tordeuse des bourgeons de l'épinette à travers les arbres subfossiles. M.Sc. thesis, Université du Québec à Chicoutimi, Saguenay, Que.

Simard, S., Elhani, S., Morin, H., Krause, K., and Cherubini, P. 2008. Carbon and oxygen stable isotop form tree-rings to identify spruce budworm outbreaks in the boreal forest of Québec. Chem. Geol. **252**: 80–87.

Simard, I., Morin, H., and Lavoie, C. 2006. A millennial-scale reconstruction of spruce budworm abundance in Saguenay, Québec, Canada. The Holocene, **16**: 31–37.

Simard, I., Morin, H., and Potelle, B. 2002. A new paleo-ecological approach to reconstruct long-term history of spruce budworm outbreaks. Can. J. For. Res. **32**: 428–438.

Su, Q., MacLean, D.A., and Needham, T.D. 1996. The influence of hardwood content on balsam fir defoliation by spruce budworm. Can. J. For. Res. **26**: 1620–1628.

Swetnam, T.W. and Lynch, A.M. 1993. Multicentury, regional-scale patterns of western spruce budworm outbreaks. Ecol. Monogr. **63**: 399–424.

Swetnam, T.W., Thompson, M.A., and Sutherland, E.K. 1985. Using dendrochronology to measure radial growth of defoliated trees. USDA For. Serv. Agric. Handb. **639**: 1–39.

Volney, W.J.A. 1988. Analysis of historic jack pine budworm outbreaks in the Prairie provinces of Canada. Can. J. For. Res. **18**: 1152–1158.

Ward, P.C., Tithecott, A.G., and Wotton, B.M. 2001. Reply: a re-examination of the effects of fire suppression in the boreal forest. Can. J. For. Res. **31**: 1467–1480.

Weber, U.M. 1997. Dendroecological reconstruction and interpretation of larch budmoth (*Zeiraphera diniana*) outbreaks in two central Alpine valleys of Switzerland from 1470–1990. Trees Struct. Funct. **11**: 277–290.

Williams, D.W. and Liebhold, A.M. 2000. Spatial scale and the detection of density dependence in spruce budworm outbreaks in eastern North America. Oecologia, **124**: 544–552.

Zhang, Q.B. and Alfaro, R.I. 2002. Periodicity of two-year cycle spruce budworm outbreaks in central British Columbia: a dendro-ecological analysis. For. Sci. **48**: 722–731.

Zhang, Q.B. and Alfaro, R.I. 2003. Spatial synchrony of the two-year cycle budworm outbreaks in central British Columbia, Canada. Oikos, **102**: 146–154.

# Chapter 8

## Forest Tent Caterpillar Outbreak Dynamics from Manitoba to New Brunswick*

*Alanna Sutton and Jacques C. Tardif*

* The authors thank the Sustainable Forest Management Network for its financial contribution to the research in Manitoba as well as Louisiana-Pacific Canada, Ltd. and Manitoba Conservation for their logistical support. Our thanks go also to Barry Cooke for his help with the research of the literature, as well as Colin Murray and Pierre-Hughes Tremblay for their suggestions and help with the text. We also thank Dr. Hubert Morin and Marie-Andrée Vaillancourt for their revisions and suggestions. The photos on this page were graciously provided by Alanna Sutton.

# 1. INTRODUCTION

Insect outbreaks are an important natural disturbance in the boreal forest region. Timber loss in the boreal zone of Canada due to insect outbreaks is estimated to be 1.3 to 2.0 times greater per year than wildfire loss (Volney and Fleming 2000). In eastern Canada, the spruce budworm (*Choristoneura fumiferana* Clem.) has been shown to negatively affect growth and survival of balsam fir (*Abies balsamea* [L.] Mill.) and white spruce (*Picea glauca* [Moench] Voss) (Blais 1958; Morin and Laprise 1990; Morin 1994; Morin et al., chapter 7). Other important conifer defoliating insects, which have caused reduced growth in North American trees, include the jack pine budworm (*Choristoneura pinus pinus* Freeman) (Simpson and Coy 1999) and the larch sawfly (*Pristiphora erichsonii* Htg.) (Girardin et al. 2001). In the case of deciduous species, the gypsy moth (*Lymantria dispar* L.) (Naidoo and Lechowicz 2001), the large aspen tortrix (*Choristoneura conflictana* F. Walker) (Ives and Wong 1988) and the forest tent caterpillar (*Malacosoma disstria* Hbn.) are important defoliators (Simpson and Coy 1999).

The forest tent caterpillar (figure 8.1) is an important defoliator of hardwood trees throughout North America, causing frequent large-scale outbreaks within the boreal forest region (Witter 1979). According to Brandt (1995), the forest tent caterpillar, which caused 25% of the total timber loss, was among the top three groups of organisms causing the largest impacts in the Prairie Provinces from 1988 to 1992. Forest tent caterpillar defoliation covered 80.6 million hectares in central Canada between 1980 and 1996 (Simpson and Coy 1999). Unlike other insects such as the spruce budworm, little information is available about forest tent caterpillar population dynamics and the consequences on affected forests. For example, very little is known about forest tent caterpillar epidemics in the Maritimes and New Brunswick.

Increased usage of the host species, such as trembling aspen (*Populus tremuloides* Michx.), by the forest industry and the necessity for a better understanding of forest disturbance dynamics other than fire in ecosystem management represent important elements justifying the present review. The objectives of this chapter are: 1) to review what is known about the life cycle and general dynamics of the forest tent caterpillar, 2) to describe forest tent caterpillar outbreak dynamics by province, from central to eastern Canada (Manitoba, Ontario, Québec and New Brunswick), with information on the number, size and duration of outbreaks, and 3) to present preliminary suggestions for forest management in areas affected by the forest tent caterpillar. We will also focus on the distribution of the insect, the factors that control outbreaks and forest impacts that are associated with this defoliating insect.

Figure 8.1
**Photos of A) larvae, B) adult, C) complete trembling aspen defoliation, and D) forest tent caterpillar cocoons in a young white spruce**

A)

B)

C)

D)

Sources: A, C, and D: A. Sutton; B: Ives and Wong (1988).

## 2. FOREST TENT CATERPILLAR ECOLOGY AND GENERAL OUTBREAK DYNAMICS

### 2.1. Life Cycle

The forest tent caterpillar is a common leaf eater in the forests of Canada, especially in the mixedwood boreal and hardwood forests (Martineau 1984). The insect is found throughout the United States and Canada (Witter 1979) and its distribution corresponds almost entirely with the range of trembling aspen, its preferred host, as well as its secondary hosts (see figure 8.2) (Hanec 1966; Sippell and Ewan 1967; Witter 1979). Although the trembling aspen is the primary host of the forest tent caterpillar, secondary species include sugar maple (*Acer saccharum* Marsh.), paper birch (*Betula papyrifera* Marsh.), balsam poplar (*Populus balsamifera* L.), red oak (*Quercus rubra* L.), bur oak (*Quercus macrocarpa* Michx.), American elm (*Ulmus americanus* L.), and basswood (*Tilia americana* L.), with little or no effects on red maple (*Acer rubrum* L.) (Hodson 1941; Sippell and Ewan 1967; Hildahl and Campbell 1975; Ives and Wong 1988).

Figure 8.2

## Distribution of the four main host species (shaded areas) and the northern geographical limit of the forest tent caterpillar (black line) in Canada (Fitzgerald 1995)

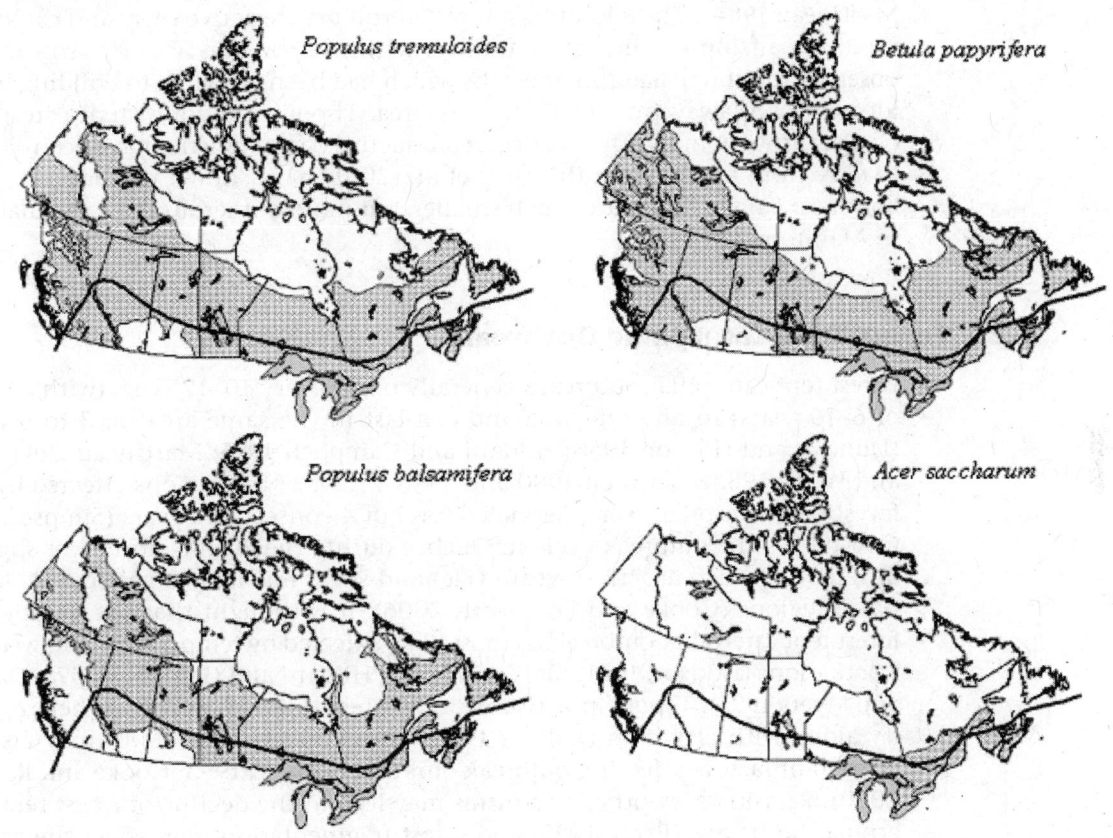

Source: <www.fs.fed.us/ne/delaware/4153/global/littlefia/species_table.html> (accessed November 19, 2007).

The forest tent caterpillar produces one generation per year, and in general, the emergence of larvae occurs in synchronization with the bud break of the host plant (Hodson 1941; Rose 1958; Sippell and Ewan 1967; Volney and Fleming 2000). Asynchrony between egg hatch and budbreak may lead to the starvation of the larvae (Hodson 1941). The larvae experiences five larval instars within one generation. During each successive stage the size of the larvae increases and the amount of food needed increases (Hodson 1941; Rose 1958; Hildahl and Campbell 1975). Young larvae feed on expanding buds and account for very little of the total defoliation, which occurs during the last larval stage (Sippell and Ewan 1967; Hildahl and Campbell 1975). Larvae begin to wander and forage for food during the 4th and 5th larval instar when food sources become depleted (Hodson 1941). Approximately 5–6 weeks elapse between the emergence of the larvae and the appearance of the first "tent" cocoons (Hodson 1941). By mid-June (in most areas), larvae have matured enough to begin forming their silk cocoons among the leaves and twigs of trees (figure 8.1D), shrubs and other vegetation, or occasionally on buildings or other man-made structures (Hodson 1941; Rose 1958; Sippell and Ewan 1967; Ives and Wong 1988). The pupa stays in the cocoon for 8–12 days; the beige/brown moth then emerges which in turn lays an egg band around young host twigs (Rose 1958; Sippell and Ewan 1967; Martineau 1984). The adult forest tent caterpillars are active fliers and can cause rapid expansion of the infested areas within a year or two. Hodson (1941) observed forest tent caterpillar moths which had been attracted to building lights located 3 km away from the outbreak area. Though the forest tent caterpillar occurs throughout North America, reproductive characteristics have been shown to vary from north to south. Parry et al. (2001) have shown that larvae in the Southern United States are much smaller, but more numerous than populations in Manitoba.

## 2.2.   Forest Tent Caterpillar Outbreaks

Forest tent caterpillar outbreaks generally occur every 10–12 years (with a range of 6–16 years) in any one area and can last in the same area for 3 to 6 years (Duncan and Hodson 1958; Hildahl and Campbell 1975; Martineau 1984; Ives and Wong 1988). Between 1980 and 1996, 21.2% of the regions affected by the forest tent caterpillar were defoliated for 3 or 4 consecutive years (Simpson and Coy 1999). The outbreak cycle in Quebec during the 20th century was slightly different between aspen- (9-year cycle) and sugar maple– (13-year cycle) dominated regions (Cooke and Lorenzetti 2006). It is thought that the duration of forest tent caterpillar outbreaks is most likely affected by temperature and weather fluctuations (Hodson 1941; Blais et al. 1955; Hildahl and Campbell 1975; Daniel and Myers 1995). Topography (Cooke and Lorenetti 2006) and forest heterogeneity along with elevation (Roland 1993; Cooke and Roland 2000) may also be important factors affecting outbreak duration (Roland 1993; Cooke and Roland 2000). Parasitism by other organisms may lead to the decline of forest tent caterpillar outbreaks (Parry 1995) and forest fragmentation may affect the ability of parasitic organisms to disperse (Cooke and Roland 2000). It has also been suggested that intrinsic factors such as larval gregariousness and adult feeding capacity play a role in all Lepidoptera[*] population behaviour (Miller 1996).

• Lepidoptera is the designation for a group of insect commonly known as butterflies.

During an outbreak, large populations of forest tent caterpillars are able to completely strip trembling aspen of their leaves before foliation is complete, as the first larvae appear while leaves are beginning to unfold (Hodson 1941; Sippell and Ewan 1967). Although refoliation by trembling aspen occurs after the initial defoliation and mortality is rare, repeated attacks by the forest tent caterpillar can make host trees susceptible to other diseases or stresses (Duncan and Hodson 1958; Sippell and Ewan 1967; Hildahl and Campbell 1975), such as drought (Hogg and Schwarz 1999; Hogg et al. 2002a) and other insects or pathogens (Hogg et al. 2002a; Brandt et al. 2003), which could eventually lead to mortality (Hildahl and Reeks 1960; Churchill et al. 1964).

Growth reduction due to forest tent caterpillar defoliation is well documented (Churchill et al. 1964; Hildahl and Campbell 1975; Hogg and Schwarz 1999), as well as branch and twig mortality (Ives and Wong 1988). Rose (1958) reported in one study that radial growth in completely defoliated trees ceased

Figure 8.3
**Figure 8.3**
**Photos of A) three consecutive white rings in trembling aspen illustrating the radial growth, and B) the anatomical magnification of a white ring (B, left side) compared to a normal ring (B, right side) showing the smaller cell size and the weaker secondary wall of woof fibres in the white ring.**

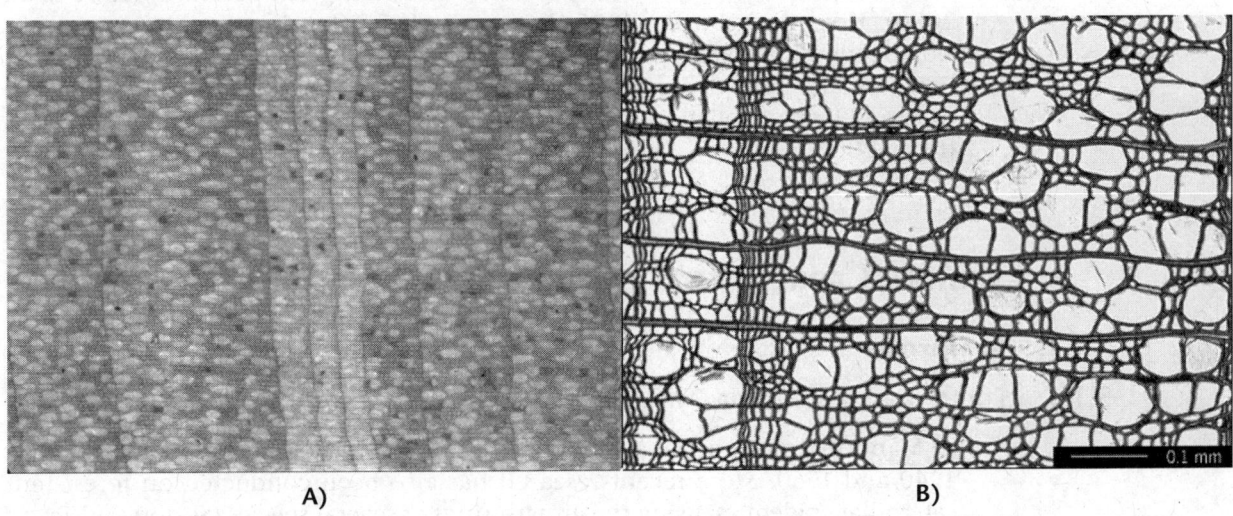

A)                                                                                    B)

Source: Sutton and Tardif 2007.

at defoliation and did not restart during the same year, even after refoliation. However, Hildahl and Reeks (1960) and Churchill et al. (1964) found that one year of severe defoliation only depressed radial growth during the year of defoliation. Trees which experienced at least two years of severe defoliation required one or two extra years to recover radial growth (Hildahl and Reeks 1960; Churchill et al. 1964). Similarly, premature growth stoppage and reduced radial growth have been observed in trembling aspen after artificial defoliation when compared to control trees (Jones et al. 2004). Severe defoliation by the forest tent caterpillar has also been associated with the formation of "white rings" (figure 8.3). These growth rings appear completely white after the wood sample has been sanded and are characterized by fibres which are small and have thinner secondary cell walls than normal fibres (Sutton and Tardif 2005) and have a reduced wood density (Hogg et al. 2002b; Sutton and Tardif 2005). White rings have already been observed in Alberta, Saskatchewan (Hogg and Schwarz 1999; Cooke 2001; Hogg et al. 2002a), and Québec (Sutton and Tardif 2005; Huang et al. 2006, 2007).

## 2.3. Outbreaks and Climate

Insects may be partly responsible for declining host stands at the southern margins of host species ranges (Volney and Fleming 2000), a factor which may become important with increasing global temperatures and climate change. Hogg and Schwarz (1999) found that defoliation was the most important factor in reducing basal area growth of trembling aspen in declining stands in Saskatchewan. Forest tent caterpillar activity was also found to be involved in dieback of aspen stands in northwestern Alberta (Hogg et al. 2002a) and sugar maple stands in southern Québec (Bauce et al. 1990; Payette et al. 1996). Trembling aspen productivity may be negatively affected if increased severity and frequency of forest tent caterpillar defoliation occurs due to increasing temperatures (Hart et al. 2000). Hogg and Hurdle (1995) have speculated that increased temperatures would reduce the extent of the forested zone due to drier conditions. An increase in temperature due to climate warming may also lead to increased mortality due to drought and defoliation in many boreal forest stands and aspen parkland areas (Hogg 2001).

## 3.  REGIONAL FOREST TENT CATERPILLAR OUTBREAK DYNAMICS

### 3.1.  Manitoba

#### 3.1.1.  Forest Tent Caterpillar Dynamics

In Manitoba, many aerial surveys of insect defoliation were conducted between 1940 and 1960. More recent research has also been conducted on forest tent caterpillar epidemics, using the growth rings of several species (Sutton and Tardif 2005, 2007). Preliminary studies have also been conducted in the boreal forest region of eastern and northern Manitoba. In general, the forest tent caterpillar is a common insect defoliator of trembling aspen in the aspen parkland forest and in agricultural wind break rows (Reeks 1971; Ives and Wong 1988). Several outbreaks have occurred in Manitoba during the 20th century (Hildahl and Reeks 1960; Sutton and Tardif 2007). Between 1951 and 1954, there was an approximate loss of 8.4% of the basal area growth in a 610,000-ha area in Manitoba and Saskatchewan (Hildahl and Reeks 1960). Between 1988 and 1992, 325,000 ha were affected by the forest tent caterpillar in Manitoba (Simpson and Coy 1999). In the Canadian prairie region (from Alberta to Manitoba), the forest tent caterpillar is one of the most important pests causing timber loss, with losses of approximately 4.1 million $m^3$ annually (Brandt 1995). Outbreaks in Manitoba are generally similar to those in Ontario (see section 3.2). In general, a severe outbreak rarely affects one region for more than two consecutive years (Hildahl and Reeks 1960; Reeks 1971; Sutton and Tardif 2007). Outbreak periodicity in Manitoba is approximately 10 years, but varies between 6 and 16 years (Hildahl and Reeks 1960). Sutton and Tardif (2007) found that the interval between outbreaks in western Manitoba was approximately 14 to 17 years. They also showed that some localised regions were sometimes subject to intense defoliation between outbreak periods.

#### 3.1.2.  Distribution and Host

The forest tent caterpillar affects a large portion of the boreal forest and aspen parkland in Manitoba (Hildahl and Reeks 1960; Reeks 1971). Outbreaks occurring in the prairie and aspen parkland area appear to be asynchronous with outbreaks in eastern Canada, especially the boreal and maritime regions (Sutton and Tardif 2005; Cooke and Lorenzetti 2006). Hildahl and Reeks (1960) suggest that trembling aspen stands are the most affected by forest tent caterpillars in Manitoba, however, growth loss and white ring formation (figure 8.3) in balsam poplar and white birch indicates that major outbreaks also affect the insect's secondary hosts (Sutton and Tardif 2007). Although agricultural pest outbreaks are generally considered more important in Manitoba than the forest tent caterpillar (Hildahl and Reeks 1960), growth loss caused by this insect has a particular significance in western Manitoba, where trembling aspen is a commercial resource.

### 3.1.3.  Outbreaks: Dynamics, Factors, and Effects

The first reports of forest tent caterpillar activity in Manitoba seem to have been recorded by Blair (1917, in Hildahl and Reeks 1960). He noted a large amount of insect activity across Canada in 1887. Dendrochronological evidences suggest a period of severe defoliation in western Manitoba during the late 1870s (Sutton and Tardif 2007). Information about the forest tent caterpillar before 1923 remains incomplete. A study completed in the Duck Mountain Provincial Forest (DMPF) identified four major outbreaks of the forest tent caterpillar after 1870: 1875–1880, 1937–1948, 1962–1965, and 1982–1985 (Sutton and Tardif 2007; see figure 8.4; also see Epp et al., chapter 12). Growth suppression (and its association with white ring formation) was observed in trembling aspen, balsam poplar, and white birch, in both older and younger stands. Growth suppression was also observed in all forest types which were dominated by trembling aspen.

Hildahl and Reeks (1960) reported four outbreaks between 1923 and 1953. During this period, the forest tent caterpillar was present each year throughout the province, with a decline in 1953 due to extreme climatic events. The outbreak of the 1960s was the most severe and caused the largest amount of radial growth reduction, with defoliation in 1962 being the most severe ever noted in Manitoba and Saskatchewan (Elliot and Hildahl 1963). Many white rings were observed in 1962 in trembling aspen and white birch samples collected in western Manitoba (figure 8.4) (Sutton and Tardif 2007). An extreme drought which occurred in 1961 (Girardin and Tardif 2005) was probably associated with the forest tent caterpillar population explosion.

Outbreaks in Manitoba are most likely affected by climate. Two outbreaks during the 1950s appear to have been affected by events such as the spring frost in 1953, which killed the young leaves and led to the starvation of the larvae (Blais et al. 1955; Hildahl and Reeks 1960). Another spring frost also appears to have been responsible for the decline of forest tent caterpillar population in the 1980s in Manitoba (Moody and Cerezke 1985; Sutton and Tardif 2007).

Although it is difficult to demonstrate the role of forest tent caterpillar defoliation in tree mortality (Hildahl and Reeks 1960), the negative effects of defoliation are evident. Jones et al. (2004) observed a premature radial growth stop and an overall reduction in radial growth in trembling aspen trees which were subjected to an artificial defoliation, compared to the control trees. In addition, severe forest tent caterpillar defoliation has been associated with white ring formation (Hogg et al. 2002b; Sutton and Tardif 2007). Hildahl and Reeks (1960) found that trembling aspen recovered quickly after a year of severe defoliation, however, more than one year was needed to recover after two consecutive years of severe defoliation. In a short-term study in Manitoba, mortality in study sites

Figure 8.4
## Chronologies of A) trembling aspen and B) white birch (continuous line and black dots), with the number of samples (dotted line), from western Manitoba

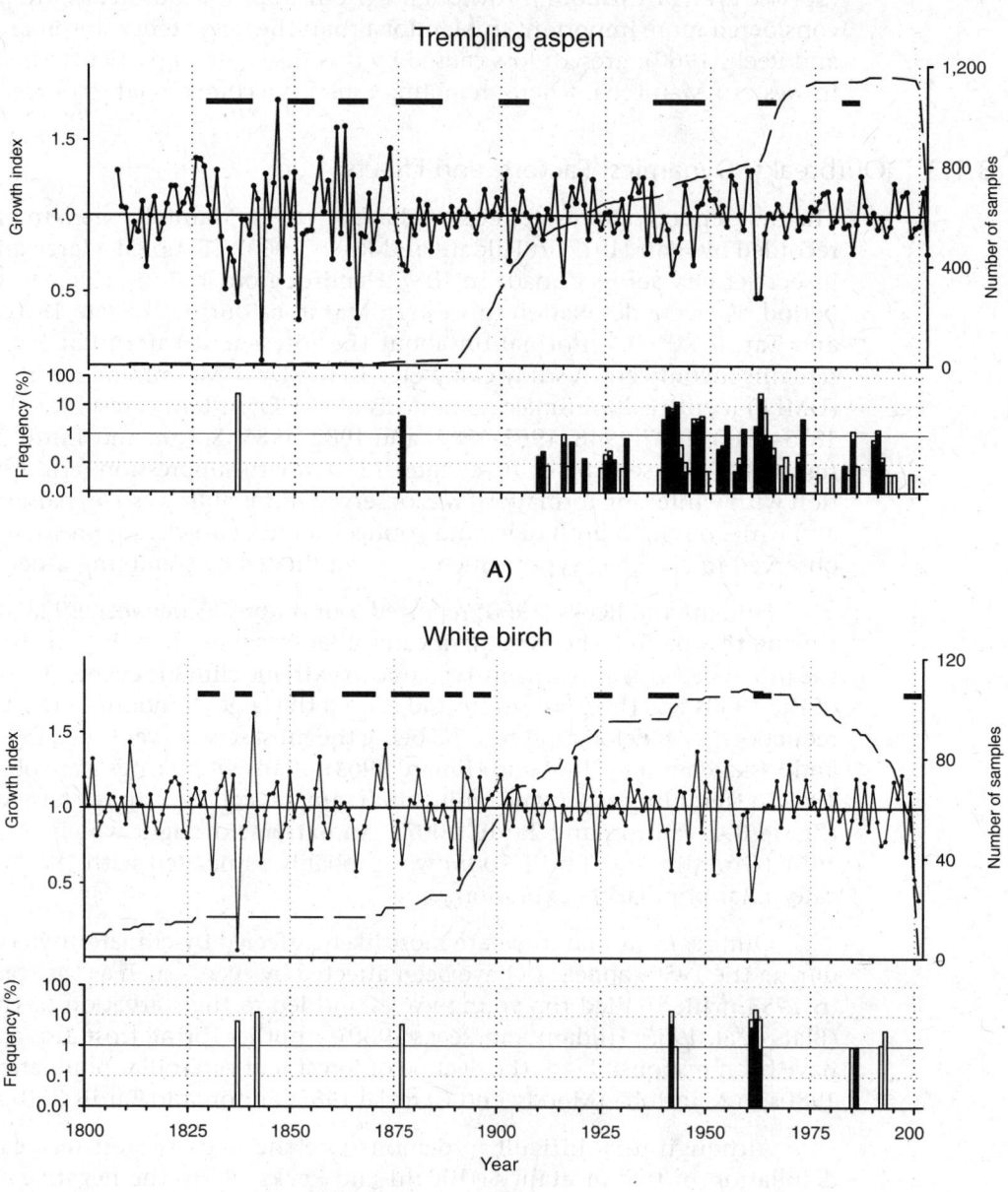

Periods identified as outbreaks are indicated by black bars in the upper part of the graph. The histogram indicates the frequency in % (on a logarithmic scale) of the samples with missing rings (grey) and white rings (black) per year. Four major forest tent caterpillar outbreak episodes were identified since 1870, during the years 1875–1880, 1937–1948, 1962–1965, and 1982–1985. From Sutton and Tardif (2007).

was negligible, however none of the sites were defoliated more than 3 consecutive years (Hildahl and Reeks 1960). Brandt et al. (2003), while studying several sites between Manitoba and Alberta, found a correlation between defoliation duration and the proportion of weakened, dead, or declining trees. Weakened trees also showed signs of other diseases in forests affected by at least two outbreaks (Brandt et al. 2003).

Sutton and Tardif (2005) suggested that white rings, which are structurally weak and associated with severe defoliation, could provide a path for pathogen dispersal and development within the tree, which could accelerate tree mortality. A synchronization of these episodes of infection of defoliated trees could have repercussions on the pattern and synchronization of tree mortality (individuals or stands of trees), which could in turn have an impact on successional processes. Brandt et al. (2003) concluded that the forest tent caterpillar was involved in the regulation of trembling aspen forest productivity and that they may play a role in mortality and general decline observed several years after defoliation. They found that trees growing in dry areas were more affected by the forest tent caterpillar than those growing in wet areas (Brandt et al. 2003). This phenomenon was also observed in western Manitoba, especially during a period of severe drought (Sutton and Tardif 2007).

## 3.2. Ontario

### 3.2.1. Forest Tent Caterpillar Dynamics

Many studies have been done on forest tent caterpillar dynamics in Ontario, but the information remains fragmented for many regions. In general, outbreaks occurring in Ontario are synchronised with those in the boreal region of Québec (Cooke and Lorenzetti 2006). The forest tent caterpillar seems to have a regular cycle of approximately 6 to 14 years with outbreaks occurring generally every 10 years (Sippell 1962). Maximum years of defoliation occur approximately every 13 years (Daniel and Myers 1995). Outbreak duration is from 3 to 9 years, with a regional average of 6 years (Sippell 1962). Cooke et al. (2009) observed that, in populated areas, outbreaks last 2.6 years on average, whereas in less populated regions, outbreaks last approximately 0.8 years, which is slightly longer than outbreaks in Québec (see section 3.3).

### 3.2.2. Distribution and Host

In Ontario after 1930, outbreaks generally had three epicentres throughout the province, in the south and the northeast of the province (Sippell 1962). Extreme cold temperatures appear to be an important factor in the distribution of the forest tent caterpillar in Ontario, more important than the natural population cycle of the insect. Cold temperatures may even affect egg hatch in certain areas (Daniel and Myers 1995). As in other provinces, outbreak duration is affected by forest fragmentation. Cooke and Roland (2000) observed that forest heterogeneity and altitude were positively associated with the duration of outbreaks. A similar association was observed between defoliation and forest fragmentation caused by roads; this pattern was also observed in Québec (Cooke et al. 2009). Roland (1993) showed that outbreaks in the mixed boreal forest of Ontario were longer in fragmented forests than in continuous forests. Forest fragmentation could favour defoliation by limiting the dispersal of parasites and pathogens that normally control forest tent caterpillar populations (Roland 1993).

### 3.2.3.  Outbreaks: Dynamics, Factors, and Effects

Between 1948 and 1988, four outbreaks were recorded in Ontario (Daniel and Myers 1995; see figure 8.5). Historical records indicate ten outbreak periods between 1867 and 1960; however, the data recorded before 1930 correspond only to agricultural areas, and are not as complete as the data collected after 1930 (Sippell 1962). Since 1948, an annual defoliation maximum has been observed every 13 years (Daniel and Myers 1995). A large portion of the province was affected by forest tent caterpillar defoliation during the 1948–1956 period, but never more than 2 consecutive years of severe defoliation have been observed in one region (Daniel and Myers 1995). Unfavourable climatic conditions during the spring of 1953 may have been responsible for the rapid forest tent caterpillar population decline in northwestern Ontario (Blais et al. 1955).

**Figure 8.5**
**Distribution of the area defoliated by the forest tent caterpillar during six outbreak periods in Ontario and Québec**

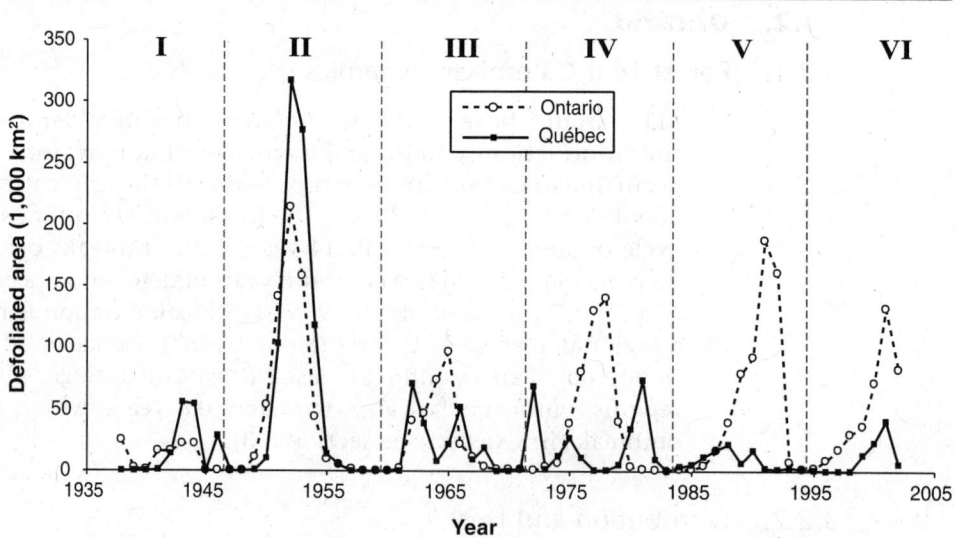

Periods are indicated by the vertical dotted lines. From Cooke et al. (2009).

Studies in Ontario have shown that forest heterogeneity has a greater effect on outbreak duration than climatic variables (Roland et al. 1998; Cooke and Roland 2000). Although there may not be a link between spring temperatures and outbreak intensity, extreme winter temperatures may influence forest tent caterpillar distribution (Daniel and Myers 1995). Roland et al. (1998) found that cold spring temperatures could regulate outbreak duration in Ontario. They showed that cold temperatures could shorten outbreaks and optimal temperature could prolong them.

Cooke et al. (2009) observed that tree mortality and eventually their decline could increase after three consecutive years of defoliation. Some have hypothesized that trembling aspen decline observed in north eastern Ontario could have been caused by frequent defoliation and the stress associated with drought (Candau et al. 2002). Sugar maple population decline in the temperate region of Ontario has also been linked to forest tent caterpillar defoliation (Gross 1995). Severe defoliation during the 1970s caused branch mortality, premature leaf colour change in the fall, leaf and wood growth reduction as well as tree mortality (Gross 1995). Gross (1995) also found that declining trees were sometimes infected by other insects that often affect weakened trees. This suggests that defoliation by the forest tent caterpillar may weaken trees, which may then become infected by other organisms which can cause mortality.

## 3.3. Québec

### 3.3.1. Forest Tent Caterpillar Dynamics

In Québec, many papers have recently been published dealing with forest tent caterpillar outbreaks (Bergeron and Charron 1994; Leblanc 1999; Jardon et al. 2002; Cooke and Lorenzetti 2006; Huang et al. 2008). The forest tent caterpillar is an important defoliator in this province, where it is associated with two important forest types, the boreal mixedwoods and the sugar maple forests. Approximately 25% of the province (384,540 km²) was defoliated at least one year between 1938 and 2002 (Cooke and Lorenzetti 2006). In Québec, it has been shown that periods of defoliation are longer and more severe in populated areas. For example, in less populated areas, only 11% of the area affected by forest tent caterpillars, between 1938 and 2002 were defoliated for more than two consecutive years, whereas the number increases to 45% in heavily populated areas (Cooke et al. 2009).

### 3.3.2. Distribution and Host

According to Fitzgerald (1995), the distribution of the forest tent caterpillar is generally limited by the 47° N (figure 8.2). However, outbreaks have been observed further north, especially during the extensive outbreak of the 1950s (Cooke and Lorenzetti 2006). The defoliation signal has also been observed in tree ring chronologies developed for trembling aspen in the Lake Duparquet area in Abitibi (48° N) (Bergeron and Charron 1994) and in other regions sampled between the 47° N and 49° N (Leblanc 1999; Jardon et al. 2002; Huang et al. 2006). Huang et al. (2007) have also reported outbreaks in poplar stands as far north as 54° N, suggesting that forest tent caterpillar outbreaks occur much further north than the normally accepted limit of 49° N.

Many factors affect the distribution and frequency of outbreaks in Québec, including topography (Cooke and Lorenzetti 2006), forest fragmentation (Fortin and Mauffette 2001), as well as the concentration of human population (Cooke et al. 2009). Fortin and Mauffette (2001) observed that forest tent caterpillar performance was higher on leaves from trees growing at the forest edge as compared to those of trees growing inside the forest. They observed that larvae fed with these leaves were heavier than those fed with leaves from trees growing inside the forest. In addition, larvae preferred leaves from sugar maple that had developed in the sun over leaves that had developed in the shade (Panzuto et al. 2001). An increase in pollution could also favour larvae performance, especially in southern Québec where the insect affects sugar maple (Fortin et al. 1997). As in the case of Ontario, increasing forest fragmentation in southern Québec could favour forest tent caterpillar activity (Fortin and Mauffette 2001; Cooke et al. in press). Outbreaks are generally longer and more frequent in disturbed rural areas (Cooke et al. 2009).

In Québec, studies have shown that outbreak frequency depends on forest type. For example, the forest tent caterpillar has an outbreak cycle of approximately 9 years in forests dominated by trembling aspen (Cooke and Lorenzetti 2006). In the broadleaf forest dominated by sugar maple, the cycle is approximately 13 years. Although priority is often given to the study of the effects of defoliation on trembling aspen, the forest tent caterpillar can also affect large areas dominated by sugar maple (Fortin and Mauffette 2001). Defoliating insects were found to be an additional stress on maple stands during their decline in 1980 (Bauce et al. 1990; Payette et al. 1996). A similar decline was observed in maple stands in New York during the 1950s (Hibben 1962).

The forest tent caterpillar can affect other secondary species in Québec. A study conducted in the Lake Duparquet area, radial growth was more suppressed in white birch than in trembling aspen during the 1950s outbreak (Bergeron and Charron 1994).

### 3.3.3.   Outbreaks: Dynamics, Factors, and Effects

Between 1938 and 2002, six distinct outbreaks have been observed in the boreal region of Québec (Charron and Bergeron 1994; Cooke and Lorenzetti 2006; see figures 8.5 and 8.6). The observed cycle was associated with natural population cycles and maximum defoliation was observed every 10 years (Cooke and Lorenzetti 2006). The same observations were made using tree ring chronologies constructed in the Abitibi region (Bergeron and Charron 1994; Leblanc 1999; Jardon et al. 2002; Bergeron et al 2002). In a reconstruction of outbreaks along a transect between the 46th and 54th parallel in western Québec, Huang et al. (2007) observed a strong regional outbreak synchronism, while outbreak initiation dates, intensity, and extent varied spatially. During the 20th century, outbreaks have affected between 11 and 97% of the province (Cooke and Lorenzetti 2006). The outbreak between 1948 and 1957 was especially intense, with 82.5% of the outbreak range affected in 1952 (Cooke and Lorenzetti 2006). In general, outbreaks in Québec are more or less synchronized, but population dynamics are different for each region (Cooke and Lorenzetti 2006). Forest tent caterpillar outbreaks appear synchronized in the temperate and maritime regions (southern Québec, the Maritimes and the northeastern United States), however they are not synchronized with the outbreaks in the boreal forest (northern Québec, Ontario and western Canada) (B. Cooke, pers. comm.).

Most outbreaks appear to be related to natural forest tent caterpillar population cycles (Cooke and Lorenzetti 2006); however, the outbreak of the 1950s seems to have also been influenced by drought conditions caused by several consecutive warm, dry summers (Bergeron and Charron 1994). Climatic

Figure 8.6
**Radial growth increment of two host species (trembling aspen and white birch) from Lake Duparquet, Québec**

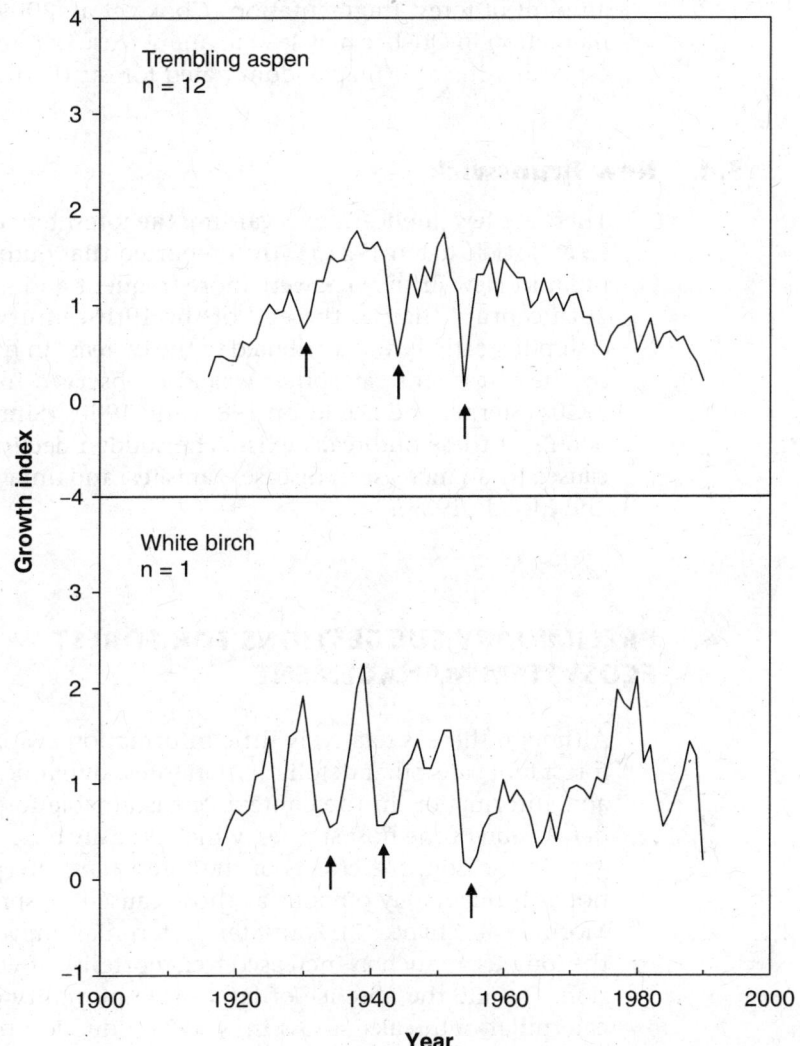

Arrows indicate three years of defoliation by the forest tent caterpillar. From Bergeron and Charron (1994).

conditions also played a major role in the decline of the 1960s outbreak. A late spring frost in 1963 (in eastern Canada) caused a widespread decline in defoliation by the forest tent caterpillar in Québec (Cooke and Lorenzetti 2006).

The effects of the forest tent caterpillar vary in Québec, as observed in other provinces. Defoliation may simply result in radial growth loss (Bergeron and Charron 1994), or may be a factor in sugar maple forest decline (Bauce et al.

1990; Payette et al. 1996), even though defoliated trees are rarely killed (Bergeron et al. 2002). Tree mortality may increase in a region submitted to defoliation in more than two consecutive years and even though it is rare, the probability of three consecutive years of defoliation occurring in one area increased with the amount of forest fragmentation (Cooke et al. 2009). An increase in forest fragmentation in Québec may lead to an increase in forest tent caterpillar defoliation, especially in sugar maple–dominated forests (Fortin and Mauffette 2001).

### 3.4.  New Brunswick

There are few publications regarding the forest tent caterpillar in New Brunswick. In an article dated 1922, Tothill reported that outbreaks of the forest tent caterpillar in New Brunswick were more frequent and severe at the beginning of the 20th century than at the end of the 19th century. The increase in forest tent caterpillar activity was attributed to the increase in human population. Defoliation by the forest tent caterpillar was also observed in New Brunswick during the 1940s (Sterner and Davidson 1981) and 1980s (Simpson and Coy 1999), but few details of these outbreaks exist. The sudden decline of the 1980s outbreak was caused by an increase in disease, parasites, and unfavourable temperatures (Kondo and Moody 1987).

## 4.  PRELIMINARY SUGGESTIONS FOR FOREST ECOSYSTEM MANAGEMENT

Although there is relatively little information available related to the impacts of forest tent caterpillar defoliation on forest dynamics, we must consider the available information in the context of sustainable forest management. Besides the defoliation of the host species, which is easily observable at the beginning of the growing season, the effects of the forest tent caterpillar on forest dynamics are not as immediately obvious as those caused by spruce budworm outbreaks (see Morin et al., chapter 7). Forest tent caterpillar impacts are however observed over the long term, such as increased tree mortality after repeated episodes of defoliation. Despite the absence of large-scale mortality directly linked to forest tent caterpillar outbreaks, a loss in wood production of the host species is directly observable after an outbreak. A better quantification of these losses is needed so that they can be taken into account during the planning stages of management and during wood allocation calculations.

Outbreaks seem to be directly or indirectly associated with host tree mortality, depending on frequency and severity. Poplar mortality, whether it is directly or indirectly caused by the forest tent caterpillar, may cause the formation of gaps in the forest canopy, which in turn favours the increase in conifer and shrub growth in the understory. Many studies indicate that trembling aspen stands are capable of maintaining themselves, even in the long term (see Epp et al., chapter 12). This suggests that delayed poplar mortality and possibly gap formation could favour species maintenance through regeneration in forest gaps. It is equally possible that, during outbreaks, understory species such as white spruce or beaked hazel could benefit from increased light penetration. Hamel and Kenkel (2001) reported that a tall, dense understory of shrubs (beaked hazel [*Corylus cornuta* Marsh.] or mountain maple [*Acer spicatum* Lam.]) in older poplar stands in the Duck Mountain Provincial Forest prevented the establishment of a second poplar cohort. The authors linked this lack of regeneration to ungulate browsing on young poplar shoots.

In any case, observations made of forest dynamics after forest tent caterpillar outbreaks indicate that outbreaks can generate a diversity of stand types, in terms of composition and structure. In terms of forest management, although clearcutting may be appropriate to ensure regeneration, not all trembling aspen stands should be managed in this way, nor should they be replaced by conifer plantations or understory plantations. In the case of non-integrated forest management where the concern is only the harvest of either coniferous trees (deciduous trees left at the cut site) or deciduous trees (coniferous trees left at the cut site), a better understanding of successional processes after cutting is needed so that ecosystem-based management may be more easily implemented.

## 5.  CONCLUSION

Myers (1998) showed that in general, insect population fluctuations in the Lepidoptera were synchronized over large expanses in the northern hemisphere during most of the 20th century. Exogenous factors such as climate could therefore be responsible for this synchronization (Daniel and Myers 1995). Although the forest tent caterpillar has generally been considered less important than other insects until recently (Reeks 1971), the effects that this defoliator has on economically valuable species should be studied further.

Knowledge of the processes that affect insect dynamics is still fragmentary, and more studies are needed so that we can better understand how to manage affected forests and to provide solutions to lessen the damage cause by defoliating insects (Cooke and Lorenzetti 2006). Understanding insect dynamics is equally important for maintaining a natural landscape and integrating timber loss and tree mortality into annual allowable cut calculations.

Our knowledge regarding the impact of the forest tent caterpillar on forest dynamics, in particular its role in gap formation after tree death, as well as the response of the understory, should be increased. For example, in Manitoba, gaps associated with forest tent caterpillar outbreaks, depending on the intensity, favour a trend towards a more coniferous stand, poplar maintenance through suckering, or an invasion of the understory by beaked hazel. A better understanding of forest dynamics in relation to forest tent caterpillar outbreaks should permit, in the medium term, the development of sustainable forest management strategies.

An adaptive management approaches will be essential so that silvicultural practices may be altered once additional information concerning the dynamics of the forest tent caterpillar are available. Adaptability will also be important in the case where climate change could affect the dynamics and extent of forest tent caterpillar outbreaks.

## REFERENCES

Bauce, E., Lachance, D., and Archambault, L. 1990. Le rôle de la livrée des forêts et de l'arpenteuse de Bruce dans le dépérissement des érablières du sud du Québec. *In* Le dépérissement des érablières, causes et solutions possibles. *Edited by* C. Camiré, W. Hendershot, and D. Lachance. C.R.B.F., Faculté de foresterie et géomatique, Université Laval, Québec, Que., pp. 39–47.

Bergeron, Y. and Charron, D. 1994. Postfire stand dynamics in a southern boreal forest (Québec): a dendro-ecological approach. Écoscience, **1**: 173–184.

Bergeron, Y., Denneler, B., Charron, D., and Girardin, M.-P. 2002. Using dendrochronology to reconstruct disturbance and forest dynamics around Lake Duparquet, northwestern Québec. Dendrochronologia, **20**: 175–189.

Blair, A.B. 1917. An historical account of the forest tent caterpillar and of the fall webworm in North America. Annual report of the Entomological Society of Ontario **47**: 73–84.

Blais, J.R. 1958. Effects of defoliation by spruce budworm (*Choristoneura fumiferana* Clem.) on radial growth at breast height of balsam fir (*Abies balsamea* (L.) Mill.) and white spruce (*Picea glauca* (Moench) Voss.). For. Chron. **34**: 39–47.

Blais, J.R., Prentice, R.M., Sippell, W.L., and Wallace, D.R. 1955. Effects of weather on the forest tent caterpillar *Malacosoma disstria* Hbn., in central Canada in the spring of 1953. Can. Entomol. **87**: 1–8.

Brandt, J.P. 1995. Forest insect- and disease-caused impacts to timber resources of west-central Canada: 1988–1992. Information Report NOR-X–341. Canadian Forest Service, Northwest Region, Northern Forestry Centre, Edmonton, Alta.

Brandt, J.P., Cerezke, H.F., Mallett, K.I., Volney, W.J.A., and Weber, J.D. 2003. Factors affecting trembling aspen (*Populus tremuloides* Michx.) health in the boreal forest of Alberta, Saskatchewan, and Manitoba, Canada. For. Ecol. Manag. **178**: 287–300.

Candau, J.-N., Abt, V., and Keatley, L. 2002. Bioclimatic analysis of declining aspen stands in northeastern Ontario. Forest Research Report no. 154. Applied Research and Development, Ontario Ministry of Natural Resources, Sault Ste. Marie, Ont.

Churchill, G.B., John, H.H., Duncan, D.P., and Hodson, A.C. 1964. Long-term effects of defoliation of aspen by the forest tent caterpillar. Ecology, **45**: 630–633.

Cooke, B.J. 2001. Interactions between climate, trembling aspen and outbreaks of forest tent caterpillars in Alberta. Ph.D. thesis, University of Alberta, Edmonton, Alta.

Cooke, B.J. and Lorenzetti, F. 2006. The dynamics of forest tent caterpillar outbreaks in Québec, Canada. For. Ecol. Manag. **226**: 110–121.

Cooke, B.J. and Roland, J. 2000. Spatial analysis of large-scale patterns of forest tent caterpillar outbreaks. Écoscience, **7**: 410–422.

Cooke, B.J., Lorenzetti, F., and Roland, J. (2009). On the duration and distribution of forest tent caterpillar outbreaks in eastcentral Canada. J. Entomol. Soc. Ont. 140 (in press)

Daniel, C.J. and Myers, J.H. 1995. Climate and outbreaks of the forest tent caterpillar. Ecography, **18**: 353–362.

Duncan, D.P. and Hodson, A.C. 1958. Influence of the forest tent caterpillar upon the aspen forests of Minnesota. For. Sci. **4**: 71–79.

Elliot, K.R. and Hildahl, V. 1963. Manitoba and Saskatchewan: forest insect and disease survey. *In* Annual report of the Forest Insect and Disease Survey 1962. *Edited by* Department of Forestry. Department of Forestry, Forest Entomology and Pathology Branch, Ottawa, Ont.

Fitzgerald, T.D. 1995. The tent caterpillars. Cornell University Press, Ithaca, N.Y., USA.

Fortin, M. and Mauffette, Y. 2001. Forest edge effects on the biological performance of the forest tent caterpillar (*Lepidoptera: Lasiocampidae*) in sugar maple stands. Écoscience, **8**: 164–172.

Fortin, M., Mauffette, Y., and Albert, P. J. 1997. The effects of ozone-exposed sugar maple seedlings on the biological performance and the feeding preference of the forest tent caterpillar (*Malacosoma disstria* Hbn.). Environ. Pollut. 97: 303–309.

Girardin, M.-P. and Tardif, J. 2005. Sensitivity of tree growth to the atmospheric vertical profile in the boreal plains of Manitoba, Canada. Can. J. For. Res. **35**: 48–64.

Girardin, M.-P., Tardif, J., and Bergeron, Y. 2001. Radial growth analysis of *Larix laricina* from the Lake Duparquet area, Québec, in relation to climate and larch sawfly outbreaks. Écoscience, **8**: 127–138.

Gross, H.L. 1995. Dieback and growth loss of sugar maple associated with defoliation by the forest tent caterpillar. For. Chron. **67**: 33–42.

Hall, J.P., Bowers, W.W., and Hironen, H. 1998. Forest insect and disease conditions in Canada, 1995. Forest Insect and Disease Survey, Natural Resources Canada, Canadian Forest Service, Ottawa, Ont.

Hamel, C. and Kenkel, N. 2001. Structure and dynamics of boreal forest stands in the Duck Mountains, Manitoba. Final Project Report 2001-4, Sustainable Forest Management Network. University of Alberta, Edmonton, Alta.

Hanec, W. 1966. Cold-hardiness in the forest tent caterpillar, *Malacosoma disstria* Hubner (*Lasiocampidae, Lepidoptera*). J. Ins. Physiol. **12**: 1443–1449.

Hart, M., Hogg, E.H., and Lieffers, V.J. 2000. Enhanced water relations of residual foliage following defoliation in *Populus tremuloides*. Can. J. Bot. 78: 583–590.

Hibben, C.R. 1962. Investigations of sugar maple decline in New York woodlands. Ph.D. thesis, Cornell University, Ithaca, N.Y., USA.

Hildahl, V. and Campbell, A.E. 1975. Forest tent caterpillar in the Prairie Provinces. Information Report NOR-X–135. Canadian Forest Service, Northern Forest Research Centre, Edmonton, Alta.

Hildahl, V. and Reeks, W.A. 1960. Outbreaks of the forest tent caterpillar, *Malacosoma disstria* Hbn., and their effects on stands of trembling aspen in Manitoba and Saskatchewan. Can. Entomol. **92**: 199–209.

Hodson, A.C. 1941. An ecological study of the forest tent caterpillar *Malacosoma disstria* Hbn., in northern Minnesota. Minnesota Technical Bulletin no. 148. University of Minnesota, Agricultural Experiment Station, St. Paul, Minn., USA.

Hogg, E.H. 2001. Modelling aspen responses to climatic warming and insect defoliation in western Canada. *In* Sustaining aspen in western landscapes: symposium proceedings. *Compiled by* W.D. Shepperd, D. Binkley, D.L. Bartos, T.J. Stohlgren, and L.G. Eskew. 13–15 June 2000; Grand Junction, Colorado. Proceedings RMRS-P-18. U.S. Department of Agriculture, Forest Service, Rocky Mountain Research Station, Fort Collins, Colo., USA, pp. 325–338.

Hogg, E.H., Brandt, J.P., and Kochtubajda, H. 2002a. Growth and dieback of aspen forests in northwestern Alberta, Canada, in relation to climate and insects. Can. J. For. Res. **32**: 823–832.

Hogg, E.H., Hart, M., and Lieffers, V.J. 2002b. White tree rings formed in trembling aspen saplings following experimental defoliation. Can. J. For. Res. **32**: 1929–1934.

Hogg, E.H., and Hurdle, P.A. 1995. The aspen parkland in western Canada: a dry-climate analogue for the future boreal forest? Water Air Soil Pollut. **82**: 391–400.

Hogg, E.H. and Schwarz, A.G. 1999. Tree-ring analysis of declining aspen stands in west-central Saskatchewan. Information Report NOR-X-359. Canadian Forest Service, Northern Forestry Centre, Edmonton, Alta.

Huang, J.-G., Bergeron, Y., and Denneler, B. 2006. Dendro-climatological analyses of trembling aspen (*Populus tremuloides* Michx.) along a latitudinal gradient in western Québec, Canada. Cultural Diversity, Environmental Variability. 7th International Conference on Dendrochronology. 11–17 June 2006, Beijing, China. Program and abstracts. *Compiled by* L. Chen, H.-Y.Qiu, X.-C. Wang, and Q.-B. Zhang. Institute of Botany, Chinese Academy of Sciences. 20 Nanxincun, Beijing, China.

Huang, J., Bergeron, Y., Denneler, B., and Tardif, J. 2007. Tree-ring reconstruction of forest tent caterpillar outbreaks along a latitudinal gradient in western Quebec, Canada. 6th International Conference on Disturbance Dynamics in Boreal Forests, "Climate Change Impacts on Boreal Forest Disturbance Regimes," Fairbanks, Alaska, USA, 30 May – 2 June 2007.

Huang, J., Tardif, J., Denneler, B., Bergeron, Y., and Berninger, F. 2008. Tree-ring evidence extends the historic northern range limit of severe defoliation by insects in the aspen stands of western Quebec, Canada. Can. J. For. Res. **38**: 2535–2544.

Ives, W.G.H. and Wong, H.R. 1988. Tree and shrub insects of the Prairie Provinces. Information Report NOR-X-292. Canadian Forest Service, Northern Forestry Center, Edmonton, Alta.

Jardon, Y., Bergeron, Y., and Morin, H. 2002. The forest tent caterpillar outbreak dynamics at the northern limit of the insect distribution, Quebec. Dendrochronology, environmental change, and human history: 6th International Conference on Dendrochronology. Québec, Que., 22–27 August 2002.

Jones, B., Tardif, J., and Westwood, R. 2004. Weekly xylem production in trembling aspen (*Populus tremuloides*) in response to artificial defoliation. Can. J. Bot. **82**: 590–597.

Kondo, E.S. and Moody, B.H. 1986. Forest insect and disease conditions in Canada 1986. Forest Insect and Disease Survey. Canadian Forest Service, Ottawa, Ont.

Leblanc, D. 1999. Performance de la livrée des forêts (*Malacosoma disstria* Hbn.) élevée sur deux hôtes près de sa limite septentrionale et reconstruction de la chronologie de ses épidémies en Abitibi-Témiscamingue au moyen d'une analyse dendrochronologique. M.Sc. thesis, Université du Québec à Montréal, Montréal, Que.

Martineau, R. 1984. Insects harmful to forest trees. Multiscience Publications Ltd.

Miller, W.E. 1996. Population behavior and adult feeding capability in lepidoptera. Environ. Entomol. **25**: 213–226.

Moody, B.H. and Cerezke, H.F. 1985. Forest insect and disease conditions in Alberta, Saskatchewan, Manitoba, and the Northwest Territories in 1984 and predictions for 1985. Information Report NOR-X-269. Canadian Forest Service, Northern Forest Research Centre, Edmonton, Alta.

Morin, H. 1994. Dynamics of balsam fir forests in relation to spruce budworm outbreaks in the boreal zone of Québec. Can. J. For. Res. **24**: 730–741.

Morin, H. and Laprise, D. 1990. Histoire récente des épidémies de la tordeuse des bourgeons de l'épinette au nord du lac Saint-Jean (Québec): une analyse dendrochronologique. Can. J. For. Res. **20**: 1–8.

Myers, J.H. 1998. Synchrony in outbreaks of forest *Lepidoptera* in the Northern Hemisphere: a possible example of the Moran Effect. Ecology, **79**: 1111–1117.

Naidoo, R. and Lechowicz, M.J. 2001. Effects of gypsy moth on radial growth of deciduous trees. For. Sci. **47**: 338–348.

Panzuto, M., Lorenzetti, F., Mauffette, Y., and Albert, P.J. 2001. Perception of aspen and sun/shade sugar maple leaf soluble extracts by larvae of *Malacosoma disstria*. J. Chem. Ecol. **27**: 1963–1978.

Parry, D. 1995. Larval and pupal parasitism of the forest tent caterpillar, *Malacosoma disstria* Hubner (Lepidoptera: Lasiocampidae) in Alberta, Canada. Can. Entomol. **127**: 877–893.

Parry, D., Goyer, R.A., and Lenhard, G.J. 2001. Macrogeographic clines in fecundity, reproductive allocation, and offspring size of the forest tent caterpillar *Malacosoma disstria*. Ecol. Entomol. **26**: 281–291.

Parry, D., Spence, J.R., and Volney, W.J.A. 1998. Budbreak phenology and natural enemies mediate survival of first-instar forest tent caterpillar (*Lepidoptera: Lasiocampidae*). Environ. Entomol. **27**: 1368–1374.

Payette, S., Fortin, M.-J., and Morneau, C. 1996. The recent sugar maple decline in southern Québec: probable causes deduced from tree rings. Can. J. For. Res. **26**: 1069–1078.

Reeks, W.A. 1971. Impact of insect outbreaks in Manitoba. Manitoba Entomol. **5**: 5–16.

Roland, J. 1993. Large scale forest fragmentation increases the duration of tent caterpillar outbreak. Oecologia, **93**: 25–30.

Roland, J., Mackey, B.G., and Cooke, B. 1998. Effects of climate and forest structure on duration of forest tent caterpillar outbreaks across central Ontario, Canada. Can. Entomol. **130**: 703–714.

Rose, A.H. 1958. The effect of defoliation on foliage production and radial growth of quaking aspen. For. Sci. **4**: 335–342.

Simpson, R. and Coy, D. 1999. An ecological atlas of forest insect defoliation in Canada 1980–1996. Information Report M-X-206E. Natural Resources Canada, Canadian Forest Service – Atlantic Forestry Centre. Fredericton, N.B.

Sippell, W.L. 1962. Outbreaks of the forest tent caterpillar, *Malacosoma disstria* Hbn., a periodic defoliator of broad-leaved trees in Ontario. Can. Entomol. **94**: 408–416.

Sippell, W.L. and Ewan, H.E. 1967. Forest tent caterpillar, *Malacosoma disstria* Hbn. *In* Important forest insects and diseases of mutual concern to Canada, the United States, and Mexico. *Edited by* A.G. Davidson and R.M. Prentice. Number Fo 47-1180. Forestry and Rural Development, Ottawa, Ont.

Sterner, T.E. and Davidson, A.G. 1981. Forest insect and disease conditions in Canada 1980. Forest Insect and Disease Survey. Canadian Forest Service, Ottawa, Ont.

Sutton, A. and Tardif, J. 2005. Distribution and anatomical characteristics of white rings in *Populus tremuloides* Michx. IAWA J. **26**: 221–238.

Sutton, A. and Tardif, J. 2007. Dendrochronological reconstruction of forest tent caterpillar outbreaks in time and space, western Manitoba, Canada. Can. J. For. Res. **37**: 1643–1657.

Tothill, J.D. 1922. Notes on the outbreaks of spruce budworm, forest tent caterpillar, and larch sawfly in New Brunswick. Proc. Acadian Entomol. Soc. **8**: 172–182.

Volney, W.J.A. and Fleming, R.A. 2000. Climate change and impacts of boreal forest insects. Agric. Ecosyst. Environ. **82**: 283–294.

Witter, J.A. 1979. The forest tent caterpillar (*Lepidoptera: Lasiocampidae*) in Minnesota: a case history review. Great Lakes Entomol. **12**: 191–197.

# Chapter 9

# Applying Knowledge of Natural Disturbance Regimes to Develop Forestry Practices Inspired by Nature in the Southern Region of the Gaspé Peninsula*

*Daniel Kneeshaw, Ève Lauzon, André de Römer, Gerardo Reyes, Jonatan Belle-Isle, Julie Messier, and Sylvie Gauthier*

\* We thank Mathieu Bouchard, Yves Bergeron, Alain Leduc, Patrick Lefort, Marie-Andrée Vaillancourt, and all the students who carried out their field work in the Gaspé region. We also thank TEMBEC and TEMREX, particularly the employees that participated in helping to produce this work. Financial support was also provided by TEMBEC, TEMREX, NSERC, and the Canadian Forest Service. The photos on this page were graciously provided by Gerardo Reyes and Jacques Morissette.

# 1. INTRODUCTION

Variations in conditions created by disturbances such as intense fire and canopy gaps generate structural and compositional diversity (Denslow 1987; Frelich 2002), two aspects critical for the maintenance of biodiversity. Given their major impacts at both stand and landscape scales, forest fires and insect outbreaks have historically been the principal disturbance agents studied in the boreal forest (Heinselman 1973; Bergeron et al. 2006; Gauthier et al., chapter 3). However, more recently, recognition that smaller, less severe disturbances can also have large impacts on the structural diversity and composition of boreal forests has been increasing (Kneeshaw 2001; McCarthy 2001). Fires produce extensive areas of relatively homogenous forest conditions in terms of composition and age-class structure. When these large, intense disturbances are less frequent or affect smaller areas, regeneration of forest cover results from partial disturbances, wherein single-tree or small-group mortality occurs (Kneeshaw 2001). This creates canopy heterogeneity within forest stands, forming a mosaic of small forest units in different stages of development (Watt 1947; Pickett and White 1985). Moreover, moderate-severity disturbances caused by windthrow or areas affected by partial mortality caused by insect outbreaks can substantially modify successional pathways and forest structure (Seymour et al. 2002; Bouchard et al. 2006).

The different characteristics of natural disturbance regimes influence forest composition and structure (Frelich 2002; Seymour et al. 2002), creating variability and complexity in habitat conditions that native species have adapted to. Haeussler and Kneeshaw (2003) proposed four questions that forest managers were required to address, to ensure that current management techniques recreate the conditions necessary to maintain biodiversity and ecological processes: 1) Do forestry practices create conditions beyond the range of natural variability found in non-managed forests? 2) Do forestry practices recreate the complete range of conditions observed naturally? 3) If differences occur between managed and natural conditions, how are biodiversity and ecological functioning affected? 4) How can forestry practices be modified to minimize these differences? This study examining a suite of disturbances affecting the boreal forest will provide the necessary information to respond to the first two questions, and provide insight and direction for the last two.

The Gaspé Peninsula is a region within the boreal forest zone, wherein a large variety of tree species are found and several disturbance types occur. Currently there is a lack of information on the relative importance of the various disturbances types across the forest landscape, while understanding of their role in forest dynamics has also remained unclear. Yet characterizing the principal natural disturbances of the boreal forest is of fundamental importance for the development of a forest ecosystem management (FEM) approach, particularly with respect to forest attributes that are profoundly affected by forest harvesting (e.g., stand age and composition, biological legacies, spatial extent of disturbances, etc.). There is urgency to acquire this knowledge as current silvicultural practices have led to a substantial reduction in the regional abundance of important species such as white pine (*Pinus strobus* L.) and spruces (*Picea* spp.) and are responsible for homogenizing the forest age-class structure and increasing the proportion of younger stands.

The primary objective of this chapter is to characterize a suite of natural disturbances that affect the boreal mixedwood region in the southern part of the Gaspé Peninsula (i.e., fire, insect epidemics, windthrow, and canopy gaps), with the goal of providing management options that aspire to reduce the differences between conditions created by forestry practices and those found naturally. We accomplished this by using a two-step approach: one, characterizing the effects of natural disturbances, and two, comparing the effects of current forest practices with natural disturbances that are most similar. For each natural disturbance type, we examined different characteristics such as the proportion of the landscape affected, disturbance frequency, composition and structure of post-disturbance stands, as well as the different anthropological and biophysical factors affecting the disturbance regime. This will help us determine threshold levels of natural variation of forest ecosystems, both in space and time. Furthermore, such information will help us describe the influence of the different disturbances on natural boreal forest ecosystem dynamics and, subsequently, let us propose guidelines to develop forestry practices that are closer to nature.

## 2.   STUDY AREA AND METHODS

The study area is located in the Chaleur Bay region of the Gaspé Peninsula, in the southeast of Québec, between 64° 22′ and 67° 42′ W and 47° 49′ to 49° 15′ N (figure 9.1). The region is within the northern limit of the temperate forest zone and southern limit of the boreal forest zone. The study area is partly located in both the balsam fir–white birch and balsam fir–yellow birch bioclimatic domains (Saucier et al. 1998). The total surface of the study area is approximately 6,480 km². The topography is composed of hills, mountains, plateaus, and valleys. Summits range between 300 and 900 m in altitude. Much of the territory is composed of colluvial and alteration deposits. The northeastern part of the territory also consists of till deposits while certain southern portions have marine deposits. The land is mostly composed of mesic sites (95%) while water represents approximately 1% of the area. Mean annual temperature is 2.5°C and mean annual precipitation varies between 900 and 1,200 mm, with 35% occurring as snowfall. The southern part of the territory is partly used for agriculture, while most of the region is subjected to forest exploitation (Robitaille and Saucier 1998).

## 2.1.   The Context of Forest Management: Characterizing a Suite of Natural Disturbances

As in the conceptual model presented Bergeron et al. (1999), we assume that intense fire returns the forest to early-successional conditions; i.e., the first cohort, with even-aged regeneration dominated by shade-intolerant, pionneer species. At this stage, small-scale, partial disturbances have little influence on forest development, as the forest is too young to be subjected to mortality associated with

**Figure 9.1**
## Location of the study area in the Gaspé Peninsula of southeastern Québec, Canada

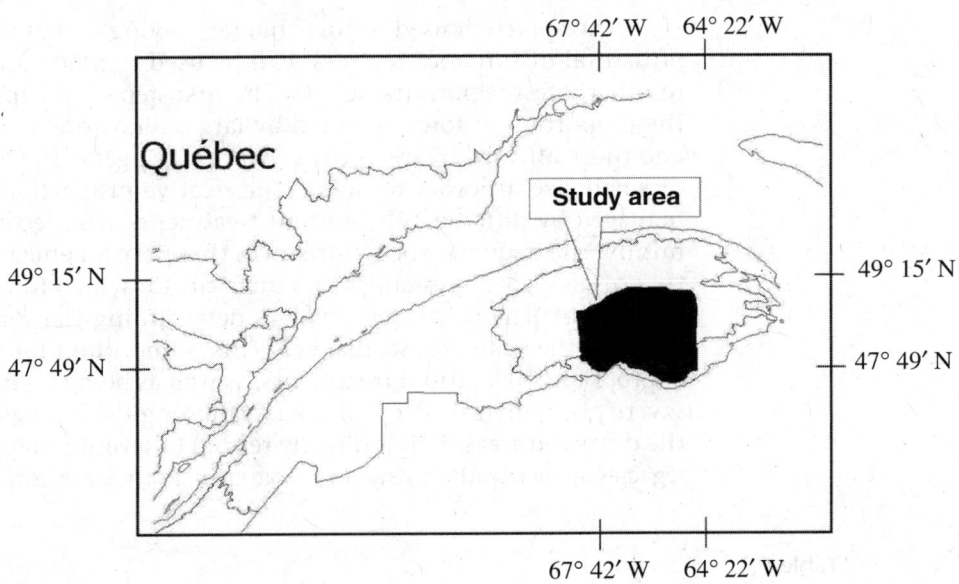

The study area covers 6,480 km² and encompasses both the balsam fir–white birch and balsam fir–yellow birch bioclimatic domains.

these disturbances. Trees are healthy and are less susceptible to windthrow when compared with mature trees. As stands get older, and in the absence of severe fire, these smaller-scale disturbances have greater effects on forest dynamics. Thus, mature and old-growth forests will be more influenced by spruce budworm (*Choristoneura fumiferana* Clem.) outbreaks, windthrow, and senescence.

Bergeron et al. (1999) proposed a conceptual model subdividing the forest into three structural types or cohorts (i.e., cohort 1: young stands; cohort 2: mature or over-mature stands; cohort 3: old-growth stands). Each cohort is subjected to a particular silvicultural treatment (also see figure 3.4, Gauthier et al., chapter 3), either clearcutting, partial cutting or selection cutting, in order to create smaller and smaller openings as stands advance through the stages. However, even if forest stands are regulated by natural disturbances creating openings of different sizes, the majority of stands do not follow a linear relationship over time; i.e., affected by large disturbances when young, moderate-severity disturbances when mature, and very small canopy openings in old-growth stands (Franklin et al. 2002). In fact, mature stands are often less vulnerable to moderate-severity disturbances than old-growth forest stands. Accordingly, canopy openings are often smaller in those stands than in older forests. For example, Kneeshaw (2001) demonstrated that the percentage of forest canopy openings, as well as the area of canopy gaps, increased with time since fire because of greater effects of disturbances such as spruce budworm outbreak in old forests than in younger forests. Consequently, we modified the initial conceptual model proposed by

Bergeron et al. (1999) in such a way that there would be a binary division of the forest, where the fire cycle can be used to determine the relative proportions of both even-aged and uneven-aged components of managed forests.

The approach used in this chapter, relying on the different characteristics of natural disturbance regimes, can be used to guide forest managers working in other forest regions (table 9.1). The first step of our approach is to determine the proportion of forest affected by large, catastrophic disturbances (i.e., fires) and those affected by other disturbances, i.e., generally less severe, of moderate or small size, in order to define the relative proportion of the territory to be managed by different silvicultural treatments. The second step requires determining the frequency of disturbances that have an effect on forest composition, thus allowing us to establish the intervals for which forestry operations can be carried out. The third step involves determining the spatial extent of affected areas for the suite of disturbances. This is important for the implementation of appropriate silvicultural treatments, as well as being an indicator of disturbance severity. A fourth stage consists of evaluating the biological legacies left within the disturbed areas. This is directly related to severity (step 3); however, biological legacies are generally considered separately, as they are critical for the maintenance

**Table 9.1**

## Important characteristics of natural disturbances and their relevance and applicability to forest management

| Characteristics of natural disturbances | Relevance and applicability to forest management |
|---|---|
| Proportion of landscape affected by different types of disturbance | The proportion of forest affected by each disturbance will serve as a guideline to determine the proportion of management units to be managed under different silvicultural systems or types of treatments. For example, the proportion of the region recently affected by fire will determine the proportion of the region to be managed by even-aged practices. |
| Frequency (return intervals) | The frequency of smaller-scale disturbances will determine intervals between interventions and, in the case of fire (as mentioned above), the percentage of the region assigned to even-aged management practices. |
| Severity | Severity at the canopy level can be determined by the proportion of trees killed within the disturbed area. This information helps us choose the silvicultural technique to be used at the stand scale. For smaller-scale disturbances, severity is often measured by size of the opening in the forest stand, as it is an indicator of the number of trees killed at this scale. |
| Size distribution | The surface area of natural disturbances allows us to suggest the size of an intervention (forest operation) in the forest region; i.e., size of clearcuts. |
| Regeneration composition after disturbance | Coupled with disturbance severity, this criterion helps to determine the silvicultural technique to employ. Coupled with disturbance frequency, regeneration composition also helps to determine the proportion of different interventions to use at the landscape scale. |
| Biological legacies (snags, coarse woody debris, remnant tree patches, etc.) | Residual structure after natural disturbance, particularly live and dead wood, is important for the maintenance of biodiversity. This criteria is linked to severity; however, it is treated separately to allow us to better refine the choice of structures to leave. |

of biodiversity and can be easily manipulated by forestry practices. Lastly, step five consists of determining the composition of the regeneration after disturbance, in order to determine if forest interventions are necessary to re-establish natural conditions.

## 3. FIRE REGIME OF THE CHALEUR BAY REGION OF THE GASPÉ PENINSULA

### 3.1. Considerations for the Determination of the Fire Regime

To evaluate the fire cycle, and therefore, the proportion of forest to be subjected to even-aged management dominated by fire-adapted or shade-intolerant species, several factors identified in the literature as influencing the variation in the fire cycle were considered (Lauzon et al. 2007). Among these, the biophysical characteristics of the territory have an important influence on the fire regime. At the regional scale, seasonal and meteorological conditions influenced fires by their effect on the precipitation regime (see Girardin et al. 2004; Girardin et al., chapter 4). At local and landscape scales, topography, altitude, surficial deposit type, vegetation type, degree of exposure, fuel availability, as well as the speed and direction of winds influenced susceptibility to fire (Dansereau and Bergeron 1993; Hély et al. 2000; Ryan 2002; Cyr et al. 2007). Climate is also responsible for spatio-temporal variations in the fire cycle (Flannigan and Van Wagner 1991; Stocks 1993; Girardin et al., chapter 4). In several regions across North America, climate change may be responsible for significant increases in fire cycles since the end of the Little Ice Age (LIA) ($\approx$1850) (Bergeron and Archambault 1993; Engelmark et al. 1994; Flannigan et al. 1998; Bergeron et al. 2001). Finally, the length of the fire cycle may also be influenced by anthropological factors. Indeed, the fire cycle has shortened in certain regions because of an insurgence of fires initiated by human activity (Johnson et al. 1990). Paradoxically, more recently, the fire cycle increased in other regions because of fire control and prevention measures (Wein and Moore 1977; Tande 1979).

We used forestry maps, aerial photos, and the age of trees (measured in the field) to determine the time since last fire in each forest stand. To estimate the fire cycle, we used survival analysis, as this takes into account that certain trees only represent a minimum time since last fire; i.e, early-succession, pioneer trees have often already disappeared (also called censured data). We also evaluated the effect of the environment, in terms of topography and time period considered, on the length of the fire cycle (Lauzon et al. 2007).

### 3.2. Fire Frequency

In the Gaspé Peninsula, the humid, coastal climate has led to the belief that fires are rare events in the region. However, according to the current age-class distribution (figure 9.2A), 52% of the boreal mixedwood forest of the Gaspé Peninsula originated from fires that occurred over the course of the last century while 48% of forests are more than 100 years old. Seventeen percent of the forest is older

### Figure 9.2
### Stand age-class distribution, and number and size of fires occurring in boreal mixedwoods of the Gaspé Peninsula

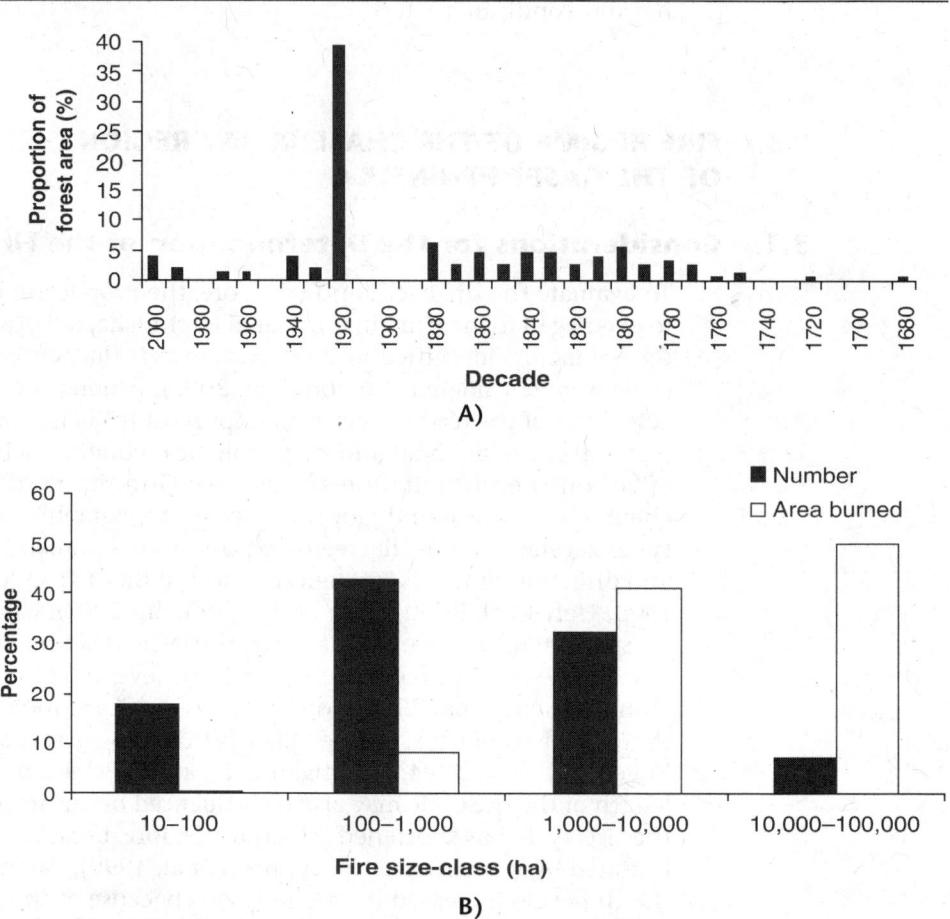

A)

B)

Legend: A) Distribution of stands age-classes according to proportion of area burned for each decade between 1680 and 2003; B) distribution of the number (*n* = 33) and size of fires (total area burned = 241,893 ha) between 1920 and 2003 in boreal mixedwoods of the Gaspé Peninsula. Note that even though more than half of the fires affected less than 1,000 ha, the majority of the area burned was the result of fires larger than 10,000 ha.

than 200 years, and 0.6% is older than 300 years. Between 1920 and 2003, 241,893 ha of forest burned, for an average of 2,914 ± 27,623 ha yr$^{-1}$, which is equivalent to a fire cycle of 250 yrs for this period (Lauzon et al. 2007).

Apart from a few intense fire episodes, the fire cycle has been relatively stable over the last three centuries. Nevertheless, survival analyses showed that a change in the length of the fire cycle occurred at the end of the LIA in our study region (Lauzon et al. 2007), which is in accordance with research from other regions of North America (Flannigan et al. 1998; Weir et al. 1999; Bergeron et al. 2006). For the period before 1850, the fire cycle is estimated at 89 years, whereas after 1850, it is estimated at 176 years. There are no significant differences between the balsam fir–white birch (96 years) and the balsam fir–yellow birch bioclimatic domains (145 years), or between the different topographical areas tested (valleys, low- and high-altitude areas). In addition, these results show that fire events are not as rare in the region as was previously believed.

A factor to consider is that since European colonization began in the 1700s, it has been difficult to gauge the effect of humans on the fire cycle. Fires of immense size have certainly affected the region, but their origin (anthropogenic or natural) remains unclear. However, during the early years of colonization, these European colonies generally lived and utilized lands close to the sea, reducing their potential impact in much of the study region, which occurs in-land. The fact that the fire cycle is much shorter than previously suggested in the scientific literature (cycles between 300 and 600 years; Lévesque 1997; Gauthier et al. 2001) suggests that humans may have had an important influence on fires in the region during recent centuries. On the other hand, an increase in fire frequency did not occur with colonization, but rather, an overall lengthening of the fire cycle can be observed as human activity increased. Climate changes towards a moister growing season that coincides with demographic growth suggest that the effect of climate is a dominant factor controlling the fire cycle.

Our analyses show that the fire cycle in the Gaspé Peninsula has varied over time, as well as with the calculation method used. In summary, the fire cycle varies between 89 and 250 years for the period between 1680 and 2003 (Lauzon et al. 2007). When calculating the fire cycle over the entire time frame, it is estimated at 161 years (Bergeron et al. 2006), which corresponds to the mean age of the territory when taking into account the data wherein only the minimum time since fire is known. This mean age has the advantage of incorporating the variations in the observed fire cycle over time, and is far less variable than the fire cycle itself. Using the mean age (or mean time since fire) to calculate the age-class distribution of the forest, based on a theoretical negative exponential curve (Van Wagner 1978), we suggest that 39% of the landscape could be submitted to even-aged silviculture if one considers 80-year rotations.

## 3.3.  Fire Severity and Size

The total surface area burned between 1920 and 2003 is 2,419 km² (241,893 ha). During the 1920s, a fire of exceptional size affected approximately 1,584 km² of the region. The second largest fire observed using aerial photos and eco-forestry maps occurred in 1995, affecting 231 km² of the territory, approximately 1/8 the size of the largest fire. These two fires were responsible for more than 50% of the area burned during this period, despite much smaller fires being far more numerous (figure 9.2B).

While we could not directly measure fire severity in all cases, we can reasonably infer that fire severities in the Gaspé Peninsula were similar to the data collected by Bergeron et al. (2002). Their data show, that on average, approximately 5% of the area affected by fire consists of preserved tree islands of varying size, while in the main areas of burned forest, various mixtures of live and burned trees are observed in different proportions. Moreover, data collected from recent fires allowed us to estimate that approximately 300 trees ha⁻¹ survived fire disturbance.

## 3.4.  Forest Composition and Regeneration after Fire

At the regeneration level, balsam fir shows a strong tendency to re-establish many years after fire. Non-commercial deciduous species such as speckled alder (*Alnus rugosa* [Du Roi] Spreng.), beaked hazel (*Corylus cornuta* Marsh.), and mountain maple (*Acer spicatum* Lam.) dominate the regeneration soon after disturbance (table 9.2). Subsequently, white birch and trembling aspen (*Populus tremuloides* Michx.) dominate the landscape 15 to 50 years after fire. Lastly, balsam fir, accompanied by black spruce (*Picea mariana* [Mill.] B.S.P.) in certain sites, dominates the stand when time since fire reaches approximately 70 years. Over the course of the last century, the proportion of white spruce (*Picea glauca* [Moench] Voss) has substantially decreased across the region because of insect epidemics (spruce bark beetle [*Dendroctonus* spp.] and the European spruce sawfly [*Diprion hercyniae*]). Black spruce is also less abundant than previously observed, possibly because of the reduction in fire frequency as fire tends to favour its regeneration. Reduced densities of white pine (*Pinus strobus* L.) and eastern white cedar (*Thuja occidentalis* L.) have also been observed in certain sites. Furthermore, current management practices such as Cutting with the Protection of Regeneration and Soils (CPRS)* tends to favour the regeneration of balsam fir (Laflèche et al. 2000), as well as trembling aspen (Fortin 1999), thus modifying the natural forest composition at the landscape scale.

• Cutting practice widely use in Québec that is similar to careful logging around advance growth (CLAAG) practised in Ontario.

Table 9.2
**Densities of various species groups according to time since fire in the study area***

| Time since fire (years) | Commercial deciduous species (stems ha⁻¹) | Non-commercial deciduous species (stems ha⁻¹) | Coniferous species (stems ha⁻¹) |
|---|---|---|---|
| 75 | 300 | 980 | 200 |
| 65 | 1,000 | 1,400 | 6,100 |
| 50 | 1,000 | 1,200 | 1,900 |
| 40 | 100 | 2,200 | 100 |
| 30 | 1,800 | 1,800 | 2,500 |
| 15 | 1,400 | 1,200 | 2,200 |

* Commercial deciduous species were primarily white and yellow birch, as well as *Populus* spp. and *Acer rubrum*; non-commercial species were composed of shrubs such as beaked hazel, alders, mountain maple, etc.

## 4.  MODERATE SEVERITY DISTURBANCES: SPRUCE BUDWORM OUTBREAK AND WINDTHROW

## 4.1.  Methodological Considerations

To estimate the proportion of the study region affected by spruce budworm outbreak and windthrow, eco-forestry maps produced in 1980 and 1990 were analyzed. Fifty aerial photographs from various locations across the study area were also analyzed. Lastly, since salvage logging operations occurred in the zones affected by natural disturbances, we adjusted our results to account for this by using data provided by the forest companies working in the region (Kneeshaw et al. 2003).

To evaluate moderate-severity disturbances, we compared aerial photographs taken before and after the last spruce budworm outbreak (1970 and 1987) for five sites. Photos were geo-referenced in order to superimpose stands over the two time periods and to subsequently determine the changes in forest composition and structure caused by the outbreak. Vegetation response, including that of the regeneration, was evaluated in 31 sites using 95, 20 × 20 m quadrats placed randomly in areas affected by windthrow and spruce budworm outbreak. Tree and shrub regeneration density were quantified according to height class (Reyes and Kneeshaw 2008).

## 4.2.  Proportion of the Study Area Affected and Disturbance Frequency

The percentage of the area affected by severe moderate-severity disturbances (i.e., disturbances between 0.5 and 60 ha and having more than 75% canopy mortality within the disturbed area, data for spruce budworm outbreak and windthrow combined) was estimated using aerial photos and eco-forestry maps. Approximately 12% of the area was affected, ranging between 5 to 15% (Kneeshaw et al. 2003). When considering light moderate-severity disturbances (i.e., those resulting in 50 to 74% canopy mortality within the disturbed area), approximately 10% of the area was affected, with size of disturbances reaching 100 ha (table 9.3). When combining light and severe moderate-severity disturbances, almost 25% of the territory has been affected. However, these disturbances do not occur very frequently. Spruce budworm outbreak occurs more frequently than windthrow, having occurred every 30 to 45 years over the course of the 20th century in the Gaspé Peninsula (de Römer et al. 2007). However, at the stand level, the most pertinent scale to silvicultural interventions, spruce budworm outbreaks only have significant impacts at 60- to 80-year intervals, essentially every other outbreak. Indeed, balsam fir needs at least 30 to 50 years to develop from the sapling to mature stages after spruce budworm outbreak (see Morin et al., chapter 6). Considering that outbreaks last approximately 15 years, there is less than 15 to 25 years between outbreaks, a time interval much too short for a new stand to be able to establish. This has also been observed in other regions of Québec (Jardon et al. 2003; Bouchard et al. 2006, 2007). Thus, silvicultural operations inspired by spruce budworm outbreak should have rotation periods of 70 to 80 years at the stand scale.

Partial and severe windthrow affect approximately 1.45% of the study area annually. Partial windthrow, which leaves many trees standing, affects approximately 1.29% of the study area, while 0.16% is affected by severe windthrow (more than 75% canopy tree mortality within the disturbed area). However, trees established in areas with steep slopes and thin soils are more susceptible to windthrow, while certain species such as balsam fir are more vulnerable to windthrow relative to other species such as yellow birch (Ruel et al. 1998). Therefore, mean area affected may not be indicative of local effects as variability in biotic and abiotic conditions across the landscape can influence susceptibility to windthrow.

Table 9.3

## Characteristics of the main natural disturbances occurring in the Gaspé Peninsula

| Category of opening size | Disturbance type | Frequency (return interval of disturbances) | Proportion of region affected | Disturbance size | Regeneration/ composition* |
|---|---|---|---|---|---|
| Large, catastrophic disturbances <br><br> Severe effects on soils and the canopy | Fire | Mean age: 161 years <br> Range: 89 to 250 years <br> 3,200 ha yr⁻¹ (interval of 80 years) <br> 0.49% yr⁻¹ | **Mean: 39%** <br> **Range: 27 to 59%** | 27 to 158,459 ha <br> Mean: 7,330 ha | BS and WB dominate <br><br> BF always present at low densities |
| Moderate-severity disturbances | Windthrow | 650 to 17,000 years (to affect all trees) <br> Partial windthrow: 1.29% yr⁻¹ <br> Severe windthrow: 0.16% yr⁻¹ | **Mean: 22%** <br> **Range: 11 to 29%** <br> These disturbances can be subdivided according to severity: <br> *Partial* (50 to 74% opening) <br> Mean: 10% of territory <br> Range: 6 to 14% | Partial windthrow 0.5 to 100 ha <br> Severe windthrow 0.5 to 60 ha | Variable, dependent on stand composition prior to disturbance 20 to 65% BF |
| | Spruce budworm outbreak | Regional scale: 30 to 45 years <br> Stand scale: 60 to 80 years | *Severe* (≥75% opening) <br> Mean: 12% of territory <br> Range: 5 to 15% | 0.5 to 45 ha | BF always dominates (71 to 89%) |
| Small disturbances | Canopy gaps (spruce budworm outbreak, windthrow, and senescence) | 40 to 140 years <br> 0.7 to 2.7% yr⁻¹ | **Mean: 39%** <br> **Range: 12 to 62%** | % canopy openness in forest stands <br><br> Mean: 42% <br> Range: 18 to 64% <br><br> Number of canopy gaps 75% < 20 m² <br><br> Canopy gap surface areas 25% > 225 m² | Large increase in BF (60 to 95%) |

* BS = black spruce (*Picea mariana* (Miller) BSP), WB = white birch (*Betula papyrifera* Marsh.), BF = balsam fir (*Abies balsamea* (L.) Mill.)

## 4.3.  Disturbance Size and Severity

Following the last spruce budworm outbreak, we observed, using aerial photos taken before and after the outbreak, an increase in canopy openness (%) in all types of forest stands (deciduous, mixed, coniferous), ranging between 25 and 60%, on average. The largest canopy openings (which represent more than 70% of the forest area disturbed) reach sizes between 29 and 45 ha. In mixed stands, canopy openness increased by 10% during the last spruce budworm outbreak, reaching 21%. The degree of openness in deciduous stands also increased from 11 to 23% during the last outbreak, then diminished to 3% during the period after the outbreak. Conifer stands were most affected, since they are composed of species with a greater vulnerability to spruce budworm outbreaks. Our results show that disturbances, forest openness and forest composition have not been

Figure 9.3
**Figure 9.3**
**Percentage of canopy openings greater than 1 ha resulting from spruce budworm outbreaks according to size class**

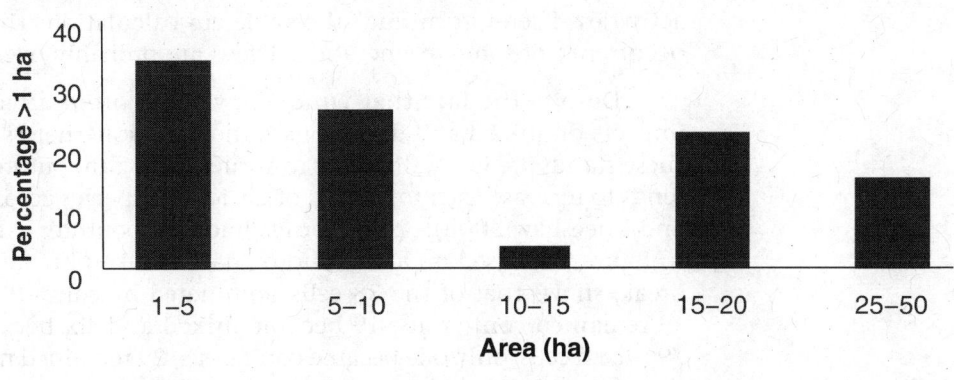

constant over time. The mean size of canopy openings was 13.4 ha, varying between 1 and 45 ha, and having substantial legacies of live trees within these openings (figure 9.3, table 9.3).

Openings caused by windthrow are similar in size relative to those created by spruce budworm outbreaks. Rarely do we observe complete canopy mortality after severe windthrow. Conifer stands are more often affected by windthrow than mixed or deciduous stands. Mean canopy openness in deciduous stands was 12% while mean canopy openness in coniferous stands was 62% after windthrow.

The size of openings and the presence of live trees within disturbed areas are characteristics common to areas disturbed by spruce budworm outbreak and windthrow. We also observed that coniferous stands were more greatly affected by these disturbances relative to mixed or deciduous stands. This is evident for spruce budworm outbreak, as balsam fir and the spruces are host species, but studies conducted in other regions of the boreal mixedwood forest observed considerable levels of mortality for certain deciduous species several years following budworm outbreaks. This is probably related to resulting changes in local environment conditions, for which these species could not adapt (Roy et al. 2001; Bouchard et al. 2005). In the region considered here, we observed an increase in canopy openness of approximately 10% in deciduous stands during the outbreak that also coincides with openings caused by windthrow in deciduous stands.

## 4.4.  Forest Composition and Regeneration after Spruce Budworm Outbreak and Windthrow

After the last budworm outbreak, the percentage of conifer cover was reduced from 77 to 10% while the proportion of mixed and deciduous cover increased from 19 to 21% and 1 to 16%, respectively. In these three stand types (deciduous, mixed, and coniferous), balsam fir is the conifer species that dominates the majority of regeneration layers, while white spruce and black spruce experienced significant declines. The proportion of white birch was maintained over time. Note that during the time period analyzed a portion of the area had not regenerated and remained open (figure 9.4). This may indicate that in some cases regeneration problems can occur after natural processes and not just after forestry activities. Therefore annual allowable cut calculations that do not consider the occurrence of some regeneration delays are probably over-optimistic.

Despite the fact that windthrow more often affects coniferous stands, impacts on mixed and deciduous stands are nonetheless important, given that these stands are less vulnerable to spruce budworm outbreak. Severe windthrow tends to increase the proportion of deciduous species across the landscape (Reyes and Kneeshaw 2008). As for spruce budworm outbreak, the example presented in figure 9.4, based on aerial photo analysis taken before and after the last outbreak, shows that of the 38 cells dominated by conifers prior to the outbreak, 13 remained coniferous, 19 became mixed and six became deciduous. Of the 79 mixed cells, only one became coniferous, 29 remained mixed, while 49 became deciduous. Lastly, of the 97 deciduous cells, none became coniferous, seven became mixed, while 90 cells remained deciduous.

**Figure 9.4**
**Changes in forest cover and composition as determined using aerial photos taken before and after the last spruce budworm outbreak for a representative forest stand**

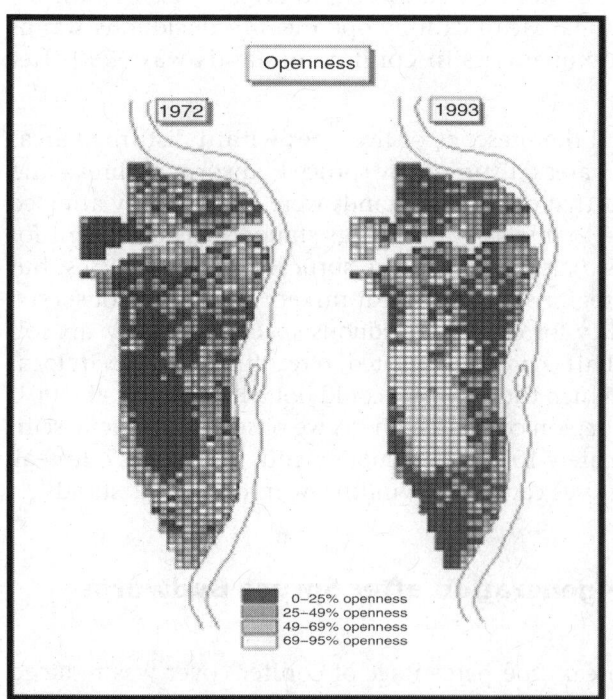 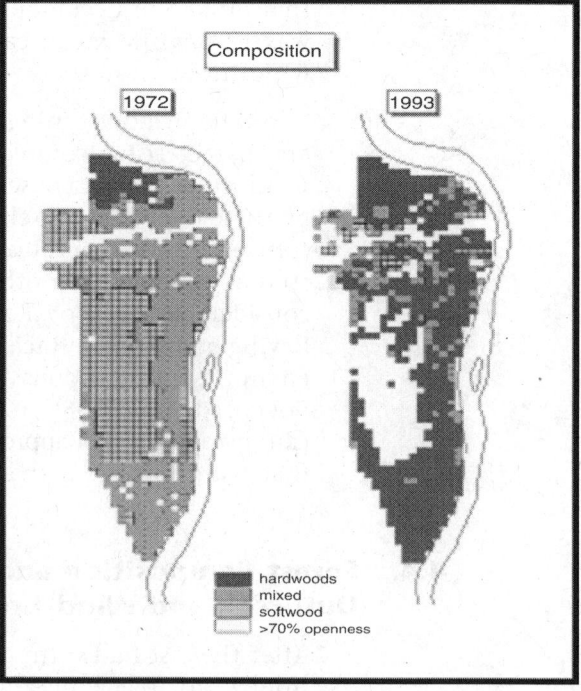

Note the increase in the openness of the forest cover as well as the reduction in conifer cover that is beneficial to the increase in deciduous cover.

Field data show that balsam fir, white birch, white spruce, rowan (*Sorbus americana*), mountain maple (*Acer spicatum*), and red maple (*Acer rubrum* L.) seedlings were more abundant after a spruce budworm outbreak than after windthrow. This is probably related to the greater quantity of coarse woody debris, which results in conditions that are not conducive for the germination and establishment of shade-intolerant species (Reyes and Kneeshaw 2008). The sudden canopy openings resulting from windthrow, compared to the gradual nature of mortality after spruce budworm defoliation, also favours shade-intolerant species. However, these changes in forest composition appear to be temporary, as over time, balsam fir dominates stands affected by windthrow.

## 5.  CANOPY GAP DISTURBANCES: SPRUCE BUDWORM OUTBREAK, WINDTHROW, AND SENESCENCE

### 5.1.  Methodology for Evaluating Canopy Gaps

The proportion of the region affected by canopy gap disturbances was only evaluated in mature and old-growth forests, given that these disturbances do not occur in young, even-aged stands (i.e., cohort 1 – Bergeron et al. 1999). Therefore, we excluded areas burned within the last 100 years from our analyses. We then identified and excluded the areas affected by moderate-severity disturbances; i.e., spruce budworm outbreak and windthrow disturbances causing more extensive canopy mortality (as described in the previous section). Subsequently, the remaining portion of the landscape was classified as undergoing a canopy gap disturbance regime (de Römer et al. 2007).

Evaluation of the canopy gap disturbance regime was based on 27 sites situated within the balsam fir–white birch and balsam fir–yellow birch bioclimatic domains (de Römer et al. 2007). These forest stands have old-growth attributes, having not been disturbed by fire or other severe events for more than 150 years. Within each site, canopy gaps traversed by transects between 150 and 500 m long were sampled to determine the proportion of canopy openness (ratio of transect under forest canopy *vs.* within a canopy gap). For each site, biotic and abiotic factors such as species composition and size of trees surrounding canopy gaps, percent slope, soil type, and exposure were quantified (de Römer et al. 2007). A sub-sample of 101 canopy gaps was more intensively examined to characterize canopy gap size, shape, as well as the composition, height, and regeneration density. Methodological details can be found in de Römer et al. (2007).

### 5.2.  Proportion of the Study Area Affected and Disturbance Frequency

Canopy gaps are omnipresent in forest stands of the Gaspé Peninsula. However, their effects are more prevalent in older forest stands. For example, Kneeshaw (2001) observed that the proportion of forest under canopy gap substantially increased with stand age. However, other research has suggested that canopy gaps also play a critical role during the transition from young to mature forest (Hill et al. 2005). Our methodology does not address this transition, thus potentially under-estimating the influence of canopy gaps within the area.

We estimate that approximately 39% of the forest area is regulated by canopy gap disturbances (table 9.3). Although individual canopy gaps affect relatively small areas (<1 ha), a large proportion of the forest is open within these old-growth forests (mean of 42%). This proportion is quite variable, ranging between 18 and 64%. According to our estimates, the rates of opening and closing of canopy gaps are rather quick, given that the majority of stands are dominated by balsam fir, a relatively short-lived species (varying between 40 and 140 years) which is highly vulnerable to insect outbreaks, windthrow, and disease. Variation in frequency is mostly dependent on stand composition (relative proportions of the various species) and local abiotic conditions. Therefore, even if individual canopy openings are not large, the frequency and number of openings (see next section) indicate that the forest is constantly changing.

## 5.3.  Disturbance Size and Severity

Canopy gaps were generally small (mean of 28 m$^2$, range between 3 and 697 m$^2$), and the size distribution followed a negative exponential (de Römer et al. 2007). The mean proportion of a forest stand under canopy gap (42%) shows that high mortality rates occur relative to other boreal and temperate forest regions (McCarthy 2001). The last spruce budworm outbreak had a large impact on all stand types. On average, 65% of canopy gaps were caused at least in part by budworm activity compared with 37% in which windthrow was observed, and 36% in which unexplained mortality was observed (one cause does not exclude the others, de Römer et al. 2007). The stands that were most disturbed were affected by windthrow, and contained a relatively important composition of black spruce before and after disturbance. Altitude and percent slope did not significantly affect disturbance severity in our study region. In contrast to results from other studies examining canopy gap dynamicss, pure stands of balsam fir were less disturbed on average by the last budworm outbreak in terms of canopy gaps compared with balsam fir stands mixed with white spruce and tolerant deciduous species (primarily yellow birch [*Betula alleghaniensis* Britton]) or black spruce (Morin 1994; Kneeshaw and Bergeron 1998). Although this result may be biased at least in part if salvage operations were concentrated in pure conifer stands following the last outbreak. Lastly, 0.7 to 2.7% of the forest area was annually disturbed by canopy gaps, with a return interval of 56 years.

## 5.4.  Forest Composition and Regeneration in Canopy Gaps

Because balsam fir is present in all forest types and canopy gaps of all dimensions, one can assert that, in contrast with other regions (Messier et al. 2005; Bouchard et al. 2006), the canopy gap disturbance regime favours balsam fir in the Gaspé Peninsula. White birch is generally maintained, but both white spruce and black spruce are clearly less abundant in the regeneration layer relative to the tree canopy layer. Some white birch saplings in gaps are more than 70 years old, which implies that this species, while considered intolerant to shade, does not necessarily require a large opening in the forest canopy for survival. This is also supported by the fact that this species was found in the smallest canopy gaps (<10 m$^2$).

Small disturbances in the canopy have important impacts on forest development as they significantly influence several parameters associated with regeneration. For example, our results indicate that when an opening is formed, the rate of sapling growth increases by a factor of eight on average, while density increases as a function of canopy gap size. White birch density was also greater in larger canopy gaps, indicating that the composition of the regeneration also varies as a function of canopy gap size.

## 6.   COMPARISON OF CPRS AND SPRUCE BUDWORM OUTBREAK

It is often suggested that different harvesting techniques emulate different natural disturbances. For example, the idea that CPRS can, to some extent, emulate the effects of spruce budworm outbreak has been raised. While several studies have presented evidence eliciting the similarities and differences between clearcutting and fire (McRae et al. 2001; Haeussler and Kneeshaw 2003; Claveau et al. 2007), comparisons have yet to be made for other disturbance types or harvesting techniques. It is important to make these comparisons to determine to what extent harvesting practices respect the variability created by natural disturbances. Essentially, we need to identify the differences in responses to harvesting practices relative to those observed after natural disturbances. We thus evaluated the similarities and differences between CPRS and severe spruce budworm outbreaks immediately after and 15 years after disturbances in terms of spatial characteristics, regeneration, and biological legacies. This comparison allowed us to predict the direction for which the forest evolves following natural and anthropogenic disturbances, and for the evaluation of the resilience of the system undergoing silvicultural treatments such as CPRS.

We made the comparison between CPRS harvesting and areas severely affected by a spruce budworm outbreak at stand and landscape scales (Belle-Isle and Kneeshaw 2007). Compared to spruce budworm outbreak, CPRS produced much larger openings in the forest cover and had a far more aggregated distribution across the landscape. In contrast, the shape of harvested areas was similar to that created by budworm outbreak. Patterns of tree mortality were also similar, although budworm outbreak left more large, live legacy trees behind, most being white spruce and white birch. The spruce budworm produced openings that were more partial and gradual in nature relative to CPRS, thus favouring shade-tolerant species such as balsam fir. Moreover, a greater number of shade-intolerant species such as white birch, pin cherry (*Prunus pensylvanica* [L.] f.) and invasive herbaceous species, were sampled in CPRS sites than in those affected by spruce budworm outbreak. Pre-commercial thinning practices (a silvicultural treatment that selects and releases young future canopy trees from competing species) used ten years after harvest have probably accelerated the process of reduced diversity, as the highest level of tree species diversity observed after CPRS seems to be a temporary condition produced before canopy closure. In fact, when examining only regenerating stems with a diameter at breast height (DBH) larger than 4 cm, we noticed a reduction in the proportion of shade-intolerant species as well as rare and infrequent species (e.g., white spruce) in the areas managed with CPRS. Tree diversity is thus relatively higher in stands affected by spruce budworm outbreak, primarily because of the greater presence of white spruce and white birch. Spruce budworm outbreak also favours a more complex stand structure, at least with respect to horizontal structure, with a greater density of snags, live legacy trees, and coarse woody debris as compared to stands affected by CPRS.

We established that important differences exist at several scales, between managed stands and those affected by spruce budworm outbreak. Modifying our harvesting practices can minimize several of these differences. For example, the use of smaller cuts dispersed throughout the forest landscape in situations wherein strategies are guided by insect outbreaks rather than intense fires could be considered. At the stand scale, a strategy aiming to maintain species diversity and residual trees will help to minimize the difference between conditions created by management practices and those found in natural forests.

## 7.  MANAGEMENT STRATEGIES AND SILVICULTURAL PRACTICES SUGGESTED FOR THE GASPÉ PENINSULA

### 7.1.  General Considerations

In the following section, we use the knowledge acquired from natural disturbance regimes to propose a FEM strategy that can be applied to the Gaspé Peninsula. While not the only possible strategy, this approach should help to re-establish the balance between current forestry practices and the natural dynamics of forests. The issues addressed here loosely follow the recommendations by Hunter (1993), as well as those from the first section of this book, which consider frequency, size, and biological legacies as key factors to emulate and integrate into forest practices in order to bring forestry closer to nature. Table 9.3 should thus serve as a basic tool for the development of FEM in other regions (Vaillancourt et al. 2009).

The first stage of our approach consists of determining the proportion of forest that should be placed under even-aged management, based on the proportion of natural forest younger than the mean regional fire return interval. We then consider the size of fires, the density of residual trees left within burned areas, and the response of the regeneration that will determine future stand composition.

A similar approach is used for older forests. However, there are several important considerations that vary according to the disturbance type and the stage of development targeted. In fact, past work related to management based on natural disturbances recognized, in an implicit or explicit manner, that there must be different scales of observation, planning, and execution (Haeussler and Kneeshaw 2003; see also Gauthier et al., chapter 1). Here we stress that the importance of each scale will change depending on the disturbance type considered. For example, the proportion of the landscape under different silvicultural regimes (even-aged *vs.* uneven-aged) will solely be determined by fire frequency. Moderate-severity disturbances affecting areas smaller than fires and intersected by contiguous forest canopy more strongly influence mature and over-mature forests (MacLean 1980; Kneeshaw 2001). Their proportion is therefore concentrated within zones of older forest, but at the stand scale they can create even-aged patches of several hectares (Watt 1947). At the temporal level, their frequency does not have a known influence, as is the case for fire frequency (however see Bouchard et al. 2008). However, they can have a key role in determining species diversity and abundance across the region (Woods 2000).

Lastly, canopy gaps have a very local impact in terms of direct effects. They are always present in forest stands wherein stand-replacing disturbances have not occurred for long periods of time (mature and old-growth forests). The fact that these forests are abundant underlines the importance of the canopy gap regime, and thus that the use of small-opening harvests is a must on a large proportion of the study area. Table 9.3, for example, shows that almost 40% (on average) of the area is naturally influenced by a canopy gap regime. The frequency of canopy gap occurrence can then be used to determine the frequency of silvicultural interventions used to emulate small gap disturbances. It should be noted that the use of canopy gap frequency differs from fire frequency estimates, since it is not to be used in determining overall management strategies; rather, the canopy gap frequency should provide silvicultural guidelines for planning forest rotations within stands.

## 7.2.   Suggested Management Guidelines Specific to the Gaspé Peninsula

### 7.2.1.   Proportion of the Landscape under Even-Aged Management

The homogenization of the forest landscape and the agglomeration of large harvest blocks encompassing tens to hundreds of hectares is a consequence of the clearcutting regime. Moreover, these practices have an amplitude of variation lying well beyond the limits of variation observed for the natural disturbance regime of the Gaspé Peninsula. This situation has caused substantial changes to the forest age-class structure, so that it is now dominated by younger stands. According to the proportion of young and old stands in natural forests, between 27 and 59% (mean of 39%) of the area should be managed with even-aged silviculture, when considering a rotation age of 80 years. We can thus determine the management target in function of this average, which is equivalent to harvesting 0.49% of the area using an even-aged silvicultural regime. This approach should however consider that disturbances such as catastrophic fire will continue to occur, and that the synergy between clearcut harvesting (CPRS) and fires will contribute to an increase in the proportion of the area occupied by first-cohort stands. Accordingly, more than half of the territory should not be managed with large clearcuts since a large proportion is subjected to moderate-severity or canopy gap disturbances.

Natural variation in the fire cycle is also an important consideration. Consequently, management objectives should not be fixed on a single threshold level, as this will result in the homogenization of the forest, but rather aimed to vary the size and frequency of interventions within the natural variability historically observed (Haeussler and Kneeshaw 2003; see Vaillancourt et al., chapter 2). In other words, spatial, temporal, and compositional differences should be considered when planning silvicultural operations. Therefore, forest managers need to have the latitude or freedom to reproduce a forest composition and structure that varies over time. In the same manner, we should not aim to achieve minimum or maximum values of management targets, in order to retain some flexibility for dealing with unexpected occurrences, such as the event a large fire, for instance.

Our data on burned area sizes indicate that 17% of the first-cohort forests should be managed using harvests ranging from 10 to 100 ha, 42% ranging from 100 to 1,000 ha, and 32% from 1,000 to 10,000 ha. We must also keep in mind that fire leaves a large quantity of residual forest cover (see Perron et al., chapter 6). Therefore, it is imperative to plan the retention of trees and residual islands into clearcut zones, as they act as seed sources for the re-colonization of clearcut areas and as refuges for fauna associated with older forests. The retention of residual trees, when well planned, contributes to the accelerated development of irregular forests.

It is also worth noting that extreme events such as large-scale fires do not have to be reproduced by clearcutting since they are extremely rare, and tend to occur despite efforts for their control. A goal of emulating these large fires would in all probability lead to a duplication of the frequency and area affected by these extreme events. Nevertheless, a reduction in fire frequency has been observed in various regions of Québec. One can thus presume that certain forest areas that would have burned under the former fire regime will remain untouched. This lengthening of the fire cycle can provide a larger degree of flexibility for forest managers (Bergeron et al. 2006; see Gauthier et al., chapter 3). However, despite the potential to replace fires with harvesting it is crucial that we consider the additive effects of natural disturbances and forest harvesting if we want to avoid creating forest conditions that lie outside the natural range of variability.

In terms of regeneration, fires are necessary for the regeneration of fire-adapted species such as black spruce and trembling aspen. In the Gaspé Peninsula, contrary to other regions of the boreal mixedwood zone (Frelich and Reich 1995; Kneeshaw and Bergeron 1998; Bouchard et al. 2006), the absence of fire leads to the homogenization of stands. In other words, small openings do not ensure tree diversity. Thus, this stresses the necessity of tailoring FEM strategies to local characteristics. Applying a uniform strategy across a forest type will not respect these differences.

In the Gaspé Peninsula, an important issue is to ensure that the natural proportions of species such as white spruce and black spruce, which have been decreasing in abundance over the course of the last century, are maintained. Data comparing CPRS with spruce budworm outbreak demonstrate that this silvicultural method does not allow for the recruitment of these fire-adapted species. At an operational scale, CPRS favours shade-tolerant species, which are present as advance regeneration. Thus, CPRS is not a technique that imitates the effects of fire at the stand scale. A strategy needs to be put in place that ensures the recruitment of these species. Other research undertaken in the region suggests that the reduced distribution of species such as white pine and eastern white cedar needs specific treatments (i.e., a fine filter approach) to ensure that they are represented according to natural historic proportions in managed forests.

## 7.2.2.   Proportion of the Landscape under Uneven-Aged Management

Our results on moderate-severity disturbances suggest that almost a quarter (22%) of the area should be managed with harvest operations that remove more than 50% of the canopy, and have sizes that vary between 0.5 and 100 ha. Half of these operations should be severe (12% of the area, >75% canopy removal) and the other half using less severe techniques (10% of the area, between 50 and 75% canopy removal). Finally, close to 40% of the area should consist of older forest, managed using a harvest regime of smaller interventions (<1 ha).

### Moderate-Severity Disturbances

Severe canopy-level disturbances, i.e. those that kill the majority of trees, are far rarer than disturbances that leave many live trees in place. To simulate the effects of severe spruce budworm outbreak, CPRS or shelterwood cutting are suggested, with the recommendation that foresters leave between 10 and 20% residual forest. These residual zones play an important role for the maintenance of bio-diversity (Eberhart and Woodward 1987; DeLong and Kessler 2000; Doyon and Sougavinski 2003). Following a severe spruce budworm outbreak, this residual forest is always composed of white birch, white cedar, and yellow birch. In order to emulate the natural outbreak regime, 30% of the harvested areas should be <10 ha, and 70% between 10 and 45 ha (table 9.3).

In terms of application in the field, one could base interventions on the range of openings associated with budworm outbreak and windthrow for forests having a large proportion of balsam fir, a species highly vulnerable to both of these disturbance agents, and on canopy gaps for forests with high densities of yellow birch, a long-lived species that naturally re-establishes in canopy gaps (Kneeshaw and Prévost 2007). If these suggestions are applied, it will then be necessary to ensure that the consequences of these actions do not lead to forest system imbalance in a direction that has never before been seen. As our results show that balsam fir dominates within canopy gaps (de Römer et al. 2007), a possible concern is that yellow birch is limited to more southern latitudes of the region. Thus, in light of maintaining species diversity of the forest, one has to ensure the maintenance of spatial and temporal heterogeneity of the forest by using a variety of silvicultural treatments of different sizes.

CPRS or shelterwood harvests could emulate windthrow, particularly in conifer stands, on the condition that an approach with variable retention is applied in order to leave a number of live trees in place. To remain within the range of openings created naturally, 60% of harvest areas should be <10 ha and 5% > 50 ha. Only a small portion of the mixed and deciduous forest should be treated with other harvesting techniques such as selection harvest, less frequent periodic harvesting (see Bouchard, chapter 13), or other types of harvesting techniques that maintain some forest cover, as severe windthrow is relatively rare. However, considering that it is more difficult to prevent windthrow than fire or insect epidemics, foresters should reflect on which factors would be appropriate to emulate. One of the guiding principles of ecosystem management is that native species are adapted to conditions found under natural disturbance regimes (Gauthier et al., chapter 1; Franklin 1993). Therefore, our goal should be to ensure the maintenance of conditions created by natural disturbances. For example, windthrow can create germination beds on mineral soils that are not (or rarely) found after CPRS. These microsites are very important for small-seeded species such as yellow birch (Smith et al. 1996; Nyland 2002; Kneeshaw and Prévost 2007). Moreover, certain animal species associated with closed forests are able to relocate more easily according to the variety of openings, and find better habitats in the structure left after windthrow than after traditional clearcut harvests (Hunter 1999; Buskirk et al. 2000; Courtois 2003).

### Canopy Gap Disturbances

Old-growth forests, where tree senescence is important, should be managed using small harvests. With respect to the canopy gap disturbance regime, natural conditions suggest that harvesting dimensions should be variable, with the majority being very small. This approach implies that the reduced size of harvested opening is compensated by more frequent harvesting intervals (table 9.3). In fact, a large proportion of the forest is naturally open and recruitment and mortality rates are very high. However, as previously mentioned, the combined effects of harvesting and natural disturbances have to be evaluated to ensure that forestry activities do not produce additive effects that lie outside the range of natural variability.

In this section, we have suggested a variety of silvicultural treatments, for which their frequency and spatial extent are inspired by the diverse natural disturbances found in our study area. It is important to stress that forest managers are not limited to our suggested guidelines, but rather need to show intuition and creativity since the objective is to develop and use forest management practices that respect natural threshold levels, and which maintain or recreate the heterogeneity found in the Gaspé Peninsula. Even if a better understanding of the variability of natural disturbances helps us to improve forest practices, this does not guarantee that a FEM strategy will have the predicted success for the maintenance of biodiversity in the system. Thus, this underlines the need to closely monitor forest management activities (see Drapeau et al., chapter 14). Other considerations, such as the restoration of species that are becoming increasingly rare, the influence of road networks (Trombulak and Frissell 2000; Bourgeois et al. 2005), conflicts with other forest uses (Kneeshaw and Gauthier 2006), as well as public perception, have to be taken into consideration. Interactions between natural disturbances, e.g., severe fire and insect epidemics (Stocks 1987), or between harvesting and natural disturbances, e.g., clearcutting and spruce budworm outbreak (Blais 1983), should also be monitored as they could generate unforeseen or novel conditions. All these factors point to the fact that an adaptive and flexible management approach is necessary in order to be able to integrate future knowledge and to appropriately respond to unforeseen events.

## 8.   CONCLUSIONS

Our work supports the establishment of forest management practices that respect natural conditions. Forestry based on only a single management strategy does not account for the variation in conditions naturally observed. Forests in the Gaspé Peninsula are shaped by a variety of natural disturbances. Furthermore, a large degree of variation is characteristic of each disturbance type, as several factors can influence disturbances and, subsequently, the different scales under consideration (stand and landscape). There are, however, certain important trends that we have observed. For example, in the absence of large catastrophic fires and windthrow, forests are dominated by balsam fir and contain a variety of structural features. Fire and windthrow are important for the establishment and recruitment of pyrogenous and shade-intolerant species. Thus, it is the variety of disturbances that ensure the compositional and structural variation found in the forests of the Gaspé Peninsula.

In this chapter, we examined disturbance frequency, the proportion of the area affected by each disturbance type, biological legacies, and information on future stand composition as represented by response of the regeneration to each disturbance. All of these factors vary within and among disturbance types. The approach that we proposed could easily be reproduced in other regions. Current forestry practices, by extensively using only a single silvicultural technique, have a tendency to homogenize the forest landscape. The comparison made in this chapter, between the effects of CPRS and spruce budworm outbreak, suggest that current silvicultural strategies and practices can be slightly modified to address the differences observed. Thus, it is imperative that changes which bring forest management closer to natural dynamics occur, while at the same time ensuring that new forest management practices respect natural processes, maintain biodiversity, and limit the combined effects of harvesting and natural disturbance. It is, therefore, important to have an understanding of the collection of natural factors if we wish to make effective, worthwhile changes.

## REFERENCES

Belle-Isle, J. and Kneeshaw, D.D. 2007. Comparison of the effects of a spruce budworm (*Choristoneura fumiferana* (Clem.)) outbreak to the combined effects of harvesting and thinning at stand and landscape scales. For. Ecol. Manag. **246**: 163–174.

Bergeron, Y. and Archambault, S. 1993. Decreasing frequency of forest fires in the southern boreal zone of Québec and its relation to global warming since the end of the "Little Ice Age." Holocene, 3: 255–259.

Bergeron, Y., Cyr, D., Drever, C.R., Flannigan, M., Gauthier, S., Kneeshaw, D., Lauzon, È., Leduc, A., Le Goff, H., Lesieur, D., and Logan, K. 2006. Past, current, and future fire frequency in Quebec's commercial forests: implications for sustainable forest management. Can. J. For. Res. **36**: 2737–2744.

Bergeron, Y., Gauthier, S., Kafka, V., Lefort, P., and Lesieur, D. 2001. Natural fire frequency for the eastern Canadian boreal forest: consequences for sustainable forestry. Can. J. For. Res. **31**: 384–391.

Bergeron, Y., Harvey, B., Leduc, A., and Gauthier, S. 1999. Stratégies d'aménagement forestier qui s'inspirent de la dynamique des perturbations naturelles: considérations à l'échelle du peuplement et de la forêt. For. Chron. **75**: 55–61.

Bergeron, Y., Leduc, A., Harvey, B., and Gauthier, S. 2002. Natural fire regime: a guide for sustainable forest management in the Canadian boreal forest. Silva Fenn. **36**: 81–95.

Blais, J.R. 1983. Trends in the frequency, extent, and severity of spruce budworm outbreaks in eastern Canada. Can. J. For. Res. **13**: 539–547

Bouchard, M., Kneeshaw, D., and Bergeron, Y. 2005. Mortality and stand renewal patterns following the last spruce budworm outbreak in mixed forests of western Quebec. For. Ecol. Manag. **204**: 297–313.

Bouchard, M., Kneeshaw, D.D., and Bergeron, Y. 2006. Forest landscape composition and structure after successive spruce budworm outbreaks. Ecology, **87**: 2319–2329.

Bouchard, M., Kneeshaw, D.D., and Bergeron, Y. 2008. Ecosystem management based on large-scale, episodic disturbances: a case study from sub-boreal forests. For. Ecol. Manag. **256**: 1734–1742.

Bouchard, M., Kneeshaw, D., and Messier, C. 2007. Forest dynamics following spruce budworm outbreaks in the northern and southern mixedwoods of central Quebec. Can. J. For. Res. **37**: 763–772.

Bourgeois, L., Kneeshaw, D.D., and Boisseau, G. 2006. Les routes forestières: le Québec doit considérer les impacts. Vertigo, **6**: 1–9.

Buskirk, S.W., Ruggiero, L.F., and Krebs, C.J. 2000. Habitat fragmentation and interspecific competition: implications for lynx conservation. *In* Ecology and conservation of lynx in the United States. *Edited by* L.F. Ruggiero et al. University Press of Colorado, Boulder, Colorado, USA, pp. 83–100.

Claveau, Y., Kneeshaw, D., and Gauthier, S. 2007. Nos pratiques s'inspirent-elles vraiment des feux? L'Aubelle, **151**: 14–21.

Courtois, R. 2003. La conservation du caribou forestier dans un contexte de perte d'habitat et de fragmentation du milieu. PhD thesis, Université du Québec à Rimouski, Rimouski, Que.

Cyr, D., Gauthier, S., and Bergeron, Y. 2007. Scale-dependent influence of topography on fire frequency in a coniferous boreal of eastern Canada. Landsc. Ecol. 22: 1325–1339.

Dansereau, P.-R. and Bergeron, Y. 1993. Fire history in the southern boreal forest of northwestern Quebec. Can. J. For. Res. **23**: 25–32.

DeLong, S.C. and Kessler, B.W. 2000. Ecological characteristics of mature forest remnants left by wildfire. For. Ecol. Manag. **131**: 93–106.

Denslow, J.S. 1987. Tropical rain forest gaps and tree species diversity. *In* The ecology of natural disturbance and patch dynamics. *Edited by* S.T.A. Pickett and P.S. White. Academic Press, New York, USA, pp. 307–323.

De Römer, A.H, Kneeshaw, D.D., and Bergeron, Y. 2007. Small gap dynamics in the southern boreal forest of eastern Canada: do canopy gaps influence stand development? J. Veg. Sci. **18**: 815–826.

Doyon, F. and Sougavinski, S. 2003. La rétention variable: un outil de sylviculture écosystémique. L'Aubelle, **144**: 13–16.

Eberhart, K.E. and Woodward, P.M. 1987. Distribution of residual vegetation associated with large fires in Alberta. Can. J. For. Res. **17**: 1207–1212.

Engelmark, O., Kullman, L., and Bergeron, Y. 1994. Fire and age structure of Scots pine and Norway spruce in northern Sweden during the past 710 years. New Phytologist, **126**: 163–168.

Flannigan, M.D., Bergeron, Y., Engelmark, O., and Wotton, B.M. 1998. Future wildfire in circumboreal forests in relation to global warming. J. Veg. Sci. **9**: 469–476.

Flannigan, M.D., and Van Wagner, C.E. 1991. Climate change and wildfire in Canada. Can. J. For. Res. **21**: 66–72.

Fortin, S. 1999. Expansion du tremble (*Populus tremuloides* Michx.), au cours du XXᵉ siècle, dans le bassin de la rivière York, en Gaspésie, Québec. M.Sc. thesis, Université du Québec à Chicoutimi, Chicoutimi, Que.

Franklin, J.F. 1993. Preserving biodiversity: species, ecosystems, or landscapes? Ecol. Appl. **3**: 202–205.

Franklin, J.F., Spies, T.A., Van Pelt, R., Carey, A.B., Thornburgh, D.A., Berg, D.R., Lindenmayer, D.B.,

Harmon, M.E., Keeton, W.S., Shaw, D.C., Bible, K., and Chen, J. 2002. Disturbances and structural development of natural forest ecosystems with silvicultural implications, using Douglas-fir forestry as an example. For. Ecol. Manag. **155**: 399–423.

Frelich, L.E. 2002. Forest dynamics and disturbance regimes: studies from temperate evergreen forests. Cambridge University Press, Cambridge, New York, USA.

Frelich, L.E. and Reich, P.B. 1995. Spatial patterns and succession in a Minnesota southern-boreal forest. Ecol. Mon. **65**: 325–346.

Gauthier, S., Leduc, A., Harvey, B., Bergeron, Y., and Drapeau, P. 2001. Les perturbations naturelles et la diversité écosystémique. Nat. Can. **125**: 10–17.

Girardin, M.P., Tardif, J., Flannigan, M.D., and Bergeron, Y. 2004. Multicentury reconstruction of the Canadian drought code from eastern Canada and its relationship with paleoclimatic indices of atmospheric circulation. Clim. Dynamics, **23**: 99–115.

Haeussler, S. and Kneeshaw, D.D. 2003 Comparing forest management to natural processes. *In* Towards sustainable management. *Edited by* P.J. Burton, C. Messier, D.W. Smith, and W.L. Adamowicz. NRC Research Press, Ottawa, Ont., pp. 307–368.

Heinselman, M.L. 1973. Fire in the virgin forest of the Boundary Waters Canoe Area, Minnesota. Quat. Res. **3**: 329–382.

Hély, C., Bergeron, Y., and Flannigan, M.D. 2000. Effects of stand composition on fire hazard in mixed-wood Canadian boreal forest. J. Veg. Sci. **11**: 813–824.

Hill, S.B. Mallik, A.U., and Chen, H.Y.H. 2005. Canopy gap disturbance and succession in trembling aspen dominated boreal forests in northeastern Ontario. Can. J. For. Res. **35**: 1942–1951.

Hunter, M.L. Jr. 1993. Natural disturbance regimes as spatial models for managing boreal forests. Biol. Conserv. **65**: 115–120.

Hunter, M.L. Jr. (Ed.). 1999. Maintaining biodiversity in forest ecosystems. Cambridge University Press, Cambridge, NY, USA.

Jardon, Y., Morin, H., and Dutilleul, P. 2003. Périodicité et synchronisme des épidémies de la tordeuse des bourgeons de l'épinette au Québec. Can. J. For. Res. **33**: 1947–1961.

Johnson, E.A., Fryer, G.I., and Heathcott, M.J. 1990. The influence of man and climate on frequency of fire in the interior wet belt forest, British Columbia J. Ecol. **78**: 403–412.

Kneeshaw, D.D. 2001. Are non-fire gap disturbances important to boreal forest dynamics? *In* Recent research developments in ecology 1. *Edited by* S.G. Pandalarai. Transworld Research Press, pp. 43–58.

Kneeshaw, D. and Bergeron, Y. 1998. Canopy gap characteristics and tree replacement in the southeastern boreal forest. Ecology, **79**: 783–794.

Kneeshaw, D.D., Bergeron, Y., Harvey, B. Grenier, D., Bouchard, M., Lauzon, E. de Romer, A., D'Aoust, V.,

Senecal, D., and Messier, J. 2003. Rapport sur le projet écosystémique pour Tembec au Témiscamingue et en Gaspésie. Grefi, UQAM, Montréal, Que.

Kneeshaw, D.D. and Gauthier, S. 2006. Accessibilité forestière accrue: panacée ou boîte de Pandore? Téoros, **25**: 36–40.

Kneeshaw, D.D., Kobe, R., Coates, D., and Messier, C. 2006. Sapling size influences shade tolerance ranking among southern boreal tree species. J. Ecol. **4**: 471–480.

Kneeshaw, D.D. and Prévost, M. 2007. Natural canopy gap disturbances and their role in maintaining mixed species forests of central Québec, Canada. Can. J. For. Res. **37**: 1534–1544.

Laflèche, V., Ruel, J.-C., and Archambault, L. 2000. Évaluation de la coupe avec protection de la régénération et des sols comme méthode de régénération de peuplements mélangés du domaine bioclimatique de la sapinière à bouleau jaune de l'est du Québec, Canada. For. Chron. **76**: 653–663.

Lauzon, È., Kneeshaw, D.D., and Bergeron, Y. 2007. Forest fire history reconstruction (1680–2003) in the Gaspesie region of eastern Canada. For. Ecol. Manag. **244**: 41–49.

Lévesque, F. 1997. Conséquences de la dynamique de la mosaïque forestière sur l'intégrité écologique du Parc national Forillon. M.Sc. thesis, Université Laval, Québec, Que.

MacLean, D.A. 1980. Vulnerability of fir-spruce stands during uncontrolled spruce budworm outbreaks: a review and discussion. For. Chron. **56**: 213–221.

McCarthy, J. 2001. Gap dynamics of forest trees: a review with particular attention to boreal forests. Environ. Rev. **9**: 1–59.

McRae, D.J., Duchesne, L.C., Freedman, B., Lynham, T.J., and Woodley, S. 2001. Comparisons between wildfire and forest harvesting and their implications in forest management. Environ. Rev. **9**: 223–260.

Messier, J., Kneeshaw, D.D., Bouchard, M., and de Römer, A. 2005. A comparison of gap characteristics in mixedwood old-growth forests in eastern and western Quebec Can. J. For. Res. **35**: 2510–2515.

Morin, H. 1994. Dynamics of balsam fir forests in relation to spruce budworm outbreaks in the Boreal Zone of Quebec. Can. J. For. Res. **24**: 730–741.

Nyland, R.D. 2002. Site preparation. *In* Silviculture: concepts and applications. Second edition. *Edited by* S. Spoolman. McGraw-Hill, NY, USA, pp. 86–116.

Pickett, S.T.A., and White, P.S., 1985. The ecology of nature disturbance and patch dynamics. Academic Press, NY, USA.

Robitaille, A. and Saucier, J.-P. 1998. Paysages régionaux du Québec méridional. Ministère des Ressources naturelles du Québec, Direction de la gestion des stocks forestiers et Direction des relations publiques, Québec, Que.

Reyes, G.P. and Kneeshaw, D.D. 2008. Moderate-severity disturbance dynamics in *Abies balsamea-Betula* spp. forests: the relative importance of disturbance type and

local stand and site characteristics on woody vegetation response. Écoscience, **15**: 241–249.

Roy, V., Jobidon, R., and Blais, L. 2001. Étude des facteurs associés au dépérissement du bouleau à papier en peuplement résiduel après coupe. For. Chron. **77**: 509–517.

Ruel, J.-C., Pin, D., and Cooper, K. 1998. Effect of topography on wind behavior in a complex terrain. Forestry, **71**: 169–173.

Ryan, K.C. 2002. Dynamic interactions between forest structure and fire behavior in boreal ecosystems. Silva Fenn. **36**: 13–39.

Saucier, J.-P., Bergeron, J.-F., Grondin, P., and Robitaille, A. 1998. Les régions écologiques du Québec méridional (3rd edition): un des éléments du système hiérarchique de classification écologique du territoire mis au point par le Ministère des Ressources naturelles du Québec. Supplement of L'Aubelle. February-March.

Seymour, R.S., White, A.S., and de Maynadier, P.G. 2002. Natural disturbance regimes in northeastern North America: evaluating silvicultural systems using natural scales and frequencies. For. Ecol. Manag. **155**: 357–367.

Smith, D.M., Larson, B.C., Kelty, M.J., and Ashton, P.M.S. 1996. The practice of silviculture: applied forest ecology. John Wiley & Sons Inc., New York, USA.

Stocks, B.J. 1987. Fire potential in the spruce budworm-damaged forests of Ontario. For. Chron. **63**: 8–14.

Stocks, B.J. 1993. Global warming and forest fires in Canada. For. Chron. **69**: 290–293.

Tandes, G.F. 1979. Fire history and vegetation pattern of coniferous forests in Jasper National Park, Alberta. Can. J. Bot. **57**: 1912–1931.

Trombulak, S.C., and Frissell, C.A. 2000. Review of ecological effects of roads on terrestrial and aquatic communities. Conserv. Biol. **14**: 18–29.

Vaillancourt, M.-A., Gauthier, S., Kneeshaw, D., and Bergeron, Y. 2009. Implementation of ecosystem management in boreal forests. Sustainable Forest Management Network, University of Alberta, Edmonton, Alta.

Van Wagner, C.E. 1978. Age-class distribution and the forest fire cycle. Can. J. For. Res. **8**: 220–227.

Watt, A.S. 1947. Pattern and process in the plant community. J. Ecol. **35**: 1–22.

Wein, R.W., and Moore, J.M. 1977. Fire history and rotation in the Acadian Forest of New Brunswick. Can. J. For. Res. **7**: 285–294.

Weir, J.M.H., Johnson, E.A., and Myanishi, K. 1999. Fire frequency and spatial age mosaic of the mixedwood boreal forest of Saskatchewan. *In* Proceedings of the sustainable management network conference. Science and practice: sustaining the boreal forest. *Edited by* T.S. Veeman, D.W. Smith, B.G. Purdy, F.J. Salkie, and G.A. Larkin. Sustainable Forest Management Network, Edmonton, Alta., pp. 81–86.

Woods, K.D. 2000. Long-term change and spatial pattern in a late successional hemlock northern hardwood forest. J. Ecol. **88**: 267–282.

# Chapter 10

# Towards an Ecosystem Approach to Managing the Boreal Forest in the North Shore Region*
## Disturbance Regime and Natural Forest Dynamics

*Louis De Grandpré, Sylvie Gauthier, Claude Allain, Dominic Cyr, Sophie Périgon, Anh Thu Pham, Dominique Boucher, Jacques Morissette, Gerardo Reyes, Tuomas Aakala, and Timo Kuuluvainen*

\* This project is a joint initiative of the Sustainable Forest Management Network (SFMN), the Canadian Forest Service, and the CRSNG-UQAM-UQAT Industrial Chair in Sustainable Forest Management, carried out in collaboration with the Ministère des Ressources naturelles et de la Faune (North Shore region), Nordbois sawmill, and Uniforêt. We thank Martin Simard and Jacques Duval for their comments on the first draft of the manuscript. We also thank Christine Simard, Martin Simard, Catherine Boudreault, André Proulx, Gilles Sauvageau, Marie-Noëlle Caron, and Marie-Andrée Vaillancourt for their assistance with the sampling and sample analyses. Thanks are extended as well to Pamela Cheers, Benoit Arsenault, and Isabelle Lamarre of Natural Resources Canada for editing the manuscript. The photos on this page were graciously provided by Martin Simard, Louis De Grandpré, Tuomas Aakala, Marie-Andrée Vaillancourt, and Michel Robert (Canadian Wildlife Service).

# 1. INTRODUCTION

Until recently the boreal forest was considered an even-aged forest and managed as such. It was assumed that succession through species replacement in the canopy and the establishment of an uneven-aged structure were unlikely to occur given the short interval between major disturbances and the ubiquity of disturbances in the boreal zone. However, as shown in the previous chapters, there is considerable variability in disturbance regimes throughout the boreal forest (Gauthier et al., chapter 3; Girardin et al., chapter 4; Morin et al., chapter 7; Sutton and Tardif, chapter 8), which leads to differences in forest composition and structure in the various boreal forest regions. Recent research has shown that barely 30% of conifer stands in the North Shore region have regular diameter structures (Boucher et al. 2003; Côté 2006). The vast majority of forest stands (70%) possess structures varying from two-storied stands to reverse J-shaped structures. Ecosystem-based management of North Shore forests centres on gaining insight into the development dynamics of old-growth stands.

When forest stands develop over very long periods of time in the absence of forest fires, their dynamics gradually become governed by other types of disturbances. Spruce budworm (*Choristoneura fumiferana* Clem.) outbreaks, windthrow and gap dynamics associated with senescence contribute to the development of structurally complex stands and forest landscapes. These disturbances create openings that allow for the establishment of new stems and growth release by suppressed stems in the understory (Doucet 1988; Ruel 1989; Kneeshaw and Bergeron 1998). The forest dynamics in certain boreal forest regions, including the North Shore, are therefore more strongly influenced by gap dynamics than by major disturbances such as fires (Kneeshaw and Bergeron 1998; McCarthy 2001; Pham et al. 2004; Bouchard et al. 2006; McCarthy and Weetman 2006).

The renewal of the forest matrix, the decline in large tracts of old forests and their increasingly simplified structural features are ecological issues of particular concern for North Shore forests. This chapter reviews what is known about the natural dynamics and the disturbance regime of productive forests in the region. Although post-fire succession is addressed, the focus will be on the development dynamics of old-growth stands with various structures, since they dominate the forest landscape on the North Shore and their dynamics are still poorly understood (De Grandpré et al. 2000; Boucher et al. 2006). Finally, a few management approaches are discussed in relation to the specific characteristics of North Shore forest dynamics.

# 2. REGIONAL CONTEXT

Bordered on the west by the Saguenay River and extending eastward along more than 1,300 km of shoreline to Labrador and northward to the 55th parallel, the North Shore is the second largest administrative region in Québec (MRNFP 2004). Although the continuous boreal forest covers less than 40% of this region, provincially it accounts for close to 30% of the volume of standing timber.

The North Shore's commercial boreal forest is influenced by a cool, maritime climate. Annual precipitation varies between 900 and 1,300 mm and mean annual temperatures range from –1.0°C to 2.5°C. The topography is generally rugged, with high hills and mountains, many of which are cut by deep north-south valleys. The dominant surficial deposits are primarily tills, with thick till deposits found in depressions and on lower slopes, and thin deposits covering slopes and some summits. Finally, fluvio-glacial deposits blanket the bottom of broad valleys (Robitaille and Saucier 1998). Ubiquitous rocky outcrops are a particular feature of the region. Thin soils and rocky outcrops are more frequent as one moves inland from the shoreline and toward the east. Thick tills are abundant in the west, from the Saguenay River to the Manicouagan River.

## 2.1.  Portrait of the Forest

The North Shore region lies within the Chibougamau-Natashquan forest area (Rowe 1972), in which black spruce (*Picea mariana* [Mill.] B.S.P.) is dominant. Two bioclimatic subdomains characterize the continuous boreal forest in the North Shore region: balsam fir–white birch, near the shoreline and on Anticosti Island, and black spruce–moss, extending north to the 52nd parallel (Thibault and Hotte 1985; Robitaille and Saucier 1998). Black spruce and balsam fir (*Abies balsamea* [L.] Mill.) are the dominant species in the region. According to Robitaille and Saucier (1998), balsam fir dominates on mesic sites in the south in association with white birch (*Betula papyrifera* Marsh.) and trembling aspen (*Populus tremuloides* Michx.). Black spruce abundance increases with latitude.

About 74% of forested areas are dominated by coniferous stands (MRNFP 2004). The remaining area is either regenerating (10%), or occupied by mixed (13%) or hardwood (3%) stands. Dominant stands consist of pure spruce, black spruce and balsam fir mixedwood, and pure balsam fir stands. Most of the stands (57%) fall in the 90-plus age class and are therefore considered mature or overmature. With the increasingly intensive forest management practices applied in recent decades, a substantial increase in regenerating stands and stands younger than 30 years has been observed. Harvesting combined with natural disturbances has meant that about one quarter of the area is occupied by stands less than 30 years old.

## 2.2.  Recent Disturbances

### 2.2.1.  Fires

Although the North Shore region is under the influence of a maritime climate, it still has forest fires, some of which can cover extensive areas. In 1991, a fire that occurred northwest of Baie-Comeau affected almost 4,000 km². Although rare, fires still have a major impact on the forest landscape (figure 10.1). The conservation branch of the Ministère des Ressources naturelles du Québec, which has tracked forest fires since the early 1920s, mapped fires on the North Shore up to the 51st parallel between 1920 and 2003. During that period, major fires (> 1,000 km²) occurred south of the 51st parallel in 1921, 1941, 1955 and 1991

**Figure 10.1**

## Overview of the magnitude and the spatial distribution of the major disturbances (fire and insect outbreaks) in the North Shore region

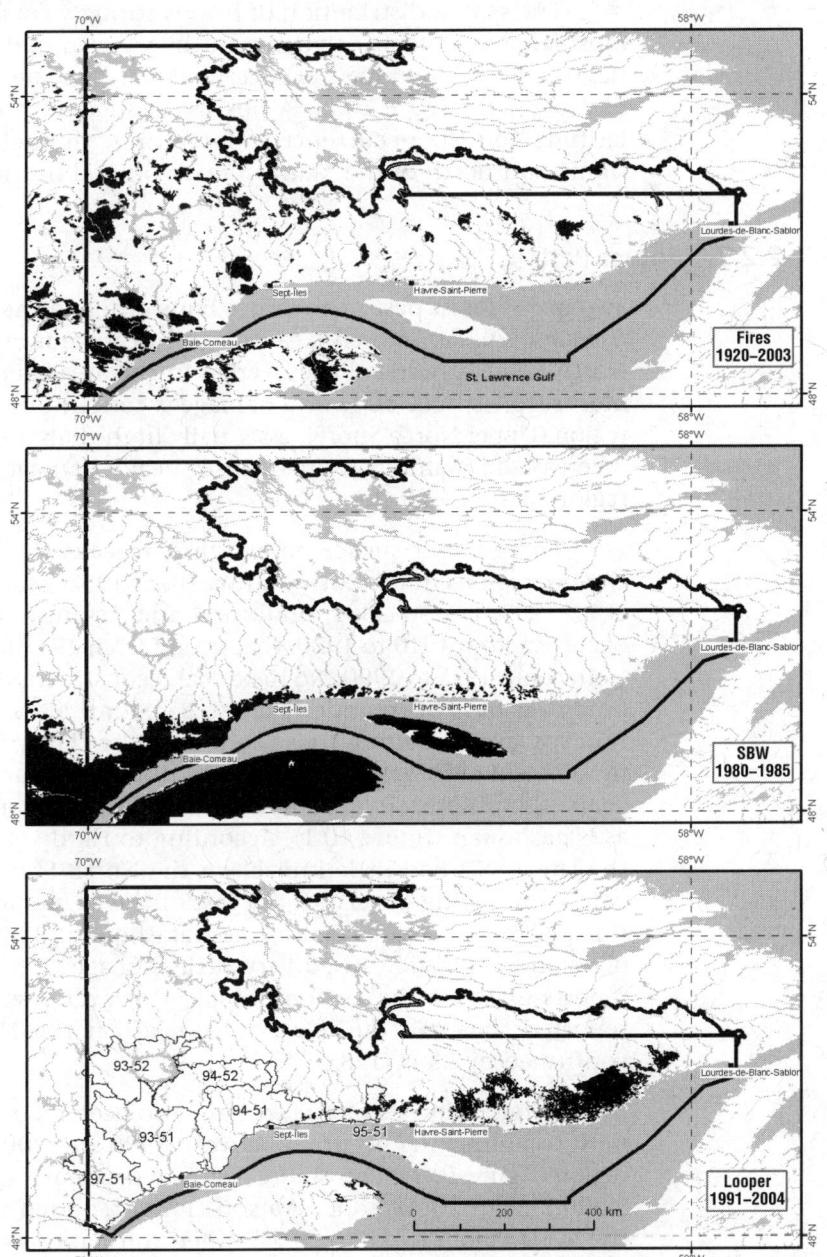

For fires, the time period covered 1920 to 2003 for the territory located south of the 51st parallel. The northern area was covered from 1940 to 2003. For SBW, the black zones correspond to areas where defoliation was observed between 1980 and 1985. Finally, for hemlock looper, black zones correspond to defoliated stands during the last outbreak (1999–2002). On this last panel, the location of six forest management units is presented.

(figure 10.2A). In most cases, a small number of fires (figure 10.2B) were involved. About 85% of the areas burned during that period resulted from fifteen fires covering more than 1,000 km$^2$.

The spatial distribution of fires is strongly correlated with longitude, with most large fire events occurring in the western part of the region. East of Sept-Îles, there was only one fire covering an area greater than 100 km$^2$ south of the 51st parallel (figure 10.1). A positive association was also observed between latitude and the occurrence of large fires; this relationship was particularly significant north of the Manic 5 reservoir and in the spruce-lichen domain.

## 2.2.2.  Insect Pests

Two major forest pests, the spruce budworm and the hemlock looper (*Lambdina fiscellaria fiscellaria* Guen.), cause significant damage to coniferous stands on the North Shore. There is little overlap in the areas affected by these insects. The most severe spruce budworm damage is concentrated in the western part of the region (Upper North Shore), essentially in the balsam fir–white birch subdomain, whereas the hemlock looper causes heavy damage in the eastern part of the region (east of Sept-Îles).

Three major spruce budworm outbreaks were recorded in a number of Quebec regions in the 20th century (Blais 1983; Morin and Laprise 1990; Morin et al. 1993; Jardon 2001; Boulanger and Arseneault 2004). The first outbreak occurred from 1910 to 1920, the second from 1947 to 1957, and the most recent from about 1975 to 1990. No historical reconstruction studies of spruce budworm outbreaks have been undertaken to date for the North Shore. However, defoliation surveys, including those conducted by Hardy et al. (1986), show that the most recent outbreak affected part of the region starting in 1974. In subsequent years, the outbreak spread eastward and infestations were found as far north as Natashquan (figure 10.1). According to Hardy et al. (1986), the outbreak in the late 1940s had little impact on the North Shore region, with most of the defoliation being observed just east of the Saguenay River. A few rare infestations were reported as far east as Natashquan. Blais (1983) reported that this defoliation episode caused very little damage to forest stands. No defoliation reports exist for the 1910 outbreak. However, Swaine and Craighead (1924) reported that this outbreak did not affect the North Shore, and this was confirmed by Blais (1983).

Hemlock looper outbreaks have also been reported on the North Shore, the most recent having occurred between 1999 and 2002 in the eastern part of the region (figure 10.1). The outbreak affected extensive areas, including more than 9,000 km$^2$ in 2000 alone. Watson (1934) reported the first looper outbreak in the North Shore region in 1927. A new infestation was reported in 1956 (Benoît and Desaulniers 1972) and in the early 1970s (Jobin and Desaulniers 1981). The impact of hemlock looper on stand dynamics is not addressed in this chapter because very little data is available on the forest dynamics associated with this insect pest in the eastern part of the North Shore region.

Figure 10.2

**Burned areas (A) and number of fires (B) for the North Shore region between 1920 and 2003**

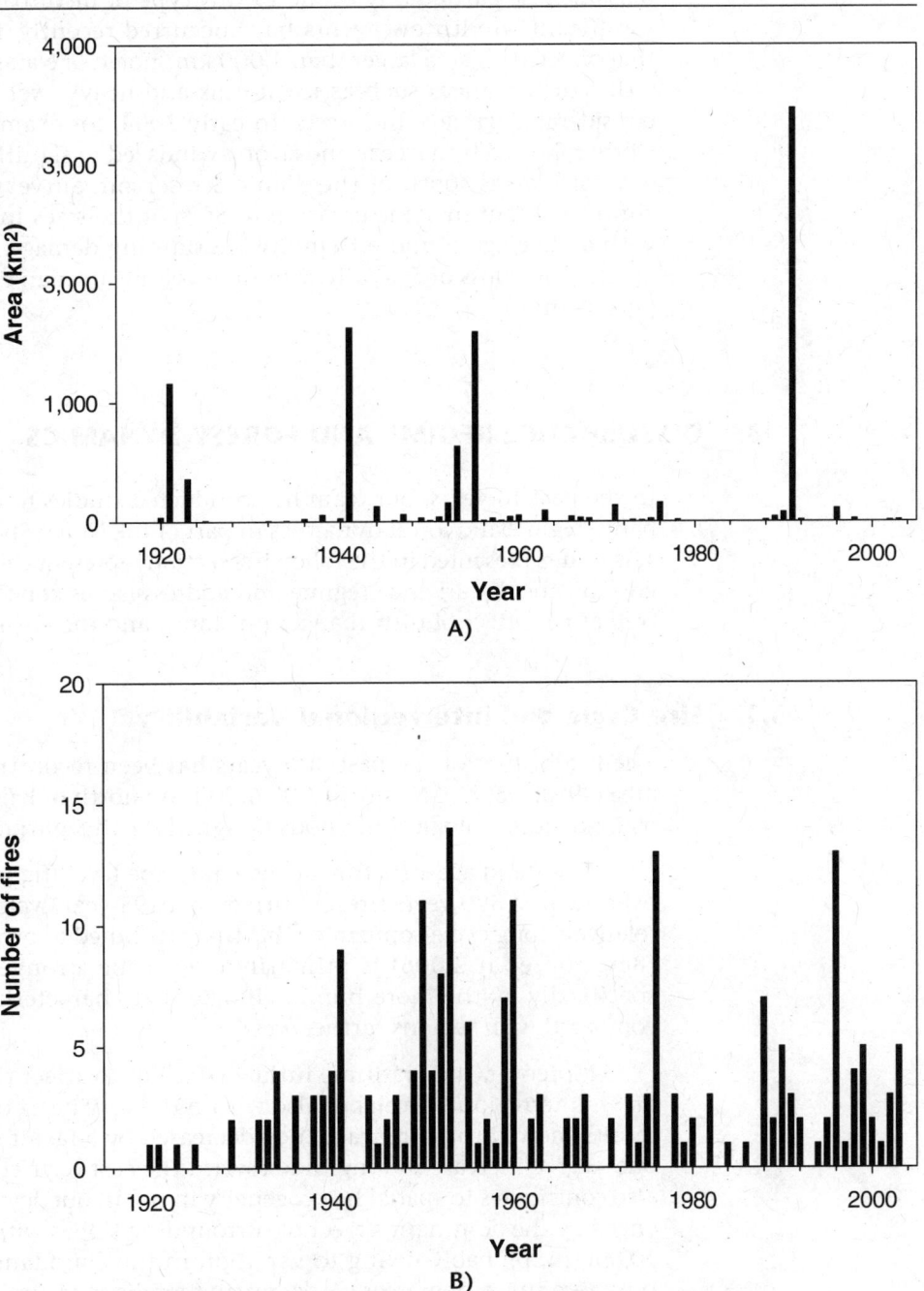

A)

B)

To cover the maximum time frame, the zone inventoried was located south of the 51st parallel only. We note that there were few years where affected areas exceeded 1,000 km². Source: Gouvernement du Québec, Ministère des Ressources naturelles, de la Faune et des Parcs (MRNFP), Direction de la conservation.

### 2.2.3.  Other Disturbance Agents

Windthrow occurs periodically, contributing to the dynamics of forest stands. Balsam fir is particularly prone to this type of disturbance (Ruel 2000). Some significant windthrow events have occurred recently, including one in 2001 that affected an area larger than 1,000 km² north of Natashquan (MRNFP 2004). Other disturbances such as ice storms and heavy, wet snowstorms can cause considerable damage in forests. In early 2004, for example, a major wet snowstorm followed by a freeze and strong winds led to significant tree breakage over extensive areas south of the Manic 5 reservoir. Surveys to assess the damage confirmed that in some cases up to 80% of the trees in a stand were affected, with an average of more than 30% sustaining damage. This event resulted in an estimated loss of 5% to 10% of fibre volume over an extensive area (J. Duval, pers. comm.).

## 3.   DISTURBANCE REGIME AND FOREST DYNAMICS

In the past 10 years, our team has conducted studies to characterize the disturbance regime and forest dynamics in part of the North Shore region (figure 10.3). The results presented in the following section represent the current state of knowledge on the disturbance regime and address issues concerning the fire cycle, as well as the effect of disturbances on stands and the ensuing forest dynamics.

### 3.1.  Fire Cycle and Interregional Variability

The fire history of the past 300 years has been reconstructed for an extensive area (49°00′–50°15′ N and 69°00′–67°00′ W) north of Baie-Comeau (figure 10.3) using dendroecological methods designed for this purpose.

The stand age structure at the landscape level (figure 10.4A) indicates that over the past 300 years fires occurred on a 295-year cycle (Cyr et al. 2007). This relatively long cycle compared with that attributed to other boreal forest regions (Bergeron et al. 2006) is primarily due to the stronger maritime influence, specifically by the more humid climate that characterizes the North Shore as compared with regions farther west.

Moreover, the maritime influence seems to affect the fire regime not only on an interregional level, but also within the very heart of the region. The study results showed that fire frequency decreased by a factor of nearly two for every one-degree increase in longitude toward the east (Cyr et al. 2007). Topography also contributes to spatial heterogeneity in fire frequency. Fire frequency is influenced by the dominant aspect of surrounding slopes within a radius of 3,500 to 10,000 m, probably owing to variation in the abundance and connectivity of the more fire-prone areas. A decreasing gradient of fire frequency is observed, from areas with a dominant south-southwest slope aspect to areas with a dominant north-northeast aspect (Cyr et al. 2007).

Figure 10.3
## Study area locations for the projects on fire history, gap dynamics, and post-fire forest succession

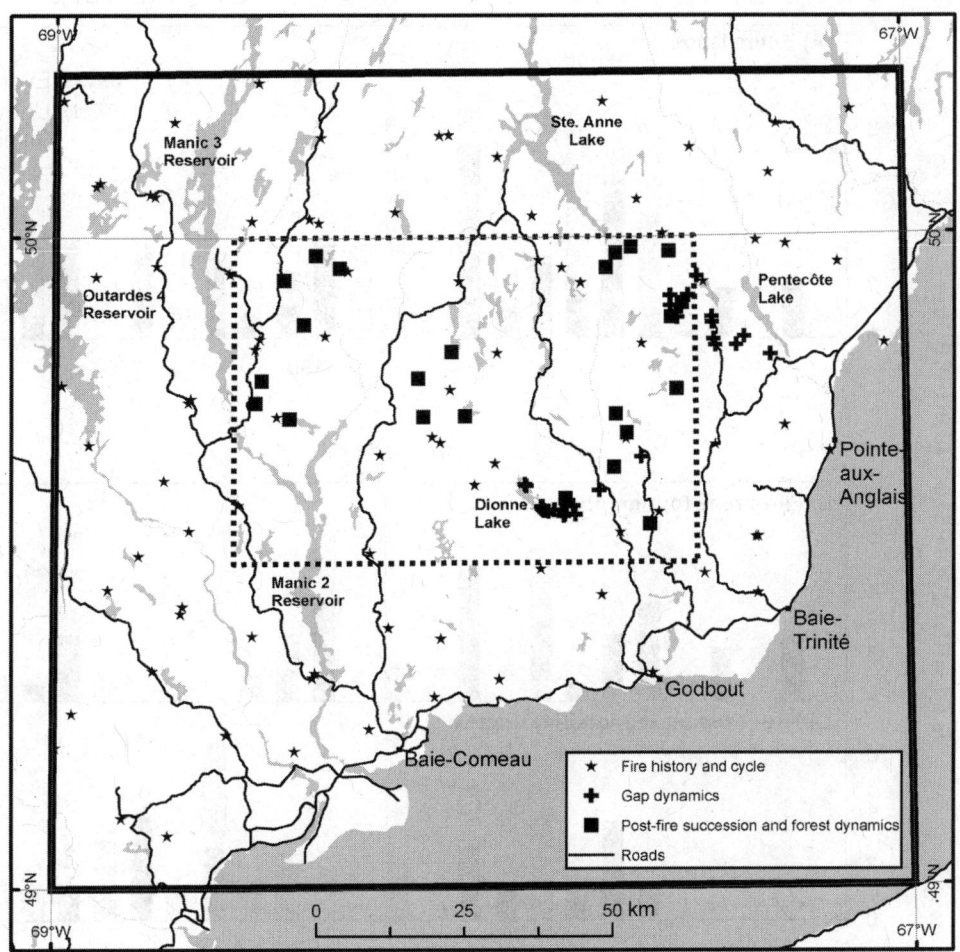

The largest square corresponds to the study area for fire history reconstruction. The territory comprised in the dotted rectangle was the one where studies on succession (photo-interpretation) and old-forest dynamics were conducted.

The percentage of annual area burned also varied considerably over time. Strong interannual (data not illustrated) and interdecadal (figure 10.4B) variability has characterized fire activity over the past 300 years. In addition, it was determined that the mean percentage of forest area burned annually was higher in the first half of the study period than in the second half. This finding, which seems to be consistent with observations in other regions of Québec (Bergeron et al. 2006), can be attributed primarily to the climate change that occurred around 1850 and resulted in less frequent periods of drought.

Figure 10.4
### Figure 10.4
## Forest area proportion according to the time elapsed since fire

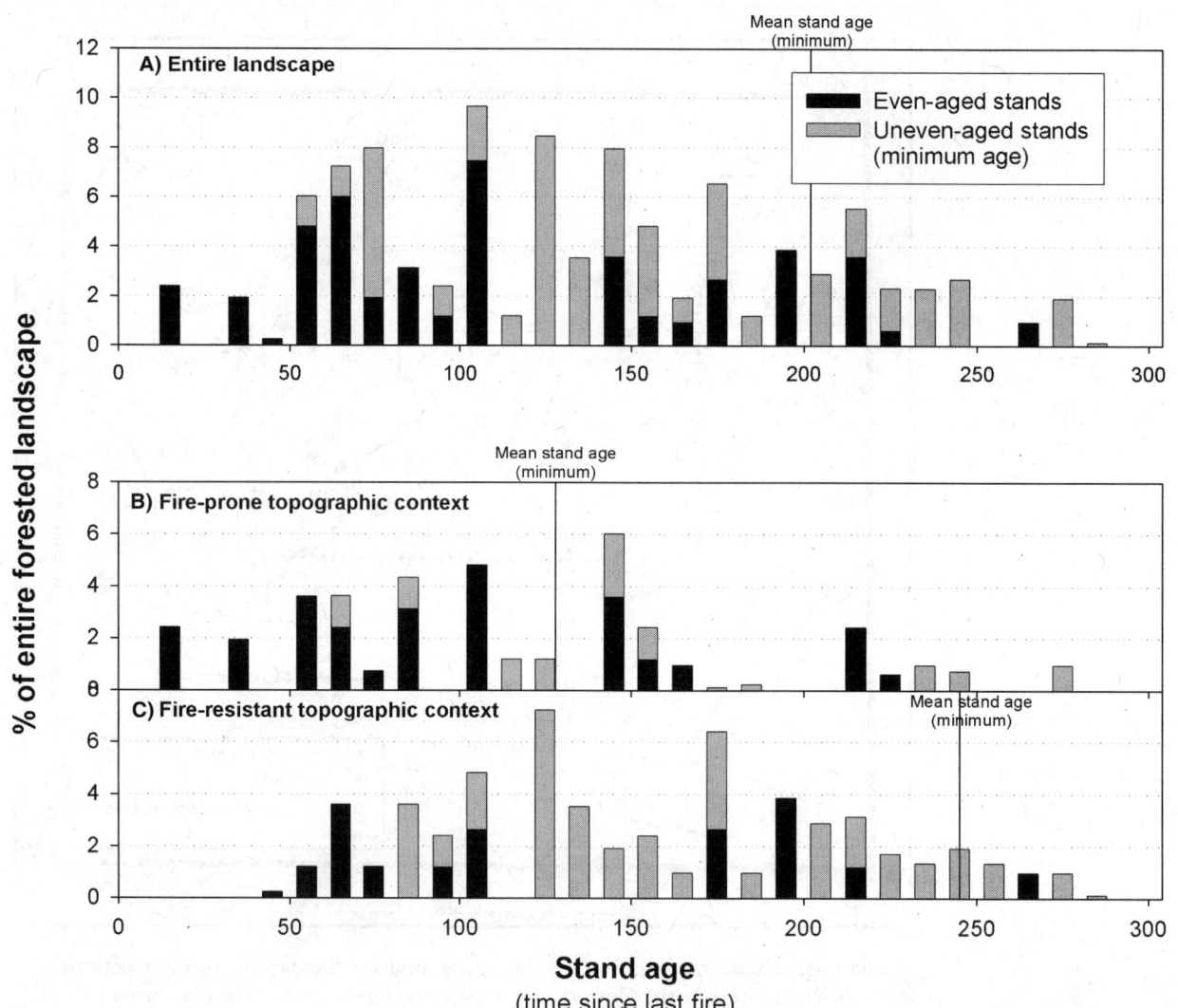

Legend: A) for the entire landscape sampled; B) for the topographic context prone to fire; C) for the topographic context that fosters fire resistance. It can be observed that mean time since fire varies as a function of the topographic context on the study area. The averages presented here differ from the one mentioned in the text (295 years) because they were obtained by means of a non-parametric calculation. Only minimal estimations are possible, however, due to the important proportion of stands with a minimum age.

## 3.2. Post-fire Succession

Archival documents (maps and aerial photographs) were used to evaluate the succession patterns that occurred in the absence of fire over a period of almost 60 years (figure 10.3). With a 295-year fire cycle, the composition and structure of many stands change over time because this average interval greatly exceeds the average life span of the constituent tree species, particularly those that become established immediately after a fire (De Grandpré et al. 2000; Gauthier et al. 2004).

Four main succession patterns, which differ in their initial post-fire composition or their successional pathway, were identified in an analysis of a series of aerial photographs (figure 10.5). Moreover, although these succession patterns are widely encountered, they are frequently found in association with certain deposit and drainage combinations.

Two of these patterns are characterized by initial dominance of shade-intolerant hardwoods (white birch and trembling aspen). The first pattern, which is associated with medium-thick, well-drained tills or thin tills with moderate

**Figure 10.5**
## Diagram representing the main successionnal pathways on the North Shore

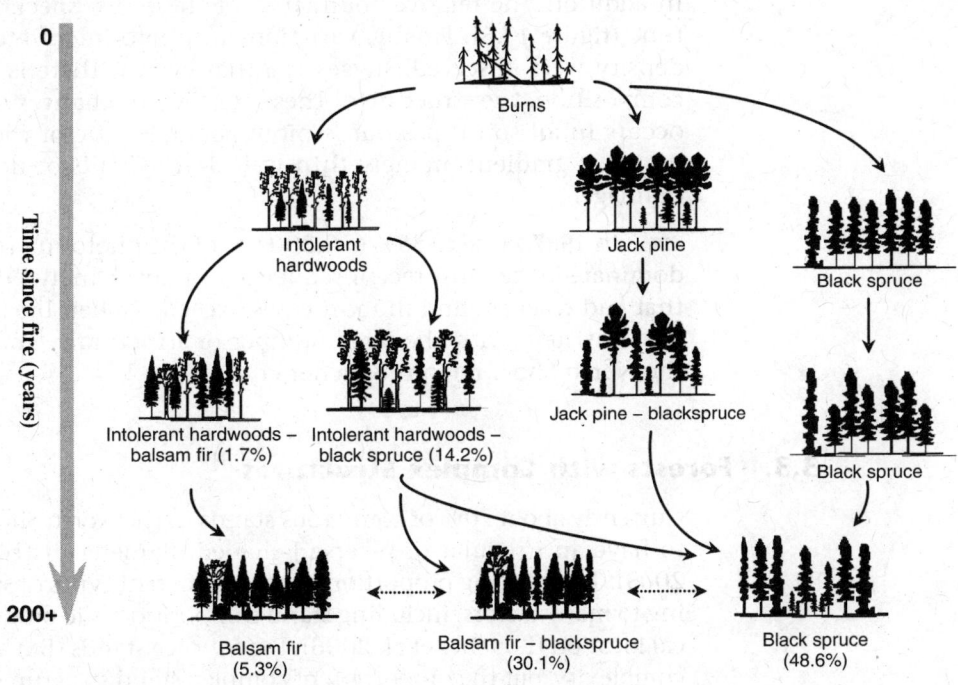

Percentages correspond to forest composition interpreted from 1987 photos. Since photo-interpreted areas from 1930 to 1987 have not undergone major disturbance during the 57-yr interval, we found very few young stands. At the landscape level and according to the fire cycle of the area, we can presume that at least 30% of the forest area consists of even-aged stands or is dominated by early-succession species.

drainage, is characterized by the strong early successional dominance of one, and sometimes both, of these species. This first cohort of trees starts to die off between 80 and 100 years post-fire and is replaced by firs that will dominate the canopy about 100 to 135 years after the fire. Thereafter, an increase in the proportion of black spruce is observed. Slightly less than 20% of the stands resulted from this succession pattern (figure 10.5). The second pattern is associated with thin to thick till deposits with moderate drainage. In this case, pioneer species are replaced by black spruce, which will dominate the stand within 100 to 140 years of the fire. Over time, an increase in the number of fir trees will be observed in the canopy. About 35% of the stands followed this successional pathway (figure 10.5).

Another pioneer species, the jack pine (*Pinus banksiana* Lamb.), sometimes becomes established immediately after a fire. In the absence of fire, black spruce will gradually replace jack pine in the canopy (De Grandpré et al. 2000). This successional pathway is observed mainly on sandy deposits in the western part of the region and in the northern area near the Manic 5 reservoir. The increased frequency of forest fires in this area has little impact on jack pine, as it is well adapted to fire (Wein and MacLean 1983; Johnson and Gutsell 1993).

The fourth, and most common, pattern of succession (accounting for more than 45% of stands) starts with black spruce and is characterized by a gradual post-fire increase in balsam fir (De Grandpré et al. 2000; Gauthier et al. 2004). In addition, the relative abundance of these two species often fluctuates over time (figure 10.5). Finally, variations in photo-interpreted age class, height, or density were observed, suggesting that even if there is little change in stand composition, the structure of these stands still changes over time. This pattern occurs in all soil types but is more characteristic of the two extremes of the drainage gradient, namely thin well-drained tills or deposits with imperfect drainage.

In 1987, close to 85% (> 3,500 ha) of the photo-interpreted forest areas were dominated (the most recent sequence of photos analyzed) by coniferous stands that had reached, and in most cases exceeded, their break-up age. These results are consistent with the large number of structurally complex stands found in the North Shore region (Boucher et al. 2003).

## 3.3.  Forests with Complex Structures

Currently about 70% of coniferous stands in the North Shore region are estimated to have an irregular or reverse J-shaped diameter distribution (Boucher et al. 2003). The existing proportion of these different types of structures varies according to many factors, including stand composition, site productivity and geographical location (Boucher et al. 2006). Pure spruce stands (BS) have the least structural complexity, but the proportion of complex stand structures increases as the abundance of balsam fir increases. In addition, a relative productivity index was developed based on the age–height relationship among dominant trees. This index provides the best explanation of variation in stand structure for all softwood composition types (pure black spruce [BS], spruce-fir [BS-BF]), balsam fir–black spruce [BF-BS] and pure balsam fir [BF]), whereas stand age seems to influence structure more in the early stages of stand development. The results suggest that productive stands become structurally complex earlier than unproductive stands and also maintain greater diameter diversity. These contrasting structural dynamics may be explained in two ways: a higher growth rate in more productive stands would likely induce earlier senescence and thus an earlier transition to an uneven-aged structure; and the maximum tree diameter in poor stands would be limited by resource scarcity, which would in turn constrain the diameter diversity of these stands even after break-up age (Boucher et al. 2006).

With a 295-year fire cycle, close to 63% of the stands can be expected to be more than 150 years old (past break-up age) and have therefore attained the status of an old forest. Although the structurally complex forest matrix could appear relatively uniform at the landscape level, at a finer resolution one might see a complex vertical structure, a two-storied structure or a single-story structure. In addition, the forest cover often features gaps of varying sizes and shapes which may or may not contain regeneration. The combination of these characteristics distributed in space and highly dynamic in time creates a structurally complex forest matrix.

## 3.4. Gap Dynamics

### 3.4.1. Characteristics of Gaps

Recent research conducted north of Baie-Comeau and Rivière-Pentecôte in three types of old-growth coniferous stands revealed a gap disturbance regime (BS, BS-BF or BF-BS, BF) (Pham et al. 2004; Périgon 2005). At the stand level, small-scale mortality events influence forest development (McCarthy 2001; Pham et al. 2004; Périgon 2005). Two studies on gap dynamics showed that on average more than 55% of the surface area of old-growth coniferous forests was in gaps. The gap fraction does not differ according to stand type. Although the gap fraction is high, it is comparable with that observed in other mature boreal forests (Lertzman and Krebs 1991; Kneeshaw and Bergeron 1998; McCarthy 2001; Bartemucci et al. 2002).

Périgon (2005) defined three types of gaps: 1) gaps with regeneration and dead trees; 2) gaps with dead trees but no regeneration; and 3) gaps with neither regeneration nor dead trees. Gaps that had not regenerated were for the most part found in BS stands (figure 10.6), where ericaceous shrubs such as *Kalmia* spp., *Ledum* spp. and *Vaccinium* spp. were present in the understory and largely dominated the gaps. The ubiquitous ericaceous shrubs in the understory of BS stands and in the gaps hinder regeneration, unlike the situation in BF stands.

The large number of unregenerated, ericaceous shrub-dominated gaps suggests that BS stands open up far more easily than they fill in. This development model has already been suggested for spruce-moss stands following one or more fires occurring in succession (Gagnon et al. 1999) or following a succession of varied disturbances (Payette et al. 2000; Payette and Delwaide 2003). Unregenerated gaps in BS stands could be caused by strong competition from ericaceous vegetation after the opening of the forest canopy. They could also be older and date from a stand-initiating fire. After a fire, microsites that are not conducive to regeneration can remain open, explaining the presence of gaps containing ericaceous vegetation and no dead trees in BS stands (Payette et al. 2000; Harper et al. 2005).

Figure 10.6

**Mean proportion of gap area within forest stands according to stand type and gap type**

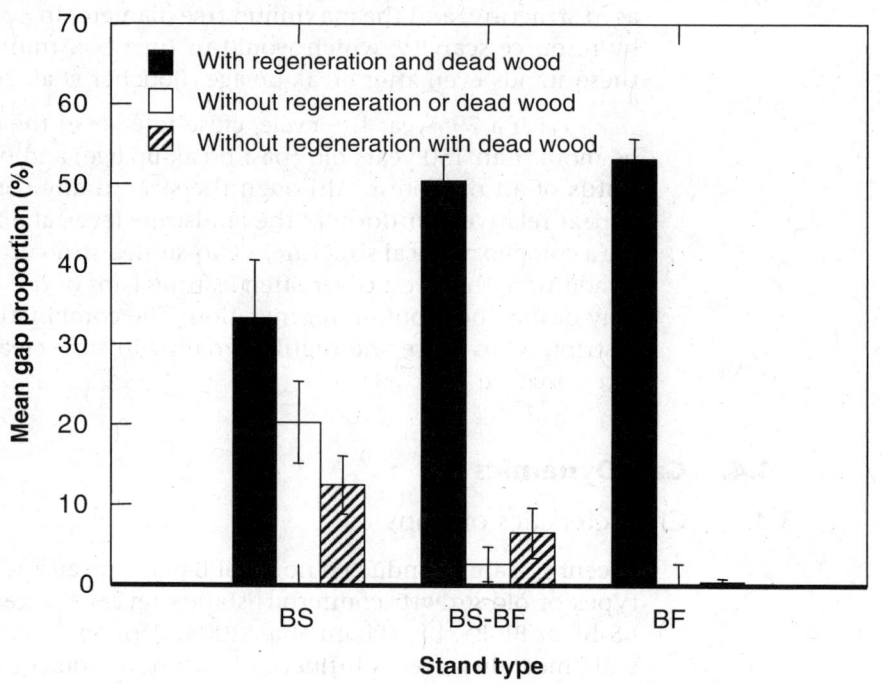

Legend: BS = black spruce stands; BS-BF = mixed black spruce–balsam fir stands; BF = balsam fir stands.

Gaps found along the transects have an average size of just over 100 m². However, more than 80% of them are smaller than 100 m² and are caused by the mortality of fewer than 10 trees (Pham et al. 2004; Périgon 2005). There is considerable variability in gap size distribution in all types of stands. The importance of large gaps in the gap size distribution (figure 10.7) is a crucial element. In BF stands, close to half of the total area covered by gaps results from openings larger than 500 m² (Perigon 2005). A significant number of large gaps are also found in other types of stands. Strong variability in gap size distribution not only has an impact on forest structure, but can also lead to marked differences in vegetation response (Bouchard et al. 2006). An increase in gap size, for example, is often associated with an increase in the establishment of prolific shade-intolerant species, such as white birch. In smaller gaps, advance softwood regeneration would have an edge in closing the forest canopy.

To gain insight into the forest dynamics associated with gap formation, Pham et al. (2004) calculated transition (replacement) probabilities within the three types of forest stands referred to above in terms of the species that created the gap and its probable successor. In the case of pure stands (BS or BF), the results show stability in long-term composition within the context of small-scale

Figure 10.7
**Gap size-class distribution according to mean area occupied in each stand type**

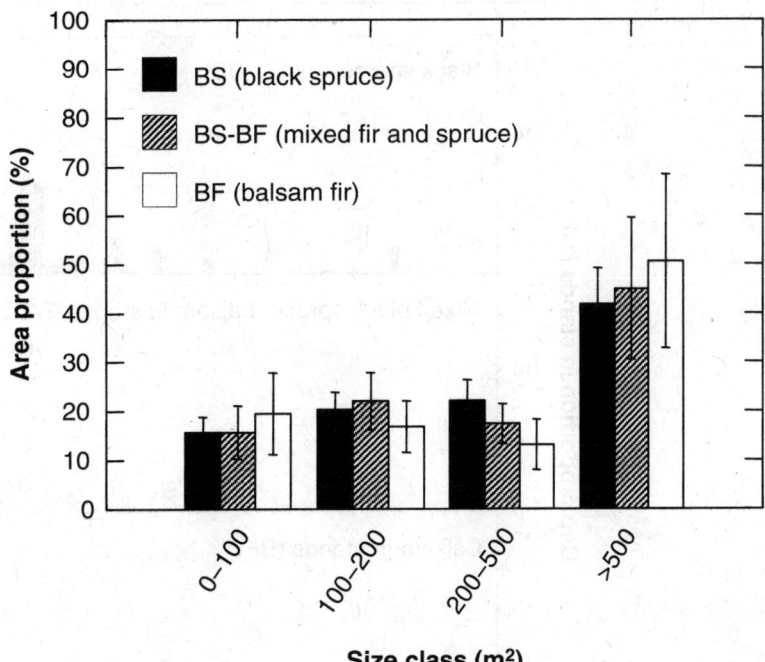

Size class (m²)

Legend: BS = black spruce stands; BS-BF = mixed black spruce–balsam fir stands; BF = balsam fir stands.

gap dynamics. Transitions in the other stand types (BS-BF or BF-BS) revealed that a change in composition at the gap level was probable. There was a strong probability of black spruce filling in gaps generated by balsam fir and of balsam fir filling in gaps resulting from black spruce mortality. These results show that changes in composition are likely at the level of individual gaps, but that globally, at the stand level, composition would remain in dynamic equilibrium.

### 3.4.2.  Gap Formation Rate

Dead trees were harvested in a number of gaps along the transects to determine the year the trees died and to measure their annual growth rate (Périgon 2005). The results were used to determine annual gap formation by stand type and to verify whether the rate of formation remained stable over time. Previous episodes of sudden declines in growth rate that could have been associated with disturbances were also characterized.

Gap dynamics plays a major role in old-growth boreal forests in eastern Canada (McCarthy 2001; Pham et al. 2004; Périgon 2005; McCarthy and Weetman 2006). Openings occur intermittently in stands, with variable spatial and temporal severity (figure 10.8). The synchronism in gap formation observed

Figure 10.8
**Figure 10.8**
**Distribution of the proportion of stands in gap according to the year of gap formation and stand type**

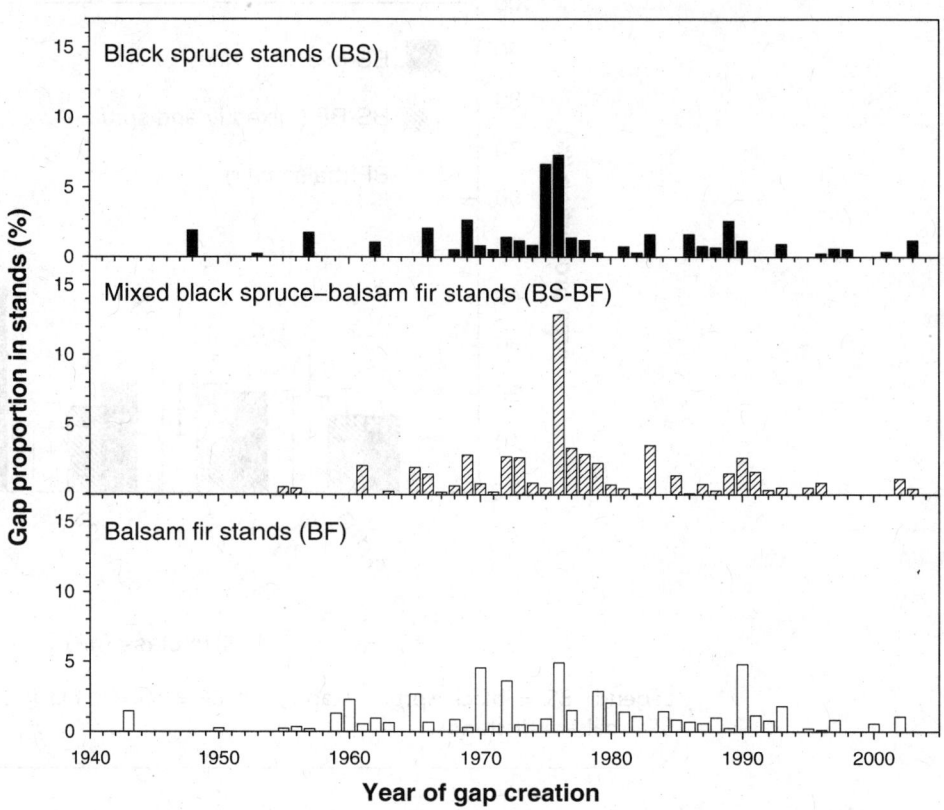

A continuous gap formation can be observed, marked by periods where gap creation is more important, consistent with the last SBW outbreak.

among stands (Périgon 2005) suggests that events at the landscape level control gap dynamics. However, stand composition and structure lead to local variability in the severity of this disturbance (MacLean 1980; Bergeron et al. 1995).

The maximum gap formation period from 1975 to 1980 (figure 10.8) coincided with the spruce budworm outbreak reported in Quebec from 1972 to 1987 (Blais 1983; Gray et al. 2000; Jardon 2001; Jardon et al. 2003). Vulnerability to the outbreak did not seem to differ between black spruce and balsam fir (figure 10.9), despite the fact that close to 40% of the gaps resulting from the death of individual spruce trees occurred during the outbreak and that black spruce is generally considered more resistant to spruce budworm attacks (MacLean 1980, 1984; Blais 1983), with less current-year and cumulative defoliation than that of balsam fir (MacLean and MacKinnon 1997).

The annual gap formation rate varies with stand composition in addition to being very heterogeneous over time. A mean annual gap formation rate for these forests would therefore not be representative of temporal gap dynamics.

**Figure 10.9**
## Gap maker frequency distribution by year of mortality for balsam fir and black spruce

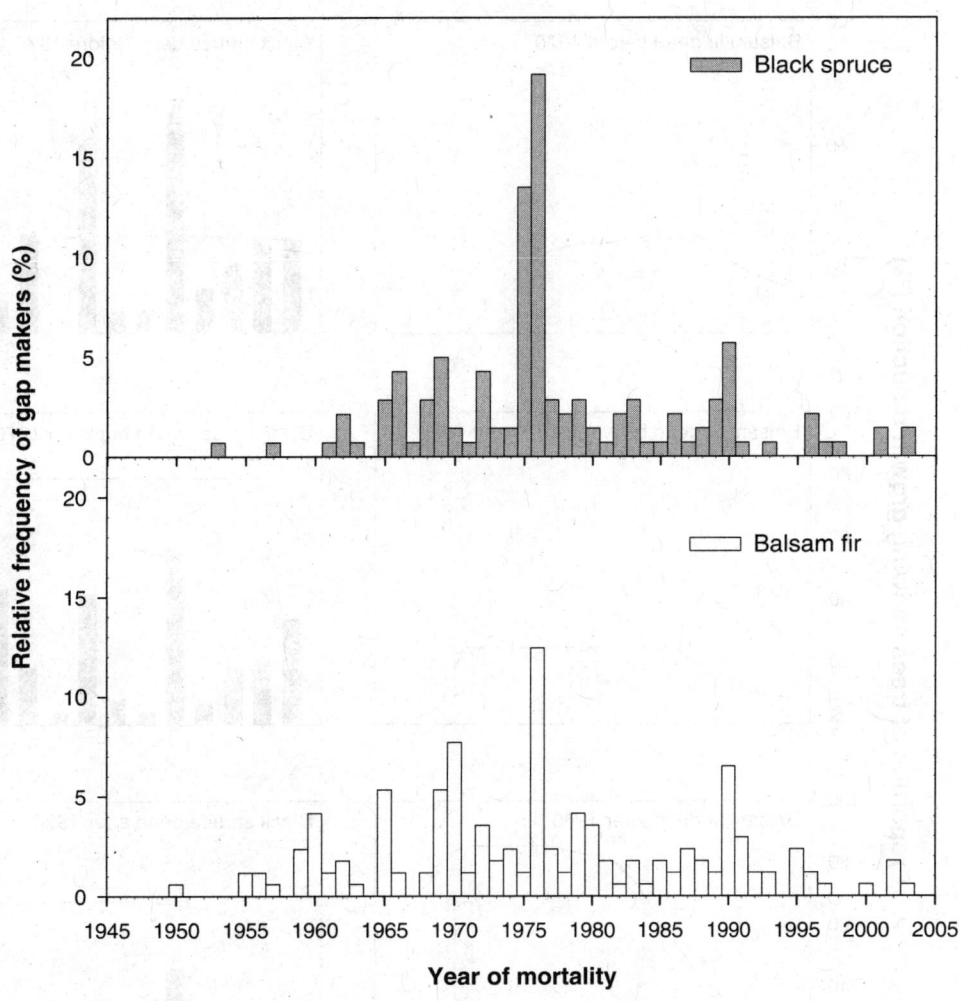

It can be observed that gap makers of the two species are more frequent during years associated to the last SBW outbreak.

The gap dynamics in old-growth coniferous forests are marked by significant gap formation events, which are probably linked to spruce budworm outbreaks. Between such events, gap formation is associated more with the mortality of single trees.

Study results concerning the number of trees exhibiting a reduction in annual growth show that, unlike balsam fir, black spruce can tolerate many periods of growth reduction during its life cycle (figure 10.10), a characteristic that has also been observed in other studies (MacLean 1980, 1984; Morin and Laprise 1990; Burns and Honkala 1990). The growth reductions observed in

### Figure 10.10
## Growth reduction frequency distribution for three periods of mortality compared between balsam fir and black spruce

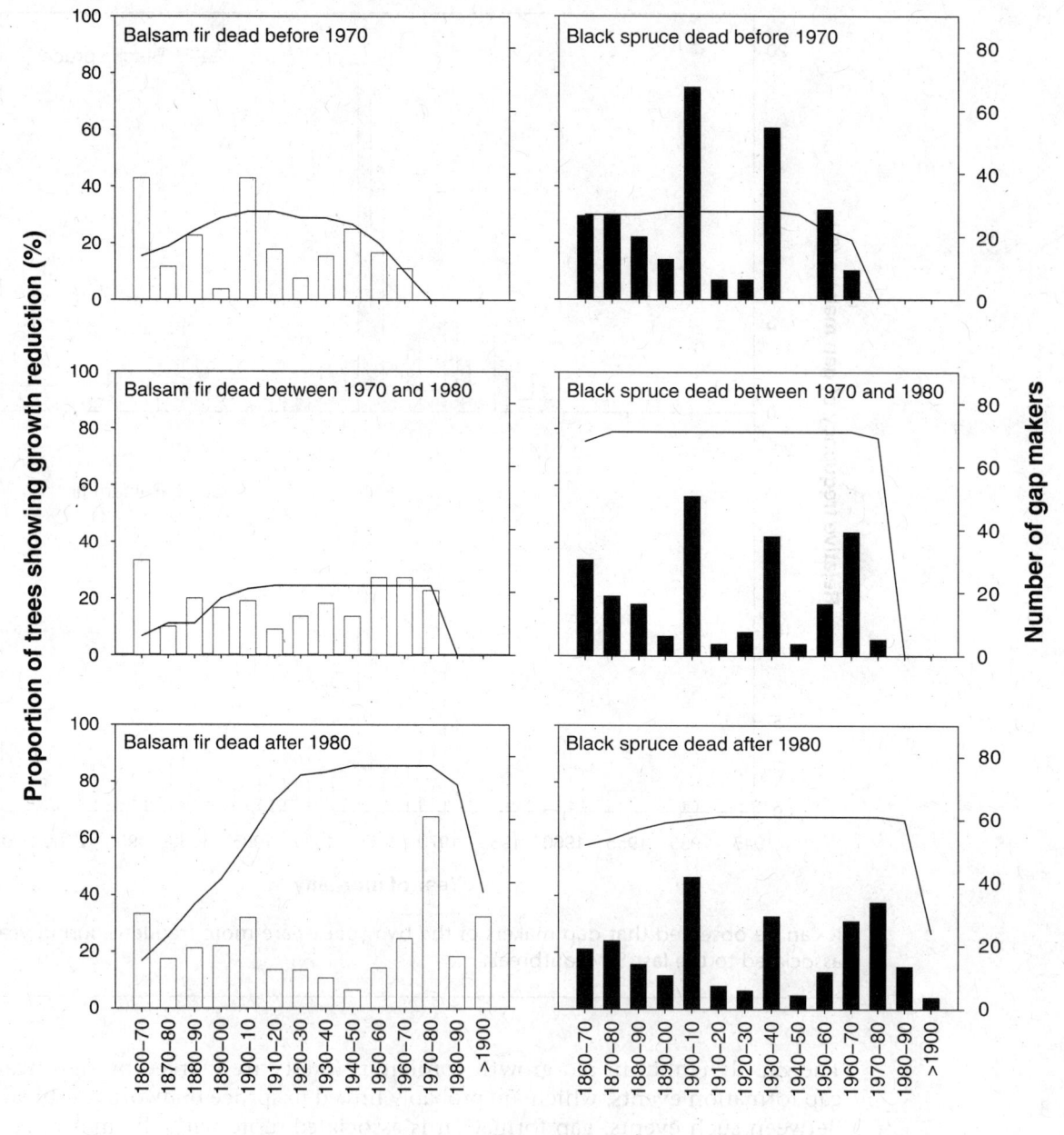

The number of gap makers analyzed is indicated by the curve. A gap maker can show several growth reduction periods. It can be observed that black spruce stems can tolerate many growth reduction periods during their life.

black spruce coincided with the beginning of the spruce budworm outbreaks recorded around 1910, 1940, and 1970 (Filion et al. 1998; Gray et al. 2000). Black spruce seems to die after being weakened by multiple periods of growth reduction, whereas balsam fir will die soon after a single such period. It is difficult to assess the impact of previous outbreak-related periods of growth reduction on mortality within stands. However, on the basis of the frequency of growth reductions that occurred in different periods, it seems clear that the ones in the early 20th century were similar to what was observed in the more recent outbreak (figure 10.10), at least in the case of the sites that were sampled. These observations are supported by the results reported by Morin et al. (chapter 7), which showed that at the provincial level the epidemic that began around 1910 was the most severe outbreak in the 20th century.

### 3.4.3.  Gap Dynamics and Dead Wood Production

Given the disturbance regime and the forest dynamics characterizing the North Shore region, the forests there produce a large quantity of dead wood. However, there are substantial differences among stands in terms of the availability of dead wood and large standing trees (Vaillancourt et al. 2008). These differences are attributable in part to the development of different types of irregular stand structures (Boucher et al. 2006), and to the type of disturbances involved. Since balsam fir–dominated stands are more prone to disturbances such as spruce budworm defoliation and windthrow, they produce a larger quantity of dead wood than black spruce–dominated stands. A study by Aakala et al. (2007) showed that estimated annual mortality for large-diameter trees varied between 0.8% for spruce stands and balsam fir–spruce mixedwood stands, and 1.2% for balsam fir stands. During outbreaks, these rates can increase appreciably.

### 3.4.4.  Large Gap Dynamics

In the balsam fir–white birch forest, mass mortalities of canopy trees are observed episodically in the southwestern part of the North Shore region (and result in the formation of larger openings). The minimum size of these large gaps is 0.2 ha, and they can reach a size of up to several hundred hectares. These disturbances tend to increase the spatial complexity of the vegetation cover at both stand and landscape levels. Disturbances such as windthrow and severe spruce budworm or hemlock looper outbreaks may contribute to the development of such structures by increasing the complexity of both stem spatial distribution within the stand and stand spatial arrangement in the landscape.

An ongoing study by Reyes (2009) in the Upper North Shore area reveals no specific regeneration patterns according to the type of disturbance (windthrow, spruce budworm or spruce budworm followed by windthrow). However, pre-disturbance stand composition as well as the ratio of uprooted to broken trees has a strong influence on the composition of post-disturbance regeneration. The study showed that balsam fir, black spruce and white birch are more abundant in regenerating disturbed stands that were originally dominated by conifers. White spruce and mountain maple (*Acer spicatum* Lamb.) are more abundant after disturbances in stands that were originally mixedwood. Sites with more uprooted woody debris are characterized by a greater abundance of raspberry

(*Rubus idaeus* L.) and dogwood (*Cornus* sp.) in the understory, whereas those with a predominance of snapped trees typically feature ericaceous shrubs, mountain-ash and fir. Understanding the mechanisms affecting vegetation response to disturbance is vital for predicting spatial and temporal changes in stand composition. In certain sectors of the North Shore region, these disturbances definitely have a greater impact on forest dynamics than fires would ever have.

## 4. CHALLENGES AND ISSUES ASSOCIATED WITH ECOSYSTEM MANAGEMENT ON THE NORTH SHORE

### 4.1. Clearcut Agglomerations and Forest Matrix Renewal

The age-class distribution of managed North Shore coniferous stands indicates that more than one quarter of the forest area is currently regenerating (table 10.1). An assessment by forest management unit (FMU) reveals strong intra-regional variability in the distribution of regeneration areas. The proportion of regeneration areas decreases as one moves eastward; however, the overall value is still higher than that estimated on the basis of the fire cycle (table 10.1). Areas in the 120-plus age class increase as one moves eastward (table 10.1). For the region as a whole, just under half of the areas fall within this age class. Based on the present fire cycle, more than 70% of the coniferous forest area is expected to belong to the 120-plus age class. In two FMUs (97-51 and 93-51), harvesting combined with natural disturbances (including the major impact of the most recent spruce budworm outbreak) have contributed significantly to the reduction in the number of old forest areas. Although the eastern part of the North Shore region is still primarily dominated by stands older than 120 years (forest management having begun only recently), a significant proportion of regeneration areas is observed in comparison with other age classes (table 10.1). Also, since harvesting is concentrated in the southern part of the FMUs, today, these sectors have a significant proportion of young even-aged forests, and are therefore very different from what they might have been in the past, with potential impact on biodiversity.

**Table 10.1**

**Distribution of the coniferous forest area (%) for timber production for each of the forest management units (FMUs) in the North Shore region for different age classes. See figure 10.1 for FMUs location.**

| Age class | FMU 97-51 | FMU 93-51 | FMU 93-52 | FMU 94-51 | FMU 94-52 | FMU 95-51 | North Shore | Natural[1] |
|---|---|---|---|---|---|---|---|---|
| 10 | 36.5 | 28.5 | 21.6 | 12.1 | 10.0 | 11.7 | 26.2 | 6.6 |
| 30 | 16.3 | 10.1 | 6.6 | 4.7 | 0.3 | 1.3 | 7.7 | 6.1 |
| 50 | 2.4 | 2.6 | 0.7 | 1.0 | 0.6 | 0.5 | 1.6 | 5.7 |
| 70 | 10.1 | 7.5 | 2.3 | 8.4 | 2.4 | 3.5 | 6.5 | 5.4 |
| 90 | 11.1 | 9.6 | 9.8 | 11.3 | 3.5 | 4.0 | 8.5 | 5.0 |
| 120+ | 23.5 | 41.7 | 59.0 | 62.5 | 83.2 | 78.9 | 49.6 | 71.2 |

1. Proportions estimated based on a 295-year fire cycle (Cyr et al. 2007).

Current knowledge of the fire regime, specifically as regards variability in the fire return interval depending on the region and aspect (Cyr et al. 2007), provides an indication of the spatial distribution that should be targeted in even-aged management. Large areas under even-aged management should be concentrated more in the western part of the region, where fire frequency is greater. In addition, at the local level, the dominant aspect of surrounding slopes needs to be considered. However, in an ecosystem-based management context, even-aged management (group clearcuts) would not be a major component of regional harvesting strategies. At any given time, tracts of even-aged forest should make up a maximum of 30% of the forest area. If the break-up age of stands exceeds 100 years, the annual harvesting rate (CPRS* type) should correspond to at most 0.3% of the area. Clustered clearcuts covering extensive areas (10 to 300 km$^2$) should be spaced fairly far apart (from 50 to 200 km), as suggested by Belleau et al. (2007).

• The CPRS (cuts with protection of regeneration and soils) is a careful-logging treatment commonly used in Québec. The CPRS is analogous to the CLAAG (careful logging around advance regeneration) used in Ontario.

Fires and other disturbances at the landscape level will continue to affect the region and create regeneration areas. In years when large fires occur, harvesting should be done primarily in the burn areas, and salvage cutting should be geared to ensuring that the legacies of the fire regime are maintained (Nappi et al. 2004). A literature review by Schmiegelow et al. (2006) on salvage logging in the boreal forest lays the groundwork for establishing an ecosystem-based salvage strategy for burn areas. The strategy is based on maintaining habitat diversity within burn areas, which reflects the variable effects of fire severity.

### 4.2.  Succession and Even-Aged Management

At the stand level, the harvesting intensity associated with CPRS is considerable compared with the effect of fire; yet the impact of this type of cutting on the organic soil layer is far less severe than that of a fire. Consequently, in addition to renewing the forest matrix, current CPRS forest management practices could lead to changes in the proportion of species present and alter succession patterns in relation to the expected sequence under a natural disturbance regime.

At the landscape (or regional) level, the increase in regeneration areas could promote the establishment of shade-intolerant species such as trembling aspen and white birch. Also, old-growth BS-BF or BF-BS stands harvested using CPRS would no longer go through an early successional stage dominated almost exclusively by black spruce, because balsam fir advance regeneration would be preserved during harvesting operations. A management approach based on CPRS could therefore also contribute to an increase in balsam fir in the landscape over the long term.

Another significant succession-related aspect of harvesting BS (pure) stands is the potential post-harvest loss of productivity, together with problems related to competition from ericaceous vegetation. Some studies show that black spruce stands that originate after harvesting or a fire characterized by little soil disturbance are less productive than stands that emerge after severe fires (Ruel et al. 2004; Lecomte et al. 2006). Canopy openings in certain types of spruce stands could promote the development of low-density stands owing to strong competition between ericaceous shrubs and black spruce regeneration (Mallik 1995). A study undertaken to characterize the ericaceous vegetation in forests in the North Shore region showed that these species were dominant in pure spruce stands with a canopy opening of at least 60% (Laberge Pelletier et al. 2007).

To minimize the impacts of CPRS, not only does harvesting intensity need to be varied, but in certain cases soil scarification is required to maintain productivity and reduce competition from ericaceous shrubs (Thiffault and Jobidon 2006). In other cases, targeted measures are needed (e.g. a decrease in the abundance of balsam fir regeneration to promote black spruce regeneration) in order to reinitiate a successional pathway.

## 4.3.  Managing a Structurally Complex Forest Matrix

Close to 70% of productive forest area in the North Shore region (with some regional variations) consists of forests that have reached or exceeded their break-up date (Boucher et al. 2003; Côté 2006). Contrary to popular perceptions, an old-growth forest is not defined as a declining stand in which mortality in the dominant cohort leads to unavoidable openings in the forest canopy. Old-growth status is, rather, associated with a development stage in which there are many age cohorts within the same stand and where canopy replacement occurs through the formation of openings of variable sizes (Oliver and Larson 1996; Frelich and Reich 2003). The challenge faced in an ecosystem-based management context is to implement approaches that will maintain and reproduce the structural diversity characterizing North Shore forests (see Bouchard, chapter 13, for examples of silvicultural treatments). To achieve this, current knowledge on gap dynamics and structural development in old-growth forests can be used as a guide.

### 4.3.1.  Reproducing Small Gaps

*Gap Size*

The results presented in the previous sections show that gap sizes range from 1 m² to several thousand square meters. These openings are caused by a variety of factors, ranging from single-tree mortality through partial windthrow to insect outbreaks. Although gaps smaller than 100 m² are more frequent in the old-growth forest matrix (Pham et al. 2004; Périgon 2005), it is totally unrealistic to think that such openings can be produced through forest management operations. However, since the larger (0.02 to 0.2 ha), but less numerous, gaps dominate in gap area calculations, they are the ones that should be reproduced through various partial cutting approaches. To prescribe the spatial distribution of such cuts, it is essential to know how these openings are spatially distributed in stands.

*Frequency*

Findings reported by Périgon (2005) on the frequency of gap formation shed light on the contagious distribution of gaps over time in relation to spruce budworm outbreak cycles (for the western part of the region). The regional periodicity of spruce budworm outbreaks is roughly 30 years (Jardon et al. 2003). Locally, there is however considerable variability in outbreak severity, which is essentially linked to stand development stage. Generally speaking, one in two outbreaks within a given stand is severe (Morin et al., chapter 7), and this can serve as a guide in management operations. However, in addition to the periodicity of disturbances, it is important to know which types of stands would show a strong and rapid growth response to interventions and allow a timber harvest every 60 years.

### The Right Intervention in the Right Place

Current knowledge on post-disturbance productivity and the development of irregular stand structures is limited. Some stands are known to develop an irregular or more complex structure far more rapidly than other stands. The presence of balsam fir in stands is associated with faster recovery of complex structure. Spruce stands, however, generally seem to take more time to develop an irregular structure following a disturbance. In light of results reported by Boucher et al. (2006), it is important to consider stand structure to be re-established after interventions, by taking stand composition and a productivity index into account.

Another limitation for the widespread application of partial cuts relates to advance regeneration and competition from ericaceous vegetation. Périgon (2005) showed that BS stands are the ones most likely to present regeneration problems after the forest canopy is opened up. Laberge Pelletier et al. (2007) have in fact identified the types of stands in which ericaceous vegetation would pose a problem once gaps have formed in the canopy. In terms of both stand productivity and ericaceous shrub competition, the presence of balsam fir in stands seems to be a determining factor for prescribing partial cuts.

## 4.3.2.  Reproducing Large Gaps

Research undertaken by Reyes (2009) shows that in addition to small gap dynamics, some sectors in the region are subject to disturbances that result in larger openings. Large gap dynamics is a factor especially in balsam fir–white birch stands in the western part of the region (FMUs 97-51 and 93-51). In addition to forest fires, large gaps resulting from insect outbreaks leave their mark on the landscape in this part of the North Shore. Cutting with retention of small merchantable stems (CPPTM) and harvesting with advance regeneration and soil protection (CPHRS) could prove to be effective treatments for reproducing the footprint of major disturbances. The intensity of the intervention needs to be varied, however (e.g. by leaving patches of dead and living trees), to mimic the spatial and temporal variability in the severity of spruce budworm outbreaks. These components are discussed in greater depth in Kneeshaw et al. (chapter 9) in relation to the Gaspé region, where the forestry context is fairly similar to that of the North Shore region.

## 4.4.  Old-Growth Forests and Biodiversity

• Aging patches are a modality adopted by Ministère des Ressources naturelles du Québec which consist in patches of forest with increased rotation length.

One of the major biodiversity challenges in the North Shore region will be to maintain large tracts of intact forest and to use different methods to reproduce the conditions characterizing the old-growth forest matrix (e.g. by using adaptive forest practices and aging patches•). As discussed in this chapter, old-growth forests are for the most part extremely dynamic systems in both space and time.

Treatments therefore need to be distributed in such a way as to reproduce the spatial patterns associated with natural forest dynamics, at various levels, from landscape to stand.

At the landscape level, maintaining tracts of intact forest is essential to protecting animal species with large home ranges, such as the woodland caribou (*Rangifer tarandus caribou*), a species that has been designated as vulnerable in Québec. Courtois et al. (2005) showed that woodland caribou avoid recently harvested areas, either taking shelter in tracts of forest specifically left intact for this purpose, or temporarily leaving the managed areas. The woodland caribou situation is a major issue on the North Shore and the species' preservation may well depend on the implementation of an ecosystem-based management strategy developed to emulate the spatial processes and patterns resulting from natural disturbances.

At the stand level, maintaining dead wood and large standing trees is another critical regional issue. These structural characteristics of stands are instrumental to sustaining diversity (Harmon et al. 1986; Hansen et al. 1991; Hunter 1999; Kuuluvainen 2002). Many wildlife species depend on the presence of large-diameter dead trees. Barrow's Goldeneye (*Bucephala islandica*), a cavity-nesting duck, is one such species. The breeding area for the Barrow's Goldeneye population in eastern North America is concentrated on the North Shore (Robert et al. 2000). Moreover, this species is likely to be designated as threatened or vulnerable in Québec owing to its limited numbers.

As Aakala et al. (2007) showed, dead wood production varies according to stand type in addition to being strongly influenced by temporal variability in disturbance severity (Périgon 2005; Reyes 2009). To maintain biodiversity, we need to adjust our management practices by taking into account the spatio-temporal variability associated with dead wood production, and we need to establish target levels for dead wood production, including decomposition class thresholds.

## 5.  CONCLUSION

The North Shore forest contradicts the long-held dogma that the boreal forest is composed primarily of even-aged stands. The large proportion of structurally complex stands in the immense North Shore forest area calls into question the way in which this forest is currently being managed. In addition, social pressure and commitments made to the international community to develop sustainable forest management practices make it necessary to review the way in which the boreal forest is managed. Another issue is whether the silvicultural treatments currently in use actually optimize production and long-term fibre yield and promote resistance to and resilience in the face of environmental change. Conventional silvicultural treatments assume and prescribe homogeneity in boreal forest stands to optimize and predict fibre yield. North Shore forest stands are however characterized by a heterogeneous structure and composition. Maintaining this heterogeneity through the use of adaptive silvicultural systems and management strategies is necessary in the context of an ecosystem-based management approach and may even prove beneficial for fibre production.

## REFERENCES

Aakala, T., Kuuluvainen, T., De Grandpré, L., and Gauthier, S. 2007. Trees dying standing in the northeastern boreal old-growth forests of Quebec: spatial patterns, rates, and temporal variation. Can. J. For. Res. **37**: 50–61.

Bartemucci, P., Coates, K.D., Harper, K.A., and Wright, E.F. 2002. Gap disturbances in northern old-growth forests of British Columbia, Canada. J. Veg. Sci. **13**: 685–696.

Belleau, A., Bergeron, Y., Leduc, A., Gauthier, S., and Fall, A. 2007. Using spatially explicit simulations to explore size distribution and spacing of regenerating areas produced by wildfires: recommendations for designing harvest agglomerations for the Canadian boreal forest. For. Chron. **83**: 72–83.

Benoît, P. and Desaulniers, R. 1972. Épidémies passées et présentes de l'arpenteuse de la pruche au Québec. Rev. Bimestr. Rech. **2**: 11–12.

Bergeron, Y., Cyr, D., Drever, C.R., Flannigan, M., Gauthier, S., Kneeshaw, D., Lauzon, È., Leduc, A., Le Goff, H., Lesieur, D., and Logan, K. 2006. Past, current, and future fire frequencies in Quebec's commercial forests: implications for the cumulative effects of harvesting and fire on age-class structure and natural disturbance-based management. Can. J. For. Res. **36**: 2737–2744.

Bergeron, Y., Leduc, A., Morin, H., and Joyal, C. 1995. Balsam fir mortality following the last spruce budworm outbreak in northwestern Québec. Can. J. For. Res. **25**: 1375–1384.

Blais, J.R. 1983. Trends in the frequency, extent, and severity of spruce budworm outbreaks in eastern Canada. Can. J. For. Res. **13**: 539–547.

Bouchard, M., Kneeshaw, D., and Bergeron, Y. 2006. Forest dynamics after successive spruce budworm outbreaks in mixedwood forests. Ecology, **87**: 2319–2329.

Boucher, D., De Grandpré, L., and Gauthier, S. 2003. Développement d'un outil de classification de la structure des peuplements et comparaison de deux territoires de la pessière à mousses du Québec. For. Chron. **79**: 318–328.

Boucher, D., Gauthier, S., and De Grandpré, L. 2006. Structural changes in coniferous stands along a chronosequence and a productivity gradient in the northeastern boreal forest of Québec. Écoscience, **13**: 172–180.

Boulanger, Y. and Arseneault, D. 2004. Spruce budworm outbreaks in eastern Quebec over the last 450 years. Can. J. For. Res. **34**: 1035–1043.

Burns, R.H. and Honkala, B.H. 1990. Silvics of North America, Vols. 1 and 2. USDA For. Serv. Agric. Handb. 654, Washington, D.C., USA.

Côté, G. 2006. Élaboration d'une typologie forestière adaptée à la forêt boréale irrégulière. M. Sc. thesis, Université Laval, Québec, Que.

Courtois, R., Sebbane, A., Gingras, A., Rochette, B., Breton, L., and Fortin, D. 2005. Changement d'abondance et adaptations du caribou dans un paysage sous aménagement. MRNF and Université Laval, Québec, Que.

Cyr, D., Gauthier, S., and Bergeron, Y. 2007. Scale-dependent determinants of heterogeneity in fire frequency in a coniferous boreal forest of eastern Canada. Landsc. Ecol. **22**: 1325–1339.

De Grandpré, L., Morissette, J., and Gauthier, S. 2000. Long-term post-fire changes in the northeastern boreal forest of Quebec. J. Veg. Sci. **11**: 791–800.

Doucet, R. 1988. La régénération préétablie dans les peuplements forestiers naturels au Québec. For. Chron. **64**: 116–120.

Filion, L., Payette, S., Delwaide, A., and Bhiry, N. 1998. Insect defoliators as major disturbance factors in the high-altitude balsam fir forest of Mount Mégantic, southern Quebec. Can. J. For. Res. **28**: 1832–1842.

Frelich, L.E. and Reich, P.B. 2003. Perspectives on development of definitions and values related to old-growth forests. Environ. Rev. **11**: S9–S22.

Gagnon, R., Morin, H., Lord, D., Krause, C., Potvin, J., Savard, G., and Cloutier, S. 1999. Nouvelles connaissances sur la dynamique naturelle des forêts d'épinette noire au Québec. L'Aubelle, **128**: 10–14.

Gauthier, S., Morissette, J., Boucher, D., and De Grandpré, L. 2004. Succession forestière dans la forêt boréale de la Côte-Nord du Québec: facteurs impliqués dans les changements de composition des espèces sur une période de près de 60 ans. *In* De Grandpré, L., Gauthier, S., Morissette, J., and Bergeron, Y. Amélioration de la précision du calcul de la possibilité forestière par une meilleure connaissance de la dynamique naturelle de la forêt boréale de la Côte-Nord. Final report presented to the Fonds forestier, MRNFP, March.

Gray, D.R., Régnière, J., and Boulet, B. 2000. Analysis and use of historical patterns of spruce budworm defoliation to forecast outbreak patterns in Quebec. For. Ecol. Manag. **127**: 217–231.

Hansen, A.J., Spies, T.A., Swanson, F.J., and Ohmann, J.L. 1991. Conserving biodiversity in managed forests. BioScience, **41**: 382–392.

Hardy, Y., Mainville, M., and Schmitt, D.M. 1986. An atlas of spruce budworm defoliation in Eastern North America 1938–1980. USDA For. Serv. Misc. Publ. No. 1449, Washington, D.C., USA.

Harmon, M.E., Franklin, J.F., Swanson, F.J., Sollins, P., Gregory, S.V., Lattin, J.D., Anderson, N.H., Cline, S.P., Aumen, N.G., Sedell, J.R., Lienkaemper, G.W., Cromack, K. Jr, and Cummins, K.W. 1986. Ecology of coarse woody debris in temperate ecosystems. Adv. Ecol. Res. **15**: 133–302.

Harper, K.A., Bergeron, Y., Drapeau, P., Gauthier, S., and De Grandpré, L. 2005. Structural development following fire in black spruce boreal forest. For. Ecol. Manag. **206**: 293–306.

Hunter Jr, M.L. (Ed.). 1999. Maintaining biodiversity in forest ecosystems, Cambridge University Press, Cambridge, UK.

Jardon, Y. 2001. Analyses temporelles et spatiales des épidémies de la tordeuse des bourgeons de l'épinette au Québec. Ph.D. thesis, Université du Québec à Montréal, Montréal, Que.

Jardon, Y., Morin, H., and Dutilleul, P. 2003. Périodicité et synchronisme des épidémies de la tordeuse des bourgeons de l'épinette au Québec. Can. J. For. Res. **33**: 1947–1961.

Jobin, L.J. and Desaulniers, R. 1981. Results of aerial spraying in 1972 and 1973 to control the eastern hemlock looper (*Lambdina fiscellaria fiscellaria* (Guen.)) on Anticosti Island. Info. Rep. LAU-X-49E. Environment Canada, Canadian Forestry Service, Laurentian Forest Research Centre, Sainte-Foy, Que.

Johnson, E.A. and Gutsell, S.L. 1993. The heat budget and fire behaviour associated with the opening of serotinous cones in two *Pinus* species. J. Veg. Sci. **4**: 745–750.

Kneeshaw, D. and Bergeron, Y. 1998. Canopy gaps characteristics and tree replacement in the southeastern boreal forest. Ecology, **79**: 783–794.

Kuuluvainen, T. 2002. Natural variability of forests as a reference for restoring and managing biological diversity in boreal Fennoscandia. Silva Fenn. **36**: 97–125.

Laberge Pelletier, C., Munson, A., and Ruel, J.-C. 2007. L'environnement des éricacées des forêts de l'est du Québec. Bulletin d'information No 6. Chaire de recherche industrielle CRSNG-Université Laval en sylviculture et faune, Université Laval, Québec, Que.

Lecomte, N., Simard, M., and Bergeron, Y. 2006. Effects of fire severity and initial tree composition on stand structural development in the coniferous boreal forest of northwestern Québec, Canada. Écoscience, **13**: 152–163.

Lertzman, K.P. and Krebs, C.J. 1991. Gap-phase structure of a subalpine old-growth forest. Can. J. For. Res. **21**: 1730–1741.

MacLean, D.A. 1980. Vulnerability of fir-spruce stands during uncontrolled spruce budworm outbreaks: a review and discussion. For. Chron. **56**: 213–221.

MacLean, D.A. 1984. Effects of spruce budworm outbreaks on the productivity and stability of balsam fir forests. For. Chron. **60**: 273–279.

MacLean, D.A. and MacKinnon, W.E. 1997. Effects of stand and site characteristics on susceptibility and vulnerability of balsam fir and spruce to spruce budworm in New Brunswick. Can. J. For. Res. **27**: 1859–1871.

Mallik, A.U. 1995. Conversion of temperate forests into heaths: role of ecosystem disturbance and ericaceous plants. Environ. Manag. **19**: 675–684.

McCarthy, J. 2001. Gap dynamics of forest trees: a review with particular attention to boreal forests. Environ. Rev. **9**: 1–59.

McCarthy, J.W. and Weetman, G. 2006. Age and size structure of gap-dynamic, old-growth boreal forest stands in Newfoundland. Silva Fenn. **40**: 209–230.

Ministère des Ressources naturelles, de la Faune et des Parcs (MNRFP). 2004. Portrait forestier, Région de la Côte-Nord. Document d'information sur la gestion de la forêt publique. Publication No 2004-3524. Ministère des Ressources naturelles, de la Faune et des Parcs, Direction régionale de la Côte-Nord, Baie-Comeau, Que.

Morin, H. and Laprise, D. 1990. Histoire récente des épidémies de la tordeuse des bourgeons de l'épinette au nord du lac Saint-Jean (Québec): une analyse dendrochronologique. Can. J. For. Res. **20**: 1–8.

Morin, H., Laprise, D., and Bergeron, Y. 1993. Chronology of spruce budworm outbreaks near Lake Duparquet, Abitibi region, Quebec. Can. J. For. Res. **23**: 1497–1506.

Nappi, A., Drapeau, P., and Savard, J.-P.L. 2004. Salvage logging after wildfire in the boreal forest: is it becoming a hot issue for wildlife? For. Chron. **80**: 67–74.

Oliver, C.D. and Larson, B.C. 1996. Forest stand dynamics, Update Edition. John Wiley and Sons, New York, USA.

Payette, S., Bhiry, N., Delwaide, A., and Simard, M. 2000. Origin of the lichen woodland at its southern range limit in eastern Canada: the catastrophic impact of insect defoliators and fire on the spruce–moss forest. Can. J. For. Res. **30**: 288–305.

Payette, S. and Delwaide, A. 2003. Shift of conifer boreal forest to lichen-heath parkland caused by successive stand disturbances. Ecosystems, **6**: 540–550.

Périgon, S. 2005. Dynamique de trouées dans de vieux peuplements résineux de la Côte-Nord, Québec. M. Sc. thesis, Université du Québec à Montréal, Montréal, Que.

Pham, A.T., De Grandpré, L., Gauthier, S., and Bergeron, Y. 2004. Gap dynamics and replacement patterns in gaps of the northeastern boreal forest of Quebec. Can. J. For. Res. **34**: 353–364.

Reyes, G. 2009. Natural disturbance dynamics in *Abies balsamea-Betula* spp. Boreal mixedwood of southern Quebec: examination of spatio-temporal factors affecting woody vegetation diversity and abundance. Ph.D. thesis. Université du Québec à Montréal, Montréal, Que.

Robert, M., Bordage, D., Savard, J.-P.L., Fitzgerald, G., and Morneau, F. 2000. The breeding range of the Barrow's Goldeneye in eastern North America. Wilson Bull. **112**: 1–7.

Robitaille, A. and Saucier, J.-P. 1998. Paysages régionaux du Québec méridional. Les Publications du Québec, Sainte-Foy, Que.

Rowe, J.S. 1972. Forest regions of Canada. Canadian Forestry Service. Ottawa, Ont., Publication No. 1300.

Ruel, J.-C. 1989. Importance de la régénération préexistante dans les forêts publiques du Québec. Ann. Sci. For. **46**: 345–359.

Ruel, J.-C. 2000. Factors influencing windthrow in balsam fir forests: from landscape studies to individual tree studies. For. Ecol. Manag. **135**: 169–178.

Ruel, J.-C., Horvath, R., Ung, C.H., and Munson, A. 2004. Comparing height growth and biomass production of black spruce trees in logged and burned stands. For. Ecol. Manag. **193**: 371–384.

Schmiegelow, F.K.A., Stepnisky, D.P., Stambaugh, C.A., and Koivula, M. 2006. Reconciling salvage logging of boreal forests with a natural-disturbance management model. Conserv. Biol. **20**: 971–983.

Swaine, J.M. and Craighead, F.C. 1924. Studies on the spruce budworm (*Cacoecia fumiferana* Clem.). Part 1. A general account of the outbreaks, injury and associated insects. Agric. Can. Bull. Tech. **37**: 3–27.

Thibault, M. and Hotte, D. 1985. Les régions écologiques du Québec méridional (2$^e$ approximation). Carte couleur à l'échelle 1: 250 000. Ministère de l'Énergie et des Ressources du Québec, Service de la recherche, Québec, Que.

Thiffault, N. and Jobidon, R. 2006. How to shift unproductive *Kalmia angustifolia – Rhododendron groenlandicum* heath to productive conifer plantation. Can. J. For. Res. **36**: 2364–2376.

Vaillancourt, M.-A., Drapeau, P., Gauthier, S., and Robert, M. 2008. Availability of standing trees for large cavity nesting birds in the eastern boreal forest of Québec, Canada. For. Ecol. Manag. **255**: 2272–2285.

Watson, E.B. 1934. An account of the eastern hemlock looper, *Ellopia fiscellaria* Gn., on balsam fir. Sci. Agric. **14**: 669–678.

Wein, R.S. and McLean, D.A. 1983. The role of fire in northern circumpolar ecosystems. John Wiley and Sons, New York, USA.

# Chapter 11

# Ecosystem Management of Québec's Northern Clay Belt Spruce Forest*
## Managing the Forest...
## and Especially the Soils

*Martin Simard, Nicolas Lecomte, Yves Bergeron, Pierre Y. Bernier, and David Paré*

*   We thankfully acknowledge the funding agencies that have supported the work presented in this chapter: the Network for Centres of Excellence in Sustainable Forest Management, the Natural Sciences and Engineering Research Council of Canada, the Fonds québécois de recherche sur la nature et les technologies, the Fonds d'action québécois pour le développement durable, the Programme de mise en valeur du milieu forestier (volet 1) of the Ministère des Ressources naturelles et de la Faune du Québec, Canada Economic Development, Natural Resources Canada, the industrial partners of the NSERC-UQAT-UQAM Industrial Chair in Sustainable Forest Management, and the Université du Québec en Abitibi-Témiscamingue. We also wish to thank Nicole Fenton, Martin Lavoie, Sonia Légaré, and Annie Belleau for their contribution to this work through insightful and productive discussions. Finally, we thank Pamela Cheers, Benoit Arsenault, and Isabelle Lamarre of Natural Resources Canada for the linguistic revision of the manuscript. Photos on this page were graciously provided by Martin Simard.

# 1. INTRODUCTION

Forests of the northern Clay Belt of Québec and Ontario (Canada) are dominated by black spruce forests. The natural landscape and stand dynamics of these forests have been widely studied (Bergeron et al. 2002; Lefort et al. 2002; Gauthier et al. 2004; Carleton and Maycock 1978, 1980). Many studies have documented disturbance (Carcaillet et al. 2001; Bergeron et al. 2004) and stand dynamics (Gauthier et al. 2000; Harper et al. 2003; Lecomte et al. 2006a), while others have compared the effects of natural disturbances with those of silvicultural treatments (Nguyen-Xuan et al. 2000; Simard et al. 2001; Fenton and Bergeron 2007; Fenton et al., chapter 15). Recent studies have focused on the paludification process (i.e. the gradual accumulation of soil organic matter ultimately leading to the development of a peatland) and its effects on forest productivity and biodiversity. Findings from these studies require us to revisit the forest management strategy previously proposed for the black spruce forest of northwestern Québec (Bergeron et al. 1999).

Forests in this region are managed commercially for their timber. Accounting for natural dynamics of these forests could help forest managers in maintaining forest productivity without compromising biodiversity. In particular, three characteristics of the regional forest dynamics must be taken into account for the successful implementation of an ecosystem management strategy. First, paludification is a pervasive and widespread phenomenon that has important consequences for the diversity and productivity of stands and landscapes. Second, soil burn severity is highly variable and affects tree regeneration, forest productivity and succession, and biodiversity. Finally, most stands show little or no tree species replacement during succession, and as a result, changes in stand structure are of particular importance. As we demonstrate in this chapter, ignoring these properties of natural forest dynamics can lead to a decrease in both biodiversity and forest productivity.

In this chapter, we describe our current understanding of natural disturbance dynamics and long-term post-fire stand succession in Québec's western black spruce–feather moss bioclimatic subdomain. Building on this foundation, we propose a new ecosystem management strategy that relies on silvicultural treatments whose effects are similar to those of the region's natural disturbance regime. We conclude by discussing societal choices regarding the regional forest management, considering the historical variability of forest and disturbance dynamics, and the global consequences of decisions taken at the regional level.

# 2. REGIONAL CONTEXT

The study area (49°00′–50°30′N; 78°30′–79°30′W) is located in northern Abitibi, a region of northwestern Québec, about 200 km south of James Bay (figure 11.1). This region of the Precambrian Shield was covered by the Laurentian ice sheet until about 9,000 years ago, when the gradual retreat of the glacier gave way

Figure 11.1
# Location of study area and of major surficial deposit types in Abitibi

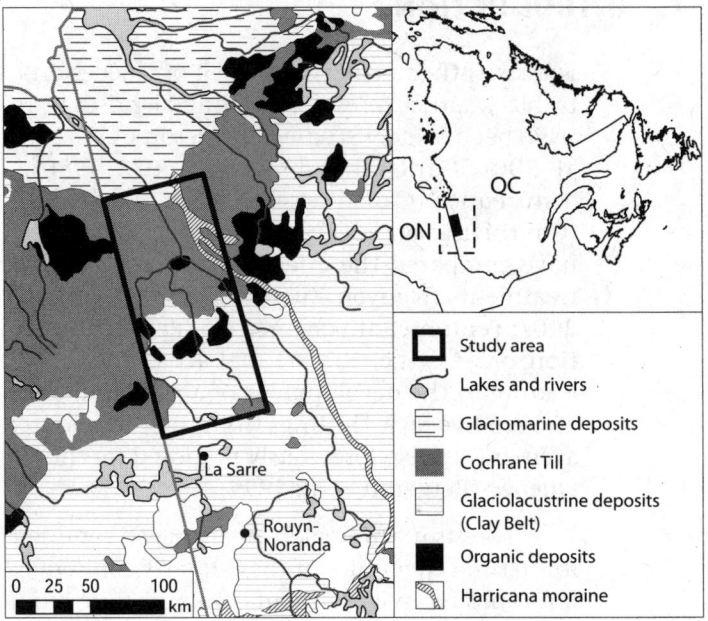

Note the location of the Cochrane Till and of organic deposits. Source: Natural Resources Canada, Geological Survey of Canada. Surficial Materials of Canada, map 1880a, <gsc.nrcan.gc.ca/map/1880a/index_e.php> and <atlas.nrcan.gc.ca/site/english/maps/environment/land/surficialmaterials>.

to the proglacial lake Barlow-Ojibway (Veillette 1994). Heavy clay deposits were left by the lake, forming the physiographic unit known today as the Clay Belt. A few centuries before lake Barlow-Ojibway drained into the Tyrrell Sea, southward glacial surges incorporated gravel into the lacustrine clay, resulting in a relatively compact surficial deposit called the Cochrane Till (figure 11.1; Veillette 1994). Most of the study area is located on this surficial deposit whose properties have a predominant effect on regional vegetation and ecological processes.

Although some important rock outcrops occur throughout the landscape, topography is generally flat (mean elevation = 250 m). On over half of this territory, organic soils deeper than 60 cm cover the Cochrane Till. Other major types of soils in the region include Luvisols and Gleysols (Soil Classification Working Group 1998). Mean annual temperature is –0.7°C and annual precipitation is 900 mm, 35% of which falls during the growing season (Environment Canada 2006). Growing season frosts are frequent, and although periglacial activity has been observed in the region (Brown and Gangloff 1980), there is no permafrost.

The study area is part of the western black spruce–feather moss bioclimatic subdomain (Robitaille and Saucier 1998) and is dominated by black spruce (*Picea mariana* [Mill.] B.S.P.) stands showing high variability in height and density. Jack pine (*Pinus banksiana* Lamb.) and trembling aspen (*Populus tremuloides* Michx.), although less abundant, form mixed stands with black spruce. Over 85% of the territory is forested, 90% of which is on public lands. Forest harvesting represents 10% of direct employment in the Abitibi-Témiscamingue administrative region, with most of the jobs in the first transformation sector (Lessard 2004).

## 3. DISTURBANCE REGIME

### 3.1. Fire Frequency

Forest wildfires are undoubtedly the primary disturbance factor in Abitibi's black spruce forests. Fire reconstruction studies have shown that the fire cycle in the region is considerably longer than it was during the Little Ice Age, increasing from 100 years before 1850, to 135 years between 1850 and 1920, to about 400 years since 1920 (Bergeron et al. 2004). Although the western black spruce–feather moss bioclimatic subdomain has the same fire cycle than the neighbouring western balsam fir–white birch (*Abies balsamifera* (L.) Mill. – *Betula papyrifera* Marsh.) bioclimatic subdomain in the south, black spruce forests are characterized by a lower number of fires that are larger in extent, compared with the boreal mixedwood forest (Gauthier et al. 2000; Bergeron et al. 2004). The numerous old paludified stands of black spruce forests do not seem to act as significant fire breaks. Therefore, they have an equal chance of burning compared with younger forests, which is corroborated by the presence of macroscopic charcoal fragments in their soils (Cyr et al. 2005).

### 3.2. Soil Burn Severity

In crown fire-dominated ecosystems, fire regimes are usually described in terms of their cycle (Vaillancourt et al., chapter 2). It is now widely recognized that fire severity (i.e. the effect of fire on vegetation and soils) is also very important to understand vegetation patterns, and many studies have described the spatial heterogeneity of fire severity within burned areas (Turner et al. 1994, 1997, 1999; Kafka et al. 2001). In boreal ecosystems that accumulate great amounts of soil organic matter, like in northern Abitibi, soil burn severity determines forest productivity and regeneration (Simard et al. 2007; Greene et al. 2007). Humus combustion stimulates nutrient cycling through the release of nutrients and the warming of the soil (MacLean 1983; Wardle et al. 1997; Smithwick et al. 2005). Organic-layer combustion also creates microsites that are essential for the regeneration of many tree species (Charron and Greene 2002; Jayen et al. 2006; Greene et al. 2007).

Ideally, soil burn severity should be measured in terms of organic matter thickness consumed by the fire (Miyanishi and Johnson 2002), but, for practical reasons, is usually quantified by the thickness of the residual layer left after the burn (Nguyen-Xuan et al. 2000; Greene et al. 2005). High severity soil burns are characterized by soil profiles in which a single charcoal layer lies directly above the mineral soil, whereas lower severity burns have a charcoal layer sitting atop a layer of residual organic matter. This residual organic layer can also contain one or several layers of charcoal produced by previous fires (figure 11.2A).

### Figure 11.2
**Residual organic-layer depth allows the separation of stands that originate from high-severity soil burns from those that establish after low-severity fires**

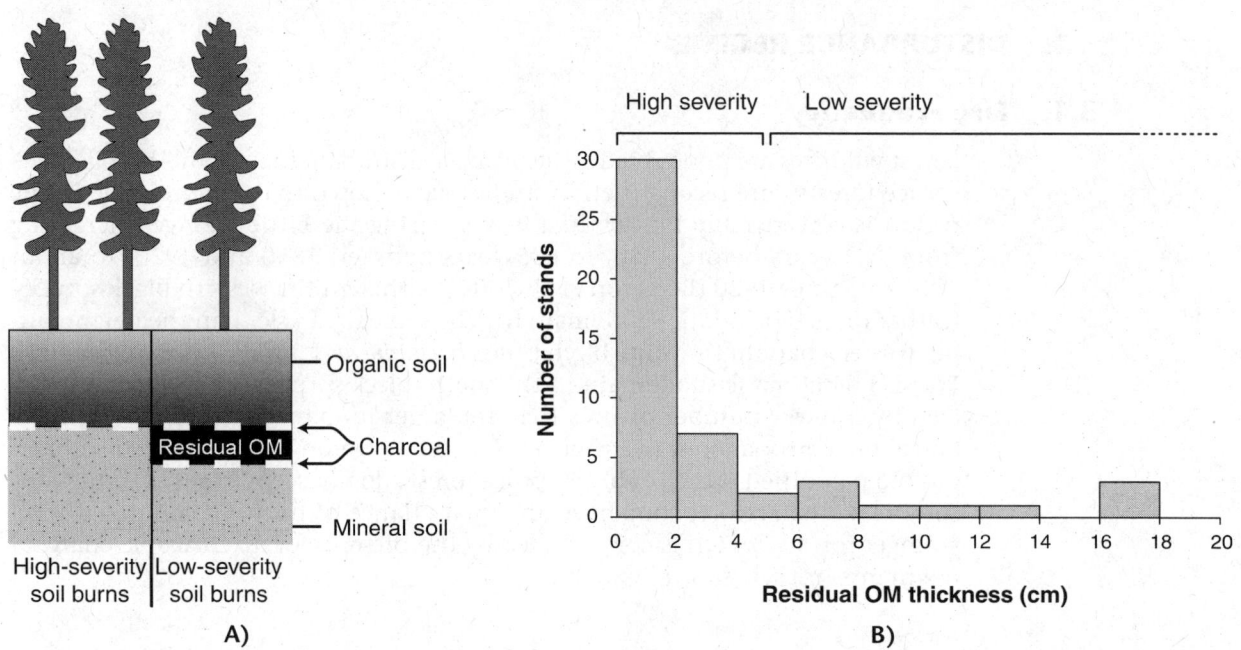

Legend: A) High-severity soil burns consume almost all the organic layer, and consequently, charcoal fragments of the last fire are found directly on the mineral soil. A low-severity soil burn, on the contrary, only partially burns the organic layer, leaving a distinct layer of residual organic matter (OM). B) Residual OM thickness makes it possible to separate stands that established after high- (residual OM < 5 cm) vs. low-severity soil burns (residual OM > 5 cm). Modified from Lecomte et al. (2005, 2006a) and Simard et al. (2007).

In black spruce forests of northern Abitibi, soil burn severity varies considerably, with residual organic layers ranging from 1 cm to more than 15 cm on average, and often exceeding 40 cm locally (figure 11.2B). Since a residual organic layer deeper than 4–5 cm significantly affects the establishment of several tree species (Johnstone and Chapin 2006; Greene et al. 2007), this thickness can be used as a threshold to distinguish high-severity soil burns (residual organic layer < 5 cm) from low-severity ones (residual organic layer > 5 cm) (Lecomte et al. 2005, 2006a). In the following text, the terms *high-* and *low-severity soil burns* essentially refer to this ecologically meaningful threshold. Our research has shown that soil burn severity is a critical factor in black spruce forest dynamics because it creates alternative successional pathways following fire (Lecomte et al. 2006a). We now know that ignoring this factor can lead to incorrect conclusions regarding their growth and successional dynamics (Lecomte et al. 2006b; see section 4).

## 3.3.  Insect Outbreaks and Windthrow

Forests become increasingly susceptible to some biotic (infestations of insect pests, pathogens, etc.) and abiotic (e.g. windthrow) disturbances as they mature. As a result, there is often a negative relationship between the frequency of these disturbances and the frequency of fires. In Québec, this can be seen as an eastward increase in spruce budworm (*Choristoneura fumiferana* Clem.) outbreak severity from Abitibi to the North Shore region (Boulet et al. 1996), concomitant with a reduction in fire frequency (Gauthier et al. 2001). Although some forest pests (the spruce budworm and the forest tent caterpillar, *Malacosoma disstria* Hbn.) have a detectable impact on tree growth (Leblanc 1999; Levasseur 2000), black spruce forests of northwestern Québec are not affected by the severe outbreaks that are more common in the southern and eastern forests of the province (Boulet et al. 1996; Gray et al. 2000; Harper et al. 2002; Morin et al., chapter 7). Windthrow actually seems to affect a larger forest area (9% for all stand age classes; 15% for the 100–250-year-old class) than does the spruce budworm (<1% for all stand age classes; about 2% for the 100–250-year-old class) (Harper et al. 2002, 2003).

## 4.  FOREST DYNAMICS

## 4.1.  Paludification

One of the most significant ecological processes occurring in the black spruce forests of northern Abitibi is the paludification of forest soils (Boudreault et al. 2002; Fenton et al. 2005; Lavoie et al. 2005; Simard et al. 2007). Paludification is the process of soil organic matter accumulation and concomitant rise in the water table that can eventually lead to the development of peatlands (Payette and Rochefort 2001). In forest sites, paludification usually begins with an excessive accumulation of forest humus followed by the development of *Sphagnum* peat (Van Cleve and Viereck 1981; Van Cleve et al. 1983; Viereck 1983). Two major types of paludification processes are recognized: "edaphic paludification" (also called paludification of wet depressions; Payette 2001), where topography plays a determinant role in keeping the water table close to the soil surface, and "successional paludification" (also called paludification of well-drained soils; ibid.), where forest succession is the main driver of the process (Simard et al. 2007). In the field, these two processes are confounded but their differences are of significance to forest management. Successional paludification is a potentially reversible biological phenomenon, whereas edaphic paludification is an intrinsic property of a site whose modification would require considerable resources (e.g. forest drainage) (Simard et al. 2009).

In northern Abitibi, on sites with a gentle slope, organic-layer accumulation (successional paludification) can be predicted fairly well if soil burn severity is taken into account (figure 11.3A; Fenton et al. 2005; Lecomte et al. 2006b). Because soil organic layers in post-fire stands develop over the residual (unburned) organic layer, their overall depth will always be greater and more variable in stands that originate from low-severity soil burns than in stands of the same age but established after high-severity soil burns. Peat accumulation is accompanied by a decrease in soil temperature (figure 11.3B) and the development of a superficial water table (figure 11.3C; Fenton et al. 2006). Water table development probably results from the low hydraulic conductivity of humified peat which slows vertical drainage, coupled to the reduction in canopy evapotranspiration caused by a reduction in stem density and leaf area.

As sites undergo paludification, the rooting zone of black spruce trees gradually migrates upwards from the mineral soil into the accumulating organic horizon (figure 11.3D) via production of adventitious roots on established trees and by the establishment by seed or layering of new stems on the forest floor. The transition of the rooting zone from the mineral to the organic horizon is a major change because peat is a poor growth substrate compared with clay (Munson and Timmer 1989; Roy et al. 1999; Lavoie et al. 2007b and c). This transition occurs mainly between 100 and 200 years after fire, which corresponds to the age at which the cohort of trees established after high-severity soil burns starts to break up (Lecomte et al. 2006a).

When the depth of accumulating organic material exceeds 60 cm, the designation of the surficial deposit changes from mineral to organic (Perron and Morin 2002). Forests growing on the fine-grained mineral soils and on the organic deposits are classified as different ecological types under the current ecological classification system of the Québec's Ministère des Ressources naturelles. However, these two ecological types are not permanent and in essence lie at both ends of the paludification gradient. The conversion from one deposit type to the other is caused by succession on the one hand (fine-grained towards organic), and by high severity soil burns on the other hand (organic to fine-grained).

The non-permanence of ecological types means that it is inappropriate to study forest succession on each deposit type because fine-grained and organic deposits are part of the same chronosequence. It is on the other hand essential to consider soil burn severity to understand post-fire succession patterns. As a result, we will emphasize studies that have quantified soil burn severity in the context of successional paludification, i.e. on fine-grained and organic deposits. These deposits represent 80% of the area occupied by productive forests (Harper et al. 2002; Lecomte and Bergeron 2005) in northern Abitibi, and an even greater percentage of the territory if unproductive forests are considered.

## 4.2. Forest Structure and Succession

Forest succession studies in the northern Abitibi region have used different approaches and data sources: chronosequence studies based on forest inventory data or photo-interpretation (Gauthier et al. 2000; Harper et al. 2002, 2003; Lecomte and Bergeron 2005; Lecomte et al. 2006a), medium-term (1970–1990)

Figure 11.3
**Soil organic matter accumulation (A), soil temperature at 10 cm below the soil surface (B), water table depth relative to the mineral-organic soil interface (C), and black spruce rooting zone (D) in relation to time since high- and low-severity fire**

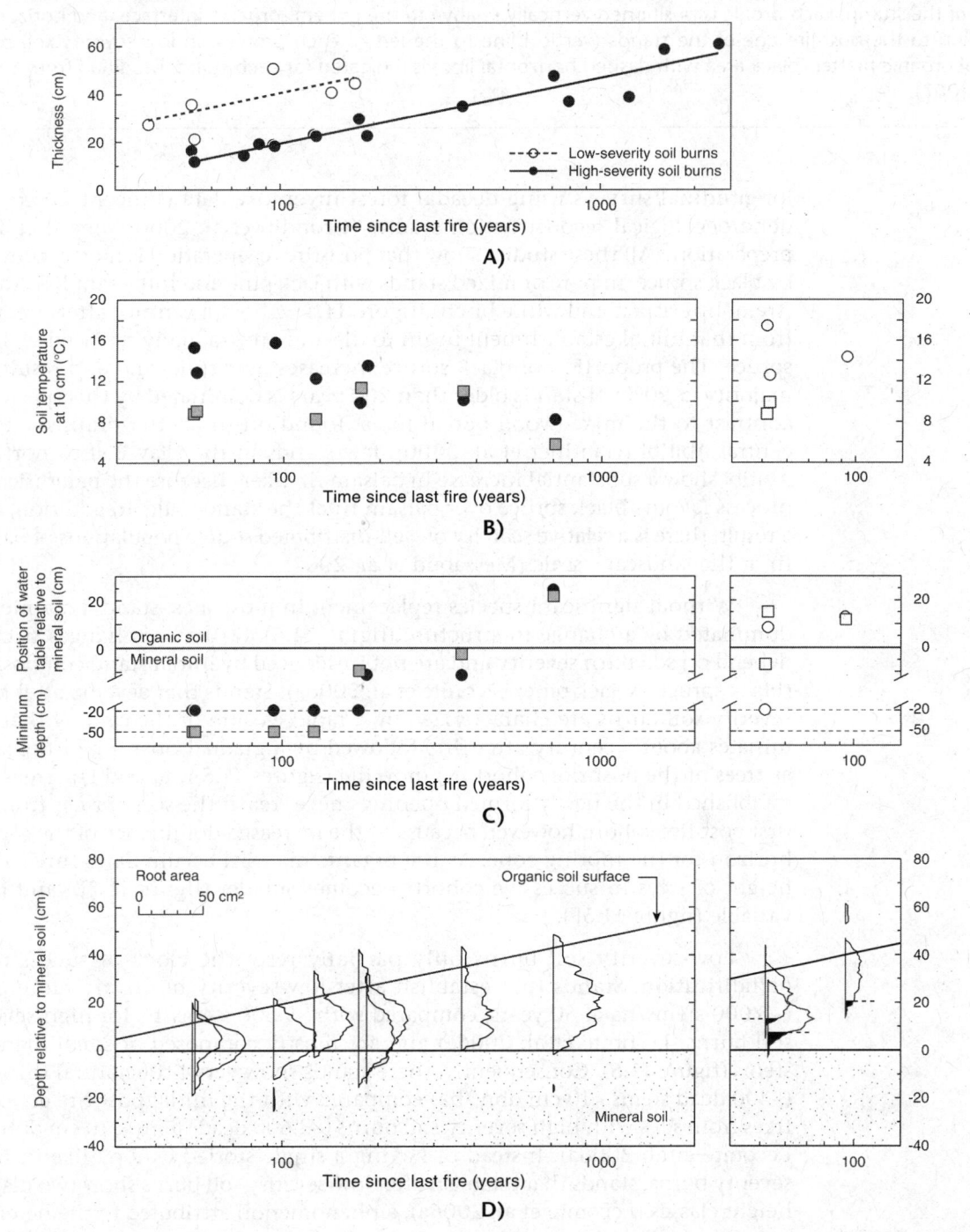

*(See legend at the top of next page.)*

Left and right panels show high- and low-severity soil burns, respectively, except in A) where high- (solid symbols) and low- (open symbols) severity soil burns are superimposed for comparison. In B) and C), each symbol represents the average of 20 measurements taken in 2003 (circles) and 2004 (squares). In C), minimal depth is indicated when the water table was not reached. In D), the oblique line represents the surface of organic horizons and each profile (mean of 3 trees of the dominant height cohort) represents the cumulative area (irregular lines to the right of each profile) occupied by black spruce roots >1 cm in diameter at a distance of 30 cm from the centre of the stump. Each profile is positioned vertically relative to the mineral-organic interface, and horizontally according to the post-fire age of the stands (vertical line to the left of each profile). In low-severity soil burns, residual organic matter (black area with dashed horizontal line) is indicated for each stand. Modified from Simard et al. (2007).

---

longitudinal surveys using decadal forest inventory data (Vincent 2004), and dendroecological reconstruction studies (Grondin et al. 2000; Simard et al. in preparation). All these studies show that post-fire regeneration is mostly provided by black spruce, in pure or mixed stands with jack pine and intolerant hardwoods (trembling aspen and white birch) (figure 11.4). About a century after fire, trees from that initial establishment begin to die and are gradually replaced by black spruce. The proportion of black spruce increases over time, and as a result, the majority (> 90%) of stands older than 200 years is dominated by this species. In contrast to the mixedwood boreal forest found on the better-drained sites of central Abitibi (Gauthier et al. 2000), few stands in the Clay Belt of northern Abitibi show a substantial increase in balsam fir, likely because the paludification process favours black spruce over balsam fir at the stand scale. In addition, or as a result, there is a relative scarcity of well-distributed source populations of balsam fir at the landscape scale (Messaoud et al. 2007).

Without significant species replacement in most sites, stand dynamics are dominated by a change in structure (figure 11.5). Structural changes strongly depend on soil burn severity and are not influenced by initial stand composition (black spruce vs. jack pine; Lecomte et al. 2006a). Stands that develop after high-severity soil burns are characterized by a rapid closure of the canopy that culminates about a century after fire, followed by a gradual opening of the stand as trees of the post-fire cohort begin to die (figures 11.5A, B, and D). Trees that established in the newly formed openings never reach the size of trees from the first post-fire cohort, however, because of the increased dominance of the organic horizon for the rooting zone. As the organic material accumulates further, the height of trees in successive cohorts becomes smaller (figure 11.5E) and more variable (figure 11.5F).

Low-severity soil burns only partially reset the clock of successional paludification. Stands that establish after low-severity burns are more open (~6,000 stems/ha at 50 years, compared with ~9,500 stems/ha for high-severity soil burns; Lecomte et al. 2006a) and are mostly composed of small-diameter stems (figure 11.5). Consequently, these stands show very little natural thinning (~300 dead stems <10 cm dbh /ha) compared with the initial post-fire cohort of trees from sites with high severity soil burns (~3,600 dead stems < 10 cm dbh /ha; Lecomte et al. 2006a). Instead of having a single-storied canopy like in high-severity burns, stands that establish after low-severity soil burns show two distinct height classes (Lecomte et al. 2006a), a phenomenon attributed to the heterogeneity in post-fire germination substrates (Greene et al. 2007; Lavoie et al. 2007a).

Figure 11.4
## Forest succession on organic and fine deposits in northwestern Québec's black spruce–moss forest according to different chronosequence studies

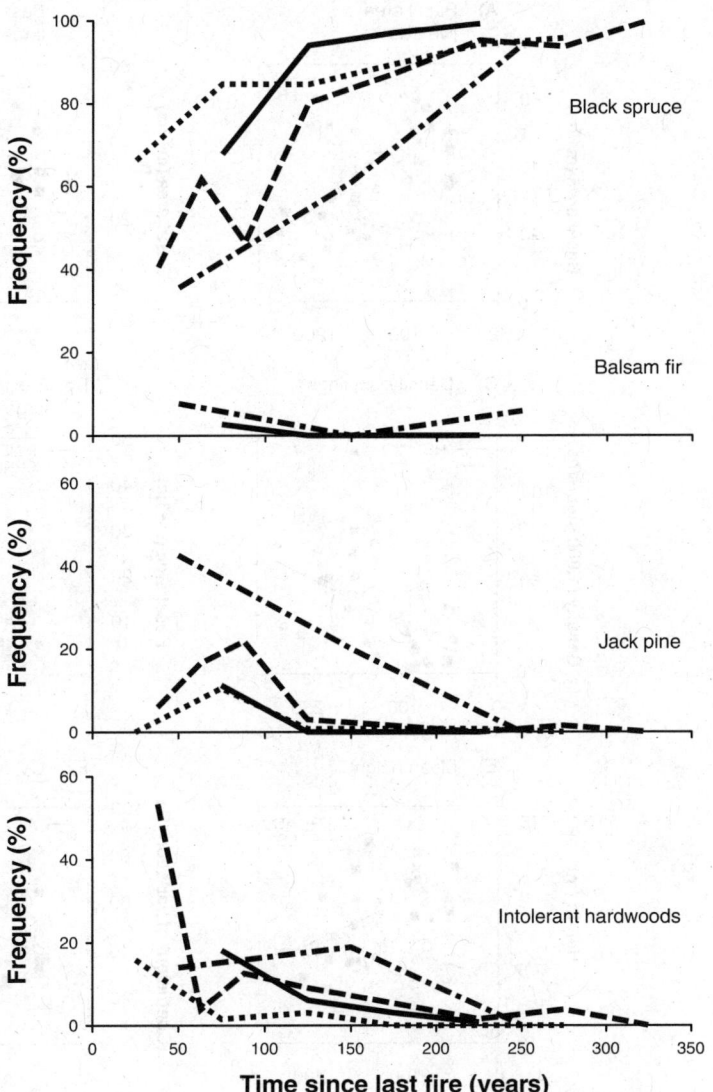

| | Source | Data type | *N* | Response variable |
|---|---|---|---|---|
| – · – · – · – | Gauthier et al. 2000 | Field surveys | 125 | % stands dominated by species |
| – – – – – | Harper et al. 2002 | Gov. ecoforestry maps | 31,033 | % area dominated by species |
| · · · · · · · | Harper et al. 2003 | Field surveys | 91 | % basal area |
| ———— | Lecomte and Bergeron 2000 | Gov. forest inventories | 781 | % stands dominated by species |

Although post-fire regeneration is provided by black spruce, jack pine, and intolerant hardwoods, old stands are dominated by black spruce.

Figure 11.5
**Changes in stand structure with time since fire, soil burn severity, and initial stand composition**

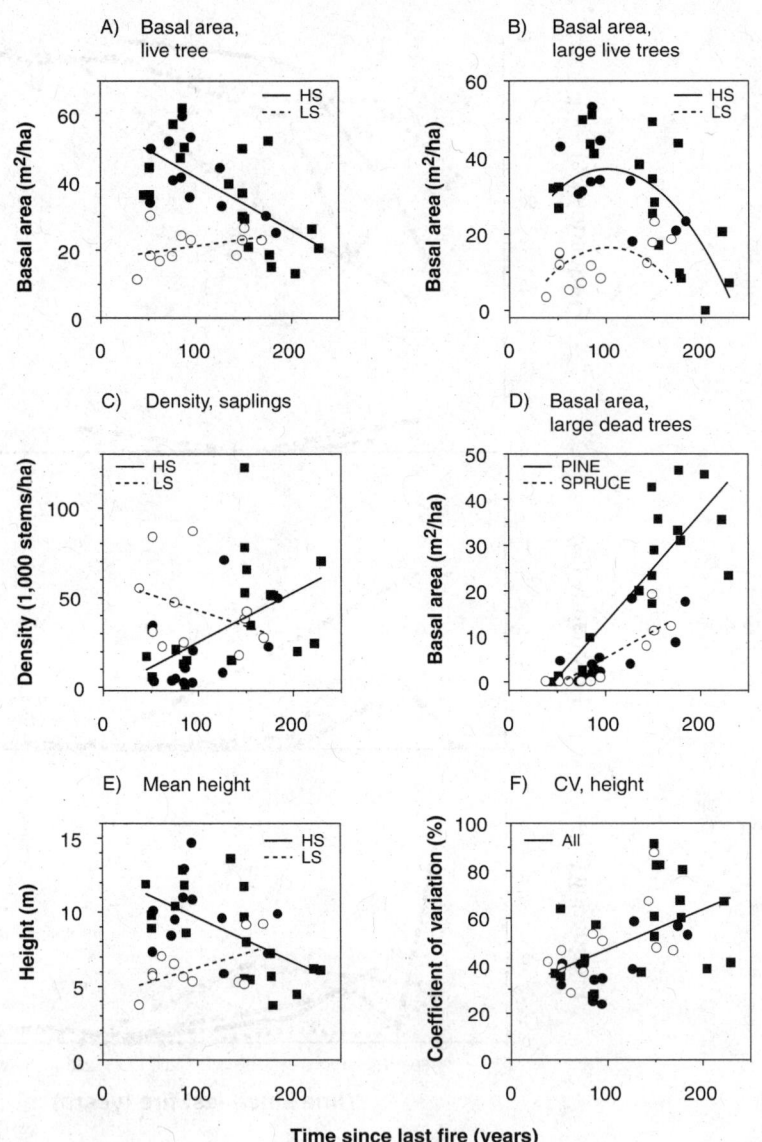

A)   Basal area, live tree
B)   Basal area, large live trees
C)   Density, saplings
D)   Basal area, large dead trees
E)   Mean height
F)   CV, height

**Time since last fire (years)**

■   Jack pine (PINE), high-severity soil burns (HS)

●   Black spruce (SPRUCE), high-severity soil burns (HS)

○   Black spruce (SPRUCE), low-severity soil burns (LS)

Note that the marked difference between stands established after high- and low-severity soil burns disappears with time. Large trees: diameter at breast height ≥ 10 cm. From Lecomte et al. (2006a).

Over time, stands originating from low-severity soil burns are subject to some infilling and their trees become larger and taller, thereby reducing the number of small-diameter stems. At about 200 years after fire, successional paludification brings about convergence in the two structural development trajectories towards a relatively open structure mostly composed of small stems.

## 4.3. Understory Succession

Understory succession patterns are similar to those of the canopy in that they are mostly affected by soil burn severity and very little by post-fire canopy composition (black spruce vs. jack pine; Lecomte et al. 2005). After a high-severity soil burn, the ground cover is dominated by feather mosses (*Pleurozium schreberi*), which are gradually replaced by *Sphagnum* moss species typical of mesic sites, then by *Sphagnum* species typical of wet sites, as the stand opens up and the water table rises (figure 11.6; Fenton and Bergeron 2006; Simard et al. 2007). Stands established after low-severity soil burns are characterized by a lower cover of *Pleurozium* and a higher colonization rate by *Sphagnum* compared with high-severity stands of the same age. Ericaceous shrubs, which show a regular increase over time following high-severity soil burns, are clearly favoured by low-severity fires that spare their roots and allow rapid recolonization by suckering.

## 4.4. Forest Productivity

Forest productivity varies over different scales of space and time. At a fine scale (1–5 years, 0.01–10 m), black spruce growth immediately following fire is affected by the interaction of microtopography and growth substrate (Munson and Timmer 1989; Roy et al. 1999; Lavoie et al. 2007a, b, c). The best substrates for black spruce rooting and growth are fibric horizons and mixtures of feather moss fibric and humic materials, whereas the worst substrates are bare mineral soil and fibric or humified *Sphagnum* horizons (Lavoie et al. 2007b, c). Peat accumulation hinders seedling growth and is negatively related to slope (Simard et al. in press and to elevation relative to average soil surface (Lavoie et al. 2007a).

At broader temporal and spatial scales (10–1000 years, 0.1–100 km), successional paludification considerably reduces potential (site index) and realized (basal area and aboveground tree biomass) forest productivity (figure 11.7; Simard et al. 2007). The reduction in forest productivity is observed not only through a post-fire chronosequence of stands, but also between successive cohorts of trees growing in the same sites. This productivity drop between successive tree cohorts occurs rapidly in stands that originate from low-severity soil burns, in which the second cohort is already particularly small (see section 4.2). Consequently, site index falls quickly (~4 m) between the first and second cohort of trees in low-severity soil burns. The effect of soil burn severity on aboveground tree biomass is considerable and reflects differences in both lower potential productivity (figure 11.7A) and the lower post-fire seedling establishment following low-severity soil burns (Lecomte et al. 2006a; Greene et al. 2007).

The presence of jack pine in black spruce–dominated stands does not change total aboveground tree biomass (Gaubert 2006; Lecomte et al. 2006a). However it seems that total tree volume in mixed stands of black spruce and trembling

**Figure 11.6**

## Changes in the abundance of major bryophyte and ericaceous shrub taxa with time since fire, soil burn severity, and initial stand composition

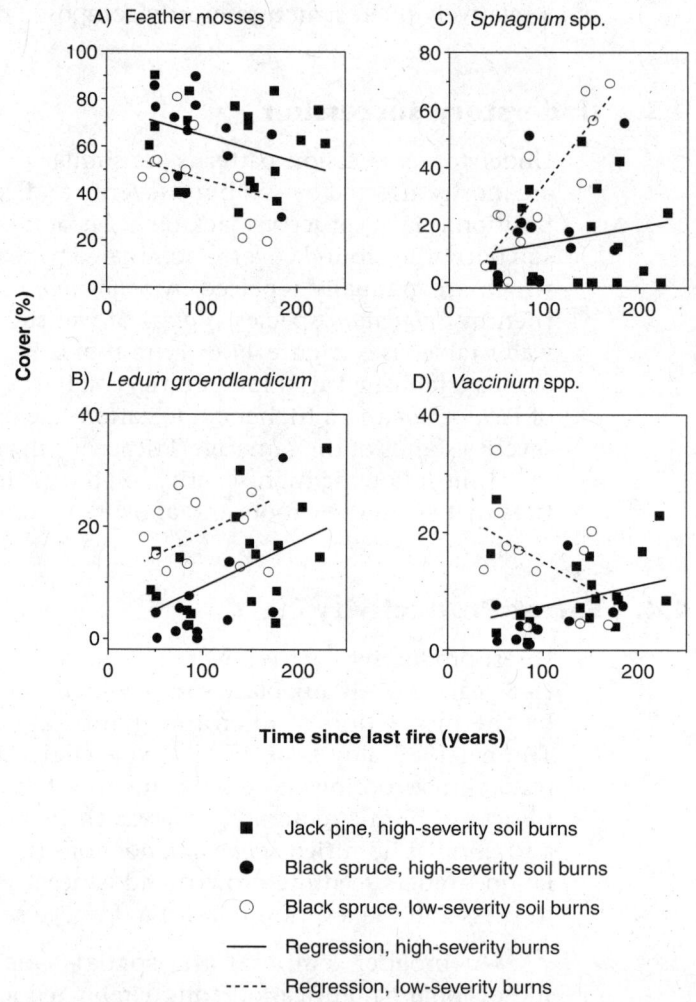

Time since last fire (years)

■ Jack pine, high-severity soil burns

● Black spruce, high-severity soil burns

○ Black spruce, low-severity soil burns

—— Regression, high-severity burns

----- Regression, low-severity burns

Note the significant increase in *Sphagnum* moss cover caused by paludification. From Lecomte et al. (2005).

aspen is greater than in pure stands of either type (Légaré et al. 2004, 2005a). This positive effect of aspen has been found for proportions of aspen up to 40%, and the extra volume that was generated was in aspen (Légaré et al. 2004). In addition, aspen litter improves the nutritional status of the stands and reduces soil organic matter accumulation rates (Légaré et al. 2005b). Although these relations do not seem constant among sites, they suggest that mixed-stand silviculture may be a win-win solution for enhancement of forest productivity and biodiversity conservation.

Figure 11.7
## Figure 11.7
### Changes in site index (A), 5-year mean basal area increment per tree (B), and total aboveground tree biomass (C) with time since fire and soil burn severity

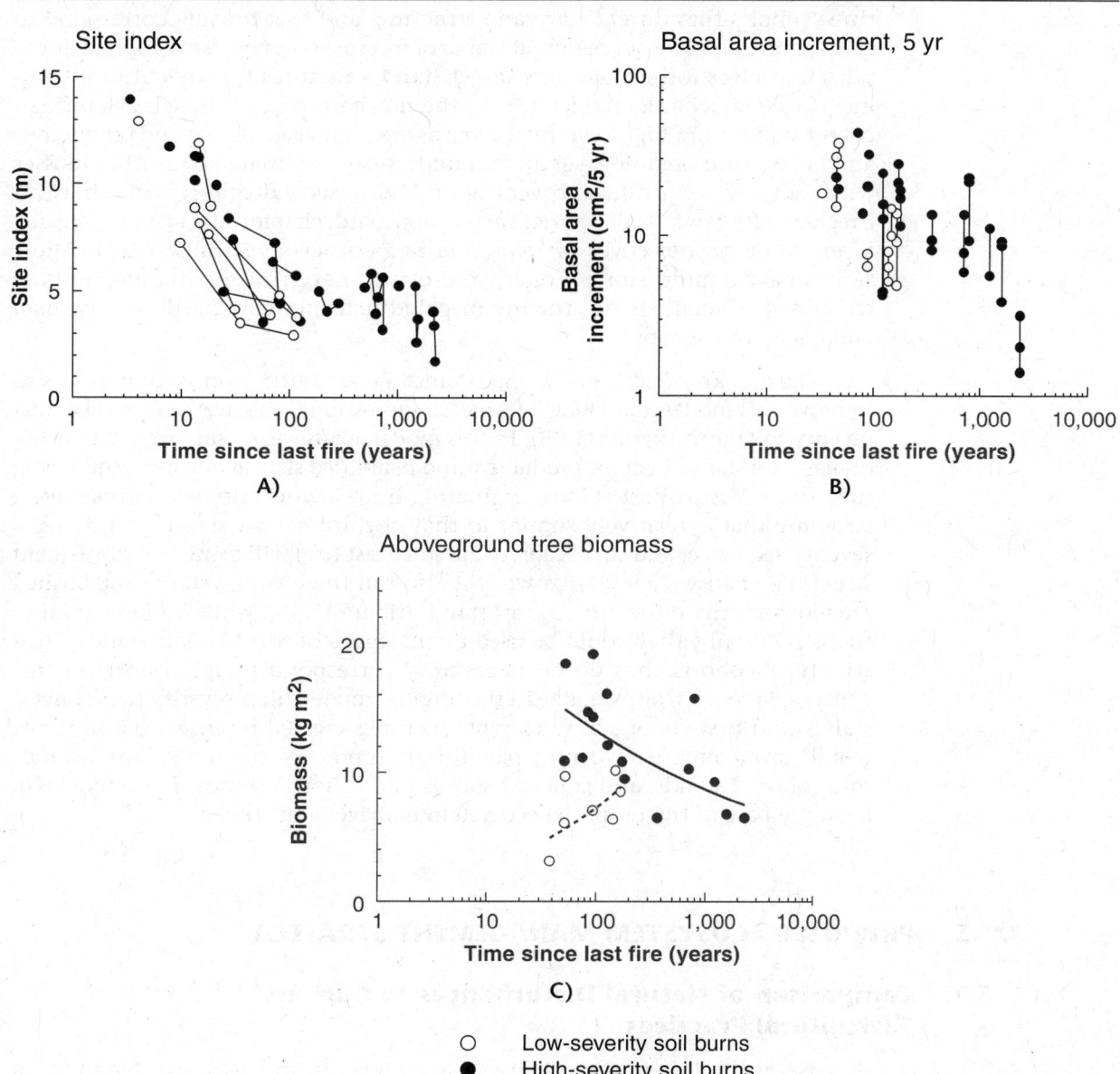

Site index represents height grown in 50 years of free growth, measured by stem analysis. In A) and B), each symbol represents a height cohort (*n* = 3 trees per cohort) growing in full light, and cohorts belonging to the same stand are connected by a line. In C), each symbol represents a stand. Note the reduction in potential (site index) and realized (basal area increment and biomass) forest productivity indices with time elapsed since last fire. From Simard et al. (2007).

## 4.5.  The Cohort Approach

The cohort approach proposed by Bergeron et al. (1999) and Harvey et al. (2002) simplifies the representation of forest succession by separating forest stands into three cohorts that differ in age and structure, and that broadly correspond to early-, mid-, and late-successional stages of forest succession. This compartmentalization gives forest managers target stand structures for silvicultural treatments (see Bouchard, chapter 13). In the northern part of the Clay Belt, first-cohort stands from high-severity soil burns are even-aged, dense, and productive, and have a thin organic layer and an understory dominated by feather mosses with a low ericaceous shrub cover (figure 11.8A). Second-cohort stands have an irregular, often two-storied structure (see Bouchard, chapter 13), a thicker organic layer, and a greater cover of *Sphagnum* and ericaceous shrubs. Third-cohort stands have a multi-storied, open, and often uneven-aged structure, and are composed of small stems growing in paludified sites colonized by *Sphagnum* and ericaceous shrubs.

Current knowledge of the importance of soil burn severity prompts us to improve this model to include alternative successional pathways created by low-severity soil burns (figure 11.8B). In this model, low-severity soil burns can bring or maintain stands in an unproductive and paludified state at any moment during succession. First-cohort stands originating from low-severity soil burns have a structure that is relatively similar to that of third-cohort stands in the high-severity sequence, and succession (time since last fire) will bring few subsequent structural changes. For clarity, we will also call these young stands established after low-severity burns first-cohort stands (figure 11.8B), while keeping in mind that structurally, they could be used as analogues of third-cohort stands. Such structural cohorts thus do no necessarily correspond to age cohorts in this expanded model. However, like in the original model, high-severity fires convert stands into first cohorts, and canopy openings created by insect outbreaks or windthrow (combined with the paludification process) convert cohort 1 stands into cohort 2 stands, and cohort 2 stands into cohort 3 stands. This model will be at the base of the proposed ecosystem management strategy.

## 5.  PROPOSED ECOSYSTEM MANAGEMENT STRATEGY

## 5.1.  Comparison of Natural Disturbances to Current Silvicultural Practices

Because of the importance of disturbance severity in the forest dynamics of northern Abitibi, it is useful to characterize the different silvicultural practices currently used in the boreal forest according to their effect (severity) on canopy and soils (figure 11.9A). Careful logging with protection of regeneration and soils or CPRS (*coupe avec protection de la régénération et des sols*, similar to Ontario's CLAAG, careful logging around advance growth) is currently the most widely used harvesting method in Québec. CPRS removes a large proportion of the canopy but, by definition, has a negligible effect on soil organic layers. In northern Abitibi, CPRS cuts are usually done during winter, which reduces even more their impact on soils. Prior to introduction of CPRS in the early 1990's, harvesting

Figure 11.8
**Stand structural development models on fine and organic deposits in northwestern Québec's spruce-moss forests**

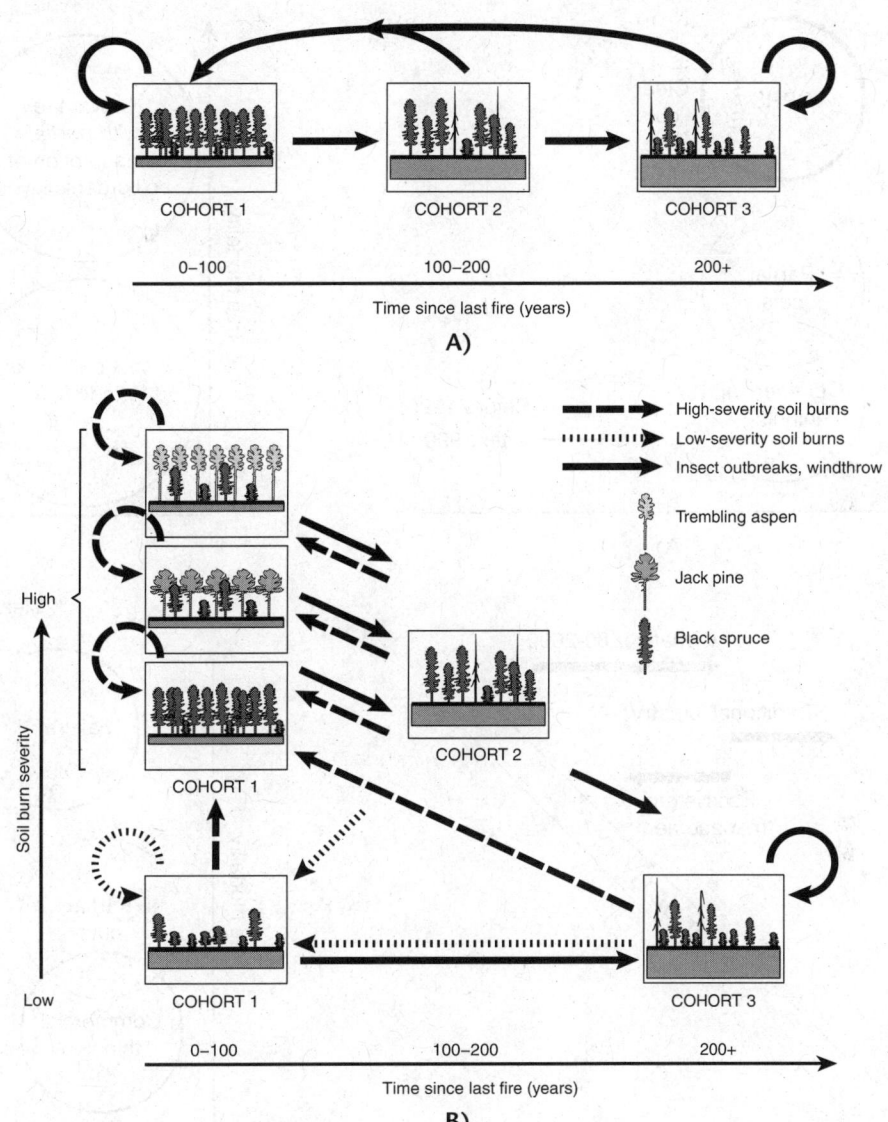

Legend: A) Structural cohort model of Bergeron et al. (1999) based on time elapsed since last stand-replacing fire; B) Model that incorporates the effects of soil burn severity and time since fire. After a high-severity soil burn, canopy composition can vary from the complete dominance of black spruce to co-dominance with jack pine or trembling aspen. All stands, irrespective of their origin, eventually converge after 200 years towards an open structure dominated by black spruce. Stands that develop after low-severity soil burns (dashed arrows) are highly paludified and unproductive and show few structural changes with succession. Organic soil thickness is shown in gray under the stands. Modified from Lecomte et al. (2006a).

**Figure 11.9**
## Characterization of forest disturbances relative to their severity on the canopy and soils

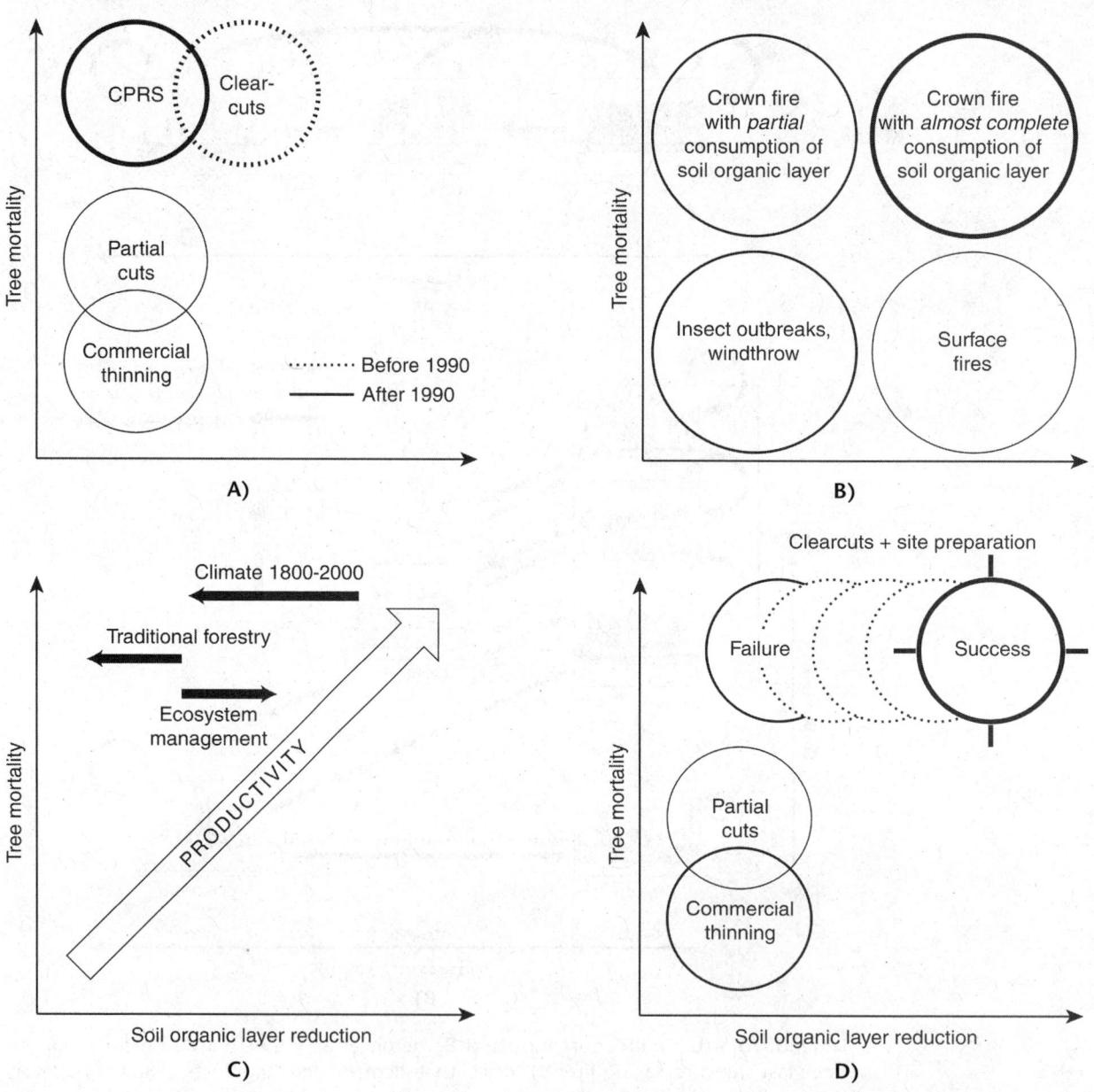

Legend: A) Traditional harvesting methods; B) Natural disturbances; C) A forest productivity gradient (open arrow) generally follows the gradient in disturbance severity (compare with B). Through their effect on the severity of disturbances at the origin of the stands, climate and forest management (solid horizontal lines) can affect forest productivity across the landscape; D) Proposed ecosystem management strategy. In A), B), and D), circle line thickness is proportional to the frequency of disturbances in the study area. CPRS: Cuts with protection of regeneration and soils. See text for more details.

involved clearcutting practices that had a greater impact on soils for a similar amount of canopy removal. Current alternative practices such as partial cuts and commercial thinning have a lower impact on the canopy and are far less frequent than CPRS, but are also done with soil protection in mind. In short, current silvicultural treatments are characterized by a low impact on soils and by a variable intensity of canopy removal.

In comparison to silvicultural practices, natural disturbances show a greater variability in severity, particularly in their impacts on soils (figure 11.9B). Insect outbreaks, windthrow, and pathogens partially open the canopy but have very little effect on accumulated organic horizons. Crown fires generally result in complete opening of the canopy, but the consumption of soil organic matter varies from nearly total (high-severity soil burns) to partial (low-severity soil burns). Surface fires have a low impact on the canopy but reduce soil organic-layer depth. However this type of disturbance is uncommon in black spruce forests and mostly occurs at the margins of large canopy burns where fire spreads sometimes exclusively on the ground (Harper et al. 2004).

This quick analysis reveals that the effects of current logging practices on soils are more similar to the effects of low-severity soil burns than those of high-severity soil burns (Bergeron et al. 2007). Studies have indeed found differences between severe fires and harvesting with respect to understory composition, organic-layer thickness, and nutrient dynamics (Nguyen-Xuan et al. 2000; Simard et al. 2001). In ecosystems that are not influenced by paludification, these differences may have less importance. We believe however that forest logging practices as they are currently implemented in the black spruce forests of northern Abitibi reduce forest productivity compared with high-severity soil burns (Fenton and Bergeron 2007) and that this effect was enhanced by the application of CPRS since the 1990s (figure 11.9C).

Currently, there are no data to determine the importance of forest productivity losses following harvesting. In addition, this anthropogenic effect is compounded by a long-term climatic trend of diminishing fire frequency (Carcaillet et al. 2001; Bergeron et al. 2004) and possibly of diminishing soil severity. This interpretation is supported by the absence of high-severity soil burns in the last 50 years in the region, whereas early-century fires were primarily severe soil burns (Lecomte et al. 2005, 2006a, b). It is thus imperative to include soil organic-layer management in forest management strategies if we want to maintain long-term forest productivity across the landscape. In northern Abitibi, this mainly means increasing usage of silvicultural practices that severely disturb soils, like site preparation techniques and prescribed burns. Use of silvicultural approaches that do not significantly disturb organic layers may reduce biodiversity at the landscape level by reducing the proportion of dense, productive stands in which species assemblages differ considerably from those of old paludified forests (Drapeau et al. 2003).

To maximize the potential of the black spruce forests of the northern Clay Belt region, we recommend intensifying forest management (figure 11.9D) by:

- increasing the frequency of commercial thinning that partially opens stand canopies; and

- using silvicultural treatments that severely disturb soil organic layers and forest canopies.

In the next two sections we describe what approaches can be used to create or maintain forest structures typical of late- and early-successional stages. Finally, in the last two sections, we discuss methods for selecting stands for treatment as well as landscape-level issues of target proportions for the different stand structures.

## 5.2.  Creating or Maintaining Late-Successional Structures Using Uneven-Aged Silviculture

Studies of stand dynamics following high-severity soil burns show that stands gradually open up by self-thinning over several decades starting about 100 years after their post-fire establishment (figure 11.5; Lecomte et al. 2006a). During the transition from a first to a second cohort, stand merchantable volume, which was distributed among many moderate-sized stems, is reduced and redistributed among a smaller number of larger stems. At longer time scales, stand development is characterized by a further opening of the canopy, resulting in third-cohort stands with few medium-sized stems and many small-sized stems (figures 11.5 and 11.8). We propose to accelerate the development of dense and even-sized stands towards uneven-sized stands by using commercial thinning implemented in dense, first-cohort stands just before inception of break-up to emulate the break-up process. Further opening up or maintenance of stand openness can be achieved by using partial cuts that protect small merchantable stems, a treatment called CPPTM in Quebec (*coupe avec protection des petites tiges marchandes*, a diameter limit cut similar to Ontario's HARP, harvesting with regeneration protection) (figure 11.10). Short-term studies have shown that the CPPTM treatment re-creates structures and characteristics of old paludified forests such as moss community composition (Fenton and Bergeron 2007; Fenton et al., chapter 15).

The objective of these operations is to maintain stands in a productive state while recreating the diversity of structures and species composition observed in forest succession. Silvicultural operations must therefore compress natural stand dynamics, which normally occur over many centuries (Lecomte et al. 2006a), into a much shorter time frame before forest productivity is significantly reduced by the paludification process. The timing of implementation of these treatments must thus not depend on a particular stand age, but rather on the achievement of silvicultural and ecological objectives (maintaining canopy cover, productivity, etc.). Stands should therefore be treated as soon as they achieve a merchantable volume sufficient for commercially viable operations.

## 5.3.  Creating Early-Successional Structures Using Even-Aged Silviculture

To convert second- and third-cohort stands to first cohorts, we propose more frequent use of intensive approaches inspired by the effects of high-severity soil burns, i.e. treatments that completely open stand canopies (clearcuts) and that disturb and reduce the depth of soil organic layers (site preparation,

**Figure 11.10**
## Proposed silvicultural treatments to recreate natural stand dynamics

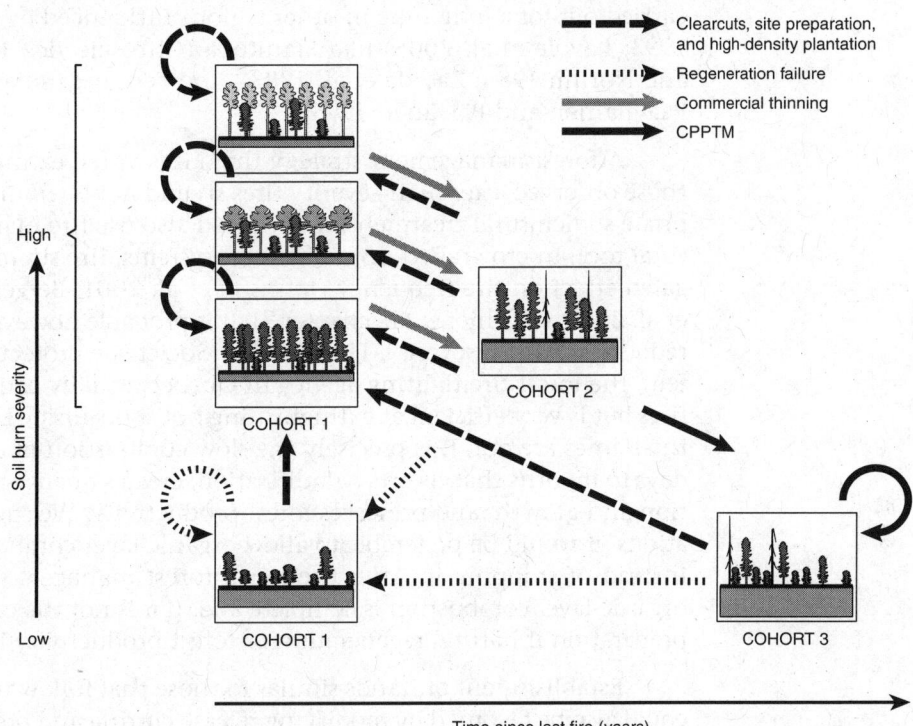

Legend: CPPTM: cuts with protection of small merchantable stems (*coupe avec protection des petites tiges marchandes*).

prescribed burns). Such treatments should be followed by artificial regeneration (plantation or seeding) to favour rapid closure of the canopy (figure 11.10). This intensive approach however represents the extreme end of a gradient of soil disturbance severity in silvicultural treatments; the decision of whether or not to carry out site preparation should be based on the degree of paludification of the stands (organic-layer thickness; see section 5.4), as estimated before and after harvesting.

In the boreal forest, scarification is a widely used site preparation technique that decreases organic-layer thickness and the abundance of competitive vegetation, increases summer soil temperature and the quantity of regeneration microsites, and improves the nutritional status of trees and site drainage (Prévost 1992, 1996; Prévost and Dumais 2003; Thiffault et al. 2004; Lavoie et al. 2005). Postharvest prescribed burn is another useful site preparation method, but is seldom used in Quebec because of technical challenges and risks of escape. This technique, however, would be particularly appropriate in sites susceptible to paludification, as it directly reduces the depth of soil organic layer and stimulates nutrient cycling in paludified sites. In northern Abitibi, recent scarification and prescribed burn trials gave encouraging but variable preliminary results (Cormier

2004; D. Paré, pers. obs.), and illustrate the importance of targeting the right stands to be treated (see section 5.4). Although the majority of the data from the northern Clay Belt are quite recent, this type of intensive operation has been carried out for a long time in other regions influenced by paludification (Sutton 1993; Lavoie et al. 2005) like Manitoba (Chrosciewicz 1976), Alaska (Dyrness and Norum 1983; Zasada et al. 1983), and Fennoscandia (Örlander et al. 1990; Paavilainen and Päivänen 1995).

A forest management strategy that aims at recreating conditions similar to those observed after high-severity fires should not be limited to the use of appropriate silvicultural treatments, but should also use fire management as a silvicultural tool. In crown fire–dominated ecosystems, fire suppression has had a mitigated effect on fire frequency (Johnson et al. 2001; Bergeron et al. 2002; Bridge et al. 2005; Gauthier et al., chapter 3). It is probable however that fire suppression reduced soil burn severity. The SOPFEU (Société de protection des forêts contre le feu), the forest fire fighting agency in Québec, usually cannot control the largest fires but is very efficient at extinguishing hot spots inside burned perimeters once the flames are out. It is precisely the slow combustion of organic soil layers over days to months that reverts paludification, creates quality microsites for regeneration and growth, and preserves forest productivity. We suggest that in most situations, it would be preferable to allow organic-layer combustion to run its course instead of stopping it. After the burn, forest managers should make sure that organic-layer combustion is complete and, if it is not the case, consider doing site preparation if natural regeneration or forest productivity is compromised.

Establishment of stands similar to those that follow high-severity soil burns could also be favoured by modifying release cutting and pre-commercial thinning procedures that currently favour conifers at the expense of hardwoods. As shown by studies undertaken in the Clay Belt, keeping a minor proportion (up to 40%) of trembling aspen in black spruce stands during release cutting and thinning operations would 1) favor the establishment of herbaceous species in the understory, 2) reduce accumulation rates of soil organic matter, 3) improve soil physical and chemical conditions, and 4) increase total merchantable volume in the stands (Légaré et al. 2001, 2005a, b). By favouring mixed silviculture of black spruce and aspen in the right proportions, there could be significant gains in hardwood volume without causing losses in coniferous volume (Légaré et al. 2005b).

Although we suggest intensive approaches for the management of soil organic matter, these treatments should be subordinated to residual forest retention objectives that aim at reproducing residual structures observed in burned sites (Kafka et al. 2001; Bergeron et al. 2002). We therefore recommend site preparation that would be associated with a variable retention of the canopy (10–50% of permanent and non-permanent retention; see Belleau and Légaré, chapter 19).

## 5.4. Targeting the Right Stands to Be Treated

In a landscape prone to paludification, it is essential to use the right treatments in the right stands. Silvicultural treatments should thus be chosen according to the stands' merchantable volume and degree of paludification as well as to the dominant type of paludification (edaphic vs. successional). Treatments like partial cuts aimed at creating or maintaining late-successional forest structures (figure 11.10) should be carried out in stands that have a low degree of paludification (shallow organic layer, low abundance of *Sphagnum* and ericaceous shrubs) and still possess a significant potential for growth. Stands that show a high degree of paludification have very little growth potential following partial cuts and should not be subjected to treatments that aim at accelerating the opening of their canopy.

Treatments that aim at creating early-successional forest structures should be limited to sites that present a high potential for productivity recovery. Stands that are primarily influenced by successional paludification could potentially recover their past productivity level using site preparation techniques, which would reduce soil organic-layer depth. Forests that are mainly influenced by edaphic paludification, however, show little potential of productivity recovery and should not be targeted by silvicultural investments. Although both types of stands possess attributes of old paludified forests and thus possibly possess an equivalent potential for maintaining biodiversity associated with this type of structure, they do not have the same potential for timber production.

Analysis of paludification dynamics currently provides a first approximation in the prediction of productivity recovery potential. We have examined how soil organic-layer thickness varied with slope (edaphic paludification) and time since last fire (successional paludification) (Simard et al. 2009). Stands growing on flat sites (0–1%) had higher paludification rates than stands of a similar age growing on slightly steeper slopes (3–7%) (figure 11.11). Consequently, 50 years after

**Figure 11.11**
## Potential for forest productivity recovery

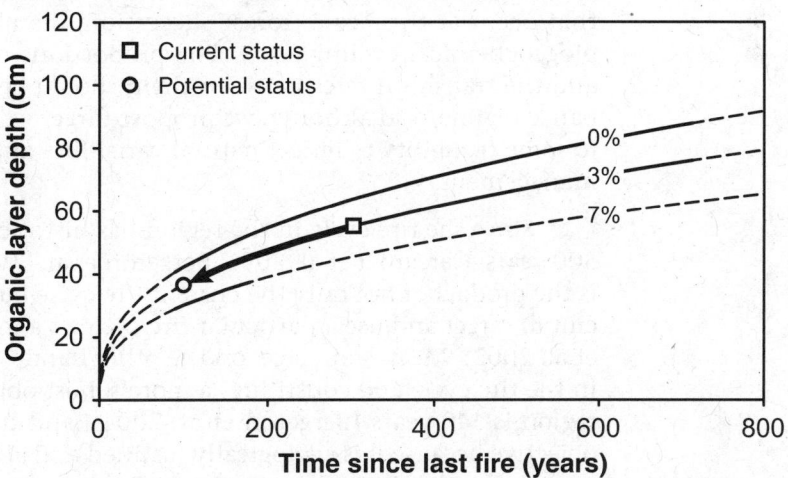

Potential for forest productivity recovery is the difference between current paludification status of a stand and the degree of paludification of a mature stand (potential), and can be predicted by slope (%; parallel curves) and stand age or by slope and current organic-layer thickness. Dashed curves indicate range of extrapolation (*n* = 260 plots). Modified from Simard et al. (2009).

disturbance (fire or harvesting with site preparation), organic-layer depth reaches about 33 cm on flat sites, but only 23 cm on sites with a 7% slope. Mature stands growing on flat sites will therefore express lower productivity than stands growing on a gentle slope because forest productivity (site index) is negatively related to organic-layer depth (Simard et al. 2009). In addition, any productivity gain by site preparation on flatter sites will be short-lived because of the accelerating effect of edaphic paludification. These observations indicate that the potential for productivity recovery increases rapidly with slope. This potential can be estimated as the difference between current organic-layer thickness and that of a mature (100 years) stand.

Productivity recovery potential can be calculated for any stand whose slope and age, or slope and peat thickness, are known, allowing the stand to be located on its paludification trajectory (figure 11.11). We are currently working at mapping the recovery potential of productivity across the black spruce forests of the northern Clay Belt. The mapping project is based on currently available spatial data (fire maps, digital elevation models) and remote sensing. This tool should allow the selection of stands that should be converted to early-successional structures (first-cohort stands), as well as stands that are currently considered unproductive but that could return to a productive state.

## 5.5.  Maintaining a Balanced Forest Mosaic

At the landscape level, one objective of ecosystem management is to approximately maintain the age structure and proportions of different stand types (cohorts) that would be observed under a natural disturbance regime (Gauthier et al. 1996; Bergeron et al. 1999; Gauthier et al. 2004; Gauthier et al., chapter 3). It is also important to favor transitions between the different cohorts (see section 3.2 in Bouchard, chapter 13) in order to maintain ecological processes that occur at this scale (forest succession, disturbances of different severity, biogeochemical cycling, etc.). The proportions of stands in different cohorts and the transition rates between them should be based on the natural disturbance regime, and although we propose target values, they should be subjected to some flexibility to reflect natural variability and to allow adaptive ecosystem management.

Since the fire cycle in the region has historically changed over time (50 to 500 years; Carcaillet et al. 2001; Bergeron et al. 2004) and that the forest mosaic is the product of not only the current fire cycle but also of past cycles, it is difficult to target and use a particular fire cycle as a management objective (Harvey et al. 2002). Mean stand age, on the other hand, shows more inertia to changes in the fire cycle and constitutes a more robust objective. Mean stand age in the region is 140 years (Bergeron et al. 2004), and this value could be used as an objective because it is ecologically justified and allows sustainable forestry to be achieved under the current fire cycle (Bergeron et al. 2006). Forest management would not aim at a strict and normalized distribution of areas by *age* classes (i.e. a rectangular distribution with no stands older than 140 years), but a negative exponential distribution of areas by *structural* classes (cohorts) (Bergeron et al. 1999; Harvey et al. 2002).

By applying the cohort approach with this distribution and a silvicultural rotation of 100 years, about 45–55% of the managed landscape (excluding conservation zones) should be composed of stands with a structure that is equivalent to the first cohort, 23–26% with a structure similar to the second cohort, and 20–30% with a structure equivalent to the third cohort. These proportions are the same as those determined by Harvey et al. (2002) for the balsam fir–white birch forest region of northwestern Québec because both zones show a similar mean stand age and fire cycle (Bergeron et al. 2004). Transition rates between the different cohorts are therefore also the same in the two zones. For example, about 55% of the first-cohort stands would be recycled in cohort 1 and 45% would be transformed in cohort 2, whereas half of the second-cohort stands would be transformed in cohort 1 and the other half would be converted in cohort 3. Finally, 67% of third-cohort stands would be converted to cohort 1 and the remaining 33% would conserve a third-cohort structure (Harvey et al. 2002; Harvey et al., chapter 18). By using these transition rates, the proportions of cohorts are relatively stable over time at the landscape scale, even when considering the effect of fire (Harvey et al. 2002). The current fire cycle (~400 years) is much longer than the historical fire cycle (= mean stand age, 140 years) that is targeted by the proposed ecosystem management. This reduced fire frequency gives some latitude that allows forest harvesting in spite of the occurrence of fire, and maintenance of landscape-scale forest structure within the limits of the natural disturbance regime (for more details, see Gauthier et al. 2004, Bergeron et al. 2006, and Gauthier et al., chapter 3).

It is also important to maintain the fraction of first-cohort stands that are dominated by jack pine (~10–20% of stands < 100 years old, figure 11.4) and by intolerant hardwoods (~10–20%), especially since trembling aspen can slow down paludification and increase stand productivity (Légaré et al. 2005a, b). Additionally, first-cohort stands can originate from high- or low-severity soil burns, and here again we should try to maintain the natural proportions of these two stand types. It is estimated that about half of the young stands in northern Abitibi established following low-severity fires, based on the proportion across the landscape of young (<100 years) open (canopy cover < 60%) stands, which typically originate from low-severity soil burns (Lecomte and Bergeron 2005). This coarse approximation should however be refined because it does not take into account the proportion of low-severity burns in older stands, which could be lower if the 19th-century fires were more severe than the 20th-century ones, in addition to being more frequent. Knowledge of the natural variability of this proportion through time would give a range to guide management decisions.

As a first approximation, first-cohort stands (which would cover 45–55% of the managed landscape) should be composed at most of 50% of stands established following low-severity disturbances (all black spruce stands because aspen- and pine-dominated stands always originate from high-severity soil burns) and at least of 50% of stands established following high-severity disturbances (15% jack pine, 15% hardwoods, and 20% black spruce). However, there are already large quantities of stands across the landscape that have a high degree of paludification (often identified on forest inventory maps as "unproductive" or "forested peatlands'), whereas productive stands in the first cohort are becoming scarce. We therefore suggest that it would be a priority in the coming decades to increase the proportion of productive forests, i.e. of young cohort 1 stands originating from high-severity disturbances both on soil and in the canopy.

The proportion of cohorts and transition rates discussed here should be flexible given the natural variability in fire frequency and severity. Even under a fire regime that is considered stable at a multi-century scale, there is interannual variability in area burned, which leads to decadal variability in the proportion of first-cohort stands across the landscape (Belleau et al. 2007; Cyr et al. in press). Such flexibility would make it possible to factor in the biophysical and economic constraints associated with reaching the target proportions, the improvement of our knowledge of landscape-scale forest dynamics, and the achievement of other forest management objectives.

## 6.  CONCLUSION

The case of the black spruce forest in northern Abitibi clearly illustrates the necessity to adjust forest management practices to regional forest dynamics. In this region, compact clay deposits (particularly in the portion located on the Cochrane Till) and flat topography favour paludification, i.e. the accumulation of soil organic matter associated with superficial water tables, colonization by *Sphagnum* mosses, low soil temperature, and reduced forest productivity. Only high-severity soil burns that consume most of the accumulated organic layer can allow establishment of dense and productive forests, whereas low-severity burns (>5 cm of residual organic layer) typically give rise to open stands composed of small stems and dominated by ericaceous shrubs.

In this region, forest management based on natural forest dynamics must therefore include silvicultural treatments that severely disturb soil organic layers in order to recreate conditions similar to those created by high-severity soil burns. In addition, partial cuts should be used to recreate or maintain old (>100 years) open forest structures, which currently represent about 60% of the landscape (Bergeron et al. 2004). We should also aim to conserve the proportions and transitions between the different structural cohorts at the landscape scale.

Certain silvicultural standards developed for the province go against the natural dynamics of northern Abitibi's black spruce forests and could have negative effects not only on biodiversity, but also on forest productivity (Bergeron et al. 2007). Furthermore, application of harvesting methods that protect the soil organic layer (e.g. CPRS) and treatments that favor conifers over hardwoods accelerate the paludification process. An ecosystem management strategy inspired by natural forest dynamics could be beneficial to the logging industry by offering approaches that would improve or at least maintain forest productivity across the territory (Bergeron et al. 2007).

Given the uncertainty associated with climate change (effects on fire frequency and severity, paludification, and forest productivity), the ecological and economic effects of novel silvicultural practices proposed here, and the emerging carbon trading markets, the practical details of the implementation of an ecosystem management strategy should constantly be re-evaluated by integrating new knowledge as it becomes available. This adaptive management approach therefore requires some flexibility in the application of silvicultural practices, within the limits set by natural variability in disturbance regimes and forest dynamics. Some of our knowledge of the northern Clay Belt's black spruce forest is still fragmentary and its improvement would allow us to refine the technical details of the ecosystem management strategy proposed here.

Natural variability in fire frequency and severity gives a certain latitude to the management of black spruce forests and consequently offers society many important choices. The challenge consists in choosing, within this variability and based on forest productivity, carbon sequestration, and multiple usage of the forest, a target proportion of stand types to conserve across the landscape. To achieve this, forest managers will have to encourage management approaches that give citizens and their forest-related values an important role in management decisions, which conforms to a more inclusive definition of ecosystem management (Grumbine 1997; Cortner and Moote 1999) and to the recommendations of the Coulombe Commission on the management of Québec's public forests (Coulombe et al. 2004). Finally, as the implementation of any vision requires the involvement of many stakeholders, the implementation of an ecosystem management strategy will depend mostly on the capacity to integrate new practices into the operational, decisional, and economic structures already in place.

## REFERENCES

Belleau, A., Bergeron, Y., Leduc, A., Gauthier, S., and Fall, A. 2007. Using spatially explicit simulations to explore size distribution and spacing of regenerating areas produced by wildfires: recommendations for designing harvest agglomerations for the Canadian boreal forest. For. Chron. **83**: 72–83.

Bergeron, Y., Cyr, D., Drever, C.R., Flannigan, M.D., Gauthier, S., Kneeshaw, D., Lauzon, È., Leduc, A., Le Goff, H., Lesieur, D., and Logan, K. 2006. Past, current, and future fire frequencies in Quebec's commercial forests: implications for the cumulative effects of harvesting and fire on age-class structure and natural disturbance-based management. Can. J. For. Res. **36**: 2737–2744.

Bergeron, Y., Drapeau, P., Gauthier, S., and Lecomte, N. 2007. Using knowledge of natural disturbances to support sustainable forest management in the northern Clay Belt. For. Chron. **83**: 326–337.

Bergeron, Y., Gauthier, S., Flannigan, M., and Kafka, V. 2004. Fire regimes at the transition between mixedwood and coniferous boreal forest in northwestern Quebec. Ecology, **85**: 1916–1932.

Bergeron, Y., Harvey, B., Leduc, A., and Gauthier, S. 1999. Forest management guidelines based on natural disturbance dynamics: stand- and forest-level considerations. For. Chron. **75**: 49–54.

Bergeron, Y., Leduc, A., Harvey, B.D., and Gauthier, S. 2002. Natural fire regime: a guide for sustainable management of the Canadian boreal forest. Silva Fenn. **36**: 81–95.

Boudreault, C., Bergeron, Y., Gauthier, S., and Drapeau, P. 2002. Bryophyte and lichen communities in mature to old-growth stands in eastern boreal forests of Canada. Can. J. For. Res. **32**: 1080–1093.

Boulet, B., Chabot, M., Dorais, L., Dupont, A., and Gagnon, R. 1996. Entomologie forestière. *In* Manuel de foresterie. *Edited by* J. Bérard and M. Côté. Les Presses de l'Université Laval, Sainte-Foy, Québec, p. 1007–1043.

Bridge, S.R.J., Miyanishi, K., and Johnson, E.A. 2005. A critical evaluation of fire suppression effects in the boreal forest of Ontario. For. Sci. **51**: 41–50.

Brown, J.-L. and Gangloff, P. 1980. Géliformes et sols cryiques dans le sud de l'Abitibi, Québec. Géo. Phys. Quat. **34**: 137–158.

Carcaillet, C., Bergeron, Y., Richard, P.J.H., Fréchette, B., Gauthier, S., and Prairie, Y.T. 2001. Change of fire frequency in the eastern Canadian boreal forests during the Holocene: does vegetation composition or climate trigger the fire regime? J. Ecol. **89**: 930–946.

Carleton, T.J. and Maycock, P.F. 1978. Dynamics of the boreal forest south of James Bay. Can. J. Bot. **56**: 1157–1173.

Carleton, T.J. and Maycock, P.F. 1980. Vegetation of the boreal forest south of James Bay: non-centered component analysis of the vascular flora. Can. J. Bot. **69**: 778–785.

Charron, I. and Greene, D.F. 2002. Post-wildfire seedbeds and tree establishment in the southern mixedwood boreal forest. Can. J. For. Res. **32**: 1607–1615.

Chrosciewicz, Z. 1976. Burning for black spruce regeneration on a lowland cutover site in southeastern Manitoba. Can. J. For. Res. **6**: 179–186.

Cormier, D. 2004. Essais comparés de diverses techniques de scarifiage dans des sites susceptibles à la paludification. Institut canadien de recherches en génie forestier (FERIC), Division de l'est, Pointe-Claire, Que.

Cortner, H.J. and Moote, M.A. 1999. The politics of ecosystem management. Island Press, Washington, D.C.

Coulombe, G., Huot, J., Arsenault, J., Bauce, É., Bernard, J.-T., Bouchard, A., Liboiron, M.A., and Szaraz, G. 2004. Rapport de la Commission d'étude sur la gestion de la forêt publique québécoise. Québec, Que. [Online] <www.mrnf.gouv.qc.ca/publications/forets/consultation/rapport-coulombe.pdf>.

Cyr, D., Bergeron, Y., Gauthier, S., and Larouche, A. 2005. Are the old-growth forests of the Clay Belt part of a fire-regulated mosaic? Can. J. For. Res. **35**: 65–73.

Cyr, D., Gauthier, S., Bergeron, Y., and Carcaillet, C. in press. Forest management is driving the eastern North American boreal forest outside its natural range of variability. Front. Ecol.

Drapeau, P., Leduc, A., Bergeron, Y., Gauthier, S., and Savard, J.P. 2003. Les communautés d'oiseaux des vieilles forêts de la pessière à mousses de la ceinture d'argile: problèmes et solutions face à l'aménagement forestier. For. Chron. **79**: 531–540.

Dyrness, C.T. and Norum, R.A. 1983. The effects of experimental fires on black spruce forest floors in interior Alaska. Can. J. For. Res. **13**: 879–893.

Environment Canada. 2006. Canadian Climate Normals 1971–2000. [Online] <www.climate.weatheroffice.ec.gc.ca> (accessed November 18, 2008).

Fenton, N.J. and Bergeron, Y. 2006. Facilitative succession in a boreal bryophyte community driven by changes in available moisture and light. J. Veg. Sci. **17**: 65–76.

Fenton, N.J. and Bergeron, Y. 2007. *Sphagnum* community change after partial harvest in black spruce boreal forests. For. Ecol. Manag. **242**: 24–33.

Fenton, N., Lecomte, N., Légaré, S., and Bergeron, Y. 2005. Paludification in black spruce (*Picea mariana*) forests of eastern Canada: potential factors and management implications. For. Ecol. Manag. **213**: 151–159.

Fenton, N., Légaré, S., Bergeron, Y., and Paré, D. 2006. Soil oxygen within boreal forests across an age gradient. Can. J. Soil Sci. **86**: 1–9.

Gaubert, C. 2006. Reconstruction dendrochronologique de l'effet de *Pinus banksiana* sur la croissance de *Picea mariana* dans la forêt boréale au nord-ouest du Québec, Canada. Master Biologie, Géosciences, Agroressources,

Environnement. Université Montpellier II, Montpellier, France.

Gauthier, S., De Grandpré, L., and Bergeron, Y. 2000. Differences in forest composition in two boreal forest ecoregions of Quebec. J. Veg. Sci. **11**: 781–790.

Gauthier, S., Leduc, A., and Bergeron, Y. 1996. Forest dynamics modelling under natural fire cycles: a tool to define natural mosaic diversity for forest management. Environ. Monit. Assess. **39**: 417–434.

Gauthier, S., Leduc, A., Harvey, B., Bergeron, Y., and Drapeau, P. 2001. Les perturbations naturelles et la diversité écosystémique. Naturaliste can. **125**: 10–17.

Gauthier, S., Nguyen, T., Bergeron, Y., Leduc, A., Drapeau, P., and Grondin, P. 2004. Developing forest management strategies based on fire regimes in northwestern Quebec. *In* Emulating natural forest landscape disturbances: concepts and applications. *Edited by* A.H. Perera, L.J. Buse, and M.G. Weber. Columbia University Press, New York, USA, pp. 219–229.

Gray, D.R., Régnière, J., and Boulet, B. 2000. Analysis and use of historical patterns of spruce budworm defoliation to forecast outbreak patterns in Quebec. For. Ecol. Manag. **127**: 217–231.

Greene, D.F., Macdonald, S.E., Cumming, S., and Swift, L. 2005. Seedbed variation from the interior through the edge of a large wildfire in Alberta. Can. J. For. Res. **35**: 1640–1647.

Greene, D.F., Macdonald, S.E., Haeussler, S., Domenicano, S., Noël, J., Jayen, K., Charron, I., Gauthier, S., Hunt, S., Gielau, E.T., Bergeron, Y., and Swift, L. 2007. The reduction of organic-layer depth by wildfire in the North American boreal forest and its effect on tree recruitment by seed. Can. J. For. Res. **37**: 1012–1023.

Grondin, P., Noël, J., Hotte, D., Tardif, P., and Lapointe, C. 2000. Croissance potentielle en hauteur et dynamique des espèces forestières sur les principaux types écologiques des régions écologiques 5a et 6a (Abitibi). Rapport interne n° 461, Gouvernement du Québec, Forêt Québec, Ministère des Ressources naturelles, Direction de la recherche forestière, Québec, Que.

Grumbine, R.E. 1997. Reflections on "What is ecosystem management?" Conserv. Biol. **11**: 41–47.

Harper, K.A., Bergeron, Y., Gauthier, S., and Drapeau, P. 2002. Post-fire development of canopy structure and composition in black spruce forests of Abitibi, Québec: a landscape scale study. Silva Fenn. **36**: 249–263.

Harper, K., Boudreault, C., De Grandpré, L., Drapeau, P., Gauthier, S., and Bergeron, Y. 2003. Structure, composition, and diversity of old-growth black spruce boreal forest of the Clay Belt region in Quebec and Ontario. Environ. Rev. **11**: S79–S98.

Harper, K.A., Lesieur, D., Bergeron, Y., and Drapeau, P. 2004. Forest structure and composition at young fire and cut edges in black spruce boreal forest. Can. J. For. Res. **34**: 289–302.

Harvey, B.D., Leduc, A., Gauthier, S., and Bergeron, Y. 2002. Stand-landscape integration in natural disturbance-based management of the southern boreal forest. For. Ecol. Manag. **155**: 369–385.

Jayen, K., Leduc, A., and Bergeron, Y. 2006. Effect of fire severity on regeneration success in the boreal forest of northwestern Québec, Canada. Écoscience, **13**: 143–151.

Johnson, E.A., Miyanishi, K., and Bridge, S.R.J. 2001. Wildfire regime in the boreal forest and the idea of suppression and fuel buildup. Conserv. Biol. **15**: 1554–1557.

Johnstone, J.F. and Chapin, F.S., III. 2006. Effects of soil burn severity on post-fire tree recruitment in boreal forest. Ecosystems, **9**: 14–31.

Kafka, V., Gauthier, S., and Bergeron, Y. 2001. Fire impacts and crowning in the boreal forest: study of a large wildfire in western Quebec. Int. J. Wildland Fire, **10**: 119–127.

Lavoie, M., Harper, K., Paré, D., and Bergeron, Y. 2007a. Spatial pattern in the organic layer and tree growth: a case study from regenerating *Picea mariana* stands prone to paludification. J. Veg. Sci. **18**: 213–222.

Lavoie, M., Paré, D., and Bergeron, Y. 2007b. Relationships between microsite type and the growth and nutrition of young black spruce on post-disturbed lowland black spruce sites in eastern Canada. Can. J. For. Res. **37**: 62–73.

Lavoie, M., Paré, D., and Bergeron, Y. 2007c. Quality of growth substrates of post-disturbed lowland black spruce sites for black spruce (*Picea mariana*) seedling growth. New For. **33**: 207–216.

Lavoie, M., Paré, D., Fenton, N., Groot, A., and Taylor, K. 2005. Paludification and management of forested peatlands in Canada: a literature review. Environ. Rev. **13**: 21–50.

Leblanc, D. 1999. Performance de la livrée des forêts (*Malacosoma disstria*) élevée sur deux hôtes près de sa limite septentrionale et reconstruction de la chronologie de ses épidémies en Abitibi-Témiscamingue au moyen d'une analyse dendrochronologique. M.Sc. thesis. Université du Québec à Montréal, Montréal, Que.

Lecomte, N. and Bergeron, Y. 2005. Successional pathways on different surficial deposits in the coniferous boreal forest of the Quebec Clay Belt. Can. J. For. Res. **35**: 1984–1995.

Lecomte, N., Simard, M., and Bergeron, Y. 2006a. Effects of fire severity and initial tree composition on stand structural development in the coniferous boreal forest of northwestern Québec, Canada. Ecoscience, **13**: 152–163.

Lecomte, N., Simard, M., Fenton, N., and Bergeron, Y. 2006b. Fire severity and long-term ecosystem biomass dynamics in coniferous boreal forests of eastern Canada. Ecosystems, **9**: 1215–1230.

Lecomte, N., Simard, M., Bergeron, Y., Larouche, A., Asnong, H., and Richard, P.J.H. 2005. Effects of fire severity and initial tree composition on understorey vegetation dynamics in a boreal landscape inferred

from chronosequence and paleoecological data. J. Veg. Sci. **16**: 665–674.

Lefort, P., Harvey, B., Parton, J., and Smith, G.K.M. 2002. Synthesizing knowledge of the Claybelt to promote sustainable forest management. For. Chron. **78**: 665–671.

Légaré, S., Bergeron, Y., Leduc, A., and Paré, D. 2001. Comparison of the understory vegetation in boreal forest types of southwest Quebec. Can. J. Bot. **79**: 1019–1027.

Légaré, S., Bergeron, Y., and Paré, D. 2005a. Effect of aspen (*Populus tremuloides*) as a companion species on the growth of black spruce (*Picea mariana*) in the southwestern boreal forest of Quebec. For. Ecol. Manag. **208**: 211–222.

Légaré, S., Paré, D., and Bergeron, Y. 2004. The responses of black spruce growth to an increased proportion of aspen in mixed stands. Can. J. For. Res. **34**: 405–416.

Légaré, S., Paré, D., and Bergeron, Y. 2005b. Influence of aspen on forest floor properties in black spruce–dominated stands. Plant Soil, **275**: 207–220.

Lessard, I. 2004. Présentation du portrait et des enjeux reliés à la forêt en Abitibi-Témiscamingue pour la Commission d'étude sur la gestion de la forêt publique du Québec. Association forestière de l'Abitibi-Témiscamingue. [Online] <www.mrnfp.gouv.qc.ca/commission-foret/pdf/Portrait_Abitibi_texte.pdf> (accessed November 11, 2007).

Levasseur, V. 2000. Analyse dendroécologique de l'impact de la tordeuse des bourgeons de l'épinette (*Choristoneura fumiferana*) suivant un gradient latitudinal en zone boréale au Québec. M.Sc. thesis. Université du Québec à Chicoutimi, Chicoutimi, Que.

MacLean, D.A. 1983. Fire and nutrient cycling. *In* The role of fire in northern circumpolar ecosystems. *Edited by* R.W. Wein and D.A. MacLean. John Wiley & Sons Ltd, Chichester, UK, pp. 111–132.

Messaoud, Y., Bergeron, Y., and Leduc, A. 2007. Ecological factors explaining the location of the boundary between the mixedwood and coniferous bioclimatic zones in the boreal biome of eastern North America. Glob. Ecol. Biogeogr. **16**: 90–102.

Miyanishi, K. and Johnson, E.A. 2002. Process and patterns of duff consumption in the mixedwood boreal forest. Can. J. For. Res. **32**: 1285–1295.

Munson, A.D. and Timmer, V.R. 1989. Site-specific growth and nutrition of planted *Picea mariana* in the Ontario Clay Belt. I. Early performance. Can. J. For. Res. **19**: 162–170.

Nguyen-Xuan, T., Bergeron, Y., Simard, D., Fyles, J.W., and Paré, D. 2000. The importance of forest floor disturbance in the early regeneration patterns of the boreal forest of western and central Quebec: a wildfire versus logging comparison. Can. J. For. Res. **30**: 1353–1364.

Örlander, G., Gemmel, P., and Hunt, J. 1990. Site preparation: a Swedish overview. FRDA report, Forestry Canada and BC Ministry of Forests, Research Branch, Victoria, BC.

Paavilainen, E. and Päivänen, J., (Eds.). 1995. Peatland forestry: ecology and principles. Ecological Studies, vol. 111. Springer-Verlag, Berlin.

Payette, S. 2001. Les principaux types de tourbières. *In* Écologie des tourbières du Québec-Labrador. *Edited by* S. Payette and L. Rochefort. Les Presses de l'Université Laval, Québec, Que., pp. 39–89.

Payette, S. and Rochefort, L. (Eds.). 2001. Écologie des tourbières du Québec-Labrador. Les Presses de l'Université Laval, Québec, Que.

Perron, J.-Y. and Morin, P. 2002. Normes d'inventaire forestier. Placettes-échantillons permanentes. Direction des inventaires forestiers, Forêt Québec, Ministère des Ressources naturelles du Québec, Gouvernement du Québec, Québec, Que.

Prévost, M. 1992. Effet du scarifiage sur les propriétés du sol, la croissance des semis et la compétition : revue des connaissances actuelles et perspectives de recherches au Québec. Ann. Sci. For. **49**: 277–296.

Prévost, M. 1996. Effets du scarifiage sur les propriétés du sol et l'ensemencement naturel dans une pessière noire à mousses de la forêt boréale québécoise. Can. J. For. Res. **26**: 72–86.

Prévost, M. and Dumais, D. 2003. Croissance et statut nutritif de marcottes, de semis naturels et de plants d'épinette noire à la suite du scarifiage : résultats de 10 ans. Can. J. For. Res. **33**: 2097–2107.

Robitaille, A. and Saucier, J.-P. 1998. Paysages régionaux du Québec méridional. Les Publications du Québec, Québec, Que.

Roy, V., Bernier, P.Y., Plamondon, A.P., and Ruel, J.-C. 1999. Effect of drainage and microtopography in forested wetlands on the microenvironment and growth of planted black spruce seedlings. Can. J. For. Res. **29**: 563–574.

Simard, D.G., Fyles, J.W., Paré, D., and Nguyen, T. 2001. Impacts of clearcut harvesting and wildfire on soil nutrient status in the Quebec boreal forest. Can. J. Soil Sci. **81**: 229–237.

Simard, M., Bernier, P.Y., Bergeron, Y., Paré, D., and Guérine, L. 2009. Paludification dynamics in the boreal forest of the James Bay lowlands: effect of time since fire and topography. Can. J. For. Res. **39**: 546–552

Smithwick, E.A.H., Turner, M.G., Mack, M.C., Chapin, F.S., III. 2005. Postfire soil N cycling in northern conifer forests affected by severe, stand-replacing wildfires. Ecosystems, **8**: 163–181.

Simard, M., Lecomte, N., Bergeron, Y., Bernier, P.Y., and Paré, D. 2007. Forest productivity decline caused by successional paludification of boreal soils. Ecol. Appl. **17**: 1619–1637.

Soil Classification Working Group, 1998. The Canadian System of Soil Classification. 3rd edition. Agriculture and Agri-Food Canada, Ottawa, Ont.

Sutton, R.F. 1993. Mounding site preparation: a review of European and North American experience. New For. **7**: 151–192.

Thiffault, N., Cyr, G., Pregent, G., Jobidon, R., and Charette, L. 2004. Régénération artificielle des pessières noires à éricacées : effets du scarifiage, de la fertilisation et du type de plants après 10 ans. For. Chron. **80**: 141–149.

Turner, M.G., Hargrove, W.W., Gardner, R.H., and Romme, W.H. 1994. Effects of fire on landscape heterogeneity in Yellowstone National Park, Wyoming. J. Veg. Sci. **5**: 731–742.

Turner, M.G., Romme, W.H., and Gardner, R.H. 1999. Prefire heterogeneity, fire severity, and early postfire plant reestablishment in subalpine forests of Yellowstone National Park, Wyoming. Int. J. Wildland Fire, **9**: 21–36.

Turner, M.G., Romme, W.H., Gardner, R.H., and Hargrove, W.W. 1997. Effects of fire size and pattern on early succession in Yellowstone National Park. Ecol. Monogr. **67**: 411–433.

Van Cleve, K., Dyrness, C.T., Viereck, L.A., Fox, J., Chapin, F.S., III, and Oechel, W. 1983. Taiga ecosystems in interior Alaska. BioScience, **33**: 39–44.

Van Cleve, K. and Viereck, L.A. 1981. Forest succession in relation to nutrient cycling in the boreal forest of Alaska. *In* Forest succession: concepts and application. *Edited by* West, D.C., Shugart, H.H., and Botkin, D.B. Springer-Verlag, New York, USA, pp. 185–211.

Veillette, J.J. 1994. Evolution and paleohydrology of glacial lakes Barlow and Ojibway. Quat. Sci. Rev. **13**: 945–971.

Viereck, L.A. 1983. The effects of fire in black spruce ecosystems of Alaska and northern Canada. *In* The role of fire in northern circumpolar ecosystems. *Edited by* R.W. Wein and D.A. MacLean. John Wiley & Sons Ltd, Chichester, UK, pp. 201–220.

Vincent, É. 2004. Succession arborée dans la forêt boréale de l'Ouest du Québec. M.Sc. thesis. Université du Québec à Montréal, Montréal, Que.

Wardle, D.A., Zackrisson, O., Hörnberg, G., and Gallet, C. 1997. The influence of island area on ecosystem properties. Science, **277**: 1296–1299.

Zasada, J.C., Norum, R.A., Van Veldhuizen, R.M., and Teutsch, C.E. 1983. Artificial regeneration of trees and tall shrubs in experimentally burned upland black spruce/feather moss stands in Alaska. Can. J. For. Res. **13**: 903–913.

Chapter *12*

# Forest Dynamics of the Duck Mountain Provincial Forest, Manitoba, and the Implications for Forest Management*

*Brock Epp, Jacques C. Tardif, Norm Kenkel, and Louis De Grandpré*

* We acknowledge Keith Knowles and Manitoba Conservation, Alanna Sutton, and France Conciatori for providing information on insect outbreaks and pathogens. We also thank Louisiana-Pacific Canada, Ltd. and Manitoba Conservation for logistic support during fieldwork. Thanks also to Alanna Sutton for the translation of the text. Funding for much of the research presented in this chapter was provided by the Sustainable Forest Management Network and the Canada Research Chair program. The photos on this page were graciously provided by Jacques C. Tardif and France Conciatori.

# 1. INTRODUCTION

In recent years the forest industry has been directed towards a more sustainable forest management approach, including ecologically based management encompassing a broad range of values over large spatial and temporal scales (Galindo-Leal and Bunnell 1995; Gauthier et al., chapter 1). Within ecological management there is an emphasis on maintaining biodiversity within the ecosystem. Several authors (Franklin 1993; Bondrup-Nielsen 1995; Gauthier et al. 1996) have suggested that one way to achieve this is to maintain a forest landscape mosaic (in terms of age, structure, and dynamics) similar to one that is maintained by natural disturbances. The preservation of mosaic elements over the landscape helps to perpetuate the natural range of habitats and ecological niches. Given the critical importance of the influence of fire and other disturbances on the composition and structure of the forest landscape, there has been interest in the adoption of natural disturbance based management (NDBM), which attempts to develop silvicultural practices that best emulate the effects of natural disturbances (Rempel 1999; Bergeron et al. 2002; Harvey et al. 2003).

Across Canada, the boreal forest exhibits a significant amount of variability in terms of species dominance, climate, geography, and the disturbance regime (Rowe 1972; Larsen 1980; Bergeron et al. 2004b). Because of this variability, it is critical that any NDBM strategy reflect the local forest conditions and historical variability in the disturbance regime. The objectives of this chapter are to examine the disturbance regime and dynamics of the mixed boreal forest of the Duck Mountain Provincial Forest (DMPF) in western Manitoba, and to discuss their implications to forest management practices within the forest.

# 2. THE DUCK MOUNTAIN PROVINCIAL FOREST

The DMPF in Manitoba is part of the western region of the mixed boreal forest in Canada (Rowe 1972). Although Canada's mixed boreal forest shares many similarities in terms of dominant tree species, there are important variations across Canada. In western Canada (Manitoba, Saskatchewan, and Alberta), the mixed boreal forest is transitional between the northern boreal forest and the drier grasslands and aspen parklands to the south (Wiken 1986; Scott 1995). The mixed tree species composition in this region is thought to be determined by a combination of factors, including moisture stress and shorter fire intervals that define the southern limit of the conifer species (Rowe and Scotter 1973; Scott 1995). In eastern Canada (eastern Ontario and Québec), the mixed boreal forest forms a transitional zone possessing environmental conditions suitable to both the northern coniferous tree species and the southern broadleaf deciduous trees (Scott 1995). As a result that region tends to be somewhat richer in tree species.

The DMPF is located in western Manitoba, about 300 km northwest of Winnipeg (between 51°N and 52°N, 100°W and 102°W), adjacent to the Manitoba-Saskatchewan border (figure 12.1). The forest covers an area of about 376,000 ha, and within the forest is Duck Mountain Provincial Park (DMPP) which covers an area of 142,400 ha. The DMPF is part of a geological feature known as the Manitoba Escarpment – a series of uplands that includes Riding Mountain to the south and Porcupine Hills to the north, all of which fall within the Mid-boreal Uplands Ecoregion and the greater Boreal Plains Ecozone (Environment Canada 2004a). The DMPF is a forest island, and the surrounding area is dominated by agricultural activity including crop production in the Swan River valley, and grazing and hay land near the base of the escarpment (Sauchyn and Hadwen 2001).

The Mid-boreal Uplands Ecoregion is part of the Subhumid Mid-boreal Ecoclimatic Region, with predominantly short, cool summers and cold winters (EWG 1989). The period of mean daily temperatures above 0°C extends from April to October, with up to 120 frost-free days and maximum precipitation occurring in July (EWG 1989). Climate data from the meteorological station located at Swan River (figure 12.1; 52°03′ N and 101°13′ W; elevation 347 m a.s.l., approx. 17 km from the NW border of DMPF) indicates that temperature ranges from a minimum of –23.0°C in January to a maximum of 24.4°C in July, and the average annual precipitation is 530 mm (Environment Canada 2004b). However, the Duck Mountain plateau is cooler than the surrounding lowlands and can receive up to 50% more precipitation (MPB 1973).

The Manitoba Escarpment is characterized by its bedrock (primarily composed of Odanah shale) and higher elevation compared to the plains to the east. The eastern boundary of the forest is made up of the Escarpment, a steep 245-m drop to the Manitoba plains (Sauchyn and Hadwen 2001). Elevation decreases more gradually to the west, at a rate of approximately 3.1 m/km (Smith et al. 1998) (figure 12.2). The current landscape of Duck Mountain emerged as the last glacier receded 12,000 years ago, and is covered by glacial deposits of clay, gravel, sand, and boulders (Manitoba Conservation 2004a), as well as stratified deposits of outwash sediments (Sauchyn and Hadwen 2001). The landscape is uneven, with slopes ranging from 10 to 15% and from 50 to more than 100 m in length (Smith et al. 1998). Soils in the region are primarily composed of well-drained Gray Luvisols associated with glacial till. There are also localized areas of Dark Gray Chernozems, and poorly drained Luvic Gleysols associated with depressions (Smith et al. 1998).

The dominant deciduous tree species of the western mixed boreal forest are trembling aspen (*Populus tremuloides* Michx.), balsam poplar (*Populus balsamifera* L.), and white birch (*Betula papyrifera* Marsh.), while the dominant conifers are white spruce (*Picea glauca* [Moench] Voss), black spruce (*Picea mariana* [Mill.] BSP), *jack* pine (*Pinus banksiana* Lamb.), eastern larch (*Larix laricina* [Du Roi] K. Koch), and some balsam fir (*Abies balsamea* [L.] Mill.) (Scott 1995; Lieffers et al. 1996a; Hamel and Kenkel 2001). Younger, post-disturbance forests occurring on finer-textured soil are usually dominated by trembling aspen and balsam poplar on well-drained slopes and richer floodplains, respectively. On coarse-textured soils jack pine community types dominate, while black spruce communities dominate on less rich organic sites (La Roi 1992; Hamel and Kenkel

Figure 12.1

## Location of the Duck Mountain Provincial Forest (DMPF) and the Mid-boreal Uplands Ecoregion (A), and boundaries of the Duck Mountain Provincial Park (DMPP) and the DMPF (B)

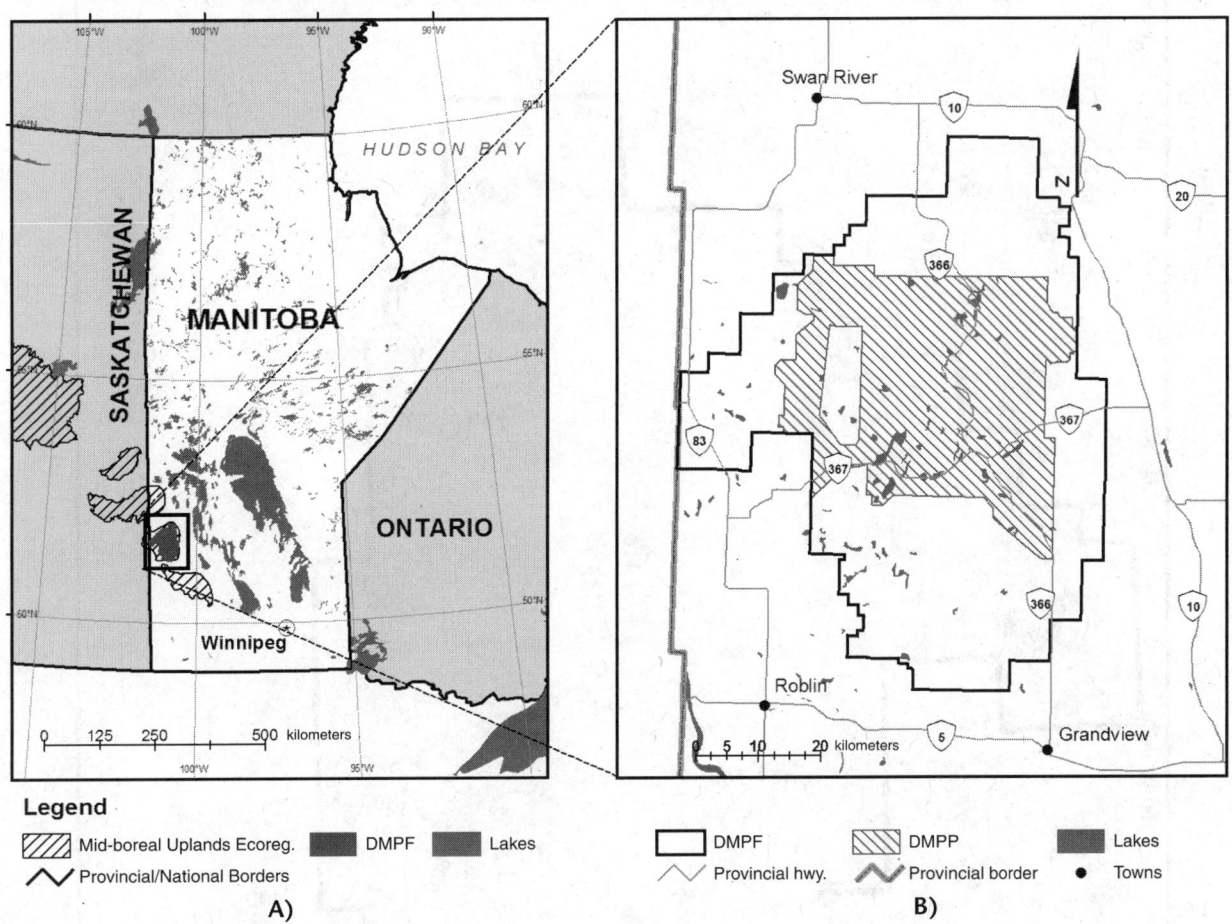

**Legend**

| | | |
|---|---|---|
| ▨ Mid-boreal Uplands Ecoreg. | ▐ DMPF | ▐ Lakes |
| ⋀ Provincial/National Borders | | |

**A)**

| | | |
|---|---|---|
| ▢ DMPF | ▨ DMPP | ▐ Lakes |
| ⋀ Provincial hwy. | ⋀ Provincial border | ● Towns |

**B)**

The meteorological station used for climate data is located at Swan River. Sources: A: from Schut (2005); B: Manitoba Conservation (2000).

2001). Post-disturbance stands dominated by white spruce are less common but also occur on finer textured soils. This post-disturbance community type relies more on the availability of a seed source and moister soil conditions (Lieffers et al. 1996b; Hamel and Kenkel 2001; Kenkel et al. 2003). Old-growth stands in this region tend to be composed of white spruce and some balsam fir with white birch on the finer textured soils, and black spruce on coarse textured and organic sites (Hamel and Kenkel 2001). In the absence of a white spruce seed source, old-growth forests may continue to be dominated by trembling aspen (Cumming et al. 2000; Hamel and Kenkel 2001).

Figure 12.2
**Topographic map of the Duck Mountain Provincial Forest (DMPF)**

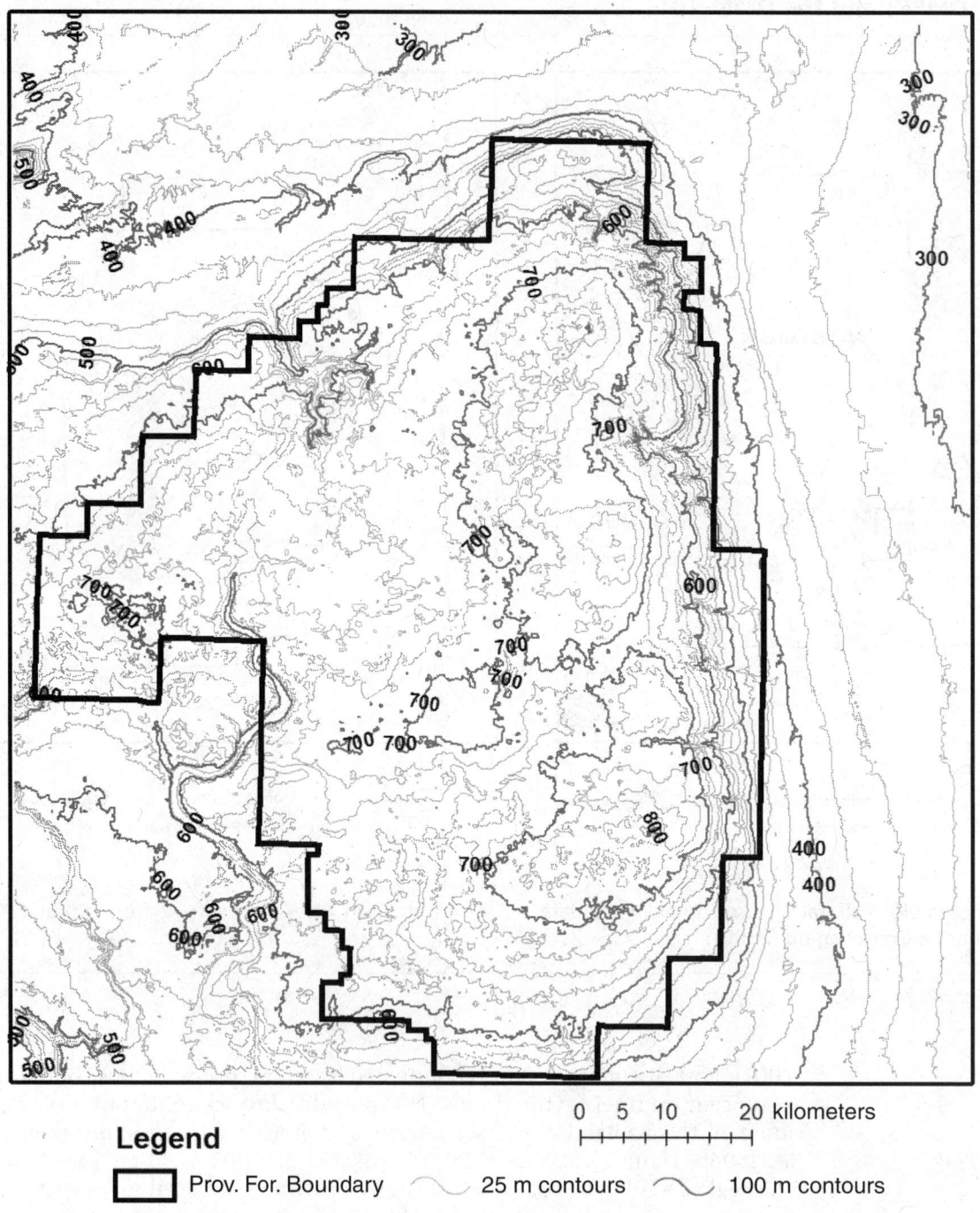

Isolines represent elevation in metres-above-sea-level at 25 m and 100 m intervals. Source: Louisiana-Pacific Canada, Ltd.

Fire is generally considered to be the most important natural disturbance in western mixed boreal forest, since it largely determines age structure at the landscape scale (Weir et al. 2000). Estimates of the fire cycle in this region for the past century range from 52 to 118 years, although current estimates revealed that the fire cycle has become much longer (>450 years in some areas) (Larsen 1997; Tardif 2004). These higher estimates may be attributable to active fire suppression and forest fragmentation resulting from human development in the period following ~1945, and/or climatic trends that are less suitable for fire (Weir and Johnson 1998; Weir et al. 2000; Tardif 2004). Despite this apparent lengthening, fire cycles are generally shorter in the western mixed boreal forest compared to the east (Larsen 1997; Weir et al. 2000; Bergeron et al. 2004b; Tardif 2004). These differences are attributable to differences in large-scale atmospheric circulation patterns which affect local climatic conditions and drought severity (Flannigan et al. 2001; Girardin et al. 2004). Secondary disturbances, such as insects, pathogens, and windthrow, also play a role in the western mixed boreal forest. Spruce budworm (*Choristoneura fumiferana* Clem.) is a primary defoliator of balsam fir, white spruce, and black spruce (Morin 1994), while forest tent caterpillar (*Malacosoma disstria* Hbn.) is a primary defoliator of trembling aspen (Peterson and Peterson 1992; Simpson and Coy 1999; Hogg et al. 2002a; Sutton and Tardif 2007; Sutton and Tardif, chapter 8). Secondary disturbances play a more important role in the eastern mixed boreal because the forests are older on average (Morin 1994; Pham et al. 2004).

The Duck Mountain area was first surveyed and mapped during the 1880s (Tyrrell, 1888). During this time the presence of Aboriginal settlements was noted in the surrounding area, as well as some agricultural activity. The DMPF was established as a forest reserve in 1906 and was considered at the time to be the most valuable timber resource in the province (Harrison 1934). In 1961 the DMPP was established with a goal of protecting the natural integrity of the park while providing recreational activities and honouring existing resource commitments (Manitoba Conservation 2003). The DMPF has provided a source of timber for European settlers as they settled the surrounding plains. Prior to 1882, when the first timber berth was given in Duck Mountain, settlers had already established a portable sawmill and harvested a considerable amount of timber for local use (Anonymous 2000). By the early 1900s, the logging industry in Duck Mountain expanded significantly with several sawmills established by the Burrows Lumber Company (Anonymous 2000). These mills continued to operate until the early 1930s when most of the timber berths in Duck Mountain were depleted (Anonymous 2000). Since 1994 the DMPF has been included in Louisiana-Pacific Canada's management license (FML #3). Louisiana-Pacific primarily utilizes hardwood fibre, including trembling aspen, balsam poplar, and white birch, for the production of oriented strand board (OSB) (LP Canada 2006a). There are also about 27 quota holders operating in the forest, primarily utilizing softwood timber and fibres. The largest quota holder operating in the region is Spruce Products Ltd., with approximately 59.6% of the total softwood quota volume (LP Canada 2006b). The forest is highly accessible, and contains several campgrounds, lakes, and trail systems that are used to support recreational activities.

# 3.  DISTURBANCE REGIMES

## 3.1.  Fires

Fires have played the primary role in shaping the age-structure of the DMPF landscape. Historically, peoples utilizing the DMPF for logging have fought fires to protect the timber resource (Stilwell 1988). Early studies in the region have documented the occurrence of large fires in the late 1800s that burned much of the forested area in the uplands (Gill 1930; Harrison 1934). Since 1961, few fires have burned in the DMPF. A 300-year fire history for the DMPF was created by Tardif (2004) based on a network of 359 sampling points, which was extended to 1,497 points using forest inventory ground truthing sites (n = 443) and plots from Louisiana Pacific-Canada Ltd. (n = 695). In 2002, the mean age of the DMPF was estimated to be 108 ± 38 years based on the 359 sampling points. Trembling aspen stands are generally the youngest, while the oldest stands, dating back to the early 1700s, are composed of black spruce and eastern larch growing on very moist sites. The time-since-last-fire (TSLF) distribution for the DMPF illustrates the impact of the late 1800s fires on the landscape (figure 12.3). Based on the distribution of stand ages and fire scars, it is estimated that approximately 83% (283,580 ha) of the DMPF area was affected by fires between 1885 and 1895. This was followed by several small fires around the southwest periphery of the forest, and then one large fire in 1961, where approximately 6% of the DMPF burned (Hamel and Kenkel 2001). After this very little fire activity was observed.

The fire history for the DMPF suggests that the fire cycle has not been constant over the last 300 years (figure 12.4), significantly lengthening after the late 1800s. The fire history of the DMPF can be divided into three distinct periods: 1) pre-European settlement (prior to 1880), 2) European settlement (1880–1961), and 3) the current, fire-absent period (1962–present). The pre-European settlement period was characterized by a short fire cycle ranging from 50 to 60 years and large fires. Growth suppression in the tree-ring data during this period suggests that these large fires were associated with prolonged periods of drought. Such prolonged drought periods were observed prior to the 1900s in Manitoba and western Canada (Sauchyn and Skinner 2001; Girardin et al. 2002; Tardif and Stevenson 2002). The large fires of the 1880s and 1890s coincided with a period of extreme drought (Girardin et al. 2004).

During the European settlement period in the early part of the 20th century, the fire cycle increased to about 200 years and numerous small fires were observed around the periphery of the DMPF (Tardif 2004). Some of these fires may have been caused by an increase in human agricultural activity, as this has been observed in other regions of the mixed boreal forest (Weir and Johnson 1998; Lefort et al. 2003). The last large fire in the DMPF occurred in 1961, which was recorded as an extremely dry year, and was brought under control after burning for about one month (Kenkel et al. 2003). This was followed by the modern, fire-absent period, in which very few fires burned, thereby dramatically lengthening the fire cycle. However, it is difficult to accurately estimate because of the short period of time (40 years). The exact reasons for the overall lengthening of the fire cycle after the 19th century are not known. It is probable that a combination of more effective fire suppression techniques and less favourable climatic

Figure 12.3
## Map of the Duck Mountain Provincial Forest showing the time-since-last-fire distribution as of 2002

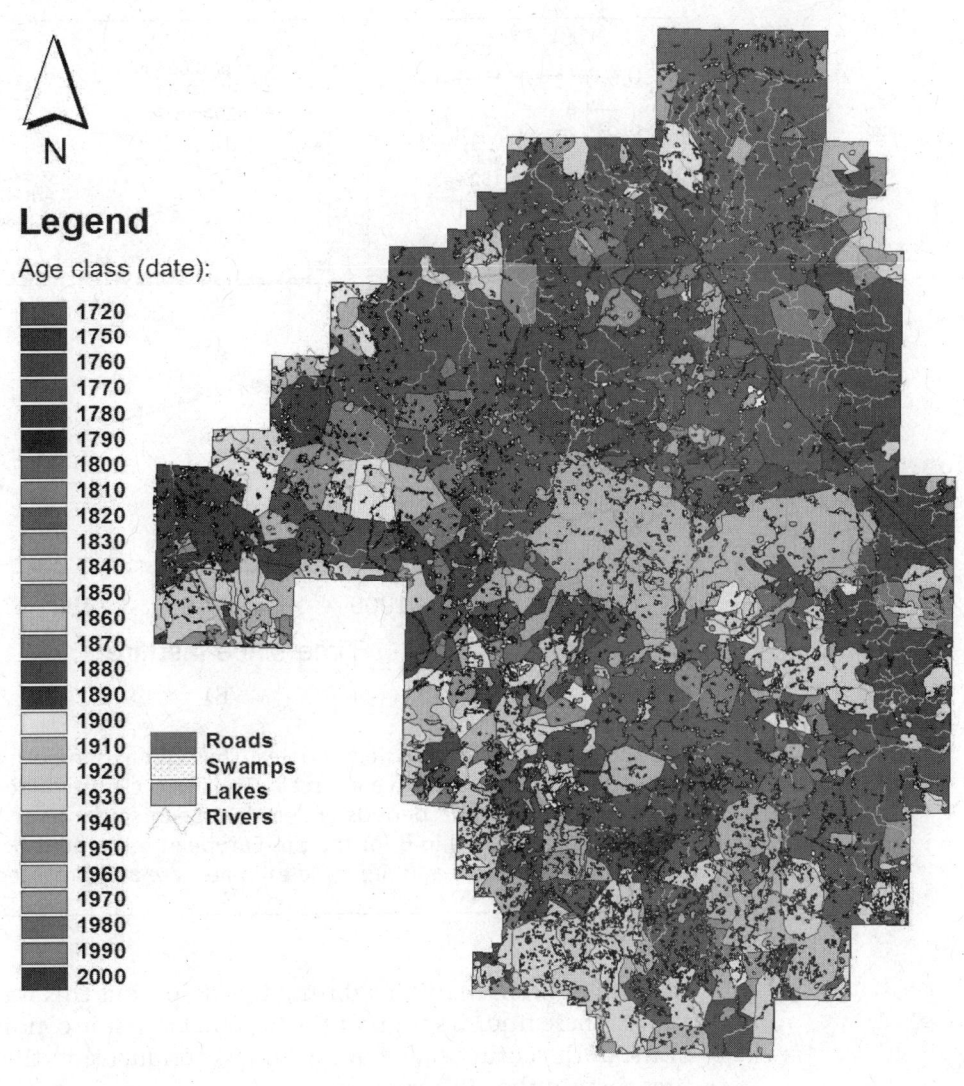

Source: Tardif (2004).

conditions for the development of large, catastrophic fires are the cause. In western Québec, Lefort et al. (2003) found that changes in the Fire-Weather Index (FWI) mostly explained changes in the fire cycle during the 20th century. They suggested that fire suppression only became effective enough to affect the fire cycle after the advent of water bombers in the 1970s. For western Canada, Sauchyn and Skinner (2001) found that the 20th century had fewer prolonged droughts compared to the 19th century. Tardif and Stevenson (2001) reported prolonged periods of growth suppression for northern white-cedar (*Thuja*

Figure 12.4

**Percent area of A) the DMPF in each age class derived from the current time-since-last-fire map and B) percent area in each age class derived from the time-since-last-fire map calculated for 1880**

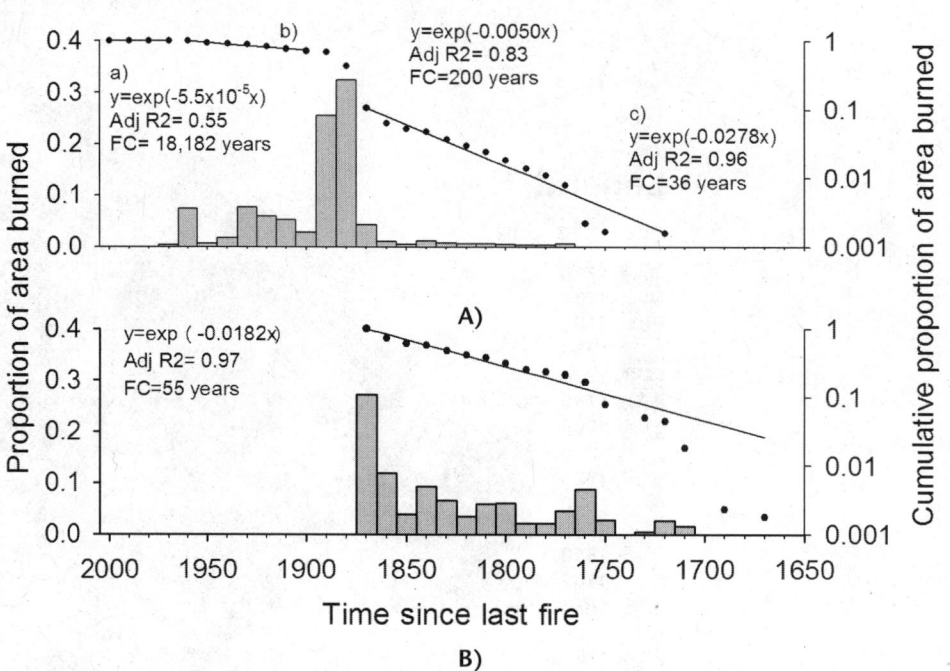

The age classes are represented as dates. The right axis in both graphs show cumulative percent area distribution on a log scale. The fire cycle estimations are presented in A for the three historical fire periods (1: pre-European settlement, 2: European settlement, 3: current period) and in B for the pre-European settlement period only. The negative exponential equations and fire cycle estimates are also provided. From Tardif (2004).

*occidentalis* L.) in Manitoba during the 1890s, but this was not observed for the 1900s. These findings support the idea that climatic conditions at the beginning of the 20th century may have been less conducive to the development of large fires than in the 19th century.

## 3.2. Insect Outbreaks and Pathogens

Insect outbreaks are an important component of the disturbance regime in the DMPF, and they are of particular concern to the regional forest industry. The four major insect pests in the region are forest tent caterpillar (FTC), spruce budworm, jack pine budworm (*Choristoneura pinus pinus* Freeman), and larch sawfly (*Pristiphora erichsonii* Hartig) (LP Canada 2006a). The primary host species for FTC is trembling aspen (Peterson and Peterson 1992; Sutton and Tardif 2007), which is the most commercially important species for Louisiana-Pacific Canada, Ltd. (LP Canada 2006a). Secondary hosts include white birch and balsam poplar

(Hildahl and Campbell 1975; Sutton and Tardif 2007). Forest tent caterpillar disturbances usually occur in cyclical outbreaks lasting for 3 to 6 years, ranging from 6 to 20 years between (Duncan and Hodson 1958; Hildahl and Campbell 1975; Martineau 1984; Ives and Wong 1988; Sutton and Tardif 2007). The insect is not necessarily a direct cause of mortality in trembling aspen but acts as a stressor that, when combined with other stressors such as drought, reduces growth and predisposes trees to damage by other pests and pathogens that may eventually lead to mortality (Hogg et al. 2002a). Drier conditions are believed to be more favourable to FTC, resulting in more severe outbreaks (Hogg et al. 2002a). More detailed information on the nature of FTC outbreaks in Manitoba is presented in chapter 8 (Sutton and Tardif).

The 20th century outbreaks of FTC in the DMPF correspond with outbreaks reported throughout the Canadian Prairie provinces (Brown 1940, 1941; Hildahl and Reeks 1960; Elliot and Hildahl 1963, 1964, 1965; Moody and Cerezke 1983, 1984, 1985). Patterns of growth suppression indicated that the outbreak of the 1940s, which was the longest, spread into the DMPF from the Porcupine Forest Reserve (PFR) to the north, and gradually intensified until about 1946 (figure 12.5; Sutton and Tardif 2007). The outbreak in the 1960s was the shortest but most intense, and was part of a larger outbreak that began after the extreme drought of 1961. The outbreak in the early 1980s also built slowly in intensity (similar to the outbreak of the 1940s), but insect activity decreased after 1984 (figure 12.5). This decrease may have been the result of severe late-spring frosts in 1984 that could have reduced the FTC population.

Spruce budworm and jack pine budworm are also important defoliators in the Prairie provinces (Simpson and Coy 1999). Spruce budworm has been observed in the DMPF but has not recently reached outbreak levels (LP Canada 2006a). Federal surveys conducted since the mid-1940s have reported low-intensity defoliation by the spruce budworm in 1951 and 1978, and extensive, moderate to severe defoliation in the mid-1980s and mid-1990s within the DMPF (FIDS 1944–1996). Jack pine budworm has also been observed (LP 2006a) but outbreaks have not been reported for the region. Regular outbreaks of the larch sawfly have been reported and studied in the DMPF (FIDS 1944–1996; Nairn et al. 1962; Turnock 1972). Moderate to severe defoliation has been observed for the periods 1911–1920, 1924–27, 1945–1953, 1961, and 1970 (FIDS 1944–1996; Nairn et al. 1962; Turnock 1972). Larch sawfly outbreaks are associated with the formation of light rings, followed by periods of growth suppression in eastern larch (Jardon et al. 1994; Girardin et al. 2001). In the DMPF, light rings (France Conciatori, pers. comm.) and prolonged periods of growth suppression (Girardin and Tardif 2005) in eastern larch corresponded with the outbreaks of the 1910s and 1940s.

While there are numerous pathogens present within the DMPF, the most widespread in terms of impact on the forest structure is armillaria root rot (*Armillaria* spp.) (Keith Knowles, pers. comm.). Armillaria is responsible for centres of tree mortality and the formation of openings in upland black spruce stands in the DMPF. Canopy gaps are created as individual trees are felled by root or butt rot (Keith Knowles, pers. comm.). Currently, volume-loss surveys are ongoing in the DMPF but surveys conducted in the Porcupine Hills to the north indicated volume losses ranging from 17% to 45% in upland black spruce stands (Pines 2006).

## Figure 12.5
## Maps of the DMPF showing the progression of three major FTC outbreaks identified for the 20th century

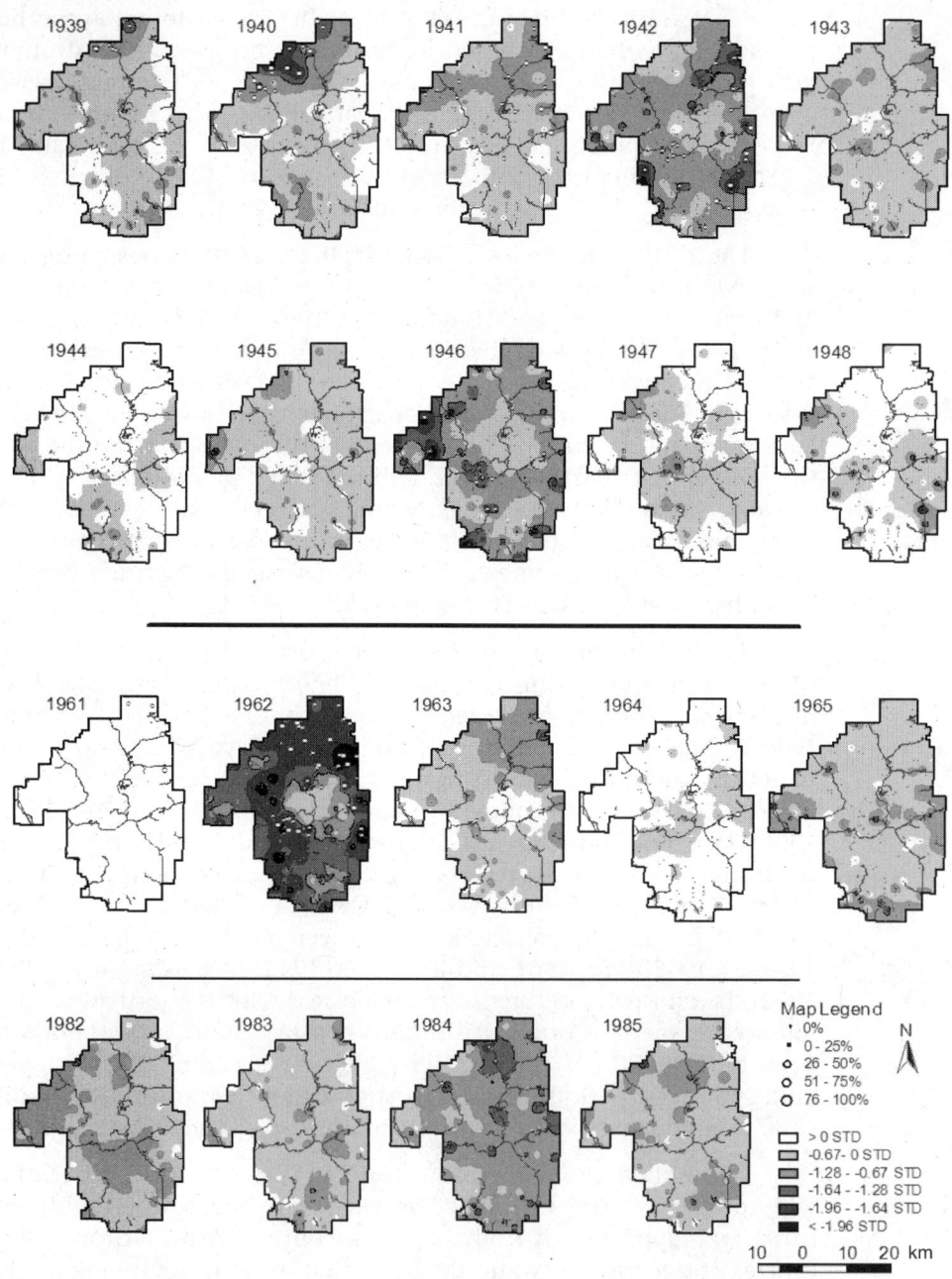

The maps present one year before and after each outbreak period. Gray shading represents mean negative growth deviation in standard deviation (STD), which expresses the importance of the reduction in radial growth compared to non-host species. Circles represent the relative frequency of white rings as a percentage. From Sutton and Tardif (2007).

## 4.  FOREST DYNAMICS

### 4.1.  Forest Succession in the Duck Mountain Provincial Forest

In the mixed boreal forest, succession is influenced by many factors, including the type and severity of disturbances, pre-disturbance conditions, and various environmental, vegetation, and soil (e.g., seedbed) characteristics (Chen and Popadiouk 2002). Generally, in the mixed boreal forest, there is a gradual shift from hardwood- to softwood-dominated stands in the absence of fire (Bergeron and Dubuc 1989; Bergeron 2000; De Grandpré et al. 2000; Hamel and Kenkel 2001).

Hamel and Kenkel (2001) characterized stand dynamics and developed a general successional model for the DMPF (figure 12.6). Five post-fire stand types were identified for the DMPF, each characterized by a dominant species; these include: 1) trembling aspen, 2) balsam poplar, 3) white spruce, 4) jack pine, and 5) black spruce. The dynamics of trembling aspen stand types are characterized by extensive root suckering, and such stands may be self-perpetuating in the absence of a nearby white spruce seed source (Hamel and Kenkel 2001). If a seed source is available, white spruce density will increase slowly over time (Kenkel et al. 2003), although recruitment of aspen, balsam poplar, and white birch may continue to occur in gaps. Balsam poplar stand types contain advance regeneration mostly composed of white spruce, however the presence of trembling aspen, white birch, balsam fir, balsam poplar, or black spruce in the understory suggests a complex successional pathway (Hamel and Kenkel 2001). As the density of the conifer component increases, trembling aspen becomes less important because light is a limiting factor in its establishment (Paré et al. 2001; Greene et al. 2002). White spruce stand types show primarily balsam fir advance regeneration when tree density is high. When density is low, white spruce dominates the advance regeneration, with some trembling aspen and white birch. These stands tend to occur adjacent to unburned areas, and tend to self-perpetuate, with balsam fir becoming more important with increasing density (Hamel and Kenkel 2001). Hamel and Kenkel (2001) described four major regeneration trends in jack pine stand types, depending on abiotic conditions and seed sources. Advance regeneration in the jack pine stand types can consist of combinations of white spruce, black spruce, trembling aspen, white birch, and balsam fir. Balsam fir, white spruce, and black spruce become more important on moist sites than on sandy, well-drained sites. Black spruce stand types tend to be very dense on mineral soil with very little advance regeneration (Hamel and Kenkel 2001). Trembling aspen or white birch may occur sporadically in gaps but are usually suppressed. In the absence of major disturbance, these stands are generally self-perpetuating. On organic substrates black spruce is much less dense and eastern larch may also occur in the initial cohort. Black spruce tends to self-perpetuate through layering in these stands.

The large fires in the late 1800s, as well as a lack of fire in the latter half of the 20th century, have resulted in a high proportion of stands in the DMPF originating from the 1880s and 1890s (figure 12.4; Tardif 2004). Stand origin estimates from a new forest lands inventory, in conjunction with the fire history data (Tardif, 2004), have been used to isolate the stands remaining on the landscape that originated from the 1885–1895 fires and characterize compositional

**Figure 12.6**
## A synoptic forest succession model for boreal mixedwood stands of the DMPF

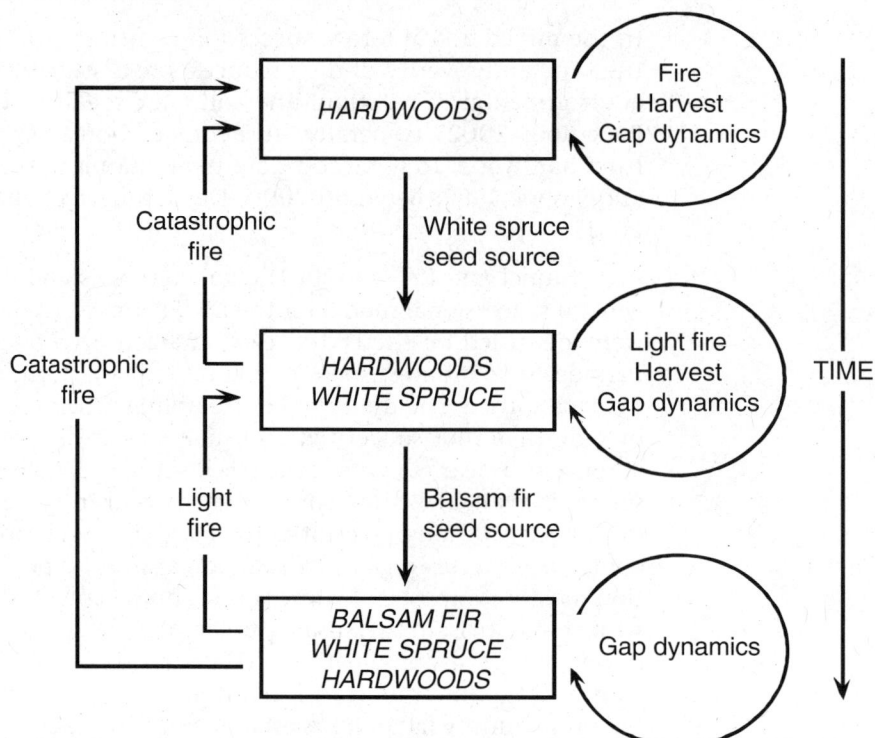

Following catastrophic fire, stands are generally dominated by hardwoods (trembling aspen, balsam poplar, and/or white birch). These stands are self-replacing under recurrent light or catastrophic fire, clear-cut harvesting, or canopy gap dynamics. Hardwood stands can develop into mixed hardwood-softwood stands only if there is a proximate white spruce seed source. Such stands may later include balsam fir if a proximate seed source is present. Selection harvesting and gap dynamics will perpetuate mixedwood stands, but catastrophic fires that remove conifer seed sources will result in reversion to hardwood dominance. Adapted from Hamel and Kenkel (2001).

and structural variability within different forest types (Epp et al. 2006). Currently about 58% of the DMPF forest area is estimated to originate from that time (figure 12.4). In our study, 73,000 ha of the 1885–1895 fire-originated forest, representing 23% of the total forest area, was retained for analysis. A high degree of variability was found in terms of tree composition and structure in the stands originating during this period (figure 12.7). Stands dominated by trembling aspen and white spruce exhibited the highest degree of variability. They ranged from nearly pure trembling aspen with single-layered even-aged canopies, to pure white spruce stands with complex, multi-layered canopies (Epp et al. 2006). Stands dominated by jack pine and black spruce, however, exhibited considerably less variability than those dominated by trembling aspen and white spruce in terms of species composition and canopy structure.

**Figure 12.7**
## Schematic diagram of species composition and structure for stands originating from the 1885–1895 fires in the DMPF

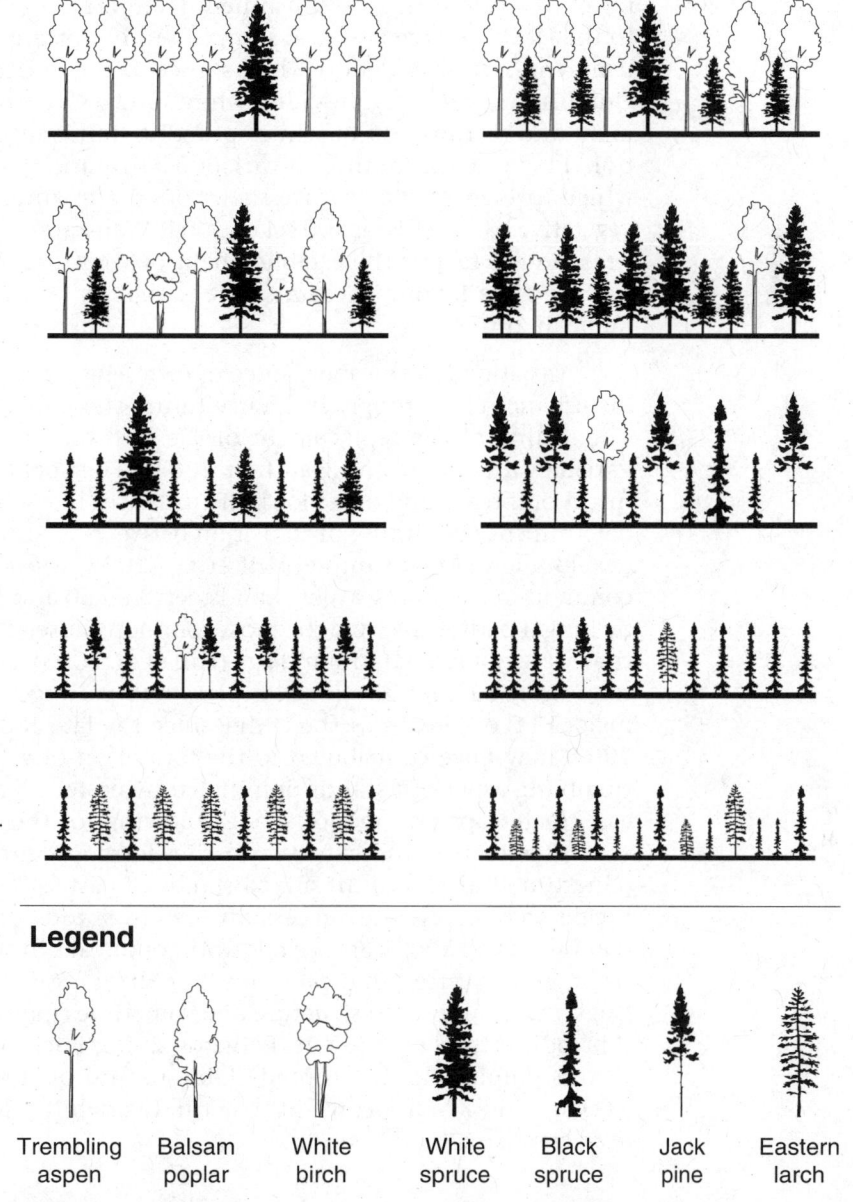

**Legend**

| Trembling aspen | Balsam poplar | White birch | White spruce | Black spruce | Jack pine | Eastern larch |

Individual stand diagrams reflect the relative mean percent cover and relative height for each tree species in groups that were classified using forest inventory data. Individual tree profiles are identified in the legend at the bottom of the figure. The relative number of individual species in each diagram reflects the relative percent cover for that species. The variation in species height for each diagram reflect the relative position and composition of the canopy layers. From Epp et al. (2006).

The structural variability observed between similarly-aged stands in the DMPF suggests that age is not the only factor determining the development of canopy structure (Epp et al. 2006). Several studies have found that the proximity of seed source plays an important role in stand development, particularly in terms of white spruce establishment (Galipeau et al. 1997; Kenkel et al. 2003). In the DMPF distance to seed source is an important determinant for the establishment of second-cohort white spruce in the understory (Kenkel et al. 2003). Unburned stands may provide a white spruce seed source for the early establishment and recruitment of white spruce in stands regenerating nearby. An additional factor contributing to variability in stand development is fire intensity, which influences the characteristics of the growing substrate and the seed and vegetative banks (Wang 2002; Lee 2004; Wang and Kemball 2005). More intense fires tend to favour the establishment of seed-regenerating conifers, while less intense fires favour deciduous trees that establish vegetatively (Wang and Kemball 2005).

Variations in the abundance of different species following stand-replacing disturbance can strongly influence future stand development (Bergeron 2000). The timing of the replacement of the post-disturbance cohort is also highly variable (Kneeshaw and Gauthier 2003). The spatial dynamics and timing of insect outbreaks and other local disturbances like windthrow may play a role in determining the timing of transition between successional stages and structural development. For example, Hill et al. (2005) found that gap development is common in trembling aspen stands between 60 and 120 years of age. Infections by fungal pathogens were the most common cause of gap-making mortality, and they were not related to stand age (Hill et al. 2005). Since FTC defoliation makes trees more susceptible to other pests and pathogens, the occurrence of three major FTC outbreaks in the DMPF since the late 1800s fires (Sutton and Tardif 2007) may have contributed to the formation of gaps in stands dominated by trembling aspen. In stands dominated by eastern larch, larch sawfly outbreaks can create gaps and reduce shade, allowing for the release of shade-intolerant seedlings in the understory and the development of structural complexity (Girardin et al. 2002). In mixed stands of eastern larch and black spruce, outbreaks may accelerate succession toward stands dominated by black spruce (Girardin et al. 2002). Spruce budworm outbreaks in stands dominated by balsam fir or white spruce can create gaps that allow for the release of understory seedlings, increasing stand structure and sometimes maintaining a mixedwood state if broadleaf trees are present (Bergeron 2000; Fricker et al. 2006). Overall, these factors emphasize the unpredictable nature of structural development, and explain why a high degree of structural variability is present in these similarly aged stands.

## 4.2. Fire and the Distribution of Species and Structural Cohorts in the DMPF

The fire history of the DMPF has had a dominating influence on the pattern of forest composition and structure observed over the landscape (Tardif 2004). Epp et al. (2006) utilized forest inventory data to classify the landscape according to species composition and structure. Forest stands were assigned to one of three cohorts similar to those defined by Bergeron et al. (2002) and Harvey et al.

(2002), representing stands in different stages of structural development (see Gauthier et al., chapter 3). Pure to mixed deciduous and coniferous stands, primarily composed of trembling aspen and white spruce, constituted about 80% of the DMPF. Stands dominated by black spruce and jack pine constitute about 14% and 6% of the area, respectively (figure 12.8). Black spruce and jack pine stands occurred most commonly in the central, higher elevation regions of the DMPF, while trembling aspen-dominated mixedwoods were most common around the perimeter of the forest adjacent to agricultural land (Epp et al. 2006). Topography and environmental conditions play a role in determining the predominant forest types on the landscape. The hummocky landforms of the central uplands result in impeded drainage and poor nutrient status, conditions conducive to the formation of black spruce–dominated forests (Hamel and Kenkel 2001). The sloping landforms around the perimeter of the DMPF are better drained and tend to have a higher nutrient status, thus providing suitable conditions for hardwoods and white spruce (Hamel and Kenkel 2001). After environmental factors, however, time since last fire is an important factor influencing the species composition of forest stands (Bergeron and Dubuc 1989; Gauthier et al. 2000). In the DMPF there tends to be an overall trend from shorter-lived, shade-intolerant hardwoods and jack pine, to shade-tolerant conifers such as black spruce, white spruce, and balsam fir (Hamel and Kenkel 2001). Visual comparison of the time-since-last-fire distribution (Tardif 2004) with the species composition of the upper canopy (Epp et al. 2006) reveals some interesting patterns (figures 12.4 and 12.8). Jack pine, which is considered an early successional species, tends to be most concentrated within the extent of the 1961 fire, while pure trembling aspen canopies tend to dominate in the southwest periphery of the forest where many small fires in the early 20th century occurred. Conversely, white spruce canopies tend to dominate in regions where the date of last fire is prior to the year 1900.

With respect to structure, approximately 60% of the DMPF is in the first cohort stage, where the canopy is dominated by trees recruited immediately post-fire, and the stands have an even-aged structure (figure 12.8; Epp et al. 2006). About 34% of the DMPF is in the second-cohort stage, in which there is evidence of break-up of the post-disturbance canopy and replacement by the second cohort. Such stands take on an uneven-aged structure. The remaining 6% is in the third cohort stage, which is characterized by gap dynamics and a complex canopy structure dominated by shade-tolerant conifers (figure 12.8). Jack pine-black spruce stand types had the highest relative proportion in the first-cohort stage while black spruce–eastern larch stand types had the highest proportion in the third-cohort stage (Epp et al. 2006). Many of the third-cohort stands are concentrated in areas originating from fires prior to the 1880s and 1890s. This distribution of cohorts is largely the result of the influence of the large, catastrophic fires of the late 1800s, and fire suppression in the latter half of the 20th century (Gill 1930; Tardif 2004). As a result, much of the forest originates from those large fires, and much of the first cohort is at or near a transitional phase toward the second cohort. The influence of large, infrequent catastrophic disturbances on the forest mosaic and succession has been studied in many forest landscapes. Studies have suggested that such disturbances show a lasting, persistent impact on the landscape and successional processes compared to smaller fires (Turner and Dale 1998; Turner et al. 1998).

**Figure 12.8**
## Map of the DMPF showing the distribution of the three structural cohorts for three major forest types

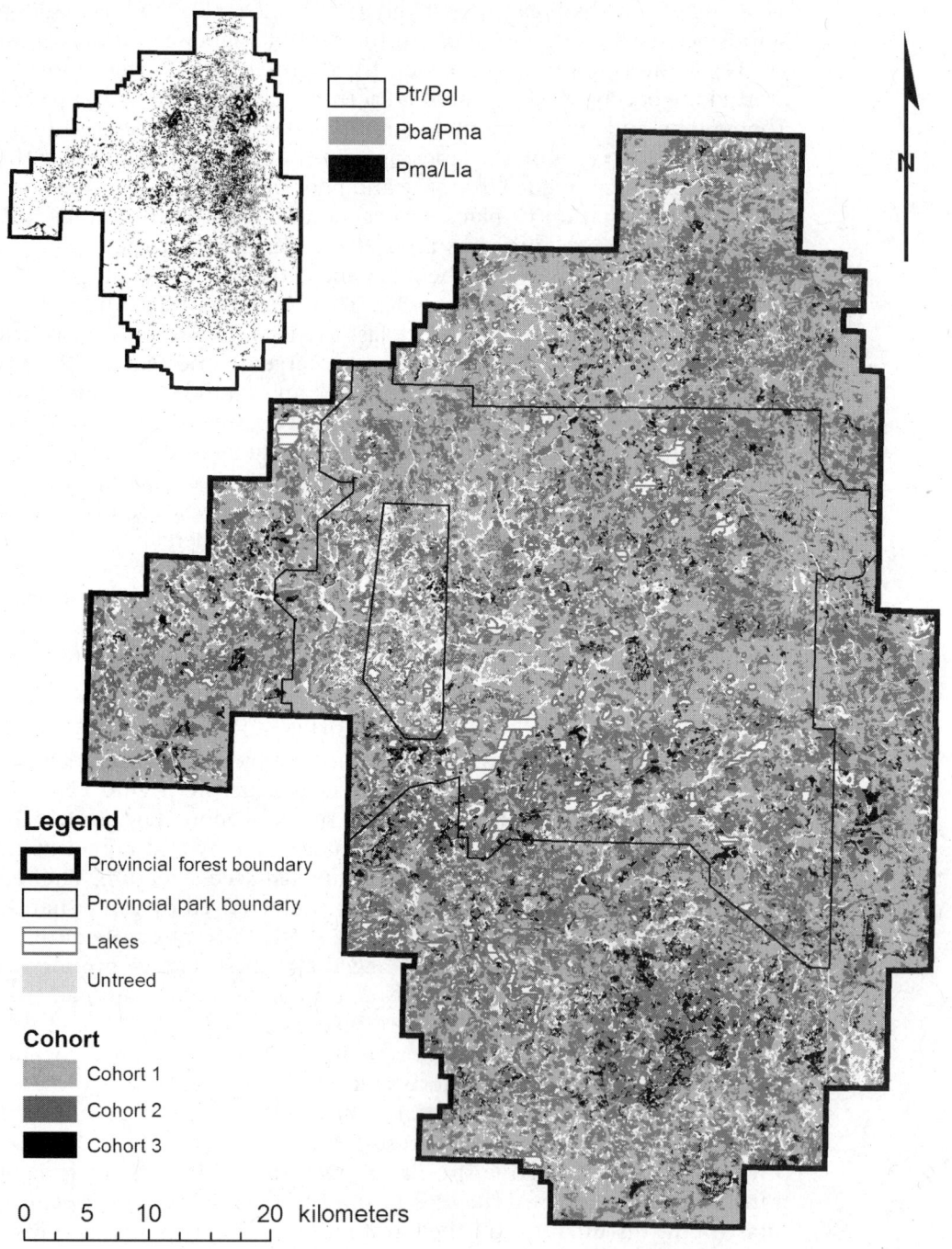

Legend: Ptr/Pgl = trembling aspen/white spruce; Pba/Pma = jack pine/black spruce; Pma/Lla = black spruce/ eastern larch. From Epp et al. (2006).

## 5.  IMPLICATIONS FOR FOREST MANAGEMENT

### 5.1.  Natural Disturbance-Based Management

Different management approaches have been proposed to maintain a natural age-class distribution similar to the one found in forests regulated by natural disturbances. For example, varying rotation lengths would allow portions of the landscape to naturally develop second- and third-cohort stand structures (Seymour and Hunter 1999).

One major difficulty with the variable rotation approach (sensu Burton et al. 1999), which would require a lengthening of the rotation for a portion of the stands, is that commercial rotation ages in the DMPF are relatively short, averaging about 75 years (Manitoba Conservation 2004b). In addition, the dominant species on the forested landscape have an average longevity that rarely surpasses 100 years. Variable rotations could potentially reduce the annual allowable cut in a fully regulated landscape and make long rotations economically unfeasible, particularly when trying to maintain a high proportion of old-growth structure (Bergeron et al. 2002). Conversely, allowing longer rotation ages may help to account for structural variability, different rates of structural development, and natural inputs of coarse woody debris that would be difficult to reproduce with traditional silvicultural practices. This would therefore permit old-growth forming successional processes for a portion of the territory.

Another approach proposed for the mixed boreal forest is to employ a combination of even-aged and uneven-aged silvicultural techniques to maintain a representative proportion of three structural cohorts in the landscape (Bergeron et al. 2001; Harvey et al. 2002; see Gauthier et al., chapter 3). Even-aged silviculture can be used to emulate the effect of a stand-replacing disturbance such as fire while uneven-aged silviculture (such as partial and selection cutting) may be used to facilitate canopy transition and emulate the gap-dynamic cycle of older stands.

The presence of a provincial park within the DMPF may provide an opportunity to maintain an appropriate portion of the landscape beyond the rotation age, by allowing the natural development of second- and third-cohort forests. However, this concentration of potential older stands would constrain the spatial distribution of second and third cohorts and may not reflect the natural distribution of age classes over the entire landscape as required in NDBM (Harvey et al. 2003). In the DMPF the tri-structural cohort approach may be advantageous from an economic standpoint because it relies primarily on structure and compositional knowledge, and all silvicultural treatments are carried out within the commercial rotation time-frame. It also provides the opportunity to control the distribution of "older" structural cohorts over the entire landscape, provided that the differences between a natural and artificially produced old-growth stand structure are acceptable from an ecological and social perspective.

There are a number of challenges to implementing the NDBM approach in the DMPF. Much of the forest is the product of the large fires of the late 1800s (Gill 1930; Tardif 2004). In addition, there is a lack of fires after 1961. Fire suppression activities and climate change make it difficult to assess the current natural length of the fire cycle in this system. Consequently, the current age structure of the landscape does not necessarily reflect that expected in the absence of these influences. The fire cycle has also been lengthening throughout the 20th century (Tardif 2004). It is therefore difficult to establish a target fire cycle length within which to develop a NDBM plan. Alternatively, Bergeron et al. (2004b) suggested that current mean time since fire be used as a benchmark on which to base forest management objectives. This strategy has an advantage over the fire cycle because it encapsulates variations in the fire cycle and their cumulative effect on the landscape (Bergeron et al. 2004b). The mean time since last fire would therefore permit the maintenance of the cohort 1, 2 and 3 forest types, while incorporating some of this variability.

The mean age of the DMPF was estimated at 108 year in 2002. Theoretically under a mean-time-since-fire management model of 110 years, approximately 49% of the forest should be within the first cohort, 25% in the second cohort, and 26% in the third cohort. Currently, the distribution of the cohorts in the DMPF, that is 60% in the first cohort, 34% in the second cohort, and 6% in the third cohort, does not reflect the distribution associated with a natural landscape submitted to a fire cycle of 110 years. Therefore, currently there is a third-cohort deficit in the DMPF. One way to remediate this deficit would be to adapt cuts in the DMPF so as to maintain a larger compositional and structural diversity by causing some forest stands to shift to the second- and third-cohort stages.

Another challenge arises from the large amount of structural and compositional variability observed in stands of similar age (originating from the 1885–1895 fires; Epp et al. 2006). As described in the introduction of this book, maintaining a representative range of structure and composition over the landscape constitutes an important ecological challenge for forest management. Reproducing such a wide variety of composition and structure through silviculture alone is unfeasible in the long term, especially since the processes that cause this variability are not fully understood. Therefore, forest managers should adopt a precautionary approach with respect to reserving a proportion of the landscape where natural processes can occur freely in order to maintain some representative range of structure. Additionally, in accordance with this approach, it would be necessary to avoid the homogenization of the stand types developing after harvest. Current legislation in Manitoba does not include prescriptions that would lead to the development of a range of representative stand structures/compositions at the landscape level. It essentially encourages an immediate return to the pre-cut tree composition and relative density (Manitoba Conservation 2001). Current standards in Manitoba do not favour forested landscape development that mimics natural forest composition and structure. In a NDBM context, these standards would need to be revised.

Finally, large, infrequent disturbance events are usually impractical to emulate in a forest management plan, yet they cannot be ignored given their historical importance in DMPF and their influence on the landscape (Turner et al. 1998; Tardif 2004). The fact that such a large portion of the DMPF burned in the 1885–1895 fires suggests that the forest extent is small relative to the potential scale of fires. However, several composition and structure types have arisen from these fire episodes. This indicates that a cautious approach would be to avoid practices that would cause an homogenization of stand composition and structure. In implementing a NDBM model that maintains the current range of structure, forest managers might consider accounting for the full range of variability present among the entire Mid-boreal Uplands Ecoregion (see Harvey et al., chapter 18 for some examples of strategies useful for small areas).

## 5.2.   Future Natural Disturbance Trends

The fire history of the DMPF has revealed that the fire return interval has increased over the last century (Tardif 2004). However, projections for the FWI based on general circulation models suggest that the western region of the mixed boreal forest will experience more extreme moisture/weather conditions as global temperature increases, and that this in turn will increase the potential fire severity (Flannigan and Van Wagner 1991; Flannigan et al. 2001). The past large fires of the late 1800s have been linked with extended periods of drought (Girardin et al. 2004), and if climatic trends in the future increase the risk of extreme drought events the risk of future catastrophic fires may also increase (Flannigan et al. 2001). In terms of silviculture, a shorter fire cycle would justify a more widespread application of even-aged silvicultural techniques (clearcuts) and accommodate shorter rotation ages. This use of primarily even-aged silviculture would be appropriate because the shorter the fire return interval is in a natural system, the greater the expected proportion of first-cohort stands in the system (Bergeron et al. 2002; Gauthier et al., chapter 3). The management towards first-cohort stands may, in fact, simplify the application of NDBM at the landscape scale in the DMPF, since Louisiana-Pacific Canada, Ltd., the primary forest manager in the region, mostly utilizes early-successional hardwoods in their operations (LP 2006a). However, with a potential future increase of catastrophic fires in this region, this strategy should be considered with caution. Unless fire suppression remains effective in extreme FWI conditions, there will be increased competition between fires and logging. This combination of disturbances could threaten biodiversity maintenance and the sustainability of the industry (Bergeron et al. 2004b). Additionally, a more widespread use of clearcuts may be less desirable from a social perspective (Lieffers et al. 2003).

Climate change will also have an impact on disturbances other than fire, particularly insect outbreaks. A shorter fire cycle in the DMPF would likely result in a lower proportion of late-successional conifer species such as balsam fir and white spruce, in favour of more fire-adapted species such as trembling aspen and jack pine (Bergeron 2000; Hamel and Kenkel 2001). A lower proportion of balsam fir and white spruce, which are the spruce budworms' primary hosts, may translate into a reduced risk for these types of outbreaks in the future (Morin 1994; Su et al. 1996; Morin et al., chapter 7). However, an increase in the proportion of trembling aspen and jack pine may increase the risk of FTC and jack pine budworm outbreaks. Evidence suggests that warmer, drier conditions are favourable for FTC (Hogg et al. 2002a). The most severe FTC outbreak of the past century in the DMPF, which took place during the early 1960s, occurred after an extreme drought event in 1961 (Sutton and Tardif 2007). If more frequent drought events are to be expected in the future (Flannigan et al. 2001), forest managers should be prepared to contend with the impact of more frequent and severe FTC outbreaks.

## 5.3.  Balancing Forest Values and Interests in the DMPF

Management of the DMPF poses many challenges, especially the inclusion of multiple interests and values associated with the forest, and in particular the presence of Duck Mountain Provincial Park (DMPP). Manitoba Conservation developed DMPP to be managed for multiple uses. There are three zones in the DMPP: a preservation zone, where timber harvesting and landscape manipulation are not permitted; a recreational zone, where timber harvest is only permitted for park maintenance; and a management zone, in which harvesting may be allowed (Manitoba Conservation 2003). The resource management area is the largest, covering 86,719 ha, while the second largest area (46,851 ha) is allocated to preservation. The protected area within the park makes up approximately 15% of the total treed area of the DMPF. If we consider that the entire population of the protected area was to develop into the third-cohort stage, this area would exceed the current population of this cohort. This would remain lower than the percentage projected in a management model based on the 110 years since last fire, where approximately 26% of the territory should be within the third cohort. This protected portion may provide managers with sufficient area for forests to develop into third-cohort stands naturally. However, this protected area is confined to a specific, continuous region. This spatial restriction would need to be considered when developing any mixed management strategies so that a representative proportion of the three cohorts could be maintained within the entire DMPF. The spatial management aspect is further complicated by limitations imposed by environmental conditions, particularly when trying to maintain representative proportions of different species. According to the draft management plan for the park (Manitoba Conservation 2003), forest planning within DMPP is primarily the responsibility of Louisiana-Pacific Canada, Ltd., but with some additional restrictions to accommodate other activities in the Park. These include operation timeframes, access limitations and road closure, and imposed buffers for certain natural features as well as historic and recreation sites (Manitoba Conservation 2003).

Much of the DMPF is accessible to the public, and many harvesting operations are in clear view of park users. Public perception also presents a challenge in terms of the distribution and application of timber harvesting and silvicultural practices within the Park. Harvey et al. (2003) outlined a strategy, in which the managed area was divided into three types of management zones depending on the major harvesting method: (1) clear-cut harvesting, (2) partial harvesting, and (3) selection harvesting. Within each zone, different silvicultural strategies were dispersed to recreate the heterogeneity of natural disturbance events. This mixed management approach (which utilizes different silvicultural treatments) would help to accommodate different values associated with forest management. Additionally, Harvey et al. (2003) suggest that this type of mixed strategy does not significantly affect the annual allowable cut. However the effect on ecological function still needs to be assessed. Such a mixed approach may be suited for the DMPF, since different management zones (including the provincial park) provide an opportunity to divide the forest into regions to be managed for different values.

## 6.   CONCLUSION

The effects of the large catastrophic fires during the late 1800s dominate the DMPF landscape. The length of the fire cycle has not been constant over the past 300 years, and appears to have increased dramatically after the 1890s. Although large fire disturbances have been relatively frequent prior to the 20th century, they have since been absent from the landscape. Forest tent caterpillar has also been an important component of the disturbance regime in the DMPF. Large outbreaks in the past century have impacted the growth of deciduous trees, particularly trembling aspen, which is the primary host of FTC and the most widespread and commercially important tree species in the DMPF. Climate change has the potential to increase the risk for both large, catastrophic fires and FTC outbreaks in this forest. Both are important factors that must be considered for sustainable forest management.

Succession in the DMPF generally proceeds from the dominance of shade-intolerant broadleaf species and jack pine to shade-tolerant conifers. There is, however, a large degree of variability in terms of structural and compositional development of stands after fire, as well as the time needed to reach the breakup of the first cohort. This variability likely originates from fire intensity, and many small-scale, unpredictable events, such as stand-scale disturbances, together with variations in the physical environment. Also, spatial factors such as proximity to seed source play a role in determining stand development. Because of the particular fire history of the DMPF, the current and future climate as well as active fire suppression, a large portion of the landscape is currently in a transitional phase from dominance of the post-disturbance cohort in the canopy, to dominance of the second cohort. In a context where the replacement of large natural disturbances by harvesting is becoming more commonplace, forest management should aim for the creation of a large diversity of stand types while maintaining the aggregation of the forest stands under regeneration. Such an approach requires not only the use of several silvicultural treatments, but also a better integration of the planning and utilization of resources between different users (deciduous vs. coniferous for example), in order to maintain successional processes. In addition, it is important to keep in mind the eventual occurrence of large natural disturbances, such as large severe fires, when creating management plans and planning harvesting operations.

The disturbance history of the DMPF provides considerable challenges for sustainable forest management and NDBM. Because the current "true" fire cycle is difficult to estimate, forest managers must choose a management scenario that would best serve ecological requirements, as well as management objectives and socio-economic needs. Potential scenarios include management based on the current distribution of cohorts in the DMPF or the larger Mid-boreal Uplands Ecoregion, or imposing a structure that reflects the "natural" cohort distribution based on the current mean forest age. These choices would also have an impact on the species composition in the forest. Furthermore, the effects of future climate change on the disturbance regime and the eventual risk of increased fire severity should be considered.

## REFERENCES

Anonymous 2000. The Lumber Industry in Manitoba. Manitoba Culture, Heritage and Tourism, Historic Resources Branch. Winnipeg, Man.

Bergeron, Y. 2000. Species and stand dynamics in the mixed woods of Quebec's southern boreal forest. Ecology, **81**: 1500–1516.

Bergeron, Y. and Dubuc, M. 1989. Succession in the southern part of the Canadian boreal forest. Vegetatio, **79**: 51–63.

Bergeron, Y., Flannigan, M., Gauthier, S., Leduc, A., and Lefort, P. 2004a. Past, current and future fire frequency in the Canadian boreal forest: implications for sustainable forest management. Ambio, **33**: 356–360.

Bergeron, Y., Gauthier, S., Flannigan, M., and Kafka, V. 2004b. Fire regimes at the transition between mixedwood and coniferous boreal forest in northwestern Quebec. Ecology, **85**: 1916–1932.

Bergeron, Y., Gauthier, S., Kafka, V., Lefort, P., and Lesieur, D. 2001. Natural fire frequency for the eastern Canadian boreal forest: consequences for sustainable forestry. Can. J. For. Res. **31**: 384–391.

Bergeron, Y., Leduc, A., Harvey, B.D., and Gauthier, S. 2002. Natural fire regime: a guide for sustainable management of the Canadian boreal forest. Silva Fenn. **36**: 81–95.

Bondrup-Nielsen, S. 1995. Forestry and the boreal forest: maintaining inherent landscape patterns. Water Air Soil Poll. **82**: 71–76.

Brown, A.W.A. 1940. Annual report of the forest insect survey: forest insect investigations 1939. Department of Agriculture Canada, Division of Entomology. Edmond Cloutier, Ottawa, Ont.

Brown, A.W.A 1941. Annual report of the forest insect survey: forest insect investigations 1940. Department of Agriculture Canada, Division of Entomology. Edmond Cloutier, Ottawa, Ont.

Chen, H.Y.H. and Popadiouk, R.V. 2002. Dynamics of North American boreal mixedwoods. Environ. Rev. **10**: 137–166.

Cumming, S.G., Schmiegelow, F.K.A., and Burton, P.J. 2000. Gap dynamics in boreal aspen stands: is the forest older than we think? Ecol. Appl. **10**: 744–759.

Dale, V.H., Lugo, A.E., MacMahon, J.A., and Pickett, S.T.A. 1998. Ecosystem management in the context of large, infrequent disturbances. Ecosystems, **1**: 546–557.

De Grandpré, L., Morissette, J., and Gauthier, S. 2000. Long-term post-fire changes in the northeastern boreal forest of Quebec. J. Veg. Sci. **11**: 791–800.

Duncan, D.P. and Hodson, A.C. 1958. Influence of the forest tent caterpillar upon the aspen forests of Minnesota. For. Sci. **4**: 71–79.

Ecoregions Working Group (EWG) 1989. Ecoclimatic regions of Canada, first approximation. Ecological Land Classification Series, No. 23. Sustainable Development

Branch, Canadian Wildlife Service, Conservation and Protection, Environment Canada, Ottawa, Ont.

Elliot, K. R. and Hildahl, V. 1963. Provinces of Manitoba and Saskatchewan: forest insect survey. *In* Annual report of the forest insect and disease survey 1962. Department of Forestry, Forest Entomology and Pathology Branch. Ottawa, Ont.

Elliot, K. R. and Hildahl, V. 1964. Provinces of Manitoba and Saskatchewan: forest insect survey. *In* Annual report of the forest insect and disease survey 1963. Department of Forestry, Forest Entomology and Pathology Branch. Ottawa, Ont.

Elliot, K. R. and Hildahl, V. 1965. Provinces of Manitoba and Saskatchewan: forest insect survey. *In* Annual report of the forest insect and disease survey 1963. Department of Forestry, Forest Entomology and Pathology Branch. Ottawa, Ont.

Environment Canada 2004a. Narrative descriptions of terrestrial ecozones and ecoregions of Canada. [Online] <www.ec.gc.ca/soer-ree/English/Framework/Nardesc/TOC.cfm> (accessed November 21, 2007).

Environment Canada 2004b. Canadian climate normals 1971–2000. [Online] <www.climate.weatheroffice.ec.gc.ca/climate_normals/results_e.html> (accessed November 21, 2007).

Epp, B.V., Tardif, J., and De Grandpré, L. 2006. Forest dynamics, and the application of a natural disturbance-based management model in Duck Mountain Provincial Forest, Manitoba, Canada. Poster presentation. 13th IBFRA Conference: New Challenges in Forest Management. August 28–30, 2006. Umeå, Sweden.

Flannigan, M., Campbell, I., Wotton, W., Carcaillet, C., Richard, P., and Bergeron, Y. 2001. Future fire in Canada's boreal forest: paleoecology results and general circulation model – regional climate modes simulations. Can. J. For. Res. **31**: 854–864.

Flannigan, M.D. and C.E. Van Wagner 1991. Climate change and wildfire in Canada. Can. J. For. Res. **21**: 66–72.

Forest Insect and Disease Survey (FIDS). 1944–1996. Forest Insect and Disease Conditions in Canada, Natural Resources Canada, Canadian Forest Service. Ottawa, Ont.

Franklin, J.F. 1993. Preserving biodiversity: species, ecosystems, or landscapes? Ecol. Appl. **3**: 202–205.

Fricker, J.M., Chen, H.Y.H., and Wang, J.R. 2006. Stand age structural dynamics of North American boreal forests and implications for forest management. Int. For. Rev. **8**: 395–405.

Galindo-Leal, C. and Bunnell, F.L. 1995. Ecosystem management: implications and opportunities of a new paradigm. For. Chron. **71**: 601–606.

Galipeau, C., Kneeshaw, D., and Bergeron, Y. 1997. White spruce and balsam fir colonization of a site in the southeastern boreal forest as observed 68 years after fire. Can. J. For. Res. **27**: 139–147.

Gauthier, S., De Grandpré, L., and Bergeron, Y. 2000. Differences in forest composition in two boreal forest ecoregions of Quebec. J. Veg. Sci. **11**: 781–790.

Gauthier, S., Leduc, A., and Bergeron, Y. 1996. Vegetation modelling under natural fire cycles: a tool to define natural mosaic diversity for forest management. Environ. Monitoring Assess. **39**: 417–434.

Gill, C.B. 1930. Cyclic forest phenomena. For. Chron. **6**: 42–56.

Girardin, M.P. and Tardif, J. 2005. Sensitivity of tree growth to the atmospheric vertical profile in the boreal plains of Manitoba, Canada. Can. J. For. Res. **35**: 48–64.

Girardin, M.P., Tardif, J., and Bergeron, Y. 2001. Radial growth analysis of *Larix laricina* from the Lake Duparquet area, Québec, in relation to climate and larch sawfly outbreaks. Écoscience, **8**: 127–138.

Girardin, M.P., Tardif, J., Flannigan, M.D., and Bergeron, Y. 2002. Reconstructing atmospheric circulation history using tree rings: one more step toward understanding temporal changes in forest dynamics. *In* Proceedings for the 3rd International Sustainable Forest Management Network Conference – Advance in forest management: from knowledge to practice, 13–15 November, Edmonton, Alta. *Edited by* T.S. Veeman et al., Edmonton, Alta. pp. 105–110.

Girardin, M.P., Tardif, J., Flannigan, M.D., Wotton, B.M., and Bergeron, Y. 2004. Trends and periodicities in the Canadian Drought Code and their relationships with atmospheric circulation for the southern Canadian boreal forest. Can. J. For. Res. **34**: 103–119.

Greene, D.F., Kneeshaw, D.D., Messier, C., Lieffers, V., Cormier, D., Doucet, R., Coates, K.D., Groot, A., Grover, G., and Calogeropoulos, C. 2002. Modelling silvicultural alternatives for conifer regeneration in boreal mixedwood stands (aspen/white spruce/balsam fir). For. Chron. **78**: 281–295.

Hamel, C. and Kenkel, N. 2001. Structure and dynamics of boreal forest stands in the Duck Mountains, Manitoba. Final Project Report 2001-4, Sustainable Forest Management Network. University of Alberta, Edmonton, Alta. [Online] <www.sfmnetwork.ca/docs/e/PR_2000-27.pdf> (accessed November 21, 2007).

Harrison, J.D.B. 1934. The forests of Manitoba. Forest Service Bulletin 85, Department of the Interior, Ottawa, Ont.

Harvey, B.D., Leduc, A., Gauthier, S., and Bergeron, Y. 2002. Stand-landscape integration in natural disturbance-based management of the southern boreal forest. For. Ecol. Manag. **155**: 369–385.

Harvey, B.D., Nguyen-Xuan, T., Bergeron, Y., Gauthier, S. and Leduc, A. 2003. Forest management planning based on natural disturbance and forest dynamics. *In* Towards sustainable management of the boreal forest. *Edited by* P.J. Burton, C. Messier, D.W. Smith, and W.L.

Adamowicz. NRC Research Press, Ottawa, Ont., pp. 395–432.

Hildahl, V., and Campbell, A.E. 1975. Forest tent caterpillar in the Prairie Provinces. Information Report NOR-X-135. Canadian Forest Service, Northern Forest Research Centre. Edmonton, Alta.

Hildahl, V. and Reeks, W.A. 1960. Outbreaks of the forest tent caterpillar, *Malacosoma disstria* Hbn., and their effects on stands of trembling aspen in Manitoba and Saskatchewan. Can. Entomol. **92**: 199–209.

Hill, S.B., Mallik, A.U., and Chen, H.Y.H. 2005. Canopy gap disturbance and succession in trembling aspen dominated boreal forests in northeastern Ontario. Can. J. For. Res. **35**: 1942–1951.

Hogg, E.H., Brandt, J.P., and Kochtubajda, B. 2002a. Growth and dieback of aspen forests in northwestern Alberta, Canada, in relation to climate and insects. Can. J. For. Res. **32**: 823–832.

Hogg, E.H., Hart, M., and Lieffers, V.J. 2002b. White tree rings formed in trembling aspen saplings following experimental defoliation. Can. J. For. Res. **32**: 1929–1934.

Hogg, E.H. and Schwarz, A.G. 1999. Tree-ring analysis of declining aspen stands in west-central Saskatchewan. Information Report NOR-X-359. Canadian Forest Service, Northern Forestry Centre. Edmonton, Alta.

Ives, W.G.H., and Wong, H.R. 1988. Tree and shrub insects of the Prairie Provinces. Information Report NOR-X-292. Canadian Forest Service, Northern Forestry Centre. Edmonton, Alta.

Jardon, Y., Filion, L. and Cloutier, C. 1994. Tree-ring evidence for endemicity of the larch sawfly in North America. Can. J. For. Res. **24**: 742–747.

Kenkel, N., Foster, C., Caners, R., Lastra, R., and Walker, D. 2003. Spatial and temporal patterns of white spruce recruitment in two boreal mixedwood stands, Duck Mountains, Manitoba. Project Reports 2003/2004, Sustainable Forest Management Network. University of Alberta, Edmonton, Alta. [Online] <www.sfmnetwork.ca/docs/e/PR_200304kenkelnrecr6.pdf> (accessed November 21, 2007).

Kneeshaw, D. and Gauthier, S. 2003. Old growth in the boreal forest: a dynamic perspective at the stand and landscape level. Environ. Rev. **11**: S99-S114.

La Roi, G.H. 1992. Classification and ordination of southern boreal forests from the Hondo – Slave Lake area of central Alberta. Can. J. Bot. **70**: 614–628.

Larsen, C.P.S. 1997. Spatial and temporal variations in boreal forest fire frequency in northern Alberta. J. Biogeogr. **24**: 663–673.

Larsen, J.A. 1980. The boreal ecosystem. Academic Press, Inc., New York, USA.

Lee, P. 2004. The impact of burn intensity from wildfires on seed and vegetative banks, and emergent understory in aspen-dominated boreal forests. Can. J. Bot. **82**: 1468–1480.

Lefort, P., Gauthier, S., and Bergeron, Y. 2003. The influence of fire weather and land use on the fire activity of the Lake Abitibi area, eastern Canada. For. Sci. **49**: 509–521.

Lieffers, V.J., Macmillan, R.B., MacPherson, D., Branter, K., and Stewart, J.D. 1996a. Semi-natural and intensive silvicultural systems for the boreal mixedwood forest. For. Chron. **72**: 286–292.

Lieffers, V.J., Messier, C., Burton, P.J., Ruel, J.-C., and Grover, B.E. 2003. Nature-based silviculture for sustaining a variety of boreal forest values. *In* Towards sustainable management of the boreal forest. *Edited by* P.J. Burton, C. Messier, D.W. Smith, and W.L. Adamowicz. NRC Research Press, Ottawa, Ont., pp. 481–530.

Lieffers, V.J., Stadt, K.J., and Navratil, S. 1996b. Age structure and growth of understory white spruce under aspen. Can. J. For. Res. **26**: 1002–1007.

Locky, D.A., Bayley, S.E., and Vitt, D.H. 2005. The vegetational ecology of black spruce swamps, fins, and bogs in southern boreal Manitoba, Canada. Wetlands, **25**: 564–582.

LP Canada, Ltd. 2006a. 20 year sustainable forest management plan (2006–2026). Louisiana Pacific Canada, Ltd., Swan River, Manitoba. [Online] <www.swanvalleyforest.ca/planning/LongTermPlan.html> (accessed November 21, 2007).

LP Canada, Ltd. 2006b. 2006–2007 annual operating plan. Louisiana Pacific Canada, Ltd., Swan River, Manitoba. [Online] <www.swanvalleyforest.ca/documents.html#AR_2005> (accessed November 21, 2007).

Manitoba Conservation 2000. Park boundaries in Manitoba – Outlines. Manitoba Conservation, Winnipeg, Manitoba. [Vector digital data, online] <web2.gov.mb.ca/mli/adminbnd.html> (accessed October, 2004).

Manitoba Conservation 2001. Forest renewal standards. Manitoba Conservation, Forestry Branch, Winnipeg, Man. [Online] <www.gov.mb.ca/conservation/forestry/forest-renewal/fr2-standards.html> (accessed November 21, 2007).

Manitoba Conservation 2003. Duck Mountain Provincial Park, Draft Management Plan, Parks and Natural Areas Branch, Winnipeg, Manitoba. [Online] <www.manitobaparks.com> (accessed November 21, 2007).

Manitoba Conservation 2004a. Duck Mountain Provincial Park, Park Information and Maps, Parks and Natural Areas Branch, Winnipeg, Manitoba. [Online] <www.gov.mb.ca/conservation/parks/popular_parks/duck_mtn/info.html> (accessed November 21, 2007).

Manitoba Conservation 2004b. Wood Supply Analysis Report for Forest Management Unit 13 and 14, Forestry Branch, Manitoba Conservation, Winnipeg, Manitoba. [Online] <www.swanvalleyforest.ca/planning/LongTermPlan.html> (accessed November 21, 2007).

Manitoba Parks Branch (MPB) 1973. Outdoor recreation master plan: Duck Mountain Provincial Park. Manitoba Department of Tourism, Recreation and Cultural Affairs, Parks Branch, Winnipeg, Man.

Martineau, R. 1984. Insects harmful to forest trees. Multiscience Publications Ltd.

Moody, B.H. and Cerezke, H.F. 1983. Forest insect and disease conditions in Alberta, Saskatchewan, Manitoba, and the Northwest Territories in 1982 and predictions for 1983. Information Report NOR-X-248. Canadian Forestry Service, Northern Forest Research Centre, Edmonton, Alta.

Moody, B.H. and Cerezke, H.F. 1984. Forest insect and disease conditions in Alberta, Saskatchewan, Manitoba, and the Northwest Territories in 1983 and predictions for 1984. Information Report NOR-X-261. Canadian Forestry Service, Northern Forest Research Centre, Edmonton, Alta.

Moody, B.H. and Cerezke, H.F. 1985. Forest insect and disease conditions in Alberta, Saskatchewan, Manitoba, and the Northwest Territories in 1984 and predictions for 1985. Information Report NOR-X-269.Canadian Forestry Service, Northern Forest Research Centre, Edmonton, Alta.

Morin, H. 1994. Dynamics of balsam fir forests in relation to spruce budworm outbreaks in the Boreal Zone of Quebec. Can. J. For. Res. **24**: 730–741.

Nairn, L.D., Reeks, W.A., Webb, F.E., and Hildahl, V. 1962. History of larch sawfly outbreaks and their effect on tamarack stands in Manitoba and Saskatchewan. Can. Entomol. **94**: 242–255.

Paré, D., Bergeron, Y. and Longpré, M. 2001. Potential productivity of aspen cohorts originating from fire, harvesting, and tree-fall gaps on two deposit types in northwestern Quebec. Can. J. For. Res. **31**: 1067–1073.

Peterson, E.B., and Peterson, N.M. 1992. Ecology, management, and use of aspen and balsam poplar in the prairie provinces. Canada. Special Report 1. Forestry Canada, Northwest region, Northern Forestry Centre, Edmonton, Alta.

Pham, A.T., De Grandpré, L., Gauthier, S., and Bergeron, Y. 2004. Gap dynamics and replacement patterns in gaps of the northeastern boreal forest of Quebec. Can. J. For. Res. **34**: 353–364.

Pines, I.L. 2006. Forest pests in Manitoba. Presented at: The Forest Pest Management Forum 2006. Natural Resources Canada, Canadian Forest Service. December 5–7, 2006, Ottawa, Ont.

Rempel, R.S. 1999. Natural disturbance analysis and planning tools. Project Report 1999-3, Sustainable Forest Management Network. University of Alberta, Edmonton, Alta. [Online] <www.sfmnetwork.ca/docs/e/PR_1999-3.pdf> (accessed November 21, 2007).

Rowe, J.S. 1972. Forest regions of Canada. Publication No. 1300, Department of the Environment, Canadian Forest Service. Ottawa, Ont.

Rowe, J.S., and Scotter, G.W. 1973. Fire in the boreal forest. Quat. Res. **3**: 444–464.

Sauchyn, D.J., and Hadwen, T. 2001. Forest ecosystems and the physical environment, Duck Mountains, west central Manitoba. Final Project Report 2001-29,

Sustainable Forest Management Network. University of Alberta, Edmonton, Alta. [Online] <www.sfmnetwork. ca/docs/e/PR_2001-29.pdf> (accessed November 21, 2007).

Sauchyn, D.J., and Skinner, W.R. 2001. A proxy PDSI record for the southwestern Canadian plains. Can. Water Res. J. **26**: 253–272.

Schut, P. 2005. A national ecological framework for Canada, GIS data. Canadian Soil Information System, Agriculture and Agri-food Canada. [Online] <sis.agr. gc.ca/cansis/nsdb/ecostrat/gis_data.html> (accessed November 21, 2007).

Scott, G.A.J. 1995. Canada's vegetation: a world perspective. McGill-Queen's University Press, Montreal, Que.

Seymour, R.S., and Hunter, M.L. Jr. 1999. Principles of ecological forestry. *In* Maintaining biodiversity in forest ecosystems. *Edited by* M.L. Hunter Jr. Cambridge University Press, Cambridge, UK, pp. 22–61.

Simpson, R. and Coy, D. 1999. An ecological atlas of forest insect defoliation in Canada 1980–1996. Information Report M-X-206E. Natural Resources Canada, Canadian Forest Service – Atlantic Forestry Centre. Fredericton, N.B.

Smith, R.E., Veldhuis, H., Mills, G.F., Eilers, R.G., Selby, C., and Santry, M. 1998. Terrestrial ecozones, ecoregions and ecodistricts of Manitoba: an ecological stratification of Manitoba's natural landscapes. Manitoba Land Resource Unit, Brandon Research Centre, Research Branch, Agriculture and Agri-Food Canada, Winnipeg, Man.

Stilwell, W.J. 1988. The Baldy Mountain cabin: the history and role of forest rangers and game wardens in the Baldy Mountain area. Regional Services Branch, Manitoba Department of Natural Resources, Winnipeg, Man.

Su, Q., MacLean, D.A., and Needham, T.D. 1996. The influence of hardwood content on balsam fir defoliation by spruce budworm. Can. J. For. Res. **26**: 1620–1628.

Sutton, A. and Tardif, J. 2005. Distribution and anatomical characteristics of white rings in *Populus tremuloides*. IAWA J. **26**: 221–238.

Sutton, A. and Tardif, J. 2007. Dendrochronological reconstruction of forest tent caterpillar outbreaks in time and space, western Manitoba, Canada. Can. J. For. Res. **37**: 1643–1657.

Tardif, J. 2004. Fire history in the Duck Mountain Provincial Forest, western Manitoba. Project Reports 2003/2004, Sustainable Forest Management Network. University of Alberta, Edmonton, Alta. [Online] <www. sfmnetwork.ca/docs/e/PR_200304tardifjfire6fire.pdf> (accessed November 21, 2007).

Tardif, J. and Stevenson, D. 2002. Radial growth – climate association of *Thuja occidentalis* L. at the northwestern limit of its distribution, Manitoba, Canada. Dendrochronologia, **19**: 1–9.

Turner, M.B., Baker, W.L., Peterson, C.J., and Peet, R.K. 1998. Factors influencing succession: lessons from large, infrequent natural disturbances. Ecosystems, **1**: 511–523.

Turner, M.G. and Dale, V.H. 1998. Comparing large, infrequent disturbances: what have we learned? Ecosystems, **1**: 493–496.

Turnock, W.J. 1972. Geographical and historical variability in population patterns and life systems of the larch sawfly (*Hymenoptera: Tenthredinidae*). Can. Entomol. **104**: 1883–1900.

Tyrrell, J.B. 1888. Notes to accompany a preliminary map of the Duck and Riding Mountains in North-Western Manitoba. Dawson Brothers, Montréal, Que.

Wang, G.G. 2002. Fire severity in relation to canopy composition within burned boreal mixedwood stands. For. Ecol. Manag. **163**: 85–92.

Wang, G.G. and Kemball, K.J. 2005. Effects of fire severity on early development of understory vegetation. Can. J. For. Res. **35**: 254–262.

Weir, J.M.H. and Johnson, E.A. 1998. Effects of escaped settlement fires and logging on forest composition in the mixedwood boreal forest. Can. J. For. Res. **28**: 459–467.

Weir, J.M.H., Johnson, E.A., and Miyanishi, K. 2000. Fire frequency and the spatial age mosaic of the mixedwood boreal forest in western Canada. Ecol. Appl. **10**: 1162–1177.

Wiken, E. 1986. Terrestrial ecozones of Canada. Ecological Land Classification Series No. 19. Lands Directorate, Ottawa, Ont.

# Part 3

# Forest Ecosystem Management Implementation

Yves Bergeron,
Sylvie Gauthier,
and Marie-Andrée Vaillancourt

Since the late 1990s, many Canadian provinces have adopted regulations that favour sustainable forest management (BCMF 1995; OMNR 2001). In Québec, the Coulombe Report (2004) that resulted from the Commission sur l'étude sur la gestion de la forêt publique québécoise• made the following recommendation: "ecosystem management must be at the heart of public forest management." Thus throughout Canada there is an increasing will to renew the traditional management model and to ensure that ecosystem management is at the centre of public forest management.

• The Québec Government mandated this Commission in 2003 in order to have an overview of the situation with regard to public forest management and to recommend changes to improve the forest management regime in the context of sustainable forest management.

Ecosystem management is based on a good knowledge of natural forest dynamics. Forest dynamics are characterized by long-term variability of disturbance regimes (fire, wind, insect outbreaks, etc.), and by stand structure and composition changes following these disturbances. Despite the interest in forest ecosystem management (FEM), the implementation of its concepts is not well developed yet and FEM plays a very limited role in the vast Canadian forest area. Consequently, the forest industry and forest managers are reluctant to move away from traditional forest management practices without any concrete alternatives based on forest dynamics.

Development and implementation of management strategies based on natural forest dynamics include several steps (Bergeron and Harvey 1997; see table below). The first step consists in reconstructing recent and historical natural disturbance regimes to define the variability range in which the ecosystem has evolved (see chapters 2 and 3). Reconstructing long-term stand changes following disturbances is also an essential component for understanding structural and compositional changes. The second step aims at comparing natural and managed landscapes with respect to key forest attributes (such as forest composition and structure, coarse woody debris, soil organic matter layer; see chapter 1) and to identify and quantify the main differences. To reduce these differences, new management objectives must be defined in a silvicultural system that can achieve these objectives (step 3). It is important to emphasize the fact that planned treatment in this type of regime is not intended to "mimic" natural disturbances but rather to preserve key ecosystem processes and the natural ecosystem mosaic (forest composition and structural diversity, etc.) at the landscape level. Thus knowledge of forest dynamics within a management unit will allow forest managers to take advantage of the natural forest potential, and to minimize constraints imposed by this dynamic and maximize forest productivity while respecting biodiversity. Different scenarios based on this ecological approach have to be implemented in the context of a management plan that takes into account social acceptability and economic feasibility (step 4). Finally, achievement of objectives must be assessed with monitoring programs in order to modify silvicultural activities in the context of an adaptative management framework (step 5).

As stressed in chapter 2 (see figure 2.1), one of the main ecosystem management objectives is to ensure that the forest management regime allows for some variability within the ecosystem's natural historical range of variability. In recent years, several studies have been conducted on boreal forest natural dynamics and on the differences between natural and managed landscapes generated by forest management (see part 2). This approach led to the identification of major differences between natural and forest management regimes. Among these, important issues for biodiversity preservation were identified such as the loss of mature and old-growth forests that dominated in natural forest landscapes, the loss of large forest landscapes to increased landscape fragmentation, the low forest retention within managed areas compared with naturally disturbed landscapes and the absence of fire as a catalyst

***Steps required for ecosystem management implementation***

1. Reconstruction of the natural disturbance regime and long-term evolution of forest stands following a disturbance

2. Comparative analysis of natural and managed landscapes and identification of the main differences

3. Development of management objectives and silvicultural activities to minimize differences between forest management and natural forest dynamics

4. Implementation of silvicultural activities in the context of a management plan that takes into account social and economic values

5. Monitoring of interventions to evaluate management objectives and modify silvicultural activities if necessary

for the nutrient recycling in some regions (Bergeron et al. 2002; see Introduction). To address these differences, solutions were proposed, for example, using adapted forestry practices in order to maintain stand composition and structure, agglomerating and increasing spacing between cutting areas in order to maintain large forest landscapes, increasing retention in cutting areas and using soil scarification and controlled burning.

Although there is still a large amount of work needed to establish natural dynamic and regional issues especially where human settlement has profoundly modified the land, current knowledge is sufficiently developed to enable us to move towards ecosystem management. However its implementation is still facing several problems. Among are: 1) the lack of development in uneven-aged silviculture within the boreal forest where the even-aged approach was traditionally practised, 2) the vast territory available for management that favours the use of a mid-intensity approach (high intensity at the canopy level, low intensity at the soil level) applied to the whole territory rather than an approach adapted to regional ecosystem particularities, and 3) the small number of forest specialists whose training is adequate enough to help move toward a management regime based on forest ecosystem dynamics.

Several experimental projects currently implement ecosystem management in Canada. This section does not review them exhaustively but its objective is to provide an overview of different tools available for the implementation of ecosystem management systems and present a few examples of projects that are now at various stages of development. It is essential to discuss first how existing silvicultural tools can be used to meet ecosystem management objectives (chapter 13). In addition, the establishment of monitoring program is an essential aspect in order to assess whether FEM objectives are reached both in terms of forest stand- and landscape-level targets and biodiversity maintenance targets (chapter 14). The first results of a partial cut experimental network implemented in Abitibi presented in chapter 15 illustrate the silvicultural and ecological evaluation of alternative forestry practices.

Uneven-aged silvicultural practices and other alternative management strategies have not been widely used so far. This makes their impact on annual allowable cut and biodiversity maintenance difficult to predict. Modelling is thus an essential tool in the development of ecosystem management strategies for assessing timber supply in uneven-aged systems (chapter 16) or developing decision support processes in which various scenarios are compared with respect to their capacity to meet management objectives using environmental and socio-economic criteria and indicators (chapter 17).

Finally, three experimental projects implemented in different ecological regions are presented: the Lake Duparquet research and teaching forest (chapter 18) and the Tembec project (chapter 19) in Abitibi, Québec, as well as variable retention strategies applied in the Upper Fraser River watershed interior wet forests in British Columbia (chapter 20). A description of these projects highlights various regional issues and proposed solutions for ecosystem management implementation.

## REFERENCES

Bergeron, Y. and Harvey, B. 1997. Basing silviculture on natural ecosystem dynamics: an approach applied to the southern boreal mixedwood forest of Quebec. For. Ecol. Manag. **92**: 235–242.

Bergeron, Y., Leduc, A., Harvey, B.D., and Gauthier, S. 2002. Natural fire regime: a guide for sustainable management of the Canadian boreal forest. Silva Fenn. **36**: 81–95.

British Columbia Ministry of Forests (BCMF). 1995. Biodiversity Guidebook. Forest Practices Code of BC. Ministry of Forests, B.C.

Coulombe, G., Huot, J., Arsenault, J., Beauce, E., Bernard, J.-T., Bouchard, A., Liboiron, M.-A., and Szaraz, G. 2004. Rapport de la Commission d'étude sur la gestion de la forêt publique québécoise. [Online] <www.mrnfp.gouv. qc.ca/commission-foret/rapportfinal.htm> (accessed November 7, 2007).

Ontario Ministry of Natural Resources (OMNR). 2001. Forest management guide for natural disturbance pattern emulation. Version 3.1. Ontario Ministry of Natural Resources, Queen's Printer for Ontario, Toronto, Ont.

Chapter *13*

# Silviculture in a Context of Forest Ecosystem Management in Boreal and Southern Boreal Forests*

*Mathieu Bouchard*

* I acknowledge S. Gauthier, A. Leduc, D. Kneeshaw, D. Pothier, Y. Bergeron, M.-A. Vaillancourt, B. Harvey, and two anonymous reviewers for their comments on the manuscript. This chapter was written in the course of a postdoctoral fellowship within the Chaire industrielle CRSNG-Université Laval en sylviculture et faune. I also thank Pamela Cheers for editing the manuscript.

# 1. INTRODUCTION

One of the main goals of Forest Ecosystem Management (FEM) is to use silvicultural techniques that are appropriate for maintaining ecosystems relatively close to their natural state, or to recover that state when they have been modified by human activity (for the differences between silviculture and forest management, see box 13.1). Because the composition and spatial structure of forest ecosystems depend strongly upon natural disturbance regimes, it is often recommended that silvicultural techniques that produce effects similar to those of natural disturbances be used (Bergeron and Harvey 1997).

In a FEM context, it is important to make sure that harvesting and other silvicultural treatments, as "natural" as they may appear at the stand level, do not create undesirable effects when they are juxtaposed in time and space. The effects of silvicultural treatments tend to be uniform compared with the variability observed following natural disturbances, and using a restricted range of silvicultural treatments could modify and simplify considerably the forest mosaic at the landscape level (Bergeron et al. 2002; Gauthier et al., chapter 3). For example, the use of clearcutting could appear compatible with ecosystem management at the local scale since, under good conditions, its impact on vegetation can be similar to that of a natural disturbance such as a severe insect outbreak or fire. At the landscape level, however, the repetitive use of such cuttings could result in a considerable drop in the average stand age, a significant increase in the abundance of hardwood species, or some tree species becoming rare (Grondin and Cimon 2003; Boucher et al. 2006). To facilitate the maintenance of forest heterogeneity at the landscape level, it becomes essential to introduce diversity in the silvicultural practices used at the stand level.

---

Box 13.1
## Silviculture and forest management

Silviculture and forest management are applied at different spatial scales, and generally refer to distinct planning levels, particularly in a North American context (Davis et al. 2001; Duchesne and Raulier 2004). Forest management involves relatively vast spatial scales, generally forest management units of several hundred hectares. It is part of the strategic planning level, which implies the regulation of forest composition and structure over long temporal scales (25 years or more). As for silviculture, it concerns the stand scale, generally 5 hectares to a few tens of hectares in boreal forests. At this tactical or operational planning levels (1–10 years), all the general objectives and concepts mentioned in the management plan are translated into concrete interventions. For example, the types of interventions to apply in specific stands or the road network layout are specified. The selection and localization of large protected areas to satisfy biodiversity protection objectives is part of the forest management planning process, but it does not concern silviculture. At a finer spatial scale, the decision to preserve or not living trees or snags for biodiversity purposes within managed stands is also part of the management process, but the implementation of this measure within individual stands will be done through silviculture.

The objective of this chapter is to examine how different silvicultural treatments can be arranged in time and space in order to provide solutions to the challenges posed by ecosystem management. We will address three aspects: 1) the similarity between the effects of silvicultural treatments and natural disturbances at the stand level, 2) the development of silvicultural systems that maintain the diversity of successional patterns characterizing natural forest dynamics at this level, and 3) the importance of considering the composition and spatial arrangement of the forest mosaic at the landscape level.

## 2.   SILVICULTURAL TREATMENTS AND NATURAL FOREST DYNAMICS AT THE STAND LEVEL

This section comprises a review of the main impacts of natural disturbances and logging on forest ecosystems at the stand level. Four important attributes of these forest ecosystems will be considered, namely the forest stand *per se* (living trees), dead wood, soils, and regeneration. For a given region, the observed differences between the effects of natural disturbances and those of silvicultural treatments, in interaction with other factors such as soils or climate, are used to identify the main shortcomings resulting from current forest management practices, to which ecosystem management will try to provide corrective solutions. In this respect, carrying out the natural disturbance/logging comparison at the stand level first will provide a better understanding of the mechanisms producing the differences observed at the landscape level.

## 2.1.   Stand Structure and Composition

Natural disturbances and silvicultural treatments are processes that influence several important features such as stand structure (diameter distribution, height) and composition (Nyland 1996; Oliver and Larson 1996). In the long term, the severity and frequency of these disturbances will also determine the stand age structure. Three types of stand age structures are commonly encountered in Québec and elsewhere in North America:

- Even-aged stands composed of one age class (figure 13.1A). This age structure results from disturbances such as severe fires, severe insect outbreaks or clearcuts. Such severe but infrequent disturbances lead to the regeneration and maintenance of stands dominated by shade-intolerant species, such as pines (*Pinus* spp.) or poplars (*Populus* spp.), or stands with a two-storied canopy structure but composed of trees of the same age (e.g., a jack pine [*Pinus banksiana* Lamb.] overstory with a black spruce [*Picea mariana* (Mill.) B.S.P.] understory, or an aspen overstory with a balsam fir understory). Tree species that tend to generate a seedling bank in the understory, such as balsam fir (*Abies balsamea* [L.] Mill.), tend to generate a dense seedling bank.

- Stands with multi-modal* age structures contain trees that can be grouped into two, or sometimes three, distinct age classes (figure 13.1B). It is easy to observe multi-modal age structures produced by natural

* These age structures are sometimes referred as "multi-cohort" (see Oliver and Larson 1996). In this chapter, the term "multi-modal" will be used to avoid the confusion with the Tree-Cohort management model (Bergeron et al. 1999), where the term "cohort" has a different meaning.

**Figure 13.1**
## Even-aged (A), multi-modal (B), and uneven-aged (C) age structures for hypothetical stands

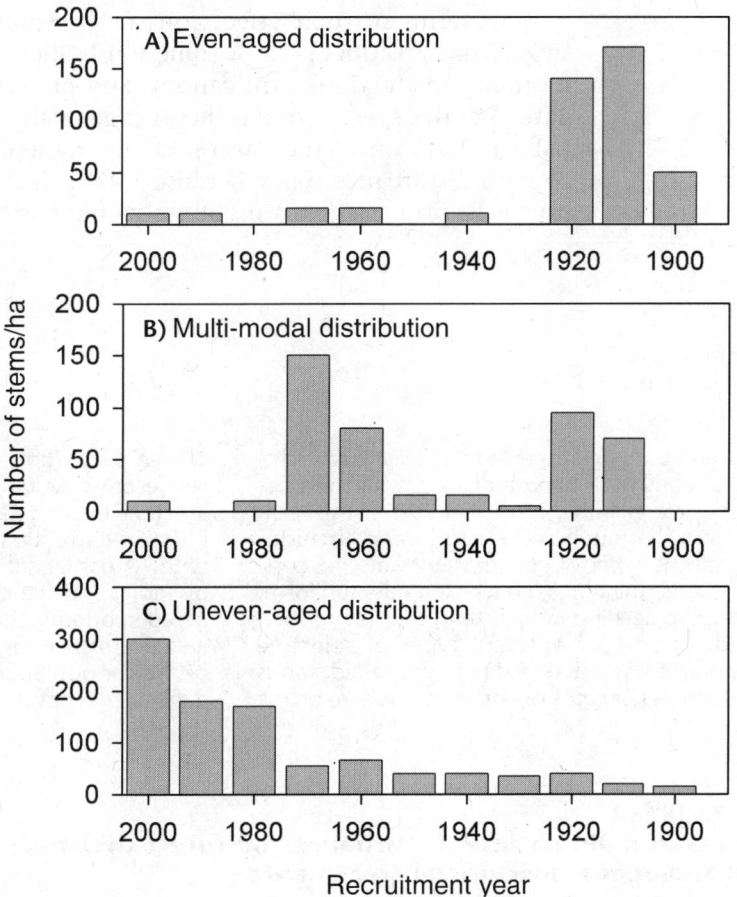

The recruitment age that is represented in the graphics does not correspond to germination year but rather to the number of growth rings at 1-m height.

disturbances such as spruce budworm (*Choristoneura fumiferana;* SBW) outbreaks or partial windthrows. For example, a stand could contain at the same time a group of trees that established following a SBW outbreak around 1910, and another group that established following a partial windthrow around 1960 (figure 13.1B). There is no specific term for the silvicultural treatments that, when used repetitively, produce stands with several distinct age classes, other than the vague term "partial cutting," which includes in fact all the types of cuttings that are not clearcuts (see box 13.2).

- Uneven-aged stands contain merchantable trees that belong to many age classes that cannot be separated from one another (figure 13.1C). Recruitment episodes producing these age structures thus occur in a quasi-continuous manner in time. Gap-scale natural disturbances (e.g., senescence) or repetitive selection cuts can create uneven-aged stand structures. Uneven-aged stands are usually dominated by shade-tolerant tree species.

In terms of their effects on stand structure, natural disturbances and silvicultural treatments show some similarities: they both create a given level of mortality in the dominant canopy, and preserve or favour the recruitment of particular tree species, thus sometimes generating changes in species composition (table 13.1). However, there also exist important differences between stands affected by both disturbance types, because a silvicultural regime that has wood production as its primary goal aims at producing optimal stand densities to favour the

---

### Box 13.2
## Partial cuts

While clearcutting clearly refers to a silvicultural treatment that consists in harvesting nearly all the trees in a stand, for example more than 90%, the term "partial cut" is much more vague and includes roughly all the types of cuts that are not clearcuts. Selection cuts, thinnings, conversion cuts, uniform, patch or strip seeding cuts (often used in the shelterwood system), or different types of selective logging (i.e., culling) that lead to high-graded stands are all examples of types of partial cuts where only a certain proportion of trees in a stand are harvested. However, these terms are generally not very precise and their use can generate some confusion, especially if similar terms are used in different regions or forest types. A partial cut can be described more precisely by indicating the species or tree sizes that are targeted in priority, the percentage of removal (in basal area, stems per hectare), or the spatial arrangement of the canopy openings that they create (localized gaps or patches that are uniform across the stand).

---

### Table 13.1
## Effects of main natural disturbances on forest dynamics at the stand level and analogous silvicultural treatments

| | Natural disturbances | | | | Analogous treatments** |
|---|---|---|---|---|---|
| **Type** | **Severity** | **Species affected*** | **Soil disturbance** | **Species favoured** | |
| SBW outbreak | High | BF, BS, WS | Light | BF, BS, WS, WB | CPR, PC |
| SBW outbreak | Moderate | BF | Light | BF, BS, WS | |
| Severe fire | High | All | Generally severe | JP, BS, TA, WB | CC + scarification |
| Surface fire | Moderate | All | Moderate | WP, RP | PC |
| Windthrow | High/moderate | All | Moderate | WS, YB, WB | CC, CPR, PC |
| Forest decline | Moderate | WB, TA, YB | Light | Tolerant | PC, SC |
| Senescence | Low | All | Light | Tolerant | PC, SC |
| Competition | Very low | All | Light | None or very tolerant | Thinning |

\* WB = white birch (*Betula papyrifera*), TA = trembling aspen (*Populus tremuloides*), YB = yellow birch (*Betula alleghaniensis*), BF = balsam fir (*Abies balsamea*), JP = jack pine (*Pinus banksiana*), BS = black spruce (*Picea mariana*), WS = white spruce (*Picea glauca*), WP = white pine (*Pinus strobus*), RP = red pine (*Pinus resinosa*).

\*\* CC = clearcut without protection of advance regeneration, CPR = final cut with protection of advance regeneration, PC = partial cut, SC = selection cut.

References consulted for natural dynamics: SBW: MacLean and Ostaff 1989; Morin 1994; Kneeshaw and Bergeron 1998; Bouchard et al. 2006a, b. Fire: Bergeron et al. 1999; Bergeron 2000. Windthrow: Ruel and Pineau 2003; Houle and Payette 1990. Forest decline: Lortie 1979; Bouchard et al. 2006a. Senescence: De Grandpré et al. 2004; Pham et al. 2004; Groot and Horton 1994.

vigour and growth of individual trees. For this reason, trees tend to be relatively well spaced out and vigorous in well managed stands compared with natural stands, which are dense and contain an important proportion of weakened trees.

Considered individually, the different disturbance types enumerated in table 13.1 provide a relatively brief and limited overview of natural forest dynamics, notably because stand age structure or stand species composition are not the product of just one silvicultural treatment or one natural disturbance event, but rather of a sequence of interventions or disturbances. The integration of treatments inside silvicultural systems that aim at maintaining a given composition or structure over the long term will be developed in section 3.

## 2.2.  Dead Wood and Residual Trees: Variable Retention Strategies

The dead wood remaining in the recently disturbed stands, either in the form of standing dead trees or coarse woody debris, is important because a significant proportion of forest organisms, plants and animals, depend on such habitats at one time or another during their life cycle (Harmon et al. 1986; Drapeau et al., chapter 14). Natural disturbances such as fire, insect outbreaks or windthrow leave large amounts of dead wood and variable proportions of surviving trees that are more or less weakened (Sturtevant et al. 1997; Kafka et al. 2001; D'Aoust et al. 2004). By comparison, logging will export wood outside the forest ecosystem. The amount of dead wood that should be left to ensure the maintenance of viable populations could be determined with sensitivity analyses based on the abundance of various indicator species depending on this resource, both at the stand and landscape levels (Drapeau et al., chapter 14). In addition to the amount of dead wood available, factors such as the species or size of dead stems should also be considered because they could be biologically important.

Even though fires are generally severe in boreal forests, it is common to observe residual islands scattered within the burn perimeter (Kafka et al. 2001). SBW outbreak can also cause significant mortality but, even in the most severe cases, it is usual to observe living trees, notably non-host species such as birches (*Betula* spp.) and eastern cedar (*Thuja occidentalis* L.) (D'Aoust et al. 2004; Bouchard et al. 2006a). These "remnant" living trees will provide seeds and shade for the future stand, create conditions that could be more favourable for the dispersion and maintenance of populations for various organisms, and will allow for dead wood recruitment in the future, thus ensuring a better continuity between pre- and post-disturbance stands (Franklin et al. 1997). The spatial distribution of surviving trees could also be important for the persistence of some animal or plant species, particularly after a clearcut. In this respect, it is sometimes recommended to concentrate the residual living trees in patches, which would be more consistent with the spatial patterns observed after a natural disturbance, instead of leaving uniformly distributed isolated trees across the stand. Moreover, some consideration must be paid to the species and health status of these residual trees because such factors could influence the probability of trees being affected by diebacks or windblown shortly after the cut.

Information on dead and remnant trees should also be integrated explicitly into silvicultural scenarios as well as annual allowable cut calculations through the adjustment of stand yield curves. These decreased yields include the wood retained in stands, as well as the anticipated growth reductions of the future stand because part of the resources (e.g., light and nutrients) will be intercepted by trees that will never be harvested (Rose and Muir 1997).

Under many provincial forest jurisdictions such as in the province of Québec, the kind of silviculture that was practised in the past has sometimes resulted in leaving important quantities of dead or dying trees within stands. This reality is often used as an argument to support the opinion that it is not necessary to devote special attention to dead wood retention. Other than the fact that this opinion neglects qualitative and quantitative differences between dead and dying trees in managed vs. unmanaged stands (with respect to stem size for instance), the fact that the actual stands contain satisfactory quantities of dead wood is of course no guarantee that it will remain the case in the future. In some jurisdictions, the dead wood issue is also susceptible to becoming more of a concern as policies aimed at maximizing the efficiency of logging activities for timber processing purposes are implemented.

## 2.3. Soils

Natural disturbances have variable effects on soils. Some disturbance types, such as severe fires, cause a reduction in the thickness of the duff layer, the exposure of mineral soils, and the elimination of understory competition, and they increase nutrient input (Nguyen-Xuan et al. 2000; Simard et al., chapter 11). By comparison, harvesting operations have a tendency to result in relatively moderate humus disturbance, especially when the operations are carried in wintertime, which can ultimately result in a thickening of the soil organic layer and a decrease in soil fertility, particularly on imperfectly drained sites in the boreal zone. Different silvicultural treatments such as soil scarification or scalping could reduce significantly the thickness of the organic layer after cutting, somewhat similarly to the effect of a severe fire. Some trials indicate that prescribed burning by itself is generally not sufficient to reduce significantly the depth of the organic layer, but it could have beneficial effects in terms of stand regeneration when used after soil scarification (Lavoie et al. 2005).

Other disturbances, such as windthrow, can lead to localized soil disturbance due to individual tree uprooting. Moderate scarification created locally by passage of heavy machinery, or small-patch scarification for the creation of germination micro-sites, are among the possible options to emulate the effect of that kind of disturbance on forest soils (Gastaldello et al. 2007). Disturbances such as insect outbreaks or forest declines (diebacks, diseases) rather tend to leave dead trees standing, which generate little soil disturbance. Types of silvicultural operations that protect the organic layer (such as cuts performed in wintertime) can succeed in reproducing these conditions.

Logging practices and particularly the passage of heavy machinery can have undesirable consequences on forest soils, an issue that is often addressed in best management practices guides (see for example the *Objectifs de protection et de mise en valeur* [Gouvernement du Québec 2007] and the *Forest Management Guidelines for the Protection of the Physical Environment* [Archibald et al. 1997]) or in forest regulations (Gouvernement du Québec 2007). For example, soil compaction and rutting can impede regeneration establishment for variable periods, and reduce the productive forest area (Archibald et al. 1997; Gouvernement du Québec 2005). These undesirable effects could be mitigated by restricting the use of heavy machinery during winter, when the soils are frozen, or by carefully planning the location of skidding trails. Obviously, these protective measures are less relevant in a context where organic matter accumulation is judged excessive (paludification context; see Simard et al., chapter 11).

## 2.4. Regeneration

The degree of canopy opening, the seed rain, as well as the amount and quality of seedbeds are all determinant influences on the germination and growth of natural regeneration. Because these different factors vary according to the type and severity of natural disturbances (table 13.1) as well as the site type, regeneration composition can vary considerably in time and space. A total lack of regeneration or drastic compositional changes can also occur following a natural disturbance, especially after intense and severe disturbances that occur in rapid succession such as insect outbreaks followed by fire (Jasinski and Payette 2005).

The use of natural regeneration is often viewed favourably in silviculture compared with artificial reforestation measures because it is the most economically profitable option, but also because it is a simple way to obtain regeneration that is both diversified and well adapted to the site. Natural regeneration established after a cut could, however, differ from what would be found after a natural disturbance. For example, silvicultural interventions could lead to the elimination of advance regeneration or to the selective removal of some seed trees. These various factors explain why some types of cuts could systematically favour certain tree species, for example balsam fir or trembling aspen in boreal forests, over species such as black spruce or jack pine (Grondin and Cimon 2003). In southern boreal forests, repeated harvests have also led in some instances to the gradual reduction of some conifer species in favour of hardwood species such as white birch (*Betula papyrifera* Marsh.), sugar maple (*Acer saccharum* Marsh.), or trembling aspen (Boucher et al. 2006). The replacement of natural disturbance regimes by forest harvesting could also increase understory competition in some site types, for example ericaceous shrubs in boreal forests (Thiffault and Jobidon 2006) or mountain maple (*Acer spicatum* Lamb.) in southern boreal forests (Lieffers et al. 2003; Grondin and Cimon 2003; Bourgeois et al. 2004). This competition can slow down or even block seedling growth during several years or decades for some tree species.

Some silvicultural treatments allow preventive control of the composition of the future stand (such as thinnings or partial cuts aimed at leaving seed trees from particular species), the conservation or the elimination of advance regeneration during harvesting operations, or the creation of appropriate seed beds. Vegetation control measures (e.g., slashing and phytocides) or site preparation techniques such as soil scarification can be considered to prevent the potential invasion by shrub species. The size of the canopy openings created by the removal of dominant trees will also have an influence on the quantity of light that reaches the understory, and on the possibility of regenerating shade intolerant species (table 13.1).

The use of artificial reforestation measures should also be considered in some situations, particularly for tree species for which natural regeneration is unlikely in the absence of natural disturbances (e.g., species with serotinous or semi-serotinous cones), or tree species that have been greatly reduced in abundance as a consequence of past forestry practices and that must be restored (e.g., red spruce [*Picea rubens* Sarg.], white pine [*Pinus strobus* L.], eastern white cedar [*Thuja occidentalis* L.]). When the use of artificial regeneration measures is required due to a deficit in natural regeneration with unsuitable composition or characteristics, it is preferable to target the areas within stands where regeneration status is critical. For example, ensuring spot planting in skid trails might be sufficient in some circumstances. The use of large-scale plantations is sometimes unavoidable, for example when it is necessary to proceed with scarification to maintain soil productivity, but this practice could lead to a greater homogenization of stand structure and composition compared with stands created by natural disturbances.

## 3. DEVELOPMENT OF SILVICULTURAL SYSTEMS THAT MAINTAIN NATURAL FOREST DYNAMICS

Reduced to its simplest expression, a silvicultural system designates a series of silvicultural treatments that follow one another in time and are generally associated with one or a few production objectives (e.g., wood products, wildlife habitats) (Nyland 1996; Smith et al. 1997). This succession of silvicultural treatments and harvestings should ensure both stand regeneration and stand quality development (Nyland 1996). The selection of the appropriate silvicultural system for a given stand could depend on stand characteristics (composition and structure), production objectives (saw logs, pulp and paper, wildlife habitats), stand accessibility, site characteristics (fertility, trafficability), and other socio-economic considerations (recreation, hunting). In an ecosystem management context, knowledge about the landscape-level structure and dynamics of preindustrial forests will also influence the choice of an adequate system or treatment for a given region or stand.

### 3.1. Traditional Silvicultural Systems

Traditionally, silvicultural systems are categorized as a function of the harvesting method used to regenerate the stand (table 13.2), as well as by the frequency and intensity of these harvestings (figure 13.2), which will determine stand structure and composition. The main silvicultural systems used in North America are uneven-aged systems, which allow for the regeneration of irregular or multi-storied stands, often dominated by shade-tolerant species, and even-aged systems, which allow the regeneration and maintenance of regular stands. Most silvicultural systems that are commonly used in boreal forests could be easily associated with one of these silvicultural systems. For example, clearcutting* or seed-tree cuttings are associated with even-aged systems (such as the uniform shelterwood system), and selection cuts are associated with uneven-aged systems (specifically the selection system). Some intermediate silvicultural systems also maintain

• Clearcutting includes different kinds of careful logging treatments that protect advance regeneration (for example CPRS in Québec or CLAAG in Ontario).

## Table 13.2
## Main silvicultural systems, grouped according to the regeneration method

| Age structure | Canopy structure | Regeneration method | Description | Example of use in an ecosystem management context |
|---|---|---|---|---|
| Even-aged | Regular | Clearcut (with or without protection of advance regeneration) | Harvest in one intervention all merchantable stems in a stand to regenerate an even-aged stand, either by natural or artificial regeneration. CPRS* and CLAAG** designate silvicultural treatments that are used to make the final harvest when advance regeneration of acceptable composition is present, or when the soils are fragile. In other circumstances, clearcuts could be followed by a scarification favouring the germination of new propagules, or by planting. | Used to emulate the effects of severe natural disturbances. In boreal forests, clearcuts leading to the recruitment of pioneer species will be analogous to a severe fire, and clearcuts maintaining mixed stands dominated by late-successional species such as balsam fir could be analogous to severe SBW outbreaks or windthrows. A minor portion of the dominant canopy can be left standing to satisfy some protection objectives (i.e., clearcuts with retention). |
| | | Regular shelterwood | One or two seeding (or education) partial cuts to establish a continuous regeneration layer at the stand scale, followed by a final cut (CPRS or CLAAG) once the regeneration is established. | Used to emulate the effects of relatively severe disturbances (fires, insect outbreaks or windthrows). The preparation of germination beds could be important for the establishment of seedlings of the desired species. |
| Semi-even-aged*** | Relatively irregular | Irregular shelterwood | A system that ensures regeneration establishment by the progressive widening of patches or strips within a stand. Contrary to the regular shelterwood, the final cut does not occur in a synchronous manner across the whole stand. | Used when a heterogeneous but relatively continuous forest cover must be maintained, while regenerating a relatively even-aged stand (or light-demanding species). Relatively analogous to severe or spatially contagious disturbances such as forest declines, or partial insect outbreaks followed by windthrows. |
| Uneven-aged | Multi-storied | Periodic partial cuts, not frequent and relatively severe (multi-modal age structure) | A system that consists of regeneration episodes separated by relatively long cutting cycles, for example 25 years or more. Each intervention must ensure stand regeneration, education, and harvesting at the same time. Regeneration establishment takes place uniformly across the stand. | Effect analogous to insect outbreaks or windthrows of moderate severity. Preferable to the "classic" selection system when site productivity is relatively low and/or road access is limited (often the case in boreal forests). |
| | Irregular | Periodic partial cuts, frequent but not very severe ("classic" selection system) | A system consisting of regeneration episodes separated by short cutting cycles (for example less than 20 years). Each intervention ("selection cut") must ensure stand regeneration, education, and harvesting at the same time. Regeneration establishment takes place in localized canopy gaps and not in a uniform manner across the whole stand. | Analogous to natural mortality by senescence and gap dynamics. May be used when stand productivity is good to regenerate shade-tolerant species. |

\* "Cut with protection of advance regeneration and soils," used in Québec.
\*\* "Careful logging around advanced growth," used in Ontario.
\*\*\* The regeneration period generally extends over at least half of the full rotation length, and thus it is not a strictly even-aged system (Smith et al. 1997).

Figure 13.2
## Frequency and intensity of regeneration cuts for the main silvicultural systems

### Silvicultural treatments

☐ Optional thinning (education)
▨ Seeding cut (regeneration)
☐▨ Multiple-use partial cut (regeneration + education)
▨■ Multiple-use partial cut (regeneration + final)
■ Final cut

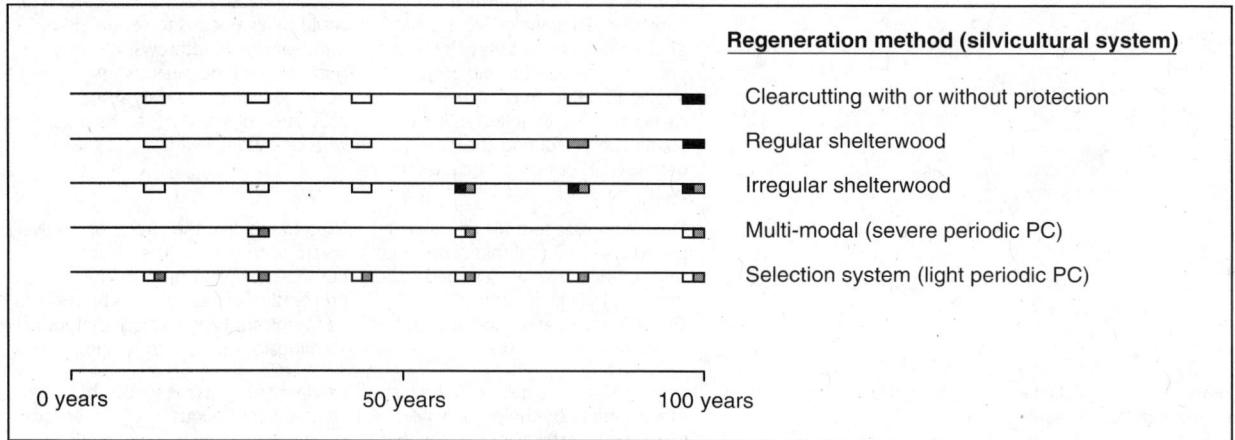

This example is based on a 100-year rotation period. Regeneration methods by clearcutting with or without protection of advance regeneration, and the shelterwood methods, include a final cut at the end of the rotation period, which is not the case for regeneration methods with periodic partial cuts.

stands with a more or less irregular structure using combinations of different harvesting techniques (e.g., progressive widening of cut patches), notably those associated with the irregular shelterwood system. Other treatments are more difficult to classify in the different systems because their primary objective is not necessarily to favour regeneration establishment or recruitment in the understory. For example, the CPPTM (cut with protection of small merchantable stems) used in Québec or the HARP (harvest with regeneration protection) used in Ontario are diameter-limit cuttings that free the existing undergrowth rather than regeneration cuttings. In some circumstances, however, the long-term use of this type of cutting may promote the maintenance of stands with multi-modal age structures and a multi-storied vertical structure. Moreover, variable retention (see section 2.2), which aims to leave living trees in harvested stands, either according to a dispersed or clumped spatial pattern, does not constitute a silvicultural system as such, because it does not generally put the emphasis on long-term stand regeneration and dynamics (Groot et al. 2005). Variable retention should be considered as an option that can be incorporated in standard silvicultural systems, for example even-aged systems with retention of a given proportion of trees after the end of the rotation period.

It is possible to establish analogies between existing silvicultural systems and the effect of natural disturbances (table 13.2), in order to adopt types of cut that maintain stands with relatively natural composition and structure. Figure 13.3 illustrates, in chronosequences, some systems that were inspired by the effect of natural disturbances. Each example shows different stand development stages after silvicultural treatments trying to reproduce the effect of specific natural disturbances. Figure 13.3A illustrates a regular shelterwood system that aims to emulate the effect of a severe fire, assuming a mixed-species initial composition

**Figure 13.3**
**Examples of silvicultural systems**

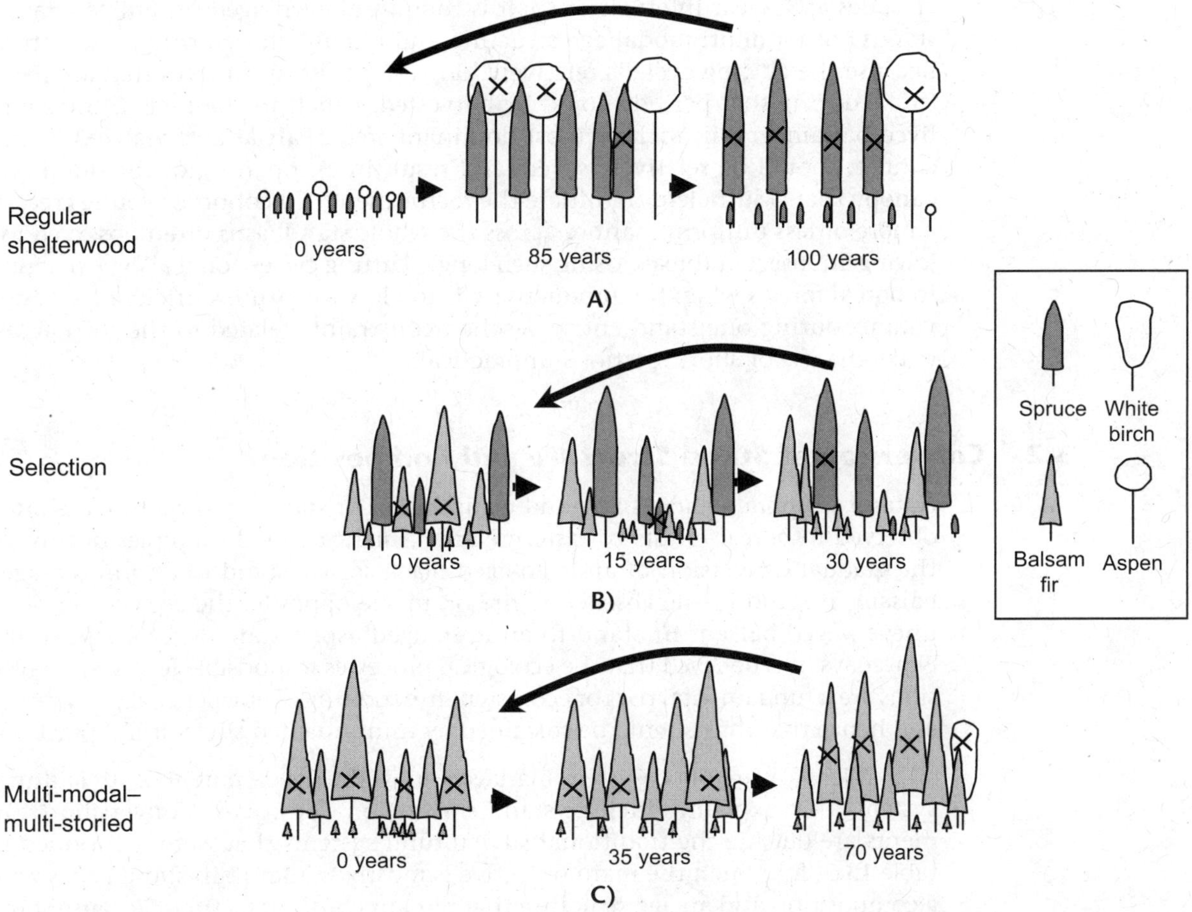

Legend: A) a regular shelterwood system resulting in an even-aged stand, B) selection system resulting in an uneven-aged stand, C) severe periodic partial cuts resulting in a multi-modal age structure and a multi-storied vertical structure. Snags and skidding roads are not illustrated in these examples, and it is assumed that the stands are relatively well stocked. Trees marked with an 'X' are those that are harvested. See text in section 4.1 for detailed explanations.

that we try to reinitiate at the end of the rotation. In this example, a partial cut is done at 85 years to promote regeneration in the understory, and a final cut at 100 years to free this regeneration from the overstory influence.

Figure 13.3B shows an uneven-aged stand that is maintained by the use of frequent but low-intensity selection cuts, emulating the effects of senescence or small-scale windthrows. Because the goal of these cuts is to favour punctual gap-scale recruitment, but not a uniform stand-scale recruitment pulse, the canopy gaps created by these cuts must be well defined. The goal of the cuts that are performed elsewhere in the stand will be to free intermediate stems that are already established rather than to initiate understory recruitment. The selection system is the one that has the effect of maximizing structural diversity at the stand level (Schütz 1997, 2001).

Finally, figure 13.3C shows a system that is analogous to the effect of several successive SBW outbreaks in a balsam fir stand, which will generate recruitment episodes at 35-year intervals or so; it is thus an uneven-aged system generating stands with a multi-modal age structure and a multi-storied canopy structure. Because the cutting cycle is relatively long (35 years), most trees that are likely to die during that period should be harvested, which in the case of the short-lived balsam fir will include most dominant trees. Partial cuts realized in this context should be relatively severe and result in an opening of the dominant canopy that is sufficient to initiate the recruitment of a cohort of young trees in a more or less uniform manner across the whole stand, as is often observed following an insect outbreak. Using such longer cutting cycles could also be practical in boreal forests when the productivity is too low to ensure sufficient harvested volume during one stand entry, or when constraints related to the road access make the use of short rotations unpractical.

## 3.2.   Conversion of Stand Structure and Composition

Natural temporal changes in stand composition or stand age-structure are often observed in boreal forests of eastern North America. Good examples of this are the gradual conversion of an even-aged black spruce stand to an uneven-aged balsam fir stand in the absence of fire or, to the opposite, the conversion of an uneven-aged balsam fir stand to an even-aged aspen stand after a severe fire. Nowadays, it is believed that the ecological processes responsible for these conversions are a fundamental part of ecosystem functioning (Gauthier et al., chapter 1), and hence that they should be sustained by using adapted silvicultural practices.

A key-feature of ecosystem management is the development of silvicultural systems that explicitly include stand conversion objectives. Conversion treatments are outside the traditional silvicultural systems that were mentioned in table 13.2, for which the main objective is mostly to maintain stand types with a composition and an age structure that remain constant in time. Because stand conversions can be associated with some losses in wood production during the transition phase, they are considered as a delicate operation that requires some patience and care in order to be carried successfully (Schütz 1997, 2001; O'Hara 2001). For example, during the conversion of a stand from an even-aged to an uneven-aged structure, a certain proportion of the stems will have to be harvested in pre-maturity or post-maturity in order to maintain some form of canopy cover during the transition phase, which can span over several decades. During the conversion of an uneven-aged stand to an even-aged structure, a proportion of the stems that have not reached target size will be harvested in pre-maturity.

The conversion of an even-aged structure (regular) to an uneven-aged one (irregular) can emulate the long-term effect of post-fire succession. This type of conversion consists in using partial cuts to gradually open the canopy and promote the implementation and growth of natural regeneration. These initial canopy openings can be dispersed uniformly across the stand or be concentrated in gaps. The most important point is maintaining a sufficient proportion of merchantable stems that are likely to survive during the entire transition phase (Schütz 1997, 2001). For this reason, it is often considered easier to initiate stand conversion in a relatively young even-aged stand (Schütz 1997, 2001; O'Hara 2001). In some conditions, silvicultural systems that favour the conversion of a stand from an even-aged to an uneven-aged structure are initially similar to the regeneration-establishment cuttings done in the shelterwood system, followed by selection-type cuttings in order to improve or maintain the irregular structure (table 13.2). In boreal forests, the conversion of the stand age-structure sometimes happens together with compositional changes, such as the replacement of even-aged stands dominated by pioneer species with uneven-aged stands dominated by balsam fir or black spruce.

In comparison, the conversion of a stand from an uneven-aged (irregular) structure to an even-aged (regular) structure is a relatively simple operation because it consists in harvesting every merchantable stem in one stand entry. For example, regeneration methods such as careful logging practices (e.g., CPPTM/ HARP) could be described as conversion treatments when they are applied in an uneven-aged stand. The conversion of a stand from an uneven-aged to an even-aged structure is often perceived unfavourably because it can have negative visual and environmental effects. However, such conversions could be useful in a context where we aim to emulate the landscape-level effects of large disturbances like fire (see subsequent sections).

## 4.   LANDSCAPE-LEVEL CONSIDERATIONS

### 4.1.   Relative Proportion of Different Treatments and Silvicultural Systems at the Landscape Level

Silvicultural treatments and systems should be integrated within an ecosystem management strategy implemented within a forest management unit. In theory, this management strategy must be based on a portrait of the preindustrial forest that is as complete as possible, including a description of natural disturbance regimes and the relative abundance of different stand types (Harvey et al. 2003). For example, the use of cuttings aimed at emulating the effects of fire, severe insect outbreaks, or senescence of individual trees should correspond to the observed frequency of these various disturbance types in preindustrial forests. Silvicultural systems that generate even-aged structures should thus be used more frequently in a region where the fire cycle is short, those to maintain multi-modal age structures should be used more frequently in a region where SBW outbreaks and windthrow are important, whereas those promoting strictly uneven-aged structures should be used where large-scale disturbances tend to be rare and where trees tend to die individually.

This regional portrait can be presented in a diagram, as shown by Gauthier et al. (2004) and Harvey et al. (2003) for a management unit located in the black spruce–feather moss bioclimatic domain in the Abitibi region (western Québec) (figure 13.4). In this region, severe stand-replacing fires are the main natural disturbance, promoting the regeneration of stands dominated by black spruce, trembling aspen, or jack pine, which gradually evolve towards exclusive black spruce dominance. According to this model, stands that belong to the first cohort (C1) are recruited immediately after fire, stands belonging to the second cohort

**Figure 13.4**
## Schematic representation of natural forest dynamics in a region of the black spruce–feather moss bioclimatic domain, western Québec

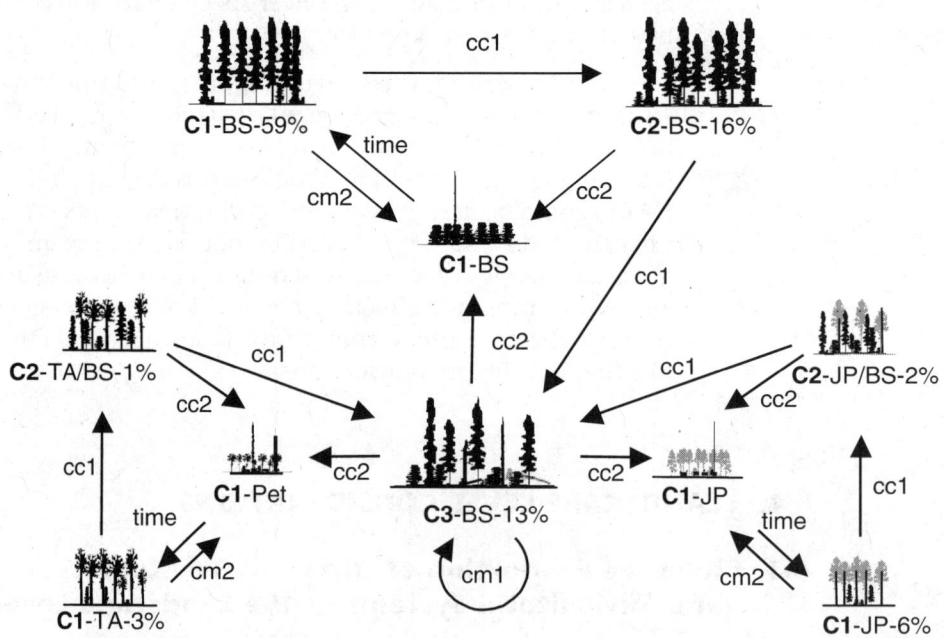

The diagram illustrates three possible successional trajectories, including initial dominance by trembling aspen (TA), black spruce (BS) or jack pine (JP). These successional stages correspond with the cohort system of Bergeron et al. (1999) (C1 to C3; see text in section 4.1), and percentages indicate the relative proportion of the different forest types within the management unit (cf. Harvey et al. 2003). The types of cuts that allow the emulation of natural processes are the following: cm1 = maintenance of uneven-aged structures, cm2 = maintenance of even-aged structures, cc1 = conversion from even-aged to uneven-aged structures, cc2 = conversion from uneven-aged to even-aged structures. Figure modified from Gauthier et al. (2004) and Harvey et al. (2003).

(C2) include a sparse overstory of senescent trees that recruited immediately after fire and a second group of trees that recruited later after the dominant canopy break-up, and stands belonging to the third cohort (C3) present a heterogeneous structure and mostly contain trees that have established long after the fire occurred (see Gauthier et al., chapter 3). These different successional stages correspond to the cohort system proposed by Bergeron et al. (1999).

The diagram presented in figure 13.4 makes it possible to place the different silvicultural systems in a precise ecosystem context for this region. For example, cuts that allow the conversion of even-aged structures to uneven-aged ones will be used to emulate the natural conversion of stands of the first cohort towards the second and third cohorts, ensuring at the same time a compositional change towards a greater dominance of black spruce (figure 13.4). Cuts that allow the conversion from uneven-aged structures to even-aged ones will be used to favour the transition of stands belonging to C2 and C3 towards C1. As for the silvicultural systems that are to maintain stable age structures and constant forest composition, they will be used to maintain first-cohort even-aged stands (clearcuts with or without protection of advance regeneration), or third-cohort uneven-aged stands (partial cuttings of variable severity such as selection cuttings). In general, the relative proportion of the different stand types in a forest management unit should thus remain relatively stable in time. This means for example that if some proportion of C3 stands are converted to C1 (through clearcutting), a corresponding number of C1 or C2 stands should be converted to C3.

As mentioned before, the extent to which the various silvicultural techniques are used to emulate natural processes is likely to vary from one region to another due to different natural disturbance regimes, different successional processes, the presence of additional tree species, or the influence of abiotic factors. Nonetheless, many authors agree on the fact that, in general, the forestry that is currently practised in boreal forests puts relatively little emphasis on silvicultural systems that favour the maintenance or conversion towards irregular structures compared with the proportions that are observed in natural ecosystems (MacDonald 1995; Bergeron et al. 1999; Lieffers et al. 2003; Groot et al. 2005). This situation probably results from the fact that even-aged systems based on clearcutting are much simpler to use, are more profitable in the short term, and do not require detailed knowledge about pre-harvest stand characteristics or the regeneration dynamics of the different tree species because there is always the possibility of using artificial reforestation measures to regenerate the stand. Nonetheless, several experiments suggest that a more prominent use of partial cuts is possible in the boreal forest, as long as stands showing suitable characteristics in terms of structure, composition, and fertility are targeted in priority (Harvey et al. 2002; Groot et al. 2005; Ruel et al. 2007; Thorpe and Thomas 2007). Taking additional factors into account such as socio-economic considerations, field experience, or availability of silvicultural tools like ecological classification guides (produced by the provincial MNRs) as well as research projects carried out in partnership with universities may also facilitate the implementation of these systems and increase the probability of success over the long term.

## 4.2.  Spatial Arrangement of the Different Harvesting Methods at the Landscape Level

One of the important characteristics of forest ecosystem management is the strategic importance that it confers on the spatial organization of harvesting areas and tracts of mature and old-growth forests. The spatial layout of silvicultural treatments and systems should be similar to the spatial pattern created by natural disturbance regimes, because this pattern plays an important role for several ecosystem processes, such as the dispersion and reproduction of animal populations, or the re-colonization by plants following a severe disturbance (Hunter 1999; Thompson and Harestad 2004). Among these disturbances, fires covering very large areas, for example 1,000 ha and more, are without doubt the most important disturbance type shaping the spatial configuration of forest ecosystems at the landscape level in boreal forest of North America. These fires create vast areas where the stands have a similar age, and are often initially dominated by pioneer species such as black spruce, aspen, birches, or pines. Clearcuts that are agglomerated over large areas, with retention of some proportion of the stands within the agglomerations, can emulate to a certain extent the effect of fires at the landscape level (Andison 2003; Gauthier et al. 2004; Belleau et al. 2007). One must however make sure that these agglomerations are well dispersed in the landscapes and that the characteristics of the residual stands (regeneration, remnant trees) are adequate (Perron et al., chapter 6). Moreover, because in nature large fires are not usually juxtaposed exactly (Belleau et al. 2007; Perron et al., chapter 6), the stands located between the clearcut clusters should be left intact or be treated with partial cuts that maintain a continuous forest cover.

In natural landscapes, the occurrence of a large forest fire often entails the conversion of a certain proportion of late-successional stands (usually with an uneven-aged or multi-modal age structure, dominated by balsam fir) towards early successional stands (even-aged, dominated by pioneer species). Emulating the effects of such disturbances can lead to silvicultural prescriptions that may appear counterintuitive if they are only considered from a stand-level perspective. For example, in order to ensure a characteristic species composition within a clearcutting agglomeration that aims to emulate the effect of a large and severe fire, it may be necessary to prescribe scarification and plantation treatments to neutralize the advance regeneration of balsam fir, a species that tends to proliferate in managed forest mosaics (Grondin and Cimon 2003). If this regeneration is not neutralized, there is a risk of seeing the regional-level abundance of balsam fir exceed its historical levels, which could have negative effects in the long term, such as a greater vulnerability to SBW outbreaks (see Morin et al., chapter 7).

Natural disturbances such as SBW outbreaks or windthrow can create severely disturbed areas that cover few or tens of hectares, which are more or less dispersed in the forest landscapes (Leblanc and Bélanger 1998; Belle-Ile and Kneeshaw 2007). These disturbances occur mainly in forest landscapes that have not been recently affected by fires, and hence they will be particularly important in regions where the fire cycle is long. The spatial pattern created by these disturbances can be emulated to a certain extent by forest practices using small clearcuts (e.g., a few hectares) dispersed throughout landscapes (even-aged stands), or else by partial cuts that maintain irregular structures and a relatively continuous canopy cover (stands with uneven-aged or multi-modal age structures).

## 4.3.  Management of the Road Network

Controlling the extent of the road network is often considered as one of the main challenges of ecosystem forest management. This control would minimize the impact caused by human presence for some sensitive animal species such as the woodland caribou (Courtois et al. 2004; Bourgeois et al. 2005). In the long term, an adequate planning of the road network also reduces the costs involved in road construction and maintenance (Andison 2003).

The size of the permanent road network can be reduced by clustering harvest operations in time and space, so that some roads could be abandoned between interventions. Silvicultural systems that do not require frequent interventions should be favoured in this context because they create stand entries that are separated by periods of 30 years or more. This refers essentially to silvicultural systems that are based on clearcutting and generate even-aged structures, or systems based on relatively intense and periodic partial cuts generating multimodal age structures. Systems featuring multi-modal age structures thus seem to hold promise for an interesting future for the boreal forests because they meet the need for maintaining a permanent forest cover and the need for minimizing the frequency of interventions and the extent of the road network (and hence financial investments). In this context, the selection system or other intensive silvicultural practices (e.g., ligniculture), which necessitate frequent interventions, should be used mainly near permanent roads (primary access roads) and close to the mills.

Overall, one could expect that a forest management strategy that includes a controlled development of the road network would result in a well developed primary road network to disperse adequately the clearcut agglomerations in the forest mosaic, and a secondary and tertiary road network less extensive than usually observed. Such a strategy will probably be simpler and less costly from an operational perspective (Richards and Gunn 2000; Andison 2003), and could also allow for some productivity gains, if for example the areas occupied by abandoned roads are reforested. In the long term however, such a strategy will generally be less productive (from a wood transformation perspective) because it implies fewer intermediate treatments such as commercial thinnings (Jamnick and Walters 1993; Andison 2003).

## 4.4.  Integration of Silviculture with Large-Scale Forest Planning

Traditionally, the silvicultural prescription was based on local stand characteristics, such as the current composition and structure, its disturbance history or site characteristics. As we have seen, this type of approach, which only considers the stand level, is not sufficient to resolve some landscape-level issues, such as the relative proportion of the different stand types, their spatial distribution, and the extent of the road network. In other words, components that were traditionally addressed at the tactical or strategic planning levels (such as the road network and the regulation of forest composition and structure) must be integrated into local-scale decision making, and vice versa (box 13.1).

The following example illustrates concretely how silviculture could be used in order to provide solutions to some ecosystem management issues. We suppose that after preliminary analysis of the preindustrial forest mosaic for a given

**Figure 13.5**
**Example illustrating the spatial distribution of different types of cuts in a fictive territory, according to an 80-year planning horizon (divided in four 20-year periods)**

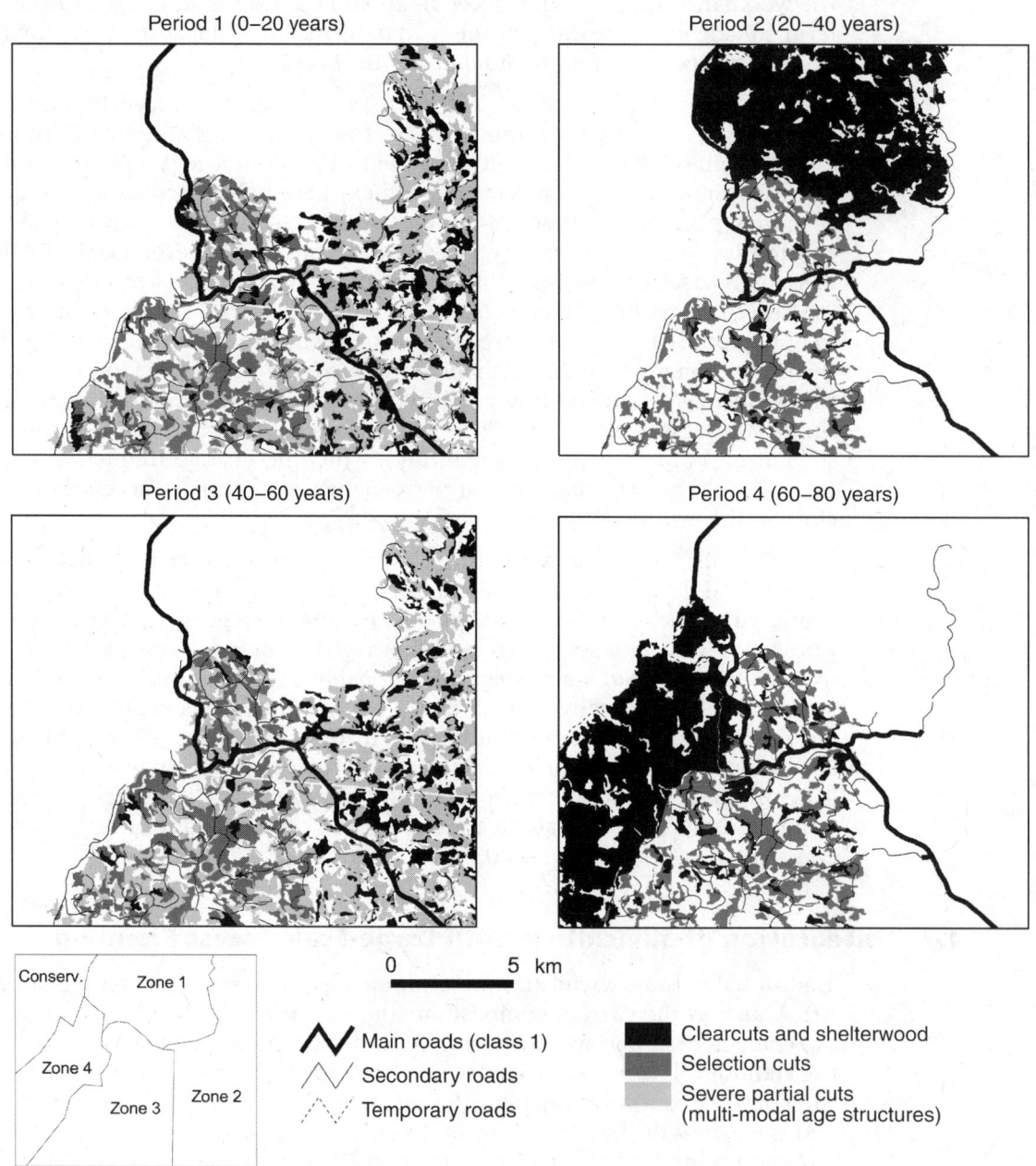

A temporary road network is used in all zones, except in zone 3 where the selection system is used, which necessitates the use of a permanent road network. The silvicultural strategies used in each zone are detailed in table 13.2. The territory also includes a block of protected forest where no silvicultural interventions are planned.

territory (disturbance regimes, forest age class structure, tree species composition), three main objectives are defined: 1) irregular stands should represent at least 50% of the forest mosaic at any time (including multi-modal and uneven-aged structures), 2) very large disturbance agglomerations that cover about 50 km² should be created in order to emulate fire effects, and 3) the extent of the road network should be minimized. Figure 13.5 presents the spatial distribution of different silvicultural treatments for a temporal horizon of 80 years in order to achieve these three objectives. Overall, it can be observed that using an appropriate zoning strategy will maintain both large agglomerations dominated by even-aged stands produced and maintained by severe disturbances (figure 13.5, zones 1 and 4) and large tracts of irregular forests where complex stand age structures predominate (figure 13.5, zones 2 and 3). This example also illustrates that silvicultural treatments resulting in even-aged or multi-modal age structures could accommodate a temporary road network (zones 1, 2 and 4), but that the presence of silvicultural systems results in uneven-aged structures, such as the selection system, that require the use of a permanent road network (zone 3). The silvicultural strategies that are used in this example are summarized in table 13.3. It must be stressed that the silvicultural prescriptions that are suggested here are very simplistic, and should be detailed by taking into consideration precise site factors and stand characteristics.

**Table 13.3**

**Examples of silvicultural strategies used in different zones of a fictive management area (figure 13.4) according to a planning period of 80 years**

| Zone | Road network | Interval between interventions | Allowed silvicultural treatment* | Resulting age structure |
|------|--------------|-------------------------------|----------------------------------|--------------------------|
| 1 | Temporary | 80 years | CC | Even-aged |
| 2 | Temporary | 40 years | CC / PC | Even-aged, multi-modal |
| 3 | Permanent | 20 years | CC / PC / SC | Even-aged, multi-modal, uneven-aged |
| 4 | Temporary | 80 years | CC | Even-aged |

* CC = clearcut, PC = severe partial cut, SC = selection cut.

## 5. CONCLUSION

Even if forest ecosystem management features the use of several silvicultural practices that are currently seldom used in North American boreal forests (partial cuts, prescribed burning), it is important to underline that its originality does not stem from the formulation of novel silvicultural treatments, but rather from a redirecting of silvicultural treatments and systems to regenerate, educate, and harvest forest stands in order to emulate or maintain some natural processes and key attributes of the forest ecosystem, particularly with respect to biodiversity. This redirection takes into account the temporal dimension, in order to constitute silvicultural systems that favour compositional and structural changes corresponding to natural forest dynamics. The treatments are also spatially distributed in order to better address ecological issues such as the rarefaction of old-growth forest landscapes, compositional changes and management of the road network. Finally, at the local level, silvicultural treatments should be modified to increase structural and compositional diversity within stands, and to keep some forest attributes that are important for biodiversity, such as dying trees and dead wood.

## REFERENCES

Andison, D.W. 2003. Tactical forest planning and landscape design. *In* Towards sustainable management of the boreal forest. *Edited by* P.J. Burton, C. Messier, D.W. Smith, and W.L. Adamowicz. NRC Research Press, Ottawa, Ont., pp. 433–480.

Archibald, D.J., Wiltshire, W.B., Morris, D.M., and Batchelor, B.D. 1997. Forest management guidelines for the protection of the physical environment. Ontario Ministry of Natural Resources, Toronto, Ont.

Belle-Isle, J. and Kneeshaw, D. 2007. A stand and landscape comparison of the effects of a spruce budworm (*Choristoneura fumiferana* (Clem.)) outbreak to the combined effects of harvesting and thinning on forest structure. For. Ecol. Manag. **246**: 163–174.

Belleau, A., Bergeron, Y., Leduc, A., Gauthier, S., and Fall, A. 2007. Using spatially explicit simulations to explore size distribution and spacing of regeneration areas produced by wildfires: recommendations for designing harvest agglomerations for the Canadian boreal forest. For. Chron. **83**: 72–83.

Bergeron, Y. 2000. Species and stand dynamics in the mixed woods of Quebec's southern boreal forest. Ecology, **81**: 1500–1516.

Bergeron, Y. and Harvey, B. 1997. Basing silviculture on natural ecosystem dynamics: an approach applied to the southern boreal mixedwood forest of Quebec. For. Ecol. Manag. **92**: 235–242.

Bergeron, Y., Harvey, B., Leduc, A., and Gauthier, S. 1999. Forest management guidelines based on natural disturbance dynamics: stand- and forest-level considerations. For. Chron. **75**: 49–54.

Bergeron, Y., Leduc, A., Harvey, B., and Gauthier, S. 2002. Natural fire regime: a guide for the sustainable management of the Canadian boreal forest. Silva Fenn. **36**: 81–95.

Bouchard, M., Kneeshaw, D., and Bergeron, Y. 2006a. Forest dynamics after successive spruce budworm outbreaks in mixedwood forests. Ecology, **87**: 2319–2329.

Bouchard, M., Kneeshaw, D., and Bergeron, Y. 2006b. Tree recruitment pulses and long-term species coexistence in mixed forests of western Quebec. Écoscience, **13**: 82–88.

Boucher, Y., Arseneault, D., and Sirois, L. 2006. Logging-induced change (1930–2002) of a preindustrial landscape at the northern range limit of northern hardwoods, eastern Canada. Can. J. For. Res. **36**: 505–517.

Bourgeois, L., Kneeshaw, D., and Boisseau, D. 2005. Les routes forestières au Québec: Les impacts environne-mentaux, sociaux et économiques. Vertigo, **6**: 164–182.

Bourgeois, L., Messier, C., and Brais, S. 2004. Mountain maple and balsam fir early response to partial and clear-cut harvesting under aspen stands of northern Quebec. Can. J. For. Res. **234**: 1049–1059.

Courtois, R., Ouellet, J.-P., Dussault, C., and Gingras, A. 2004. Forest management guidelines for forest-dwelling caribou in Québec. For. Chron. **80**: 598–607.

D'Aoust, V., Kneeshaw, D., and Bergeron, Y. 2004. Characterization of canopy openness before and after a spruce budworm outbreak in the southern boreal forest. Can. J. For. Res. **34**: 339–352.

Davis, L.S., Johnson, K.N., Bettinger, P.S., and Howard, T.E. 2001. Forest management: to sustain ecological, economic, and social values. McGraw-Hill, New York, USA.

De Grandpré, L., Simard, M., and Gauthier, S. 2004. Dynamique de développement de peuplements résineux dans la forêt boréale de l'Est du Québec: une analyse dendroécologique. Rapport remis au Fonds Forestier, Direction de la recherche forestière, Ministère des Ressources naturelles, de la Faune et des Parcs, Québec, Que.

Duchesne, L. and Raulier, F. 2004. Description sommaire et analyse critique de l'approche de modélisation de l'aménagement forestier pour fins d'évaluation de la possibilité forestière. *In* Rapport détaillé du comité chargé d'examiner le calcul de possibilité forestière. *Edited by* R. Jobidon. Ministère des Ressources naturelles, de la Faune et des Parcs, Québec, Que, pp. 5–49.

Franklin, J.F., Berg, D.R., Thornburgh, D.A., and Tappeiner, J.C. 1997. Alternative silvicultural approaches to timber harvesting: variable retention harvest systems. *In* Creating a forestry for the 21st century. *Edited by* K.A. Kohm and J.F. Franklin. Island Press, Washington, D.C., USA, pp. 111–139.

Gastaldello, P., Ruel, J.-C., and Paré, D. 2007. Micro-variations in yellow birch (*Betula alleghaniensis*) growth conditions after patch scarification. For. Ecol. Manag. **238**: 244–248.

Gauthier, S., Nguyen, T.-X., Bergeron, Y., Leduc, A., Drapeau, P., and Grondin, P. 2004. Developing forest management strategies based on fire regimes in northwestern Quebec, Canada. *In* Emulating natural forest landscape disturbances. *Edited by* A.H. Perera, L.J. Buse, and M.C. Weber. Columbia University Press, New York, USA, pp. 219–229.

Gouvernement du Québec. 2007. Regulation respecting standards of forest management for forests in the domain of the State, R.S.Q. c. F-4.1, r.1.001. Replaced, D. 498-96, 1996 G.O. 2, 2570; eff. 96-05-23; see c. F-4.1, r. 1.001.1. Éditeur officiel du Québec, Que.

Grondin, P. and Cimon, A. 2003. Les enjeux de biodiversité relatifs à la composition forestière. Ministère des Ressources naturelles, de la Faune et des Parcs, Gouvernement du Québec, Québec, Que.

Groot, A. and Horton, B.J. 1994. Age and size structure of natural and second growth peatland *Picea mariana* stands. Can. J. For. Res. **24**: 225–233.

Groot, A., Lussier, J.-M., Mitchell, A.K., and MacIsaac, D.A. 2005. A silvicultural systems perspective on changing Canadian forestry practices. For. Chron. **81**: 50–55.

Harmon, M.E., Franklin, J.F., Swanson, F.J., Sollins, P., Gregory, S.V., Lattin, J.D., Anderson, N.H., Cline, S.P., Aumen, N.G., Sedell, J.R., Lienkaemper G.W., Cromack, K.J., and Cummins, K.W. 1986. Ecology of coarse woody debris in temperate ecosystems. Adv. Ecol. Res. **15**: 133–302.

Harvey, B.D., Leduc, A., Gauthier, S., and Bergeron, Y. 2002. Stand-landscape integration in natural disturbance-based management of the southern boreal forest. For. Ecol. Manag. **155**: 369–385.

Harvey, B.D., Nguyen-Xuan, T., Bergeron, Y., Gauthier, S., and Leduc, A. 2003. Forest management planning based on natural disturbance and forest dynamics. *In* Towards sustainable management of the boreal forest. *Edited by* P.J. Burton, C. Messier, D.W. Smith, and W.L. Adamowicz. NRC Research Press, Ottawa, Ont., pp. 395–432.

Houle, G. and Payette, S. 1990. Seed dynamics of *Betula alleghaniensis* in a deciduous forest in northeastern North America. J. Ecol. **78**: 677–690.

Hunter, M.L. Jr. (Ed.). 1999. Maintaining biodiversity in forest ecosystems. Cambridge University Press, Cambridge, UK.

Jamnick, M.S. and Walters, K.R. 1993. Spatial and temporal allocation of stratum-based harvest schedules. Can J. For. Res. **23**: 402–413.

Jasinski, J.P.P. and S. Payette. 2005. The creation of alternative stable states in the southern boreal forest, Québec, Canada. Ecol. Monogr. 75: 561–583.

Kafka, V., Gauthier, S., and Bergeron, Y. 2001. Fire impacts and crowning in the boreal forest: study of a large wildfire in western Quebec. Int. J. Wildland Fire, **10**: 119–127.

Kneeshaw, D.D., and Bergeron, Y. 1998. Canopy gap characteristics and tree replacement in the southeastern boreal forest. Ecology, **79**: 783–794.

Lavoie, M., Paré, D., Fenton, N., Groot, A., and Taylor, K. 2005. Paludification and management of forested peatlands in Canada: a literature review. Environ. Rev. **13**: 21–50.

Leblanc, M. and Bélanger, L. 1998. La sapinière vierge de la Forêt Montmorency et de sa région: une forêt boréale distincte. Ministère des Ressources naturelles du Québec, Mémoire de recherche forestière #136.

Lieffers, V.J., Messier, C., Burton, P.J., Ruel, J.-C., and Grover, B.E. 2003. Nature-based silviculture for sustaining a variety of boreal forest values. *In* Towards sustainable management of the boreal forest. *Edited by* P.J. Burton, C. Messier, D.W. Smith, and W.L. Adamowicz. NRC Research Press, Ottawa, Ont., pp. 481–530.

Lortie, M. 1979. Arbres, forêts et perturbations naturelles au Québec. Les Presses de l'Université Laval, Sainte-Foy, Que.

MacDonald, G.B. 1995. The case for boreal mixedwood management: an Ontario perspective. For. Chron. **71**: 725–734.

MacLean, D.A. and Ostaff, D.P. 1989. Patterns of balsam fir mortality caused by an uncontrolled spruce budworm outbreak. Can. J. For. Res. **19**: 1087–1095.

Morin, H. 1994. Dynamics of balsam fir forests in relation to spruce budworm outbreaks in the boreal zone of Quebec. Can. J. For. Res. **24**: 730–741.

Nguyen-Xuan, T., Bergeron,Y., Simard, D., Fyles, J., and Paré, D. 2000. The importance of forest floor disturbance in the early regeneration patterns of the boreal forest of western and central Quebec: a wildfire versus logging comparison. Can. J. For. Res. **30**: 1353–1364.

Nyland, R.D. 1996. Silviculture: concepts and applications. McGraw-Hill, Inc., New York, USA.

O'Hara, K.L. 2001. The silviculture of transformation: a commentary. For. Ecol. Manag. **151**: 81–86.

Oliver, C.D. and Larson, B.C. 1996. Forest stand dynamics. John Wiley & Sons, New York, USA.

Pham, A., De Grandpré, L., Gauthier S., and Bergeron, Y. 2004. Gap dynamics and replacement patterns in gaps of the northeastern boreal forest of Quebec. Can. J. For. Res. **34**: 353–364.

Richards, E.W. and Gunn, E.A. 2000. A model and Tabu search method to optimize stand harvest and road construction schedules. For. Sci. **46**: 188–203.

Rose, C.R. and Muir, P.S. 1997. Green-tree retention: consequences for timber production in forests of the western Cascades, Oregon. Ecol. Appl. **7**: 209–217.

Ruel, J.-C. and Pineau, M. 2003. Windthrow as an important process for white spruce regeneration. For. Chron **78**: 732–738.

Ruel, J.-C., Roy, V., Lussier, J.-M., Pothier, D., Meek, P., and Fortin, D. 2007. Mise au point d'une sylviculture adaptée à la forêt boréale irrégulière. For. Chron. **83**: 367–374.

Schütz, J.P. 1997. Sylviculture 2: La gestion des forêts irrégulières et mélangées. Presses Polytechniques et Universitaires Romandes, Lausanne, Switzerland.

Schütz, J.P. 2001. Opportunities and strategies of transforming regular forests to irregular forests. For. Ecol. Manag. **151**: 87–94.

Smith D.M., Larson, B.C., Kelty, M.J., and Ashton, P.M.S. 1997. The practice of silviculture: applied forest ecology. 9th edition. Wiley, New York, USA.

Sturtevant, B.R., Bissonette, J.A., Long, J.M., and Roberts, D.W. 1997. Coarse woody debris as a function of age, stand structure, and disturbance in boreal Newfoundland. Ecol. Appl. **7**: 702–712.

Thiffault, N. and Jobidon, R. 2006. How to shift unproductive *Kalmia angustifolia – Rhododendron groenlandicum* heath to productive conifer plantation. Can. J. For. Res. **36**: 2364–2376.

Thompson, I.D. and Harestad, A.S. 2004. The ecological and genetic basis for emulating natural disturbance in forest management: theory guiding practice. *In* Emulating natural forest landscape disturbances. *Edited by* A.H. Perera, L.J. Buse, and M.C. Weber. Columbia University Press, New York, USA, pp. 29–42.

Thorpe, H.C. and Thomas, S.C. 2007. Partial harvesting in the Canadian boreal: success will depend on stand dynamic responses. For. Chron. **83**: 319–325.

# Chapter 14

# An Adaptive Framework for Monitoring Ecosystem Management in the Boreal Black Spruce Forest*

*Pierre Drapeau, Alain Leduc, Daniel Kneeshaw, and Sylvie Gauthier*

\* We thank the numerous graduate students and colleagues whose contributions to stimulating discussions have helped shape our reflections on ecosystem management. Particular thanks go out to Virginie Arielle Angers, Yves Bergeron, Suzanne Brais, Catherine Boudreault, Chris Buddle, Marianne Cheveau, Louis De Grandpré, Réjean Deschênes, Karen Harper, Brian Harvey, Christian Hébert, Louis Imbeau, Jean-Pierre Jetté, Émilie Lantin, Patrick Lefort, Antoine Nappi, Marcel Paré, Michel Saint-Germain, Jean-Pierre Savard, Marie-Andrée Vaillancourt, and Tim Work. We would equally like to thank our contributing editors, Pierre Larue and Rhéaume Courtois, for their pertinent and constructive commentary, which served to greatly improve the final version of this chapter. This chapter was made possible by the financial support of the Sustainable Forest Management Network (SFMN), the National Research Council of Canada, the Chaire industrielle CRSNG UQAT-UQAM en aménagement forestier durable, the Fonds FQRNT – Programme action concertée du Gouvernement du Québec, the Canadian Wildlife Service, and the Canadian Forest Service.

# 1. INTRODUCTION

The development of forest management strategies that strive to maintain the processes, structures, and functions of forest ecosystems, all the while permitting the commercial extraction of wood, is an idea that has been widely debated internationally over the past fifteen years (Hunter 1990; Franklin 1993; Attiwill 1994; Haila et al. 1994; Angelstam 1998; Bergeron et al. 1999; Harvey et al. 2002). Forest Ecosystem Management (FEM) is founded on the idea of reproducing on managed landscapes the range of forested conditions present under natural disturbance regimes within the limits of historic variation of these forests (see Gauthier et al., chapter 1). Its premise is that species inhabiting such ecosystems are adapted to the range of natural disturbances that may take place and hence to the extent of resulting forest conditions. In this way, a central tenet of FEM is that landscapes managed such that they closely embody natural forest conditions should mitigate the negative impacts of conventional forestry on ecosystem biodiversity, though this hypothesis remains to be tested.

A monitoring program is a prerequisite for assessing the outcome of management practices and its uncertainties. This requires an approach comprising both 1) a measure of the degree to which site-level management interventions meet ecosystem-based management targets and 2) a measure of the capacity of these targets to attain their ultimate objective, i.e. the maintenance of biological diversity on managed landscapes. The rationale behind such a program was presented by Kneeshaw et al. (2000). In parallel, the concepts of *effectiveness monitoring* and *validation monitoring* introduced by Mulder et al. (1999) were recently adopted by Rempel et al. (2004), who speak of evaluative and prescriptive indicators. Those associated with monitoring the implementation of management targets (*effectiveness* or evaluative) determine if management plan objectives were attained. Indicators developed to measure the capacity of management targets to maintain biological diversity (*validation* or prescriptive) verify the hypothesis that a relationship exists between the management action and its effect on biological diversity. This type of monitoring program falls intrinsically within an adaptive management framework as it is subject to a process of continual improvement (*sensu* Holling 1978; Walters 1997; Wilhere 2002). Knowledge acquired through the use of such indicators (organisms or ecological processes) can lead to the refinement of management actions to maintain required forest conditions.

In this chapter, we propose an approach to ecosystem management monitoring which can be implemented at the operational level of the forest management unit (*sensu* Kneeshaw et al. 2000). The method identifies the principal management targets to achieve and proposes indicators measuring the capacity of FEM targets to maintain biodiversity in managed forests. In the context of such a monitoring approach, FEM is considered an experiment which, at the landscape scale, is oriented toward specific questions where management targets (e.g. proportion of mature forest, abundance of critical habitat attributes), and indicators related to the capacity of these targets to maintain biological diversity, are defined. The indicators are an integral part of an iterative process of knowledge acquisition pertaining to ecosystems and their components (including species) which may lead to a redefining of original management targets.

## 2. MANAGEMENT TARGETS AS THE BASIS OF A MONITORING PROGRAM

The first stage of monitoring is to determine the extent to which FEM targets have been met on the ground. Though conventional even-aged forestry may in part be analogous to forest fire effects (Bergeron et al. 2002), the reconstruction of natural disturbance regimes has clearly shown that at the scale of the forest management unit (several thousand square kilometres), it does not reproduce the full range of conditions generated by natural disturbances. At this scale, the maintenance of older forests or those containing structural attributes inherent to such forests is now recognized as the primary condition to be met in order to bridge the gap between natural and managed forests (Spies et al. 1994; Bergeron et al. 1999, 2002; Seymour and Hunter 1999). This becomes evident given that natural forested landscapes are older on average than managed forested landscapes (Bergeron et al. 2001; Harper et al. 2002; Cyr et al. 2009). Various strategies have been proposed to alleviate this situation (see Gauthier et al., chapter 3 for more details); in general they involve lengthening rotation age on part of the landscape or varying harvesting practices, notably by favouring partial cuts, which remove part of the forest cover while conserving structural attributes (dead wood, large stems) characteristic of older forests (Bergeron et al. 1999, 2002, 2007; Bergeron 2004). In accordance with fire regimes, the amount of structurally "old" forest desired may vary geographically (Bergeron et al. 2004a; Gauthier et al., chapter 3).

In an operational context, Gauthier et al. (2004) proposed a forest management plan for an area of northwestern Quebec (Matagami ecoregion) involving clearcutting and partial cutting in such a way as to maintain the naturally occurring variety of forest types in the proportions in which they would be found under natural fire regimes within that region (Bergeron et al. 2004b). This example illustrates how the diverse array of forest conditions needing to be maintained under a FEM regime may be spatially distributed in a real applied setting. By definition a proportion of the territory is subject to clearcutting and, consequently, consists of regenerating areas meant to reproduce conditions similar to those that would be caused by a severe disturbance event. Clearcuts should be located within a larger forest matrix whose imperative is to ensure a continuum of forest cover. This matrix may be subject to partial cutting practices insofar as it assures a continuity of forest cover that is representative of older forests. Both partial and clearcutting treatments present particular issues which fall within three strategic objectives: 1) assuring the quality and configuration of residual forest in areas subject to clearcutting, 2) maintaining continuity in the availability of key habitat attributes in areas subject to partial cutting, and 3) assuring connectivity between the two types of harvesting areas at the scale of the forest management unit.

## 2.1.  Quality and Configuration of Residual Forest in Clearcut Areas

Several studies have shown that within areas affected by wildfire, a significant proportion of living forest remains in the form of residual "islands" that escaped fire and are embedded within a surrounding matrix of less severely burned forest with a significant proportion of live trees (Turner and Romme 1994; Kafka et al. 2001; Bergeron et al. 2002; Perron et al., chapter 5). For the boreal forest of Québec, Leduc et al. (2000) determined that 5 to 18% of burned areas are spared in the form of small islands of unburned residual forest. In the Saguenay–Lac-Saint-Jean region, a study by Perron (2003) estimated that burned landscapes in the boreal forest consist of 10 to 35% surviving residual forest (from isolated trees to remnant patches of various size), 1 to 8% of which consist of island remnants. For a series of fires that took place in 1995 and 1996, Bergeron et al. (2002) found that 30 to 50% of total burned area was either mildly affected or spared entirely by fire, a finding which is by no means negligible.

Retention strategies currently in place fail to adequately integrate the complex reality of residual forests within fire-burned landscapes (Cissell et al. 1999). Bergeron et al. (2007) point out that aside from riparian buffers (permanent retention in the case of partial cutting) and temporary cutblock separators (harvested when adjacent regeneration attains a height of 3 metres), current forest management practices in Québec produce remnant forest conditions quite different from naturally burned landscapes. With this in mind, in clearcut areas, new retention targets should be established which strive to maintain similar proportions of live trees, clusters of live trees, and unburned islands as are found naturally after fire (see Bouchard, chapter 13). Guidelines for retention within clearcut areas should address the spatial configuration of residual forest, the size and form of residual patches, and the distance between residual elements (Kafka et al. 2001; OMNR 2001; Perron 2003; Perron et al., chapter 6).

Detailed analysis of the spatial configuration of residual forest spared by a fire near Lake Crochet, Québec, showed that lightly affected zones were distributed throughout the larger burned area (Kafka et al. 2001). The authors noted that the size of unburned residual stands was 52 ha on average and that these areas were often surrounded by lightly burned forest. Finally, Kafka et al. (2001) found that, contrary to common practice in managed forests, the shape of residual forests on naturally burned landscapes is not systematically linear; they are therefore less susceptible to edge effects and comprise a greater quantity of interior habitat than riparian buffers or cutblock separators (Darveau et al. 1995; Ruel et al. 2001; Mascarúa-López et al. 2006; Boudreault et al. 2008). Consequently, although it converts substantial portions of the landscape to an early seral stage of succession, a major disturbance agent like fire nevertheless maintains a certain proportion of mature and older forest within its boundaries. In turn, these residual habitats may serve as biological refuges while the disturbed matrix regenerates (Schieck and Hobson 2000).

In order to assure the maintenance of biological diversity associated with older forests, it is important that residual habitats be able to persist on managed landscapes so that linkages exist between similar habitats, and in particular with large tracts of forest that may act as sources of recruitment (recruitment rate > mortality rate) (Villard et al. 1993; Burke and Nol 2000; Bennett 2003). The significant proportion of living stems left behind by natural wildfire assures an inherent connectivity between residual habitats (Bennett 2003); thus forest management planning should carefully consider the spatial configuration of forest retention. To this effect, cutover areas should include diverse forms of retention (from individual trees to groups of trees or linear strips (riparian corridors)) in order to ensure connectivity within areas subject to clearcutting. The maintenance of such residual habitat networks may secure not only biological refuges but also linkages important to the persistence and reestablishment of some biological populations (Bennett 2003). Species associated with mature forest can be spatially confined by regenerating areas interpreted as "hostile." As conditions become more favourable, these species may disperse more freely, and use of residual habitats increases until such time as the adjacent forest is of sufficient quality to allow population reestablishment. In order for such species to make use of biological refuges, travel corridors and eventually source habitats permitting population reestablishment, residual habitats need to be permanently maintained on managed landscapes. This is presently not the case in Québec, where by virtue of current harvesting regulations (Règlement des normes d'intervention (RNI) dans les forêts du domaine public) residual habitats are to be partially or completely harvested once cutovers have attained 3 metres in height (Gouvernement du Québec 2007a).

The need to develop strategies that guarantee ongoing sources of residual habitats applies equally to mosaic cutting, the checkerboard style of dispersed cutting recently adopted for Québec's Crown lands of boreal forest. Indeed, though "first-pass" cutblocks are required to attain 3m in height before "second-pass" blocks can be harvested, for all intents and purposes the resulting landscape is comparable to that of traditional one-pass clearcuts in terms of canopy openness and dominant seral stage. Therefore issues of permanent retention of remnant habitats are by no means resolved under a mosaic cutting regime. It is often argued with respect to the clearcutting method that remnant habitats are included in calculations of the annual allowable cut (AAC) and therefore should be harvested in order to keep the forest industry viable. It should be noted, however, that the systematic removal of older forest from harvested areas can lead to the local extirpation of numerous species. Such practices could indirectly harm companies in the voluntary process of forest certification, which by necessity entails the maintenance of biological diversity. Ecological arguments aside, the harvest of residual habitats can have important economic impacts given the increasing interest of wood buyers to ecologically certified forest products.

Failure to maintain residual habitats clearly contravenes the principles and criteria of sustainable forest management entrenched in many provincial regulations (Québec's *Forest Act*, Ontario's *Crown Forest Sustainability Act*, etc.). Scientific knowledge on habitat requirements of wildlife species in the boreal forest has greatly advanced in recent years (Darveau et al. 1995, 2001; Schmiegelow et al. 1997; Schieck et al. 1995; Schieck and Hobson2000; Bayne and Hobson 1997, 2001, 2002; Hobson and Schieck 1999; Hobson and Bayne 1999; Drolet et al. 1999; Imbeau et al. 1999, 2001, 2003; Potvin et al. 1999, 2000; Samson and Huot 1999; Drapeau et al. 2000, 2002, 2003, 2005; Hannon et al. 2002; Imbeau and Desrochers 2002; Boulet et al. 2003; Courtois 2003; Nappi et al. 2003, 2004; Potvin and Bertrand 2004; Dussault et al. 2005a, b; Etcheverry et al. 2005a, b; Hannon and Drapeau 2005; Courtois et al. 2007; Simon et al. 2000, 2002). To satisfy the criteria of sustainable forest management, and more particularly that of biodiversity maintenance, new knowledge must be taken into account with up-to-date standards for best practices. With the perspective of adaptive management, one way for legislators to integrate recently acquired scientific knowledge into current policy would be to remove residual habitats from the AAC.

Finally, remnant habitats on post-fire landscapes consist, among other things, of older forest comprising structural attributes (large live trees, standing dead wood and woody debris in diverse stages of decay) critical for an important variety of species (Imbeau et al. 2001; Schmiegelow and Mönkkönen 2002; Drapeau et al. 2003). Consequently, the structural quality of remnant habitats (abundance and variety of dead wood and living stems of large diameter) in managed forests is an important consideration as it can play a functional role in not only species dispersal (Machtans et al. 1996; Hannon and Schmiegelow 2002) but also species reproduction (Imbeau and Desrochers 2002; Leboeuf 2004). This facilitates as much the recolonization of cutovers as it does the long-term maintenance of biodiversity. The quality of remnant forest (composition and structure) is therefore another factor to consider when planning retention in cutblock areas (Scott et al. 2001; Drapeau et al. in press).

## 2.2.   Continuity of Key Habitat Attributes in Partially Cut Areas

Forest management based on even-aged silviculture reproduces only a portion of the natural diversity of stand conditions found in the boreal forest (Gauthier et al. 2004; Gauthier et al., chapter 3). The elimination of heterogeneous or uneven-aged forest structures in favour of uniform stands with little or no dead wood may have serious implications for the maintenance of biodiversity, in particular for those species adapted to a continuous supply of dead wood (Angelstam and Mikusinski 1994; Franklin et al. 2000; Imbeau et al. 2001; Drapeau et al. 2002, 2009; Bütler et al. 2004). Furthermore, silvicultural practices that eliminate trees of large diameter compromise the future recruitment of substrates for large cavity-nesting birds on managed landscapes (Vaillancourt et al. 2008). An important challenge where partial cutting is concerned consists in maintaining continuity of structural habitat attributes associated with older forests. Dead wood recruitment in naturally aging forests can vary from the asynchronous mortality of individual trees to the synchronous mortality of numerous trees (windthrow, insect infestation). This variability results in a wide range of decay stages presenting conditions quite different from those which characterize stands at the age of commercial maturity (Harper et al. 2002, 2003; Périgon 2006).

A fine understanding of these patterns and their underlying dynamics (Kruys et al. 2002; Aakala et al. 2008) will improve harvesting and retention methods in areas subject to partial cutting. It is not merely a question of putting into practice a new arsenal of silvicultural practices (cuts with protection of small merchantable stems, shelterwood cutting, commercial thinning) but of adapting such practices to the reference conditions provided by older forests in order to assure the continued availability of deadwood in these areas, both standing and on the forest floor.

That being said, we still have little knowledge as to the number, quality, and distribution of dead wood that needs to be maintained in order to assure the conservation of species and processes associated with this key habitat attribute (Drapeau et al. in press). Descriptive studies on the distribution of dead wood in irregular uneven-aged forests (see Harper et al. 2003, 2005) offer some preliminary management targets, but such knowledge needs to be integrated into simulation models capable of projecting structural attributes in time in the context of partial cutting scenarios. Sound data on the decay rate of tree species and their transition times from one decay stage to another should serve as a basis for such management scenarios (see Kruys et al. 2002; Aakala et al. 2008).

## 2.3. Connectivity at the Scale of Management Units

The boreal forest, like other ecosystems driven by large-scale disturbances, consists of species tolerant to these important landscape-scale changes in composition and structure of the forest cover (Hunter 1992; Drapeau et al. 2000; Schmiegelow and Mönkkönen 2002). In this ecosystem, the issue for biodiversity may not be anthropogenic landscape fragmentation *per se*, but whether the scale and magnitude of such fragmentation is within the range of variation of natural disturbances, and likewise, within habitat thresholds of species of concern (see Jansson and Angelstam 1999; Angelstam et al. 2003; Radford et al. 2005). At the scale of the forest management unit (from several hundreds to thousands of square kilometres), consideration of the spatial configuration of clearcut areas (clearcut or partial cut areas) becomes essential to the maintenance of ecological functions (e.g. dispersion, colonization, population dynamics) and processes (e.g. trophic interactions, woody debris decomposition and nutrient cycling). The size and spacing of clearcut areas should thus be informed by knowledge of the spatio-temporal distribution of natural disturbances. Dendroecology studies provide information on the size, frequency, and severity of fire events, but they are limited by tree longevity and offer little information about the long-term variability of these fire regime characteristics (Johnson 1992; Bergeron et al. 2001, 2004a, b; Lefort et al. 2003). A promising approach for measuring variability at the scale of the forest management unit is to model natural disturbance behaviour based on the size and frequency of disturbance events (Boychuck et al. 1997; Wimberly et al. 2000).

Belleau et al. (2007) used this approach to evaluate the temporal variation in patterns created by fire throughout the boreal forest of Canada; they then proposed strategic forest management objectives applying to the even-aged portion of FEM units. Basing their models on numerous fire-history studies (Suffling et al. 1982; Foster 1983; Larsen 1997; Weir et al. 2000; Bergeron et al. 2001, 2004; Gauthier et al. 2001; Lesieur et al. 2002; Lefort et al. 2003), they were able to incorporate empirical parameters for fire size and frequency. The results of simulations spanning several hundred years show that the proportions of the territory to treat with clearcutting vary as a function of fire cycle and average fire size (figure 14.1). For example, in order to reproduce conditions similar to fires in northwestern Québec, simulations revealed that the size of regenerating areas could vary between 100 and 1,000 km². Likewise, over a planning horizon of 25 years, the minimal spacing between wildfires varies from 6 to 31 km depending on size (figure 14.2). Hence, clearcut areas should be

**Figure 14.1**
## Estimates of the average proportion of forest less than 100 years on the landscape and its temporal variation for three mean fire size classes

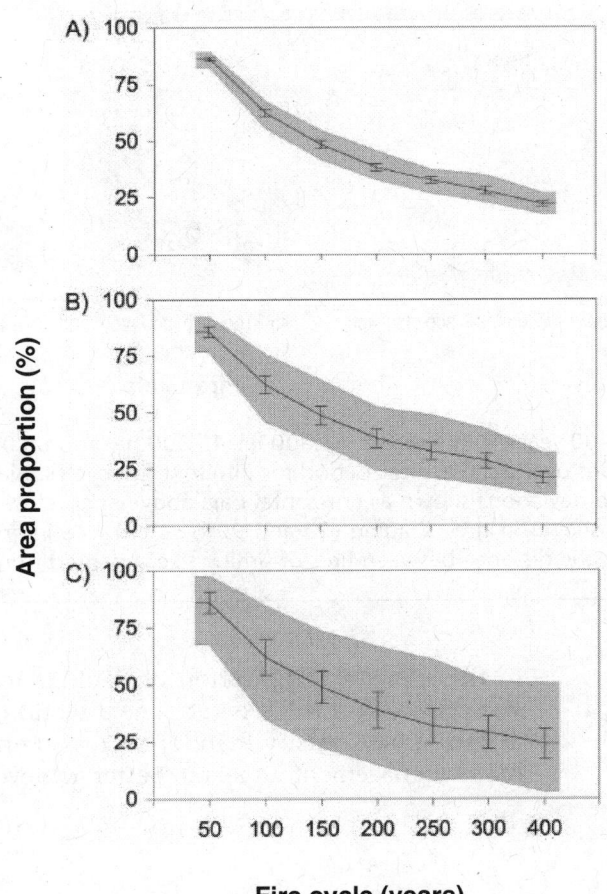

Legend: grey shading: temporal variability; A) 3,000 ha; B) 15,000 ha; C) 60,000 ha mean fire size. These estimations were obtained through simulations over a 1,000-year period for different fire cycles. Note that the proportion is similar regardless of fire size, but the induced variation is much larger for large fires. Adapted from Belleau et al. (2007).

distant from one another within the same range of distance and bordering clearcut areas should be large tracts of mature and older forest. These forests, in turn, could be the object of partial cutting.

Putting such recommendations into practice should return the forest mosaic to a more natural state and thereby attenuate the cumulative negative impacts of large-scale clustering of clearcut blocks on biological diversity (Kouki and Väänänen 2000; Smith et al. 2000; Leboeuf 2004). Although this approach is in a first step ecosystem-based, using wildfires' spatial distribution at a very large scale, it does not exclude the integration of knowledge on organisms' response

**Figure 14.2**
## Estimates of the distance between regenerating areas (≤25 years) for different fire regimes

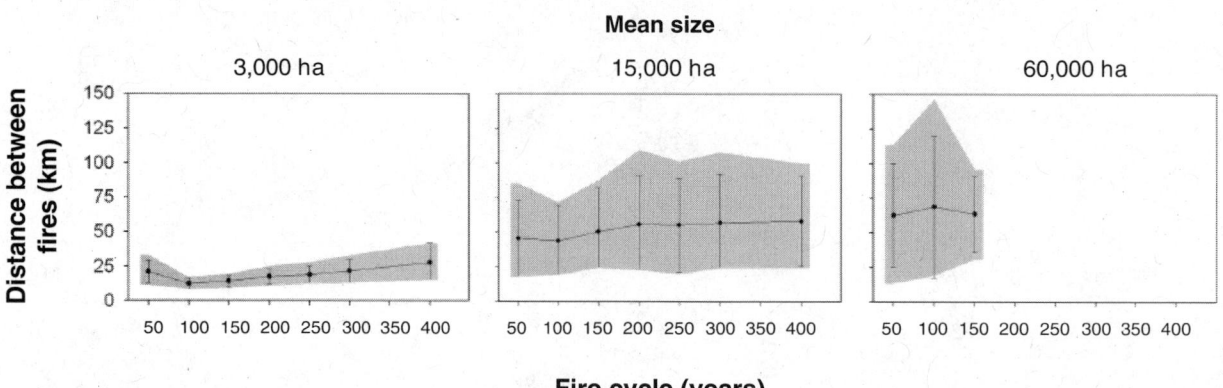

Frequency = 50–400 years. Average size = 3,000 ha, 15,000 ha, and 60,000 ha. These simulations were obtained through simulations over a 1,000-year period for different fire cycles. Black lines represent average distance values, with standard deviations shown as horizontal bars above and below points. Temporal variation surrounding average values is shown in grey shading (10 and 90% confidence intervals). The figure shows that as fire size increases, so does the distance between fires of similar size. Adapted from Belleau et al. (2007).

to the spatial configuration of habitats to guide harvesting areas dispersion strategies. As we will see in the next section, the use of indicators as a means for assessing biodiversity maintenance also enables us to refine these ecosystem-based management targets to better achieve the goals of FEM.

## 3.   INDICATORS FOR ASSESSING BIODIVERSITY MAINTENANCE

The maintenance of forest conditions within the range of natural variation is a coarse filter management strategy (*sensu* Hunter 1990) where the diversity of habitat types present at the scale of the forest management unit is maintained in similar proportions to those under a natural disturbance regime. Evaluating the success of such an approach requires direct measure of organisms' response to management targets (Kneeshaw et al. 2000). For areas subject to clearcutting, this means evaluating the ability of residual habitats to assure the ecological functions (e.g. post-reproduction dispersion, migration, growth, and reproduction) necessary for the maintenance of species associated with these habitats. Within partial harvesting blocks, evaluation should be based on the capacity of residual forest structure (e.g. canopy remaining, large trees both dead and living) to provide quality habitat for associated organisms.

Since the second half of the 1990s, numerous publications have taken on the question of which biodiversity indicators to adopt in the environmental monitoring of sustainable forest management practices and strategies (Block et al.

1995; McClaren et al. 1998; Lindenmayer et al. 1999; Noss 1999; Carignan and Villard 2002; Rempel et al. 2004; Thompson 2006). In Québec, an interministerial Integrated Resource Management (IRM) project was put in place in 1991; its goal among others was to use the habitat requirements of a suite of species associated with diverse forest types over a complete range of successional stages to develop Habitat Suitability Indices (HSI) (Blanchette and Ostiguy 1996). Both game and non-game species were used to evaluate the potential changes to their habitat caused by harvest-induced transformation of the forested landscape. Following the example of McLaren et al. (1998) for Ontario, indicator species designated to monitor the state of Québec's forests in the context of IRM were selected at the scale of the dominant forest regions of the province (balsam fir–white birch, balsam fir–yellow birch, and black spruce–feather moss bioclimatic domains). As reported by Thompson (2006), the rationale for selection of these biological indicators was based on their natural history, habitat associations, and scales of spatial and temporal response as well as considerations of technical and economic feasibility with regards to implementing monitoring programs (Blanchette and Ostiguy 1996; McClaren et al. 1998; Crête 2002). An important additional issue with the selection of indicators is how these can be used at the operational scale of forest management units, which can cover areas of over 1,000 km$^2$ (Kneeshaw et al. 2000).

In parallel with the selection of indicator species, much work has been done on the concepts of keystone species and umbrella species, which have been criticized for using a limited number of species to make generalized conclusions about the state of the environment and its biological diversity (Frankel and Soulé 1981; Wilcox 1984; Landres et al. 1988; Mills et al. 1993; Lambeck 1997; Fleishman et al. 2000; Roberge and Angelstam 2004). Recently, a shift to a more multi-species approach (Block et al. 1995; Roberge and Angelstam 2004; Huggett 2005) centered on groups of species sensitive to habitat alteration was proposed (Lambeck 1997; Imbeau et al. 2001; Drapeau et al. 2003, 2009; Guénette and Villard 2005). This latter approach implies that the evaluation of the capacity of target forest conditions to maintain biodiversity relies on the functional response of such groups of sensitive species.

Up until now, indicators have been used to monitor the general state of change in forested environments rather than actually gauging the success of a given management approach at an operational level. In what follows, we propose indicators which, by their response to habitat conditions, allow us to evaluate the capacity of FEM to maintain elements important for the biological diversity in boreal forests (table 14.1). For evaluating how the FEM approach is able to maintain biodiversity, the choice of functional species groups should be based on species 1) who exhibit a strong and quick response to changes in forested landscapes and who are most affected by forestry activities, 2) that collectively imply coverage of the full range of spatial and temporal scales impacted by landscape modification, and 3) for which a significant body of ecological knowledge exists.

We base our approach on five taxonomic groups of organisms: 1) the forest-dwelling woodland caribou (*Rangifer tarandus caribou*), a threatened ungulate of the boreal forest whose home range more or less matches the scale of the forest management unit (several hundreds if not thousands of km$^2$), 2) the American

Table 14.1
## Proposed multi-scale biodiversity indicators for a monitoring program that assesses the effectiveness of several FEM targets

| Indicators | Response Scale | FEM targets examined |
|---|---|---|
| Woodland caribou | Forest Management Unit (hundreds to thousands of km²) | – Persistence of large older forest tracks at FMU scale<br>– Connectivity between clearcut areas and partial cut areas |
| American marten | Landscape (tens to hundreds of hectares) | – Amount and spatial organization of remnant forests within clearcut areas<br>– Habitat quality (forest cover and key structural attributes) in partial cut areas |
| Forest birds and deadwood associates | Landscape, Stand (several hectares to tens of hectares) | – Habitat quality within remnants in clearcut areas<br>– Forest cover within partial cut areas<br>– Amount and Quality of standing deadwood in both partial and clearcut areas |
| Epiphytic lichens | Stand | – Edge effects and Interior habitat conditions in both remnants of clearcut areas and standing forest cover in partial cut areas |
| Saproxylic insects | Stand, tree | – Amount and Quality of standing and downed deadwood |

marten (*Martes americana*), a medium-sized carnivore with a strategic role in the trophic chain of the boreal ecosystem, whose home range can vary from several hundred to a few thousand hectares, 3) bird species associated with older forests (>100 yrs) or with dead wood (representing habitat conditions expressed at the scale of several hectares to several tens of hectares), 4) epiphytic lichens, which are sensitive to openings in the forest canopy and therefore to the configuration of residual habitat at scales varying from several tens to several hundred square metres, and 5) saproxylic insects (associated with dead wood) which act at the scale of the individual tree as a function of its decomposition stage. For the boreal forest, many studies have shown these taxonomic groups to be good indicators of changes in forest cover (Chapin et al. 1998; Drapeau et al. 2000; Potvin et al. 2000; Imbeau et al. 2001, 2003; Courtois 2003; Drapeau et al. 2003; Nappi et al. 2003; Rheault et al. 2003; Jacobs et al. 2007; Saint-Germain et al. 2007a, b, c; Boudreault et al. 2008).

### 3.1.   Woodland Caribou

The woodland caribou is generally present in small herds throughout the boreal forest as well as in mountainous parts of both eastern and western Canada. It is highly vulnerable to habitat modification and suffers increased mortality in the presence of natural or anthropogenic disturbance (Courtois et al. 2007). It is classified as a species at risk throughout its range, in large part due to habitat loss and fragmentation associated with human activities (Thomas and Gray 2002). A national caribou recovery strategy is currently in place (COSEWIC 2002).

In Québec, woodland caribou populations have declined over the course of the last few decades to the point that they were designated as vulnerable in March of 2005 under *An Act respecting threatened or vulnerable species* (Gouvernement du Québec 2007b). A vulnerable species is a species which is likely to become endangered unless circumstances threatening its survival and reproduction improve. A woodland caribou recovery committee has been working on a plan for Québec's woodland caribou. The committee will soon propose a strategy for the species founded on the conservation of a network of large forested areas currently being used by caribou within a broader sector subject to industrial forest management. The strategy will also predicate the adoption of forest management practices which favour the long-term integrity of the boreal forest (Courtois 2003; Courtois et al. 2004). This combination of elements follows the ecosystem approach by diversifying forest harvesting practices (Bergeron et al. 1999; Bergeron 2004) in order to reproduce the full range of natural variability in forest cover at the scale of the forest management unit.

The presence and long-term maintenance of herds in areas subject to partial cutting could reflect the functional capacity of management targets (contiguity and structure of forest cover) to accommodate this species in particular, but also for a suite of other species associated with mature and overmature forest conditions. The woodland caribou could therefore assume the role of an umbrella species (Lambeck 1997; also see Roberge and Angelstam 2004 for an exhaustive review). Theoretically, with greater retention of both living and dead trees than is currently required by Québec's regulations (RNI), presence of woodland caribou in residual habitats (remnant patches, linear strips in cutover areas) should increase. This could in turn allow assessing the functional capacity of such landscapes to enable dispersion and habitat use by caribou. The maintenance of woodland caribou in FMUs subject to ecosystem management is therefore an indicator of the adequacy of connectivity between partial cut and clearcut areas at a large spatial scale.

## 3.2. American Marten

The American marten has often been identified as an indicator species representing late stages of forest succession (Larue 1992; Thompson 1994; McClaren et al. 1998; Smith and Schaefer 2002). Potvin et al. (2000) studied marten habitat selection using radio-telemetry in a black spruce–dominated area of the Abitibi region of Quebec, 60% of which had been recently cut in an agglomerated fashion typical of RNI standards. Their study revealed significant avoidance of recent cutblocks: marten winter home ranges contained a larger proportion of closed canopy forest (>30 yrs) and a smaller proportion of open regenerating areas than what were randomly available within the larger study area. The authors concluded that martens are unable to tolerate more than 35% of their home range as cutovers (figure 14.3). Hargis et al. (1999) found a similar threshold response for marten in the coniferous forest of Utah and recommended that no more than 25% of landscapes $\geq 9$ km$^2$ in size be comprised of harvested or natural openings combined.

The American marten may therefore be monitored on the basis of the proportion as well as the spatial configuration of residual habitat on managed landscapes. Furthermore, given its smaller home range size, it responds to changes in forest composition and structure at a finer scale than woodland caribou (i.e. several hundred hectares vs. several thousand square kilometers) and is therefore an appropriate meso-level indicator of habitat fragmentation.

Figure 14.3
**Home range distribution of four American martens on a managed landscape in Abitibi, Québec**

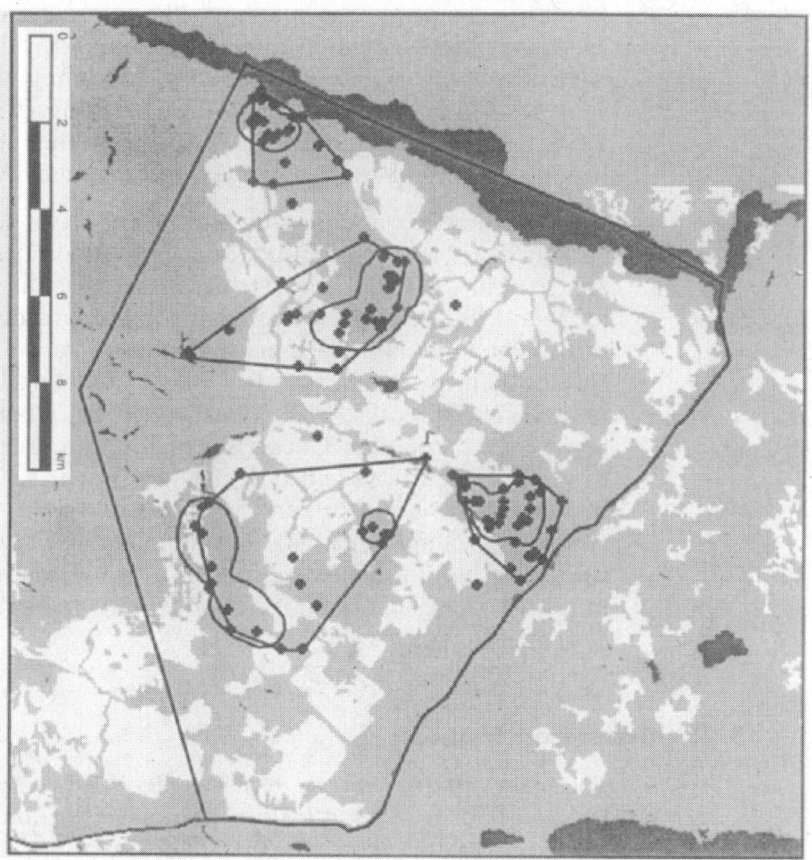

White areas correspond to cutover areas. Note that for the most part, martens occupy forested parts of the managed landscape. Adapted from Potvin et al. (2000).

## 3.3.  Birds Associated with Older Forests and Dead Wood

The composition of bird communities is known to change significantly along a gradient extending from early post-fire stages to those of old-growth forests (Schieck et al. 1995; Imbeau et al. 1999; Drapeau et al. 2000, 2003; Simon et al. 2000). This phenomenon is strongly related to changes in forest structure (height and density of forest canopy) and composition (particularly in mixedwood forests) associated with natural forest succession and the influence of the surrounding landscape. For example, 100–120-year-old stands are propitious for species associated with closed canopy forests, such as the Swainson's Thrush (*Catharus ustulatus*), Golden-crowned Kinglet (*Regulus satrapa*), and the Bay-breasted Warbler (*Dendroica castanea*). As forests age there is a gradual die-off in the first cohort, creating habitat conditions suitable for species that feed on dying and recently dead wood such as the Brown Creeper (*Certhia americana*) and the Red-breasted Nuthatch (*Sitta canadensis*), which reach their highest abundance in old-growth forest conditions (figure 14.4). Coincidentally, Imbeau et al. (1999) identified this collection of species as a cause for concern with regards to even-aged management of black spruce forest in the Saguenay–Lac-Saint-Jean region. Proportional net reduction of old stands thereby impinges on the functional role of these forest types in the convergence of an important assemblage of forest birds. Old forests are dynamic and the process of aging results in a structural complexity that supports numerous niche habitats to which individual species are closely associated. As forest structure increases in complexity it accommodates an increasing number of species, notably those associated with dead woody material (Imbeau et al. 2001; Drapeau et al. 2002).

**Figure 14.4**
## Occupancy rate of the Red-breasted Nuthatch in old forests (>100 years) as a function of the quantity of old forest surrounding sampled sites

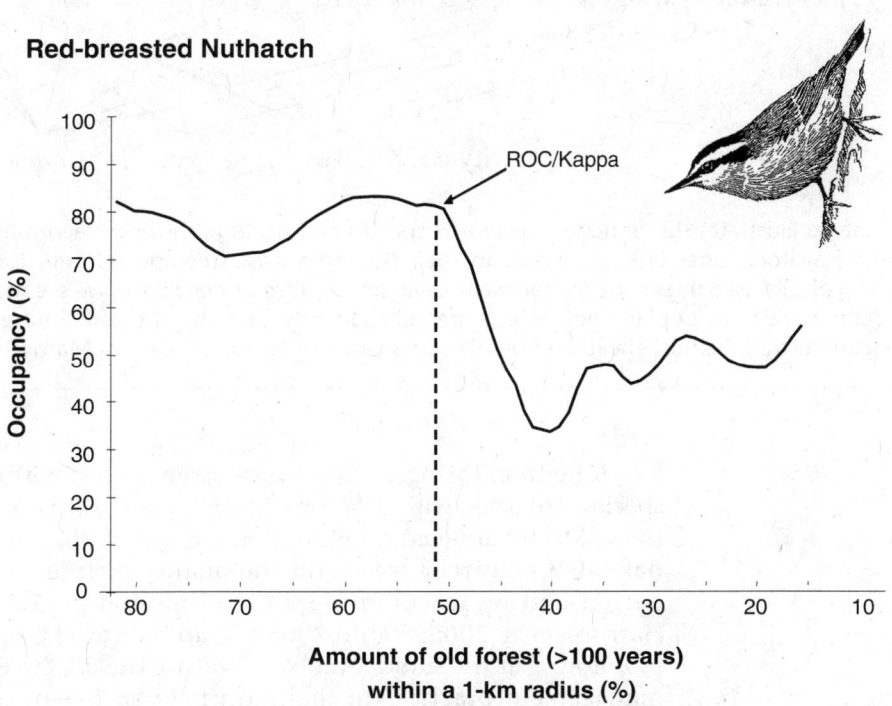

The dotted line indicates a threshold in that quantity where the probability of occurrence this species drops significantly. The threshold was determined via performance analyses (Receiver Operating Characteristics [ROC]; Manel et al. [2001], and the Kappa statistic, Fielding and Bell [1997]) of observed vs. predicted presence-absence values using the logistic model. When the proportion of forest older than 100 years is less than 50%, we note a significant drop in the occurrence rate of this species (P. Drapeau, unpublished data).

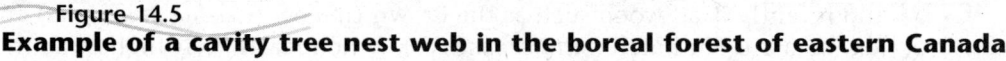

## Figure 14.5
## Example of a cavity tree nest web in the boreal forest of eastern Canada

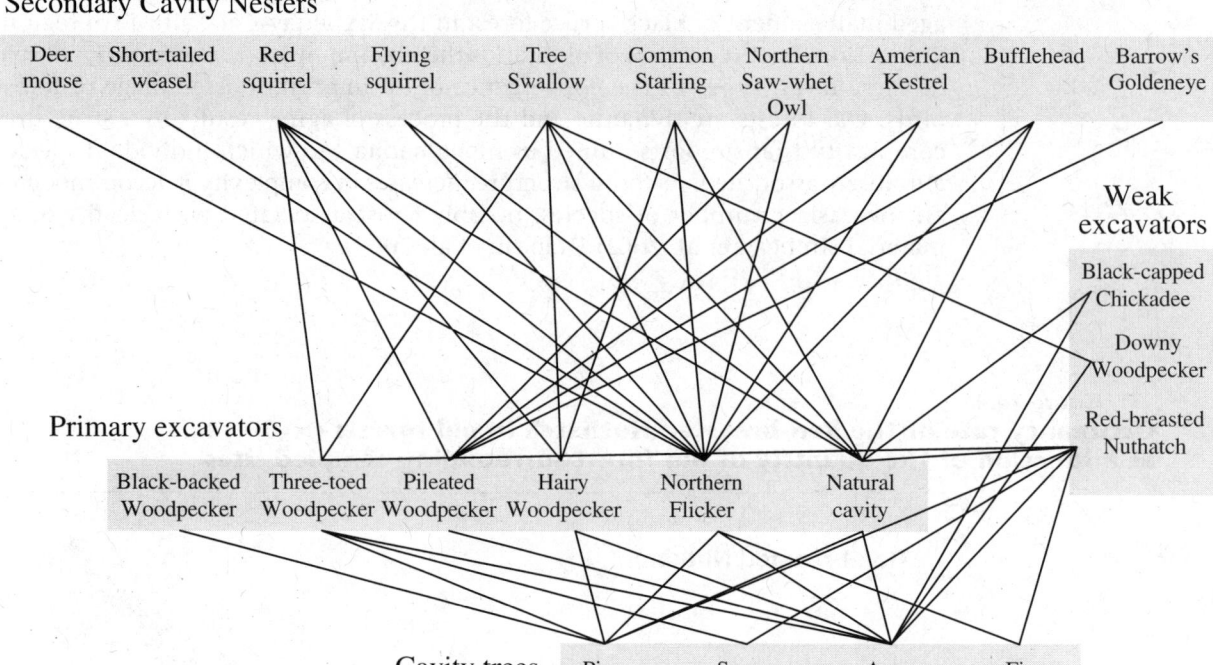

The schema illustrates the principal cavity tree species and distinguishes between primary, secondary, and weak cavity excavators. Lines link excavators to their preferred host tree species and, likewise, illustrate the cavity type(s) typically used by secondary and weak cavity excavators. For example, we see that the Pileated Woodpecker generates cavities in poplar trees, which are subsequently used by the red squirrel, the Saw-whet Owl, the American Kestrel, the Bufflehead and the Barrow's Goldeneye. Adapted from Martin et al. (2004).

The monitoring of forest bird assemblages, with particular emphasis on species associated with dead wood which collectively form nest webs (figure 14.5) (*sensu* Martin and Eadie 1999), should permit us to gauge the quality of residual habitats in clearcut areas and the quality of forest conditions generated by partial cutting in adjoining areas (Simon et al. 2000; Leupin et al. 2004; Harrison et al. 2005; Poulin 2005). As an indicator, it contributes to biodiversity monitoring at the stand level by reflecting the effectiveness of forest ecosystem management practices in maintaining a sufficient quality and quantity of standing dead wood.

### 3.4. Epiphytic Lichens and Edge Effects

In a recent article about monitoring in circumboreal forests, Thompson (2006) noted disapprovingly that biodiversity indicators selected for monitoring programs in the boreal forest of Canada include few invertebrates and nonvascular plants, which in Scandinavian boreal forests have proven to be important indicators of biological diversity (Esseen et al. 1996; Jonsson and Jonsell 1999; Berglund and Jonsson 2001). Recent studies in the boreal forest of Québec have been published which address these target species groups (Boudreault et al. 2002; Saint-Germain et al. 2004a, b, c). Boudreault et al. (2002) documented changes in the distribution of nonvascular plants along a structural and compositional gradient represented by numerous stages of forest succession, from post-fire stands to those that had not burned for over 200 years. In managed black spruce forests of northwestern Québec, Rheault et al. (2003) demonstrated that epiphytic or arboreal lichens are sensitive to edges along interfaces between cutblocks and residual forests (figure 14.6). This is in keeping with what has been documented

## Figure 14.6
## Biomass distribution of three types of epiphytic lichen as a function of distance from forest-cutover interfaces (edges) in clearcut areas of northwestern Québec

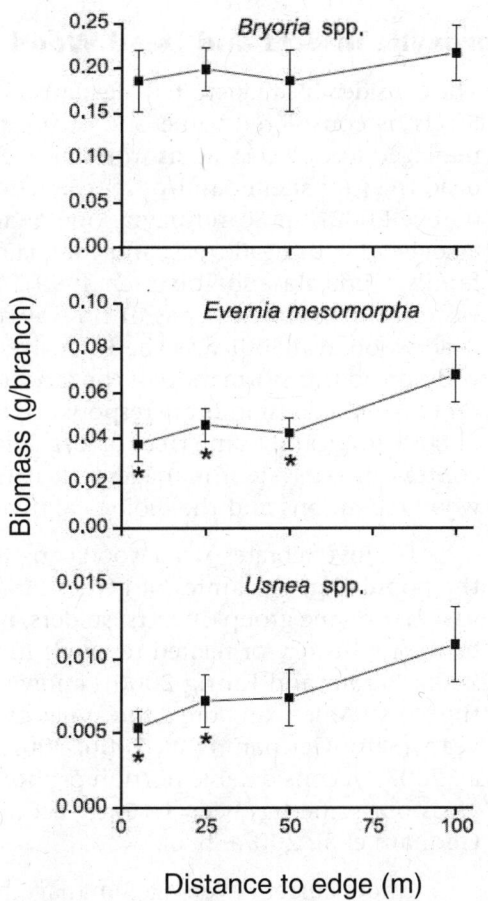

Both *Usnea* spp. and *Evernia mesomorpha* are susceptible to edge effects. Adapted from Rheault et al. (2003).

in Europe (Esseen and Renhorn 1998; Kivistö and Kuusinen 2000; Hilmo and Holien 2002) and elsewhere in North America (Sillett 1994; Campbell and Coxson 2001; Hylander et al. 2005). Within areas characterized by large agglomerated clearcuts, Boudreault et al. (2008) have shown that the integrity of epiphytic lichen communities is at risk if linear corridors are the only form of habitat retention. In order to sustain residual populations of epiphytic lichens in such landscapes, the authors emphasize the need to maintain interior habitat conditions using wider linear strips and non-linear forms of retention (e.g. spheric, square, rectangular, or irregular). This finding corroborates with Mascarúa-López et al. (2006), who concluded that habitat structure in the form of linear retention is significantly different from that found in continuous old-growth forest. Elsewhere, a study currently underway (C. Boudreault, pers. comm.) will allow us to determine whether partial harvesting creates conditions sufficient for the maintenance of epiphytic lichen populations via measures of colonization and biomass accumulation. The degree to which partial cutting maintains closed canopy forest conditions can therefore be evaluated using this functional species group as a biodiversity indicator.

## 3.5.  Saproxylic Insects and Dead Wood

The considerable reduction of dead woody debris, at both the stand and landscape levels, is considered to be one of the principal causes of biodiversity loss in managed forest ecosystems worldwide. Saproxylic insects, which feed on or use dead trees for shelter and/or reproduction, are particularly affected by this loss (Grove 2002). In Scandinavia, one quarter of species listed as endangered are associated with dead trees, and the majority of these belong to the invertebrate family (Virkkala and Toivonen 1999). While extensive forest management in eastern North America has in no way as systematic or important an effect on dead wood availability as the intensive forestry practised in Northern Europe, widespread transformation of the landscape to early seral forest and the concurrent loss of older stands are responsible for a significant reduction in the density of both snags and living trees of large diameter (Vaillancourt et al. 2008). In the context of ecosystem management it is therefore important to monitor dead wood conditions and the biological diversity with which they are associated.

For invertebrates, dead wood constitutes a substrate essential for life, driving the population dynamics of a very large diversity of species belonging to this vast taxonomic group (insects, spiders, millipedes, centipedes). While there may be a long history of related research in boreal forests of Europe (Jonsson et al. 2005; Ranius and Fahrig 2006), knowledge of saproxylic insect distribution in the North American boreal forest has greatly improved over the course of recent years (Saint-Germain et al. 2006, 2007a, b, c; Vanderwel et al. 2006; Jacobs et al. 2007). A considerable portion of this knowledge comes from recently burned areas but some has been recently acquired for post-rotation age forests (Saint-Germain et al. 2007a, b, c).

Under current even-aged management practices, very little dead wood and very few large trees remain after forest operations (Spies et al. 1994; Vaillancourt et al. 2008). In recent years variable retention of live trees and dead wood has been introduced to managed forests of western Canada and the United States

(Franklin et al. 1997; Mitchell and Beese 2002). Such practices have been ratified in the forest management guidelines of certain government agencies (BC Ministry of Forests 1998). With a view to biodiversity conservation, variable retention is intended to maintain a certain proportion of the structural variability that is inherent to irregular or uneven-aged forests in even-aged management systems (*sensu* Bergeron et al. 1999). Québec's current regulations (RNI) contain no provision for the maintenance of pre-harvest forest structure in managed cutblocks. Under RNI standards, retention is not meant to address organisms associated with large trees or dead wood, an essential element in the maintenance of forest biodiversity. Knowledge already acquired on the use of dead wood by saproxylic insect communities in naturally older forests (Jacobs et al. 2007; Saint-Germain et al. 2007b) should serve as a baseline against which comparisons can be made on the use of dead wood in partial cuts and in residual habitats (riparian strips, cutblock separators, or remnant islands) within clearcut areas (Webb 2006; Webb et al. 2008).

When considered collectively, the indicators proposed provide an opportunity to evaluate complementary responses to ecosystem-based dispersion and retention strategies at numerous spatial scales. This monitoring approach allows us to effectively evaluate the influence of ecosystem management practices on the biological diversity of boreal forests.

## 4. ECOSYSTEM MANAGEMENT AND SALVAGE LOGGING IN BURNED FORESTS

In recent years, numerous studies have been conducted in burned parts of the mixed and coniferous boreal forest (Hutto 1995; Saab and Dudley 1998; Schieck and Hobson 2000; Drapeau et al. 2002, 2005; Hoyt and Hannon 2002; Morissette et al. 2002; Nappi et al. 2003, 2004; Saint-Germain et al. 2004a, b; Hannon and Drapeau 2005; Larrivée et al. 2005; Saab et al. 2005; Smucker et al. 2005; Boulanger and Sirois 2007; Koivula and Schmiegelow 2007; Kotliar et al. 2007). These studies have shown that recently burned ecosystems possess certain qualities that are absent in managed areas subject to clearcutting. Among others, the density of standing dead trees of commercial value (>10 cm in diameter) is significantly higher in recent burns than in clearcuts. These trees provide important substrates for saproxylic insect populations (Saint-Germain et al. 2004a) and woodpeckers, their predators (Nappi et al. 2003) (figure 14.7). Concurrently, they tend to be an increasing source of wood supply in the form of salvage logs for the forest industry (Nappi et al. 2004). This reduces the availability of natural attributes associated with burned forests and exercises pressure on that portion of biological diversity which depends on these resources (Hannon and Drapeau 2005). Given the assumption that managed forests must incorporate the maintenance of biodiversity as a sustainable management objective in their planning process, the scientific community raises several concerns as to the validity of pervasive salvage logging in burned forests (Nappi et al. 2004; Lindenmayer et al. 2004; Lindenmayer and Noss 2006; Schmiegelow et al. 2006; Lindenmayer et al. 2008). Recent burns represent an important reference point for comparing the response of boreal forest ecosystems and associated fauna to natural versus

**Figure 14.7**

## The probability of use of dead standing trees by the Black-backed Woodpecker in burned forests as a function of tree diameter and degradation stage

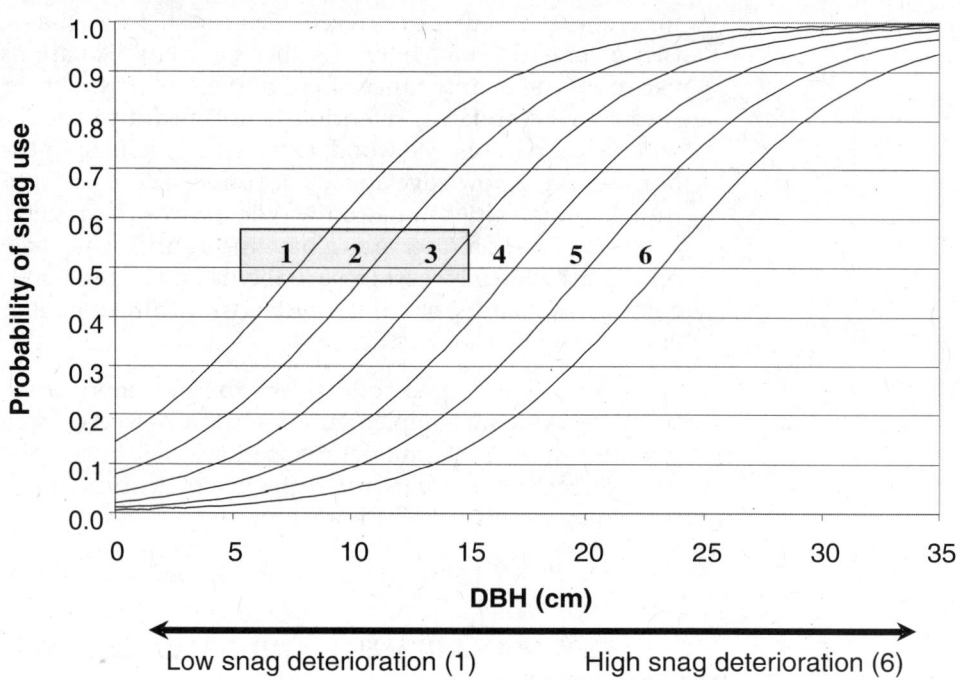

Probability of use for foraging diminishes as tree condition declines. Weakly degraded trees (categories 1 to 3) of large diameter (>15 cm) are most likely to be used as a foraging substrate. Adapted from Nappi et al. (2003).

anthropogenic disturbances (Nappi et al. 2004; Schmiegelow et al. 2006). It would therefore be prudent to maintain a portion of burned landscapes in their natural state in order to permit the maintenance of ecological processes inherent to these ecosystems.

The importance of burned landscapes in natural forest dynamics cannot be understated. As a habitat source for an important component of boreal forest biodiversity, they play a strategic role in the context of FEM, though this necessarily implies major changes to current salvage logging practices. These changes must serve to meet ecosystem management targets that are founded on traits of unmodified burned landscapes, portions of which are to be conserved within the larger managed burned matrix. That portion which is conserved must encompass the natural range of post-fire forest conditions by incorporating previous knowledge on variation in fire severity, fire intensity, and heterogeneity of the resulting forest mosaic. This assessment needs to be made at a much finer scale than what is provided by provincial fire impact maps, which only offer a general idea of the extent of fire damage. Once management targets are established, their effectiveness can be measured via monitoring of species most likely to be affected by post-fire salvage logging (*resource-limited species, sensu* Lambeck 1997), as was recently done by the U.S. Forest Service (Saab and Dudley 1998). Relative success in maintaining biodiversity should therefore be measured using indicator species recognized as being users of dead wood, such as the Black-backed Woodpecker (*Picoides arcticus*) (Hutto 1995; Nappi et al. 2003) or saproxylic insects (Jacobs 2004; Saint-Germain et al. 2004a, b; Boulanger and Sirois 2007).

## 5. CONCLUSION

Implicit to forest ecosystem management is a fundamental change in the way we conceive "management": for here what we plan to leave behind becomes as important as what we propose to harvest. Forest retention (partially cut land-scapes, residual stands in clearcut blocks) must no longer be considered a loss of wood supply but an effective means of maintaining biological diversity. Beyond this matter, it is imperative that we are able to gauge the attainment of management objectives ultimately underlying such an approach. Forest eco-system management (FEM) strategies must therefore be validated through moni-toring programs in which acquired knowledge on indicator behaviour permits us to make quantitative predictions and, if necessary, informed adjustments to management targets.

The monitoring approach that has been proposed in this chapter is in line with what Block et al. (1995) and Kneeshaw et al. (2000) have previously put forward, in that it is founded on both 1) the monitoring of forest ecosystem conditions put in place and 2) the monitoring of elements of biodiversity (species groups) which inherit or are otherwise affected by these conditions, with a view to examining the relationship between management actions and their effect on biological diversity (*sensu* Mulder et al. 1999; Rempel et al. 2004). With regard to choice of indicators, our approach distinguishes itself from others in that selection of species or groups of species is less generic (focusing on species sensi-tive to forest cutting) and more precise, as each indicator is associated with a particular management target at a particular spatial scale. For example, epiphytic lichens allow us to evaluate the quality of residual habitat in terms of the pro-portion that can be considered interior versus edge habitat, particularly within clearcuts. Likewise, relative occurrence of woodland caribou in areas subject to partial cutting serves as an indicator of functional habitat quality. In this sense our approach is in line with Rempel et al. (2004) and Thompson (2006) in that biodiversity monitoring programs must be structured as long-term experiments complete with particular test hypotheses, the results of which permit us to respond in an adaptive way in our choice of forest management actions. Monitoring programs of this nature can lead to the refinement and redefinition of management targets via this dynamic iterative process. To this end, recent developments in the study of ecological thresholds (Andrén et al. 1997; Jansson and Angelstam 1999; Imbeau et al. 2001; Radford and Bennett 2004; Guénette and Villard 2005; Radford et al. 2005; Drapeau et al. 2009) appear promising for setting conservation benchmarks and predicting the response level of organisms to management interventions.

In the absence of naturally disturbed landscapes as a point of comparison, the overall performance of the ecosystem approach on managed landscapes cannot be ascertained. Consequently, zoning on forested landscapes must include an acceptable proportion of protected areas. These "natural controls" serve as reference points which enable us to evaluate the ability of managed landscapes to maintain the structure, functions, and comprising elements (specific diversity) of forest ecosystems. This proves particularly relevant in the context of climate change, as natural forest dynamics are likely to be affected by this phenomenon (Bergeron et al. 2004; also see Hannah et al. 2007; Girardin et al., chapter 4), exacerbating the benchmark function of protected areas as we monitor the differences between natural and managed landscapes. In an ecosystem management perspective, the function of protected areas must be expanded to incorporate their role as the reference conditions on which forest management is based. Viewed in this light, protected areas are, in fact, a critical component of any forest ecosystem management strategy.

In conclusion, the monitoring approach we have proposed in this chapter was developed with an understanding that application of Forest Ecosystem Management (FEM) at the landscape scale is an experiment aligned toward specific questions for which particular management targets are defined (e.g. proportion of mature forest, abundance or density of critical habitat attributes), as are indicators that measure the capacity of these targets to maintain biological diversity. The indicators form an integral part of an iterative process of knowledge acquisition pertaining to ecosystems and their essential components (notably species), which, in the spirit of adaptive management, can lead us to redefine initial management targets in order to better achieve the objectives of FEM.

## REFERENCES

Aakala, T., Kuuluvainen, T., Gauthier, S., and De Grandpré, L. 2008. Standing dead trees and their decay-class dynamics in the northeastern boreal old-growth forests of Quebec. For. Ecol. Manag. 255: 410–420.

Andrén, H., Delin, A., and Seiler, A. 1997. Population response to landscape changes depends on specialization to different landscape elements. Oïkos, 80: 193–196.

Angelstam, P. 1998. Maintaining and restoring biodiversity in European boreal forests by developing natural disturbance regimes. J. Veg. Sci. 9: 593–602.

Angelstam, P., Bütler, R., Lazdinis, M., Mikusinski, G., and Roberge, J.-M. 2003. Habitat thresholds for focal species at multiple scales and forest biodiversity conservation: dead wood as an example. Ann. Zool. Fenn. 40: 473–482.

Angelstam, P. and Mikusinski, G. 1994. Woodpecker assemblages in natural and managed boreal and hemiboreal forest: a review. Ann. Zool. Fenn. 31: 157–172.

Attiwill, P.M. 1994. The disturbance of forest ecosystems: the ecological basis for conservative management. For. Ecol. Manag. 63: 247–300.

Bayne, E.M. and Hobson, K.A. 1997. Comparing the effects of landscape fragmentation by forestry and agriculture on predation of artificial nests. Conserv. Biol. 11: 1418–1429.

Bayne, E.M. and Hobson, K.A. 2001. Effects of habitat fragmentation on pairing success of ovenbirds: importance of male age and floater behavior. Auk, 118: 380–388.

Bayne, E.M. and Hobson, K.A. 2002. Apparent survival of male ovenbirds in fragmented and forested boreal landscapes. Ecology 83: 1307–1316.

Bélisle, M., Desrochers, A., and Fortin, M.J. 2001. Influence of forest cover on the movements of forest birds: a homing experiment. Ecology, 82: 1893–1904.

Belleau, A., Bergeron, Y. Leduc, A., Gauthier S., and Fall, A. 2007. Using spatially explicit simulations to explore size distribution and spacing of regenerating areas produced by wildfires: recommendations for designing harvest agglomerations for the Canadian boreal forest. For. Chron. 83: 72–83.

Bennett, A.F. 2003. Linkages in the landscape: the role of corridors and connectivity in wildlife conservation. IUCN, Gland, Switzerland.

Bergeron, Y. 2004. Is regulated even-aged management the right strategy for the Canadian boreal forest? For. Chron. **80**: 458–462.

Bergeron, Y., Drapeau, P., Gauthier, S., and Lecomte, N. 2007. Using knowledge of natural disturbances to support sustainable forest management in the northern Clay Belt. For. Chron. **83**: 326–337.

Bergeron, Y., Flannigan, M., Gauthier, S., Leduc, A., and Lefort, P. 2004b. Past, current, and future fire frequency in the Canadian boreal forest: implications for sustainable forest management. Ambio, **33**: 356–360.

Bergeron, Y., Gauthier, S., Flannigan, M., and Kafka, V. 2004a. Fire regimes at the transition between mixedwood and coniferous boreal forest in northwestern Québec. Ecology, **85**: 1916–1932.

Bergeron, Y., Gauthier, S., Kafka, V., Lefort P., and Lesieur, D. 2001. Natural fire frequency for the eastern Canadian boreal forest: consequences for sustainable forestry. Can. J. For. Res. **31**: 384–391.

Bergeron, Y., Gauthier, S., Leduc, A., and Harvey, B. 2002. Natural fire regime: a guide for sustainable management of the Canadian boreal forest. Silva Fenn. **36**: 81–95.

Bergeron, Y., Harvey, B., Leduc, A., and Gauthier, S. 1999. Forest management guidelines based on natural disturbance dynamics: stand- and forest-level considerations. For. Chron. **75**: 49–54.

Berglund, H. and Jonsson, B.G. 2001. Predictability of plant and fungal species richness of old-growth boreal forest islands. J. Veg. Sci. **12**: 857–866.

Blanchette, P. and Larue, P. 1993. Développement d'un indice de qualité de l'habitat pour la paruline couronnée (*Seiurus aurocapillus*) au Québec. Ministère du Loisir, de la Chasse et de la Pêche, Direction générale de la ressource faunique, Gestion intégrée des ressources. Technical document 93/2.

Blanchette P. and Ostiguy. D. 1996. Méthode de sélection des espèces représentatives utilisée dans le cadre du projet de développement de la gestion intégrée des ressources. Gouvernement du Québec, Ministère de l'Environnement et de la Faune, Ministère des Ressources naturelles, Gestion intégrée des ressources. Technical document 96/1.

Block, W.M., Finch, D.M., and Brennan, L.A. 1995. Single species vs. multiple species approaches for management. *In* Ecology and management of neotropical migratory birds: a synthesis and review of critical issues. *Edited by* T.E. Martin and D.M. Finch. Oxford University Press, New York, USA, pp. 461–476.

Boudreault, C., Bergeron, Y., Drapeau, P., and Mascarúa-López, L. 2008. Edge influence on epiphytic lichens in remnant forests (cutblock separators, riparian buffers, and large forest remnants) of the eastern boreal forest in Canada. For. Ecol. Manag. **255**: 1461–1471.

Boudreault, C., Bergeron, Y., Gauthier S., and Drapeau, P. 2002. Bryophyte and lichen communities in mature to old-growth stands in eastern boreal forests of Canada. Can. J. For. Res. **32**: 1080–1093.

Boulanger, Y. and Sirois, L. 2007. Postfire succession of saproxylic arthropods, with emphasis on Coleoptera, in the North boreal forest of Québec. Environ. Entomol. **36**: 128–141.

Boulet, M., Darveau, M., and Bélanger, L. 2003. Nest predation and breeding activity of songbirds in riparian and nonriparian black spruce strips of central Québec. Can. J. For. Res. **33**: 922–930.

Boychuk, D., Perera, A.H., Ter-Mikaelian, M.T., Martell D.L., and Li. C. 1997. Modelling the effect of spatial scale and correlated fire disturbances on forest age distribution. Ecol. Model. **95**: 145–164.

British Columbia Ministry of Forests. 1998. Field manual for describing terrestrial ecosystems: tree attributes for wildlife. Land management handbook, no. 25. British Columbia Ministry of Environment, Lands, and Parks, Victoria, B.C.

Burke, D.M. and Nol, E. 2000. Landscape and fragment size effects on reproductive success of forest-breeding birds in Ontario. Ecol. Appl. **10**: 1749–1761.

Burton, P.J., Kneeshaw, D.D., and Coates, K.D. 1999. Managing forest harvesting to maintain old growth in boreal and sub-boreal forests. For. Chron. **75**: 623–631.

Bütler, R., Angelstam, P., Ekelund, P., and Schlaepfer, R. 2004. Dead wood threshold values for the Three-toed Woodpecker presence in boreal and sub-alpine forest. Biol. Conserv. **119**: 305–318.

Campbell, J. and Coxson, D.S., 2001. Canopy microclimate and arboreal lichen loading in subalpine spruce–fir forest. Can. J. Bot. **79**: 537–555.

Carignan, V. and Villard, M.-A. 2002. Selecting indicator species to monitor ecological integrity: a review. Environ. Monit. Assess. **78**: 45–61.

Chapin, T.G., Harrison D.J., and Katnik, D.D. 1998. Influence of landscape pattern on habitat use by American marten in an industrial forest. Conserv. Biol. **12**: 1327–1337.

Cissel, J.H., Swanson, F.J., and Weisberg, P.J. 1999. Landscape management using historical fire regimes: Blue Rive, Oregon. Ecol. Appl. **9**: 1217–1231.

Courtois, R. 1993. Description d'un indice de qualité d'habitat pour l'orignal (*Alces alces*) au Québec. Ministère du Loisir, de la Chasse et de la Pêche, Service de la faune terrestre, Gestion intégrée des ressources. Technical document 93/1.

Courtois, R. 2003. La conservation du caribou forestier dans un contexte de perte d'habitat et de fragmentation du milieu. Ph.D. thesis, Université du Québec à Rimouski, Rimouski, Que.

Courtois, R., Ouellet, J.-P., Breton, L., Gingras, A., and Dussault, C. 2007. Effects of forest disturbance on density, space use, and mortality of woodland caribou. Écoscience, **14**: 491–498.

Courtois, R., Ouellet, J.-P., Dussault C., and Gingras, A. 2004. Forest management guidelines for forest-dwelling caribou in Quebec. For. Chron. **80**: 598–607.

COSEWIC 2002. COSEWIC assessment and update status report on the woodland caribou *Rangifer tarandus caribou* in Canada. Committee on the Status of Endangered Wildlife in Canada. Ottawa, Ont.

Courtois, R., Ouellet, J.-P., Gingras, A., Dussault, C., Breton, L., and Maltais, J. 2003. Historical changes and current distribution of caribou, *Rangifer tarandus*, in Québec. Can. Field-Nat. 117: 399–414.

Crête, M. 2002. Proposition pour la mise en place d'un réseau de placettes permanentes visant à suivre l'évolution de la biodiversité dans les forêts du Québec. Société de la faune et des parcs du Québec, Québec, Que.

Crête, M., Brais, S., Campagna, M., Darveau, M., Desponts, M., Déry, S., Drapeau, S., Drolet, B., Jetté, J.-P., Maisonneuve, C., Nappi, A., and Petitclerc, P. 2004. Pourquoi et comment maintenir du bois mort dans les forêts aménagées du Québec: avis scientifique. Société de la faune et des parcs du Québec, Direction du développement de la faune et ministère des Ressources naturelles, Direction de l'environnement forestier, Québec, Que.

Darveau, M., Beauchesne, P., Bélanger, L., Huot, J., and Larue, P. 1995. Riparian forest strips as habitat for breeding birds in boreal forest. J. Wild. Manag. 59: 67–78.

Darveau, M., Labbé, P., Beauchesne, P., Bélanger, L., and Huot, J. 2001. The use of riparian forest strips by small mammals in a boreal balsam fir forest. For. Ecol. Manag. 143: 95–104.

Desrochers, A. and Hannon, S.J. 1997. Gap crossing decisions by dispersing forest songbirds. Conserv. Biol. 11: 1204–1210.

Drapeau, P., Leduc, A., and Bergeron, Y. 2009. Bridging ecosystem and multiple species approaches for setting conservation targets in managed boreal landscapes. *In* Setting conservation targets in managed forest landscapes. *Edited by* M.-A. Villard and B.-G. Johnson. Cambridge University Press, Cambridge, UK, pp. 129–60.

Drapeau, P., Leduc, A., Bergeron, Y., Gauthier, S., and Savard, J.-P. 2003. Les communautés d'oiseaux des vieilles forêts de la pessière à mousses de la ceinture d'argile: problèmes et solutions face à l'aménagement forestier. For. Chron. 79: 531–540.

Drapeau, P., Leduc, A., Giroux, J.-F., Savard, J.-P., Bergeron, Y., and Vickery, W.L. 2000. Landscape-scale disturbances and changes in bird communities of boreal mixed-wood forests. Ecol. Mon. 70: 423–444.

Drapeau, P., Nappi, A., Giroux, J.-F., Leduc, A., and Savard, J.-P. 2002. Distribution patterns of birds associated with snags in natural and managed eastern boreal forests. *In* Proceedings of the Symposium on the ecology and management of dead wood in Western forests. *Edited by* W.F. Laudenslayer, P.J. Shea, B.E.,Valentine, C.P. Weatherspoon, and T.E. Lisle. Reno, Nev. USDA Forest Service general technical report PSW-GTR 181, USDA Forest Service Pacific Southwest Research Station, Albany, Calif., USA, pp. 193–205.

Drapeau, P., Nappi, A., Imbeau, L., and Saint-Germain, M. (in press). Standing deadwood for keystone bird species in the eastern boreal forest: managing for snag dynamics. For. Chron.

Drapeau, P., Nappi, A., Saint-Germain, M., and Angers, V.-A. 2005. Les régimes naturels de perturbations, l'aménagement forestier et le bois mort dans la forêt boréale québécoise. *In* Bois mort et à cavités, une clé pour des forêts vivantes. *Edited by* D. Vallauri, J. André, B. Dodelin, R. Eynard-Machet, and D. Rambaud. WWF/Tec & Doc, Lavoisier, Paris, France, pp. 45–55.

Drolet B., Desrochers, A., and Fortin, M.-J. 1999. Effects of landscape structure on nesting songbird distribution in a harvested boreal forest. Condor, 101: 699–704.

Dussault, C., Courtois, R., Ouellet, J.-P., and Girard, I. 2005a. Space use of moose in relation to food availability. Can. J. Zool. 83: 1431–1437.

Dussault, C., Ouellet, J.-P., Courtois, R., Huot, J., Breton, L., and Jolicoeur, H. 2005b. Linking moose habitat selection to limiting factors. Ecography, 28: 619–628.

Esseen, P.-A. and Renhorn, K.-E. 1998. Edge effects on an epiphytic lichen in fragmented forests. Conserv. Biol. 12: 1307–1317.

Esseen, P.-A., Renhorn, K.-E., and Pettersson, R.B. 1996. Epiphytic lichen biomass in managed and old-growth boreal forests: effect of branch quality. Ecol. Appl. 6: 228–238.

Etcheverry, P., Crête, M., Ouellet, J.-P., Rivest, L.-P., Richer, M.-C., and Beaudoin, C. 2005a. Population dynamics of snowshoe hares in relation to furbearer harvest. J. Wild. Manag. 69: 771–781.

Etcheverry, P., Ouellet, J.-P., and Crête, M. 2005b. Response of small mammals to clear-cutting and precommercial thinning in mixed forests of southeastern Quebec. Can. J. For. Res. 35: 2813–2822.

Fielding, A.H. and Bell, J.F. 1997. A review of methods for the assessment of prediction errors in conservation presence/absence models. Environ. Conserv. 24: 38–49.

Fleishman, E., Murphy, D.D., and Brussard P.F. 2000. A new method for selection of umbrella species for conservation planning. Ecol. Appl. 10: 569–579.

Foster, D.R. 1983. The history and pattern of fire in the boreal forest of southeastern Labrador. Can. J. Bot. 61: 2459–2471.

Frankel, O.H. and Soulé, M.E. 1981. Conservation and evolution. Cambridge University Press, Cambridge, UK.

Franklin, J. 1993. Preserving biodiversity: species, ecosystems or landscapes? Ecol. Appl. 3: 202–205.

Franklin, J.F., Berg, D.R., Thornburgh, D.A., and Tappeiner, J.C. 1997. Alternative silvicultural approaches to timber harvesting: variable retention harvest systems. *In* Creating a forestry for the 21st century. *Edited by* K.A. Kohm and J.F. Franklin. Island Press, Washington, D.C., USA, pp. 111–140.

Franklin, J.F., Lindenmayer, D., MacMahon, J.A., McKee, A., Magnuson, J., Perry, D.A., Waide, R., and Foster, D.

2000. Threads of continuity. Conserv. Biol. Pract. **1:** 9–16.

Gauthier, S., Leduc, A., Harvey, B., Bergeron, Y., and Drapeau, P. 2001. Les perturbations naturelles et la diversité écosystémique. Nat. Can. **125:** 10–17.

Gauthier, S., Nguyen, T., Bergeron, Y., Leduc, A., Drapeau, P., and Grondin, P. 2004. Developing forest management strategies based on fire regimes in northwestern Quebec, Canada. *In* Emulating natural forest landscape disturbances: concepts and applications. *Edited by* A.H. Perera, L.J. Buse, and M.G. Weber. Columbia University Press, New York, USA, pp. 219–229.

Gobeil, J.-F. and Villard, M.-A. 2002. Permeability of three boreal forest landscape types to bird movements as determined from experimental translocations. Oikos, **98:** 447–458.

Gouvernement du Québec. 2007a. Règlement sur les normes d'intervention dans les forêts du domaine public. Chapitre F-4.1, r.1.001. Remplacé, D. 498-96, 1996 G.O. 2, 2570; eff. 96-05-23; voir c. F-4.1, r. 1.001.1. Éditeur officiel du Québec, Que.

Gouvernement du Québec. 2007b. An Act respecting threatened or vulnerable species. R.S.Q., c. E-12.01. Éditeur officiel du Québec, Que.

Gouvernement du Québec. 2007c. Forest Act. R.S.Q., c. F-4.1. Éditeur officiel du Québec, Que.

Grove, S.J. 2002. Saproxylic insect ecology and the sustainable management of forests. Annu. Rev. Ecol. Syst. **33:** 1–23.

Guénette, J.-S. and Villard, M.-A. 2005. Thresholds in forest bird response to habitat alteration as quantitative targets for conservation. Conserv. Biol. **19:** 1168–1180.

Haila, Y., Hanski, I.K., Niemelä, J., Punttila, P., Raivo, S., and Tukia, H. 1994. Forestry and the boreal fauna: matching management with natural forest dynamics. Ann. Zool. Fenn. **31:** 187–202.

Hannah, L., Midgley, G., Andelman, S., Araújo, M., Hugues, G., Martinez-Meyer, E., Pearson, R., and Williams, P. 2007. Protected area needs in a changing climate. Frontiers Ecol. Environ. **5:** 131–138.

Hannon, S.J. and Drapeau, P. 2005. Bird responses to burning and logging in the boreal forest of Canada. Studies Avian Biol. **30:** 97–115.

Hannon S.J, Paszkowski, C.A, Boutin, S., Degroot, J., MacDonald, S.E., Wheatley, M., and Eaton, B.R. 2002. Abundance and species composition of amphibians, small mammals, and songbirds in riparian buffer strips of varying widths in the boreal mixedwood of Alberta. Can. J. For. Res. **32:** 1784–1800.

Hannon, S.J. and Schmiegelow, F.K.A. 2002. Corridors may not improve the conservation value of small reserves for most boreal birds. Ecol. Appl. **12:** 1457–1468.

Hargis, C.D., Bissonnette, J.A., and Turner, D.L. 1999. The influence of forest fragmentation and landscape pattern on American martens. J. Appl. Ecol. **36:** 157–172.

Harmon, M.E., Franklin, J.F., Swanson, F.J., Sollins, P., Gregory, S.V., Lattin, J.D., Anderson, N.H., Cline, S.P., Aumen, N.G., Sedell, J.R., Lienkaemper, G.W., Cromack,

K.J., and Cummins, K.W. 1986. Ecology of coarse woody debris in temperate ecosystems. Adv. Ecol. Res. **15:** 133–302.

Harper, K.A., Bergeron, Y., Drapeau, P., Gauthier, S., and De Grandpré, L. 2005. Structural development following fire in black spruce boreal forest. For. Ecol. Manag. **206:** 293–306.

Harper, K.A., Bergeron, Y., Gauthier, S., and Drapeau, P. 2002. Structural development of black spruce forests following fire in Abitibi, Québec: a landscape scale investigation. Silva Fenn. **36:** 249–263.

Harper, K.A., Boudreault, C., De Grandpré, L., Drapeau, P., Gauthier, S., and Bergeron, Y. 2003. Structure, composition and diversity of old-growth black spruce boreal forest of the Clay Belt region in Québec and Ontario. Environ. Rev. **11:** S79–S98.

Harrison, R.B., Schmiegelow, F.K.A., and Naidoo, R. 2005. Stand-level response of breeding forest songbirds to multiple levels of partial-cut harvest in four boreal forest types. Can. J. For. Res. **35:** 1553–1567.

Harvey, B.D., Leduc A., Gauthier S., and Bergeron, Y. 2002. Stand-landscape integration in natural disturbance-based management of the southern boreal forest. For. Ecol. Manag. **155:** 369–385.

Hilmo, O. and Holien, H. 2002. Epiphytic lichen response to the edge environment in a boreal *Picea abies* forest in central Norway. The Bryologist, **105:** 48–56.

Hobson, K.A. and Bayne, E. M. 1999. Breeding bird communities in boreal forest of western Canada: consequences of "unmixing" the mixedwoods. Condor, **102:** 759–769.

Hobson, K.A. and Schieck, J. 1999. Changes in bird communities in boreal mixedwood forests: harvest and wildfire effects over 30 years. Ecol. Appl. **9:** 849–863.

Holling, C.S. 1978. Adaptive environmental assessment and management. John Wiley, New York, USA.

Hoyt, J.S. and Hannon, S.J. 2002. Habitat associations of Black-backed and Three-toed Woodpeckers in the boreal forest of Alberta. Can. J. For. Res. **32:** 1881–1888.

Huggett, A.J. 2005. The concept and utility of ecological thresholds in biodiversity conservation. Biol. Conserv. **124:** 301–310.

Hunter Jr., M.L. 1990. Wildlife, forests, and forestry: principles of managing forests for biological diversity. Prentice-Hall, Englewood Cliffs, N.J., USA.

Hunter Jr., M.L. 1992. Paleoecology, landscape ecology, and conservation of neotropical migrant passerines in boreal forests. *In* Ecology conservation of neotropical migrant landbirds. *Edited by* J.M. Hagan and D.W. Johnston. Smithsonian Institution Press, Washington, D.C., USA, pp. 511–523.

Hutto, R.L. 1995. Composition of bird communities following stand-replacement fires in northern Rocky Mountain (USA) conifer forests. Conserv. Biol. **9:** 1041–1058.

Hylander, K., Jonsson, B.G., and Nilsson, C. 2005. Substrate form determines the fate of bryophytes in riparian buffers strips. Ecol. Appl. **15:** 674–688.

Imbeau, L. and Desrochers, A. 2002. Area sensitivity and edge avoidance: the case of the Three-toed Woodpecker (*Picoides tridactylus*) in a managed forest. For. Ecol. Manag. **164**: 249–256.

Imbeau, L., Drapeau, P., and Mönkkönen, M. 2003. Edge species in forest patches within agricultural landscapes: are we confusing response to edges and successional status? Ecography, **26**: 514–520.

Imbeau, L., Mönkkönen, M., and Desrochers, A. 2001. Long-term effects of forestry on birds of the eastern Canadian boreal forests: a comparison with Fenno-scandia. Conserv. Biol. **15**: 1151–1162.

Imbeau, L., Savard, J.-P., and Gagnon, R. 1999. Comparing bird assemblages in successional black spruce stands originating from fire and logging. Can. J. Zool. **77**: 1850–1860.

Jacobs, J.M. 2004. Saproxylic beetles assemblages in the boreal mixedwood of Alberta : succession, wildfire, and variable retention forestry. M.Sc. thesis, University of Alberta, Edmonton, Alta.

Jacobs, J.M., Spence, J.R., and Langor, D.W. 2007. Influence of boreal forest succession on dead wood qualities on saproxylic beetles. Agric. For. Entomol. **9**: 3–16.

Jansson, G. and Angelstam, P. 1999. Threshold levels of habitat composition for the presence of the Long-tailed Tit (*Aegithalos caudatus*) in a boreal landscape. Landsc. Ecol. **14**: 283–290.

Johnson, E.A. 1992. Fire and vegetation dynamics: studies from the North American boreal forest. Cambridge University Press, Cambridge, UK.

Jonsson, B.G. and Jonsell, M. 1999. Exploring potential indicators in boreal forests. Biodiver. Conserv. **8**: 1417–1433.

Kafka, V., Gauthier, S., and Bergeron, Y. 2001. Fire impacts and crowning in the boreal forest: study of a large wildfire in western Quebec. Int. J. Wildland Fire, **10**: 119–127.

Kivistö, L. and Kuusinen, M. 2000. Edge effects on the epiphytic lichen flora of *Picea abies* in middle boreal Finland. Lichenologist, **32**: 387–398.

Kneeshaw, D.D., Leduc, A., Drapeau P., Gauthier, S., Paré, D., Doucet, R., Carignan, R., Bouthillier, L., and Messier, C. 2000. Development of integrated ecological standards of sustainable forest management at an operational scale. Forestry Chronicle **76**: 481–493.

Koivula, M.J. and Schmiegelow, F.K.A. 2007. Boreal woodpecker assemblages in recently burned forested landscapes in Alberta, Canada: effects of post-fire harvesting and burn severity. For. Ecol. Manag. **242**: 606–618.

Kotliar, N.B., Kennedy, P.L., and Ferree, K. 2007. Avifaunal responses to fire in southwestern montane forests along a burn severity gradient. Ecol. Appl. **17**: 491–507.

Kouki, J. and Väänänen, A. 2000. Impoverishment of resident old-growth forest bird assemblages along an isolation gradient of protected areas in eastern Finland. Ornis Fenn. **77**: 145–154.

Kruys, N., Jonsson, B.G., and Stähl, G. 2002. A stage-based matrix model for decay-class dynamics of woody debris. Ecol. Appl. **12**: 773–781.

Lafleur, P.-E. and Blanchette P. 1993. Développement d'un indice de qualité de l'habitat pour le Grand Pic (*Dryocopus pileatus*) au Québec. Ministère du Loisir, de la Chasse et de la Pêche, Direction générale de la ressource faunique, Gestion intégrée des ressources. Technical document 93/3.

Lambeck, R.J. 1997. Focal species: a multi-species umbrella for nature conservation. Conserv. Biol. **11**: 849–856.

Landres, P.B., Verner, J., and Thomas, J.W. 1988. Ecological uses of vertebrate indicator species: a critique. Conserv. Biol. **2**: 316–328.

Larrivée, M., Fahrig, L., and Drapeau, P. 2005. Effects of a recent wildfire and clearcuts on ground-dwelling boreal forest spider assemblages. Can. J. For. Res. **35**: 2575–2588.

Larsen, C.P.S. 1997. Spatial and temporal variations in boreal forest fire frequency in northern Alberta. J. Biogeogr. **24**: 663–673.

Larue, P. 1992. Développement d'un indice de qualité de l'habitat pour la martre d'Amérique (*Martes americana*) au Québec. Ministère du Loisir, de la Chasse et de la Pêche, Direction générale de la ressource faunique, Gestion intégrée des ressources. Technical document 92/7.

Leboeuf, M. 2004. Effets de la fragmentation générée par les coupes en pessière noire à mousses sur huit espèces d'oiseaux de forêt mature. M.Sc. thesis, Université du Québec à Montréal, Montréal, Que.

Leduc, A., Bergeron, Y., Drapeau, P., Harvey, B., and Gauthier, S. 2000. Le régime naturel des incendies forestiers: un guide pour l'aménagement durable de la forêt boréale. L'Aubelle, **135**: 13–22.

Lefort, P., Gauthier, S., and Bergeron, Y. 2003. The influence of fire weather and land use on the fire activity of the Lake Abitibi area, Eastern Canada. For. Sci. **49**: 509–521.

Lesieur, D., Gauthier, S., and Bergeron, Y. 2002. Fire frequency and vegetation dynamics for the south-central boreal forest of Quebec, Canada. Can. J. For. Res. **32**: 1996–2009.

Leupin, E.E., Dickinson, T.E., and Martin, K. 2004. Resistance of forest songbirds to habitat perforation in a high-elevation conifer forest. Can. J. For. Res. **34**: 1919–1928.

Lindenmayer, D.B., Burton, P., and Franklin, J.F. 2008. Salvage logging and its ecological consequences. Island Press, Washington, D.C.

Lindenmayer, D.B., Cunningham, R.B., Pope, M., and Donnelly, C.F. 1999. The Tumut fragmentation experiment in south-eastern Australia: The effects of landscape context and fragmentation of arboreal marsupials. Ecol. Appl. **9**: 594–611.

Lindenmayer, D.B., Foster, D.R., Franklin, J.F., Hunter, M.L., Noss, R.F., Schmiegelow, F.A., and Perry, D. 2004.

Salvage harvesting policies after natural disturbance. Science, **303**: 1303.

Lindenmayer, D.B and Noss, R.F. 2006. Salvage logging, ecosystem processes, and biodiversity conservation. Conserv. Biol. **20**: 949–958.

Machtans, C.S., Villard, M.-A., and Hannon, S.J. 1996. Use of riparian buffer strips as movement corridors by forest birds. Conserv. Biol. **10**: 1366–1379.

Manel, S.H., Williams, C., and Ormerod, S.J. 2001 Evaluating presence–absence models in ecology: the need to account for prevalence. J. Appl. Ecol. **38**: 921–931.

Martin, K., Aitken, K.E.H., and Wiebe, K.L. 2004. Nest sites and nest webs for cavity-nesting communities in Interior British Columbia, Canada: nest characteristics and niche partitioning. Condor, **106**: 5–19.

Martin, K. and Eadie, J.M. 1999. Nest webs: a community-wide approach to the management and conservation of cavity-nesting birds. For. Ecol. Manag. **115**: 243–257.

Mascarúa-López, L., Harper, K., and Drapeau, P. 2006. Edge influence on forest structure in large forest remnants, cutblock separators, and riparian buffers in managed black spruce forests. Écoscience, **13**: 226–233.

McComb, W.C., Spies, T.A., and Emmingham, W.H. 1993. Douglas-fir forests: managing for timber and mature-forest habitat. J. For. **91**: 31–42.

McLaren, M.A., Thompson, I.D., and Baker, J.A. 1998. Selection of vertebrate wildlife indicators for monitoring sustainable forest management in Ontario. For. Chron. **74**: 241–248.

Mills, L.S., Soulé, M.E., and Doak, D.F. 1993. The keystone-species concept in ecology and conservation. BioScience, **43**: 219–224.

Mitchell, S.J. and Beese, W.J. 2002. The retention system: reconciling variable retention with the principles of silvicultural systems. For. Chron. **78**: 397–403.

Morrissette, J.L., Cobb, T.P., Brigham, R.M., and James, P.C. 2002. The response of boreal forest songbird communities to fire and post-fire harvesting. Can. J. For. Res. **32**: 2169–2183.

Mulder, B.S., Noon, B.R., Spies, T.A., Raphael, M.G., Palmer, C.J., Olsen, A.R., Reeves, G.H., and Welsh, H.H. 1999. The strategy and design of the effectiveness monitoring program for the Northwest Forest Plan. USDA Forest Service, Pacific Northwest Forest Station, General Technical Report PNW-GTR-437.

Nappi A., Drapeau, P., Giroux, J.-F., and Savard, J.-P.L. 2003. Snag use by foraging Black-backed Woodpeckers in a recently burned eastern boreal forest. Auk, **120**: 505–511.

Nappi, A., Drapeau P., and Savard, J.-P.L. 2004. Salvage logging after wildfire in the boreal forest: is it becoming a hot issue for wildlife? For. Chron. **80**: 67–74.

Noss, R.F. 1999. Assessing and monitoring forest biodiversity: a suggested framework and indicators. For. Ecol. Manag. **115**: 135–146.

Périgon, S. 2006. Dynamique de trouées dans de vieux peuplements irréguliers de la Côte-Nord, Québec: analyse temporelle. M.Sc. thesis. Université du Québec à Montréal, Montréal, Que.

Perron, N. 2003. Peut-on et doit-on s'inspirer de la variabilité naturelle des feux pour élaborer une stratégie écosystémique de répartition des coupes à l'échelle du paysage? Le cas de la pessière noire à mousse de l'Ouest au Lac-Saint-Jean. Ph.D. thesis, Université Laval, Québec, Que.

Potvin, F., Bélanger, L., and Lowell, K. 2000. Marten habitat selection in a clearcut boreal landscape. Conserv. Biol. **14**: 844–857.

Potvin, F. and Bertrand, N. 2004. Leaving forest strips in large clearcut landscapes of boreal forest: a management scenario suitable for wildlife? For. Chron. **80**: 44–53.

Potvin, F., Courtois, R., and Bélanger, L. 1999. Short-term response of wildlife to clear-cutting in Quebec boreal forest: multiscale effects and management implications. Can. J. For. Res. **29**: 1120–1127.

Poulin, M. 2005. Effets des coupes partielles sur les oiseaux en forêt de pessière à mousses de l'Est du Canada. M.Sc. thesis, Université du Québec à Montréal, Montréal, Que.

Radford, J.Q. and Bennett, A.F. 2004. Thresholds in land-scape parameters: occurrence of the White-browed Treecreeper (*Climacteris affinis*) in Victoria, Australia. Biol. Conserv. **117**: 375–391.

Radford, J.Q., Bennett, A.F., and Cheers, G.J. 2005. Landscape-level thresholds of habitat cover for wood-land-dependent birds. Biol. Conserv. **124**: 317–337.

Ranius, T. and Fahrig, L. 2006. Targets for maintenance of dead wood for biodiversity conservation based on extinction thresholds. Scan. J. For. Res. **21**: 201–208.

Rempel, R.S., Andison, D.W., and Hannon, S.J. 2004. Guiding principles for developing an indicator and monitoring framework. For. Chron. **80**: 82–90.

Rheault, H., Drapeau, P., Bergeron, Y., and Esseen, P.-A. 2003. Edge effects on epiphytic lichens in managed black spruce forests of eastern North America. Can. J. For. Res. **33**: 23–32.

Roberge, J.-M. and Angelstam, P. 2004. Usefulness of the umbrella species concept as a conservation tool. Conserv. Biol. **18**: 76–85.

Ruel, J.-C., Pin, D., and Cooper, K. 2001. Windthrow in riparian buffer strips: effect of wind exposure, thinning, and strip width. For. Ecol. Manag. **143**: 105–113.

Saab, V.A. and Dudley, J.G. 1998. Responses of cavity-nesting birds to stand-replacement fire and salvage logging in Ponderosa pine/Douglas-fir forests of south-western Idaho. U.S. Department of Agriculture, Forest Service Research Paper RMRS-RP-11.

Saab, V.A., Powell, H.D.W., Kotliar, N.B., and Newton, K.R. 2005. Variation in fire regimes of the Rocky Mountains: implications for avian communities and fire management. Studies Avian Biol. **30**: 76–96.

Saint-Germain, M., Buddle, C., and Drapeau, P. 2006. Sampling saproxylic Coleoptera: Scale issues and the importance of behavior. Environ. Entomol. **35**: 478–487.

Saint-Germain, M., Buddle, C., and Drapeau, P. 2007a. Primary attraction and random landing in host-selection by wood-feeding insects: a matter of scale? Agr. For. Entomol. **9**: 227–235.

Saint-Germain, M., Drapeau, P., and Buddle, C. 2007b. Occurrence patterns of aspen-feeding wood-borers (Coleoptera: Cerambycidae) along the wood decay gradient: active selection for specific host types or neutral mechanisms? Ecol. Entomol. **32**: 712–721.

Saint-Germain, M., Drapeau, P., and Buddle, C. 2007c. Host-use patterns of saproxylic wood-feeding Coleoptera adults and larvae along the decay gradient in standing dead black spruce and aspen. Ecography, **30**: 737–748.

Saint-Germain, M., Drapeau, P., and Hébert, C. 2004a. Habitat use of pyrophilous coleoptera in recently burned black spruce forest of Central Quebec. Biol. Conserv. **118**: 583–592.

Saint-Germain, M., Drapeau, P., and Hébert, C. 2004b. Species composition and substratum use of xylophagous insects on fire-killed trees in a burned black spruce forest of Quebec. Can. J. For. Res. **34**: 677–685.

Saint-Germain, M., Drapeau, P., and Hébert, C. 2004c. Landscape-scale habitat selection patterns of *Monochamus scutellatus* (Say) (Coleoptera: Cerambycidae) in a recently burned black spruce forest. Environ. Entomol. **33**: 1703–1710.

Samson, C. and Huot, J. 1998. Movements of female black bears in relation to landscape vegetation type in southern Québec. J. Wildl. Manag. **62**: 718–727.

Schieck, J., Nietfeld M., and Stelfox. J.B. 1995. Differences in bird species richness and abundance among three successional stages of aspen-dominated boreal forests. Canadian Journal of Zoology **73**: 1417–1431.

Schieck, J. and Hobson, K.A. 2000. Bird communities associated with live residual tree patches within cut blocks and burned habitat in mixedwood boreal forests. Can. J. For. Res. **30**: 1281–1295.

Schmiegelow, F.K.A., Machtans, C.S., and Hannon, S.J. 1997. Are boreal birds resilient to forest fragmentation? An experimental study of short-term community responses. Ecology, **78**: 1914–1932.

Schmiegelow, F.K.A., and Mönkönnen, M. 2002. Habitat loss and fragmentation in dynamic landscapes: avian perspectives from the boreal forest. Ecol. Appl. **12**: 375–389.

Schmiegelow, F.K.A., Stepnisky, D.P., Stambaugh C.A., and Koivula, M. 2006. Reconciling salvage logging of boreal forests with a natural-disturbance management model. Conserv. Biol. **20**: 971–983.

Seymour, R. and Hunter Jr., M. 1999. Principles of ecological forestry. *In* Maintaining biodiversity in forest ecosystems. *Edited by* M.L. Hunter Jr. Cambridge University Press, Cambridge, UK, pp. 22–61.

Sillet, S.C. 1994. Growth rates of two epiphytic cyanolichen species at the edge and in the interior of a 700-year-old Douglas fir forest in the western Cascades of Oregon. The Bryologist, **97**: 321–324.

Sillett, S.C., McCune, B., Peck, J.E., Rambo, T.R., and Ruchty, A. 2000. Dispersal limitations of epiphytic lichens result in species dependent old-growth forests. Ecol. Appl. **10**: 789–799.

Simon, N.P.P., Schwab, F.E., and Diamond, A.W. 2000. Patterns of breeding bird abundance in relation to logging in western Labrador. Can. J. For. Res. **30**: 257–263.

Smith, K.G., Ficht, E.J., Hobson, D., Sorensen, T.C., and Hervieux, D. 2000. Winter distribution of woodland caribou in relation to clear-cut logging in west-central Alberta. Can. J. For. Res. **78**: 1433–1440.

Smith, A.C., and Schaefer, J.A. 2002. Home-range size and habitat selection by American marten (*Martes americana*) in Labrador. Can. J. Zool. **80**: 1602–1609.

Smucker, K.M., Hutto, R.L., and Steele, B.M. 2005. Changes in bird abundance after wildfire: importance of fire severity and time since fire. Ecol. Appl. **15**: 1535–1549.

Spies, T.A., Ripple, W.J., and Bradshaw, G.A. 1994. Dynamics and pattern of a managed coniferous forest landscape in Oregon. Ecol. Appl. **4**: 555–568.

Suffling, R., Smith, B., and Dal Molin, J. 1982. Estimating past forest age distributions and disturbance rates in North-western Ontario: a demographic approach. J. Environ. Manag. **14**: 45–56.

Thomas, D.C., and D.R. Gray. 2002. Update COSEWIC status report on the woodland caribou *Rangifer tarandus caribou* in Canada. In COSEWIC assessment and update status report on the woodland caribou *Rangifer tarandus caribou* in Canada. Committee on the Status of Endangered Wildlife in Canada. Ottawa, Ont.

Thompson, I.D. 1994. Marten populations in uncut and logged boreal forests in Ontario. J. Wild. Manag. **58**: 272–280.

Thompson, I.D. 2006. Monitoring of biodiversity indicators in boreal forests: a need for improved focus. Environ. Monit. Assess. **121**: 263–273.

Turner, M.G. and Romme, H.W. 1994. Landscape dynamics in crown fire ecosystems. Lands. Ecol. **9**: 59–77.

Vaillancourt, M.-A., Drapeau, P., Gauthier, S., and Robert, M. 2008. Availability of standing trees for large cavity-nesting birds in the eastern boreal forest of Québec, Canada. For. Ecol. Manag. **255**: 2272–2285.

Vanderwel, M.C., Malcolm, J.R., Smith, S.M., and Islam, N. 2006. Insect community composition and trophic guild structure in decaying logs from eastern Canadian pine-dominated forests. For. Ecol. Manag. **225**: 190–199.

Villard, M.-A., Martin, P.R, and Drummond, C.G. 1993. Habitat fragmentation and pairing success in the Ovenbird, *Seiurus aurocapillus*. Auk, **110**: 759–768.

Virkkala, R. and Toïvonen, H. 1999. Maintaining biological diversity in Finnish forests. Finnish Environment Institute, Helsinki, Finland.

Walters, C.J. 1997. Adaptive policy design: thinking at larger spatial scales. *In* Wildlife and landscape ecology. *Edited by* J.A. Bissonnette. Springer-Verlag Inc., New York, USA, pp. 386–394.

Webb, A. 2006. The effects of alternative harvesting practices on saproxilic beetles in eastern mixedwood boreal forest of Quebec. M.Sc. thesis, McGill University, Montréal, Que.

Webb, A., Buddle, C.H., Drapeau, P., and Saint-Germain, M. 2008. Use of remnant boreal forest habitats by saproxylic beetle assemblages in even-aged managed landscapes. Biol. Conserv. **141**: 815–826.

Weir, J.M.H., Johnson, E.A., and Miyanishi, K. 2000. Fire frequency and the spatial age mosaic of the mixedwood boreal forest in western Canada. Ecol. Appl. **10**: 1162–1177.

Wilcox, B.A. 1984. *In situ* conservation of genetic resources: determinants of minimum area requirements. *In* National parks: conservation and development. *Edited by* J.A. McNeely and K.R. Miller. Smithsonian Institution Press, Washington, D.C., USA, pp. 638–647.

Wilhere, G.F. 2002. Adaptive management in habitat conservation plans. Conserv. Biol. **16**: 20–29.

Wimberly, M.C., Spies, T.A., Long, C.J., and Whitlock, C. 2000. Simulating historical variability in the amount of old forest in the Oregon Coast Range. Conserv. Biol. **14**: 167–180.

Chapter *15*

# Silvicultural and Ecological Evaluation of Partial Harvest in the Boreal Forest on the Clay Belt, Québec*

*Nicole Fenton, Hervé Bescond, Louis Imbeau, Catherine Boudreault, Pierre Drapeau, and Yves Bergeron*

* In addition to the co-authors many researchers and students have been involved in this project. In alphabetical order: Jean Bégin (U. Laval), Ingrid Cea (UQAT), Marianne Cheveau (UQAM), Louis De Grandpré (SCF), Jean Ferron (UQAR), Art Groot (SCF), Daniel Kneeshaw (UQAM), Alain Leduc (UQAM), Philippe Meek (FERIC), Johanne Morasse (Collège de l'Abitibi-Témiscamingue), Delphin Ruché (UQAT), Jean-Claude Ruel (U. Laval), Annick St-Denis (UQAM), Osvaldo Valeria (UQAT), Stéphane Valois (UQAR), Tim Work (UQAM), and Saliha Zouaoui (UQAM). The project is financed by the UQAT-UQAM NSERC Chair in Sustainable Forest Management, the FQRNT Fonds forestiers, and the Canadian Forest Service/NSERC Forest Research Partnership Program with the collaboration of the industrial partners of the SFM Chair. Thanks also to the operators and other professionals who contributed to the success of this project in the field. The photos on this page were graciously provided by Brock Epp, Antoine Nappi, and Darwyn Coxson.

# 1. INTRODUCTION

• CPRS (careful logging with protection of regeneration and soils) is the main harvest technique in Québec.

•• Partial cutting in this chapter refers to the specific techniques and retention levels within the experimental network. For a broader definition see Bouchard, chapter 13.

The almost exclusive use of even-aged harvest systems (e.g., clearcuts or CPRS•) in boreal Canada has significantly altered the landscape of regions with a long fire cycle and subsequently a dominance of old forest (second- and third-cohort stands where dominant trees reach the canopy many years after fire; Gauthier et al., chapter 3). As a result young, even-aged stands are overrepresented on the landscape. One potential solution to this problem is the increased use of partial cutting•• as a harvest system in landscapes where natural disturbance–based management is the objective. Partial cutting may be used as either a conversion cut to gradually transform the structure of stands from even to uneven (i.e. from cohort 1 to cohort 2 or from cohort 2 to cohort 3) or to conserve the uneven structure of a cohort 2 or 3 stand (see Bouchard, chapter 13).

The fire cycle on the Clay Belt of Québec and Ontario is relatively long (mean age of forests is approximately 150 years; Bergeron et al. 2001), and a significant proportion of stands develop uneven and irregular structure. As stands with complex structures (uneven and irregular) are more abundant on the landscape than those with a regular structure (Simard et al., chapter 11), they represent an important habitat for species such as mosses (particularly liverworts; Fenton and Bergeron 2006), lichens (Boudreault et al. 2003), and birds (Drapeau et al. 2003) that are rarely found in young or mature forests (i.e., cohort 1). They are therefore important for the maintenance of biodiversity.

A diversification of silvicultural practices used within a managed landscape including both even- and uneven-aged management systems would be appropriate in this region (Bergeron et al. 1999; Bergeron et al. 2001). Partial cuts [e.g., CPPTM (only stems >9 cm in diameter are harvested) and CAMC (adapted cuts maintaining canopy cover)] can be viewed as succession cuts that would maintain the irregular structure of old-growth forests within this managed landscape. Interestingly, the use of partial cuts meets the objectives of both conservation and silviculture as it favors the growth of the small stems present in uneven stands, while they would be harvested under the conventional low-retention system (CPRS). Therefore, there are many reasons to practice partial cuts in the boreal forest. However, there are few experimental or operational trials underway, and the relatively short time span (<10 years) of this work makes it difficult to determine whether the desired results are achieved. This is because the simple creation or preservation of irregular structure does not guarantee that these stands will be used by old-growth specialist species. There are therefore two levels of uncertainty: 1) Are we able to recreate irregular and uneven structure using partial harvesting techniques? 2) Do species use the created habitat?

Since 1998 we have developed an experimental network of partial cuts in order to evaluate the economic (operational costs and fibre production) and ecologic efficiency of partial cuts in the forests of the Clay Belt of Québec.

Initially, only CPPTM was used, but CAMC was used in subsequent years. Our objectives are to:

1. evaluate silvicultural and operational aspects of the partial harvest techniques;

2. evaluate whether these partial harvest techniques permit the creation or maintenance of old-growth structure;

3. evaluate, using indicators, the effect of these techniques on biological diversity (vascular and non-vascular plants, birds, and small mammals); and

4. evaluate the effect of partial harvest on a few target game species (Spruce Grouse [*Falcipennis canadensis*], snowshoe hare [*Lepus americanus*]).

## 2. METHODOLOGY

Since establishment in 1998, ten experimental sites have been installed across Abitibi-Témiscamingue, on the Clay Belt of Québec and Ontario (figure 15.1). Both black spruce–feather moss and boreal mixedwood regions were targeted. Most of the harvested stands had an irregular structure and were composed of pure black spruce (*Picea mariana* [Mill.] B.S.P.). A few of the harvested stands were mixed, with trembling aspen (*Populus tremuloides* Michx.) and balsam fir (*Abies balsamea* [L.] Mill.) as co-dominants.

In order to ensure that a broad spectrum of on-the-ground operations were included, sites were selected across the region with different industrial partners, each using different contractors to complete the harvests. Each site is made up of three blocks (minimum of 25 ha), one each of low-retention cut (CPRS), partial cut, and unharvested control. The first two sites (Muskuchii and Dufay) were harvested using CPPTM, but all of the following sites were harvested with CAMC. CPPTM left a residual stand that was very heterogenous, with sectors with very high retention and others with none, because it is a diameter limit cut. CAMC, where a certain proportion of stems in all diameter classes are retained, is more dependent on the cooperation and ability of machinery operators, but generally results in a more even cover and the retention of larger stems (table 15.1).

Within each block (partial cut, CPRS, and control), 17 nested plots were established. Within the largest 11.28 m radius plots all trees (>9 cm) were measured (DBH, height, vigour) and enumerated before and immediately after harvest. During the summer following harvest, nested plots were established inside the 11.28 m plots in which regeneration was quantified and the understory vegetation was characterized. Transects were used to evaluate the abundance of coarse woody debris (DBH > 5 cm). Canopy openness and organic matter thickness were also evaluated for each plot. The entire set of variables was measured on all sites immediately after harvest, and will be re-measured five and ten years post-harvest (four sites have already been re-measured). Tree vigour, mortality, and incline are recorded at the end of each growing season in order to follow growth and blowdowns in the partial cuts and the controls.

## Figure 15.1
## Map indicating the experimental partial cut network sites (triangles) in Abitibi

Note: The Muskuchii site has three experimental blocks.

## Table 15.1
## Intensity of harvest in all the sites of the partial cut network

| Site | Living trees | | | | | | Snags | | |
|---|---|---|---|---|---|---|---|---|---|
| | Basal area before harvest (m²/ha)[3] | Basal area after harvest (m²/ha) | Proportion of stems harvested (%) | Number of stems >9 cm before harvest (#/ha) | Number of stems >9 cm after harvest (#/ha) | Number of stems >9 cm harvested (%) | Basal area before harvest (m²/ha) | Basal area after harvest (m²/ha) | Proportion of stems harvested (%) |
| Muskuchii 1[1] | 23.80 | 6.79 | 71.47 | 1 084 | 482 | 55.54 | – | – | – |
| Muskuchii 2[1] | 24.57 | 9.50 | 61.33 | 1 119 | 632 | 43.52 | – | – | – |
| Muskuchii 3[1] | 12.54 | 4.12 | 67.15 | 829 | 407 | 50.90 | – | – | – |
| Dufay[1] | 17.67 | 5.74 | 67.52 | 823 | 250 | 69.62 | – | – | – |
| Maïcasagi[2] | – | 7.51 | 45.00 | – | 404 | – | – | 2.24 | – |
| Gaudet[2] | 12.55 | 2.13 | 83.03 | 826 | 197 | 76.13 | – | 0.81 | – |
| Fénélon[2] | 22.66 | 3.47 | 84.69 | 971 | 224 | 76.97 | 3.47 | 3.07 | 11.53 |
| Puiseaux[2] | 19.68 | 4.87 | 75.25 | 1 018 | 375 | 63.17 | 3.92 | 1.06 | 72.96 |
| Surimau[2] | 10.62 | 1.11 | 89.55 | 667 | 100 | 85.00 | 3.28 | 0.59 | 82.01 |
| Villars[2] | 17.33 | 9.71 | 43.97 | 1 052 | 602 | 42.78 | 1.10 | 0.75 | 31.82 |

1. Sites harvested by CPPTM.
2. Sites harvested by CAMC.
3. Basal area includes all stems.

## 3. RESULTS

### 3.1. Objective 1: Evaluate Silvicultural and Operational Aspects of Partial Harvest Techniques

The silvicultural success of the partial cuts was evaluated immediately post-harvest, in terms of the amount of stems removed, and the health and stability of the remaining trees. Overall, contractors successfully maintained part of the standing timber (table 15.1) and the basal area removed varied from 44.0% to 89.5% across the network. In all sites a significant number of merchantable stems (DBH > 9 cm) were retained within the cutover.

Two concerns generally associated with partial cuts are 1) the amount of wounding to the residual trees caused by harvest and 2) the increased susceptibility of the residual stand to blowdown, due to increased wind fetch and compaction damage to roots. As blowdowns occur during several years post-harvest, five-year growth results are necessary in order to accurately evaluate the impact of partial cuts on blowdowns. As only data from Muskuchii and Dufay is available, only these sites are presented here.

Five years after harvest, results varied widely among sites. In the first site established (Muskuchii 1), basal area and volume of merchantable trees diminished, mainly because of blowdowns (table 15.2). However, this blowdown rate is comparable to the rates in other CPPTM in Abitibi (Bégin and Riopel 2001;

### Table 15.2

### Change in the basal area and volume of merchantable stems after five growing seasons

| | Site | Treatment | Growth of residuals[1] | Recruitment[2] | Gross growth[3] | Snags[4] | Mortality[5] | Net growth[6] |
|---|---|---|---|---|---|---|---|---|
| Basal area (m²/ha) | Muskuchii 1 | Control | 2.89 | 0.83 | 3.72 | 1.04 | 2.02 | 0.66 |
| | Muskuchii 1 | CPPTM | 0.04 | 0.04 | 0.08 | 0.21 | 0.03 | −0.16 |
| | Muskuchii 1 | CPRS | 0.00 | 0.00 | 0.00 | 0.00 | 0.00 | 0.00 |
| | Dufay | Control | 3.04 | 0.90 | 3.95 | 0.26 | 1.08 | 2.61 |
| | Dufay | CPPTM | 0.85 | 0.58 | 1.43 | 0.47 | 0.82 | 0.14 |
| | Dufay | CPRS | 0.00 | 0.00 | 0.00 | 0.00 | 0.00 | 0.00 |
| Volume (m³/ha) | Muskuchii 1 | Control | 25.08 | 1.99 | 27.07 | 8.10 | 14.45 | 4.51 |
| | Muskuchii 1 | CPPTM | 3.96 | 0.98 | 4.93 | 10.34 | 1.68 | −7.09 |
| | Muskuchii 1 | CPRS | 0.00 | 0.00 | 0.00 | 0.00 | 0.00 | 0.00 |
| | Dufay | Control | 24.68 | 1.97 | 26.65 | 1.72 | 3.71 | 21.22 |
| | Dufay | CPPTM | 7.19 | 1.69 | 8.89 | 2.17 | 4.76 | 1.96 |
| | Dufay | CPRS | 0.00 | 0.00 | 0.00 | 0.00 | 0.00 | 0.00 |

1. Growth of residuals: increase in the basal area or in the volume of trees with a DBH > 9 cm that were still alive at the end of the 5-year measuring cycle.
2. Recruitment: trees that have grown to a DBH > 9 cm during the 5-year measuring cycle.
3. Gross growth: increase in volume or basal area during the 5-year measuring cycle (Growth of residuals + Recruitment).
4. Blowdowns: basal area or volume of trees blown down during the 5-year measuring cycle.
5. Mortality: basal area or volume of trees that died during the 5-year measuring cycle.
6. Net growth: increase in basal area and merchantable volume over the 5-year measuring cycle [Gross growth − (Blowdowns + Mortality)].

Miron et al. 2006). This is encouraging as Muskuchii is the site with the most topographic variation and therefore the highest risk of blowdown within the network. At Dufay, there was a slight increase in basal area and volume of merchantable stems, but it was inferior to that of the control. Several factors could explain these variable results, including different reaction times of the residual trees to the opening of the canopy, different degrees of exposure to wind in the different sites, as well as the small number of stems that remain in the CPPTM. That these results are preliminary as they include only five seasons and two sites.

The amount of wounding to residual merchantable stems occasioned during harvest varied between 11 and 17% in Villars, Fénélon and Puiseaux. The rate did not vary between sites, but was very variable from one plot to the next. The types of machinery used as well as the experience of the operator in these three sites do not explain the wounding rate. The initial structure of the stand, at a small scale, is the most likely explanation for the variable wounding rate, as it may be difficult to avoid wounding trees in dense thickets. After one year of growth, 11.5% of wounded stems were dead or blown over, a rate that does not differ from that of stems that were not wounded.

CAMC and CPRS were also compared, for the same area, in terms of productivity of operations. A study was completed by the Forest Engineering Institute of Canada (FERIC) during the harvest of the Gaudet site in order to determine the variation in operational productivity between CPRS and partial cut (Hillman 2003). During the study the productivity (m³/machine hours [MH]) was greater in the partial cuts (13.1 m³/MH) than in the CPRS (10.4 m³/MH). This difference may be explained by the increase in the average stem size in the partial cut compared to the CPRS. While productivity in the partial cut was greater, it is important to keep in mind that partial cuts also imply greater costs, as less fibre is harvested (Hillman 2003; Meek and Cormier 2004). A separate study, begun in 2006, will evaluate the costs associated with both types of harvest and will develop a model in order to minimize costs associated with skidding (O. Valeria, pers. comm.)

## 3.2. Objective 2: Evaluate Whether Partial Harvest Techniques Permit the Creation or Maintenance of Old-Growth Structure

Beyond the silvicultural objectives, the partial cuts were applied on the landscape in an attempt to maintain the structure of old-growth boreal forests while harvesting some merchantable timber. The success in meeting this objective varied among the sites, as different companies and contractors were involved in each site. Figure 15.2 illustrates the before and post-harvest diameter class curves of three sites, chosen to represent the variability observed across the network (Muskuchii 2, Gaudet, and Villars). In all cases merchantable trees were left post-harvest, however at Gaudet the distribution of the trees among the diameter classes differed from the pre-harvest distribution. In this case it is important to keep in mind that the objective of the cuts was not only to maintain the pre-harvest structure but to simulate changes that would occur in a naturally ageing forest, thus creating the structure of an old-growth forest. Therefore, a difference between the pre-harvest and post-harvest structure is the desired effect if the

**Figure 15.2**
**Diameter class distribution (2-cm classes) of stems before and after harvest in Muskuchii 2, Gaudet, and Villars**

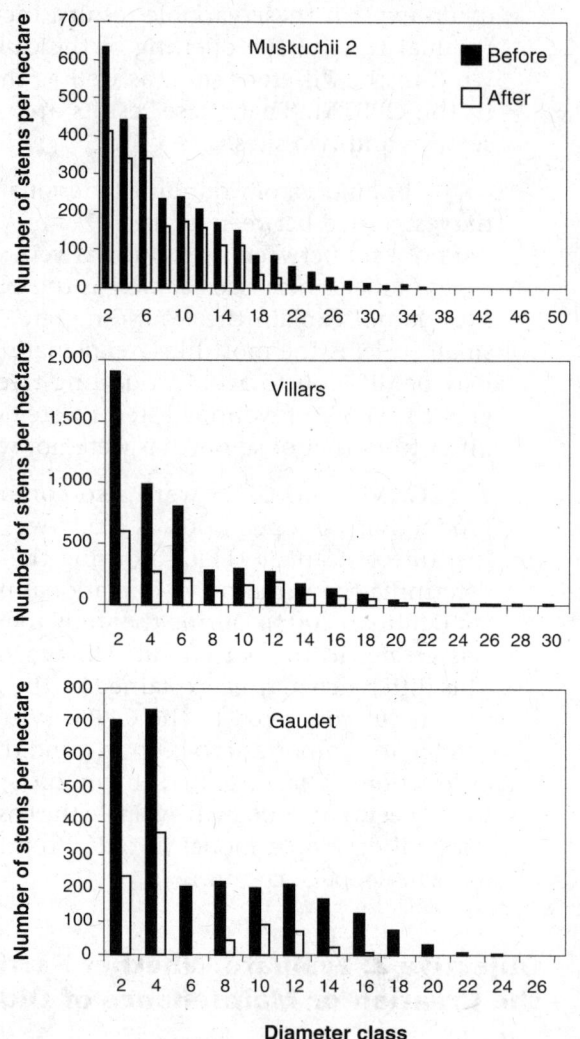

pre-harvest structure was not that of an old-growth forest. In this light, the result at Gaudet is the desired one, as the largest stems are lost from very old stands (Harper et al. 2002; Lecomte and Bergeron 2005).

Studies are currently underway to compare the post-harvest structures to natural post-fire structures that are both at the same stage of development, and older (Lecomte and Bergeron 2005; Lecomte et al. 2006). Simulations will also be performed to project the current diameter structure into the future, in order to estimate the long-term development of the post-harvest stands.

### 3.3.  Objective 3: Evaluate, Using Indicators, the Effect of Partial Harvest on Biological Diversity

A series of studies were undertaken to determine whether the stand structure maintained by partial cuts resulted in at least a community more similar to control stands or old-growth forests than traditional CPRS. Groups that have been demonstrated to be sensitive to changes in local stand structure were chosen as indicator species for the old-growth ecosystem; these were forest birds, small mammals, understory vegetation, bryophytes (*Sphagnum* spp.), and epiphytic lichens (see Drapeau et al., chapter 14). Studies on lichens and arthropods are still underway.

### 3.3.1.  Forest Birds

Despite the fact that most boreal forest birds are adapted to large-scale disturbance on the forest landscape, some species are more strictly associated with old-growth forests and their abundances decline when the landscape is dominated by clear-cuts (Imbeau et al. 1999; Drapeau et al. 2003). The composition of the community and the abundance of certain species of forest birds can be considered as old-growth forest indicators. More specifically, the presence of certain species in partial cuts would indicate that an old-growth structure had been maintained. Studies across North America, especially in the west, indicate that forest bird species richness (i.e., number of species present) increases after partial harvest, however the results are variable for old-growth indicator species, with short term effects being negative or non-existent (see Steventon et al. 1998; Young and Hutto 2002).

The abundance and richness of the forest bird community was assessed before and one year after CAMC at Maïcasagi (Poulin 2005). This site was selected as the partial cut area is large enough for effects to be attributable to forest harvest and not the surrounding landscape. A distinction was made during sampling between sections of the partial cut blocks that were actually harvested (i.e., the machinery trails and the reach of the machinery arm) and the areas within the cut block that were unharvested (between the machinery trails beyond the reach of the machinery arm).

The overall richness and abundance of the community increased at point counts that had been harvested compared to point counts within the control and the unharvested areas of the partial cuts (table 15.3). This increase in richness was driven by an increase in the number of generalist species and species associated with open areas. Overall, old-growth forest species did not decrease in richness or abundance one year after partial harvest. At the level of individual species only one species, the Golden-crowned Kinglet (*Regulus satrapa*), decreased in abundance after partial harvest (table 15.4). The remaining old-growth forest species observed both before and after harvest (Brown Creeper [*Certhia americana*], Swainson's Thrush [*Certhia americana*], Boreal Chickadee [*Poecile hudsonicus*], American Three-toed Woodpecker/Black-backed Woodpecker [*Picoides dorsalis/ P. articus*], Winter Wren [*Troglodytes troglodytes*], Red-breasted Nuthatch [*Sitta canadensis*]) did not decrease in abundance after harvest, and in fact the abundance of the Winter Wren increased significantly after harvest. This increase cannot however be linked to partial harvest since no harvest took place in the control sites.

**Table 15.3**

**Mean abundance (standard error) and richness of different elements of the forest bird community before and after harvest in control (n = 50), unharvested portions of partial cuts (n = 17), and harvested portions of partial cuts (n = 29; Poulin 2005)**

| | Control | | Unharvested areas in the partial cut | | Harvested areas in the partial cut | |
|---|---|---|---|---|---|---|
| | **Before** | **After** | **Before** | **After** | **Before** | **After** |
| Total richness | 9.78 (1.82) | 7.62 (2.11) a | 9.18 (2.24) | 8.35 (1.90) ab | 8.69 (2.77) | 9.14 (2.13) b |
| Old-growth forest species richness | 3.00 (1.07) | 2.42 (1.20) | 2.94 (1.09) | 2.24 (1.89) | 2.55 (1.24) | 2.10 (1.14) |
| Generalist species richness | 6.78 (1.34) | 5.20 (1.74) a | 6.24 (1.89) | 6.12 (1.65) b | 6.14 (2.07) | 7.03 (1.78) b |
| Total abundance | 12.10 (2.80) | 9.16 (2.80) a | 11.24 (3.21) | 9.64 (2.34) ab | 10.17 (3.37) | 10.79 (2.46) |
| Abundance of old-growth forest species | 3.70 (1.36) | 2.70 (1.42) | 3.47 (1.94) | 2.29 (1.21) | 2.59 (1.30) | 2.17 (1.20) |
| Abundance of generalist species | 8.40 (2.19) | 6.46 (2.41) a | 7.76 (2.91) | 7.35 (2.06) ab | 7.58 (2.68) | 8.62 (2.06) b |

1. The data is based on 50 listening stations systematically distributed in the partial cut block and the control. Each station was sampled for 15 minutes between 5:30 am and 9:00 am in June.

Note: Different letters indicate significant differences among the different treatments after harvest (p < 0.05; ANOVA).

**Table 15.4**

**Mean abundance (standard error) before and after harvest of old-growth forest species in control (n = 50), unharvested portions of partial cuts (n = 17), and harvested portions of partial cuts (n = 29; Poulin 2005)**

| | Control | | Unharvested areas in the partial cut | | Harvested areas in the partial cut | |
|---|---|---|---|---|---|---|
| | **Before** | **After** | **Before** | **After** | **Before** | **After** |
| Brown Creeper | 0.48 (0.54) | 0.36 (0.56) | 0.71 (0.69) | 0.41 (0.51) | 0.28 (0.45) | 0.28 (0.45) |
| Swainson's Thrush | 0.32 (0.47) | 0.46 (0.54) | 0.76 (0.66) | 0.59 (0.51) | 0.38 (0.49) | 0.66 (0.61) |
| Boreal Chickadee | 0.34 (0.52) | 0.26 (0.53) | 0.29 (0.59) | 0.24 (0.44) | 0.34 (0.48) | 0.21 (0.41) |
| American Three-toed Woodpecker/Black-backed Woodpecker | 0.08 (0.27) | 0.28 (0.50) | 0.06 (0.24) | 0.12 (0.33) | 0.14 (0.41) | 0.14 (0.44) |
| Winter Wren | 1.02 (0.50) | 0.68 (0.68) a | 1.00 (0.79) | 0.35 (0.61) ab | 0.79 (0.49) | 0.07 (0.26) b |
| Red-breasted Nuthatch | 1.42 (0.52) | 0.52 (0.54) a | 0.59 (0.51) | 0.47 (0.51) b | 0.52 (0.51) | 0.69 (0.47) b |

Different letters indicate significant differences among the different treatments after harvest (p < 0.05; ANOVA).

These results suggest that the partial cuts were successful in maintaining a structure and density that fulfilled the habitat requirements of the old-growth forest species. However, two points need to be considered before the widespread generalization of these results. The forest bird community was only assessed one year after harvest, and it is possible or even likely that population level effects would not appear until a later point after harvest, as has been suggested by previous studies (Schmieglow et al. 1997; Chambers et al. 1999). As a consequence it is premature to state that partial harvest had no impact on the majority of old-growth forest birds. Secondly, the site used in this study experienced a fairly light level of extraction, with only 33% of the stems removed, compared to the rest of the partial harvest network (table 15.1). While several studies indicate that harvest levels around 1/3 do not effect old-growth forest species abundance (Chambers et al. 1999; Simon et al. 2000), it is not certain that higher levels of extraction (2/3 – 3/4 in most of the network) will have the same effect. However, mature and overmature stands in this landscape are fairly open by nature (Lecomte et al. 2006), and the opening of the canopy at higher levels of extraction (i.e, 66%) may be closer to an extraction level of 33% in a more closed stand. In order to verify whether these results apply at higher harvesting levels, further work should be undertaken.

### 3.3.2.   Small Mammals

Small mammals are generally considered as keystone species in forest ecosystems as they are both the prey of terrestrial and avian predators (Potter 1978; Hanski et al. 1991) and the consumers of seeds, mushrooms, lichens, plants, and invertebrates (Martell 1981; Martell and Macaulay 1981; Ure and Maser 1982; MacCay and Storm 1997). The response of the small mammal community to partial cuts compared to control and CPRS was studied in both the black spruce–feather moss (site Muskuchii) and boreal mixedwood regions (Dufay site; Cheveau 2003). Two species were examined in more detail: a forest specialist, the red-backed vole (*Clethrionomys gapperi*), and a generalist frequently associated with open areas, the deer mouse (*Peromyscus maniculatus*). The small mammal community was assessed during trapping sessions that lasted eight to twelve days, using live and pitfall traps.

Overall, species composition in partial cuts was more similar to the uncut control than to CPRS in both the black spruce–feather moss and mixedwood ecoregions. However, in the black spruce region there was little difference in species composition among the three treatment types, perhaps because of the relatively open character of the old forest in this region. The populations of the two indicator species varied the most among the treatments. Red-backed vole (old-growth forest specialist) abundance in partial cuts was intermediate to its abundance in the control and the CPRS in the black spruce–feather moss region (figure 15.3), while in the mixedwood region red-backed vole abundance was highest in the partial cuts and control. Deer mouse abundance was highest in the CPRS in the black spruce–feather moss and mixedwood regions.

Overall, these results suggest that partial cuts offer an intermediate level of disturbance compared to CPRS and unharvested forests. While falling short of maintaining the same levels of small mammal populations, which a precise imitation of old-growth structure would aim to do, the intermediate level of disturbance of partial cuts may permit the maintenance of small mammal populations on the managed landscape.

**Figure 15.3**

**Red-backed vole and deer mouse abundance in spruce (Muskuchii) and mixed boreal (Dufay) forests in the three treatment types**

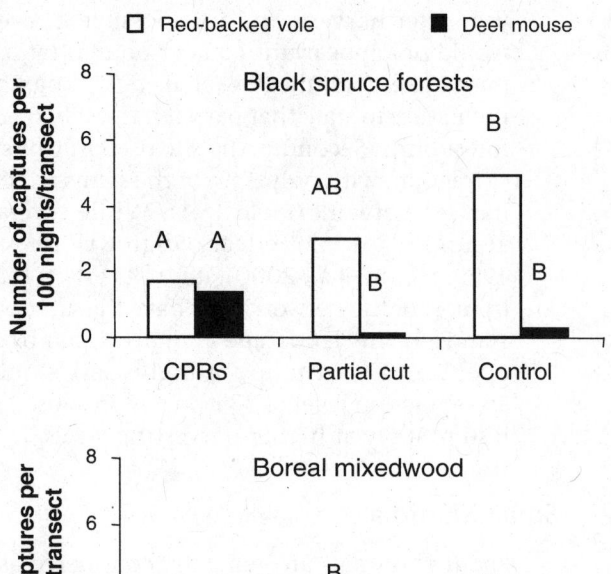

Different letters indicate significant differences among treatments for each species. The abundance of red-backed voles in the partial cut treatment is intermediate to its abundance in the control and the CPRS treatments in the spruce forest. However in the boreal mixedwood red-backed vole abundance was higher in the partial cut and control blocks. Deer mouse abundance was higher in the CPRS in both ecoregions.

### 3.3.3.    Understory Vegetation•

• For simplicity plant names are only listed in Latin, following the nomenclature of Marie-Victorin (1995).

Vegetation on the forest floor (i.e. understory) is a good indicator of the status of a forest, as it is directly influenced by the canopy (Macdonald and Fenniak 2007; Roberts 2007). The trees control the amount of light and water available to the understory as well as the quality and quantity of litter that falls on the plants. Therefore partial cuts will have both a direct effect (linked to mechanical disturbance during harvest) and an indirect effect (linked to the removal of trees and opening of the canopy) on the composition of the understory. For these reasons, the understory vegetation (including seedlings of tree species) was assessed in CPRS, partial cut and unharvested control, five years after treatment in sites Muskuchii 1, 2 and 3 (table 15.5).

Table 15.5
## Frequency (absolute) and mean abundance and standard error (%) of understory plants, five years after harvest in the three Muskuchii sites

| | Control | | Partial cut | | CPRS | |
|---|---|---|---|---|---|---|
| | Freq. | Abundance | Freq. | Abundance | Freq. | Abundance |
| **Tree seedlings** | | | | | | |
| *Abies balsamea* | 11 | 5.09 ± 5.24 | 5 – | 5.40 ± 2.51 | 19 + | 10.37 ± 8.78 |
| *Betula papyrifera* | 7 – | 1.86 ± 1.57 | 20 | 4.23 ± 5.45 | 42 + | 4.74 ± 4.76 |
| *Picea mariana* | 72 | 9.03 ± 7.29 | 90 + | 7.24 ± 6.05 | 45 – | 7.69 ± 6.34 |
| *Pinus banksiana* | 0 – | 0 | 2 – | 1.00 | 42 + | 6.38 ± 5.00 |
| *Populus tremuloides* | 2 – | 3.00 ± 2.83 | 9 | 3.44 ± 3.43 | 19 + | 6.58 ± 5.98 |
| *Prunus pensylvanica* | 2 – | 1.00 | 2 – | 3.50 ± 2.12 | 31 + | 5.26 ± 3.44 |
| **Pioneer species** | | | | | | |
| *Amelanchier* spp. | 1 – | 1.00 | 7 | 5.43 ± 3.82 | 13 + | 6.15 ± 3.85 |
| *Carex* spp. | 2 – | 5.50 ± 6.36 | 8 | 3.88 ± 3.40 | 13 + | 4.69 ± 6.29 |
| *Epilobium angustifolium* | 11 – | 4.27 ± 4.92ab | 24 | 3.63 ± 2.87a | 53 + | 8.06 ± 5.91b |
| *Gramineae* | 3 – | 5.00 ± 5.29a | 16 | 4.06 ± 5.96a | 41 + | 11.46 ± 8.95b |
| *Pteridium aquilinum* | 1 – | 7.00 | 2 – | 7.00 ± 8.49 | 7 + | 18.14 ± 9.60 |
| *Rosa acicularis* | 4 | 2.00 ± 1.41a | 11 | 2.73 ± 1.61a | 14 | 10.21 ± 6.56b |
| *Rubus idaeus* | 1 – | 2.00 | 0 – | 0 | 17 + | 11.47 ± 19.15 |
| *Salix* spp. | 11 | 7.09 ± 7.75 | 19 + | 6.74 ± 6.37 | 11 | 12.18 ± 10.88 |
| **Forest species** | | | | | | |
| *Aralia nudicalis* | 14 | 5.00 ± 4.06ab | 7 | 7.29 ± 4.15b | 13 | 3.00 ± 1.41a |
| *Clintonia borealis* | 36 | 6.67 ± 6.02ab | 37 | 4.51 ± 3.83a | 42 | 8.36 ± 6.22b |
| *Coptis groenlandica* | 24 | 7.21 ± 6.53 | 14 | 10.00 ± 8.14 | 29 | 10.34 ± 6.89 |
| *Cornus canadensis* | 81 – | 8.30 ± 6.90a | 107 | 7.85 ± 6.70a | 131 + | 12.77 ± 8.15b |
| *Diervilla lonicera* | 7 | 3.86 ± 3.76a | 5 | 4.60 ± 4.83a | 5 | 18.00 ± 9.46b |
| *Epigea repens* | 3 – | 7.67 ± 9.07a | 11 + | 3.36 ± 2.73a | 2 – | 19.50 ± 3.54b |
| *Gaultheria hispidula* | 9 | 11.22 ± 8.15b | 6 | 2.83 ± 1.84a | 6 | 3.67 ± 3.01ab |
| *Kalmia angustifolia* | 28 – | 10.61 ± 6.91a | 81 + | 14.28 ± 7.49b | 10 – | 9.40 ± 7.93a |
| *Rhododendron groenlandicum* | 14 | 8.21 ± 6.51 | 22 | 10.45 ± 7.92 | 16 | 8.50 ± 6.00 |
| *Linnaea borealis* | 25 | 8.20 ± 7.12 | 28 | 4.43 ± 4.65 | 36 | 8.47 ± 7.28 |
| *Lycopodium* spp. | 11 | 5.91 ± 5.24 | 11 | 3.36 ± 2.46 | 11 | 7.64 ± 5.57 |
| *Maianthemum canadense* | 65 | 8.66 ± 7.00 | 55 | 6.67 ± 4.77 | 63 | 6.30 ± 6.54 |
| *Rubus pubescens* | 1 | 2.00 | 5 | 15.00 ± 8.72 | 2 | 1.50 ± 0.71 |
| *Trientalis borealis* | 18 – | 2.72 ± 1.72a | 23 | 7.39 ± 8.62b | 39 + | 12.21 ± 8.95c |
| *Vaccinium angustifolium* | 88 – | 9.84 ± 8.03a | 137 + | 15.77 ± 9.08b | 105 | 10.51 ± 7.97a |
| *Vaccinium myrtilloides* | 54 | 12.59 ± 9.72 | 75 | 12.63 ± 8.68 | 61 | 9.97 ± 6.72 |
| *Viburnum cassinoides* | 1 – | 1.00 | 14 + | 5.36 ± 3.23 | 4 | 3.75 ± 3.10 |
| **Lichens** | | | | | | |
| *Cladina mitis* | 23 | 12.00 ± 8.48 | 33 + | 9.82 ± 7.25 | 2 – | 2.00 ± 1.41 |
| *Cladina rangiferina* | 55 | 13.95 ± 9.82b | 101 + | 14.30 ± 9.11b | 23 – | 6.83 ± 6.51a |
| *Cladina stellaris* | 22 | 12.59 ± 8.25 | 42 + | 10.05 ± 7.76 | 2 – | 1.50 ± 0.71 |
| **Bryophytes** | | | | | | |
| *Dicranum fuscescens* | 26 | 3.65 ± 3.42 | 23 | 2.65 ± 1.43 | 26 | 2.46 ± 1.96 |
| *Dicranum polysetum* | 45 – | 3.40 ± 3.06 | 63 | 4.48 ± 3.48 | 77 + | 4.64 ± 3.69 |
| *Hylocomium splendens* | 3 | 10.33 ± 9.71 | 5 | 3.60 ± 4.16 | 2 | 1.50 ± 0.71 |
| *Pleurozium schreberi* | 123 – | 20.62 ± 6.40c | 170 + | 18.31 ± 7.97b | 140 | 15.46 ± 7.80a |
| *Polytrichum* spp. | 8 – | 4.88 ± 3.64 | 14 | 5.64 ± 6.78 | 22 + | 5.77 ± 5.92 |
| *Ptilium crista-castrensis* | 36 + | 9.00 ± 7.63 | 18 – | 6.94 ± 7.43 | 27 | 6.67 ± 6.60 |
| *Ptilidium ciliare* | 25 – | 5.04 ± 3.71 | 58 + | 6.12 ± 5.23 | 37 – | 6.11 ± 5.23 |

Note: Tree species seedlings are included. The frequency of each species is compared among the treatments with chi-square analysis ("–" is less frequent, "+" more frequent). Abundances were also compared (ANOVA) and the letters indicate significant differences.

Overall, the differences in community composition were restricted to an increased importance of species common to disturbed areas (e.g. *Amelanchier* spp., *Carex* spp., *Epilobium angustifolium, Rubus ideaus*) in the CPRS, compared to the unharvested control. Seedlings of early successional tree species (trembling aspen, [*Populus tremuloides*], jack pine [*Pinus banksiana*], pin cherry [*Prunus pensylvanica* L.]) were almost exclusively restricted to the CPRS. Their colonization in these sites was probably due to the direct effects of harvest, i.e. the exposure of mineral soil and increased light intensity.

Some pioneer species did establish in the partial cuts, however to a lesser degree than in the CPRS. For example, *Amelanchier* spp., *Epilobium angustifolium, Carex* spp. and the grasses were more abundant in partial cuts than in the controls, but less abundant than in the CPRS. This could be due to reduced soil disturbance in the partial cut compared to the CPRS. Among the pioneer species, only the *Salix* spp. were more frequent in the partial cuts than in the CPRS.

No forest species were restricted to control or partial cut sites, and in fact many species generally considered "forest species" increased in frequency and or abundance in partial cuts and CPRS (e.g., *Cornus canadensis, Trientalis borealis*). A few species were more frequent and abundant in the partial cuts, particularly black spruce, *Vaccinium myrtilloides* and *Kalmia angustifolia*. These species were able to take advantage of the increased light availability in the partial cuts compared to the control. A greater impact of the cuts was seen in the lichens and mosses, as lichen and moss cover decreased from control to partial cut to CPRS.

These results suggest that partial cuts have an intermediate effect on the understory community compared to CPRS and unharvested areas. The reduced abundance and frequency of weedy species suggest that the long-term impact of this treatment on the community may be limited. Particularly the maintenance of lichen cover in the partial cuts is interesting as it suggests that these treatments may maintain crucial woodland caribou (*Rangifer tarandus*) habitat. Finally, the abundant establishment of black spruce seedlings after partial harvest suggests that this treatment type could favour regeneration and create multilayer structured stands (see Bouchard, chapter 13) in the future.

### 3.3.4.  *Sphagnum* Mosses

In the black spruce–feather moss region many old-growth sites are paludified, i.e. over time they have developed a thick waterlogged layer of organic material that overlays the mineral soil. This transition to a treed peatland can be seen in changes in the species of *Sphagnum* (generally water-loving peatland mosses) on the forest floor (Fenton and Bergeron 2006; Simard et al., chapter 11). A study was undertaken to see if the *Sphagnum* spp. present after CAMC were more similar to those found in old-growth paludified forest than the *Sphagnum* spp. found in the control or the CPRS sites. The three sites chosen for this study were paludified before harvest.

After partial harvest, species common to old-growth paludified sites increased in abundance (figure 15.4) compared to control and CPRS sites (Fenton and Bergeron 2007). While this indicates that the structure and conditions created by partial cut had the desired effect on this component of the ecosystem,

Figure 15.4
## Cover (%) of three *Sphagnum* species in the three treatments (control, partial cut, and CPRS) in Fénélon, Puisseaux, and Gaudet

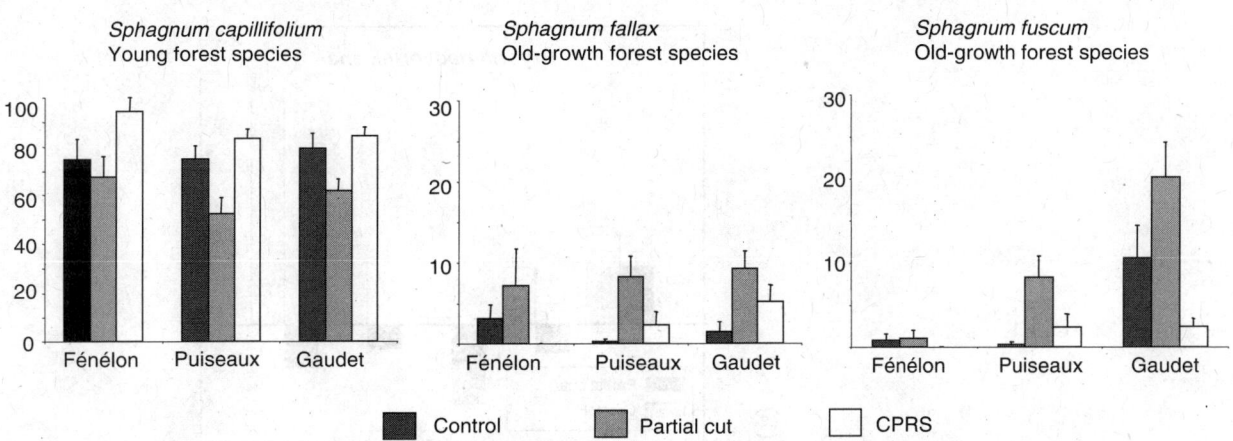

The abundance of species associated with paludified old-growth species increased after partial cut. From Fenton and Bergeron (2007).

it may also indicate a problem. Paludified forests have a lower productivity than non-paludified forests, and if partial cuts promote paludification in the black spruce–feather moss region in the Clay Belt the long-term productivity of these sites will be compromised (see Simard et al., chapter 11). However, this study was completed only one to two years after harvest and this trend needs to be verified in a longer-term study.

### 3.3.5. Epiphytic Lichens

Lichens are one of the most species-rich groups in old-growth forests (Boudreault et al. 2003). As microclimatic changes (increased temperature and wind, decreased humidity) associated with changes in stand structure have a significant impact on epiphytic lichens (lichens growing on trees), they are an important indicator of the conservation value of partial cuts. Two species of epiphytic lichens were chosen for study: *Evernia mesomorpha*, which is common in open forests, and *Bryoria nadvomikiana*, which is common in closed humid forests. Both species were installed in a growth experiment in the partial cut and control blocks of three sites: Maïcasagi, Fénélon, and Muskuchii. The lichens were weighted twice annually for two years.

Overall both species grew at a higher rate in the control than in the partial cut (10.0% vs. –1.8% for *Bryoria* and 7.3% vs. 2.6% for *Evernia*). During the first year a seasonal effect was observed (figure 15.5): during the summer, both species grew as well or better in the partial cuts than in the control, but the situation changed during winter when the lichens in the partial cuts grew significantly slower than those in the control, and even lost biomass in some cases. During the second year (summer and winter) growth of both species was lower in the

**Figure 15.5**
## Winter and summer growth rates of two epiphytic lichen species in control and partial cut blocks

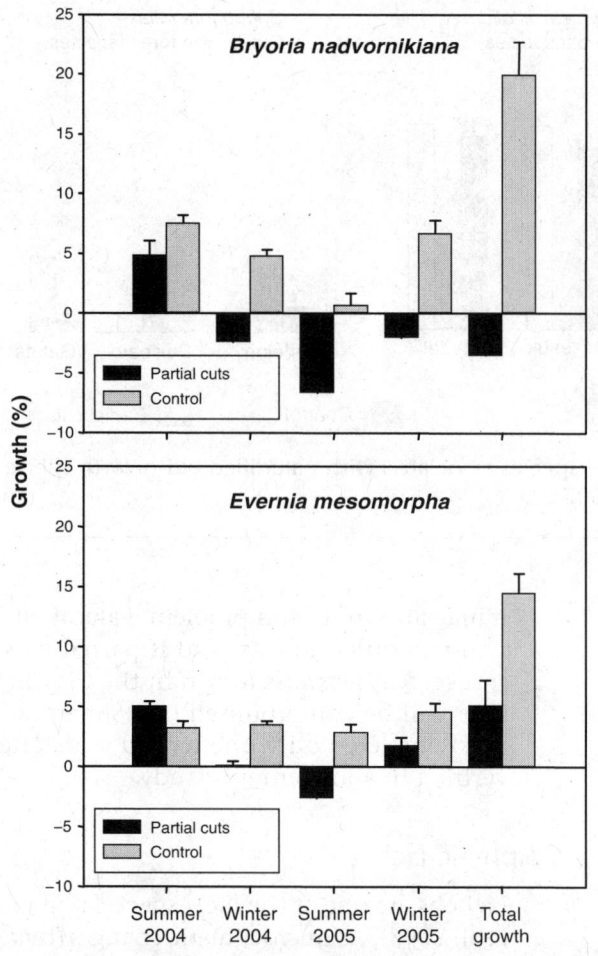

During the first growing season the summer growth rates in the partial cut blocks were equivalent to or faster than the control. However, during winter growth was significantly slower in the partial cuts than in the controls. During the second growing season (winter and summer) the growth rates of both species was significantly lower in the partial cuts than in the controls.

partial cuts than in the control, and in the partial cuts a greater number of individuals lost mass. Furthermore *Bryoria*, the closed-forest species, lost more biomass than *Evernia*, the open-forest species. The more open canopy in the partial cuts probably explains the higher summer growth rate of the lichens during the first growing season. However, in the following seasons, the different microclimatic conditions in the two treatment types resulted in very different growth rates. The lower growth rate in the partial cut could be due to greater penetration of sunlight and resultant drying, which can damage the lichens, and also greater wind speed, which can fragment the lichens. This is particularly true of *Bryoria* with its hanging filamentous shape. While partial cuts maintain populations of epiphytic lichens by maintaining their habitat, the quality of this habitat is reduced, at least in the short term.

## 3.4. Objective 4: Evaluate the Effect of Partial Harvest on Target Game Species (Spruce Grouse and Snowshoe Hare)

Game species can be affected by partial cuts in a variety of ways, notably the reduction in horizontal cover, which is necessary for predator avoidance. In northern spruce forests the most common small-game species are the Spruce Grouse and the snowshoe hare. The short-term effects of partial cuts on these species was evaluated at Maïcasagi and in the surrounding cut blocks that had been harvested with a seed-tree retention cut (Ruché 2005; Valois 2005). Seed-tree retention is a type of partial cut generally used in mature even-aged stands where the dominant trees are gradually harvested (see Bouchard, chapter 13). The dominant trees left after harvest serve as seed sources, favouring natural regeneration (MRNF 2005). These sites were chosen as they were the only ones with a treated surface area large enough to effectively study the impact of harvest type, as the game species have home ranges over ten hectares in size.

In ground-nesting birds such as the Spruce Grouse, the probability of predation is strongly influenced by nest visibility. Vegetation around the nest reduces the transmission of auditory, olfactory, and visual signs to predators. Partial cuts may then increase the vulnerability of this species to predation by simplifying the vegetative structure and reducing the tree density around the nest. On the other hand, the females cover the eggs with leaves when they are laid, a behavioural adaptation that reduces the visibility of the nest and predation risk. Artificial nests similar to Spruce Grouse nests were used to compare, during the egg-laying period, the predation risk in stands treated with CAMC, seed-tree retention (24 and 34% of surface area harvested, respectively), and control. The effect of covering both plasticine and chicken eggs with leaves on clutch survival was also assessed. The results were interpreted using three definitions of predation: 1) the classic definition, generally used with artificial nests (i.e., a nest has been predated when at least one egg has been destroyed, displaced, removed or bitten); 2) a definition that does not take into account the predation of mammals too small to consume Spruce Grouse eggs; and 3) a new definition that takes into account events that might realistically affect the survival of the eggs. The primary result of this study is that the predation rate did not significantly differ between harvested and unharvested stands, regardless of the definition of predation used (figure 15.6). Other studies are currently under way to determine whether partial cuts provide adequate habitat for Spruce Grouse during the spring displaying season, and the adult male summer molt.

The short-term effect of partial cutting of retention blocks within a mosaic harvest system on the snowshoe hare was also examined. Snowshoe hares are considered a keystone species in boreal spruce forests, as they are the prey of several aerial and terrestrial predators. Partial cuts were completed in the spring of 2002 in residual blocks, leaving some clusters of trees intact (Maïcasagi site).

Figure 15.6
**Percentage of predated and non-predated nests in the control and partial cut blocks***

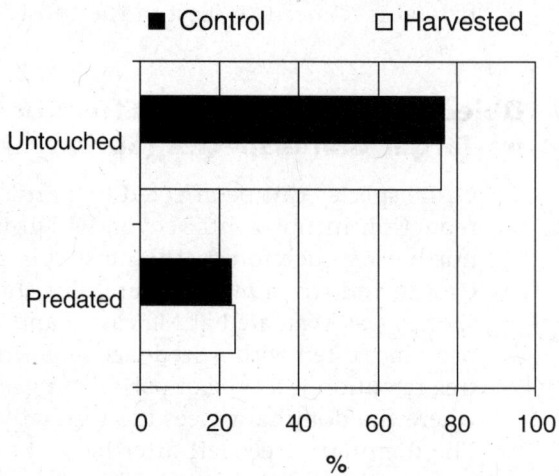

*For Maïcasagi site, partial cuts and seed-tree cuts were pooled.
Predated nests had at least one egg that was destroyed, moved or bitten. Predation rate did not vary among the partial cut and control blocks.

Surveys of tracks, hare pellets, and browsed stems (vegetation consumed by hare) were performed to determine the use by snowshoe hares of the different habitat types. During the year following harvest snowshoe hares infrequently used the harvested habitat, while they continued to be present in unharvested habitat. Similar results were found in the seed-tree harvest, one to three years after harvest (figure 15.7).

Snowshoe hares appear to be negatively affected by partial cuts, due to a reduction in vegetative cover. Within the variably harvested sites, hares were primarily found in untouched patches where vertical cover and the surface area of deciduous stems were higher. While our results indicate that partial cuts have a negative impact on snowshoe hares, they also suggest that if partial cuts are applied in an irregular (heterogeneous) way across the cut block (leaving some uncut patches) they might permit the short-term maintenance of hares and possibly other small mammal populations. The retained patches should be targeted at areas with particularly suitable habitat for this species, i.e. where there is considerable lateral cover and available food, to provide the best possible habitat within the sub-optimal landscape. The size and distribution of these retention patches within partial cuts should be determined via further studies in order to determine the best combinations to maintain this species within harvested blocks.

**Figure 15.7**
**Presence (%) of snowshoe hare scat and browsed vegetation in partial cut and control blocks in Maïcasagi**

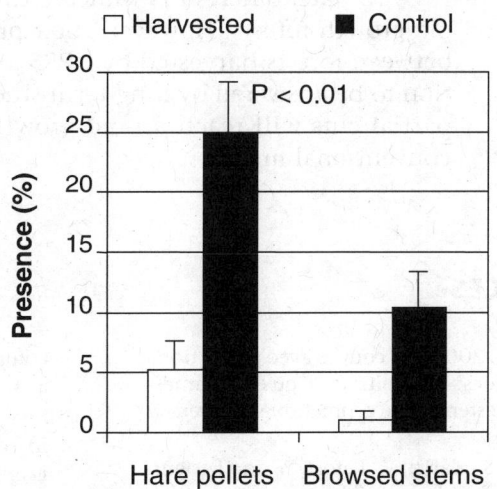

Hares were significantly less present in the partial cut than in the control blocks.

## 4.  CONCLUSION

The silvicultural results of the partial cut trials are encouraging. Partial cuts are operationally viable, at reasonable cost, using machinery that is currently in use. Mortality due to blowdowns was significant in one site with considerable topographic variation but seems to be at a more acceptable level in spruce stands on clay, the dominant forest type in this region. It is however too early to fully assess the effect of the partial cuts on the growth of the residual stands, and a definitive answer will not be available until all of the sites have been remeasured. Paludification remains an important problem on this landscape, and the possibility of it being accelerated by partial cuts is an important element to consider. In paludified stands, the choice between a partial cut and a regeneration cut followed by site preparation (to diminish the thickness of the organic layer) should be made in function of the state of each stand, taking into account both the structure and the actual and potential levels of paludification (see Simard et al., chapter 11). A relatively limited understanding of the bryophyte community would allow forest managers to make an informed choice.

Preliminary results indicate that it is difficult to reproduce old-growth structure and habitat in a short period after harvest. Partial cuts represent an intermediate level of disturbance compared to CPRS. However, as indicated by the different taxa studied, the structure created by partial cuts does not permit, in the short term, to recreate an old-growth habitat. These results are preliminary and it is too early after treatment to observe significant changes in community composition. The long-term monitoring of these sites is necessary to determine whether this type of treatment truly reproduces the types of habitat conditions required to maintain biodiversity and game species.

To date, our results indicate that ecologically partial cuts do not create old-growth forests. However, their presence on the landscape reduces the gap between forests harvested by CPRS and old-growth forests. The ultimate question to be answered by long-term studies is to what degree stands harvested by partial cuts will reach an old-growth status faster than stands harvested by conventional means.

## REFERENCES

Bégin, J. and Riopel, M. 2001. Les coupes avec protections des petites marchandes: 4 ans plus tard, où en sommes-nous? Faculté de foresterie et de géomatique, Université Laval, Québec, Que.

Bergeron, Y., Gauthier, S., Kafka, V., Lefort, P., and Lesieur, D. 2001. Natural fire frequency for the eastern Canadian boreal forest: consequences for sustainable forestry. Can. J. For. Res. **31**: 384–391.

Bergeron, Y. and Harvey, B. 1997. Basing silviculture on natural ecosystem dynamics: an approach applied to the southern boreal mixedwood forest of Quebec. For. Ecol. Manag. **92**: 235–242.

Boudreault, C., Bergeron, Y., Gauthier, S., and Drapeau, P. 2003. Bryophyte and lichen communities in mature to old-growth stands in eastern boreal forests of Canada. Can. J. For. Res. **32**: 1080–1093.

Chambers, C.L., McComb, and Tappeiner II, W.C. 1999. Breeding bird responses to three silvicultural treatments in the Oregon coast range. Ecol. Appl. **9**: 171–185.

Cheveau, M. 2003. Dynamique naturelle des petits mammifères et effets des coupes partielles sur la structure de leurs populations en forêt boréale de l'est de l'Amérique du Nord. M.Sc. thesis, Université du Québec à Montréal, Montréal, Que.

Drapeau, P., Leduc, A., Bergeron, Y., Gauthier, S., and Savard, J.-P.L. 2003. Bird communities in old lichen-black spruce stands in the Clay Belt: problems and solutions regarding forest management. For. Chron. **79**: 531–540.

Fenton, N.J. and Bergeron, Y. 2006. Facilitative succession in a boreal bryophyte community driven by changes in available moisture and light. J. Veg. Sci. **17**: 65–76.

Fenton, N.J. and Bergeron, Y. 2007. *Sphagnum* community change after partial harvest in black spruce boreal forests. For. Ecol. Manage. **242**: 24–33.

Hanksi, I., Hansson, L., and Henttonen, H. 1991. Specialist predators, generalist predators, and the microtine rodent cycle. J. Ani. Ecol. **60**: 353–367.

Harper, K., Bergeron, Y., Gauthier, S., and Drapeau, P. 2002. Post-fire development of canopy structure and composition in black spruce forests of Abitibi, Quebec: a landscape scale study. Silva Fenn. **36**: 249–263.

Hillman, D. 2003. Harvesting with the protection of small merchantable stems: costs and implementation. Forest Engineering Research Institute of Canada (FERIC), Pointe-Claire, Que. Internal Report RI-2003-04-24.

Imbeau, L., Savard, J.-P.L., and Gagnon, R. 1999. Comparing bird assemblages in successional black spruce stands originating from fire and logging. Can J. Zool. **77**: 1850–1860.

Lecomte, N. and Bergeron, Y. 2005. Successional pathways on different surficial deposits in the coniferous boreal forest of the Quebec Clay Belt. Can. J. For. Res. **35**: 1984–1995.

Lecomte, N., Simard, M., and Bergeron, Y. 2006. Effects of fire severity and initial tree composition on stand structural development in the coniferous boreal forest of north western Quebec, Canada. Écoscience, **13**: 152–163.

Macdonald, S.E. and Fenniak, T.E. 2007. Understory plant communities of boreal mixedwood forests in western Canada: natural patterns and response to variable-retention harvesting. For. Ecol. Manag. **242**: 34–48.

Marie-Victorin, F. 1995. Flore laurentienne. 3rd edition. Les Presses de l'Université de Montréal, Montréal, Que.

Martell, A.M. 1981. Food habits of southern red-backed vole (*Clethrionomys gapperi*) in northern Ontario. Can. Field. Nat. **95**: 319–324.

Martell, A.M. and Macaulay, A.L. 1981. Food habits of deer mice (*Peromyscus maniculatus*) in northern Ontario. Can. Field. Nat. **95**: 325–328.

McCay, T.S. and Storm, G.L. 1997. Masked shrew (*Sorex cinereus*) abundance, diet, and prey selection in an irrigated forest. Amer. Mid. Nat. **138**: 268–275.

Meek, P. and Cormier, D. 2004. Studies of the first entry phase in a shelterwood harvesting system. Advantage, **5**(43).

Ministère des Ressources naturelles (MRN). 2002. Normes d'inventaire forestier: placettes échantillons permanentes.

Direction des inventaires forestiers, Ministère des Ressources naturelles du Québec.

Ministère des Ressources naturelles et de la Faune (MRNF). 2005. Instructions relatives à l'application du règlement sur les valeurs des traitements sylvicoles admissibles en paiement des droits (Exercice 2005–2006). Ministère des Ressources naturelles et de la Faune.

Miron, S., Riopel, M., and Bégin, J. 2006. Expérimentation de coupes avec protections des petites tiges marchandes. Résultats du remesurage 5 ans après une CPPTM. Université Laval, Québec, Que.

Potter, G.L. 1978. The effect of small mammals on forest ecosystem structure and function. *In* Populations of small mammals under natural conditions. *Edited by* D.P. Snyder. Pymatuning Laboratory of Ecology, Special Publication No. 5., University of Pittsburg, Pittsburg, USA, pp. 181–191.

Poulin, M. 2005. Effets des coupes partielles sur les oiseaux en forêt de pessière à mousses de l'Est du Canada. M. Sc. thesis, Université du Québec à Montréal, Montréal, Que.

Roberts, M.R. 2007. A conceptual model to characterize disturbance severity in forest harvests. For. Ecol. Manag. **242**: 58–64.

Ruché, D. 2005. Influence de la dispersion des coupes totales et du traitement en coupes partielles sur la qualité du Tétras du Canada dans la pessière noire à mousses de l'Ouest du Québec. M.Sc. thesis, Université du Québec en Abitibi-Témiscamingue, Rouyn-Noranda, Que.

Schmiegelow F.K.A., Machtans, C.S., and Hannon, S.J. 1997. Are boreal birds resilient to forest fragmentation? An experimental study of short-term community responses. Ecololgy, **78**: 1914–1932.

Simon, N., Schwab, F.E., and Diamond, A.W. 2000. Patterns of breeding bird abundance in relation to logging in western Labrador. Can. J. For. Res. **30**: 257–263.

Steventon, J.D., MacKenzie, K., and Mahon, T. 1998. Response of small mammals and birds to partial cutting vs clearcutting in northwest British Columbia. For. Chron. **74**: 703–713.

Ure, D.C. and Maser, C. 1982. Mycophagy of red-backed voles in Oregon and Washington. Can. J. Zool. **60**: 3307–3315.

Valois, S. 2005. Influence à court terme de la coupe partielle sur des mammifères de la forêt boréale. M.Sc. thesis, Université du Québec à Rimouski, Rimouski, Que.

Young, J.S. and Hutto, R.L. 2002. Use of landbird monitoring database to explore effects of partial-cut timber harvesting. For. Sci. **48**: 373–378.

# Chapter 16

## Modelling Complex Stands and the Effects of Silvicultural Treatments*

*Jean-Pierre Saucier and Art Groot*

* Part of the work presented in this case study was supported by the Lake Abitibi Model Forest and the Sustainable Forest Management Network. We would like to thank the field staff, including Leanne McKinnon and Michael Adams. We would also like to thank Alain Leduc, Sylvie Gauthier, and Marie-Andrée Vaillancourt, whose comments contributed to the final version of the manuscript. The photos on this page were graciously provided by Jean-François Bergeron (MRNF) and Claude Morneau (MRNF).

# 1. INTRODUCTION

Once the principles of forest management based on natural ecosystem dynamics have been established, we must be able to identify the appropriate interventions for maintaining or recreating the natural structures observed in multi-cohort stands. To achieve this, siviculurists will likely have to either apply silvicultural treatments which were rarely used in the past and under certain conditions, or devise new treatments. How the stands will respond to these treatments is therefore unknown. Forest managers, who must predict the future evolution and availability of the forest resource over a long time frame in order to ensure sustainability of the resource, will also need to know how stands subjected to this new silvicultural regime will behave and how the forest landscape will evolve. Since there is no long-term experience with management based on the Three-cohort Model (*sensu* Bergeron et al. 1999), modelling is an essential component for implementation of this management approach.

For forest management purposes, growth models have traditionally been developed for monospecific stands containing a single age cohort. As our knowledge of natural disturbances increases, these models appear to be increasingly inadequate for the multi-aged stands that frequently result from these disturbances. Furthermore, management of these types of stands using treatments designed to maintain forest cover over long periods of time will create highly heterogeneous stands, often containing more than one tree species and with an irregular diameter distribution.

Reproducing the dynamics of these structurally complex stands, whether of natural or anthropogenic origin, and predicting their reactions to silvicultural interventions requires the development of flexible models that accurately reproduce the growth of each tree in the stand and their interactions. Single-tree-level modelling is particularly necessary given the lack of long-term data for calibrating whole-stand models for all the ranges of density and treatment intensity with different originating structures and compositions. However, in order to ensure the accuracy of a tree model while providing a better estimate of error propagation, it is useful to work at both the individual-tree and stand levels by introducing site quality and resource availability constraints into the model.

Landscape-level models are also required in order to reproduce the mosaic of ecosystems at various successional stages that will result from forest management actions. By using these tools, we can ensure longer-term attainment of ecosystem-based forest management objectives and visualize the distribution of multi-cohort stands in the landscape.

In this chapter, we briefly describe different possible modelling approaches at the stand level with a quick review of their limitations and advantages. We then present a model currently being developed by our team, and briefly describe the landscape-level approaches and discuss the qualities desired for this type of modelling. We conclude that the choice of model depends on the questions that we want answered and that a change in scale frequently comprises a change in underlying processes. The challenges posed by each of these scales (stand and landscape) may therefore require the use of different models.

## 2. STAND-LEVEL MODELLING

### 2.1. Possible Modelling Approaches

Although modelling approaches form a continuum, ranging from the simplest to the most complex, they can be classified into three major categories: 1) whole-stand models, 2), single-tree distance-independent models, and 3) single-tree distance-dependent models (Munro 1974; Vanclay 1995). Each of these categories can also be further subdivided depending on whether the model is adjusted empirically using statistical relationships (empirical-based models) or is based on the simulation of growth-related processes (process-based models) (Mäkelä et al. 2000; Johnsen et al. 2001). For more details on the approaches applicable to mixed or irregular stands, see Vanclay (1995), Peng (2000), Franc et al. (2000), as well as Porté and Bartelink (2003).

### 2.1.1. Whole-Stand Models

Whole-stand models are commonly used in forest planning to generate empirical (Plonski 1974; Boudoux 1978) or normal yield tables (Vézina and Linteau 1968; Pothier and Savard 1998). They are generally designed to estimate forest yield as a function of site richness and incorporate natural mortality due to self-thinning. Most were developed for monospecific even-aged stands and very few incorporate the effect of tree senescence on stand characteristics (Pothier and Savard 1998). These models are therefore unsuitable for modelling yields in stands with a more complex structure (second and third cohorts for example). They are also not well adapted for simulating the effect of silvicultural treatments such as partial cutting treatments on the growth of residual stems or for taking treatment-related mortality into consideration. However, they are easy to use and require little computational power, which can be an advantage in fitting a landscape-level model. Some process-based models exist for whole stands, for example the 3-PG model (Landsberg and Waring 1997; Fournier et al. 2000), but they are used to estimate productivity rather than to model yield or succession (Groot et al. 2004).

Whole-stand models meet most management planning needs for even-aged forest stands. However, the lack of data for the full range of structures and silvicultural treatments limits the ability to calibrate these models. Consequently, they cannot adequately meet the planning needs of multi-aged stands (Groot et al. 2004). Furthermore, since these models are generally only adjusted using data from stands aged 20 to 100 years, the stand senescence phase, during which the stand goes through the transition from an even-aged structure (first cohort) to a stand with more than one tree cohort (second and third cohorts), is not covered. Since a large proportion of the area in the boreal zone may be occupied by stands over 100 to 150 years old, depending on the disturbance and cutting regime, whole-stand models may prove inadequate for the purposes of ecosystem-based forest management. It is therefore necessary to develop a type of model that can express the dynamics of stand structure and composition. One possible solution is to subdivide stand growth into elemental growth units, i.e. the trees.

### 2.1.2.  Single-Tree Distance-Independent Models

Single-tree distance-independent models estimate the growth of trees in a stand as a function of their individual characteristics, stand characteristics and site characteristics. However, since the exact location of the tree relative to its neighbours is not known, the stand and site characteristics are similar for all trees, while the tree characteristics are identical for all trees of the same size. Mortality is generally expressed as a function of tree and stand parameters. In some cases, regeneration and understory vegetation growth as well as susceptibility to insect and disease attacks may be included in the model (Groot et al. 2004).

The Forest Vegetation Simulator (FVS), developed by the USDA Forest Service, is an example of a single-tree distance independent model. It incorporates a number of regional variants applicable to eastern North America, including N-Twigs for New England (Bankowski et al. 1996) and TWIGS/STEMS for the Great Lakes region (Miner et al. 1988). However, the applicability of the various models included in the FVS to a specific region must be verified. A version, called FVS[ontario], was calibrated for Ontario using permanent plot data from the boreal zone and from the Great Lakes–St. Lawrence zone (Lacerte et al. 2006). This recalibration for Ontario was necessary since the validation of the existing FVS versions with Ontario boreal data had produced unsatisfactory results (Lacerte et al. 2006).

The difficulty in using singe-tree distance-independent models to estimate the effect of partial disturbances or silvicultural treatments lies in the fact that inter-stem competition is considered similar for stems of the same diameter, regardless of their social position. In multi-aged stands, where openings are created in the stand, stems of the same size can have very different competitive environments. Moreover, Hökkä and Groot (1999), who developed a single-tree distance-independent model for peatland black spruce (*Picea mariana* [Mill.] B.S.P.) stands, were uncertain about their model's ability to predict stem development for stands with clumped stem distribution, even though they were able to obtain accurate basal area growth results at the stand level over a 40-year horizon. Diameter class models constitute a variant, intermediate between whole-stand models and single-stand distance-independent models (Vanclay 1995), and are more similar to one or the other depending on the range and the number of classes considered.

### 2.1.3.  Single-Tree Distance-Dependent Models

Single-tree distance-dependent models were developed in order to explain the growth of each stem more accurately, taking into account its specific neighbourhood and competition relationships. If the position of each tree in the stand is known, an empirical model of this type can calculate a competition index specific to each tree (Peng 2000; Mailly et al. 2003). This often provides a better estimate of individual stem growth than distance-independent competition indices (Mailly et al. 2003). However, the question of whether competition indices that require knowing the location of the trees in the stand yield better results at the stand level is a matter of ongoing debate.

In this type of model, the crucial element is the selection of the factor or factors that will be used as the model engine. The amount of light intercepted by the stems is directly related to tree growth (Bartelink et al. 1997). Light is also used as the basic factor in the SORTIE process-based model, originally developed to estimate growth and recruitment of understory trees in mixed stands in eastern North America (Pacala et al. 1993, 1996) and adapted to Canada by Coates et al. (2003). However, these versions of the model do not take inter-site differences into consideration, since light is the only factor considered (Groot et al. 2004). Models that combine physical site factors and the amount of light that is intercepted by all the crowns of the stand during the growing season should yield accurate growth estimates (of both the tree and the stand) while taking into account both site quality differences and regional differences (Groot and Saucier 2008).

Single-tree models therefore appear to be better adapted to ecosystem-based forest management. In fact, relationships between stem growth and certain variables describing their environmental conditions and their social status can be used to modulate the stand's reaction to various degrees of openings created by natural disturbances, silvicultural interventions, or mortality of senescent trees. It therefore appears that single-tree distance-dependent models should be able to better reproduce the variations in the stems' reactions to changes in stand structure. This is so because stems of the same size will react differently depending on whether they are at the edge of an opening or in direct competition with larger neighbouring trees. However, distance-dependent models, which require knowing the position of each tree, cannot be applied directly to forest inventory data, where individual stems are not mapped. Development of these models requires the collection of very detailed data, which is very costly. Consequently, their use could be limited to research applications (Vanclay 1995). The use of various modelling scales must therefore be considered in order to meet all needs. For example, the average response of a number of simulations ($n > 1,000$) could be used to estimate the response of a stand population in a management unit to a particular type of silvicultural treatment. In addition, as Vanclay (2003) pointed out, new data acquisition methods, such as airborne LIDAR, which can provide the coordinates, crown width, and height of all the trees in a stand, are opening up the path to the development of models based on completely new principles.

## 2.2. Essential Model Qualities

In order to select a modelling approach capable of meeting ecosystem-based forest management needs, it is necessary to draw up a list of essential and desirable functions in models based on the characteristics of the stands for which they will be used and on the ecosystem mosaics that comprise these stands. The goal of the modelling exercise may differ: for instance, some model families will be aimed mainly at providing a better understanding of forest succession dynamics and stand response to new silvicultural practices, while others will have the sole objective of predicting stand yield. Depending on the objective in question, the emphasis placed on accuracy will vary significantly, as well as the field of application. It is also necessary to ensure model robustness, i.e. the ability of the model to provide reasonable results in a wide variety of situations, and to represent the effect of site characteristics on growth.

### 2.2.1. Characteristics of Multi-Aged Stands

With the aim of maintaining forest ecosystem biodiversity and productivity, various management methods based on an understanding of the mechanisms of the natural disturbance regime have been proposed (Attiwill 1994; Bergeron and Harvey 1997; Angelstam 1998; Hunter 1999; Bergeron et al. 1999, 2002; Gauthier et al., chapter 1). Ecosystem-based forest management requires considering both the landscape level, in order to preserve a mosaic of stands of ages and structures similar to that created by natural disturbances (Gauthier et al. 1996; Hunter 1999), and the stand level, in order to create the diversity of structure and composition attributes characteristic of natural stands (Franklin 1993; Seymour and Hunter 1999).

Since natural disturbance regimes, including those driven by fire, allow some stands to outlive the normal lifespan of the trees that compose them (Bergeron et al. 2001; Kneeshaw and Gauthier 2004), the trees from the post-disturbance cohort begin to die and are gradually replaced by other cohorts of trees in the canopy. This results in a transition from an even-aged, frequently monospecific structure to a more irregular structure comprised of trees from different cohorts and usually with the coexistence of several different species (De Grandpré et al. 2000; Gauthier et al. 2000; see Gauthier et al., chapter 3 for more details on the succession mechanisms that drive the transition from an even-aged stand to a multi-aged stand).

In addition, natural disturbances often leave groups of residual trees that will contribute to the irregularity of the regenerated stands (Bergeron et al. 1999). It is this diversity of structures and compositions that ecosystem-based management seeks to preserve (Bergeron et al. 1999, 2002) through a variety of silvicultural treatments (Harvey et al. 2002; Gauthier et al. 2002, 2004). This management approach will produce a mosaic composed of even-aged stands, thinned stands with a more or less regular structure, as well as multi-aged stands, which will result in heterogeneity in both stem height and spatial distributions, due to the gaps of varying sizes or successive interventions. This internal stand complexity poses a challenge to modellers and it is therefore imperative that the modelling approaches, at the stand level, be able to take these factors into consideration.

In order to determine the choice of modelling approach or type of model, it is also important to ensure that the data required to calibrate and validate it are available. Otherwise, the result would be an accurate and realistic model, but with very limited application (Vanclay 1995). In order to express ecosystem complexity, an individual, distance-dependent model developed at the tree scale could be used to explore and understand stem reaction to various treatments. However, this type of model would be impossible to apply to conventional inventory data. That is why, once properly adjusted and tested, this model could be generalized to a single-tree distance-independent model. This generalization may reduce accuracy, but not in all cases (Zeide 2003; Busing and Mailly 2004). However, it could be applied to inventory data representing larger areas and thus fit into a landscape-level model.

### 2.2.2. Model Robustness

In modelling, the aim is to express the effect of complex processes occurring at the stand, tree, leaf, and even cellular level on one or more variables. However, establishing very good mathematical relationships at the level of one or more of these processes is no guarantee of success at the stand scale. Each equation has a margin of error and sometimes a bias, and when they are combined in the course of producing longer-term projections, this can result in drift. Because modelling necessarily constitutes a simplification of reality, a projection will also have a certain margin of error. The goal is therefore to obtain a forecast that is, insofar as possible, free of bias, so that any errors in the forecast for one stand will be offset by opposite errors from another stand and so that the resulting long-term forest planning can therefore be used with confidence (Groot et al. 2004).

Model robustness means the ability of a model to provide an unbiased response, in a broad range of compositions and structures, with different interventions, and over a fairly long time frame. Robustness is an essential quality of a model, since a forecast that does not accurately describe reality could reduce management strategy options in the future (Groot et al. 2004). In order to avoid having a tree model that is accurate only at the individual tree level and to enable the model to be used for longer-term projections at the stand level, it is necessary to incorporate interactions between these two levels in the model (Mäkelä 2003). For example, a tree model based on light interception by individual crowns could include a global constraint on the amount of light captured by the whole stand.

### 2.2.3. Light, the Engine of Growth

The basic process that provides energy to plants is photosynthesis, i.e. the conversion of light energy to sugars. Hence, light can be considered the engine of growth. It follows that an accurate estimate of the amount of light captured by the crown of a tree is directly related to its growth (Pacala et al. 1993, 1996; Bartelink et al. 1997; Canham et al. 2004; Groot 2004).

A number of studies have demonstrated a direct correlation between the amount of light intercepted and biomass production, in fields (Monteith 1977) and in forests, at both the stand (Cannell et al. 1987; Will et al. 2002; Allen et al. 2005) and tree levels (Kaufman and Ryan 1986; Brunner and Nigh 2000; MacFarlane et al. 2002). The slope of this relationship is generally referred to as radiation use efficiency or light use efficiency. Bartelink et al. (1997) and Groot (2004) have suggested that this relationship could be used as a basis for tree-level modelling. We have seen that the amount of light or radiation useful for photosynthesis can be estimated at the tree or stand level (Brunner 1998; Groot 2004). The efficiency of conversion of this energy into biomass or wood thus remains to be determined.

The term "light use efficiency" has been used broadly to designate the relationships between intercepted light and various values, such as gross total production, net total production or net aboveground biomass production (Groot and Saucier 2008). Even intercepted light has been estimated in several ways, from global shortwave radiation to photosynthetically active radiation, estimated annually or on a growing-season basis.

Mäkelä (2003) suggested various constraints that can be applied at a more general level, such as: 1) constraints related to environmental resources; 2) functions limiting optimal use of resources; and 3) allometric ratios. Resource availability constraints have clear advantages since they are based on the principle of conservation of matter and energy (Groot et al. 2004). Each site on which a stand is growing is capable of supplying the stand with a finite amount of water and nutrients, and the resources consumed by one tree are no longer available for the others. Similarly, the amount of light above the forest canopy is not captured equally by those trees in the stand that are in competition to capture a larger share of that light. The trees that have succeeded in developing a broader crown and that occupy a better social position will capture more light, leaving less for the others. The introduction of global constraints concerning intercepted light and environmental resources appears to be one way of ensuring model robustness (Battaglia and Sands 1998; Meinzer 2003). Groot et al. (2004) demonstrated, for a dense black spruce stand, that a model without constraints might generate unrealistic volume increments if crown size were modified, even slightly.

## 2.2.4. Growth Response to Site Characteristics

The conversion of light to biomass by a stem may be limited by competition for other resources such as water and nutrients (Mäkelä 2003). In addition, the reaction of a tree that, as a result of thinning or natural mortality of a competitor, has access to more light, may differ depending on the availability of water (Dhôte 1992) or other resources. However, interactions in the soil are more difficult to analyze than the effect of competition for light (Mäkelä 2003). For that reason, incorporating site characteristics into a growth model can make it possible to take into account to some degree the availability of these resources.

In addition to water supply and nutrient availability in the soil, other characteristics can vary depending on the site, such as local climate (resulting in differences in the length of the growing season) or the distribution of precipitation over the growing season. A number of studies have examined the relationships between growth and environmental variables, either by establishing direct relationships between a site index (SI) and environmental variables (Ung et al. 2001; McKenney and Pedlar 2003; Hamel et al. 2004; Seynave et al. 2005) or between the SI and an ecological classification (Bélanger et al. 1995; Larocque et al. 1996; Bédard 2002; Saucier et al. 2006).

From a perspective of spatialization of the results of a landscape-level growth model, the use of detailed soil variables is sometimes problematic since these variables are rarely mapped over large areas. Hence, models linking the SI to environmental variables, while often yielding good results, are applicable only to the plots where these variables were measured. For example, Seynave et al. (2005)

included in their model variables such as soil pH and C:N ratio (among others). This limits the use of their model to inventory plots containing these variables. To generalize to the landscape level, other relationships must be established, between these variables and the geology, for example, or with an ecological classification. However, more general physical variables, such as type of surface deposits or drainage class, can be used when mapped. If climatic variables alone are chosen as independent variables, then they must be estimated spatially using interpolation models, as McKenney and Pedlar (2003) and Ung et al. (2001) did.

However, Wang et al. (2005) caution users about synoptic environmental variables since they frequently exhibit collinearity. For example, temperature and precipitation are frequently correlated with altitude and latitude. These authors demonstrated that the SI of lodgepole pine (*Pinus contorta* Dougl. ex Loud) in Alberta was more strongly associated with geographic variables, particularly altitude, latitude, and longitude, and less strongly associated with the nutrient or moisture regime of the plots in their study.

There is an ongoing debate concerning the use for growth modelling purposes of 1) SIs derived from temporary or permanent inventory plots rather than stem analysis (Raulier et al. 2003; Mailly et al. 2004; 2) SIs derived from biophysical variables (biophysical indices, Ung et al. 2001), rather than from trees measured in plots (phytometric indices); and 3) SIs in continuous values using equations or in discontinuous values related to an ecological classification (Wang et al. 2005; Pojar et al. 1987; Bélanger et al. 1995; Saucier et al. 2006). Even direct estimation of SI using study trees can be problematic since suitable trees, i.e. trees that have been dominant their entire lives, are not always available in the field to estimate the SI (Wang et al. 2004; Mailly et al. 2004).

## 2.3.   Case Study: Development of the IVY Model

To illustrate the process of designing a growth model adapted to complex stands such as multi-cohort stands, we will use the example of the IVY model. This model is an interface which incorporates all components of the model into a user-friendly tool so that the results of a model can be used by forest managers.

The IVY model, which is being developed by a team led by Art Groot (Canadian Forest Service), incorporates various components and uses an iterative mechanism (figure 16.1). The starting point is a list of trees measured in a plot, with their diameters, heights, and, ideally, crown characteristics. To estimate crown light interception at the tree level, it is necessary to know the location of the stems relative to one another. Since the geographic location of the plot is known, the length of the growing season and the above-canopy radiation characteristics can then be calculated. The CORONA model is then used to calculate the amount of light intercepted during the growing season, taking into consideration direct radiation and diffuse radiation. Since the type of site and its associated volume increment efficiency are known, the amount of intercepted light is converted to volume increment for each tree and is summed for whole stand scale results. The height increment of each stem can then be calculated as a function of its height and social status, which can also be expressed as a function of the light intercepted since this data includes competition. Using the height and volume increments, the diameter increment can be determined

**Process for calculating growth as a function of intercepted light in the IVY model**

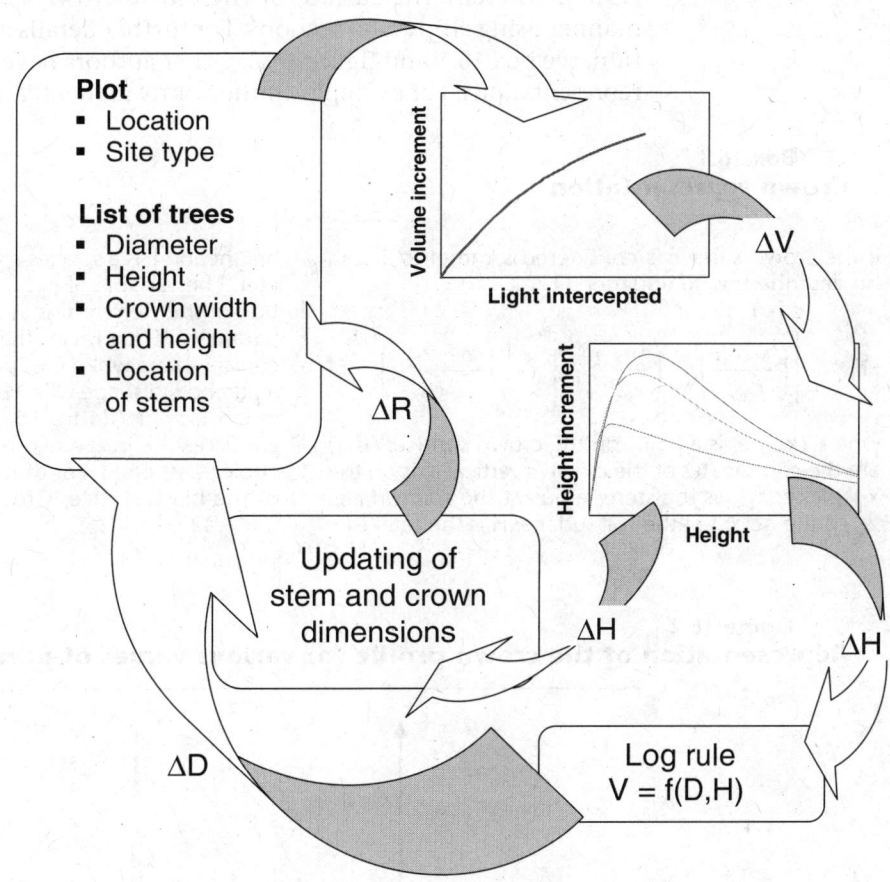

Note:  ΔV: Volume increment;  ΔH: Height increment;
ΔD: Stem diameter increment;  ΔR: Crown radius increment;
V: Stem volume;  D: Diameter at breast height;  H: Stem height.

iteratively using a volume equation. Crown characteristics are then updated using crown width increment equations, which are a function of the stem height increment. The end result is a list of trees whose diameters, heights, and crown widths have been updated. A new iteration can be begun if necessary.

## 2.3.1. IVY Growth Module

The IVY growth module uses a multi-stage process. The amount of radiation intercepted by the crown of a tree at a given location is determined using the CORONA model (Groot 2004), which is a spatially explicit model for individual trees which determines how the various light rays strike the tree crowns and uses functions to describe the transmission of light through each crown.

### Crown Representation

To represent crown surface, CORONA uses the crown model proposed by Koop (1989), whereby the surface of the entire crown is described in a simplified manner using ellipsoidal sections. For further details concerning this representation, see box 16.1 and figure 16.2. Other authors have used much simpler crown representations. For example, in the SORTIE model (Pacala et al. 1996), the crowns

---

**Box 16.1**
## Crown representation

If the crown surface is considered symmetric, it can be described using equation [1].

[1]
$$\left[ \left( \frac{x-h}{r_{stem}} \right)^2 + \left( \frac{y-r}{r_{stem}} \right)^2 \right]^{E/2} + \left( \frac{z - z_{surf}}{H} \right)^E = 1$$

where $(x, y, z)$ is a point on the crown surface; $(h, r)$ are the coordinates of the crown's vertical axis on the x-y plane; $r_{stem}$ is the stem radius at the soil surface; $z_{surf}$ is the height of the soil surface; H is the total tree

height and E is a species-specific crown profile parameter. The variable of $r_{stem}$ can be expressed as a function of the stem radius at breast height. The value of parameter E expresses the shape of the crown, i.e. a conical shape when E = 1, a neloid shape when E < 1, a paraboloid shape when E > 1 and a cylindrical shape when E >> 1 (figure 16.2). Setting the parameter E = 2 results in stem profiles that conform with a cubic paraboloid. For the upper portion of the crown of the black spruce, Groot (2004) found a value of E = 1.43.

---

**Figure 16.2**
## Representation of the crown profile for various values of parameter E

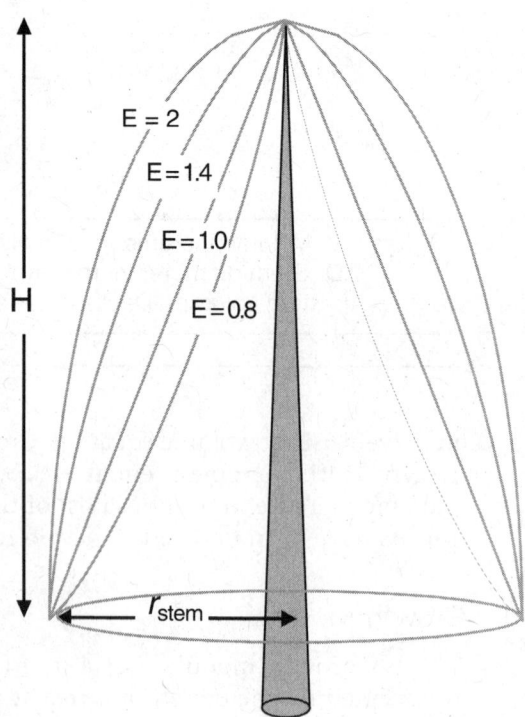

From Groot (2004).

of individual trees, from the seedling stage to the adult age, are represented by cylinders with a crown radius estimated using an empirical function of the DBH (Canham et al. 2004). In addition, the interception of the trunks below the base of the crown is ignored in this model.

In CORONA, in order to simplify and optimize the calculation of the radiation striking the crowns, the number of rays with a low angle of incidence is minimized and the rays strike the crowns and trunks in a geometric pattern in which the distance between rays is constant regardless of the horizontal or vertical angle. It is very difficult to separate the amount of light absorbed from the amount of light that is reflected upward, or the amount of light that is transmitted from the amount of light that is reflected downward. CORONA therefore uses functions to globally estimate the light intercepted by the crown. Of the two models tested by Groot (2004), the simple intersection model, where a fraction of the light is intercepted at each point of entry in the crown, was preferred since it generally had a lower margin of error. Canham et al. (1994) noted that this appears to be generally consistent with the fact that tree foliage is more densely distributed around the periphery of the crown.

CORONA assumes that no light is transmitted through tree stems. Groot (2005) demonstrated the importance of taking into consideration the interception of light by the canopy in order to prevent bias and overestimation of light interception by the trunks. According to this study, failing to consider interception by the trunks would result in a bias of 3.5% for black spruce and 14.9% in the case of trembling aspen (*Populus tremuloides* Michx.).

### Estimation of Intercepted Light

To estimate the amount of intercepted light, it is necessary to estimate first the amount of available light above the stand, and then the amount that is intercepted by each individual crown.

The amount of light is estimated using shortwave radiation and has two components: direct radiation and diffuse radiation. The total radiation intercepted by a stand can be estimated at a given point in time by comparing the above-canopy measurements with the below-canopy measurements at the same time under overcast skies or at sunset. To relate the amount of light intercepted to an annual increment (volume, basal area, height, or diameter), the total incident radiation for the entire growing season must be determined. To do this, CORONA and the methods described by Groot (2004, 2005) and by Groot and Saucier (2008) are used.

First, the mean daily temperatures from weather stations located near the site of interest are calculated in order to determine the start and end of the growing season, i.e. the photosynthetically active period. Incident shortwave radiation is then calculated on a daily basis for the study site, separating the direct and diffuse components, which are then combined for the entire growing season and distributed among the stems using CORONA.

For conifers, the growing season is considered to begin after three consecutive days during which the mean daily temperature exceeds 5°C and to end when there are no longer three consecutive days above 5°C at the end of the season. In the case of deciduous trees, the leaf development period at the start of the growing season and the leaf drop date must be considered. For example, for trembling aspen, Groot and Saucier (unpubl.) considered the growing period to start once the cumulative total of degree-days reached 70, as suggested by Parry et al. (1988). For a study in northeastern Ontario, this resulted in an average growing season of 168 days for black spruce and 139 days for trembling aspen.

### Relationship between Intercepted Light and Volume Increment

When we consider the relationship between the amount of light intercepted by an individual crown on a growing season basis and the stem volume increment, it is generally observed that the volume increment increases with the amount of light intercepted. In a study of peatland black spruce stands in northeastern Ontario, Groot and Saucier (2004) observed a relationship that can be described by a monomolecular equation (figure 16.3):

$$[2] \qquad\qquad Iv = C_0[1 - \exp(-C_1 l_{int})]$$

where $Iv$ is the wood volume increment of a stem expressed in $dm^3\ year^{-1}$;

exp is the exponential;

$l_{int}$ is the amount of light intercepted by the crown in $GJ\ year^{-1}$;

and $C_0$ and $C_1$ are parameters.

**Figure 16.3**
**Relationship between gross volume increment of black spruce stems and amount of light intercepted for two experimental plots in northeastern Ontario**

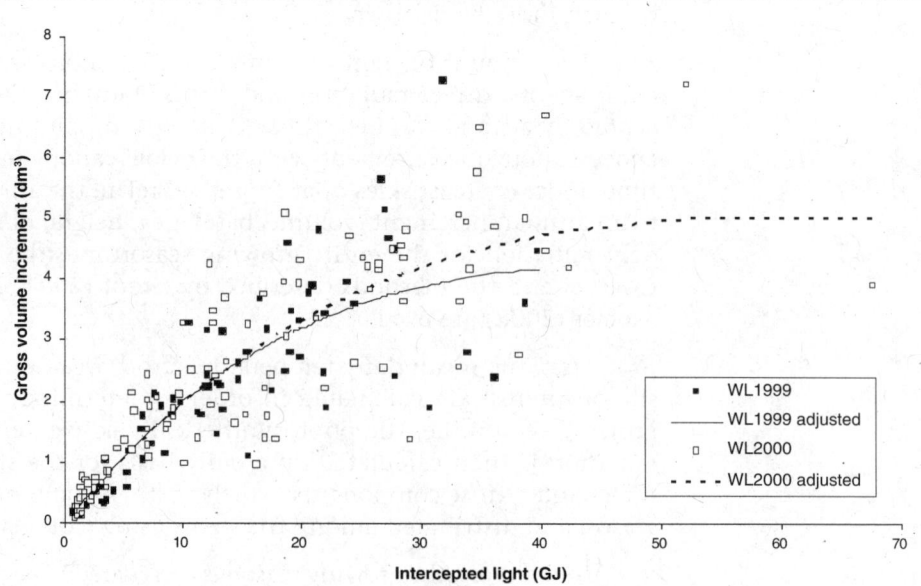

From Groot and Saucier (2004).

The stem volume increment increases non-linearly with the increase in light captured by the crown. This indicates that the efficiency of conversion of light to wood volume decreases as light capture increases. This example shows a plateau in the volume increment between 40 and 50 GJ of intercepted radiation. This is consistent with observations made in other species, such as Douglas fir (*Pseudotsuga menziesii* [Mirb.] Franco) (Brunner and Nigh 2000), loblolly pine (*Pinus taeda* L.) (Macfarlane et al. 2002), ponderosa pine (*Pinus ponderosa* Dougl. ex Laws), and Colorado white fir (*Abies concolor* [Gord. & Glend. Lindl.] ex. Hildebr.) (Gersonde and O'Hara 2005).

The decrease in efficiency of use of intercepted radiation for trunk growth could be attributed to a greater allocation of resources to the branches (Raulier et al. 1996) or to the roots. Other theories suggest that the decrease in productivity observed with increasing tree height is attributable to hydraulic limitations, to greater use of energy for respiration or even to senescence (Ryan et al. 1997; Seymour and Kenefic 2002).

## 2.3.2. Concept of Volume Increment Efficiency

Ivy forecasts stem volume increment, since this is typically the production measure of interest in forest management. This also facilitates comparisons with published growth models. This approach differs from the usual application of the concept of light-use efficiency, which generally considers the total biomass and then distributes it among the various compartments of the trees (trunks, branches, roots, etc.). The rate of conversion of light to gross merchantable volume is called volume increment efficiency (VIE) by Groot and Saucier (2008). VIE is calculated, at the stand level, by dividing the sum of the stem volume increment by the total amount of global shortwave radiation intercepted by tree crowns during the growing season. VIE is expressed in $dm^3$ $GJ^{-1}$ and can also be calculated for a stand or for a single tree.

According to Runyon et al. (1994), the bulk of the variation in net primary production between forest types is attributable to the amount of light intercepted by the vegetation cover during the growing season rather than to differences in radiation use efficiency. However, the yield tables generally used in forestry (Plonski 1974; Pothier and Savard 1998) show large differences in volume increment within a region that has a homogenous climate. These differences can be explained using site index (SI). For stands with a closed canopy, it could be presumed that these variations in wood production are due more to differences in VIE than to variations in the amount of light intercepted (Groot and Saucier 2008). Using the concept of volume increment efficiency would thus be a practical method for replacing the traditional SI. It remains to be determined how the VIE varies as a function of species, site quality, or stand characteristics.

In their research, Groot and Saucier (2008) demonstrated the relationships between VIE and certain site and stand parameters. They calculated VIE by dividing the mean annual gross volume increment of a stand by the amount of light intercepted by the stand during the growing season. The gross volume increment is calculated for permanent sample plots using the following equation (Husch et al. 1972):

[3] $$G_{g+i} = S_2 + I + D_2 - S_1 - D_1$$

where

$G_{g+i}$ is the gross volume increment including ingrowth (m³ ha⁻¹) of trees with a diameter of more than 2.5 cm;

$S_1$ and $S_2$ represent, respectively, the initial and final volume of the survivor trees;

$D_1$ and $D_2$ represent, respectively, the initial and final volume of trees recorded as dead or missing during the measurement period; and

I represents the ingrowth volume (i.e. that exceeded the 2.5-cm diameter threshold during the measurement period).

The periodic increment is then divided by the number of growing seasons between the measurements of the permanent plots, taking into consideration the dates of the measurements in the case of incomplete seasons. The result is an average annual increment expressed in m³ ha⁻¹ year⁻¹. The amount of light intercepted by the stand was estimated, for the entire growing season, by the difference in the radiation measurements taken in the absence of tree cover and under the forest tree cover at 20 points systematically distributed in the permanent plots (Groot and Saucier 2008).

These studies enable comparison of volume increment efficiency based on site characteristics determined according to the applicable ecological classification (Taylor et al. 2000), for black spruce and trembling aspen stands in northeastern Ontario. For black spruce stands, light interception, gross volume increment, VIE and SI were all significantly different (p = 0.01) as a function of site type (table 16.1). Inter-site differences were more pronounced for volume increment and for VIE than for SI. VIE generally increases from wet, nutrient-poor sites (ES14, ES13, ES11, and ES12), with values ranging from 0.071 to 0.168 dm³ GJ⁻¹, to mesic sites (ES6), where it reaches 0.345 dm³ GJ⁻¹. This means that for the same amount of intercepted light, the stand produces nearly five times more wood on a mesic site than on a wet, nutrient-poor site. A linear relationship is observed between VIE and SI (figure 16.4), but VIE more effectively expresses inter-site differences than SI according to Newman-Keuls multiple-comparison tests (table 16.1).

Similar results were observed for trembling aspen, for which VIE and gross volume increment were significantly higher for fine-textured soils (ES7f), with a VIE of 0.497 dm³ GJ⁻¹, compared to coarse-textured soils (ES3 and ES7c), with a VIE of 0.330 dm³ GJ⁻¹, despite the fact that there were no significant differences in intercepted radiation between sites.

The differences observed in VIE among site types are generally consistent with the logic of the ecological classification (Taylor et al. 2000) and the gradients in soil moisture and nutrient regimes. Although stands on poor sites tend to intercept less radiation than those on richer sites, they have lower VIE values. Total cumulative stem wood per unit of intercepted light is therefore lower on wet, nutrient-poor sites. Runyon et al. (1994) observed that the allocation of resources to the below-ground components of the tree increased by 20% to 60% as constraints related to environmental variables increased. VIE may thus reflect the combined effect of differences in resource allocation as a function of the site.

### Table 16.1
### Effect of site type on gross volume increment, radiation interception, volume increment efficiency (VIE), and site index (SI) for black spruce

| Site type[1] | Number of plots | Gross volume increment (m³ ha⁻¹ year⁻¹) | Photosynthetic season radiation interception (GJ m⁻² year⁻¹) | Volume increment efficiency[2] (dm³ GJ⁻¹) | Site index (at breast height, age 50) |
|---|---|---|---|---|---|
| ES14 | 3 | 1.16[a] | 1.56[a] | 0.071[a] | 7.9[a] |
| ES13 | 2 | 2.21[ab] | 1.88[ab] | 0.118[ab] | 7.7[a] |
| ES11 | 4 | 2.65[ab] | 2.11[ab] | 0.122[ab] | 6.8[a] |
| ES5 | 2 | 2.94[ab] | 1.46[a] | 0.197[bc] | 9.5[ab] |
| ES12 | 3 | 4.14[abc] | 2.44[b] | 0.168[b] | 10.0[ab] |
| ES2 and ES4 | 5 | 5.55[bcd] | 2.47[b] | 0.222[bcd] | 12.9[ab] |
| ES8 | 2 | 7.10[cd] | 2.21[ab] | 0.317[cd] | 10.9[ab] |
| ES6 | 3 | 8.97[cd] | 2.53[b] | 0.345[d] | 15.5[b] |
| $F_{7,16}$ | | 8.14 | 6.32 | 11.65 | 4.24 |
| p > F | | <0.001 | 0.001 | <0.001 | 0.009 |
| MSE | | 2.48 | 0.0760 | 0.0029 | 5.69 |

1. Site type according to Taylor et al. (2000).
2. Analysis was on square-root transformed values and site type means are transformed back to original units.

Legend: ES2: jack pine–xeric coarse soil; ES4: black spruce–jack pine–xeric coarse soil; ES5: black spruce–mesic fine soil; ES6: black spruce–trembling aspen–mesic fine soil; ES8: black spruce–feather moss–sphagnum–subhydric soil; ES11: black spruce–Labrador tea–organic soil; ES12: black spruce–larch–Labrador tea–organic soil; ES13: black spruce–larch–speckled alder–minerotrophic organic soil; ES14: black spruce–leather leaf–ombrotrophic organic soil.

Note: Values within a column followed by the same letter are not significantly different according to the Newman-Keuls multiple comparison test.

Adapted from Groot and Saucier (2008).

### Figure 16.4
### Relationship between black spruce site index (SI) and volume increment efficiency (VIE) for various experimental plots in northeastern Ontario

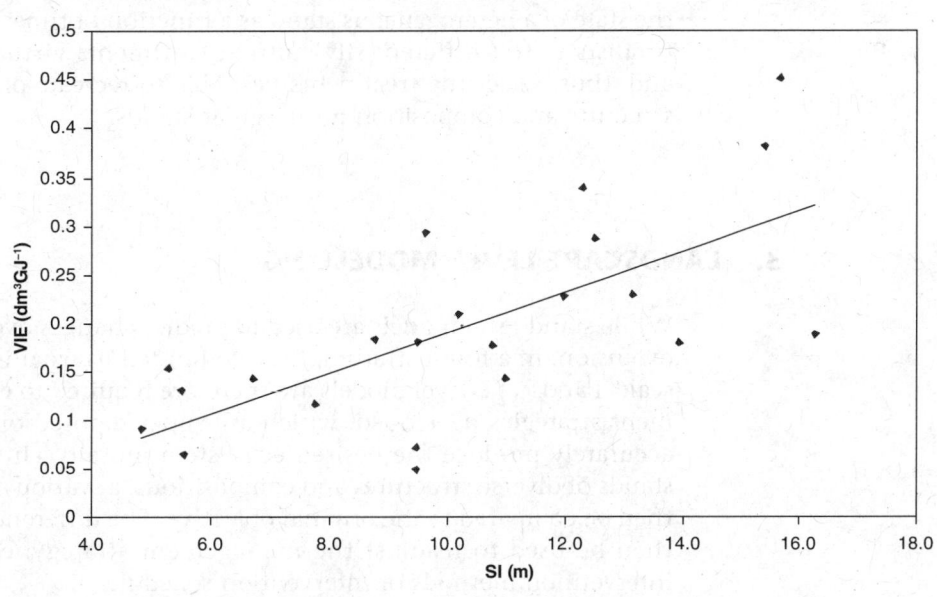

From Groot and Saucier (2008).

### 2.3.3.  Development of Ivy

The methods already described are used in the increment module of the Ivy model. Increment models appear to be incomplete for estimating stand yield, since mortality and recruitment modules for small stems, as well as establishment modules, must still be added to the models. However, the amount of intercepted light is also a good yardstick for expressing competition and should be used to estimate competition-induced mortality. The use of knowledge about the shade tolerance of different species, in conjunction with the amount of intercepted light, is also a good basis for predicting the type and quantity of regeneration under the forest canopy (Stadt et al. 2005). In addition, the structure of Ivy will enable changes in environmental conditions to be taken into consideration. For example, the change in the amount of available light following thinning will result in a favourable environment for the residual stems; their growth should improve and their mortality will no longer be a consequence of competition for light. Using light as the model engine also makes it possible to realistically represent certain disturbances. For example, insect defoliation could be represented by an increase in light transmission through crowns. However, certain disturbances that induce mortality following these openings (windthrow for example) should also be taken into consideration.

Once the various modules have been individually calibrated and all the model components interlinked, the model will have to be validated using permanent sample plots, by incorporating data from other regions of the boreal forest, in order to verify that the model accurately represents stem and canopy development. Any bias or error that may result from successive use of the submodules must also be calculated. A simplified version of the model, operating at the stand level, could also be tested over longer time sequences in order to ensure that its behaviour remains within the area's natural range of variability.

The result will be a flexible model that will be able to simulate changes in the state of a heterogeneous stand as a function of time. This will allow silviculturalists to test different silvicultural treatments virtually, assess their effects and, thus, select the treatments best able to recreate or perpetuate the desired structure and composition for irregular stands.

## 3.   LANDSCAPE-LEVEL MODELLING

While stand-level models are used to predict changes over time in a stand or, by extension, in a forest stratum, they do not tell us what is happening on a larger scale. Landscape-level models are therefore required to ensure that the management strategies developed, which are superimposed on natural disturbances, accurately produce the desired ecosystem mosaic. This mosaic, comprised of stands of diverse structures and compositions, at various successional stages, can then be compared to the original objective. The differences or convergences will then be used to readjust the management strategy, either by changing the intervention methods or intervention schedule.

## 3.1.    Approach

Forest managers have long made use of models for the purpose of calculating annual allowable cut. These models are generally based on yield tables or other stand-level models to predict the trajectory of forest strata. They simulate the effect of management strategies, i.e. the schedule of interventions, the planned silvicultural work, and the successes or delays in regeneration, on the volumes of wood available. The results of long-term simulations, generally more than 100 years, thus make it possible to estimate annual sustained yield, i.e. the amount of wood that can be harvested in perpetuity without jeopardizing the resource. One example of this type of system is the SYLVA II software, which is used in Quebec (MER 1988). Although the systems used calculate the parameters of the various strata for each period, knowledge about the state of the forest, in terms of composition, structure, or successional stage, is generally not the objective of the simulation. Since these models are rarely spatially explicit, it is very difficult to visualize the resulting landscape after a given number of years. Yet, the services that an ecosystem can provide are to a large extent dependent on the spatial arrangement and interaction of the landscape elements (Pretzsch et al. 2007). In addition, only the interventions planned in the strategies are simulated, since the effect of natural disturbances, whose risk of occurrence increases with time, is generally completely disregarded.

Various simulation tools developed since the 1990s, such as the WOODSTOCK modelling software, now make it possible to perform the same work while providing spatially explicit results (MacLean et al. 1999). The addition of an optimization system enables a number of values of interest to be followed over time, such as the volume of certain secondary species or the proportion of populations exceeding a certain age, for instance, and to set upper and lower limits for these values (Lappi 1992). Hence, the system is no longer controlled only by wood production objectives, but can ensure that other objectives related to quality and to the juxtaposition of ecosystems at the landscape scale are met.

## 3.2.    Desired Qualities

This section analyzes the desirable qualities in a model that will be capable of evaluating the achievement of ecosystem-based forest management objectives over a longer period and at the landscape scale.

### 3.2.1.    Spatially Explicit Results

For the purposes of ecosystem-based forest management, it is necessary to preserve in the landscape a mosaic of stands of varied ages and structures whose proportions are within the range of historical variations that can be created by stand dynamics and natural disturbances. This requires a landscape-level model that can spatially express the simulation results. There are two main spatialization approaches. In the first approach, the forest landscape is divided into cells of equal size and geometric shape that can be followed over time. This is the approach used in the LANDIS model (Gustafson et al. 2000) and in SELES (Fall and Fall 2001). This method is particularly well suited to data from pixel-based satellite images. In the second approach, a forest map or, better yet, an ecosite map is used as the basis. In this approach, forest stands are the cells to which the yield models and management strategies will apply. Regardless of the choice of basic cell, ideally they must be homogeneous in terms of the current characteristics of the vegetation as well as its natural dynamics and growth potential. That is why a fairly fine sectioning is desirable in order to take into consideration the characteristics of both the physical environment and the vegetation.

### 3.2.2.  Introducing Natural Disturbances

One of the essential uses of landscape-level models is to estimate the effects and risks associated with phenomena that occur at this scale, such as fire, windthrow or insect epidemics (Pretzsch et al. 2007). One of the recognized weaknesses of models for calculating annual allowable cut is their inability to consider the effect of natural disturbances, which is quite different from that of forest management interventions. While harvesting is regulated over time by constraints such as accessibility, composition, successional stage, and schedule, natural disturbances have a more stochastic behaviour. Obviously, certain factors can help determine the vulnerability of a stand to windthrow or defoliation, but if high winds and epidemics do not materialize, no damage will occur. Although it is possible to estimate a fire return interval, it is the random occurrence or combination of several climatic factors, for example, a lightning storm occurring after an extended period of drought, that will determine whether a forest fire occurs. In addition, the disturbances vary considerably in severity. For example, at the local scale, an insect epidemic can create lower-density stands with few residual trees belonging to species vulnerable to this insect, while in other areas there may not be a single living tree that survives. In traditional models, it is always possible to include provision for a certain proportion of volume lost to fire, but there is no way to localize these effects. In fact, while optimization is possible with a spatially explicit model such as WOODSTOCK, introducing stochastic factors into these types of models is difficult; these factors therefore can only be added afterwards in a simulation mode based on the optimal solution. In order to explore the effect of natural disturbances, stochastic factors must be introduced into the model and a number of simulations must be carried out in order to estimate the average effects at the landscape level (Burnett et al. 2003). The result is a model suitable for guiding ecosystem-based forest management.

### 3.2.3.  Meeting Management Objectives

The goal of ecosystem-based forest management is to ensure the preservation of multiple ecosystem functions while generating benefits for society. However, the wood production function is subordinate to maintaining certain key characteristics of stand structure and composition and a distribution of different stand age classes in the landscape. The management objectives are therefore more complex than simply ensuring an uninterrupted supply of wood. To manage these multiple objectives over a large area and long period of time, it is absolutely essential to have tools capable of testing several strategies, relatively quickly, and evaluating their ability to attain most of the objectives. These tools should also incorporate sufficient flexibility to consider both the critical thresholds of a particular resource (for example, a proportion of forests more than 100 years old) and objectives for which a wide range of variation may be acceptable.

For that reason, the ideal landscape-level model must necessarily include not only spatialization functions, but also optimization functions in order to manage a large number of constraints or objectives, given the growing diversity of forestry interventions, such as harvesting designed to maintain forest covers of different intensities and different distribution patterns.

### 3.2.4.  Model Example

Of the various landscape-level models, the LANDIS model (Gustafson et al. 2000) includes a number of desirable functions for meeting ecosystem-based forest management needs. This model has functions that enable it to perform spatial simulations of forest succession as well as of silvicultural interventions. The landscape is broken down into a grid of equal-sized individual cells (10 m × 10 m to 500 m × 500 m) which can be stratified into land types or ecoregions. The forest stand within each cell is described using a list of the species and age cohorts that are present or absent. A succession module is used to modify the species composition over time according to regeneration and mortality rules. The effect of human interventions and of certain disturbances can be incorporated. The harvesting module can simulate six silvicultural systems ranging from clearcutting to periodic-entry systems for forest cover maintenance and can be applied to a stand or to an area. Disturbances such as violent winds and fires are introduced stochastically by specifying a return period. Finally, the model time step is 10 years. It is presented by its developers not as a tool for calculating the annual allowable cut or for harvest scheduling, but rather as a tool for estimating the forest's response, at the landscape level, to a management strategy (Gustafson et al. 2000). Other capabilities can be added. For example, LANDIS was used as a basis for the LANCLIM model developed to estimate the effects of topography, climate, and land use on forest structure and forest dynamics (Pretzsch et al. 2007). When used over a fairly long time frame, this type of model could help predict the effects of climate change on the forest landscape.

### 4.  CONCLUSION

An heterogeneous forest can be considered a mosaic of nested scales (Franc et al. 2000). That is why the choice of model depends on the questions we want answered. A change in scale frequently requires that we use a different process. At the individual tree level, forest managers are interested in the rate of growth and in stem quality. At the stand level, predicting mortality and recruitment becomes essential to estimating yield. Finally, including disturbances is essential to estimating annual allowable cut and to obtaining realistic forecasts on the long-term evolution of ecosystem characteristics at the landscape level. The challenges posed by each of these scales may therefore require the use of different models.

To test silvicultural interventions virtually or predict the reaction of a stand to a treatment or disturbance, a stand-level model, using the tree as the basic element, is necessary. However, it is also essential to validate silvicultural strategies at the landscape level, where the stand is used as the basic element, in order to measure both their long-term effects and their effects over large areas.

The approach to modelling forest stand dynamics that uses light as the engine of growth linked to a site typology meets the needs of complex forest stand management. The amount of light intercepted by a crown can be considered not only in relationship with its volume and height increment, but also with the evolution of its crown characteristics, with competition-induced mortality and with regeneration under the forest canopy. We can therefore project over time the composition and structure of stands with considerable internal heterogeneity, such as multi-aged stands. Using a single-tree distance-dependent model enables us to use simulations to test the effect of different silvicultural treatments. These treatments can vary in intensity, but also in terms of the structure of the resulting stands. Forest managers can then select, as a function of the range of possible dynamics, the treatments that most effectively achieve sustainable development objectives, both those that promote maintenance of forest cover and complex structures and those that promise increased production.

Models applicable to a larger scale can be used to simulate the effect of interventions or natural disturbances at the forest landscape level, generally composed of a mosaic of sites occupied by stands of varying compositions and structures that vary significantly in resilience and dynamics. Finally, new models, at different scales, are constantly being developed to explore increasingly complex issues, such as natural cycles, for example the carbon, water, and nutrient cycles.

## REFERENCES

Allen, C.B., Will, R.E., and Jacobson, M.A. 2005. Production efficiency and radiation use efficiency of four tree species receiving irrigation and fertilization. For. Sci. **51**: 556–569.

Angelstam, P.K. 1998. Maintaining and restoring biodiversity in European boreal forests by developing natural disturbance regimes. J. Veg. Sci. **9**: 593–602.

Attiwill, P.M. 1994. The disturbance of forest ecosystems: the ecological basis for conservative management. For. Ecol. Manag. **63**: 247–300.

Bankowski, J., Dey, D., Boysen, E., Woods, M., and Rice, J. 1996. Validation of NE_TWIGS 3.0 for tolerant hardwoods stands in Ontario. Ontario Ministry of Natural Resources, Ontario Forest Research Institute Information Paper, no. 130.

Bartelink, H.H, Kramer, K., and Mohren, G.M.J. 1997. Applicability of the radiation-use efficiency concept for simulating growth of forest stands. Agric. For. Meteorol. **88**: 169–179.

Battaglia, M. and Sands, P.J. 1998. Process-based forest productivity models and their application in forest management. For. Ecol. Manag. **102**: 13–32.

Bédard, S. 2002. L'estimation du potentiel de croissance des stations forestières: exemple du sous-domaine de la sapinière à bouleau jaune de l'est du Québec. Ministère des Ressources naturelles, Direction de la recherche forestière, Forêt-Québec.

Bélanger, L., Paquette, S., Morel, S., Bégin, J., Meek, P., Bertrand, L., Beauchesne, P., Lemay, S., and Pineau, M. 1995. Indices de qualité de station du sapin baumier dans le sous-domaine écologique de la sapinière à bouleau blanc humide. For. Chron. **71**: 317–325.

Bergeron, Y., Gauthier, S., Kafka, V., Lefort, P., and Lesieur, D. 2001. Natural fire frequency for the eastern Canadian boreal forest: consequences for sustainable forestry. Can. J. For. Res. **31**: 384–391.

Bergeron, Y. and Harvey, B. 1997. Basing silviculture on natural ecosystem dynamics: an approach applied to the southern boreal mixedwoods of Quebec. For. Ecol. Manag. **92**: 235–242.

Bergeron, Y., Harvey, B., Leduc, A., and Gauthier, S. 1999. Forest management guidelines based on natural disturbance dynamics: stand forest level considerations. For. Chron. **75**: 49–54.

Bergeron, Y., Leduc, A., Harvey, B.D., and Gauthier, S. 2002. Natural fire regime: a guide for sustainable management of the Canadian boreal forest. Silva Fenn. **36**: 81–95.

Boudoux, M. 1978. Tables de rendement empiriques pour l'épinette noire, le sapin baumier et le pin gris au Québec. Gouvernement du Québec, Ministère des Terres et Forêts, COGEF, 101 p.

Brunner, A. 1998. A light model for spatially explicit forest stand models. For. Ecol. Manag. **107**: 19–46.

Brunner, A. and Nigh, G. 2000. Light absorption and bole volume growth of individual Douglas-fir trees. Tree Phys. **20**: 323–332.

Burnett, C., Fall, A., Tomppo, E., and Kalliola, R. 2003. Monitoring current status of and trends in boreal land use in Russian Karelia. Conserv. Ecol. **7**: 8 [Online] <www.consecol.org/vol7/iss2/art8> (accessed December 3, 2007).

Busing, R.T. et Mailly, D. 2004. Advances in spatial, individual-based modeling of forest dynamics. J. Veg. Sci. **15**: 831–842.

Canham, C.D., Finzi, A.D., Pacala, S.W., and Burbank, D. H. 1994. Causes and consequences of resource heterogeneity in forests: interspecific variation in light transmission by canopy trees. Can. J. For. Res. **24**: 337–349.

Canham, C.D., LePage, P.T., and Coates, K.D. 2004. A neighbourhood analysis of canopy tree competition: effects of shading versus crowding. Can. J. For. Res. **24**: 778–787.

Cannell, M.G.R., Milne, R., Sheppard, L.J., and Unsworth, M.H. 1987. Radiation interception and productivity of willow. J. Appl. Ecol. **24**: 261–278.

Coates, K.D., Canham, C.D., Beaudet, M., Sachs, D.L., and Messier, C. 2003. Use of a spatially explicit individual-tree model (SORTIE/BC) to explore the implications of patchiness in structurally complex forests. For. Ecol. Manag. **186**: 297–310.

De Grandpré, L., Morissette, J., and Gauthier, S. 2000. Long-term post-fire changes in the northeastern boreal forest of Quebec. J. Veg. Sci. **11**: 791–800.

Dhôte, J.-F. 1992. Hypotheses about competition for light and water in even-aged common beech (*Fagus sylvatica* L.). For. Ecol. Manag. **69**: 219–232.

Fall, A. and Fall, J., 2001. A domain-specific language for models of landscape dynamics. Ecol. Model. **141**: 1–18.

Fournier, R.A., Guindon, L., Bernier, P.Y., Ung, C.-H., and Raulier, F. 2000. Spatial implementation of models in forestry. For. Chron. **76**: 929–940.

Franc, A., Gourlet-Fleury S., and Picard, N., 2000. Une introduction à la modélisation des forêts hétérogènes. École Nationale du Génie Rural des Eaux et des Forêts, Nancy, France.

Franklin, J.F. 1993. Preserving biodiversity: species, ecosystems, or landscapes? Ecol. Appl. **3**: 202–205.

Gauthier, S., De Grandpré, L., and Bergeron, Y. 2000. Differences in forest composition in two ecoregions of the boreal forest of Québec. J. Veg. Sci. **11**: 781–790.

Gauthier, S., Leduc, A., and Bergeron, Y. 1996. Vegetation modelling under natural fire cycles: a tool to define natural mosaic diversity for forest management. Environ. Monit. Assess. **39**: 417–434.

Gauthier, S., Lefort P., Bergeron Y., and Drapeau, P. 2002. Time since fire map, age-class distribution and forest dynamics in the Lake Abitibi Model Forest. Information Report, Laurentian Forestry Centre, Québec Region, Canadian Forest Service, No. LAU-X-125E.

Gauthier, S., Nguyen, T.-X., Bergeron, Y., Leduc, A., Drapeau, P., and Grondin, P. 2004. Developing forest management strategies based on fire regimes in northwestern Quebec, Canada. *In* A.H. Perera, L.J. Buse and M.G. Weber. Emulating natural forest landscape disturbances: Concepts and Applications. Columbia University Press, New York, USA, pp. 219–229.

Gersonde, R.F. and O'Hara, K.L. 2005. Comparative tree growth efficiency in Sierra Nevada mixed-conifer forests. For. Ecol. Manag. **219**: 95–108.

Groot, A. 2004. A model to estimate light interception by tree crowns, applied to black spruce. Can. J. For. Res. **34**: 789–799.

Groot, A. 2005. Biases in LI-COR Plant Canopy Analyzer estimates of seasonal light interception by black spruce and trembling aspen canopies. Can. J. For. Res. **35**: 2664–2670.

Groot, A., Gauthier, S., and Bergeron, Y. 2004. Stand dynamics modelling approaches for multicohort management of eastern Canadian boreal forests. Silva Fenn. **38**: 437–448.

Groot, A. and Hökkä, H. 2000. Persistence of suppression effects on peatland black spruce advance regeneration after overstory removal. Can. J. For. Res. **30**: 753–760.

Groot, A. and Saucier, J.-P. 2004. Towards a light-capture based stand dynamics model. Conference presented at IUFRO "Meeting the Challenge: Silvicultural Research in a Changing World", Montpellier, 14–18 June 2004.

Groot, A. and Saucier, J.-P. 2008. Volume increment efficiency of *Picea mariana* in northern Ontario, Canada. For Ecol. Manag. **255**: 1647–1653.

Gustafson, E.J., Shifley, S.R., Mladenoff, D.J., Nimerfro, K.K., and He, H.S. 2000. Spatial simulation of forest succession and timber harvesting using LANDIS. Can. J. For. Res. **30**: 32–43.

Hamel, B., Bélanger, N., and Paré, D. 2004. Productivity of black spruce and jack pine stands in Quebec as related to climate, site biological features and soil properties. For. Ecol. Manag. **191**: 239–251.

Harvey, B.D., Leduc, A., Gauthier, S., and Bergeron, Y. 2002. Stand-landscape integration in natural disturbance-based management of the southern boreal forest. For. Ecol. Manag. **155**: 369–385.

Hökkä, H. and Groot, A. 1999. An individual-tree basal area growth model for second-growth peatland black spruce. Can. J. For. Res. **29**: 621–629.

Hunter, M.L., Jr. (Ed.). 1999. Maintaining biodiversity in forest ecosystems. Cambridge University Press, Cambridge, UK.

Husch, B., Miller, C.I., and Beers, T.W. 1972. Forest mensuration. The Ronald Press Co., New York, USA.

Johnsen, K., Samuelson, L., Teskey, R., McNulty, S., and Fox, T. 2001. Process models as tools in forestry research and management. For. Sci. **47**: 2–8.

Kaufmann, M.R. and Ryan, M.G. 1986. Physiographic, stand and environmental effects on individual tree growth and growth efficiency in subalpine forests. Tree Phys. **2**: 47–59.

Kneeshaw, D.D. and Gauthier, S. 2004. Old growth in the boreal forest: a dynamic perspective at the stand and landscape level. Env. Rev. **11**: S99–S114.

Koop, H. 1989. Forest dynamics. Springer-Verlag, Berlin, Germany.

Lacerte, V., Larocque, G.R., Woods, M., Parton, W.J., and Penner, M. 2004. Calibration of the forest vegetation simulator (FVS) model for the main forest species of Ontario, Canada. Ecol. Model. **199**: 336–349.

Lacerte, V., Larocque, G.R., Woods, M., Parton, W.J., and Penner, M. 2006. Testing the Lake States variant of FVS (Forest Vegetation Simulator) for the main forest types of northern Ontario. For. Chron. **80**: 495–506.

Landsberg, J.J. and Waring, R.H. 1997. A generalised model of forest productivity using simplified concepts of radiation use efficiency, carbon balance, and partitioning. For. Ecol. Manag. **95**: 209–228.

Lappi, J. 1992. JLP, a linear programming package for management planning. The Finish Forest Institute, Research paper 414, Suonenjoki, Finl.

Larocque, G.R, Parton, W.J., and Archibald, D.J. 1996. Polymorphic site productivity functions for black spruce in relation to different ecological types in northern Ontario. Natural Resources Canada, Canadian Forest Service, Laurentian Forestry Centre, Information Report 18 LAU-X-119E.

MacFarlane, D., Green, E.J., Brunner, A., and Burkhart, H.E. 2002. Predicting survival and growth rates for individual loblolly pine trees from light capture estimates. Can. J. For. Res. **32**: 1970–1983.

MacLean, D.A., Ethridge, P., Pelham, J., and Emrich, W. 1999. Fundy Model Forest: partners in sustainable forest management. For. Chron. **75**: 219–227.

Mailly, D., Turbis, S., Auger, I., and Pothier, D. 2004. The influence of site tree selection method on site index determination and yield prediction in black spruce stands in northeastern Québec. For. Chron. **80**: 134–140.

Mailly, D., Turbis, S., and Pothier, D. 2003. Predicting basal area increment in a spatially explicit, individual tree model: a test of competition measures with black spruce. Can. J. For. Res. **33**: 435–444.

Mäkelä, A. 2003. Process-based modelling of tree and stand growth: towards a hierarchical treatment of multiscale processes. Can. J. For. Res. **33**: 398–409.

Mäkelä, A., Landsberg, J., Ek, A.R., Burk, T.E., Ter-Mikaelian, M., Ågren, G.I., Oliver, C.D., and Puttonen, P. 2000. Process-based models for forest ecosystem management: current state of the art and challenges for practical implementation. Tree Phys. **20**: 289–298.

McKenney, D.W. and Pedlar, J.H. 2003. Spatial models of site index based on climate and soil properties for two boreal tree species in Ontario, Canada. For. Ecol. Manag. **175**: 497–507.

Meinzer, F.C. 2003. Functional convergence in plant responses to the environment. Oecologia, **134**: 1–11.

Miner, C.L., Walters, N.R., and Belli, M.L. 1988. A guide to the TWIGS program for the north central states.

USDA Forest Service General Technical Report NC-125.

Ministère de l'Énergie et des Ressources (MER). 1988. Sylva, logiciel de simulation de la possibilité annuelle de coupe (version 2.00) : manuel de l'usager. Ministère de l'Énergie et des Ressources, Québec, Que.

Monteith, J.L. 1977. Climate and the efficiency of crop production in Britain. Phil. Trans. R. Soc. Lond. B. **281**: 277–294.

Munro, D.D. 1974. Forest growth models – a prognosis. *In* Growth models for tree and stand simulation. *Edited by* J. Fries. Royal College of Forestry, Research Notes 30, Stockholm, Sweden.

Pacala, S.W., Canham, C.D., Saponara, J., Silander, J.A., Jr., Kobe, R.K., and Ribbens, E. 1996. Forest models defined by field measurements: estimation, error analysis and dynamics. Ecol. Monogr. **66**: 1–43.

Pacala, S.W., Canham, C.D., and Silander, J.A., Jr. 1993. Forest models defined by field measurements: I. The design of a northeastern forest simulator. Can. J. For. Res. **23**: 1980–1988.

Parry, D., Spence, J.R., and Volney, W.J.A. 1988. Budbreak phenology and natural enemies mediate survival of first-instar forest tent caterpillar (*Lepidoptera: Lasciocampidae*). Environ. Ent. **27**: 1368–1374.

Peng, C. 2000. Growth and yield models for unevenaged stands: past, present and future. For. Ecol. Manag. **132**: 259–279.

Plonksi, W.L. 1974. Normal yield tables (metric). Ontario Ministry of Natural Resources.

Pojar, J., Klinka, K., and Meidinger, D.V. 1987. Biogeoclimatic ecosystem classification. For. Ecol. Manag. **22**: 119–154.

Porté, A. and Bartelink, H.H. 2002. Modelling mixed forest growth: a review of models for forest management. Ecol. Model. **150**: 141–188.

Pothier, D. and Savard, F. 1998. Actualisation des tables de production pour les principales espèces forestières du Québec. Ministère des Ressources naturelles, Gouvernement du Québec, Québec, Que.

Pretzsch, H., Grote, R., Reineking, B., Rötzer, T.H., and Seifert, S.T. 2007. Models for ecosystem management: a European perspective. Ann. Bot. **2007**: 1–23.

Raulier, F., Lambert, M.-C., Pothier, D., and Ung, C.-H. 2003. Impact of dominant tree dynamics on site index curves. For. Ecol. Manag. **184**: 65–78.

Raulier, F., Ung, C.-H., and Ouellet, D. 1996. Influence of social status on crown geometry and volume increment in regular and irregular black spruce stands. Can. J. For. **26**: 1742–1753.

Runyon, J., Waring, R.H., Goward, S.N., and Welles, J.M. 1994. Environmental limits on net primary production and light-use efficiency across the Oregon transect. Ecol. Appl. **4**: 226–237.

Ryan, M.G., Binkley, D., and Fownes, J.H. 1997. Age-related decline in forest productivity: pattern and process. Adv. Ecol. Res. **27**: 213–262.

Saucier, J.-P., Gagné, C., and Bernier, S. 2006. Comparing site index by site types along an ecological gradient in southern Québec using stem analysis. *In* Workshop "Eastern Canusa: Les sciences forestières au-delà des frontières", Québec, 19–21 October 2006. *Edited by* G.R. Larocque, M. Fortin, and N. Thiffault. Conference manual, second edition, pp. 136–143.

Seymour, R.S. and Hunter, M.L., Jr. 1999. Principles of ecological forestry. *In* Maintaining biodiversity in forest ecosystems. *Edited by* M.L. Hunter Jr. Cambridge University Press. Cambridge, UK.

Seymour, R.S. and Kenefic, L.S. 2002. Influence of age on growth efficiency of *Tsuga canadensis* and *Picea rubens* trees in mixed-species, multi-aged northern conifer stands. Can. J. For. Res. **32**: 2032–2042.

Seynave, I., Gegout, J.-C., Hervé, J.-C., Dhôte, J.-F., Drapier, J., Bruno, É., and Dumé, G. 2005. *Picea abies* site index prediction by environmental factors and understorey vegetation : a two-scale approach based on survey databases. Can. J. For. Res. **35**: 1669–1678.

Stadt, K.J., Lieffers, V.J., Hall, R.J., and Messier, C. 2005. Spatially explicit modeling of PAR transmission and growth of *Picea glauca* and *Abies balsamea* in the boreal forests of Alberta and Quebec. Can. J. For. Res. **35**: 1–12.

Taylor, K.C., Arnup, R.W., Merchant, B.G., Parton, W.J., and Nieppola, J. 2000. A field guide to forest ecosystems of northeastern Ontario. Ontario Ministry of Natural Resources, NEST Field Guide FG-001.

Ung, C.-H., Bernier, P.-Y., Raulier, F., Fournier, R.A., Lambert, M.-C., and Régnière, J. 2001. Biophysical site indices for shade tolerant and intolerant boreal species. For. Sci. **47**: 83–95.

Vanclay, J.K. 1995. Growth models for the tropical forests: a synthesis of models and methods. For. Sci. **41**: 7–42.

Vanclay, J.K. 2003. Realising opportunities in forest growth modelling. Can. J. For. Res. **33**: 536–541.

Vézina, P.E. and Linteau, A. 1968. Growth and yield of balsam fir and black spruce in Québec. Canadian Forest Research Laboratory, Information Report no. Q-X-2.

Wang, Y., Raulier, F., and Ung, C.-H. 2005. Evaluation of spatial prediction of site index obtained by parametric and nonparametric methods: a case study of lodgepole pine productivity. For. Ecol. Manag. **214**: 201–211.

Will, R.E., Munger, G.T., Zhang, Y., and Borders, B.E. 2002. Effects of annual fertilization and complete competition control on current annual increment, foliar development, and growth efficiency of different aged *Pinus taeda* stands. Can. J. For. Res. **32**: 1728–1740.

Zeide, B. 2003. The U-approach to forest modeling. Can. J. For. Res. **33**: 480–489.

# Chapter 17

# Scenario Planning and Operational Practices within a Sustainable Forest Management Plan*
## An Approach Developed by LP Canada, Manitoba

*Margaret Donnelly, Laird Van Damme, Tom Moore, Rob S. Rempel, and Paul Leblanc*

* The authors acknowledge the contributions of Dr. Norm Kenkel, University of Manitoba; Dr. Jacques Tardif, University of Manitoba, and LP Swan Valley Forest Resource Division staff. Funding for research was provided by LP Canada Ltd. and the Sustainable Forest Management Network. We also gratefully acknowledge the contribution of the many stakeholders, organizations, and interested individuals who contributed time and energy to developing the management scenarios and attending LP's Advisory Committee meetings. The photos on this page were graciously provided by Margaret Donnelly and Marie-Andrée Vaillancourt.

# 1. INTRODUCTION

The conservation of biodiversity is a fundamental aspect of a sustainable forest management approach. Research suggests that biodiversity can be maintained through the development of planning and operational forest practices designed to maintain forest conditions similar to those created by natural disturbance. This results in a forest mosaic that is more similar to what is observed under natural conditions, and habitat to which boreal forest species have adapted (Hunter 1990).

Natural disturbances, including fire, insects, disease, and wind, are prevalent in boreal forest ecosystems and are key drivers of many ecosystem processes. As the principal element of both forest and stand level change, natural disturbances contribute to landscape heterogeneity in terms of species composition, structure, and pattern. Research into natural disturbances has become more prevalent in the last 20 years, with particular attention paid to the differences between forests initiated by natural disturbances compared to those resulting from forest management practices (see Jetté et al., Introduction). This research has further reinforced the need to adapt current forest management practices to maintain landscape and stand level diversity.

In many regions, when the outcome of natural disturbance is compared with the outcome of current forest management practices around the three factors that characterize a disturbance regime: frequency, size, and severity (see Vaillancourt et al., chapter 2), three concerns often emerge:

1. the maintenance of over-mature and old-growth forests;

2. the maintenance of landscape structure;

3. the maintenance of remnant habitats in regenerating areas.

Several jurisdictions have developed forest management guidelines based on the natural disturbance model (BCMF 1995; OMNR 2001) as part of a sustainable forest management approach. Likewise, many forest companies have developed practices designed to emulate natural disturbance patterns including variable retention and the maintenance of forest structure at both stand and landscape scales (Serrouya and D'Eon 2004; Sougavinski and Doyon 2005).

In order to more closely approximate natural disturbances it is important that alternative approaches to forest management practices are compared and explored, including forest planning, silvicultural systems, and other operational practices. This chapter describes a strategic planning process developed by LP Canada that incorporated the Natural Disturbance Model (NDM) and associated principles (Bergeron et al. 2002) into the 20-Year Sustainable Forest Management Plan for Forest Management License # 3 in west-central Manitoba (figure 17.1). Scenario planning was used to explore and compare alternative strategies to promote the conservation of biodiversity and develop a landscape planning framework while being operationally feasible. Within this chapter, we intend to illustrate 1) the process used to explore and compare different management scenarios, and 2) the process of assessing the effectiveness of scenarios in reaching biodiversity conservation objectives.

The primary goal of the scenario planning exercise was to test alternative management strategies and implementation options to achieve a suite of sustainable forest management (SFM) objectives. The conservation of biological diversity was one of the key goals of the Plan. The planning team was to develop scenarios that would test various arrangements of harvest areas or disturbance patches,• harvest systems, and regeneration strategies to "design" a landscape that would conserve biodiversity while harvesting the provincially set annual allowable cut (AAC). Alternative management or silvicultural approaches including variable retention, softwood understory protection, and a range of patch sizes were tested in PATCHWORKS (PW) and evaluated with our assessment tools. Specific objectives of the planning process were to:

1. develop a series of forest management scenarios for the Duck Mountain Provincial Forest (DMPF) including a natural disturbance management scenario;

2. evaluate the effectiveness of the management scenarios to conserve biodiversity in the future forest in terms of the maintenance of:

   ▪ forest composition (species and age class);

   ▪ landscape pattern (patch size and distribution); and

   ▪ landscape structure (remnant habitats and residual forest).

• Disturbance patches describe the latent effect of harvest operations, usually referring to those lasting for more than one planning period, and is typically characterized as the time until a "free-to-grow" condition has occurred.

## 2.   STUDY SITE

LP Canada Ltd operates an oriented strand board (OSB) facility near the Town of Minitonas, located in west-central Manitoba, Canada. LP's Swan Valley Forest Resources Division is located nearby in Swan River and is responsible for the management of the Company's operations. The OSB mill utilizes aspen, balsam poplar, and white birch harvested from boreal forest located nearby in the Duck Mountains and Porcupine Hills. LP is responsible for forest management and renewal activities on crown lands within the Duck Mountains under Forest Management License (FML) Area #3, including both hardwood and softwood quota holders (figure 17.1).

The Duck Mountain Ecodistrict in Manitoba is one of a series of highlands that rises above the prairie across Manitoba and Saskatchewan on Canada's Boreal Plains Ecozone (Smith et al. 1998). Located in west-central Manitoba, the study area includes the DMPF and Park, an area of approximately 3,770 km². The complex physiography of the region gives rise to many different types of forest communities, interspersed with lakes and wetlands. Mixedwood stands composed of trembling aspen (*Populus tremuloides* Michx.), balsam poplar (*Populus balsamifera* L.), jack pine (*Pinus banksiana* Lamb.), white birch (*Betula papyrifera* Marsh.), white spruce (*Picea glauca* [Moench] Voss) and/or balsam fir (*Abies balsamea* [L.] Mill.) dominate the uplands of the Duck Mountains. Trembling aspen stands

Figure 17.1
**Forest Management License Area #3 and Duck Mountain Provincial Park (DMPP) limits in west-central Manitoba**

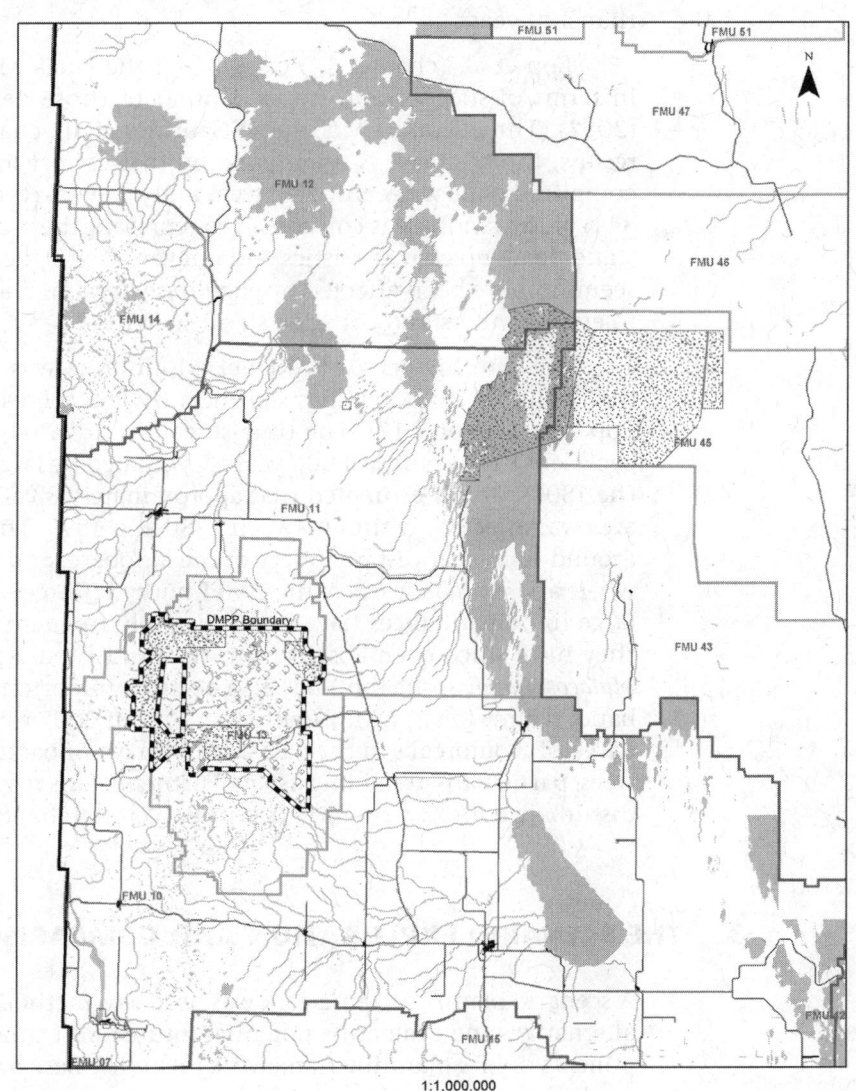

1:1,000,000

occur in well-drained sites at lower elevations, often mixed with balsam poplar in richer, moister sites. Black spruce is common in poorly drained, nutrient-deficient areas, often in association with tamarack. White spruce often occurs in the understory of aspen stands and may remain in a suppressed state for extended periods. Recruitment of white spruce advance regeneration into the upper canopy is also common (Hamel and Kenkel 2001).

Recent inventories indicate a large portion (84%) of the DMPF is forested. This includes 7% of the area that is comprised of treed fens/bogs and other wetlands dominated by black spruce and lowland conifer species. The other 77%

is upland hardwood, mixedwood, and softwood forest. Non-forested wetlands comprised of bogs, fens, swamps, and marsh account for 9% of the total unit. Only 16% of the forest is young (1–60 yrs old), while almost 80 % of DMPF is mature forest (61–120 yrs old). The remaining 4% of the forested area is older than 120 years.

Epp et al. (chapter 12) categorized the Duck Mountain forested landbase in terms of successional stages, similar to those described by Bergeron et al. (2002) (Three-Cohort Model, see Gauthier et al., chapter 3). According to their results, the landbase is dominated by first-cohort forest (60%) with a canopy comprised of post-fire initiated trees and an even-aged structure. Approximately 34% of the landbase is considered to represent the second cohort demonstrating stand break-up characteristics and uneven-aged structure. Only 6% of the forest seems to have been affected by gap phase dynamics and is dominated by shade-tolerant conifers and complex stand structure.

The fire history of the Duck Mountains was studied by Tardif (2004), indicating the fire cycle has significantly lengthened since the late 1800's (see Epp et al., chapter 12). The time-since-last-fire analysis revealed a variable fire cycle length over the last 300 years dominated by large, catastrophic fires during the 1800s. Tardif estimated that approximately 83% (283,580 ha) of the DMPF area was affected by fires between 1880 and 1899. There were several small fires around the southwest periphery of the landbase, and then one large fire in 1961, where approximately 6% of the DMPF burned (Hamel and Kenkel 2001). Although large fire disturbances have been relatively frequent prior to the 20th century, they have since been absent from the landscape. Forest tent caterpillar (FTC; *Malacosoma disstria* Hbn.) has also been an important component of the disturbance regime in the DMPF (Sutton and Tardif 2007; Sutton and Tardif, chapter 8). Large FTC outbreaks in the past century have impacted the growth of deciduous trees, particularly trembling aspen. Further details regarding the Duck Mountains disturbance history are provided in Epp et al. (chapter 12).

## 3.  THE SCENARIO EXPLORATION AND COMPARISON FRAMEWORK

A scenario planning framework was used as a method of exploring management alternatives and evaluating potential future forest conditions. The principal components of LP's planning framework including base data and other model inputs, predictive or forecasting models, and assessment tools are linked through a series of analyses and evaluations (figure 17.2). In this chapter, we will focus on the biodiversity effects assessment using different management scenarios.

The LP Plan had two main SFM objectives: 1) maintain forest ecosystem health and function, and 2) provide forest-based goods and services for present and future generations. Since forest ecosystem function is reliant upon species composition, structure, and pattern, several key strategies were designed to address these objectives with indicators to evaluate the ability of a given scenario to meet them. The principles of natural disturbance management suggest that species composition, structure and pattern in managed forests should be kept similar to those found in natural systems. These characteristics can provide a useful guide and were considered throughout the planning process.

**Figure 17.2**
**Overview of LP's scenario planning framework**

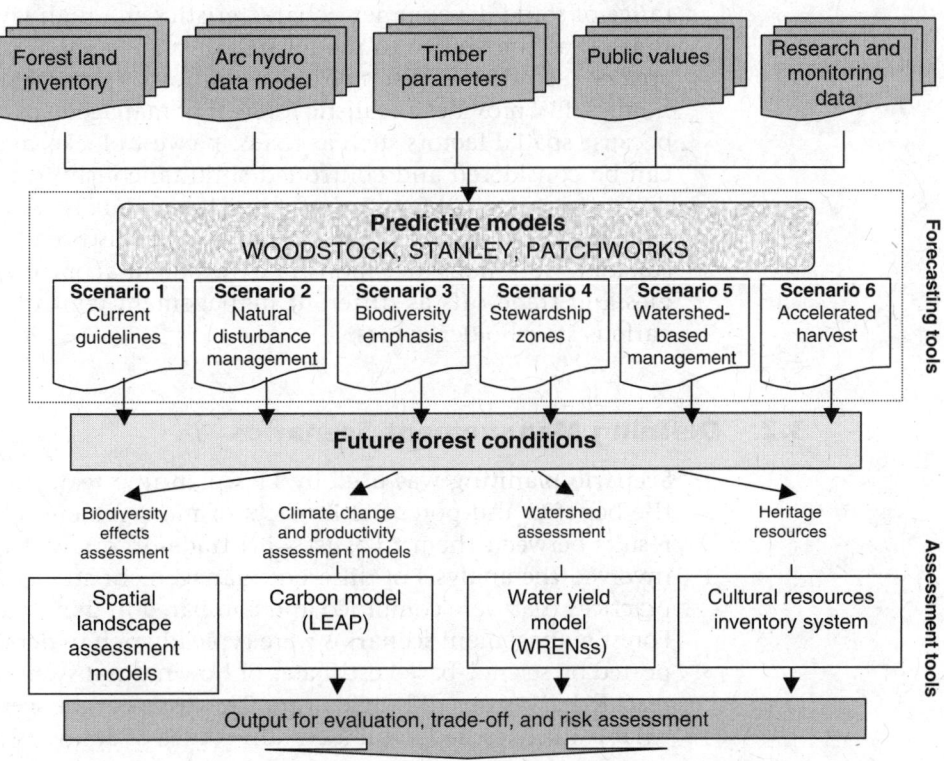

This framework includes input data, predictive models, and assessment tools used.

## 3.1.  Model Input and Modelling Tools

During the implementation of LP's 10-Year Forest Management Plan (1996–2005), LP actively developed a program of research, monitoring, inventory, and data collection to address knowledge and information gaps and prepare for the next strategic level plan, which was due in 2006. The extensive datasets compiled by LP through research and monitoring activities, in addition to an enhanced ecological forest inventory for terrestrial and aquatic ecosystems, were key inputs to the planning process and were used in a suite of predictive models and assessment tools for use in the Plan (figure 17.2). In particular, the spatial planning tool PATCHWORKS (PW) (Lockwood and Moore 1993) provided the ability to build a robust data environment to model and test a suite of forest management scenarios and create digital maps of the predicted future forest landscapes. PW has an open data approach which made it easy to incorporate research findings into the management scenarios.

PW is a spatially explicit planning tool that maintains forest area information and relationships to the managed landbase throughout the planning exercise. LP Swan Valley selected PW as a planning tool for its ability to track a wide range of stand development characteristics at a high level of spatial resolution. It also has the ability to model spatial management policies, such as harvest area or disturbance patch size distributions, transportation logistics, and economics. PW provides a realistic forecast of management alternatives and impacts because spatial factors such as roads, harvest blocks, and adjacency constraints can be considered and controlled simultaneously in the same model used to determine sustainable wood supply. The model also functions well as a growth simulation and trade-off analysis tool, and was used by LP during public consultation workshops to demonstrate the potential outcome of management strategies and trade-offs as different management controls were adjusted to meet various stakeholder values.

## 3.2.  Defining Management Scenarios

Scenario planning was used by LP's planning team to quantitatively estimate the benefits and potential impacts of management alternatives and compare results between them to assist with trade-off analysis and decision-making. It involved the analysis of different management strategies, goals, activities, and practices that were combined and compared in dynamic computer simulations. Forest management scenarios were typically rich in detail, with predictions supported by science-based estimates of how natural systems change over time and react to harvesting and silviculture. The results of the scenario modelling describe what is likely to happen if a certain course of action is pursued.

The development of management scenarios, including one based on the Natural Disturbance Model, was conducted and assessed within PW through a series of analysis "rounds." The objective was to use the first round of analysis to study the "corners of the management space" (i.e., wood fibre on one extreme to absolute conservation on the other) to better understand system responses and sensitivity to managing a few critical objectives prior to including all management objectives. This first step included several different scenarios (table 17.1). The three rounds included three different "sets" of analyses followed by public consultation and discussions about trade-offs and preferences.

## 3.3.  Predicting Future Forest Conditions

Long-term planning based on natural disturbance approaches required the ability to predict natural patterns of forest change over time. Present forest stand conditions were used as a starting point, and future stand and landscape conditions were modelled according to our understanding of how forests develop and change in the absence of stand-replacing disturbances like fire and harvesting. Different harvesting and habitat maintenance scenarios were compared over long-time horizons using models, and results used to set operational objectives in the short term. In the course of the scenario planning exercise several key activities were undertaken in order to forecast future forest landscapes. These included:

- the development of habitat element curves. These curves are similar to yield curves traditionally used in forestry, but also contain habitat attributes important for biodiversity assessment such as snag density, coarse woody debris content, and green tree retention volumes. These curves enabled the merging of ideas, data, and relationships about stand development, structural attributes, operational practices, and stand regeneration into a system useful for PW and modelling future forest conditions (LP Canada 2006);

- stand dynamic and succession models for Duck Mountain ecosystems (Hamel and Kenkel 2001; Kenkel et al. 2003);

- the development of a successional stage index as a metric for landscape composition and structure in scenario planning (LP Canada 2006);

- an ecosystem representation analysis using ecosites (LP Canada 2006);

- spatially explicit, multiple-scale songbird habitat models to evaluate scenario results in terms of biodiversity objectives (Rempel et al. 2006).

These various models were integrated into the process through PATCHWORKS. The PW model considers many spatial objectives and constraints directly in the determination of each scenario's sustainability. Other aspects of analysis, including some of the biodiversity assessment tools, were completed outside the PW environment.

## 3.4. Assessing the Effectiveness of Predicted Outcomes to Achieve the Objectives

PATCHWORKS results include reports on a suite of targets and indicators which are used to evaluate the ability of a given scenario to attain the objectives and values it was designed to achieve (e.g., variable retention practices to provide snags in future forest). Some indicators such as the proportion of snags are readily interpreted directly from the output while other indicators need further analysis or processing for interpretation. Biodiversity is an example of a value we wish to maintain in future forest landscapes, but the ability to directly measure or report on it is difficult.

The natural processes of disturbance and succession result in a forest structure that has inherent diversity. The targets and limits for structural diversity (e.g., age class, cover type, and interspersion) are determined in part by the predicted response of a set of focal species and the desire to keep sufficient habitat in place for all species of the group, over time. This structural diversity occurs at multiple spatial scales because structuring processes also occur at multiple scales. Stand-replacing disturbance events dominate structural effects at the landscape scale, while gap-phase disturbance dominate at the stand scale. Three dominant variables capture much of this structural diversity: cover type, age class or structure, and interspersion of age classes and cover types. They are directly influenced by forest disturbance, whether it is the result of natural disturbance processes or forest management practices. Therefore the principal objective of a coarse filter approach is to maintain the natural diversity of forest structure and pattern on the landscape.

Of particular interest to biodiversity conservation and our scenario assessments was the development of a method to assess the effectiveness of predicted future landscapes to provide wildlife habitat, especially a diverse range of ecosystem types for both migratory and resident bird species. Songbirds are a diverse group of organisms involving species adapted to the full range of naturally occurring forest conditions which make it an interesting group to use to assess forest practice effects at a landscape scale (see Drapeau et al., chapter 14). Songbirds are a particularly important component of biodiversity in the Duck Mountains and had been extensively monitored by LP over a 7-year period, providing an excellent database and baseline condition for modelling. Data from LP's monitoring program was used to develop multiple scales, spatially explicit models of resource selection for a range of bird species in the Duck Mountains.

The Spatial Landscape Assessment Model (SLAM) is an independent, spatially explicit computer assessment tool that allows for reporting of indicators of biodiversity at multiple scales using spatial songbird habitat models applied to PW outputs of forest conditions at different time intervals (Rempel et al. 2006). Specifically, it is a biodiversity assessment tool based on bird species response to habitat to evaluate the scenarios for the LP Plan.

## 4.  COMPARISON OF MANAGEMENT SCENARIOS FOR THE DUCK MOUNTAIN PROVINCIAL FOREST

Initially, several scenarios were defined to compare combinations of different management strategies and operational practices necessary to achieve a suite of management objectives (table 17.1). LP's scenario planning process deviated from the traditional approach to wood supply analysis in that Manitoba Conservation Base Case levels (including AAC volume forecasts) were cast as one of several competing and alternative scenario targets. Thus, for each scenario, wood supply level was treated critically as a hypothesis, and the target level was considered reasonable and viable but was not always the key objective. For each scenario, the analysis conducted in PW attributes different weight to each target, depending on their importance. For example, the weight attributed to retention was higher in the Biodiversity Emphasis Scenario than for the Accelerated Harvest Scenario. For several scenarios, harvest objectives (in volume) were never considered more important than biodiversity objectives. Overall, this provided the ability to assess scenarios in terms of multiple indicators in addition to wood supply.

Four examples of scenarios that were initially explored are presented in table 17.1. Management controls are the elements within the scenarios that are varied or "turned on," and reflect the various strategies or practices utilized by the Company during planning and operations. That is, the management controls reflect the elements of forest management that LP controls or effects in the Duck Mountains. For instance, the size of the harvest area or block can easily be

Table 17.1
## Table 17.1
## Examples of management scenarios developed for analysis during the first round

| Management control | Accelerated harvest | Biodiversity emphasis | Stewardship zones | Natural disturbance management |
|---|---|---|---|---|
| | Ramp up harvest, even flow, no guidelines | Broad range of variability, high retention older canopy | Triad zoning (20-60-20)* | Natural disturbance management |
| Natural disturbance patterns | n/a | Broad | Depends on zone | Resemble natural |
| Retention | No retention | 16-24 stems/ha | Depends on zone | 8-12 stems/ha |
| Opening size limit | n/a | < 100 ha | Depends on zone | > 100 ha |
| Road budget | Minimal cost | n/a | n/a | n/a |
| Cover type constraints | n/a | Stability | Depends on zone | Stability |
| Silviculture | Conventional | Cohort management | Depends on zone | Cohort management |

\* 20% intensive management, 60% ecosystem management, 20% conservation.
The scenarios were developed to meet different objectives and management controls varied to achieve the various objectives.

controlled through planning and layout, all blocks can be specified 15 hectares in size, or, if desired, a range of sizes from 25 to 500 hectares can be achieved. Ecosystem-based management and the conservation of biodiversity were desired elements of LP's SFM approach. Different strategies were incorporated into the various scenarios to conserve biodiversity including ecosystem representation, stewardship and management zoning concepts, the Three-Cohort Model (Bergeron et al. 2002), and old-forest retention strategies.

Following the first round of analysis and public consultation, the scenarios were refined to four variants based on comparisons between the modelling results and the objectives, as well as preferences and concerns raised by the public and scientific advisory committees. The Accelerated Harvest Scenario was eliminated since it was designed primarily to explore a timber-based emphasis, which was not feasible due to regulatory constraints as well as the lack of biodiversity objectives. The Biodiversity Emphasis Scenario was varied to include aspects of the Natural Disturbance Management Scenario. The Stewardship Zones Scenario was eliminated since the landbase already had different management zones due to the presence of the Duck Mountain Provincial Park.

Finally, following the same process, two scenarios were retained for the last round: the Preferred Scenario and the Manitoba Base. The Manitoba Base Case is a scenario based on current provincial forest management regulations and requirements, and the Province's Wood Supply Analysis and AAC determination. The Preferred Scenario is a combination of the Biodiversity Emphasis, Water Conservation, and Current Guidelines scenarios since it was considered the most feasible in terms of the current regulatory framework, the ability to achieve the biodiversity objectives, and was socially and economically acceptable. Details on those two scenarios retained for the third round of analysis are presented in table 17.2.

Table 17.2

## Patchwork scenario descriptions for Round 3 scenario comparisons

|  | Manitoba Base Case | Preferred Scenario |
|---|---|---|
| Scenario description | This simulation used Patchworks (PW) to replicate the MC Woodstock analysis as closely as possible. This produces results similar to those derived from traditional wood supply analyses with more management controls. Current regulatory requirements and LP Environment License conditions have been applied. The MC analysis was aspatial and converted to a spatial output in PW. | Stakeholder values, operational considerations, and a spatial sustainability analysis were incorporated into this simulation using PW and computer assessment tools. The maintenance of biodiversity and the current forest composition over time were key objectives. Enhanced operational and economic feasibility. Water conservation objectives more robust. |
| Assumptions | MC derived yield curves, standard guideline silviculture (clear-cut harvests), succession rules re-set at 130 yrs, etc. | Habitat element curves provide stand age, volume, development, and successional dynamics including silvicultural treatments and post-harvest responses. |
| Annual allowable cut/ volume targets[1] | 387,582 m³/yr hardwood harvest<br>162,533 m³/yr softwood harvest<br>Even flow volumes | 387,582 m³/yr hardwood harvest<br>162,533 m³/yr softwood harvest<br>Minimum harvest levels |
| Harvest patch size targets[2] | 10 – 50 ha: 70%<br>50 – 100 ha: 30% | 10 – 50 ha: 30%<br>50 – 100 ha: 70% |
| Disturbance patch size targets[3] | 0 – 25 ha: 30%<br>25 – 50 ha: 15%<br>50 – 100 ha: 10%<br>100 – 250 ha: 15%<br>250 – 500 ha: 30% | 0 – 25 ha: 25%<br>25 – 50 ha: 15%<br>50 – 100 ha: 15%<br>100 – 250 ha: 15%<br>250 – 500 ha: 30% |
| Age class distribution | Even-aged, normalized forest. No old-forest retention targets. | Old-forest retention targets to conserve structural diversity. |
| Forest cover stability | No attempt to maintain same forest unit stability. No upper limit on softwood harvest. | Forest cover stability targets set and enhanced renewal strategies to maintain mixedwood stands in future forest. |
| Ecosystem representation | No consideration given to ecosystem representation or rare sites. | Ecosystem representation analysis completed; targets set to conserve rare ecosites and enhance existing protected areas network. |
| Post-harvest stand development[4] | No realization of harvest treatment including variable retention supplemental planting or understory protection. | Habitat element curves enable forecast of treatment responses including habitat structural elements. |

1. Annual allowable cut determined by Province as part of wood supply analysis; however LP did not complete a wood supply analysis and used the AAC as a target and assessed sustainability of a suite of indicators.
2. Harvest patches describe an area of contiguous forest that is harvested within the same planning period and within a specified search radius from other harvested areas. All neighbours within this search radius that are harvested during the same time period are considered to be part of the same harvest patch, as well as all of their respective connected neighbours, etc.
3. Disturbance patches describe the latent effect of harvest operations, usually referring to those lasting for more than one planning period, and is typically characterized as the time until a "free to grow" condition has occurred (i.e. green-up: softwood height = 2 m (~15 years) or hardwood height = 3 m (~10 years). All polygons that were younger than green-up and were within the search threshold of each other were considered to be part of the same disturbance patch.
4. Post-harvest stand development reflects additional plot data on 100–200-year-old stands including species composition, age, and other structural attributes resulting from gap phase dynamics and successional processes.

## 5.  EFFECTIVENESS OF MANAGEMENT SCENARIOS

The effectiveness of the management scenarios to conserve biodiversity in the future forest was assessed using a coarse filter approach based on three indicators that gave information on the maintenance of:

- forest composition (species and age class);

- landscape pattern (patch size and distribution); and

- landscape structure (remnant habitats and residual forest).

The effectiveness of the various management scenarios in terms of the ability to achieve targets and indicators were assessed using the framework presented in figure 17.3. All scenarios were evaluated for a suite of indicators in

**Figure 17.3**
**Indicator assessment framework depicting hierarchy of goals, objectives, strategies, management controls, and targets**

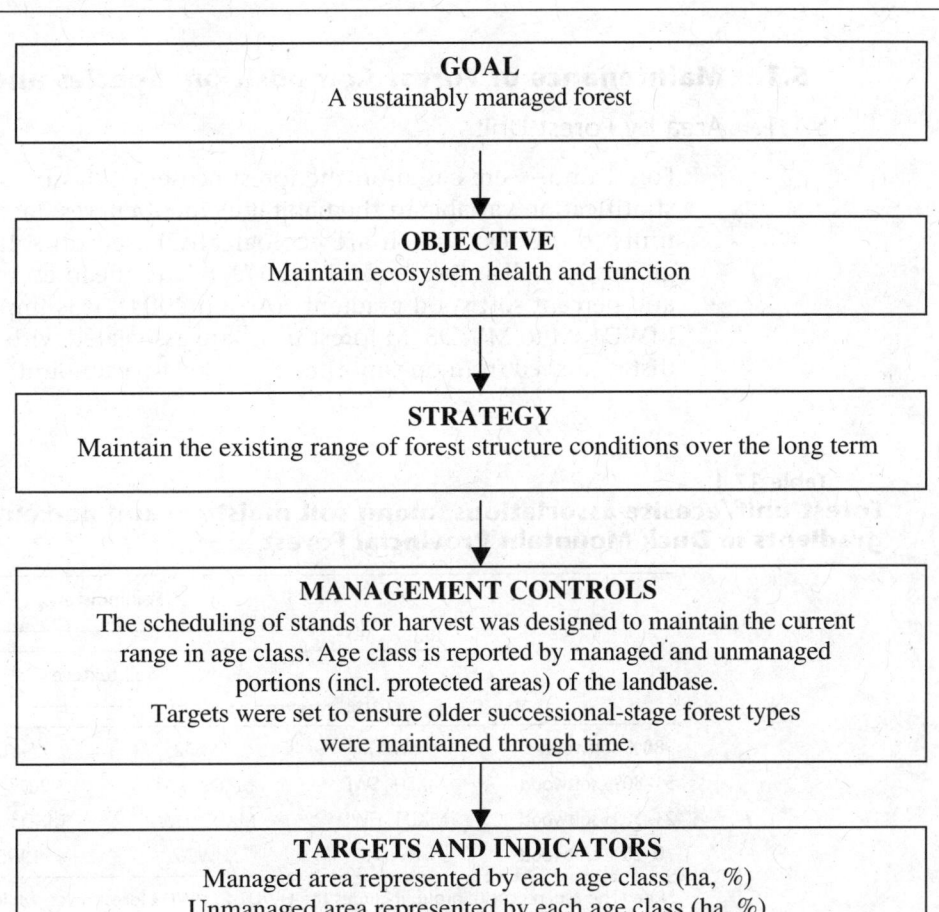

**GOAL**
A sustainably managed forest

**OBJECTIVE**
Maintain ecosystem health and function

**STRATEGY**
Maintain the existing range of forest structure conditions over the long term

**MANAGEMENT CONTROLS**
The scheduling of stands for harvest was designed to maintain the current range in age class. Age class is reported by managed and unmanaged portions (incl. protected areas) of the landbase.
Targets were set to ensure older successional-stage forest types were maintained through time.

**TARGETS AND INDICATORS**
Managed area represented by each age class (ha, %)
Unmanaged area represented by each age class (ha, %)

Table 17.3

**Strategies and indicators retained to evaluate scenario effectiveness in maintaining biodiversity in future forest landscapes**

| Strategies | Indicators |
|---|---|
| Maintain existing range of forest structure conditions over the long term | – Area by forest unit (ha)<br>– Managed area by age class (ha) |
| Maintain a functional landscape pattern of forest cover and habitat types consistent with principles of natural disturbance pattern emulation | – Number of disturbance patches by size class (ha)<br>– Disturbance patches by size (ha) as percent of total disturbance<br>– Harvest patch frequency distribution |
| Maintain representation of the current range of wildlife habitat associations | – Habitat by bird species based on spatial habitat models |

terms of the ecological, social, and economic aspects of sustainable forest management; however, only the results for a subset of indicators related to biodiversity conservation strategies are presented here for the two scenarios retained for the third round of analysis (table 17.3).

## 5.1.   Maintenance of Forest Composition: Species and Age Class

### 5.1.1.   Area by Forest Unit

Forest units were based on the forest ecoseries classification that was used as a stratification variable in the habitat element curves. LP scenarios have 13 forest units (ecoseries), which are ecologically based on soil moisture and texture groups, as well as tree cover. Table 17.4 relates the forest units along soil moisture and percent softwood gradients (Arnup 2004). It is important to note that the MWD1_M to MWD3_M forest units are associated with similar ecosites but are distinguished from one another by the relative amounts of softwood within the

Table 17.4

**Forest unit/ecosite associations, along soil moisture and percent softwood gradients in Duck Mountain Provincial Forest**

| | Soil moisture<br>dry ——————————————→ wet | | | |
|---|---|---|---|---|
| | Soil texture<br>coarse ——————————————→ organic | | | |
| >80% softwood | SWD1 | SWD2 | SWD3 | SWD4 |
| 51–80% softwood | MXD1_SWD | MXD2_SWD | MXD3_SWD | |
| 21–50% softwood | MXD1_HWD | MXD2_HWD | MXD3_HWD | |
| 0–20% softwood | HWD1 | HWD2 | HWD3 | |

Note: The aspen ecosite group includes the HWD1 to HWD3 forest units, while the black spruce ecosite group is analogous to the SWD 4 forest unit. The SWD1 to SWD3 forest units are associated with the conifer ecosite group, including white and black spruce and jack pine. The remainder of forest units are associated with the mixedwood ecosite group.

Legend: SWD = softwood, HWD = hardwood, MXD = mixedwood.

mixedwoods that were found to influence successional pathways in permanent sampling plots analysis. Note that the Manitoba Base Case uses sampling strata, based strictly on the tree canopy, and are somewhat different from the LP forest units. Many ecological and economic indicators are related to different combinations of forest unit and age class, making the forest unit and age class combination a cornerstone for multiple indicators of biodiversity in the scenario planning approach.

Figure 17.4 illustrates the areas of forest units over 200 years for the Manitoba Base Case and Preferred scenarios. In the case of the Manitoba Base Case (figure 17.4A), the light grey represents a hardwood mixedwood forest unit that becomes hardwood within 60 years due to a reliance on natural regeneration. The Preferred Scenario shows that the hardwood mixedwood forest unit can be maintained when understorey protection and planting treatments are applied and PW assigns weights to favour forest unit stability (figure 17.4B). This feature is desirable from a forest biodiversity perspective.

One of the distinguishing features of the Preferred Scenario, with its full suite of environmental management objectives and targets, is that forest unit stability can be achieved without significantly compromising other important economic (i.e. AAC) and ecological objectives (figure 17.4B). Given the importance of forest unit stability as it related to social, environmental, and ecological values, LP planners and analysts consider this result among the most significant outcomes of the scenario development process.

## 5.1.2.   Area by Age Class

Central to LP's Plan was the development of a strategy to conserve and maintain biodiversity in both present and future forests. Public values relating to the conservation of plant and animal species, forest and wetland habitats, and older forest areas were ranked high in surveys conducted and at public consultation meetings. The need to maintain forest areas for both recreation and cultural values were also highly ranked, by both local and provincial level advisory committees (KBM Forestry 2005).

The Duck Mountain Provincial Park is located within the boundaries of the Provincial Forest and presents interesting opportunities and challenges from a management perspective given the presence of two zones where commercial harvesting is not permitted (figure 17.1; see Epp et al., chapter 12). Of particular interest are the areas of contiguous forest within the Park zones that will become old-growth forest through successional processes. Approximately 14% of the Duck Mountain Provincial Forest is represented in the two management zones with no resource development activities and will continue to follow natural succession and stand dynamic trajectories providing for older forest habitats in the future. However, LP recognized the need to maintain older forest patches and structural elements (e.g. snags, down woody debris, open or closed canopy) across the landscape in addition to older forest habitats concentrated within the Provincial Park boundaries. Therefore, plan scenarios were developed with an objective to maintain older forest elements in the managed as well as unmanaged areas of the Provincial Forest.

**Figure 17.4**
**Area by forest units for the Manitoba Base Case (A) and the Preferred Scenario (B)**

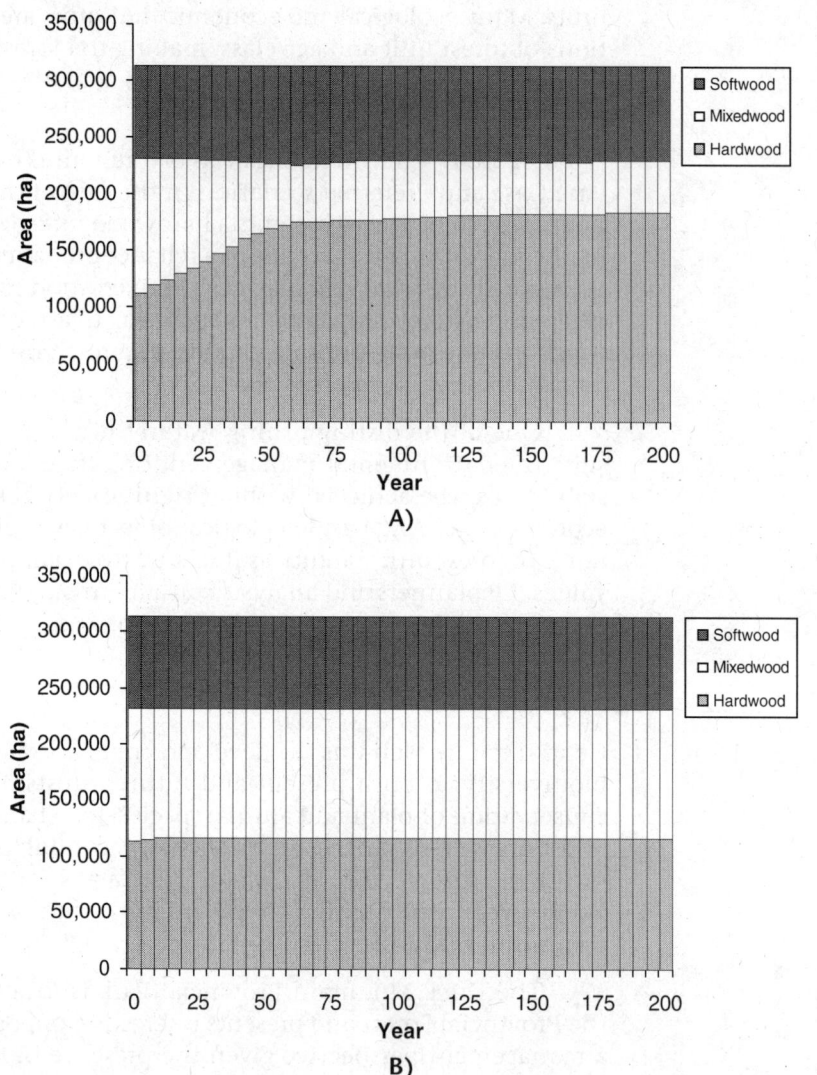

A)

B)

The amount of areas by forest unit varies from the Manitoba Base Case (A) but stays stable for the Preferred Scenario (B) since forest cover stability was a priority and assigned a greater weight.

Old forests are particularly important from a biodiversity perspective and provide structural habitat elements that are not found at younger forest stages. Preliminary analysis of field collected bird habitat data and the forest land inventory found stand age to be a very weak indicator of habitat structural elements. A more robust, process-based approach was needed that reflected stand dynamics as well as age. Habitat elements are required for the spatial landscape assessment models and are important biodiversity indicators.

The Successional Stage Index (SSI) was developed as a method of tracking the availability of older successional stands in future forests (figure 17.5). In order to capture stand structural elements particular to older-forest age classes the SSI was added to the habitat element curves for input to PW. The SSI was developed based on the Three-Cohort Model of Bergeron et al. (2002) and results of research activities to develop succession models for the Duck Mountains and Riding Mountain National Park (Hamel and Kenkel 2001; Kenkel et al. 2003; see Epp et al., chapter 12). The Successional Stage Index has five phases which reflect stand successional dynamics and structural attributes, including:

1. Regenerating phase

2. Stem exclusion phase

3. Canopy opening phase

4. Canopy closing phase

5. Gap dynamics phase

The addition of SSI data to the habitat element curves enabled planners to set retention targets for SS4 and SS5 stands to retain older forest areas and the desired habitat structural attributes associated with them.

In order to maintain the balance of age classes required for biodiversity conservation LP established objectives to maintain a proportion of older forest at all times on the landbase. Targets were set in PATCHWORKS to retain older age

**Figure 17.5**

## The Successional Stage Index depicted on a growth curve representing stand volume versus stand age relationship

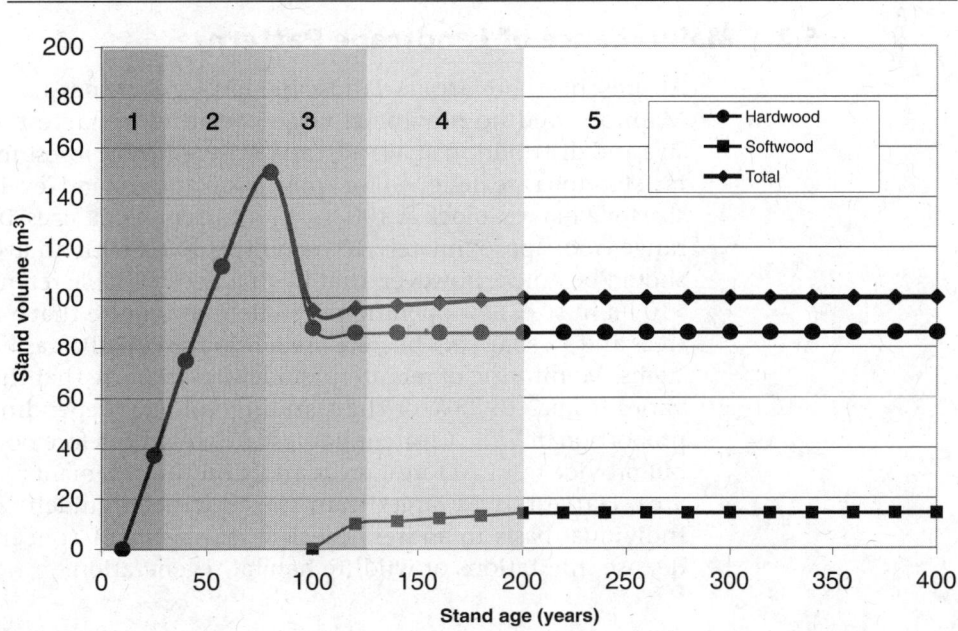

classes (SS4, SS5) in the forest over time. The Preferred Scenario ensured that sufficient areas remained in older age classes and that the relative amounts of young, mature, and older ages were within reasonable bounds given the fire cycles and other natural disturbance events in the forest, based upon the fire history work by Tardif (2004) in this forest and from other related boreal studies (e.g., Bergeron et al. 1999).

An unavoidable consequence of forest harvest is that older forests are removed and replaced by younger forest. Therefore, old forest becomes less abundant over time in each scenario. The older forests which currently dominate the Provincial Forest are being replaced by younger even-aged forests, typical of areas where forests are harvested by the forest industry in all parts of Canada. There are slight differences in forest age class structure between the two scenarios (figure 17.6A and B). There appears to be a reasonable level of age class diversity expected to persist over long periods of time at the harvest rate and patterns being tested. Although LP's harvest practices retain structure in the harvest patches and through a network of reserves, further analysis and consultation with government and stakeholders is required to determine how acceptable these patterns of age classes are.

The spatial distribution of older forest areas through time is presented in figure 17.7 for the Preferred Scenario. Age class targets were set to ensure the distribution of older forest areas outside the DMPP in addition to the larger contiguous areas that will occur naturally in the Park. Therefore, targets were set in PW to maintain specified amounts of SS4 and SS5 habitat during scenario analysis. Using this approach, the Preferred Scenario chosen following Round 3 analysis provided 20,000–30,000 ha of older forest within the managed (potentially harvestable) landbase, in addition to the older forest contained within the backcountry and recreation park zones (figure 17.7).

## 5.2.   Maintenance of Landscape Patterns

At present, there are no landscape level guidelines for forest management in Manitoba and no provincial targets relating to harvest and disturbance patch size and distribution at a landscape scale. Current forest management guidelines for Manitoba are designed for application at the stand level and set the maximum size for a harvest block at 100 ha. LP's current block sizes, based on this guideline, range from approximately 10 ha up to 100 ha with an average size of ~40 ha. It should be noted, however, that LP practices variable retention on harvest blocks >10 ha in size. LP's operational guidelines require that a minimum of 12 green trees and 12 snags per hectare are left in harvested areas in patches or individual stems. Monitoring of retention guidelines indicate that the retention proportion varies from 2 to 38% of the standing volume, depending on sites, for a mean proportion of 12%. Clearcut harvest areas are therefore not necessarily "cut clear" but provide residual forest areas and structural elements within the regenerating area. Harvest areas larger than the provincial guidelines are approved on an individual basis to address specific management concerns such as insect and disease infestations or wildlife habitat regeneration.

Figure 17.6
**Managed area by age-class results for the Manitoba Base Case (A) and the Preferred Scenario (B)**

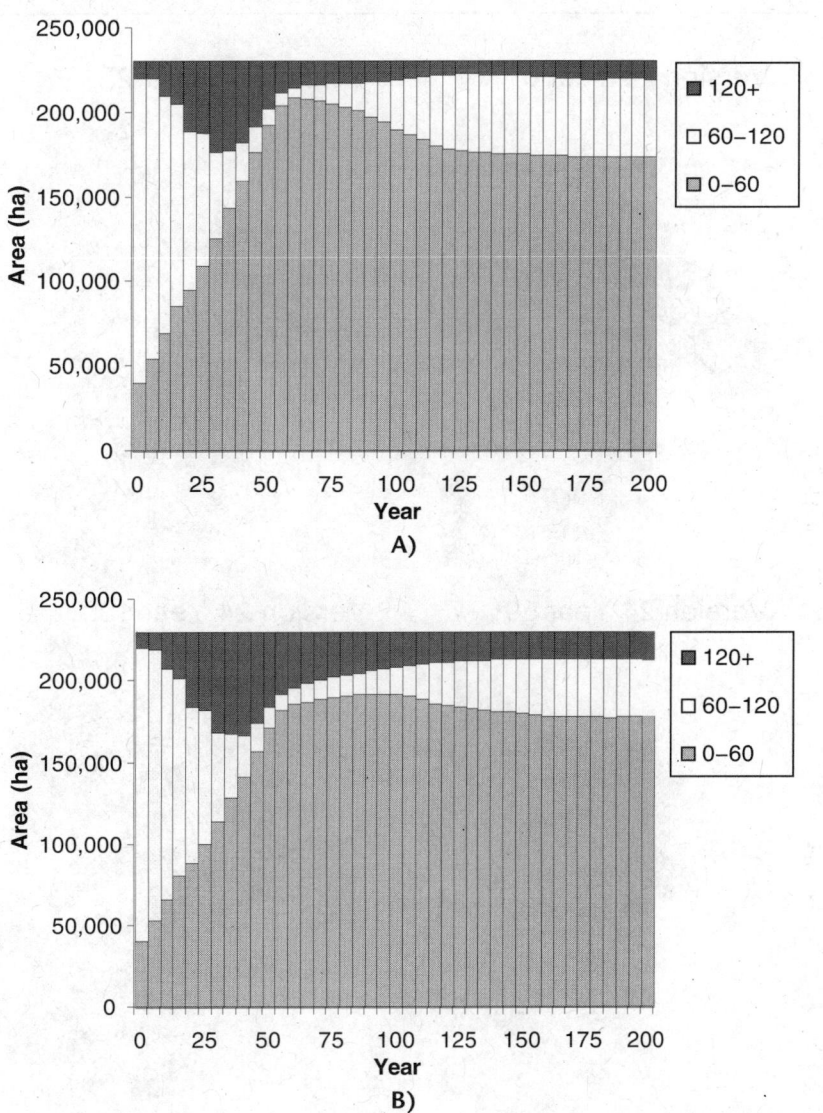

The preferred scenario maintained 20,000–30,000 ha more old-growth forest within the managed forest land base compared to the Manitoba base case.

For the purpose of LP's analysis, harvest patches are defined as contiguous areas of harvest that occurred within a single planning period (5 years). Continuity was assessed based on harvest polygons that were within 100 m of each other. Disturbance patches describe the latent effect of harvest operations, usually referring to those lasting for more than one planning period, which is typically characterized as the time until a "free to grow" condition has occurred (i.e., green-up:

Figure 17.7
**Map of Duck Mountain Provincial Forest displaying predicted pattern of age-class distribution assessed using SLAM through time (from year 0 to 100) for the Preferred Scenario**

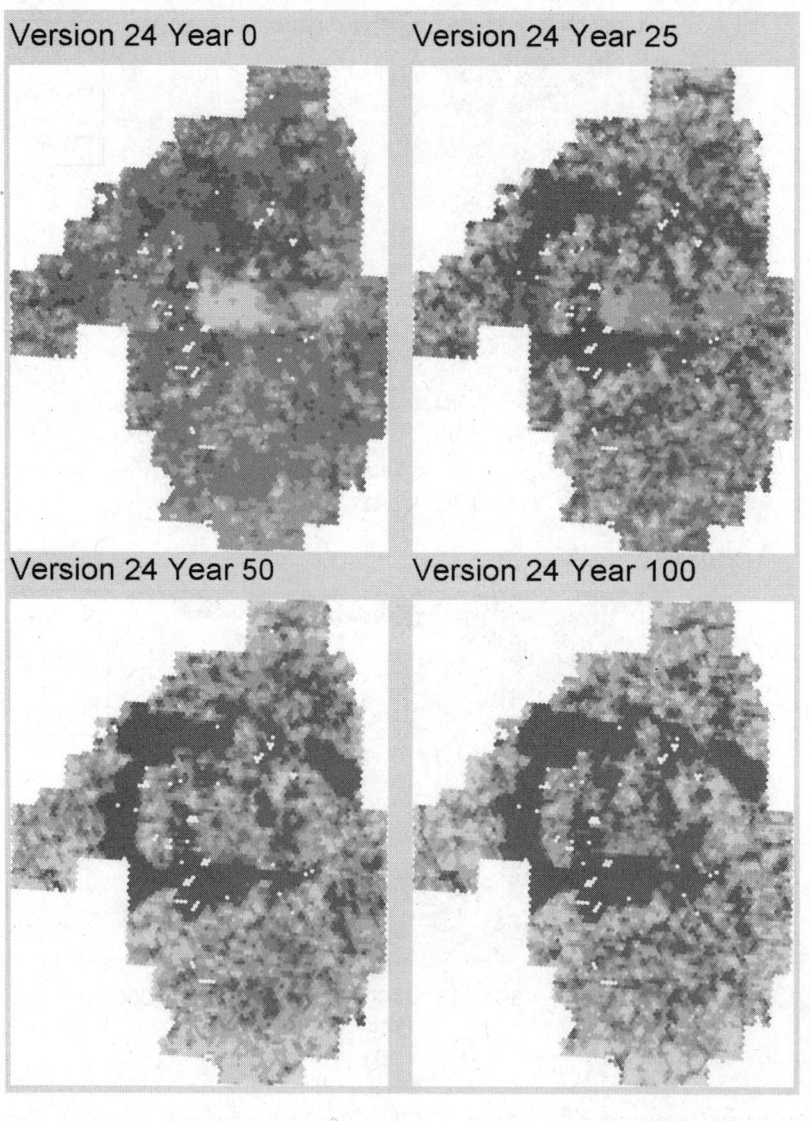

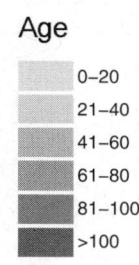

Legend: pale zones: young stands; dark zones: old stands.

softwood height = 2 m [~15 years] or hardwood height = 3 m [~10 years]). All polygons that were younger than green-up and were within the search threshold of each other were considered to be part of the same disturbance patch.

Stakeholder concerns about fragmentation and habitat connectivity, plus guidance from the natural disturbance-based forest management (NDM), contributed to the move toward larger harvest block sizes in the Plan. Harvest patch size distribution targets were designed to more closely resemble historical fire patch size distributions for the region. Compared to previous plans in which harvest patch sizes were limited to a narrow range and large cuts were avoided, the approach used now calls for a wider range of sizes that includes some larger cutovers. The anticipated result is a landscape configuration of harvest patch sizes, forest types, and ages that mere closely resemble the landscape patterns expected in fire-driven boreal ecosystems. However a closer fit to the NDM was not accomplished due to regulatory constraints regarding harvest block sizes in Manitoba.

Tables 17.5 and 17.6 present the harvest patch and disturbance patch size as percentage of total number of disturbances and the harvest patch frequency distribution for the Preferred Scenario for the first 20 years of the planning horizon. The public showed some acceptance of incorporating larger disturbances into the landscape to better emulate fire patterns, but very large harvest areas were not popular with the public and were a concern due to the potential for

**Table 17.5**
**Harvest patches by size (ha) as a percentage of total number of disturbances for the Preferred Scenario**

| Year | Size (ha) | | | |
|---|---|---|---|---|
| | 0–10 | 10–50 | 50–100 | 100+ |
| 5 | 18.45 | 38.38 | 36.53 | 6.64 |
| 10 | 18.07 | 44.88 | 32.23 | 4.82 |
| 15 | 20.23 | 43.59 | 31.05 | 5.13 |
| 20 | 20.00 | 45.07 | 30.42 | 4.51 |

**Table 17.6**
**Disturbance patches by size (ha) as a percentage of total number of disturbances for the Preferred Scenario**

| Year | Size (ha) | | | | | | |
|---|---|---|---|---|---|---|---|
| | 0–25 | 25–50 | 50–100 | 100–250 | 250–500 | 500–1,000 | 1,000+ |
| 0 | 54.44 | 29.29 | 13.17 | 2.51 | 0.44 | 0.15 | 0.00 |
| 5 | 49.39 | 22.41 | 19.21 | 5.49 | 3.35 | 0.15 | 0.00 |
| 10 | 51.64 | 11.31 | 25.18 | 6.57 | 4.56 | 0.73 | 0.00 |
| 15 | 59.94 | 10.17 | 21.13 | 5.32 | 2.97 | 0.31 | 0.16 |
| 20 | 60.99 | 10.61 | 18.24 | 6.28 | 3.59 | 0.30 | 0.00 |

adverse effects on water yield. Regulatory requirements relating to harvest block size, adjacency rules and regeneration standards were also a concern while planning. Therefore only a modest change to the broad range of harvest blocks was proposed. Although there is a shift from a high representation of very small harvest patches to a more representative distribution, approximately 92% of the total area harvested is made of patches less than 100 ha in size, and the two largest size classes (500 ha or larger) account for an average of 5.2% of the total area harvested.

## 5.3. Maintenance of Landscape Structure: Remnant Habitats and Residual Forest

The range of variation associated with natural patterns can never be exactly matched through management, so decisions must still be made to set targets and limits for such factors as age-class structure, cover-type distribution and clearcut size and distribution. From a biodiversity perspective, the question remains as to whether these NDM-based targets and limits will effectively meet biodiversity objectives. To address this question (or hypothesis) we developed a focal species assessment approach (Hannon and McCalllum 2004), where we determined through research which songbird species in the Duck Mountains were associated with the extremes of conifer/deciduous, young/old, or contiguous/fragmented forest conditions. Conceptually, this was viewed in terms of niche-space, which we termed the "biodiversity box" (figure 17.8; Rempel et al. 2006). By maintaining available habitat for the species that utilize the "corners of the box" we should conserve habitat through time by ensuring the box does not "shrink," leaving species without suitable habitat or "outside the box."

**Figure 17.8**
**Bird associations with coarse filter attributes such as forest type diversity, age-class diversity and spatial patterns**

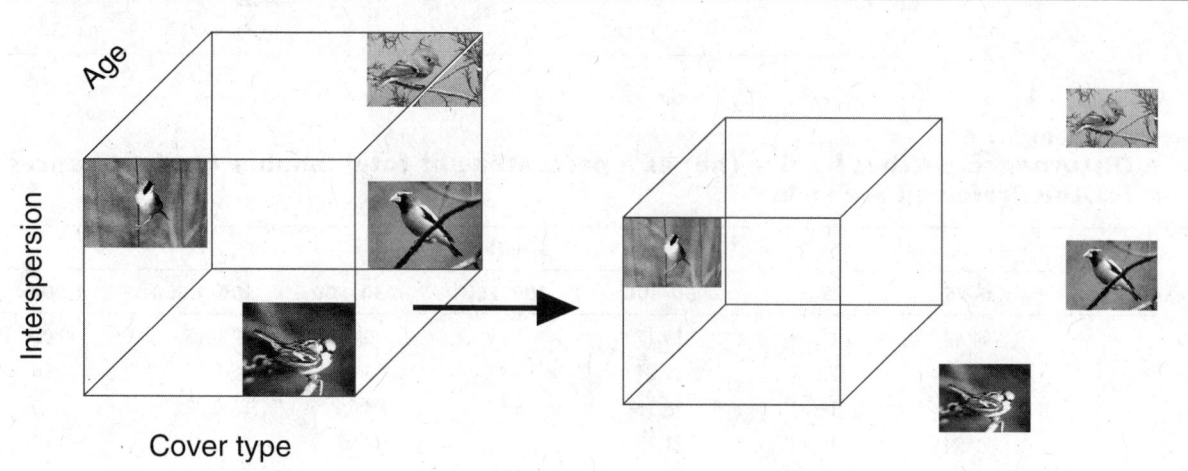

Using the predicted future forest landscapes from the PW model, logistic regression models were used to predict probabilities of future habitat occupancy based on both stand- and landscape-scale conditions. The models assess relative rates of habitat occupancy, and cannot be used to estimate absolute numbers of individuals. Based on niche-space analysis, factor associations from the habitat models, and model-accuracy performance, spatial habitat models for a focal species group were selected to represent the extremes of the three major axes (age class, cover type, and interspersion). The management goal is to not "shrink the box" in terms of structural diversity, thereby maintaining a full range of plants and animals on the landscape. If management reduces the range of variability associated with the forest structural attributes, it may also reduce the species diversity associated with this range.

Future forest landscapes were assessed using spatial landscape assessment models (SLAMs) developed for 17 bird species. The probability of habitat occupancy was determined for predicted future forest landscapes for the Duck Mountain Provincial forest for all scenarios over a 200-year planning period. Results from the SLAMs are presented as surface density maps of the predicted probability of habitat occupancy by species for the various scenarios through time. Stand and landscape level strategies designed to maintain landscape structure and conserve wildlife habitat and residual forest were evaluated using scenarios to predict future forest landscapes in PW for a 200-year planning horizon. This coarse filter scenario assessment revealed that forest structure characteristics (e.g., canopy height, species composition, snag density) and landscape pattern characteristics (e.g., interspersion and edge density) varied among forest management scenarios, but these variations tended to be relatively minor in the context of the entire forest landscape.

Model results predict that several species of birds responded favourably to forest management (e.g., Veery [*Catharus fuscescens*], Hermit Thrush [*Catharus guttarus*], Chestnut-sided Warbler [*Dendroica pensylvanica*]) because management created early successional habitat with higher levels of age class interspersion (figure 17.9). Others responded in a neutral fashion (e.g., Ovenbird [*Seiurus auroocapilla*], American Redstart [*Setophaga ruticilla*]) because the conditions required by these species (mature hardwood) were being sustained in a relatively stable manner over time. The habitat occupancy of several species was predicted to be negatively affected (e.g. Blue-headed Vireo [*Vireo solitarius*], Boreal Chickadee [*Poecile hudsonica*], Brown Creeper [*Certhia americana*]), because of changes in age class structure and interspersion in larger contiguous areas of older softwood forest in the interior of the Duck Mountains (figure 17.10). LP does not harvest softwood, but the planning process requires LP to consider the consequences of logging by secondary quota holders. An example of the output from the spatial landscape assessment models is presented in figures 17.9 and 17.10. The maps below depict habitat occupancy at year 50 for the Hermit Thrush, which, over the planning horizon, responds favourably to forest management practices, and for the Brown Creeper, which responds negatively. Overall, it appears that sufficient habitat is being provided to maintain the songbird community over time, but species associated with softwood in the Duck Mountains interior should be monitored to ensure their maintenance over the long run.

Figure 17.9
**Map of the Duck Mountain Provincial Forest with predicted probability of habitat occupancy for the Hermit Thrush at year 50**

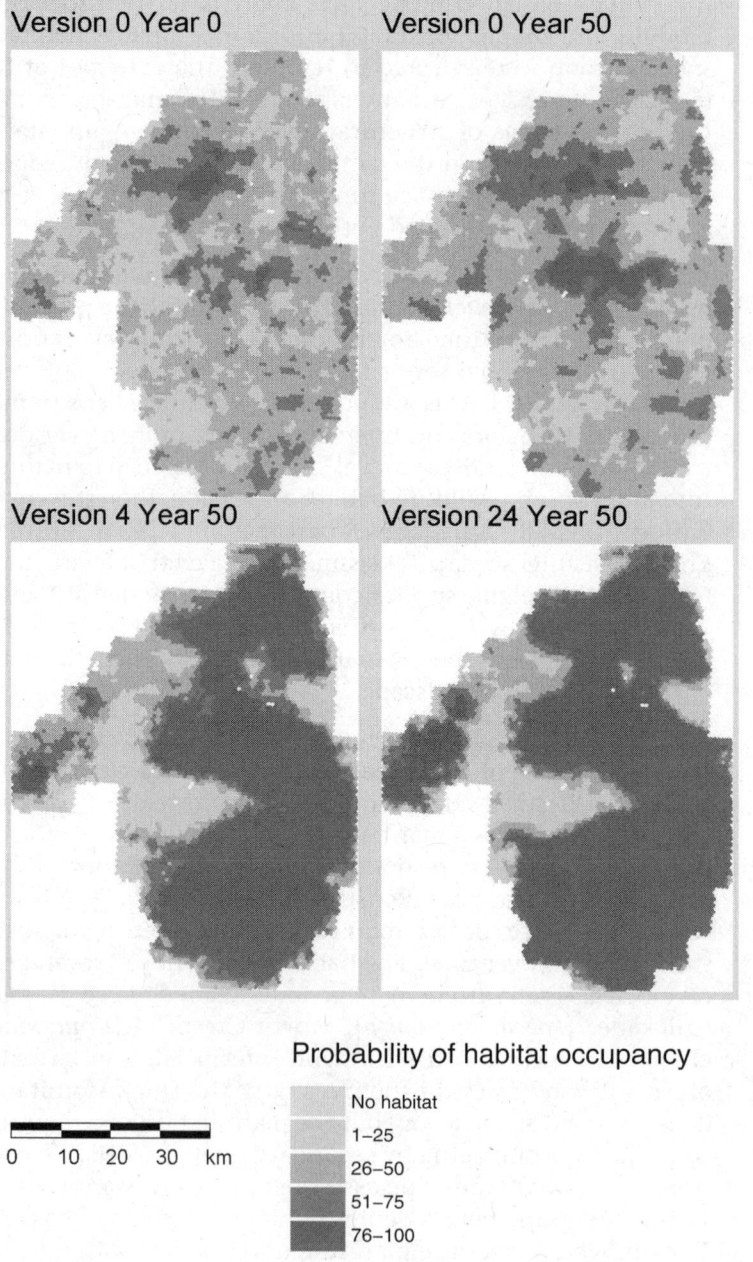

Probability of habitat occupancy

No habitat
1–25
26–50
51–75
76–100

0  10  20  30  km

Legend: Version 0 = No Harvest Scenario, Version 4 = Accelerated Harvest Scenario, Version 24 = Preferred Scenario; analyses predict a positive response for the Preferred Scenario (lower left).

**Figure 17.10**
**Map of the Duck Mountain Provincial Forest with predicted probability of habitat occupancy for the Brown Creeper at year 50**

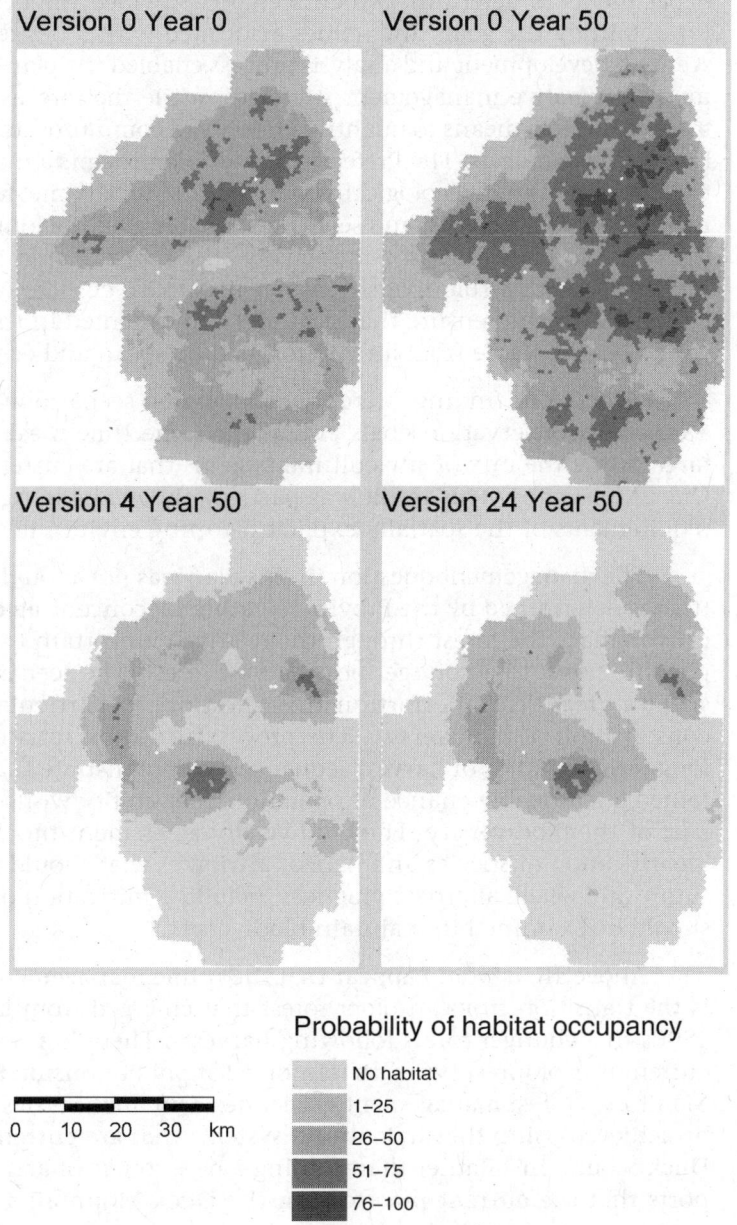

Probability of habitat occupancy

No habitat
1–25
26–50
51–75
76–100

0  10  20  30  km

Legend: Version 0 = No Harvest Scenario, Version 4 = Accelerated Harvest Scenario, Version 24 = Preferred Scenario; analyses predict a negative response for the Preferred Scenario (lower left) as well as for the Accelerated Harvest Scenario.

## 6. CONCLUSION

The scenario planning exercise was an iterative process, involving rounds of analysis and consultation with the public, stakeholders, and various government departments to determine which scenario provided the best potential solution for obtaining the goals and values associated with a desired future forest. The scenario development and analysis process enabled the planning team to explore many alternative management strategies, while the forecasting and assessment tools provided a means to quantify effects and communicate results during stakeholder consultations. The Preferred Scenario represents a management approach that balanced values (ecological, economic, and social), met federal and provincial regulatory requirements, and seems sustainable over the long term. In particular, the scenario planning process focused on a range of non-timber and environmental indicators (biodiversity, water resources, climate change impacts, and cultural values) to ensure that strategies implemented in the Preferred Scenario would be acceptable from an environmental, social and economic perspective.

Overall our findings were that the chosen scenario will be able to achieve a range of conservation goals and at the same time meet the provincial AAC target using the mix of silvicultural systems that are currently practised on the Duck Mountains. This result was particularly encouraging given the stringent requirements of the spatially explicit planning environment.

The management question in this plan was not about how much to harvest (that is determined by the Province); rather LP concentrated on how to strategically manage the forest through the relatively uncertain transition from an old post-fire forest to a younger post-harvest forest. The scenario planning process spent a great deal of effort on this question, in particular managing for the conservation of biodiversity. The process also substantiated all forecasts with long-term forecasts of harvest sequences to demonstrate that the strategy maintained a reasonable chance of operational feasibility while maintaining a large part of the biodiversity. The biodiversity assessment process also enabled the identification of species and habitat attributes that should be monitored in the future and where alternate practices including mitigation or adaptation options should be examined to maintain biodiversity.

Above all, it would appear that the prime management issue on this forest is the transition from an older forest that emerged from large fires in the late 1800s to a younger forest following harvests. There is a desire to maintain the current mix of forest types, as reinforced at public consultation meetings and in Manitoba's forest management guidelines. Our analysis has shown that this can be achieved using the silvicultural systems that are currently practised on the Duck Mountains. Rather than finding a new course of action, our analysis supports that the current practices on the Duck Mountains are sustainable and appropriate and can be improved upon with subtle changes to the harvest schedule, silviculture practices, and the development of a landscape-based approach to planning.

Some of the approaches that were examined during the scenario planning process were outside the range of conventional thinking (e.g., three-cohort management silviculture, and stewardship zoning), but it was a worthwhile exercise to investigate the potential outcomes these relatively new concepts have to offer.

The preferred alternative represented a balanced outcome that achieved the AAC targets while having a satisfactory response to a number of coarse and fine filter indicators. Although the Natural Disturbance Management Scenario was ultimately not chosen as the preferred scenario, several elements of the Natural Disturbance Model were included in the Preferred Scenario including targets to retain older successional forest elements, silviculture treatments including variable retention and understory protection, and harvest areas that encompass a range in patch size distribution. An important aspect of LP's 20 years FMP is the inclusion of an adaptive management context as their part of their approach to SFM. Following the implementation of the plan, LP's monitoring program will collect data and conduct research to assess the suite of biodiversity indicators selected in the plan to evaluate the landscape pattern and conditions resulting from the Preferred Scenario compared to those based on the natural disturbance model. Results from the effectiveness monitoring program will provide feedback to refine or adjust LP's planning and operational practices and report to the government and public on their progress towards SFM and the implementation of an ecosystem approach.

## REFERENCES

Arnup, R. 2004. Terrestrial and Wetland Ecosite Classification System for the Duck Mountains, Man. Unpublished.

Bergeron, Y., Drapeau, P., Gauthier, S., and Lecomte, N. 2007. Using knowledge of natural disturbances to support sustainable forest management in the northern Clay Belt. For. Chron. **83**: 326–337.

Bergeron, Y., Harvey, B., Leduc, A., and Gauthier, S. 1999. Forest management guidelines based on natural disturbance dynamics: stand- and forest-level considerations. For. Chron. **75**: 49–54.

Bergeron, Y., Leduc, A., Harvey, B.D., and Gauthier, S. 2002. Natural fire regime: a guide for sustainable management of the Canadian boreal forest. Silva Fenn. **36**: 81–95.

British Columbia Ministry of Forests (BCMF). 1995. Biodiversity Guidebook. Forest practices code of British Columbia. British Columbia Ministry of Forests, Victoria, B.C.

Bunnell, F.L., Dunsworth, B.G., Huggard, D.J., and Kremsater, L.L. 2003. Learning to sustain biological diversity on Weyerhaeuser's coastal tenure. The Forest Project. Weyerhaeuser, Nanaimo, B.C.

Gauthier, S., Nguyen, T., Bergeron, Y., Leduc, A., Drapeau, P., and Grondin, P. 2004. Developing forest management strategies based on fire regimes in northwestern Quebec, Canada. *In* Emulating natural forest landscape disturbances: concepts and applications. *Edited by* A.H. Perera, L.J. Buse, and M.G. Weber. Columbia University Press, New York, USA, p. 219–229.

Hamel, C. and Kenkel, N. 2001. Structure and dynamics of boreal forest stands in the Duck Mountains, Manitoba. Final Project Report 2001-4, Sustainable Forest Management Network. University of Alberta, Edmonton, Alta. [Online] <www.sfmnetwork.ca/docs/e/PR_2001-4.pdf> (accessed November 14, 2007).

Hannon, S.J. and McCallum, C. 2004. Using the focal species approach for conserving biodiversity in landscapes managed for forestry. Sustainable Forest Management Network. University of Alberta, Edmonton, Alta. [Online] <www.sfmnetwork.ca/docs/e/SR_200405hannonsusin_en.pdf> (accessed August 26, 2008).

Hunter, M.L. Jr. 1990. Wildlife, forests and forestry: principles of managing forests for biological diversity. Prentice-Hall Inc., Englewood Cliffs, N.J., USA.

KBM Forestry Consultants. 2005. Report on public values survey for LP Canada forest management plan. *In* Louisiana-Pacific Canada, Ltd. (LP). 20-year sustainable forest management plan (2006–2026). Louisiana Pacific Canada, Ltd., Swan River, Man. [Online] <www.swanvalleyforest.ca/planning/LongTermPlan.html> (accessed November 14, 2007).

Kenkel, N., Foster, C., Caners, R., Lastra, R., and Walker, D. 2003. Spatial and temporal patterns of white spruce recruitment in two boreal mixedwood stands, Duck Mountains, Manitoba. Project Reports 2003/2004, Sustainable Forest Management Network. University of Alberta, Edmonton, Alta. [Online] <www.sfmnetwork.ca/docs/e/PR_200304kenkelnrecr6.pdf> (accessed November 14, 2007).

KPMG. 1995. Manitoba's Forest Plan... towards ecosystem based management. Report to Manitoba Natural Resources.

Louisiana-Pacific Canada (LP) Ltd. 1995. 10-year forest management plan (1996–2005) for FML # 3. Louisiana-Pacific Corporation, Swan River, Man.

LP Canada, Ltd. 2006a. 20-year sustainable forest management plan (2006–2026). Louisiana Pacific Canada, Ltd., Swan River, Man. [Online] <www.swanvalley forest.ca/planning/LongTermPlan.html> (accessed November 14, 2007).

LP Canada, Ltd. 2006b. 2006–2007 annual operating plan. Louisiana Pacific Canada, Ltd., Swan River, Man. [Online] <www.swanvalley forest.ca/documents.html #AR_2005> (accessed November 14, 2007).

Manitoba Clean Environment Commission. 1996. Report on public hearings: Louisiana-Pacific Canada Ltd. 10-year forest management plan (1996–2005). Manitoba Clean Environment Commission. Steinbach, Man. [Online] <www.cecmanitoba.ca/index.cfm?pageID=9 &function=Step2&CatID=4> (accessed November 14, 2007)

Ontario Ministry of Natural Resources (OMNR). 2001. Forest management guide for natural disturbance pattern emulation. Version 3.1. Ontario Ministry of Natural Resources, Queen's Printer for Ontario, Toronto, Ont.

Rempel, R.S., Donnelly, M., Van Damme, L., Gluck, M., Kushneriuk, R., and Moore, T. 2006. Patterns and processes in forest landscapes. *In* Consequences of human management. Proceedings of the 4th meeting of IUFRO working party 8.01.03, September 26–29, 2006, Locorotondo, Bari, Italy. *Edited by* R. Lafortezza and G. Sanesi. Accademia Italiana di Scienze Forestali, Tipografia La Bianca, Bari, Italy, pp. 161–167.

Serrouya, R. and D'Eon, R. 2004. Variable retention forest harvesting: research synthesis and implementation guidelines. Sustainable Forest Management Network, University of Alberta, Edmonton, Alta. [Online] <www.sfmnetwork.ca/docs/e/SR_200405serrouyarvari_en.pdf> (accessed August 26, 2008).

Smith, R.E., Veldhuis, H., Mills, G.F., Eilers, R.G. Fraser, W.R., and Lelyk, G.W. 1998. Terrestrial ecozones, ecoregions and ecodistricts: an ecological stratification of Manitoba's natural landscapes. Techical Bulletin 98-9E. Land Resource Unit. Brandon Research Centre, Research Branch of Agriculture and Agri-Food Canada, Winnipeg, Man.

Sougavinski, S. and Doyon, F. 2002. Variable retention: research findings, trial implementation, and operational issues. Sustainable Forest Management Network, University of Alberta, Edmonton, Alta. [Online] <www.sfmnetwork.ca/docs/e/Variable%20Retention-final%20 english.pdf> (accessed August 26, 2008).

Sutton, A. and Tardif, J. 2007. Dendrochronological reconstruction of forest tent caterpillar outbreaks in time and space, western Manitoba, Canada. Can. J. For. Res. **37**: 1643–1657.

Tardif, J. 2004. Fire history in the Duck Mountain Provincial Forest, western Manitoba. Project Reports 2003/2004, Sustainable Forest Management Network. University of Alberta, Edmonton, Alta. [Online] <www.sfmnetwork.ca/docs/e/PR_200304tardifjfire6fire.pdf> (accessed November 14, 2007).

TetrES. 1996. LP's environment impact statement. Vol. 1. Louisiana-Pacific Canada Ltd. Forest Management License #3. 10-Year Forest Management Plan (1996–2005). TetrES Consultants Inc. Winnipeg, Man.

TetrES. 2003. LP Performance Evaluation. TetrES Consultants Inc. Winnipeg, Man.

Chapter *18*

# Forest Ecosystem Management in the Boreal Mixedwood Forest of Western Québec*
## An Example from the Lake Duparquet Forest

*Brian D. Harvey, Yves Bergeron, Alain Leduc, Suzanne Brais, Pierre Drapeau, and Claude-M. Bouchard*

\* This chapter is a contribution of our participation in the Centre of Excellence in Sustainable Forest Management and of the CRSNG/UQAM/UQAT Industrial Chair in Sustainable Forest Management and the Centre for Forest Research. In addition, this work would not have been possible without the financial support of the Quebec Ministry of Natural Resources and Wildlife. Thanks to Mark Fox for translating the first draft of this chapter. Photos on this page were graciously provided by Suzanne Brais. We thank Jacques Morissette (CFS, Laurentian Forestry Centre), who produced the figure 18.3.

# 1. INTRODUCTION

## 1.1. Forest Ecosystem Management: Only Possible for Large Forest Areas?

We have seen in previous chapters that an ecosystem approach to forest management requires a good understanding of the natural disturbance regime at a regional level, if not on a much larger spatial scale. In the boreal forest, probably more than any other biome, the area affected by major disturbances, particularly fires, surpasses the usual annual harvested area by forest industry (Belleau et al. 2007). If the aim of ecosystem management or Seymour and Hunter's (1999) "ecological forestry" is to understand and work in harmony with ecological patterns and processes, the spatial attributes of fires – especially with respect to the size of affected areas and their distribution in time – represent a major challenge to forest management. It becomes even more critical when your entire area of interest can potentially go up in smoke in one large fire! Does this mean that forest ecosystem management cannot be applied to small land areas? In short, the answer is no. In fact, this chapter will try to demonstrate that the integral reproduction of all patterns and processes is not necessarily required in an ecological approach to forest management. In effect, knowing that it is not possible to incorporate as much of the natural variability as in larger areas, especially in terms of disturbance or cutblock size, managers of small forest areas have two choices: 1) manage their forests as a component of the larger surrounding forest area or 2) try to achieve landscape-level objectives comparable to those set by an ecosystem management approach applied over larger areas but on a much smaller scale (see box 18.1). These two approaches have both advantages and drawbacks. In this chapter we present the second approach as it applies to the Lake Duparquet Research and Teaching Forest (LDF). Using a larger reference area allows us to examine the forest management objectives in a perspective that is in equilibrium with the regional disturbance regime – the big picture.

## 1.2. Objective of the Chapter

The objective of this chapter is to provide a real life example illustrating how an understanding of the natural processes and patterns, acting at both regional and stand scales, can serve as a reference for developing a forest ecosystem management approach for a relatively small forest area. The specific area in question is the Lake Duparquet Research and Teaching Forest, a forest management unit of 80 km² located in the southern boreal mixedwood forest of western Quebec, halfway between the cities of Rouyn-Noranda and La Sarre. We will begin with a brief history of the events that led to the creation of the LDF (which were relevant to the development of the ecosystem management approach) and a biophysical description of the area. We will then describe the underlying principles and important elements of the ecosystem management approach applied in the LDF and how we interpret ecological knowledge of the disturbance regime and forest dynamics at the landscape and stand levels. From the establishment of strategic, forest-level composition objectives to the development of an ecosystem-

Box 18.1

## What is important to maintain in an ecosystem management approach applied to a small forest?

If it is out of question to mimic the spatial patterns of large disturbances like forest fires in small forest units, what should forest ecosystem management incorporate at this scale? First, remember that the natural forest mosaic at the regional level remains the reference for forest age structure and composition attributes, for any ecosystem approach. By aiming to create conditions close to this regional reference, a small forest area can be managed as a sort of microcosm of regional forest diversity. Establishing management objectives to maintain a mix of stand ages and compositions that reflects the natural range – young, mature, old, hardwoods, mixedwoods, softwoods – is a good start to forest ecosystem management on any scale. Obviously, this kind of forest diversity is difficult, if not impossible, to attain through even-aged, clearcut and plantation management. A variety of silviculture approaches is needed to produce this diversity, with particular attention needed to be paid to attributes associated with old stands (mixed compositions, irregular or uneven-aged structures, down and standing woody debris).

Second, if the size of large forest fires exceeds the area of your forest unit, this is not necessarily the case for small fires (which affect the forest mosaic less but are more frequent than large fires) or disturbances such as insect outbreaks, windthrow, and individual tree senescence. So even for a small forest unit, there is an interest in varying the size of cuts to better reflect the variability of natural openings (see Kneeshaw et al., chapter 9; Bouchard, chapter 13). While keeping in perspective the total area of the forest unit, this means that relative proportions of large cuts and agglomerations of cutovers (hundreds of hectares), of intermediate openings (tens of hectares) and of small gaps (tens and hundreds of m²) should find some underpinnings in the natural disturbance regime. In the region, what is known historically about the relative importance of forest fires,

disturbances cause by insect outbreaks, windthrow, or mortality caused simply by senescence? The answer, even if not supported by specific studies, will provide an indication concerning the size distribution of cuts to target and silvicultural approaches to adopt or experiment.

While varying cutblock size, even in a small forest unit, it is worth thinking about how the spatial distribution of harvesting (or regeneration) areas affects the forest matrix. Small harvesting areas distributed over the whole forest and an unlimited road access certainly have advantages for the forest manager and other land users. But are there benefits to other users, to certain animal or plant species or for protection reasons to keep a part of the forest relatively inaccessible, *a little bit wilder*? In a small forest unit, do small, intact forest tracts have an intrinsic ecological value that we don't find in a landscape fragmented by 2- or 3-pass patch cutting and the well-developed road network needed to support it? When planning on mid- and long-term horizons, it is important to anticipate how the distribution of forest interventions will affect the size and configuration of the residual forest. Clustering some harvesting sectors in order to create relatively large regeneration areas can create openings close to that of small fires while maintaining larger forest tracts intact elsewhere. Another aspect that should be considered is the connectivity between mature forest tracts and zones set aside for conservation, habitat protection, or for uses other than fibre production. A cartographic approach using, among other landscape features, natural corridors offered by riparian buffers and wetlands as well as highlands and abrupt topography can provide a good first sketch in planning connectivity. Finally, all forest ecosystem management approaches, even those applied to small forest units, should favour integral conservation of rare or fragile habitats or stands.

based silviculture, it should become apparent that, even for relatively small forest areas, it is possible to develop management approaches that strive to incorporate natural forest dynamics.

## 2. A BRIEF HISTORY OF THE LAKE DUPARQUET FOREST

Research in the area surrounding Lake Duparquet started in the late 1970s when Yves Bergeron began his doctoral studies with André Bouchard at the Université de Montréal. This early work resulted in an ecological classification that integrated information on physiography, geomorphology, soils, and vegetation dynamics in the western part of Roquemaure and Hébécourt townships (Bergeron et al. 1983). This ecological classification, which has since been aligned with the classification system developed by the Ministère des Ressources naturelles et de la Faune du Québec (MRNF, Blouin et Berger 2002), has been key to structuring and organising ecological information and to management of this forest area over the last 25 years.

The Université du Québec à Montréal (UQAM) undertook its first research in forest ecology in the area in 1985 and, two years later, the Université du Québec en Abitibi-Témiscamingue (UQAT) established a forest research and development unit, the Unité de recherche et de développement forestier de l'Abitibi-Témiscamingue. Around this time, the level of research at Lake Duparquet and the surrounding area now occupied by the LDF increased considerably and involved scientists and students working in a number of fields of study, most notably in forest ecology but also in forest nutrition, plant physiology, entomology, wildlife ecology, paleoecology, conservation biology, forestry genetics, and silviculture (Harvey 1999). This mixture of students and researchers generated a tremendous amount of scientific knowledge about the forest ecosystems of the Abitibi region. Moreover, while individual studies may have initially appeared to be disparate and unconnected to one another, figuratively, they formed the pieces of a huge ecological puzzle, that has come together to form a revealing picture of the southeastern boreal mixedwood ecosystem (Bergeron and Harvey 1997).

The Quebec *Forest Act* contains provisions for the creation of research and teaching forests on public land. In order to ensure protection of research and monitoring sites and with the underlying idea of creating a demonstration forest to showcase the the integration of ecological knowledge in forest management, UQAT and UQAM submitted a proposal to the MRNF to create a research and teaching forest in 1993. The Lake Duparquet Research and Teaching Forest was officially created two years later, after two forestry companies, Norbord Industries and Tembec, gave up their wood volume rights to the area. Both companies maintain a privileged relationship with the LDF and, with representatives from the MRNF, sit on the Lake Duparquet Forest management committee. Like the other 15 research and teaching forests in Quebec, management of the forest is legally structured by a management convention signed between the two universities and the Quebec government. This convention describes the rights and responsibilities of all the parties with regard to forest and land management of the area.

## 3.  DESCRIPTION AND ZONING OF THE TERRITORY

### 3.1.  Climate and Biophysical Characteristics

Located in the Western balsam fir–white birch bioclimatic subdomain (48°30′N, 79°20′W) and in the ecological region of the Abitibi Lowlands (5a) (Saucier et al. 1998), the LDF covers an area of 8,045 ha. The climate is relatively cold and dry. The average annual temperature varies between 0°C and 2.5°C with a growing season of 150–160 days. Annual precipitation is relatively low, between 800 and

900 mm. The LDF has a slightly hilly topography (average altitude of 289 m) with some hills up to 400 m in altitude. The territory is mainly characterized by mesic and subhydric clay deposits (56% of the landscape). Organic hydromorphic deposits associated with depressions and plains, and rock outcrops and till complexes on mid- to upper-slopes and summits cover most of the remaining areas. While covering a smaller area, lacustrine plains and levees are characteristic of riparian sites and are of interest from an ecological viewpoint.

## 3.2. Natural Disturbances

The Lake Duparquet Forest is situated within a larger 15,000 km² area, extending from 48° to 50° N, for which a forest fire history reconstruction study was undertaken by Bergeron et al. (2001, 2004). The area spans two bioclimatic subdomains, namely the Western balsam fir–white birch and the Western black spruce–feather moss. Analysis of the fire history map indicates that the area has been affected by two important fire periods, before 1850 and between 1910 and 1930 (figure 18.2). While fires that have affected the spruce–feather moss subdomain have, on average, been larger than those in the balsam fir–white birch subdomain, the fire cycle (the time required to burn an area equal to that of the study area) is not significantly different between the two zones. From this study, we can draw three interpretations relevant to the ecosystem management approach: 1) the fire cycle has increased over the last 150 years (table 18.1); 2) the mean age of the forest (time since the last fire, excluding cutovers and other human disturbances) is about 140 years; and 3) approximately 57% of forest stands are older than 100 years and more than 20% are older than 200 years (figure 18.2). This information is particularly important in establishing strategic objectives for the conservation of ecosystem diversity.

The Lake Duparquet Forest has been affected by a dozen fires over the last 300 years (Dansereau and Bergeron 1993). The two biggest fires, which occurred in 1760 and 1923 (figure 18.3), have had an important effect on the age structure and composition of the forest now under management (figure 18.3; see section 3.4). While fires have had a major influence in shaping the landscape mosaic, the forest has also been affected by insect outbreaks. According to Morin et al. (1993), three spruce budworm (*Choristoneura fumiferona* Clem., SBW) outbreaks occurred during the 20th century, in 1919–1929, 1930–1950, and 1970–1987. Old balsam fir stands originating from the 1760 fire are now characterized by a high proportion of canopy gaps (Kneeshaw and Bergeron 1998) and an accumulation of snags and downed woody debris, mainly resulting from balsam fir mortality during the last outbreak (Bergeron and Leduc 1998; Hély et al. 2000). Mixedwood and intolerant hardwood forests of the region have been affected as well by up to six forest tent caterpillar (*Malacosoma disstria* Hübner) outbreaks between 1938 and 2002 (Cooke and Lorenzetti 2006). These disturbances, along with mortality of single or groups of trees caused by windthrow, stem senescence, or other factors, influence the structure and composition of forest stands that have been spared by fire. The specificity of these disturbances, in terms of stand ages and compositions affected, has resulted in a range of canopy opening sizes (Kneeshaw and Bergeron 1998; D'Aoust et al. 2004), and their dynamics, often associated with mature and over-mature stand development stages, contribute to the complexity of structure and composition of forest stands.

Figure 18.1
**Figure 18.1**
**Time-since-fire map for western Abitibi and northern Québec**

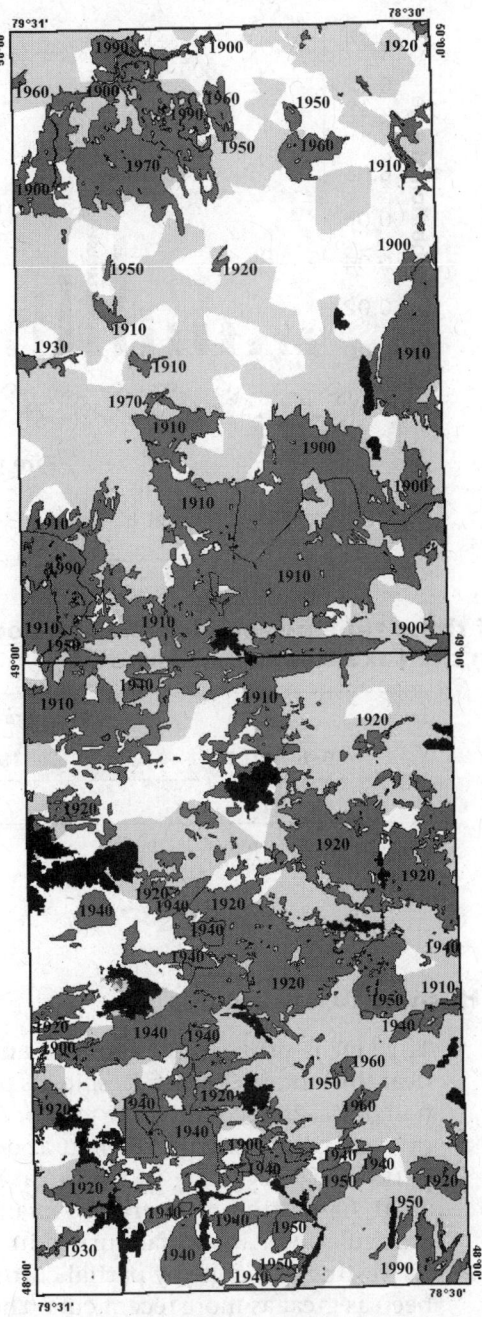

Areas that burned after 1901 are indicated in 10-year classes (dark gray). Light gray and white zones represent areas where stand initiation occurred between 1801 and 1900, and before 1800, respectively (from Bergeron et al. 2004). Note the importance of forest older than 200 years in the study area.

**Figure 18.2**
**Proportion of the western Abitibi and northern Québec territory initiated by wildfires indicated on the map presented in figure 18.1**

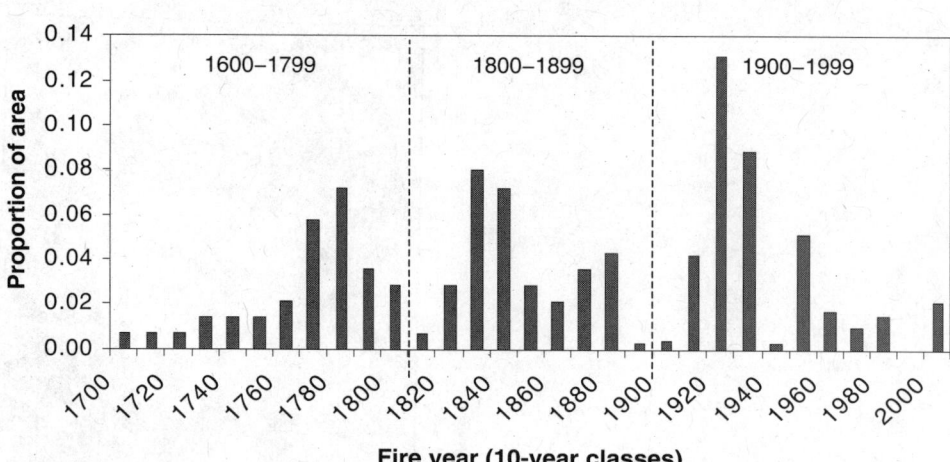

Adapted from Bergeron et al. (2001).

**Table 18.1**
**Fire cycle of three time periods for the territory of western Abitibi surrounding the Lake Duparquet Forest**

| | Fire cycle (years) | |
|---|---|---|
| **Before 1850** | **1850 to 1920** | **Since 1920** |
| 83 | 146 | 325 |

Adapted from Bergeron et al. (2001).

## 3.3.　Human Disturbances

The LDF is situated just south of an agricultural area surrounding Lake Abitibi near the Ontario border. Despite its proximity to areas colonized during the latter half of the 20th century, the territory west of Lake Duparquet was never colonized or cleared for agriculture or rural development. However, there have been two periods of harvesting in this area. During the first period, between 1920 and 1950, harvesting, primarily by diameter limit, occurred in some of the forest, particularly in areas that burned in 1760. According to Bescond (2002), as most harvesting was done by partial cutting, the effect on forest composition has not been as great as more recent cuts. The second period, between the late 1970s and 1987, consisted of mechanized salvage logging undertaken during the last spruce budworm outbreak in the western and southern parts of the forest, again in the area affected by the 1760 fire. Since the end of the 1990s and under the management of the universities, a little less then 60 ha have been harvested per year using diverse partial cut and clearcut treatments.

Figure 18.3
**Fire map and zoning of Lake Duparquet Forest**

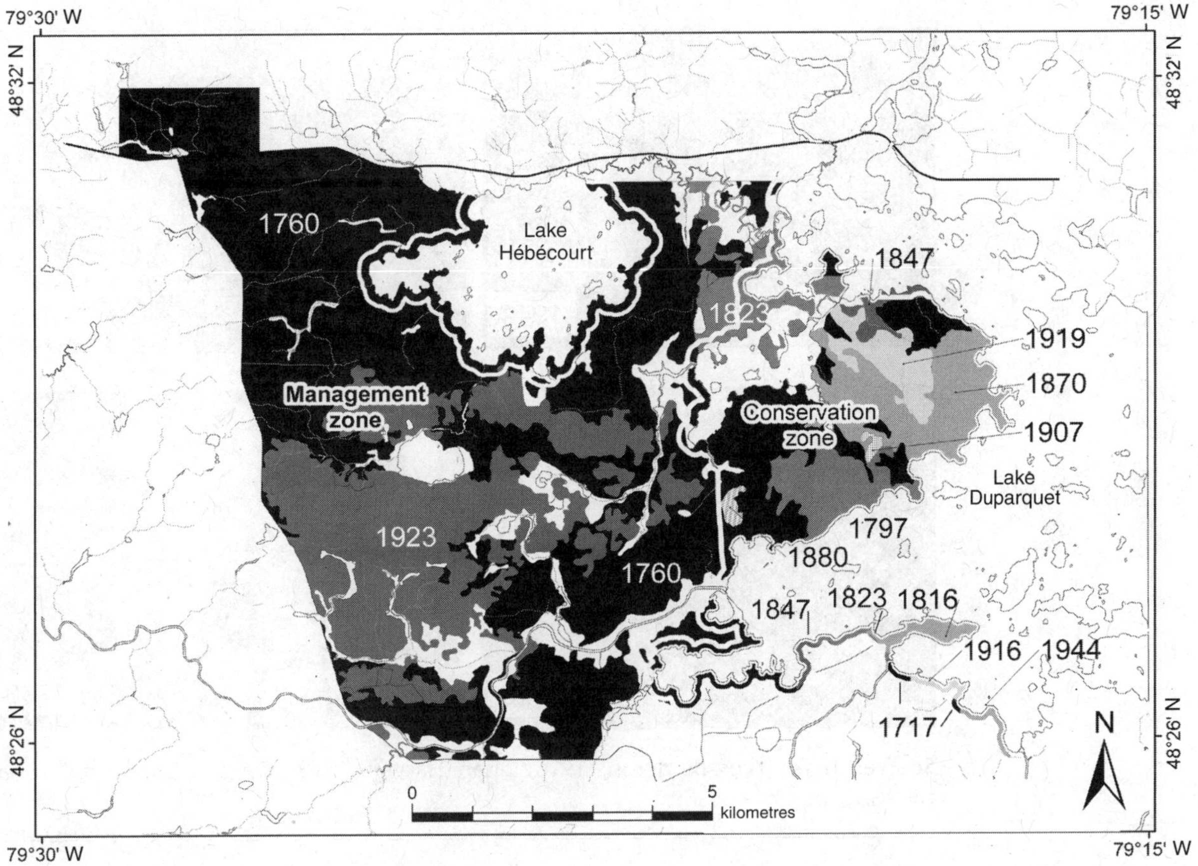

## 3.4.  Forest Composition

Balsam fir (*Abies balsamea* [L.] Mill.) is the dominant species of mature forests in the area, and is associated with white spruce (*Picea glauca* [Moench] Vosh), black spruce (*Picea mariana* [Mill.] B.S.P.), and white birch (*Betula papyrifera* Marsh.). Eastern white cedar (*Thuja occidentalis* L.) is a late successional species associated with balsam fir on mesic sites and is also found on riparian sites, rich organic deposits, and rocky outcrops. Cedars aged from 500 to 1,000 years old have been found rooted in rocky outcrops of islands on Lake Duparquet (Archambault and Bergeron 1992). After fire, jack pine (*Pinus banksiana* Lamb.), trembling aspen (*Populus tremuloides* Michx.), balsam poplar (*Populus balsamifera* L.), and white birch form mixed or pure stands on mesic sites. In the absence of fire and when seed sources of shade tolerant species (fir, spruces, and cedar) are present, these stands of pioneer species develop into mixed compositions with layered canopies (figure 18.4). Eastern larch (*Larix laricina* [Du Roi] K. Koch.) grows mainly on hydric organic sites in pure stands or with black spruce. Because red pine (*Pinus resinosa* Ait.) is not adapted to large crown fires that characterize

**Figure 18.4**
**Four stands located on mesic glaciolacustrine clays at different periods following the last wildfire**

Source: 1944: Yves Bergeron; 1916: Brian Harvey; 1823: Dave Coates; 1760: Brian Harvey.

the boreal forest, its distribution is largely limited to isolated populations, often mixed with white pine (*Pinus strobus* L.), on the islands and shores of Lake Duparquet where fires are typically less severe (Bergeron 1991; Bergeron et al. 1997). Scattered white pines, remnants of extensive populations that existed between 6800 and 2200 BP (Bergeron et al. 1998), are associated with the old black spruce stands on summits and outcrops.

## 3.5.  Zoning

The zoning approach known as the Triad was originally proposed by Seymour and Hunter (1992). The concept aims at reconciling wood production and biodiversity maintenance using a zoning strategy according to three land-use categories: conservation, extensive, *close to nature* forest management, and intensive forest management (see Gauthier et al., chapter 1). When the LDF was officially created, the territory was divided into two functional zones, a conservation zone and a management zone. The conservation zone covers 25% of the area and contains a mosaic of forests that originated from a dozen fires that occurred

Box 18.2
## The conservation zone: why 25%?

When the Lake Duparquet Forest was created in 1995, a contiguous portion covering about 2,000 ha in the eastern part of the forest was designated as a "conservation zone" (figure 18.3). This zone is relatively inaccessible, except by water, and has been affected by about ten forest fires between 1717 and 1944. The limits of this zone were not established to obtain 25% of the LDF but simply to incorporate a natural reference of the forest mosaic of the Western balsam fir–white birch bioclimatic subdomain or region. Because numerous ecological studies have been conducted in this area (and published) and several monitoring programs are ongoing, its scientific and heritage value is priceless – all the more, in fact, because of the constant threat of mining development!

between 1717 and 1944 (see box 18.2). This natural area has been set aside for ecological monitoring and assessment and serves as a reference area for evaluating the effects of management practices in the remaining 75% of the forest. About 65% of the LDF is dedicated to silviculture adapted to natural stands in the management zone. Because this zone covers most of the forest, certain measures are taken to maintain key habitat attributes (coarse filter) within the managed forest matrix. Finally, an intensive forest management zone (within the management zone) covers roughly the remaining 10% of the territory and is aimed at increasing wood production to compensate for possible decreases associated with the two other zones. In this zone, plantations established during the 1980s to 1990s are intensively tended. This includes at least one manual (brushsaw) cleaning and one fill planting to attain near full stocking and, over a large area, pruning and a manual sanitary treatment of stems infected by white pine weevil (*Pissodes strobi* Peck). On a small part of the intensive management zone, we are also evaluating the potential of highly intensive silviculture, or *ligniculture*, using poplar and larch hybrids and genetically improved white and Norway spruce (*Picea abies* [L.] Karst.) and intensive plantation tending to maximize fibre production (figure 18.5). This three-tiered vocation approach to zoning – conservation (25%), ecosystem management (65%), and intensive/ligniculture management (10%) –, while not necessarily representing the proportions generally associated with the triad, does represent the essence of the concept.

## 4.  FOREST ECOSYSTEM MANAGEMENT APPLIED TO THE LAKE DUPARQUET FOREST

### 4.1.  Four Main Principles

The specific mission of a research and teaching forest, particularly its focus on experimentation, provides a flexibility that is lacking within the current regulatory framework of public forest management. It is this flexibility that has allowed the development of a management plan that has placed ecological objectives at the heart of the approach. It is interesting to note that, a decade later, the Commission on the management of Quebec's public forest (Coulombe et al. 2004) recommended that ecosystem management form the central tenet of the province's management policy. The main goals of this approach are to maintain

**Figure 18.5**
**Practising ligniculture on small areas is an integral component of the ecosystem management approach**

A)                                                                                    B)

Legend: A) 3-year hybrid poplar plantation; B) mixed plantation of hybrid poplars and genetically improved white spruce.
Source: A) Simon Laquerre; B) Brian Harvey.

key attributes of the natural forest mosaic within managed areas and integrate knowledge on stand- and landscape-level ecological processes into management and silviculture activities. In this sense, natural disturbance-based management –, an analogue of the footprint of the natural disturbance regime on the natural forest mosaic – constitutes a variant of the coarse filter approach and a strategy for maintaining the functional diversity of ecosystems.

## 4.1.1. Using the Forest Ecological Classification Framework

Forest ecological classifications are an essential tool for structuring knowledge about the diversity and dynamics of forest ecosystems (Grondin et al. 2003). At the regional level, forest ecosystem frameworks provide basic information about climatic and biophysical factors that influence forest composition. As well, the proportion of different surface deposit types in a region provides a good indication of the potential vegetation and relative productivity of forests. Ecological type identification guides, developed by the MRNF for each ecological region of Quebec, are indispensable references for forest managers and silviculturists. Moreover, ecological classification and stand dynamics studies based on ecological types provide a framework for associating vegetation communities with different successional stages and estimating forest composition for large areas based on the proportion occupied by different sites and the time elapsed since the last major disturbance.

## 4.1.2. Integrating Fire Regime Characteristics at the Forest Level

When considering natural disturbance regimes, we necessarily refer to an area that extends over hundreds of square kilometres. Thus, even if our primary concern is the management of a small forest unit, we have to be aware that our forest is part of a much larger system in which processes associated with climate, hydrology, disturbances, dispersion of species, etc., act on both a large scale as well as on a smaller or more local scale. As previously mentioned, one of our objectives is to use the LDF as a demonstration area for an ecosystem management approach that uses knowledge about the regional natural-disturbance regime and natural dynamics for a relatively small forest unit.

### *Mean Stand Age: A Reference for the Fire Cycle*

So where do we begin? Regional fire history studies (Bergeron et al. 2004; figure 18.1) provide a revealing picture of the influence of disturbance regime on forest age structure (figure 18.2). It is clear that the natural regional landscape is characterized by a forest that is considerably older than previously thought. Despite the changes in fire cycle during the Holocene period (Carcaillet et al. 2001) and over the last 300 years (Bergeron et al. 2004), the mean age (time since the last fire) of the forest mosaic in western Abitibi has remained relatively constant for the past three centuries, at around 140 years. As highlighted by Gauthier et al. (chapter 3), one of the consequences of large-scale industrial forestry has been the rejuvenation of the forest (*i.e.* an increasing proportion of young forests). If old-growth forests compose an important part of the natural biodiversity, at least at the ecosystem level, it is reasonable to believe that any sustainable forest management approach should include a strategy to maintain these elements, even within managed areas.

Current forest composition has been influenced by permanent site factors and current disturbance regime and by climatic factors and disturbances that have occurred over the previous centuries. For this reason, the idea of establishing a forest age structure target based on a particular fire cycle is hard to justify (Armstrong 1999). On the other hand, mean forest age incorporates, at least partially, the historical variability of fire cycles and is relatively stable (Gauthier et al. 2002; Bergeron et al. 2004). Using mean forest age as a reference for the fire cycle provides a target that is relatively constant and ensures the maintenance of older stands as part of a forest management strategy.

### *Establishing Composition and Age Structure Objectives*

After establishing a fire cycle target based on mean time since fire in the region, we were able to establish a theoretical age structure (based on an inverse exponential curve) of the forest matrix at equilibrium with the fire cycle (Van Wagner 1978). The method used to establish forest composition objectives has been described in detail by Leduc et al. (1995) and Gauthier et al. (1996). Using the ecological classification framework, knowledge on forest dynamics and age structure modelling, this approach provides a means to develop forest composition models as a function of a variety of fire cycles and regional conditions of ecological types (relative proportion of different surface deposit-drainage types). From the biophysical portrait of the Lake Duparquet forest (the relative proportion occupied by the main ecological or site types), we associated forest dynamics or dominant successional pathways with each ecological type. The 140-year fire cycle then gives us a reference with regard to the proportion of the forest that should be covered by different forest "cohorts" (see Bergeron et al. 1999 and the next section).

### The Notion of "Cohorts"

The notion of cohorts, as presented by Bergeron et al. (1999), simplifies the establishment of forest-level composition objectives by referring to three broad stand development stages (see box 18.3). This approach allows us to respond to the following question: What percentage of my forest should be covered by 1) stands that are generally even-aged, and of equivalent or younger than the commercial rotation age (cohort 1); 2) transition stands, which are generally mixed and contain some first-cohort trees as well as shade-tolerant species that were recruited in the understory of the pioneer stands (cohort 2); and 3) stands considered as over-mature or old-growth that have a relatively complex vertical and horizontal structure and where pioneer species are almost completely absent, except perhaps as dead wood or in canopy gaps (cohort 3)? Figure 18.8 illustrates stand types associated with each cohort and with the seven dominant ecological types in the Lake Duparquet Forest. The percentage of forest area that is targeted for each cohort depends on the reference fire cycle and the age of stand break-up: the shorter the fire cycle and the older stand break-up age, the higher the proportion of first-cohort stands in the forest. Using a mean fire cycle of 140 years for the LDF with the a break-up age of 80–110 years, approximately 45–55% of the managed zone should be covered with first-cohort stands, 23–26% with second-cohort stands and 20–30% with third-cohort stands (figure 18.7; Harvey et al. 2002).

## 4.1.3.    Integrating Natural Dynamics in Silviculture

The "cohort approach" is essentially a coarse filter approach aimed at maintaining the diversity of forest types that characterize the natural forest mosaic. Using previous studies on regional natural forest dynamics (e.g. Bergeron et al. 1983, 2002; Grondin et al. 1999; Gauthier et al. 2000; Bergeron 2000), the concept allows us to classify the range of forest types associated with the three cohorts and with dominant ecological types and to roughly illustrate the development of forest stands through time. Once the notion of cohorts is well understood, it becomes clear that, in contrast with the "simple" stands associated with the first cohort, the structure and composition characterizing stands of the second and third cohorts cannot be "constructed" during a single harvesting cycle. Naturally, it takes time for an even-aged, first-cohort stand to develop into a more structurally complex stand that contains a high proportion of late-successional species. One aspect that is directly related to the forest age structure problematic (and

### Box 18.3
## Cohorts: clarifying the notion

In demographic terms, cohort generally designates a group of individuals that have the same origin or that were born at the same time and which evolve together through time. However, in our forest ecosystem management approach, we give another meaning to the notion of cohort. In this case, the term cohort refers to three broad stages of stand development: the early, mid- and late successional stages (figure 18.6). For example, in the Western balsam fir–white birch bioclimatic subdomain, even-aged stands composed of pioneer, fire-adapted species generally constitute first-cohort stands. These include post-fire stands of jack pine, trembling aspen, balsam poplar, and white birch that dominate during approximately 100 years following a fire. The first cohort thus corresponds to the both definitions of cohort. On mesic sites, white spruce and balsam fir that regenerate from seed immediately after fire or that are gradually recruited afterwards are more abundant in the understory than in the overstory established following the disturbance. Second-cohort stands generally comprise two components: residual first-cohort stems and tolerant softwood stems that have recruited into the overstory, and/or various species (shade intolerant to tolerant) that establish in gaps. The second cohort is thus a transition stage occurring approximately 75 to 175 years after fire; species composition is often mixed and stand structure tends to be less even-aged than in first-cohort stands. The third cohort, composed of "old-growth" stands, is characterized by an uneven-aged or irregular structure, the presence of specific species such as eastern white cedar, balsam fir, and spruce, an important amount of dead wood, and the quasi-absence of first-cohort living stems (except perhaps a for few sparse birches).

To this simple model of successional dynamics, other disturbances (for example, insect outbreaks or windthrow) can induce important modifications to stand structure and composition and thus complicate the association of stands to one of the three cohorts. When trying to associate a stand with a cohort, particularly following secondary disturbances, it helps to consider its future development. Depending on the residual structure and the regeneration present following a secondary disturbance, a stand can be associated with the first cohort (few residual standing trees and dominated by intolerant hardwood recruitment), the second cohort (many residual trees and mixed species regeneration), or the third cohort (residual trees are dominant and regeneration is mainly composed of shade-tolerant conifers).

#### Figure 18.6
## Simplified representation of composition and structure for forest types associated to the Three-Cohort Model located on mesic sites in the western balsam fir–white birch bioclimatic subdomain

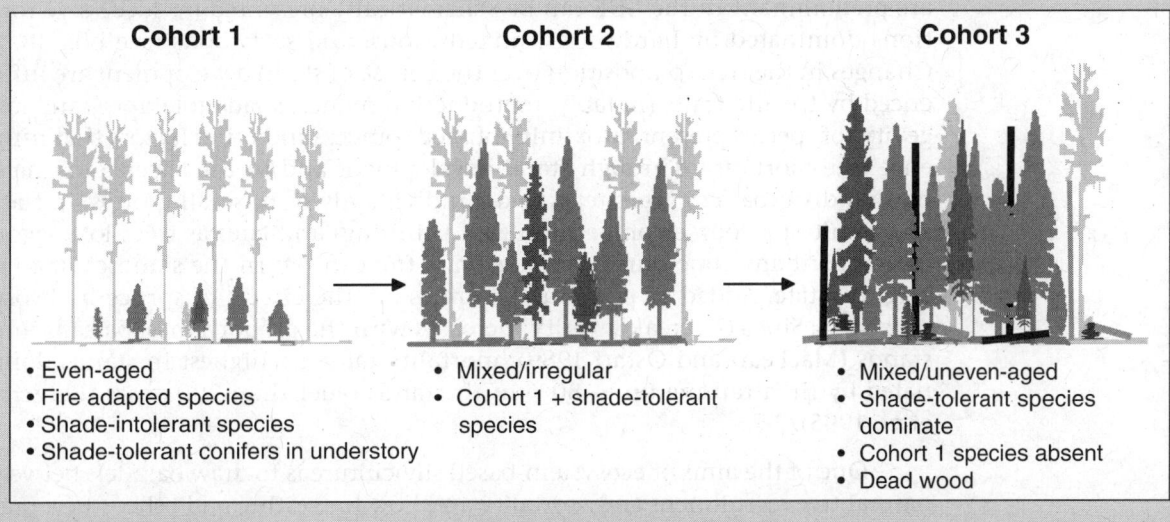

| Cohort 1 | Cohort 2 | Cohort 3 |
|----------|----------|----------|
| • Even-aged<br>• Fire adapted species<br>• Shade-intolerant species<br>• Shade-tolerant conifers in understory | • Mixed/irregular<br>• Cohort 1 + shade-tolerant species | • Mixed/uneven-aged<br>• Shade-tolerant species dominate<br>• Cohort 1 species absent<br>• Dead wood |

**Figure 18.7**
## Conceptual model of cohorts and fluxes induced by natural dynamics and by forest management at the landscape level

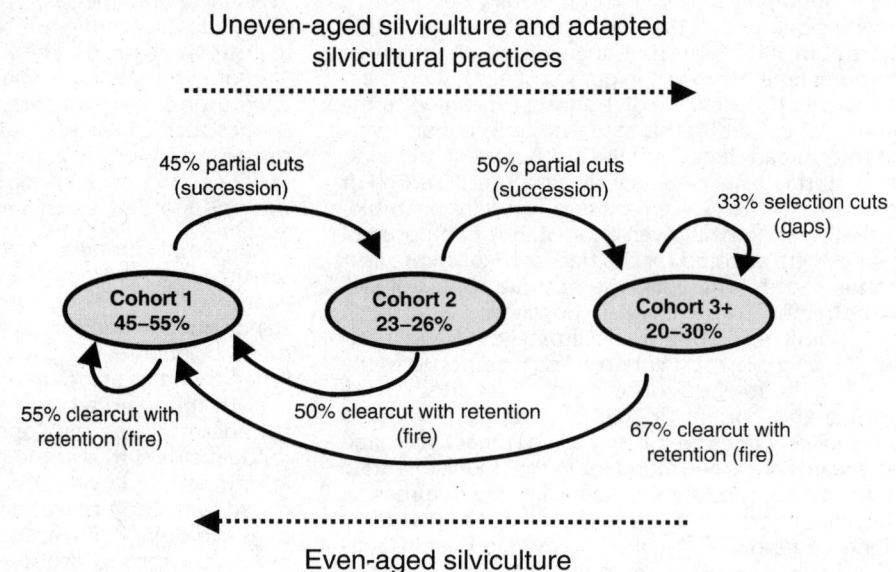

See figure 18.8 for forest types associated to the three cohorts and different site types. Adapted from Harvey et al. (2002).

the structure of older stands) is the importance of adapting silviculture practices to better incorporate elements of the natural disturbance regime and natural dynamics (figure 18.7).

Dynamics on rich, mesic sites (glaciolacustrine clays and deep tills) that are predominant in the LDF can be schematically presented by successive rotations dominated by hardwoods, mixedwoods, and softwoods (see box 18.3). Changes in the tree composition over the course of stand development are influenced by the life traits (notably, reproduction modes, shade tolerance, and longevity) of species present, proximity of seed sources, and other factors that influence tree mortality. Through stand development and in the absence of major natural disturbances, trees tend to die individually or in small groups, initially as a result of resource competition (self-thinning) and later as trees lose vigour in aging. At any time during succession, a fire can return the stand close to its original state. Added to post-fire dynamics are the effects of spruce budworm outbreaks. Since outbreak severity increases with the proportion of balsam fir in stands (MacLean and Ostaff 1989), mortality rates are highest in stands dominated by fir, attaining up to 80% in fir stands older than 200 years (Bergeron et al. 1995).

One of the aims of ecosystem-based silviculture is to draw parallels between natural tree recruitment and mortality and how harvesting and other forest practices affect these processes (Harvey and Brais 2007). For example, the transition of aspen-dominated stands towards mixed stands (aspen-fir-spruce) often occurs

**Figure 18.8**

## Representation of dominant forest types according to the three cohorts and the site types found in the Lake Duparquet Forest

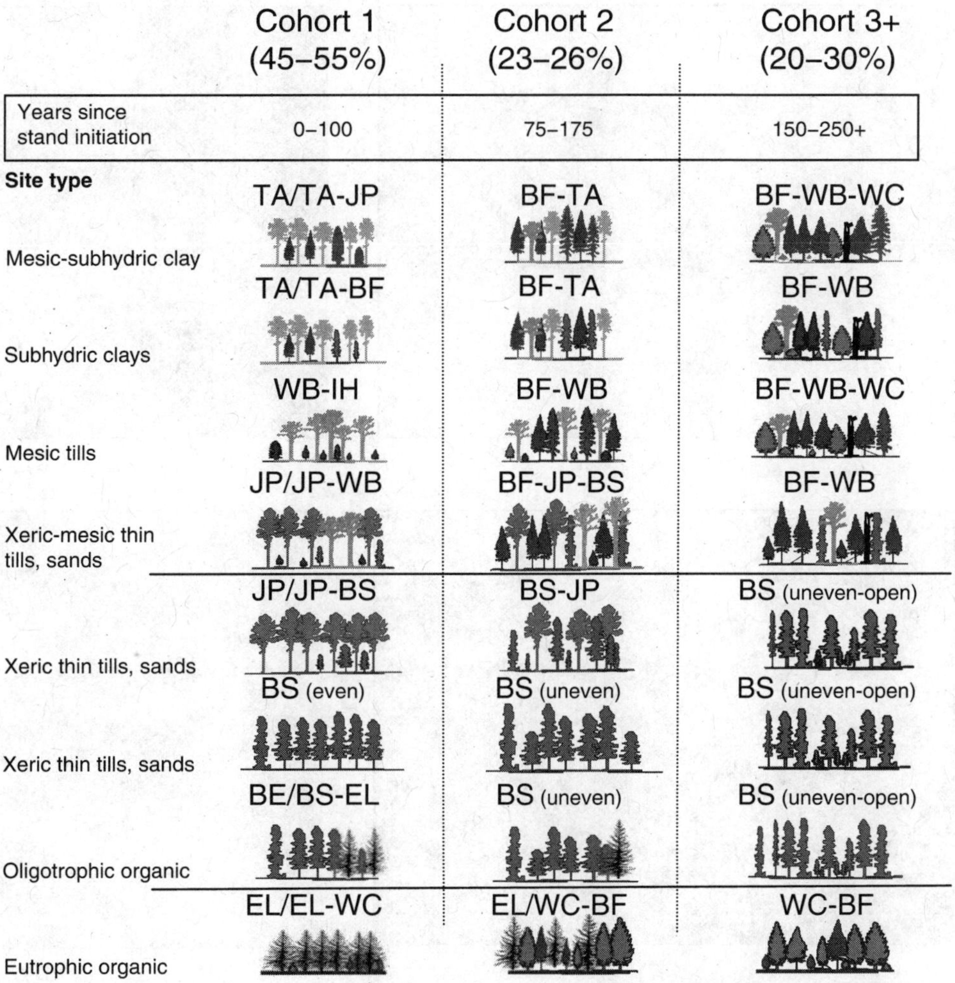

For each site type, stands first develop from types found on the left to those on the right as time since last fire increases. Stand development and uneven-aged silviculture on a given site type progresses from left to right whereas severe disturbance and even-aged silviculture revert stands to cohort 1 on left. See table 18.3 for stand type abbreviations. Adapted from Harvey et al. (2002).

naturally through gradual decline of aspen which, in turn, favours growth of suppressed conifers (Chen and Popadiouk 2002). Thus partial cutting of aspen or a "succession cut" would better produce this dynamic than simply clearcutting with regeneration protection (figure 18.9B). In the same manner, tree mortality after several years of budworm defoliation can create canopy gaps of varying size and dimension, depending on the amount of balsam fir in stands (D'Aoust et al. 2004; Bouchard et al. 2005, 2006). The silvicultural analogue to this natural

Figure 18.9
**Examples of silvicultural practices aiming at maintaining forest type diversity of the natural forest mosaic**

A)

B)

C)

D)

Legend: A) clearcut with dispersed and aggregated retention within a jack pine–aspen–birch even-aged stand (cohort 1 → cohort 1); B) succession cut within an even-aged aspen stand harvest in 2/3 of aspen stems (cohort 1 → cohort 2); C) cut with disperse white spruce seed trees followed by scarification within an open post-SBW outbreak balsam fir stand (treatments conducted just before 2006 seed year); D) selection cut within an uneven-aged black spruce stand (cohort 3 → cohort 3).
Sources: A) Philippe Duval; B) Brian Harvey; C) Philippe Duval; D) Claude Bouchard.

mortality could consist of a gradient of treatments, from careful logging to variably-sized gap cuts to recreate similar mortality patterns in mixed stands. Because tree mortality processes and the biological legacies left by natural disturbances and harvesting are inevitably different, the objective is not to perfectly imitate natural disturbances but rather to reduce the magnitude of differences generated by natural dynamics and different silviculture treatments (Haeussler et al. 2007).

### 4.1.4. Implementing a Monitoring Program and an Adaptive Management Approach

If forest ecosystem management is founded, at least partially, on cueing on the natural disturbance regime, it is important to understand and to quantify the various aspects of the natural regime and to determine the extent to which different management corollaries reflect or depart from the regime. Even management "inspired by nature" will generally differ enormously in terms of size, distribution, disturbance frequency, and biophysical impacts from the natural regime that has to be quantified.

Without any reference to natural forests, it is virtually impossible to evaluate the degree to which the composition, the productivity, the diversity, and the ecosystem functions are altered by harvesting and other management practices. The conservation zone of the LDF (see figure 18.3 and box 18.2) acts as such a reference landscape, and numerous studies aimed at understanding the disturbance history, dynamics, and ecology of natural forests have been undertaken in the area. Coincidentally, following field studies in the conservation zone, the MRNF recently recognized four "exceptional forest ecosystems" within this zone. The MRNF's *Écosystèmes forestiers exceptionnels* program is part of the government's biodiversity protection strategy and recognizes the inherent ecological value of natural ecosystems that are either old or rare or habitat of rare species.

Environmental monitoring activities are conducted on several fronts in the LDF. First, since the natural forest ecosystem constitutes the reference base for the evaluation of changes directly or indirectly induced by human interventions, a network of forest monitoring sites was established in natural forests in the conservation zone in the early 1990s. One-hectare forest plots have been mapped to characterize and monitor dynamics in forest stands originating from different fire years (figures 18.3 and 18.4). As well, two meteorological stations were installed on an island in Lake Duparquet and within the LDF. Long-term weather monitoring is useful in and of itself and for use in dendroecological studies. As well, a network of 200 permanent plots (1/40 ha) has been established in the two zones for long-term monitoring.

Over the past five years, 500 nesting tree cavities have been monitored in the conservation zone to determine characteristics of trees used by cavity nesters and to study the succession of species that create and use cavities. This protocol is providing information relevant to variable retention strategies applied in the management zone (see figure 18.9A).

Monitoring of the SAFE project (see box 18.4) and in sectors of variable retention harvesting provides a framework for evaluating management practices in the forest. One of the working hypotheses of the forest ecosystem management approach is that it is possible to generate old-growth stand attributes using adaptive silviculture practices. Mid- to long-term monitoring of these treatments is therefore central to testing this hypothesis.

Box 18.4
## Adaptive management and the SAFE project

The management approach proposed for the Western balsam fir–white birch bioclimatic subdomain is based on a conceptual model (Bergeron and Harvey 1997) which was built on several ecological studies conducted on the natural disturbance regime and forest dynamics of the region. The working hypothesis of this model may be stated as follows: By diversifying silvicultural approaches and by targeting, in particular, certain attributes of old forests, it is possible 1) to maintain an age structure and forest composition close to that of the mosaic created by the natural disturbance regime, and 2) to better maintain indigenous biodiversity and the ecological functions of forest ecosystems. Like any model, the passage from theory to practice requires an experimental framework that allows hypothesis testing in the field, in order to assess if, in practice, the model reflects reality, and if there are important gaps between anticipated and obtained effects, to modify the practices (or the model). This is the essence of adaptive management, and even if measuring real effects in

forestry can take years, it has to be done! To this end, the SAFE project (*Sylviculture et aménagement forestier écosystémique* or Silviculture and Forest Ecosystem Management) includes a series of silviculture experiments conducted in stands of different ages, compositions and structures that aims at testing the hypothesis mentioned above and the operational aspects of adaptive silviculture practices (figure 18.10; Brais et al. 2004). Several related questions are associated with each experience. For example: Does partial cutting in even-aged stands accelerate the development of attributes associated with older stands? What is the response of residual trees, advance regeneration, and the understory following partial cuts of different intensities? Does group selection or gap cutting generate more structural complexity than diffuse commercial thinning treatments? What are the effects of different intensities and configurations of treatments on dead wood dynamics and on associated living organisms and ecological processes? (<web2.uqat.ca/safe>)

Figure 18.10
## Aerial view of two of the three SAFE 1 project blocks

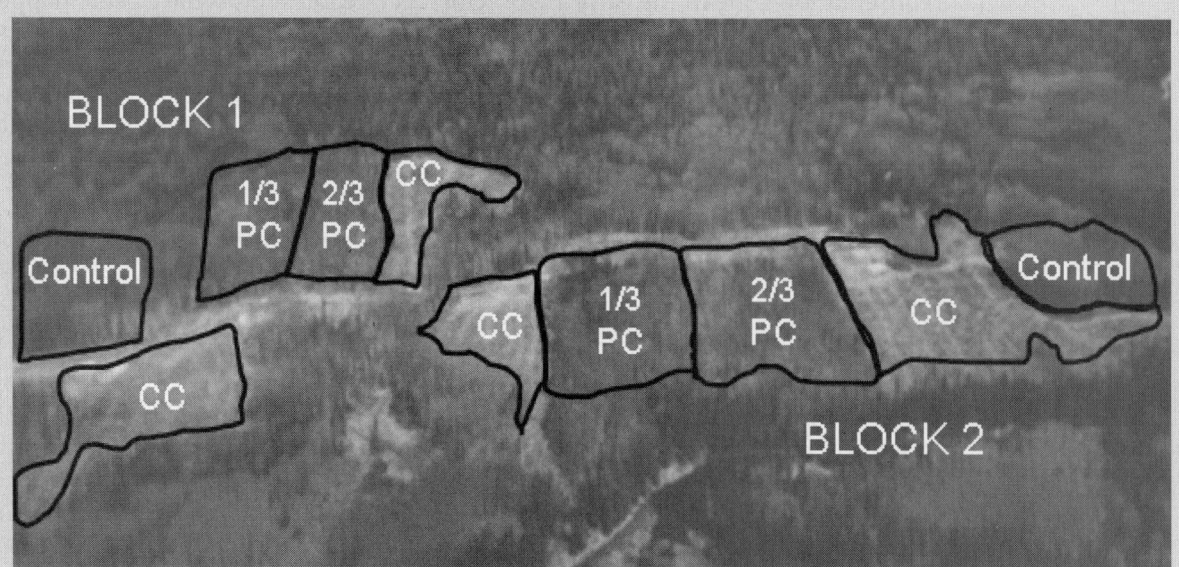

Source: René Beaudouin.

## 4.2. Other Important Elements of the Forest Ecosystem Management Approach

### 4.2.1. Recognizing the Value of Forest Ecological Functions

A landscape of healthy plantations can be a thing of beauty that needs no justification in managed areas. It should be recognized, however, that these forests cannot furnish all the ecological functions provided by a more diversified forest. It is generally accepted that the apparent disorder of natural stands, which are less homogeneous and Cartesian than plantations, contains greater structural diversity and offers shelter to a diversified array of flora and fauna. While recognizing the quality of wood as a construction material and primary resource for pulp and paper fabrication, it must be recognized that a tree, whether alive or dead, also plays an important role in the functioning of forest ecosystems. Think about the insects that depend on dead wood and which provide a food source for a huge variety of other organisms, or about the importance of dead trees for nest sites and, eventually, as inputs of soil nutrients. If we perceive the forest and trees for values other than their commercial and product values, it may be easily acknowledged that the natural character of a forest, and even patches of trees left in a harvest block, fill "non-utilitarian" – *but useful* – ecological functions. Reduced to its simplest expression, a tree that dies naturally in the forest should not be considered as a "wasted" tree. This is not a new concept; 50 years ago, the great American forester and conservationist Aldo Leopold (1966) recognized the intrinsic value of maintaining all of the components of an ecosystem as vast and complex as the boreal forest:

> *A land ethic of course cannot […] prevent the alteration, management, and the use of […] "resources", but it does affirm their right to continued existence, and, at least in spots, their continued existence in a natural state (A Sand Country Almanac, 240).*

### 4.2.2. Dead Wood Management

The Fennoscandian forest experience has exposed the critical relationship between the reductions of dead wood (DW) in managed forests and the loss of biodiversity (Niemelä 1997; Sipola and Renvall 1999; Nilsson et al. 2001). The retention of DW in managed forests is a major challenge for foresters, not only because it appears to oppose maximum fibre yield objectives, but also because our understanding of the contribution of dead wood to various ecosystem processes still remains fragmentary. Studies undertaken in the LDF have shown that harvesting prescriptions can be modulated in order to accelerate or decrease stem mortality and thus "produce" more or less dead wood in the short to medium term (Harvey and Brais 2007). However, in order to establish retention and production targets for dead wood, relationships between DW attributes (abundance, species, size, state of decay) and ecosystem processes involved need to be understood. The dead wood research program conducted in the LDF is designed to: 1) characterize the dynamics of dead wood in natural forests and in the managed stands of the SAFE project (see box 18.4); 2) understand the functional relationships between the attributes of DW and the organisms that depend on it; and 3) assess the contribution of dead wood to carbon and nitrogen cycles.

Variable retention strategies used in the Pacific Northwest (Franklin *et al.* 1997) and in British Columbia (Mitchell and Beese 2002) have been proposed as a means of increasing forest structural complexity and of providing a sort of archipelago of residual habitats (Fischer et al. 2006) in a system that remains essentially even-aged. The major difference between variable retention and the "cohort" management approach applied in the LDF is that the latter includes silvicultural practices aimed at transforming some even-aged stands into more complex structures and using silviculture to maintain a portion of the managed landscape in forests with structures and compositional attributes similar to those of over-mature and old-growth forests. This being said, the two approaches are complementary in an ecosystem management perspective. Dispersed and aggregated retention measures have been applied since 2002 to all harvest blocks where the objective is to convert a stand to the first-cohort stage (figure 18.9A), and a monitoring program has been established to study the evolution of residual trees over time. This program will allow us to answer questions with regard to the length of time residual trees remain alive, the susceptibility of the various species to breakage or uprooting, the use of the trees by wildlife, etc.

## 4.2.3.   Agglomerating Regeneration Zones

Compared to the size of wildfires (Bergeron et al. 2002) and the cumulative area of industrial harvest blocks in the region, the relatively small area of the management zone of the Lake Duparquet Forest considerably limits the possibility of varying size of cutovers or regeneration areas to reflect natural patterns. In a study of landscape-level changes in forest composition conducted in the LDF, using four series of aerial photographs (1965, 1972, 1983, 1994), Latrémouille (2008) observed that the size of first-cohort stands was relatively constant at around 42 ha. A simple interpretation of this result would be to suggest fixing harvest block size at about this area. However, the author concludes that the approach used for this study (focusing on a stand level perspective) does not capture the variability acting on a larger scale, such as that presented by the historical fire map (figure 18.3). In fact, even the variation in the size of fires that have occurred within the LDF over the past three centuries (table 18.2) suggests that the principle of varying the size of forest openings to better reflect the variability in disturbance size remains relevant. Excluding the 1923 and 1760 fires that extended outside the LDF limits (especially 1760), of the eight fires that occurred within the forest, four burned between 10 and 54 ha and the four others between 142 and 392 ha.

**Table 18.2**
**Areas of wildfires entirely located within the Lake Duparquet Forest, in ascending order**

| Wildfire year | 1907 | 1880 | 1847 | 1816 | 1919 | 1797 | 1823 | 1870 | 1923[1] |
|---|---|---|---|---|---|---|---|---|---|
| Area (ha) | 10 | 16 | 52 | 54 | 142 | 205 | 267 | 392 | 1 630 |

1.  In fact, limits of the 1923 fire exceed the LDF perimeter. The burn area was almost the double of its area within the LDF.

Source: Dansereau and Bergeron (1993).

### 4.2.4. Limiting the Permanent Road Network and Ensuring Connectivity between Large Forest Tracts

Even at the scale of the LDF, access is perceived as a major issue for forest managers and residential hunters (i.e. those having a camp in the forest) for whom the quality of the hunting experience depends in part on low hunter density and more particularly on low circulation of patrolling hunters.

To date, limited development of the permanent road system has been accomplished by agglomerating regeneration areas and concentrating harvesting activities during the winter freeze-up. However, because a certain access to the area is important for research and demonstration activities, as well as for conducting inventories and silvicultural treatments, the temporary roads are of a higher standard than winter roads normally built in Abitibi and are generally suitable for trucks in dry conditions. This is done by ditching road beds and installing culverts to standards close to those of permanent forest roads and constructing an adequate but non-gravelled road surface. In addition, these "semi-permanent" roads are generally seeded in the spring following harvest to stabilize and revegetate the surface in order to minimize runoff and erosion. This approach also allows the possibility of gravelling and upgrading roads in the future.

As access to the management zone of the LDF is increased in the future, the proportion of the forest that is not influenced by our activities will decrease. The conservation zone, situated in the eastern part of the forest (figure 18.3), should remain relatively wild, and for this reason there is no interest in developing access close to this part of the forest. Furthermore, the function of the conservation zone as a biological refuge for the rest of the LDF and neighbouring areas depends in part on the degree of connectivity between this zone and other major forest tracts. Forests associated with riparian and wet lands are well suited to this end and provide a landscape-level template to maintain a minimum continuous cover between the conservation zone located near Lake Duparquet, the Hébécourt Hills (a protected sector within the management zone), and a 160-m buffer around Lake Hébécourt.

### 4.2.5. Implementing Ligniculture on a Restricted Portion of the Area

Although still in an embryonic stage in Québec, *ligniculture*, a form of forest production modelled after intensive agriculture, constitutes a promising option to meet an increasing demand for wood resources, particularly when interests other than those of the forest processing industry are increasingly vying for the same land base. The experimental framework for testing ligniculture in the Lake Duparquet Forest (see Section 3.5 and figure 18.4) and elsewhere in Abitibi-Témiscamingue is aimed at determining which poplar and larch clones, genetically improved families of spruce, and planting and tending regimes are best adapted to the regional climatic and edaphic conditions.

## 5.  ANNUAL ALLOWABLE CUT CALCULATIONS AND STRATEGIC PLANNING

### 5.1.  Area-Based Allowable Cut Calculation

Allowable cut calculations for the first general management plan of the LDF were strictly area-based, rather than volume-based, primarily because the area approach is better suited for meeting forest-level composition objectives that reflect the natural forest mosaic. The area-based approach also has its advantages for relatively small forest areas. The allowable cut calculation was thus relatively simple. First, all areas affected by provincial forest environmental protection regulations (riparian zones, unproductive sites, steep slopes) were subtracted from the total area of the management zone. The resulting total productive forested area not subject to regulatory restrictions was then divided by an estimated average harvesting age of 70 years. This approach essentially corresponds to the *contenance* (capacity) method (Bérard and Côté 1996). While average harvest age of 70 years is probably not appropriate for all the forest stations, the fact that productive mesic clay sites dominate the forest suggests that it is generally appropriate. An annual allowable cut of about 69 ha, which correspond to 1/70 of the total exploitable area, was thus attributed to the Lake Duparquet Forest. In practical terms, the annual area of clearcutting and partial cutting is limited to 69 ha although the actual annual area harvested is usually somewhat less. Figure 18.7 illustrates the relative mix of harvesting regimes (even-aged silviculture, partial harvesting, and selection) to be applied in order to maintain stand- and forest-level structural diversity. The targeted proportions are presented on table 18.3. Areas treated by commercial thinning are not integrated into annual allowable calculations until the final cut. The mean annual volume harvested is about 11,200 m³ but has varied from year to year.

### 5.2.  General Management Planning

Similar to forest planning for large management units, strategic planning for the Lake Duparquet Forest has been a laborious task involving multiple inputs (figure 18.11). General management planning did not strictly adhere to the planning process described in the *Manual d'aménagement forestier* (Forest Management Manual) (MRNFP 2003). For example, rather than working with the "priority production groups" (such as "SEPM," which includes fir, spruce, jack pine, and larch), we adopted an approach that recognized ecological differences between species and allowed more than one management objective per "management stratum" or working group.

The ecosystem management approach applied at the LDF requires an evaluation of the proportion of the forest that should be occupied by each working group, based on fire cycle and the different site conditions that occur in the management zone. The exercise consists of estimating the "equilibrium" or balanced forest composition for each site type (or broad geomorphological type), then determining a global objective for the forest by taking into account the relative area occupied by each of these site types. The results of this exercise are presented in the form of percentages of the forest area that should be maintained in the different working groups in order to reflect the forest mosaic (forest composition and age structure) that theoretically characterizes the territory under the natural disturbance regime (table 18.3). The current state of the forest, although largely influenced by site conditions and the time elapsed since the last major disturbance, is also affected by historical factors (e.g. composition and age structure of the stands occupying the territory at the time of the last disturbance). For this reason the area objectives for each working group should be considered as approximations rather than rigid targets. For example, compromises could be made in the case of targeted areas for jack pine and trembling aspen working groups, given that both are first-cohort stand types and found on mesic clay sites.

The objective of regulating the age structure of the working groups is to ensure sustained production of each group. The simplest manner to attain this objective is to equally distribute the area allocated for each working group over all age classes up to commercial age. For example, for a commercial rotation of 70–75 years, there are fifteen 5-year age classes in which the managed area should be uniformly distributed, or about 6.7% (1/15) of the total area of the working group per age class. Obviously, this principle may prove difficult to respect if current age structure is very different from the normal or fully regulated structure. As regularization of the LDF is achieved by regularizing each working group, the first step of the modelling process consists of calculating the standardized age structure for each working group. This distribution is calculated simply by dividing the total area targeted for the group by the number of age classes up to commercial age of maturity. For example, the standardized and desired age structure of aspen stands in the LDF would be of 0.67% (10.1%/15) of the management zone in each 5-year class. Thus we ensure a sustained yield in this working group while maintaining its area at an average occupation level of 10.1% of the management unit. It should be clarified that 0.67% refers to the 5-year recruitment rate of the working group rather than the harvest rate. (Aspen stands may potentially be recruited from harvesting stands other than those of pure aspen.)

Like most working groups associated with the first cohort, aspen and jack pine stands can be naturally or artificially returned to their original composition after harvesting. Although this strategy can be used to regenerate certain types of stands, it is inadequate for others. For example, we anticipate the need to recruit (or to favour succession of) some working groups from a different group in the case where the working group is under- or over-represented within the forest mosaic, or if the internal dynamics of the working group tend to promote its conversion towards an advance successional stage.

The following example shows how it is possible to establish flows between different working groups (see figures 18.7 and 18.8). Mixedwood stands dominated by shade-tolerant softwoods (fir, spruce, cedar) with a hardwood component of aspen (TS•-TA, cohort 2) can be treated with partial cutting (or succession cutting) of aspen stems in order to enhance the softwood component (towards a composition of TS or TS-IH, cohort 3). They can also be returned to a dominant composition of shade intolerant species (TA or IH, cohort 1) or a mixed hardwood (TA-TS, cohort 1 or 2) by clearcutting or careful logging. In the same manner, aspen-dominated mixed stands (TA-TS) can be treated with succession cuts (towards TS-TA) or could serve as a recruitment group for the TA working group by clearcutting.

• TS: tolerant softwoods;
  TA: trembling aspen;
  IH: intolerant hardwoods.

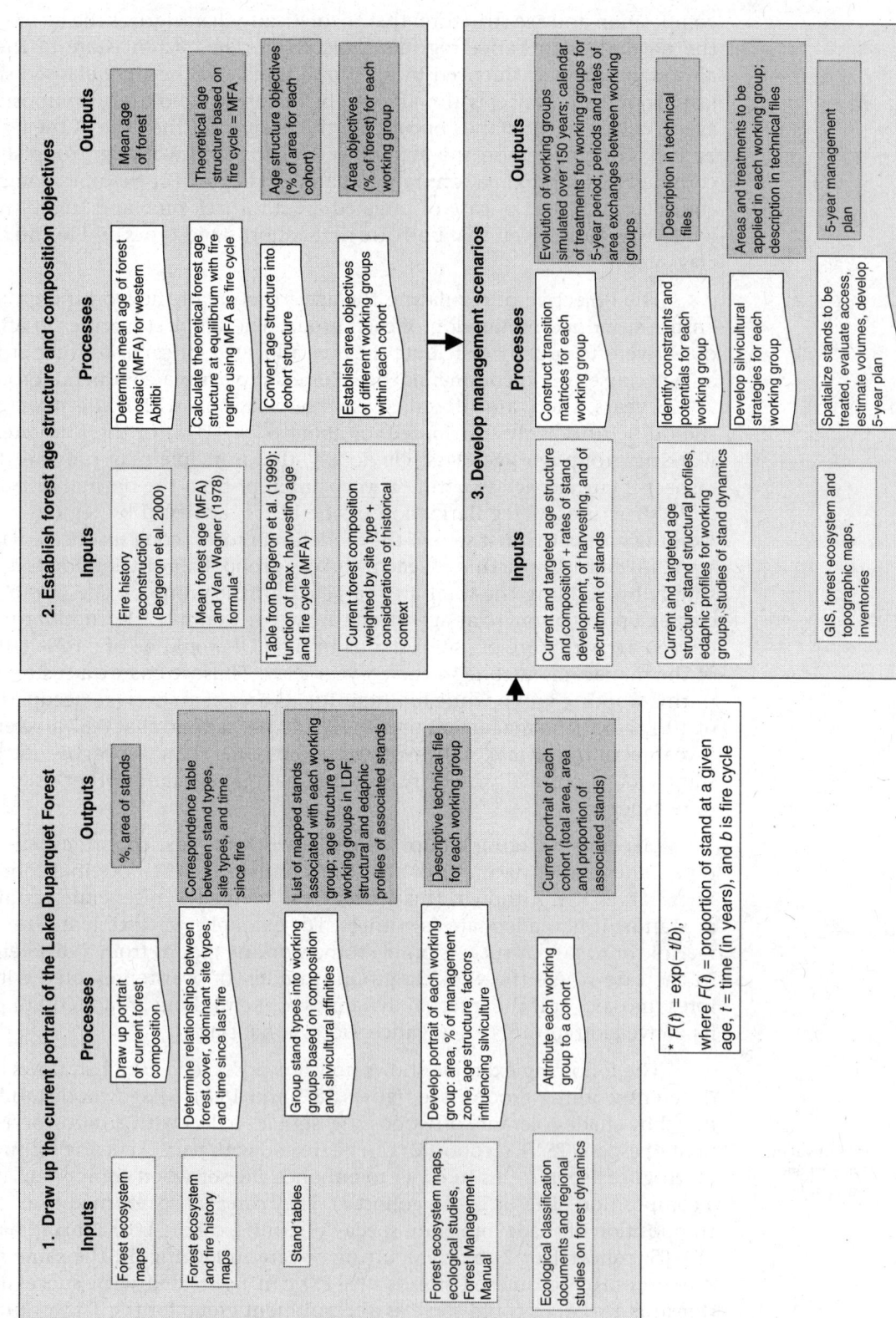

**Figure 18.11**
**Planning process for the first general management plan of the Lake Duparquet Forest**

### Table 18.3
## Summary of current* and targeted areas for working groups in the Lake Duparquet Forest

| Forest working group | Abbreviations | Cohort | Current area (ha) | Current area (%) | Targeted area (ha) | Targeted area (%) | Change (%) |
|---|---|---|---|---|---|---|---|
| Pure aspen stands or aspen/jack pine | TA, TA-JP | 1 | 427 | 8.2 | 526 | 10.1 | +1.9 |
| Jack pine stands | JP, JP-X, BS-JP | 1 | 889 | 17.1 | 889 | 17.1 | 0.0 |
| White birch stands and intolerant hardwoods | WB, IH | 1 | 397 | 7.6 | 397 | 7.6 | 0.0 |
| Even-aged black spruce stands | BS, BS-BF | 1 | 531 | 10.2 | 531 | 10.2 | 0.0 |
| Larch stands | EL, EL-BS, BS-EL | 1 | 224 | 4.3 | 134 | 2.6 | –1.7 |
| **Total cohort 1** | | | **2,468** | **47.4** | **2,477** | **47.6** | **+0.2** |
| Aspen stands with tolerant softwoods | TA-TS | 2 | 406 | 7.8 | 306 | 5.9 | –1.9 |
| White birch or intolerant hardwood stands with tolerant softwoods | WB-TS, IH-TS | 2 | 906 | 17.4 | 106 | 2.0 | –15.4 |
| Tolerant softwood stands with aspens | TS-TA | 2 | 102 | 2.0 | 350 | 6.7 | +4.8 |
| Uneven-aged black spruce stands | BS, BS-BF | 2, 3 | 384 | 7.4 | 708 | 13.6 | +6.2 |
| **Total cohort 2** | | | **1,798** | **34.6** | **1,470** | **28.2** | **–6.3** |
| Tolerant softwoods stands with birches or intolerant hardwoods | TS-WB, TS-IH | 3 | 682 | 13.1 | 909 | 17.5 | +4.4 |
| Balsam fir stands | BF, BF-BS | 3 | 159 | 3.1 | 159 | 3.1 | 0.0 |
| Eastern cedar stands | EC, TS-EC, EC-BS, EC-EL | 3 | 97 | 1.9 | 189 | 3.6 | +1.8 |
| **Total cohort 3** | | | **938** | **18.0** | **1,257** | **24.2** | **+6.1** |

*At the time of first general plan development.

Legend: BS: black spruce, BF: balsam fir, EC: eastern cedar, EL: eastern larch, JP: jack pine, IH: intolerant hardwoods, TA: trembling aspen, TS: tolerant softwoods, WB: white birch.

Management simulations were conducted using transition matrices. The evolution of each working group was simulated from current age structure to the desired normal or fully regulated structure over 150 years. Matrices incorporated, in 5-year increments, the development or aging of each working group, the areas affected by harvesting and those recruited into the 0–5-year age class following harvest. It also integrated constraints related to age structure and operating windows for harvesting of working groups. For periods when recruitment into one working group occurs by harvesting another group, the matrices of the two groups indicate these flows. Using these parameters, we established the areas to be harvested for 5-year periods in each of the 12 working groups of the Lake Duparquet Forest. A technical note of each group summarizes parameters relevant to management such as current and targeted area base and age structure, current age structure profile of stands over 70 years, and edaphic profile (current area distribution on different site types). Each file also contains a description of management objectives, silviculture strategies, and anticipated challenges as well as the list of the mapped stands composing the working group.

# 6. CONCLUSION

## 6.1. Social Acceptability of the Approach

Forest ecosystem management, particularly the notion of natural disturbance-based forest management, has been falsely criticized for failing to recognize humans as an integral part of Nature. Presented or interpreted as a strict orthodoxy in which the ultimate goal is to mimic nature or natural processes as faithfully as possible, this approach simply does not make sense. The social acceptability of the forest ecosystem management approach practised in the Lake Duparquet Forest was evaluated by a committee called the *Groupe de travail sur l'analyse multicritère de la Forêt du lac Duparquet* (Lake Duparquet Forest Multicriteria Analysis Working Group) (see Hébert et al. 2003). This group was established in 2000 to act as a sort of integrated resource management advisory council, to survey concerns of stakeholders (users of the forest) and to take these concerns into consideration in forest planning. Curiously, despite the fact that taking other forest users' concerns into account imposes additional constraints on management, it has become evident that, when well explained, the forest ecosystem management approach is well perceived by all concerned parties. Admittedly, the research forest does present certain particularities including the size of harvest areas and the designation of a quarter of the area as a conservation zone. Nevertheless, when the use of clearcutting and of partial cutting and limiting the permanent forest road network are presented and perceived as part of a global strategy to maintain at least part of the natural character of the forest within a managed landscape, the approach is overwhelmingly approved by stakeholders. Certainly, the scale of interventions is very different from that used on larger management units of public forest, where the spatial aspects of disturbances could be better integrated into management.

## 6.2. Can We Apply This Approach Elsewhere?

Every management unit is different, and the management approach applied to an area depends on the particularities of the forest, the stakeholders and communities that live nearby, the industrial structure, as well as management constraints associated with the land base. Even if other forests do not have the same degree of involvement of forest scientists, most aspects found in the ecosystem management approach applied in the Lake Duparquet Forest can still be applied elsewhere. Obviously it is important to start from the regional ecological information base of interest, and, to this end, Quebec's forest ecological framework covers all commercial forests in the province. Apart from these tools, specific studies on natural disturbance regimes, stand dynamics, and silviculture trials, if they exist, can provide the basis for an ecosystem management approach that includes the objective of maintaining the natural character of the forests.

## REFERENCES

Archambault, S. and Bergeron, Y. 1992. A 802-year tree ring chronology from the Quebec boreal forest. Can. J. For. Res. **22**: 674–682.

Armstrong, G.W. 1999. A stochastic characterisation of the natural disturbance regime of the boreal mixedwood forest with implications for sustainable forest management. Can. J. For Res. **29**: 424–433.

Belleau, A., Bergeron, Y., Leduc, A., Gauthier, S., and Fall, A. 2007. Using spatially explicit simulations to explore size distribution and spacing of regeneration areas produced by wildfires: recommendations for designing harvest agglomerations for the Canadian boreal forest. For. Chron. **83**: 72–83.

Bérard, J.A. and Côté, M. (Eds.) 1996. Manuel de foresterie. Ordre des ingénieurs forestiers du Québec. Les Presses de l'Université Laval, Québec, Que.

Bergeron, Y. 1991. The influence of island and mainland lakeshore landscapes on boreal forest fire regimes. Ecology, **72**: 1980–1992.

Bergeron, Y. 2000. Species and stand dynamics in the mixed-woods of Quebec's southern boreal forest. Ecology, **81**: 1500–1516.

Bergeron, Y., Bouchard, A., Gangloff, P., and Camiré, C. 1983. La classification écologique des milieux forestiers d'une partie des cantons d'Hébécourt et de Roquemaure. Études écologiques no. 9, Université Laval, Québec, Que.

Bergeron, Y., Denneler, B., Charron, D., and Girardin, M.-P. 2002. Using dendrochronology to reconstruct disturbance and forest dynamics around Lake Duparquet, northwestern Quebec. Dendrochronologia, **20**: 175–189.

Bergeron, Y., Gauthier, S., Flannigan, M., and Kafka, V. 2004. Fire regimes at the transition between mixedwoods and coniferous boreal forests in Northwestern Quebec. Ecology, **85**: 1916–1932.

Bergeron, Y., Gauthier, S., Kafka, V., Lefort, P., and Lesieur, D. 2001. Natural fire frequency for the eastern Canadian boreal forest: consequences for sustainable forestry. Can. J. For. Res. **31**: 384–391.

Bergeron, Y. and Harvey, B. 1997. Basing silviculture on natural ecosystem dynamics: an approach applied to the southern boreal mixedwoods of Québec. For. Ecol. Manag. **92**: 235–242.

Bergeron, Y., Harvey, B., Leduc, A., and Gauthier, S. 1999. Forest management guidelines based on natural disturbance dynamics: stand and forest level considerations. For. Chron. **75**: 39–53.

Bergeron, Y. and Leduc, A. 1998. Relationships between change in fire frequency and mortality due to spruce budworm outbreak in the southeastern Canadian boreal forest. J. Veg. Sci. **9**: 492–500.

Bergeron, Y., Leduc, A., Harvey, B., and Gauthier, S. 2001. Natural fire regime: a guide for sustainable management of the Canadian boreal forest. Silva Fenn. **36**: 81–95.

Bergeron, Y., Leduc, A., and Li, T.X. 1997. Explaining the distribution of *Pinus* spp. in a Canadian boreal insular landscape. J. Veg. Sci. **8**: 37–44.

Bergeron, Y., Morin, H., Leduc, A., and Joyal, C. 1995. Balsam fir mortality following the last spruce budworm outbreak in northwestern Quebec. Can. J. For. Res. **25**: 1375–1384.

Bergeron, Y., Richard, P.J.H., Carcaillet, C., Flannigan, M., Gauthier, S., and Prairie, Y. 1998. Variability in Holocene fire frequency and forest composition in Canada's southeastern boreal forest: a challenge for sustainable forest management. Conserv. Ecol. **12**: art. 6. [Online] <www.consecol.org/Journal/vol2/iss2/art6>.

Bescond, H. 2002. Reconstitution de l'historique de l'exploitation forestière sur le territoire de la Forêt d'enseignement et de recherche du lac Duparquet au cours du 20ième siècle et influence sur l'évolution des peuplements forestiers. M.Sc. thesis, Université du Québec à Montréal, Montréal, Que.

Blouin, J. and Berger, J.-P. 2002. Guide de reconnaissance des types écologiques de la région écologique 5a – Plaine de l'Abitibi. Ministère des Ressources naturelles du Québec, Forêt Québec. Direction des inventaires forestiers, Division de la classification écologique et productivité de stations.

Bouchard, M., Kneeshaw, D., and Bergeron. Y. 2005. Mortality and stand renewal patterns following the last spruce budworm outbreak in mixed forests of western Quebec. For. Ecol. Manag. **204**: 297–313.

Bouchard, M., Kneeshaw, D., and Bergeron. Y. 2006. Forest dynamics after successive spruce budworm outbreaks in mixedwood forests. Ecology, **87**: 2319–2329.

Brais, S., Harvey, B.D., Bergeron, Y., Messier, C., Greene, D., Belleau, A., and Paré, D. 2004. Testing forest ecosystem management in boreal mixedwoods of northwestern Quebec: initial response of aspen stands to different levels of harvesting. Can. J. For. Res. **34**: 431–446.

Carcaillet, C., Bergeron, Y., Richard, P.J.H., Fréchette, B., Gauthier, S., and Prairie, Y.T. 2001. Change of fire frequency in the eastern Canadian boreal forests during the Holocene: does vegetation composition or climate trigger the fire regime? J. Ecol. **89**: 930–946.

Chen, H.Y.H. and Popadiouk, R.V. 2002. Dynamics of North American boreal mixedwoods. Environ. Rev. **10**: 137–166.

Cooke, B.J. and Lorenzetti, F. 2006. The dynamics of forest tent caterpillar outbreaks in Québec, Canada. For. Ecol. Manag. **226**: 111–121.

Coulombe, G., Huot, J., Arsenault, J., Beauce, E., Bernard, J.-T., Bouchard, A., Liboiron, M.-A., and Szaraz, G. 2004. Rapport de la Commission d'étude sur la gestion de la forêt publique québécoise. Que. [Online] <www.mrnfp.gouv.qc.ca/commission-foret/rapportfinal.htm> (accessed November 7, 2007).

Dansereau, P. and Bergeron, Y. 1993. Fire history in the southern boreal forest of northwestern Quebec. Can. J. For. Res. **23**: 25–32.

D'Aoust, V., Kneeshaw, D., and Bergeron, Y. 2004. Characterization of canopy openness before and after a spruce budworm outbreak in the southern boreal forest. Can. J. For. Res. **34**: 339–352.

Fischer, J., Lindenmayer, D.B., and Manning, A.D. 2006. Biodiversity, ecosystem function, and reslience: ten guiding principles for commodity production landscapes. Front. Ecol. Environ. **2**: 80–86.

Franklin, J.F., Berg, D.R., Thornburg, D.A., and Tappeiner, J.C. 1997. Alternative silvicultural approaches to timber harvesting: variable retention harvest systems. *In* Creating a forestry for the 21st century: the science of ecosystem management. *Edited by* K.A. Kohm and J.F. Franklin. Island Press, Washington, D.C., USA, p. 111–139.

Gauthier, S., De Grandpré, L., and Bergeron, Y. 2000. Differences in forest composition in two ecoregions of the boreal forest of Québec. J. Veg. Sci. **11**: 781–790.

Gauthier, S., Leduc, A., and Bergeron, Y. 1996. Forest dynamics modelling under a natural fire cycle: a tool to define natural mosaic diversity in forest management. Environ. Monitor. Assess. **39**: 417–434.

Gauthier, S., Lefort P., Bergeron Y., and Drapeau, P. 2002. Time since fire map, age-class distribution and forest dynamics in the Lake Abitibi Model Forest. Information Report, Laurentian Forestry Center, Québec Region, Canadian Forest Service, No. LAU-X-125E.

Grondin, P., Blouin, J., and Racine, P. 1999. Rapport de classification écologique du sous-domaine bioclimatique de la sapinière à bouleau blanc de l'ouest. Ministère des Ressources naturelles du Québec, Direction des inventaires forestiers, Québec, Que.

Grondin, P., Saucier, J.-P., Blouin, J., Gosselin, J., and Robitaille, A. 2003. Information écologique et planification forestière au Québec, Canada. Direction de la recherche forestière, ministère des Ressources naturelles, de la Faune et des Parcs. Note de recherche forestière n° 118.

Haeussler, S., Bergeron, Y., Brais, S., and Harvey, B. 2007. Natural dynamics-based silviculture for maintaining plant biodiversity in *Populus tremuloides*–dominated boreal forests of Canada. Can. J. Bot. **85**: 1158–1170.

Harvey, B. 1999. The Lake Duparquet research and teaching forest: building a foundation for ecosystem management. For. Chron.**75**: 389–393.

Harvey, B.D. and Brais, S. 2007. Partial cutting as an analogue to stem exclusion and dieback in aspen (*Populus tremuloides*) dominated boreal mixedwoods: implications for deadwood dynamics. Can. J. For. Res. **37**: 1525–1533.

Harvey, B.D., Leduc, A., Gauthier, S., and Bergeron, Y. 2002. Stand-landscape integration in natural disturbance-based management of the southern boreal forest. For. Ecol. Manag. **155**: 369–385.

Hebert, D., Harvey, B.D., Wasel, S., Roberts, J., Hamersley Chambers, F., and Burton, P.J. 2003. Implementing sustainable forest management: some case studies. *In* Towards sustainable management of the boreal forest. *Edited by* P.J. Burton, C. Messier, D.W. Smith, and W.L. Adamiwicz. NRC Research Press, Ottawa, Ont., p. 893–952.

Hély, C., Bergeron, Y., and Flannigan, M.D. 2000. Coarse woody debris in the southeastern Canadian boreal forest: composition and load variations in relation to stand replacement. Can. J. For. Res. **30**: 674–687.

Kneeshaw, D.D. and Bergeron, Y. 1998. Canopy gap dynamics and tree replacement in the southeastern boreal forest. Ecology, **79**: 783–794.

Latrémouille, C. 2007. Impacts d'un aménagement forestier écosystémique sur la mosaïque forestière et sur les coûts d'approvisionnement en bois: étude de cas de la Forêt d'enseignement et de recherche du lac Duparquet. M.Sc. thesis, Université du Québec en Abitibi-Témiscamingue, Rouyn-Noranda, Que.

Leduc, A., Gauthier, S., and Bergeron, Y. 1995. Prévision de la composition d'une mosaïque forestière naturelle soumise à un régime des feux: proposition d'un modèle empirique pour le nord-ouest du Québec. *In* Méthodes et réalisations de l'écologie du paysage pour l'aménagement du territoire. *Edited by* G. Domon and J. Falardeau. Polyscience publication, Morin Heights, Que., p 197–205.

Leopold, A. 1966. A Sand County almanac with essays on conservation from Round River. Ballantine Books, New York, USA.

MacLean, D.A. and Ostaff, P.A. 1989. Patterns of balsam fir mortality caused by an uncontrolled spruce budworm outbreak. Can. J. For. Res. **19**: 1987–1095.

Ministère des Ressources naturelles, de la Faune et des Parcs (MRNFP). 2003. Le manuel d'aménagement forestier, 4e édition. Gouvernement du Québec, Québec, Que.

Mitchell, S.J. and Beese, W.J. 2002. The retention system: reconciling variable retention with the principles of silvicultural systems. For. Chron. **78**: 397–403.

Morin, H., Laprise, D., and Bergeron, Y. 1993. Chronology of spruce budworm outbreaks in the lake Duparquet region, Abitibi, Québec. Can. J. For. Res. **23**: 1497–1506.

Niamelä, J. 1997. Invertebrates and boreal forest management: a review. Conserv. Biol. **11**: 601–610.

Nilsson, S., Hedin, J., and Niklasson, M. 2001. Biodiversity and its assessment in boreal and nemoral forests. Scand. J. For. Suppl. **3**: 10–26.

Saucier, J.-P., Bergeron, J.-F., Grondin, P., and Robitaille, A. 1998. Les régions écologiques du Québec méridional (3e version): un des éléments du système hiérarchique de classification écologique du territoire mis au point par le ministère des Ressources naturelles du Québec. Supplément de L'Aubelle, no. 124, 12 p.

Seymour, R.S. and Hunter, Jr., M.L. 1992. New forestry in eastern spruce-fir forests: principles and application to Maine. Maine Agricultural and Forestry Experiment Station, Misc. Publication 716.

Seymour, R.S. and Hunter, Jr., M.L. 1999. Principles of ecological forestry. *In* Maintaining biodiversity in forest ecosystems. *Edited by* M.L. Hunter, Jr. Cambridge University Press, Cambridge, UK, p. 22–61.

Sippola, A.-L. and Renvall, P. 1999. Wood-decomposing fungi and seed-tree cutting: a 40-year perspective. For. Ecol. Manag. **115**: 183–201.

Van Wagner, C.E. 1978. Age-class distribution and the forest fire cycle. Can. J. For. Res. **8**: 220–227.

# Chapter 19

## Project Tembec
### Towards the Implementation of a Forest Management Strategy Based on the Natural Disturbance Dynamics of the Northern Abitibi Region[*]

*Annie Belleau and Sonia Légaré*

[*] We thank the Ministère des Ressources naturelles et de la Faune for their collaboration, particularly Elaine Cyr, Vincent Nadeau, and Marcel Paré. In addition, we thank all those at Tembec who worked on the project, especially Louis Dumas. Finally we acknowledge the considerable support and contributions from the Industrial Chair NSERC-UQAT-UQAM in Sustainable Forest Management.

# 1. INTRODUCTION

• CSA: Canadian Standard Association, SFI: Sustainable Forestry Initiative, FSC: Forest Stewardship Council

The increasing interest for forestry practices based on natural disturbance patterns and the arrival of forestry certification programs (CSA, SFI, and FSC[*]) that open new markets have encouraged Tembec to invest in a new forest management project. Originally coordonated by the NSERC-UQAT-UQAM Industrial Chair in Sustainable Forest Management, a three-year research project focused on "the development of a forest management strategy based on the natural disturbance regime of the northern Abitibi region." In collaboration with the Ministère des Ressources naturelles et de la Faune du Québec (MRNF), Tembec, and Norbord, this first project was undertaken on a 4,750 km² area situated in the boreal forest of western Québec. Tembec, who was interested in implementing the approach within a general forest management plan and then applying it at a large scale, took the lead of the project in 2002 and organized a multidisciplinary committee composed of forestry scientists and engineers from the Université du Québec en Abitibi-Témiscamingue (UQAT), the MRNF, and both Tembec and Norbord. The first objective of this committee was to develop a five-year forest management plan based on the region's natural forest dynamics and disturbance regime.

With the experience gained through the NSERC-UQAT-UQAM Industrial Chair in Sustainable Forest Management, Tembec wishes, through its committee, to adapt the approach to the constraints of the industry and ensure that it continually responds to both the goals for the region and the certification requirements of the company. Based on the concept of coarse and fine filters (Hunter 1999), the forest management strategy developed here should ensure the maintenance of biodiversity and the resilience of disturbed ecosystems (see Gauthier et al., chapter 1). Over the long term, Tembec wishes to restore and maintain the forest mosaic within its regional historical range of variability.

The current project is located between 49°00′ and 51°30′ N and 78°30′ and 79°31′ W and covers an area of 10,826 km² (figure 19.1). The Forest Management Unit (FMU) 085-51 is situated in the boreal forest within ecological region 6a–Plaine du lac Matagami, which is located at the extreme west of the western black spruce–feather moss bioclimatic subdomain. The main tree species found in the region, listed in order of abundance, are black spruce (*Picea mariana* [Mill.] B.S.P.), jack pine (*Pinus banksiana* Lamb.), and trembling aspen (*Populus tremuloides* Michx.). The mean annual temperature is –0.7°C and the average precipitation is 905.5 mm (Environment Canada 2006). The surficial geology of the region is characterized by clay deposits left behind following the retreat of proglacial Lake Ojibway (Vincent and Hardy 1977).

To date, the project's approach has been divided into four steps. The first step consisted of constructing a pre-industrial snapshot of the FMU that documents the disturbance regime and describes the forest dynamics of the forest management unit. Part of this work had already been accomplished by the NSERC-UQAT-UQAM Industrial Chair in Sustainable Forest Management during the first phase of the project (Nguyen 2000, 2002; Gauthier et al. 2004). Second, management goals and objectives were identified by comparing the snapshot

Figure 19.1
**Location of the Forest Management Unit (FMU) 085-51, in Abitibi**

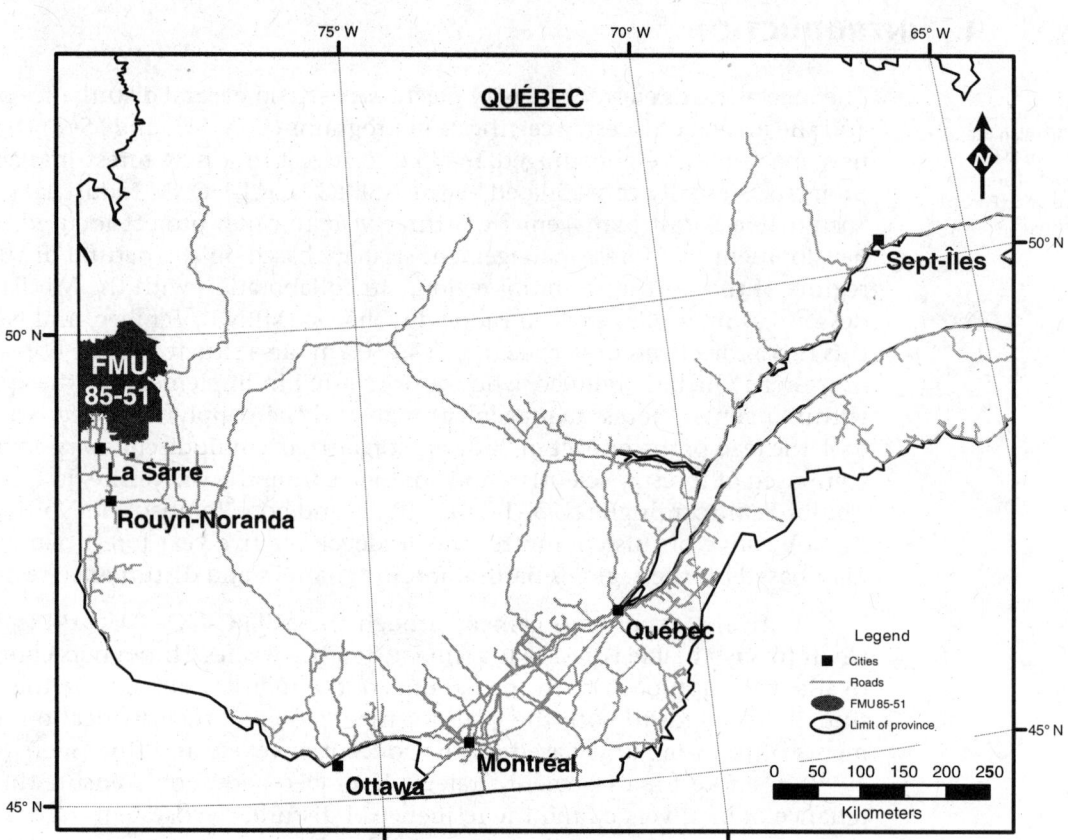

The FMU is located in the Western black spruce–feather moss bioclimatic subdomain and covers 10,826 km².

established during step one with the current state of the FMU. Third, methods and strategies were suggested in order to respond to the diverse goals and objectives. Finally, the methods and strategies elaborated were integrated into the annual allowable cut calculation in order to evaluate the socioeconomic impacts of the approach and to identify parameters that have to be optimized in order to develop the most acceptable ecological, social, and economic approach.

## 2.   AN APPROACH BASED ON LOCAL KNOWLEDGE

### 2.1.   Disturbance Regime

Excluding forestry, fire is the principal large-scale disturbance that affects the region (see Simard et al., chapter 11). Fire is recognized as the factor that has historically created the landscape and age structure of the forest. The fire cycle of the region has varied historically and possessed a 101-year fire cycle before 1850, a 135-year cycle between 1850 and 1920, and a 398-year cycle since the 1920s. The fire history reconstruction of the landscape indicates a mean forest age of around 148 years (Bergeron et al. 2004). While most of the fires were of small size, large, rare fire events (>1,000 ha) have had the most impact on the region and have burned more than 90% of its area (Bergeron et al. 2004). Over the last 60 years, 55% of the fires in the territory were of a size between 950 and 20,000 ha (Bergeron et al. 2002). While normally considered severe, fires do leave a certain quantity of green forests (unburned and partially burned patches). Leduc et al. (2000) estimate that fires leave around 5% of the forest intact in the form of islands within the burned area and that partially burned areas can make up to 30 to 50% of the affected area. To the east of the study area, Perron (2003) estimates that boreal forest fires leave between 10 and 35% intact residual forest with 1 to 8% in the form of islands. Few studies have been conducted regarding the spatial distribution of fire events (see Perron et al., chapter 6). According to simulations projecting the regional fire regime (with a random fire distribution and agglomeration of burned areas over a 25-year period), Belleau et al. (2007) suggest a minimal 6 km spacing of regeneration areas.

Although of less importance, two other disturbances also play roles in forest stand dynamics, particularly between fire episodes. These are insect outbreaks and windthrows. Three major spruce budworm (SBW) (*Choristoneura fumiferana* Clem.) outbreaks have occurred in western Québec in the last century: from 1919 to 1929, 1930 to 1950, and 1970 to 1987 (Morin et al. 1993). Furthermore, 1 to 5 forest tent caterpillar (*Malacosoma disstria* Hbn.) outbreaks have also occurred (MNRF 2003). However, the frequency and severity of these outbreaks were low in the area targeted by the ecosystem management plan. The relative abundance of balsam fir (*Abies balsamea* [L.] Mill.) and trembling aspen, which are the preferred hosts for the SBW and forest tent caterpillar respectively, are low within the region (<10%), with the stands being greatly dispersed throughout the forest mosaic. Stands aged between 100 and 300 years appear to be the most affected by windthrows. In total, around 15% of the area occupied by the stands is affected by partial windthrows (Harper et al. 2002). Finally, the proportion of gaps created following tree senescence increases with stand age. In forests less than 150 years old, the majority of gaps (80%) are less than 30 m². These gaps are generally created by the mortality of 10 or less mature trees. The regeneration within the gaps is composed primarily of black spruce. Given the cold soils, the accumulation of organic material, and the rising of the water table, tree growth is slow and does not permit the gaps to close. In absence of severe fire, areas currently possessing closed-canopy forests will therefore likely open with time (Lecomte and Bergeron 2005; Lecomte et al. 2006; Simard et al., chapter 11).

## 2.2. Stand Dynamics

The western most part of the western black spruce–feather moss subdomain corresponds physiographically to the Clay Belt (northwestern and northeastern Ontario). This area is known to be subject to paludification (Taylor et al. 1987). Paludification is a phenomenon where a significant layer of organic material accumulates on top of the mineral soil. This accumulation is caused by several factors and limits the long-term productivity of sites by favouring high water tables, decreasing microbial activity, and reducing nutrient availability (see Simard et al., chapter 11; Van Cleve and Viereck 1981; Taylor et al. 1987; Fenton et al. 2006; Simard et al. 2007). This phenomenon also influences forest succession by promoting the development of open stands of irregular structure (Harper et al. 2002, 2005; Lecomte and Bergeron 2005; Lecomte et al. 2006). In the absence of fire, stands in the region typically converge towards open black spruce stands of low productivity. The ability of black spruce to maintain themselves at these paludified sites, along with their shade tolerance, are factors contributing to their dominance in these stands (Harper et al. 2002; Lecomte and Bergeron 2005). Jack pine and trembling aspen are typically found only on well-drained sites following fire, due to their intolerance to shade and paludification.

In order to simplify the forest dynamics in the study area, the stands were classified into 11 developmental stages during the first phase of the project (Gauthier et al. 2004). These stages were divided into three successional pathways (trembling aspen, jack pine, and black spruce) and three developmental stages (cohorts; *sensu* Bergeron et al. 1999) (figure 19.2). The first cohort corresponds

**Figure 19.2**
**Model illustrating the natural succession pathway in the western black spruce–feather moss bioclimatic subdomain**

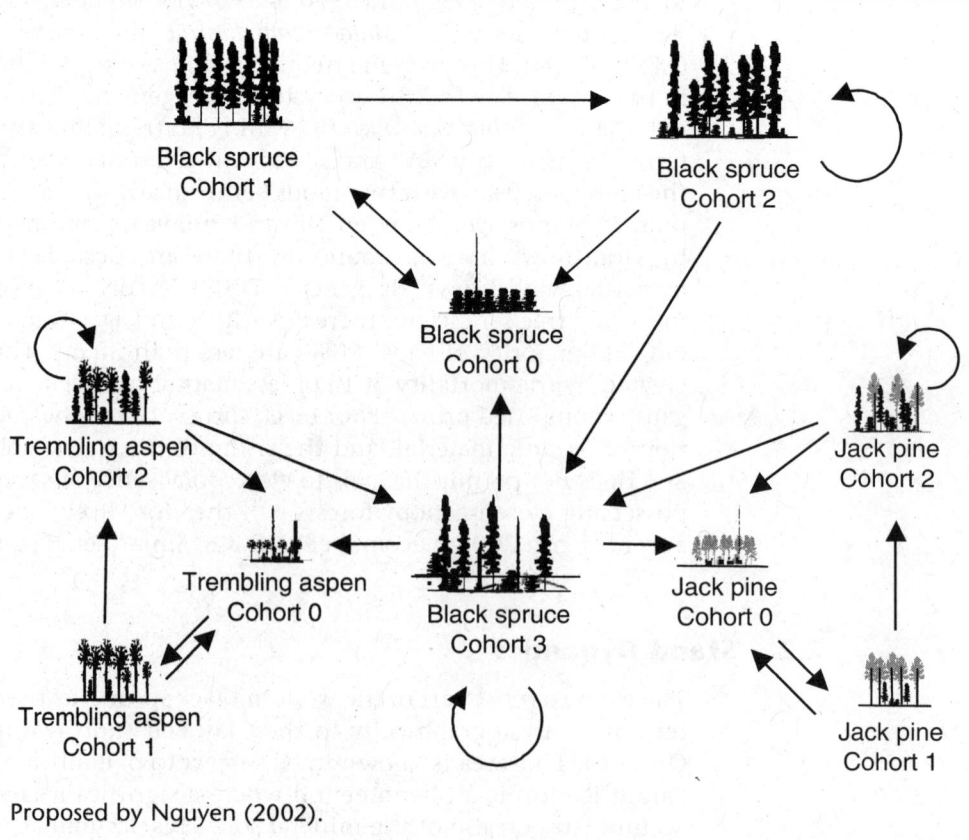

Proposed by Nguyen (2002).

with stands initiated by fire. The stands associated with this cohort are typically dense and closed with very little vertical structure. In the prolonged absence of fire, the stands of cohort 1 evolve into the stands of cohort 2, which are semi-open and possess a more irregular structure. During this developmental stage, shade-intolerant species (jack pine and trembling aspen) are gradually replaced by black spruce, which is shade tolerant and capable of reproducing by layering in the sub-canopy. Finally, in the prolonged absence of fire, gap dynamics become the main factor responsible for changes in forest structure. Gap dynamics allows for the complete replacement of the trees established during the first and second cohort stages by a third cohort, which is comprised primarily of black spruce. The stands of this developmental stage are relatively open with a well-developed vertical structure and uneven-aged structures. At sites susceptible to paludification, species succession and the arrival of the third cohort may be accelerated. In addition, if a fire is not sufficiently severe to consume the organic material to the mineral soil, the post-fire stands may develop characteristics typically found only in cohort 3, i.e. open stands with an uneven-aged structure (see Simard et al., chapter 11; Lecomte et al. 2006).

## 2.3. Fauna and Flora: Conservation Needs

A multitude of animal and vegetation species have been reported in the region targeted by the project. Some of these species are in a precarious state and are sensitive to forestry (Dallaire and Légaré 2006). As it is impossible to monitor the conditions of all of the species within the study area, four "focal" species have been selected for monitoring. Focal species are species that are both abundant within the study area and sensitive to anthropic disturbances (Lambeck 1997; Fleishman et al. 2001). The four targeted species are the woodland caribou (*Rangifer tarandus caribou*), the snowshoe hare (*Lepus americanus*), the American marten (*Martes americana*), and the Red-breasted Nuthatch (*Sitta canadensis*). These four species are particularly sensitive to anthropic disturbance, the simplification of both forest structure and composition, habitat fragmentation, and the disappearance of forest characteristics associated with mature and old-growth forests (see Drapeau et al., chapter 14).

The forest-dwelling woodland caribou, a species considered vulnerable in Québec, is found principally in the northern part of the study area. This species requires a home range that is sufficiently large for its migrations and that contains coniferous forests. The caribou is affected by the destruction and fragmentation of its habitat. In addition to being sensitive to anthropic disturbances, it is also sensitive to increases in the landscape's proportion of aspen-dominated stands, which favour predators such as the wolf (Courtois et al. 2002). The woodland caribou is currently the object of a Québec-wide reestablishment plan, and specific management plans have been developed for each sector in which it is found. For the sector corresponding with the study area, the management plan is based largely on an ecosystem management strategy and the preservation of habitat important to the caribou (rutting, calving, and summer feeding grounds).

The American marten is a predator species that requires large and healthy prey populations (field voles, snowshoe hare, etc.) in order to survive. The marten occupies the southern part of the study area. It is sensitive to forest fragmentation and prefers the interior of mature or old-growth conifer-dominated forests where it can find old trees, snags, and abundant woody debris on the forest floor (Potvin 1998; Potvin et al. 2000). These structural elements serve as habitat for birthing and feeding and as protective cover for both the marten and its prey. Although old forests may not necessarily possess the largest prey populations, it is within this type of habitat that the marten has the most predatory success (Alvarez 1996).

The snowshoe hare is considered a key species, as it is prey for a large number of predators including lynx, marten, wolve, fox, and the Great Horned Owl (Guay 1994; Brugerolle 2003). In order to survive, the hare requires a small area (less than 10 ha) with a mosaic of habitats that include variously aged coniferous cover and open environments. Summer habitat, where the hare feeds primarily on herbaceous plants, is rarely a limiting factor (Wolff 1980). Winter habitat is more constraining, as a sufficient number of branches, preferably of less than 3 mm in diameter, must be available above the snow cover (Guay 1994). The hare chooses its habitat more as a function of the environment's structure than of its species composition. A lateral obstruction of at least 40% between 1 and 3 m above the ground is, among others, a key factor for good shelter and abundant food (Guay 1994). In addition, the hare prefers coniferous stands of less than 12 m in height with a cover less than 60%, as this allows for dense vegetation to develop below the canopy. The edges between coniferous stands and logged areas also seem to be utilized for a distance up to 100 m from the cut area (Guay 1994). The reduction in lateral obstructions following logging also directly affects hare habitat. These obstructions are not found for 10 years after logging in mixed stands and for between 20 to 30 years in coniferous stands (Potvin et al. 2005).

The Red-breasted Nuthatch is a species usually associated with mature coniferous forests, particularly those susceptible to the spruce budworm (SBW, Marchand and Blanchette 1995). In the summer, the Red-breasted Nuthatch feeds primarily on insects found on tree trunks while in autumn it feeds on seeds found in the cones of balsam fir, spruce, and pine. It is a cavity-dwelling bird that requires snags or moribund trees for nesting. The trees chosen are generally >20 cm at breast height and between 2 and 30 m tall (12 m on average, Marchand and Blanchette 1995). The Red-breasted Nuthatch may be sensitive to fragmentation and its absence in cut areas is likely related to the scarcity of large trees and snags.

The main conservation goals for biodiversity within the project area are related to the maintenance of specific faunal habitat along with the conservation of old forests and a reduction in their fragmentation. Several strategies could be considered to address these goals, including the protection of habitats or the maintenance of old-growth forest attributes (snags and woody debris). Currently, the Québec government is working to set up a network of protected areas and biological refuges. These currently cover around 6% of the Forest Management Unit (FMU). The protected areas should ensure the long-term protection of certain habitats and their recruitment over time. However, as the proportion of the area being protected remains modest (<12%), the concept of maintaining large mature and old-growth forests zones should be integrated into our management strategy.

## 3. MANAGING THE FOREST WITH CLEAR OBJECTIVES

The acquisition of knowledge concerning the region allowed a pre-industrial snapshot (before 1970) of the state of the forest to be reconstructed. This snapshot was then used as a guideline for establishing targets that should be maintained or recreated through long-term forestry management. Furthermore, the snapshot was compared with the current state of the forest (table 19.1) in order to identify goals and management objectives (table 19.2). The principal goals related to the study area concern the maintenance of biodiversity, including old-growth forest and vertical structure (cohort), improving the distribution of disturbed areas (cuts, fires, windthrows), and the retention of structural elements in disturbed areas.

The general management strategy is inspired largely from the region's forest dynamics and disturbance regime and proposes management compromises that address not only environmental objectives, but also social and economic ones. Several methods were proposed to attain each identified goal and objective (table 19.2). For example, with regard to the maintenance of caribou habitat, the strategy proposes to limit the number of active logging areas in order to limit habitat fragmentation and to conserve the sites frequently visited by the woodland caribou (sites of interest include rutting, calving, and grounds). This mixed approach includes the coarse filter, which is the ecosystem management strategy that aims to keep the forests within their historical natural variability, as well as the fine filter, which relates to the identification and conservation of sites of interest. The retention of structural elements and residual forests within the cut areas is an ecological issue that became apparent when the quantity and distribution of structural elements and residual forests were compared with those of natural disturbances. The retention of structural elements and forest residuals also facilitates multiple uses including hunting, trapping, recreation, and forestry activities, in addition to improving the visual appearance of the cuts and serving as habitat refuges. Inside our logging areas, at least 25% of the area should be covered by residual forests in the form of corridors (500-m minimum width), peninsulas (500-m minimum width), and insular blocks (50 ha minimum). In addition to these residual forests, variable retention harvesting (e.g., clearcutting or careful logging treatments with retention of small patches or individual trees) should be conducted on at least 20% of the harvested area.

An evaluation of the success of attaining the objectives following the initiation of the region's ecosystem management plan will serve as a baseline for a follow-up monitoring system (see Drapeau et al., chapter 14). The monitoring system uses the adaptive management approach (Holling 1978), that will assess each strategy and method put forward to meet our objectives. When necessary, the methods and strategies may be revised to better respond to the objectives. In certain cases the objectives themselves may be revised in order to better align them with the identified management goals.

Table 19.1

## Comparison between the current forest state and the pre-industrial forest characteristics

| Characteristics | Categories | Natural variability | Current state |
|---|---|---|---|
| Developmental stage (corresponding to age for the spruce forests)[1] | No age class (very young) | 5% | 13% |
| | Regeneration (20 years or less) | 15% | 18% |
| | Young (up to 80 years and young uneven-aged stand) | 23% | 41% |
| | Mature and old-growth (81 years or more and old uneven-aged stand) | 57% | 28% |
| | *(More than 200 years)* | *(20%)* | |
| Structure[2] | Even (cohort 1) | 62% | 72% |
| | Irregular (cohort 2) | 21% | 16% |
| | Uneven open (cohort 3 or more) | 17% | 12% |
| Composition[1] | Coniferous | 83% | 76% |
| | *(black spruce forests)* | *(71%)* | *(68%)* |
| | Mixed | 12% | 16% |
| | Deciduous | 5% | 8% |
| Mosaic[3] | Disturbance size (fires) | 900 to 20,000 ha (90% < 1,000 ha) Fires with size greater than 1,000 ha accounted for the majority of the burned area | Area affected – 10–10,000 ha – (~98% < 2,000 ha) |
| | Distribution of disturbances (fires) | Random, the distance between fires is variable Minimum distance between the disturbed areas: 6 km | Minimum distance between the disturbed areas: 0.3 km |
| Severity[3] | Residual forest | 10 to 35% with 1 to 8% in the form of islands of 1 to 3 ha | Temporary residual forests within the cuts: 19–62% – Riparian buffers |
| | Mortality within the disturbance | Total mortality: 45 to 65% Partial mortality: 30 to 50% | – Cutblock separators – Forest buffer along road borders – Inaccessible or non-commercial stands |
| Frequency of disturbance[3] | Fires | Variable over the last 300 years (from the fire cycle of 1850–1920: estimated at 135 years) Mean forest age: 148 years | Current fire cycle about 398 yr Mean harvest rotation cycle of 80 years |
| | Gap disturbances | Not documented | Mean age target in regard to a forest full even-aged exploitation: 40–45 years |

1. Data derived from the 1st (1971) and 3rd (1994) 10-year provincial forest inventories, Ministère des Ressources naturelles et de la Faune, Direction des inventaires. The 1st 10-year period served to estimate the natural variability (given the low exploitation of the region during this time period). The current state of the region was derived from the 3rd inventory.
2. The natural variability was estimated using the age distribution and theoretical thresholds of forest with an average age of 150 years (Nguyen 2000); the current values were estimated using inventory plots from the 3rd inventory with their cohort designation being based on the structure of the forest.
3. The values used to characterize the natural variability were taken from Leduc et al. 2000, Bergeron et al. 2002, 2004, Perron 2003 and Belleau et al. 2007. The current states were characterized with the aid of spatial analysis tools (McGarigal and Marks 1995), manually using ArcView 3.3, and with the data derived from the 3rd inventory.

Table 19.2
## Management issues and objectives defined for FMU 085-51

| Issues | Objectives | Strategies proposed to address the objectives |
|---|---|---|
| The maintenance of old-growth forests | • Maintain between 28.5 to 38% of old-growth forests (FSC 2004)<br><br>Definition of old-growth forest: "a forest that has attained or passed its maximum silvicultural rotation time." | 1. Establish protected areas and respect the various designations assigned with the area (ex: moose containment area, fragmented forests).<br>2. Fully implement the MRNF strategy of maintaining old-growth forests as stated in the *objectifs de protection et de mise en valeur des ressources du milieu forestier* (OPMV) (MRNF 2005):<br>– create biological refuges on 2% of the area;<br>– implement the use of adapted silvicultural practices on 7% of the productive land area;<br>– establish aging patches on 10% of the productive land area.<br>3. Recognize the old-growth forest components contained within:<br>– riparian buffers both eligible and ineligible for partial cutting;<br>– stands less than 50 m³/ha that are ineligible for harvesting;<br>– forests ineligible for harvesting. |
| | • Return to the estimated historical cohort proportions established with a 150-year fire cycle (cohort 1: 62%; cohort 2: 21%; cohort 3: 17%) (Nguyen 2000)<br><br>Definition of cohort: "cohort 1 includes even-aged stands created by a severe disturbance; cohort 2 includes a regular stands created by a partial disturbance; cohort 3 includes uneven-aged stands created by several partial disturbances along with the gradual opening caused by gap formation." | 1. Contribution of strategies envisioned for the maintenance of old-growth forests.<br>2. Ensure a better distribution of cohorts through time by using partial cut treatments that allow for a quantity of wood to be removed, but also allows for an extension of the cutting cycle for productive stands.<br>3. Identify advanced paludified stands to cohorts 2 and 3 (based on the work of Simard et al. 2007) These stands are considered unproductive by the MRNF and therefore should not be exploited.<br>4. Over the long term, cohort 2 and 3 stands will be created through:<br>– the use of silviculture treatments that structure the stands in a manner to accelerate their passage towards cohorts 2 and 3;<br>– the contribution of long-term strategies that maintain old forests in which harvesting is not possible or predicted;<br>– the contribution of old burned areas with low post-fire productivity that will never be harvested. |
| Improvement of the distribution pattern of logging areas | • Recreate the natural distribution pattern of regeneration areas over the long term, as proposed by Belleau et al. (2007) | 1. Increasing the size of the logging areas in order to create large regeneration areas (in the order of 1,000 to 50,000 ha, including unproductive areas and a certain proportion of residual forest).<br>2. Maximise the space between logging clusters (6 to 12 km) in order to favour the conservation of large undisturbed forest.<br>3. Vary the dimensions and shapes of logging areas. |
| The retention of structural elements in logging areas | • Maintain from 10 to 50% of residual forest in each logging areas throughout the cutting cycle (Leduc et al. 2000; Perron 2003; FSC 2004) | 1. Implement clearcutting or careful logging (CPRS) on 45 to 65% of the logging area.<br>2. Maintain 5% of residual forest within 5% of the cutblocks using:<br>– CPRS with the retention of small patches 200 to 400 m² (OPMV);<br>– retention of individual trees (25 stems/ha).<br>3. Maintain individual standing living trees on 30 to 50% of the logging areas with the aid of:<br>– residual forest blocks;<br>– forest peninsulas;<br>– variable retention cuts (cut with protection of small merchantable stems (CPPTM), diameter limit harvesting, and harvesting with protection of advance regeneration).<br>4. Preserve snags and trees valuable to fauna within the interior of each retained residual forest proposed. |

**Table 19.2** *(continued)*
## Management issues and objectives defined for FMU 085-51

| Issues | Objectives | Strategies proposed to address the objectives |
|---|---|---|
| The retention of structural elements in logging areas *(continued)* | • Ensure an economic compromise between residual forest preservation and wood production needs | 1. Use of both permanent and non-permanent retentions:<br>  – *Permanent retention*: diameter limit harvesting, harvesting with small patch retention, harvest with advance regeneration protection, cut with protection of small merchantable stems, preservation of intact or partially cut riparian buffers and the use of adaptive silviculture practices. Average approximate contribution of 19%.<br>  – *Temporary retention (10, 15, or 20 years based on the forest composition)*: implementing peninsulas (10%), insular blocks (10%), and aging patches (harvesting delay period: rotation time +15 years). Average approximate contribution of 20%. |
| Maintaince of recently disturbed habitats | • Conserve between 10 to 50% of forests recently affected by natural disturbances (Leduc et al. 2000; Perron 2003) | 1. At the time of salvage logging, preserve around 30% of each forest type (age × composition) that was present before the disturbance.<br>2. Preserve all intact riparian buffers. |
| Protection of riparian environments | • Limit the erosion of sediment into water bodies and the loss of aquatic habitats by maintaining at all times a 20-m-wide riparian buffer along all permanent water bodies (RNI) | 1. Maintain at all times a minimum of 20-m-wide unharvested riparian buffer or a maximum harvest of one third of the stems for at most one third of the buffer (around 80% of the riparian buffers in the FMU are theoretically unharvestable because of a low stem density).<br>2. At important sites, an additional 45 m or more should be applied. These sites may be important for the conservation of habitat or cultural values. |
| Preserve forest of high conservation value | • Ensure, using specific intervention methods or protection methods to maintain the identified conservation values | 1. Apply the old-growth forest conservation strategy of the MRNF, i.e. the identification of biological refuges and temporary conservation of aging patches along with the implementation of adaptive silviculture practices.<br>2. Apply the habitat protection guidelines for threatened species and vulnerable forest environments of the MRNF by applying the MRNF's prescribed protection methods for the identified habitats and by integrating within the general forest management plan specific management plans for threatened or vulnerable species with large home ranges, such as the woodland caribou.<br>3. Apply MRNF's guidelines that target the maintenance of the quality of the landscape and that aim to respect public and First Nation needs by identifying sites of interest and sensitive landscapes, consulting other users of the study area, and signing First Nations harmonization agreements with regard to the methods and modalities applied. |
| Conservation of woodland caribou | • Maintain woodland caribou habitat and ensure its renewal through time for the northern section of the forest management unit | 1. Integrate the measures agreed upon in the caribou management plan into the FMU's management plan including the application of ecosystem management on the majority of the unit, the establishment of protected areas, and the protection of sectors of interest for the caribou (areas used for rutting, birthing, and summer feeding).<br>2. Use partial cutting at the edges and within the spruce lichen woodlands in order to favour their expansion.<br>3. Agglomerate harvest activities to limit fragmentation of the habitat and to concentrate disturbance within a single sector. |
| Conservation of the American marten | • Maintain American marten habitat and ensure its renewal through time for the southern section of the forest management unit | 1. Apply the old-growth forest conservation strategy of the MRNF, i.e. the identification of biological refuges and temporary conservation of aging patches along with the implementation of adaptive silviculture practices.<br>2. Favour the creation of interior habitats by increasing measures to maintain old forests.<br>3. Agglomerate harvest activities in order to limit habitat fragmentation. |

Table 19.2 (continued)
**Management issues and objectives defined for FMU 085-51**

| Issues | Objectives | Strategies proposed or envisioned that address the objectives |
|---|---|---|
| Forestry roads planning | • Limit habitat fragmentation by better managing the development of the road network | 1. Implement a complete access road management plan:<br>  – avoid the construction of roads within or near protected areas;<br>  – define maintenance and abandonment strategies for all road types;<br>  – preserve the isolation of sensitive cultural or biological zones;<br>  – aim for a just and equitable balance between landscape isolation for ecological, social, and economic issues and the desire to open access to vehicular traffic for recreational and operational purposes. |
| Forest productivity | • Minimize paludification after intervention | 1. In paludified stands, limit the use of partial cuts, promote the passage of machinery throughout the logging area or the use of a heavy ground preparation following by planting. The prescribed burning of organic material could also be envisioned as a ground preparation.<br>2. Limit the opening of the forest cover at 30% at the time of partial cutting. |
|  | • Increase the production of wood fibre | 1. Develop a silvicultural intensification strategy that, among other factors:<br>  – accommodates a maximum of 5% of the territory with rapid-growth, high-yield species;<br>  – ensures the full reforestation of disturbed zones and restores roads and slash pile areas into productive forest;<br>  – ensures the close monitoring and silvicultural education of regenerating stands. |

## 4. PRELIMINARY TRIALS BEFORE FULL IMPLEMENTATION

Until now, little work has been conducted in Québec regarding the initiation and operation of an ecosystem management plan or possible problems pertaining to its implementation (Brais et al. 2004; <www.projettriade.ca>, <www.fm.ulaval.ca>). From a theoretical standpoint, the approach has become more refined. However, several points have still not found concrete applications. The broad strategic outlines for ecosystem management listed in table 19.2 correspond with those that Tembec wishes to apply over the long term in FMU 085-51. However, several elements of the strategy must first be tested experimentally in order to evaluate their impacts on production costs, planning and management efficiency, operational feasibility, and above all, environmental performance. Since 2007, two strategies, among others, have been the subject of experimentation before their large-scale application: 1) the implementation of a logging area trial where part of the harvest treatments aimed at maintaining residual forests and 2) trials using adapted silvicultural practices that aim to conserve the forest cover and attributes of old-growth forests. Furthermore, other trials will be conducted in the region using partial cut treatments that aim to accelerate the succession of stands towards cohorts 2 and 3, in addition to trials that have already been completed within a partial cut research framework (see Fenton et al., chapter 15). For each trial conducted in the area, a direct follow-up with regard to the costs, outputs, and environmental impacts will be made. The achievement of the initial objectives will also be evaluated and the strategies adjusted so that the global strategy more fully addresses the landscape management issues of the FMU.

The natural disturbance-based management trials will be conducted in the eastern-central part of FMU 085-51 in the heart of the Rainboth Township. The area targeted for 2007–2008 occupies around 900 ha, including unproductive stands and non-forest areas (water). The field trial is located within the trapping grounds of the Cananasso family, who actively trap and hunt in the area. The Plamondon River is located at the southern limit of the 2007–2008 trial site and is widely used as a canoe route by the region's First Nation peoples. In this sector, the forest composition is dominated, in order of importance, by mixed stands of spruce and jack pine, spruce and deciduous trees, and deciduous trees. The majority of stands are aged between 70 and 120 years old. Surficial deposits are primarily composed of imperfect to poorly drained glaciolacustrine clays and poorly drained organic soil.

• The CPRS is a careful logging treatment commonly used in Québec, The CPRS is analogous to the CLAAG (careful logging around advance regeneration) used in Ontario.

The 2007–2008 trials (figure 19.3) plan around 400 ha of cuts with protection of regeneration and soils (CPRS)• with 133 ha with the retention of small patches (patches of 200 to 400 m²) and around 42 ha with the retention of individual standing trees (25 stems/ha). The remaining areas are organized into permanent and temporarily retained blocks and peninsulas. After harvesting, and without taking into consideration additional measures that may be imposed

## Figure 19.3
## Natural disturbance-based management field trial 2007–2008, located in Rainboth Township, Abitibi, Québec

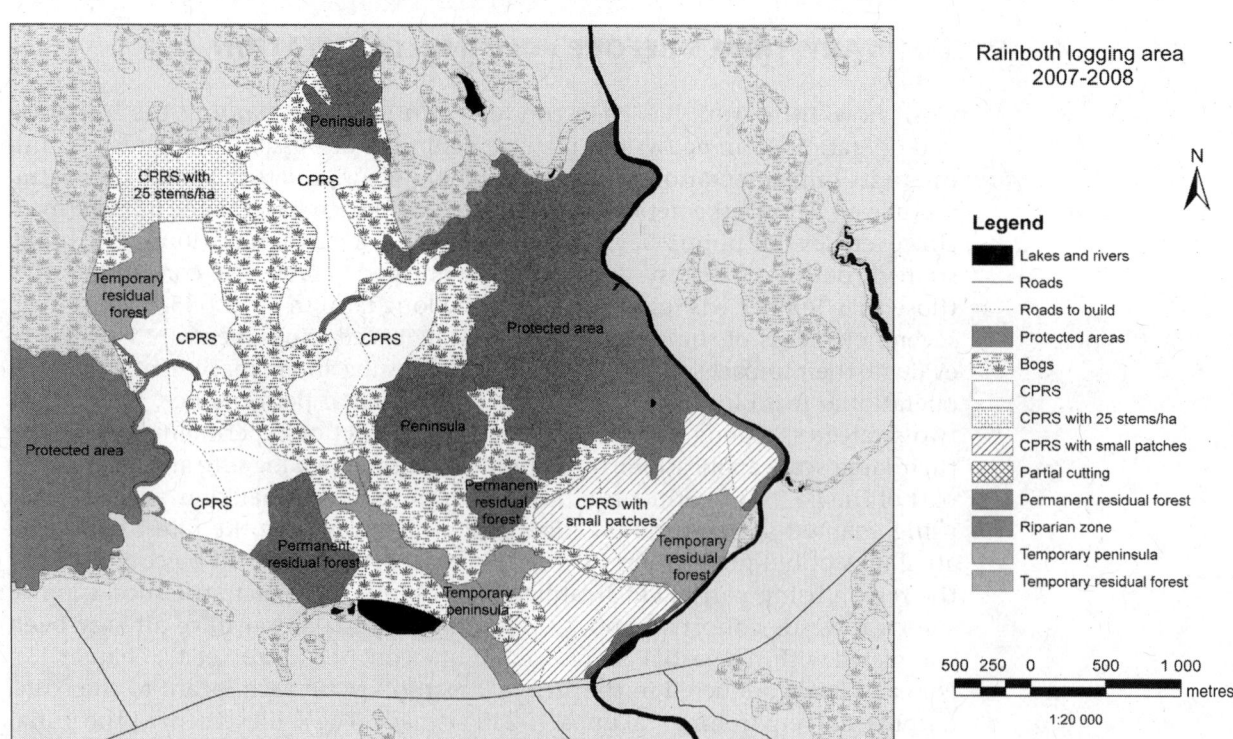

Source: From the 3rd inventory program, map sheet 32E08NE.

by the MRNF to achieve First Nation needs, a retention with around 37% of residual forest is envisioned for 2007–2008. The anticipated retention is representative of the pre-disturbance forest and is divided into two classes, with 22% being organized into permanent retention and 15% into temporary retentions (for a 10-to-20-year period, based on the nature of the stands). Areas designated for temporary retention will be located in accessible areas near roads in order to limit damage to advance regeneration at the time of return for the final cut. Using both permanent and temporary retentions is a compromise measure that aims to minimize the impact on the annual allowable cut and to ensure a certain economic profitability to the industry, while still maintaining for a time a quantity of high-quality faunal habitat.

## 5. THE CHALLENGE OF APPLYING LARGE-SCALE ECOSYSTEM MANAGEMENT GIVEN THE CURRENT LANDSCAPE

Over the long term, Tembec would like to distribute its cutblocks throughout the landscape following a pattern inspired by those created by natural disturbances. Using simulations, Belleau et al. (2007), proposed a minimum average spacing and size distribution of clustered logging areas under different fire regimes. The proposed distribution is based on the realistic premise that fire events in the region occur randomly and that the stands burn independently of their age. Table 19.3 summarizes, for the fire regime in the study area, the dispersion and distribution characteristics proposed for the proportion of forests managed under an even-aged system. The proposed logging areas are continuous surfaces that include the presence of unproductive stands and residual forests. For each logging area, Belleau et al. (2007) recommend retaining a minimal constraining space of 6 km up until the time when the average height of the stands within the logging area reach 4 m. At this point the regenerated forest should become functional for fauna that require forest cover (Jacqmain 2003; Imbeau et al. 1999).

**Table 19.3**
**Desired characteristics for the size distribution and dispersion of logging areas with regard to the forest proportion managed under an even-aged system[1]**

| Size (ha) of logging areas | Mean number (max.) of logging areas over 25 years | Minimum mean spacing (km) between logging areas based on their size (absolute min. value) |
|---|---|---|
| 0–2,000 | 1 (2) | 39.3 (21.8) |
| 2,000–5,000 | 1 (2) | |
| 5,000–10,000 | 1 (2) | 50.4 (24.5) |
| 10,000–20,000 | 1 (3) | 12.4 (7.2) |
| 20,000–50,000 | 3 (3) | |
| All sizes | | 8.6 (6.1) |

1. Under the regional fire regime, 48.4% of the forest should be at least 100 years old and managed using even-aged systems.

Using updated georeferenced ecoforestery maps of the study area (including harvests conducted up until March 2006), we evaluated at what point the current landscape differs from both the desired landscape, particularly the targeted forest retention percentage (10 to 50%) in logging areas. To achieve this, we used ArcGis 9.1 to merge all of the regeneration areas of less than 4 m in height and located at least 1,000 m from each other (Perron 2003). We then evaluated the size distribution of these harvest clusters and the distance to the closest neighbour (border to border) according to size class (McGarigal and Marks 1995). The portions of residual productive and unproductive forest within the clusters were also evaluated. Table 19.4 displays the results of the analysis. We observed that, contrary to what was desired (table 19.3), the study site contains a large number of small clusters (less than 100 ha) and, at the other extreme, one very large cluster (630,449 ha) of harvests (including residual and unproductive forests). In addition, the minimal spacing is on average 3 to 10 times smaller than the theoretical spacing of the natural-disturbance regime. Consequently, the forest matrix between logging areas is becoming more restricted and fragmented.

It is important to avoid further fragmentation of the forest matrix that is located between the already existing logging areas until the regenerating forests attain heights and densities that permit their utilization by fauna. For the moment, we wish to limit the creation of new logging areas within the undisturbed and minorly disturbed zones. Increasing the size of already existing logging areas or agglomerating logging areas is recommended. The increased use of partial cut treatments would allow for the expedited exploitation of certain types of stands within the forest matrix while still allowing habitat and forest cover to be maintained. The current lack of tools allowing for the tactical and operational simulation of various silvicultural treatments limits our capacity to optimally plan the logging areas.

**Table 19.4**
**Characteristics of the disturbed agglomerations already present in the FMU 085-51[1]**

| Size (ha) of disturbed areas | Number of disturbed areas | Percentage (%) of retention within disturbed areas (excluding unproductive stands) | Minimum mean spacing (km) between disturbed areas based on their size |
|---|---|---|---|
| 0–2,000 | 83[2] | 43 (34) | 2.6 |
| 2,000–5,000 | 2 | 88 (37) | |
| 5,000–10,000 | 0 | N/A | N/A |
| 10,000–20,000 | 0 | N/A | N/A |
| 20,000–50,000 | 0 | N/A | |
| >50,000 | 1 | 84 (35) | |
| All sizes | | | 1.5 |

1.  Disturbed areas less than 1,000 m from each other were clustered.
2.  68 disturbed areas were less than 100 ha in size.

## 5.1.  Regulation Challenges

In order to apply the ecosystem management strategies described above within the study area, requests for the exemption of several existing regulations must be made under section 25.3 of the *Forest Act* (Gouvernement du Québec 2007a). The project requires the exemption, among others, of sections 47, 74, 75, 79.1, 79.2, 79.3, 79.4, 79.5, 79.6, 79.7, 79.8, 80, 87, 88, 89, and 95 of the *Regulation respecting standards of forest management for forests in the domain of the State* (RNI) (Gouvernement du Québec 2007b). Generally, these sections specify the size of area that should be cut in a single block, the utilization of a strategy distributing cuts in a mosaic pattern, the use of linear separators between harvested areas, the harvesting of all trees possessing a commercial diameter if they are healthy and dry, the utilization of cutting treatments that assure the protection of soils and regeneration, and the prohibition of operations within spruce-lichen woodlands. The area of cut patches permitted provides a concrete example where an exemption will be necessary. Currently, regulations limit the harvested area to 150 ha within the boreal forest and require a linear separator 60 to 100 metres wide between each cut block. However, in the field trial design (figure 19.3), harvest parcels can reach sizes greater than 150 ha with separators taking the form of either clusters of retained blocks or peninsulas. For each regulation exemption request, Tembec must propose a replacement method, an explanation of the circumstances justifying the exemption, and the objectives targeted by the replacement method. The objectives proposed here will be used as a base by the MRNF to evaluate the new methods envisioned in the plan that addresses the landscape management issues.

Beyond the exemption requests concerning the forestry management regulations, no study or simulated scenario allows for the exact prediction of the impacts that forest ecosystem management will have on wood production in the region managed by Tembec. Several factors contribute to this fact: 1) many uncertainties remain regarding the application of ecosystem management theory and the response of stands to the different silvicultural treatments; 2) few simulation tools take into consideration many aspects of ecosystem management (spatial and temporal resolution, planning at several scales, social and economic constraints, etc.) or operational constraints currently exist; and 3) few experiments document the effectiveness of the mitigation measures, including intensive management. Many of these uncertainties could be resolved in the short term. A number of tests initiated several years ago at the Lake Duparquet research and teaching forest (Lake Duparquet Forest, LDF) are now ready to deliver information with regard to the response of stands to certain treatments (see Fenton et al., chapter 15; Harvey et al., chapter 18). Combined with the development of new modelling and monitoring tools (a current research project being conjointly conducted with UQAT), the continuing acquisition of knowledge concerning stand dynamics and their productivity (see Simard et al., chapter 11), and the development of new silvicultural treatments, Tembec endeavours to document a general ecosystem management plan and implement concrete applications by the year 2013.

## 5.2. Towards Meeting Our Objectives with Adaptive Management and Monitoring Systems

Given that the ecosystem management approach is experimental, it is important to have an adaptive management approach and to improve the system that monitors the environmental impacts of forest management on the different values within the study area. The disposal of dangerous residuals and impacts related to the quality of air, water, soil, and species are relatively well protected by the environmental regulations listed in Québec's *Environment Quality Act* (Gouvernement du Québec 2007c) and the RNI. In addition, Tembec's current management system conforms to the standards of ISO 14 000 (CSA 2005), which ensure that these regulations, as well as procedures for harvest and haulage, are followed. However, biological diversity is only slightly considered in these procedures. While the provincial government has recently imposed best management practices guidelines (*Objectifs de protection et de mise en valeur des ressources du milieu forestier* [OPMV]) (MRNF 2005) on the forest industry, which serve as a basis for the evaluation of management practices in order to ensure sustainable forest resources, there are still no concrete follow-up measures in the field nor are there any established monitoring indicators. The maintenance of biodiversity is an important goal, but is difficult to evaluate. The identification of indicators and focal species should be based on scientific knowledge that permits the evaluation of our impacts in order to readjust and minimize our interventions on the landscape. To do this, a number of significant indicators that are economically feasible to monitor over the long term should be developed. Given the complexity of the task and its importance in achieving sustainable forest management, a rigorous monitoring system should be developed by integrating information coming from, among others, civil organizations (hunting and fishing clubs, birding clubs), government ministries (MRNF, Ministère du Développement durable, de l'Environnement et des Parcs [Ministry of Sustainable Development, Environment, and Parks]), as well as university research centres (Centre for Forest Research and the Industrial Chair NSERC-UQAT-UQAM in Sustainable Forest Management).

The perception of the new management strategy by the public and First Nations should also be assessed. In this regard, some indicators have already been put in place by the Forever Green / Impact Zero[*] program, but an official analysis framework should be established. A functional analysis framework would link the analysis of the indicators with the preparation of different plans and reports. Currently, only the five-year forest management plan requires a public consultation. However, as this program is, along with all of the landscape management strategies, detailed within the region's general management plan, the integration of environmental indicators at the landscape level (including economic, social, and First Nation indicators) during the preparation phase of the general plan would improve the adaptive management strategy. Environmental indicators corresponding to the scale of the disturbance could be analysed during the preparation phases of the annual plan and annual report. Furthermore, collaboration between universities, government researchers, and Tembec is essential to long-term adaptive management. This collaboration would make it possible to evaluate the relevance of the monitoring system's indicators with new knowledge and issues set forth by the researchers on a year-to-year basis.

[*] For more information visit <www.tembec.com>.

## 6.  CONCLUSION

Although the theoretical concepts are well integrated within the general forest management plan (including the 2008–2013 plan) and certain trials have already been implemented in 2007, much work still remains to be done at the operational level to fully implement the ecosystem management approach. In fact, the integration of natural stand dynamics and ecological types has only been touched upon since these discussions, as the silvicultural prescriptions were not written into the general management plan. In order to better intervene within the undisturbed forest matrix and limit the increase of clear-cut logging areas, silvicultural practices at the stand level should be developed and implemented in the 2008–2013 annual plans. In these plans, we envision a silviculture that applies the right treatment at the right location and that responds not only to wood fibre production objectives, but also to the conservation of faunal habitats and other forest resources. Likewise, with the aim of reducing pressure on large tracts of intact forest and to facilitate both the implementation of the strategy on the majority of the region's area and the creation of conservation areas, an intensification strategy should be envisioned. This strategy, applicable within productive sectors and near mills, would ensure the optimal regeneration of stands and maintain productivity throughout full and partial cutting cycles. The use of heavy ground preparation, education regarding stand techniques (spacing and thinning), and fertilization could also be used. The feasibility of implementing a silviculture with improved, rapid-growth tree species on a small portion of the territory should also be evaluated to further support the intensification strategy.

Finally, the intensification and diversification of management strategies require substantial investments from the forest industry, which are sometimes difficult to get due to the mode of land tenure that offers few guarantees to investors concerning the return on their investments. An increased sharing of environmental responsibilities along with better guarantees or greater investment on the part of the government would greatly facilitate the development of an economically, socially, and ecologically sustainable forestry. Finally, a more integrated management that includes all facets of the forest, along with the development of long-term planning tools, is required to ensure the full passage of theoretical knowledge to practical application.

## REFERENCES

Alvarez, E. 1996. La forêt mosaïque: une alternative d'aménagement pour le maintien de la martre dans la sapinière boréale ? M. Sc. thesis, Université Laval, Québec, Que.

Belleau, A., Bergeron, Y., Leduc, A., Gauthier, S., and Fall, A. 2007. Using spatially explicit simulations to explore size distribution and spacing of regenerating areas produced by wildfires: recommendations for designing harvest agglomerations for the Canadian boreal forest. For. Chron. **83**: 72–83.

Bergeron, Y., Gauthier, S., Flannigan, M., and Kafka, V. 2004. Fire regimes at the transition between mixed-woods and coniferous boreal forest in northwestern Quebec. Ecology, **85**: 1916–1932.

Bergeron, Y., Harvey, B., Leduc, A., and Gauthier, S. 1999. Forest management guidelines based on natural disturbance dynamics: stand- and forest-level considerations. For. Chron. **75**: 49–54.

Bergeron, Y., Leduc, A., Harvey, B.D., and Gauthier, S. 2002. Natural fire regime: a guide for sustainable management of the Canadian boreal forest. Silva Fenn. **36**: 81–95.

Brais, S., Harvey, B.D., Bergeron, Y., Messier, C., Greene, D., Belleau, A., and Paré, D. 2004. Testing forest ecosystem management in boreal mixedwoods of northwestern Quebec: initial response of aspen stands to different levels of harvesting. Can. J. For. Res. **34**: 431–446.

Brugerolle, S. 2003. Caractérisation de l'habitat du lièvre d'Amérique à différentes échelles spatiales: une étude en forêt mélangée. M.Sc. thesis, Université Laval, Québec, Que.

Canadian Standards Association (CSA). 2005. Plus 14 000. The ISO 14000 Essentials – A Practical Guide to Implementing the ISO 14000 Standards. 2nd edition. Canadian Standard Association, Mississauga, Ont.

Courtois, R., Ouellet, J.-P., de Bellefeuille, S., Dussault, C., and Gingras, A. 2002. Lignes directrices pour l'aménagement forestier en regard du caribou forestier. Société de la faune et des parcs du Québec, Direction de la recherche sur la faune, Université du Québec à Rimouski. Rimouski, Que.

Dallaire, S. and Légaré, S. 2006. Évaluation de la présence de forêts à haute valeur pour la conservation telles que définies par le principe 9 du Forest Stewardship Council. Unités d'aménagement forestier 85-51 et 85-62. Gestion des ressources forestières, Abitibi. Version 3.

Environment Canada. 2006. Canadian Climate Normals 1971–2000. [Online] <www.climate.weatheroffice. ec.gc.ca./climate_normals/index_e.html> (accessed November 7, 2007).

Fenton, N., Légaré, S., Bergeron, Y., and Paré, D. 2006. Soil oxygen within boreal forests across an age gradient. Can. J. Soil Sc. **86**: 1–9.

Fleishman, E., Murphy, D.D., and Blair, R.B. 2001. Selecting effective umbrella species. Conserv. Biol. Pract. **2**: 17–23.

Forêt modèle du Bas-Saint-Laurent. 2002. Indices de qualité d'habitat. Extension ArcView, version 2.0. Réseau de forêts modèles, Université du Québec à Rimouski, Rimouski, Que.

Forest Stewardship Council (FSC). 2004. National Boreal Standard. Version of August 6, 2004. Forest Stewardship Council, Canada Working Group, Toronto, Ont.

Gauthier, S., Nguyen, T., Bergeron, Y., Leduc, A., and Drapeau, P. 2004. Developing forest management strategies based on fire regimes in northwestern Quebec. *In* Emulating natural forest landscape disturbances: Concepts and applications. *Edited by* A.H. Perera, L.J. Buse, and M.G. Weber. Columbia University Press, New York, USA, pp. 219–229.

Gouvernement du Québec. 2007a. Forest Act. R.S.Q., c. F-4.1. Éditeur officiel du Québec, Que.

Gouvernement du Québec. 2007b. Regulation respecting standards of forest management for forests in the domain of the State. R.S.Q., c. F-4.1, r.1.001. Replaced, D. 498-96, 1996 G.O. 2, 2570; eff. 96-05-23; see c. F-4.1, r. 1.001.1. Éditeur officiel du Québec, Que.

Gouvernement du Québec. 2007c. Environment Quality Act. R.S.Q., c. Q-2, Éditeur officiel du Québec, Que.

Guay, S. 1994. Modèle d'indice de qualité d'habitat pour le lièvre d'Amérique (*Lepus americanus*) au Québec. Gouvernement du Québec, Ministère des Ressources naturelles, Ministère de l'Environnement et de la Faune, Gestion intégrée des ressources, technical document 93/6.

Harper, K.A., Bergeron, Y., Drapeau, P., Gauthier, S., and De Grandpré, L. 2005. Structural development following fire in black spruce boreal forest. For. Ecol. Manag. **206**: 293–306.

Harper, K.A., Bergeron, Y., Gauthier, S., and Drapeau, P. 2002. Post-fire development of canopy structure and composition in black spruce forest of Abitibi, Quebec: a landscape scale study. Silva Fenn. **36**: 246–263.

Holling, C.S. 1978. Adaptive environmental assessment and management. John Wiley and Sons, New York, USA.

Hunter Jr, M.L. (Ed.). 1999. Maintaining biodiversity in forest ecosystems. Cambridge University Press, Cambridge, UK.

Imbeau, L., Savard, J.-P.L., and Gagnon, R. 1999. Comparing bird assemblages in successional black spruce stands originating from fire and logging. Can. J. Zool. **77**: 1850–1860.

Jacqmain, H. 2003. Rabbit Habitat Project: Analyse biologique et autochtone de la restauration de l'habitat du lièvre d'Amérique après coupe sur la terre des Cris de Waswanipi. Unpublished M.Sc. thesis, Université Laval, Québec, Que.

Jenkins, S.H. 1980. A size-distance relation in food selection by beavers. Ecology, **61**: 740–746.

Lambeck, R.J. 1997. Focal species: a multi-species umbrella for nature conservation. Conserv. Biol. **11**: 849–857.

Lecomte, N. and Bergeron, Y. 2005. Successional pathways on different surficial deposits in the coniferous boreal forest of the Quebec Clay Belt. Can. J. For. Res. **35**: 1984–1995.

Lecomte, N., Simard, M., and Bergeron, Y. 2006. Effects of fire severity and initial tree composition on stand structural development in the coniferous boreal forest of northwestern Quebec, Canada. Écoscience, **13**: 152–163.

Leduc, A., Bergeron, Y., Drapeau, P., Harvey, B., and Gauthier, S. 2000. Le régime naturel des incendies forestiers: un guide pour l'aménagement durable de la forêt boréale. L'Aubelle, novembre-décembre: 13–22.

Marchand, S. and Blanchette, P. 1995. Modèle d'indice de qualité de l'habitat pour la Sitelle à poitrine rousse (*Sitta canadensis*) au Québec. Gouvernement du Québec, Ministère de l'Environnement et de la Faune, Direction générale de la ressource faunique et des parcs, Gestion intégrée des ressources, technical document 92/6.

McGarigal, K. and Marks, B.J. 1995. FRAGSTATS: spatial pattern analysis program for quantifying landscape structure. USDA Forest Service Gen. Tech. Rep. PNW-351.

Morin, H., Laprise, D., and Bergeron, Y. 1993. Chronology of spruce budworm outbreaks near Lake Duparquet, Abitibi Region, Québec. Can. J. For. Res. **23**: 1497–1506.

Ministère des Ressources naturelles et de la Faune (MNRF). 2003. Historique des épidémies d'insectes en milieu forestier depuis 1938. Ministère des Ressources naturelles, de la Faune et des Parcs du Québec. [Online] <www.mrn.gouv.qc.ca/forets/fimaq/insectes/fimaq-insectes-histoire.jsp> (accessed November 7, 2007).

Ministère des Ressources naturelles, de la Faune et des Parcs (MNRFP). 2005. Objectifs de protection et de mise en valeur des ressources du milieu forestier. Plans généraux d'aménagement forestier 2007–2012. Document de mise en œuvre. Ministère des Ressources naturelles, de la Faune et des Parcs.

Naiman, R.J., Melillo, J.M., and Hobbie, J.E. 1986. Ecosystem alteration of boreal forest streams by beaver (*Castor canadensis*). Ecology, **67**: 1254–1269.

Nguyen, T. 2000. Développement d'une stratégie d'aménagement forestier s'inspirant de la dynamique des perturbations naturelles pour la région Nord de l'Abitibi. Rapport de recherche effectuée dans le cadre du Volet I du programme de mise en valeur des ressources du milieu forestier. Université du Québec en Abitibi-Témiscamingue, Rouyn-Noranda, Que.

Nguyen, T. 2002. Développement d'une stratégie d'aménagement forestier s'inspirant de la dynamique des perturbations naturelles pour la région Nord de l'Abitibi. Rapport de recherche effectuée dans le cadre du Volet I du programme de mise en valeur des ressources du milieu forestier. Université du Québec en Abitibi-Témiscamingue, Rouyn-Noranda, Que.

Perron, N. 2003. Peut-on et doit-on s'inspirer de la variabilité naturelle des feux pour élaborer une stratégie écosystémique de répartition des coupes à l'échelle du paysage? Le cas de la pessière noire à mousses de l'ouest au Lac-Saint-Jean. Ph.D. thesis, Université Laval, Québec, Que.

Potvin, F. 1998. La martre d'Amérique (*Martes americana*) et la coupe à blanc en forêt boréale: une approche télémétrique et géomatique. Ph.D. thesis, Université Laval, Québec, Que.

Potvin, F., Bélanger, L., and Lowell, K. 2000. Marten habitat selection in a clearcut boreal landscape. Cons. Biol. **14**: 844–857.

Potvin, F., Breton, L., and Courtois, R. 2005. Response of beaver, moose and snowhoe hare to clear-cutting in a Quebec boreal forest: a reassessment 10 years after cut. Can. J. For. Res. **35**: 151–160.

Simard, M., Lecomte, N., Bergeron, Y., Paré, D., and Bernier, P.-Y. 2007. Forest productivity decline caused by successional paludification of boreal forests. Ecol. Appl. **17**: 1619–1637.

Taylor, S.J., Carleton, T.J., and Adams, P. 1987. Understory vegetation change in a chronosequence. Vegetatio, **73**: 63–72.

Van Cleve, K. and Viereck, L.A. 1981. Forest succession in relation to nutrient cycling in the boreal forest of Alaska. *In* Forest succession: concepts and application. *Edited by* D.C. West, H.H. Shugart, and D.B. Botkin. Springer-Verlag, New York, USA.

Vincent, J.-S. and Hardy, L. 1977. L'évolution et l'extension des lacs glaciaires Barlow et Ojibway en territoire québécois. Géogr. Phys. Quat. **31**: 357–372.

Wolff, J.O. 1980. The role of habitat patchiness in the population dynamics of snowshoe hares. Ecol. Monogr. **50**: 111–130.

Chapter **20**

# Old-Forest Conservation Strategies in Wet-Trench Forests of the Upper Fraser River Watershed, British Columbia*

*Darwyn Coxson and David Radies*

\* Discussions on old-forest management policies with Craig DeLong and Shannon Carson are gratefully acknowledged. Previous collaborations with Susan Stevenson, Trevor Goward, Shelly Benson, and Jocelyn Campbell have helped shape our understanding of lichen response to old-forest habitats in B.C.'s wet trench environments. Funding support from the Sustainable Forest Management Network, B.C. Forest Science Program, B.C. Forest Investment Account (T.R.C. Cedar Ltd.) and Mountain Equipment Co-op is gratefully acknowledged.

# 1. INTRODUCTION

Historically, old-forest ecosystems in the upper Fraser River watershed were regarded as non-merchantable or decadent parts of the timber harvesting land-base. Conversion of old forests to more productive younger stands was seen as an appropriate management objective (Sloan 1956). With greater appreciation of the biological values represented by old forest ecosystems, however, management for old-forest values must now deal with the legacy of past forest harvesting practices. In the upper Fraser River watershed these legacy issues fall into three major categories: 1) the loss of old forests within regional landscapes; 2) the spatial configuration of remaining mature and old-growth forests within regional landscapes; and 3) the simplification of the internal structure of previously harvested forest stands. Among major contributing factors that we must consider is that past harvesting pressures have not been evenly distributed within regional landscapes. Rather there have been disproportionate logging impacts on sites that have high biodiversity values. Implementation of policies that recognize the biological value of these sites poses immediate challenges for land managers.

A first step in this direction has been taken with the designation of landscape level targets for retention of old forests, using the historic natural range of variability (NRV) as a guide for setting threshold limits. However, a major gap in public policy persists, in that designated stands which meet the age class threshold for old-forest retention are treated as essentially similar entities. Current policies do not readily distinguish between old forests with high biodiversity values and old forests with low biodiversity values or the differing availability of these forest types in present-day landscapes.

Overriding this discussion is the consideration of how ecosystem processes in the upper Fraser River watershed, which historically shaped landscapes dominated by a range of old-forest communities, will function in future managed landscapes. What trade-offs should we be making between the intensity of forest harvesting practices and the biological attributes desired within future forests? Can alternative forest harvesting practices (e.g., variable retention or partial-cut harvesting) retain aspects of old-forest values within managed forests? Or does the widespread adoption of alternative forest harvesting practices result in fragmentation of a wider landscape, if assumptions of timber supply remain unchanged? Although these questions are faced by forest managers across Canada, the high proportion of old forests within B.C.'s wet trench landscapes creates unique challenges for management of these forests in B.C.

In this chapter, we will address how the proportion and configuration of old forests in regional landscapes has shaped past logging practices and the challenges that this creates for present-day forest managers. We will discuss the growing importance of alternative forest harvesting techniques and their potential to maintain key elements of canopy structure normally associated with old-growth forests. Our assessment will focus primarily upon the response of canopy lichen communities, which represent significant biodiversity indicators in these ecosystems. Finally, we will discuss regulations adopted in the region and the difficulties of reconciling forest management policies – as currently practised – with the preservation of old-growth forests.

## 2. UPPER FRASER WATERSHED REGIONAL CONTEXT

In north-central British Columbia the upper Fraser River is contained within the Rocky Mountain Trench: a broad valley paralleling the steep western face of the Rockies and separating them from the interior (and much older) ranges of the Columbia Mountains. This area of the upper Fraser River watershed is characterized by high precipitation, up to 480 mm during the summer period alone (out of 1,030 mm annually), as prevailing westerlies carry Pacific storm systems over the interior mountain ranges (Reynolds 1997). The proximity of the upper Fraser to the continental divide and the consequent incursion of arctic air masses during the winter period, however, results in a long lasting snow pack, with up to 2 m of settled snow accumulating in mid-elevation cedar-hemlock stands by late winter.

These conditions of high summer precipitation, prolonged snowmelt into late spring, and summer fog in valley bottom locations, have historically promoted the development of landscapes that were dominated by old forests. DeLong (1998) estimated that fire return intervals in these montane forests ranged from 244 years to over 1,600 years – depending on slope position, precipitation levels, and aspect. Many of the wet toe-slope positions on north-facing slopes show no previous history of stand-destroying fires (from buried charcoal horizons), although individual trees (or clusters of trees) have shown evidence of past lighting strikes.

Forest communities in the Rocky Mountain Trench of the Upper Fraser headwaters include three biogeoclimatic ecosystem community (BEC) zones: the Sub-boreal Spruce (SBS), the Interior Cedar Hemlock (ICH), and the Engelmann Spruce Subalpine Fir (ESSF) zones (Meidinger and Pojar 1991). SBS and ESSF forests are found in valley bottom and upper elevations respectively. Both of these zones are primarily dominated by hybrid white spruce (*Picea engelmanni* Parry × *P. glauca* [Moench] Voss) and subalpine fir (*Abies lasiocarpa* [Hook] Nutt.) (though stand structural characteristics and understory forb composition are noticeably different between these two forest types). Between these two spruce-fir-dominated BEC zones lies a band of ICH forests at mid-elevations. These forests are dominated by western red cedar (*Thuja plicata* Donn.) and western hemlock (*Tsuga heterophylla* [Raf.] Sag.), with the presence of Douglas fir (*Pseudotsuga menziesii* [Mirb.] Franco.), subalpine fir, and hybrid white spruce (depending on site conditions).

## 3.   LOSS OF OLD-GROWTH FORESTS WITHIN REGIONAL LANDSCAPES

Human impacts in the wet portions of the Rocky Mountain Trench have occurred mainly along valley bottoms (SBS) and mid-elevation slopes (ICH), commonly in parallel with the development of modern transportation corridors. Extensive forest harvesting (and accidentally set fires) accompanied the development of the railroad along the south-facing slopes of the Rocky Mountain trench in the early twentieth century. Similarly, development of a highway corridor in the mid-60s along the north-facing slopes of the Rocky Mountain Trench resulted in extensive fragmentation of ICH stands. Although nearly 80% of the higher elevation mountain forests in the wet trench area (figure 20.1) can still be classified as mature or old-growth forest cover (>140 years in age), only 40% of valley bottom spruce forests in the wet-trench are still greater than 140 years in age (Anonymous 2005). A major priority of landscape level forest management policies in the upper Fraser River watershed has been one of minimizing risks to biodiversity. This area contains provincially significant populations of mountain caribou (*Rangifer tarandus caribou*), grizzly bears (*Ursus arctos*), and several species of Pacific salmons.

**Figure 20.1**
**Location of wet-trench mountain and wet-trench valley natural disturbance units in the Prince George Timber Supply Area, central-interior B.C.**

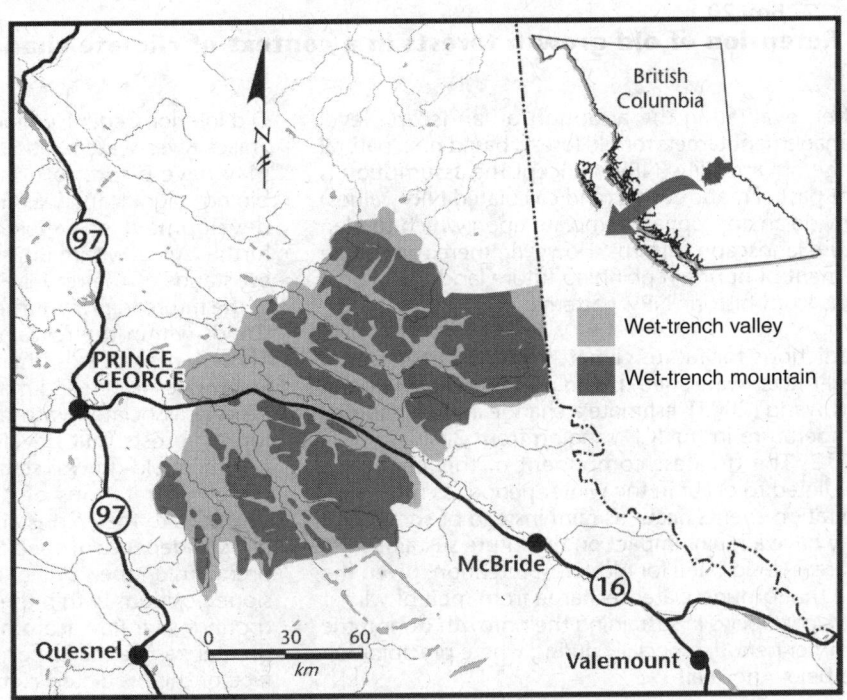

Table 20.1
**Table 20.1**
**Old-forest cover in wet-trench landscapes**

| | Natural disturbance unit | |
| --- | --- | --- |
| | **Wet-trench valley** | **Wet-trench mountain** |
| Total (ha) | 457,333 | 312,471 |
| Current old forest (ha) | 228,554 | 244,929 |
| % current old forest | 50.0 | 78.3 |
| Current interior forest (ha) | 144,154 | 212,106 |
| % current interior forest | 31.5 | 67.8 |

## 3.1. Measures Based on the Natural Range of Variability

In a comprehensive review of landscape level management objectives, the Prince George Timber Supply Analysis (Anonymous 2004) adopted the concept of managing landscapes within "a natural range of variability" (NRV). NRV-based management tries to predict the natural range of variability that has occurred in a given ecological attribute over time and then set management practices that will maintain that ecological attribute within "naturally occurring" boundaries (see Vaillancourt et al., chapter 2). Under a NRV concept we can estimate, for instance, what proportion of regional landscapes would historically have been covered in old forests and how much variation would have occurred in this proportion over long time periods (see box 20.1).

**Box 20.1**
**Retention of old-growth forests in a context of climate change**

When evaluating the adoption of landscape level management targets for old forests based on a natural range of variability (NRV) concept the assumption is that past climatic events (and calculated NRV values) provide an appropriate template upon which to plan future landscape patterns. However, there may be an element of hubris in planning future land use allocations from historic NRV patterns.

Predictions for future climate change in the upper Fraser River watershed, based on models of Hamann and Wang (2005), estimate a change in mean annual temperature in the ICH ranging from 2.9°C to 5.7–6.9°C. The greatest component of this change is predicted to occur in the winter period, as more precipitation events occur as rain, instead of snow. This may have a major impact on the future sustainability of areas designated for old-forest retention; given the role that ground water recharge from melt of winter snowpack plays in sustaining their growth during the summer period, especially during where precipitation are below normal.

Old Interior-Cedar Hemlock (ICH) stands in the upper Fraser River watershed (the wet-trench landscapes) may have disproportionate national conservation biology significance, as the climate envelope for the development of wet ICH forests shrinks in areas further south within British Columbia. Inland rainforest stands of the upper Fraser River watershed may, in the future, represent the sole examples of old-forest stands within the remaining ICH climate envelope. Though new areas to the north and east of the upper Fraser river watershed may develop characteristics of climate associated with the present-day inland rainforest, forests that develop on these new sites will not attain old-growth status for many centuries. Additionally, limitations of dispersal may pose serious obstacles to the colonization of new habitats by old-growth-dependent species. This also places a greater premium on the value of retaining forests in wet toe-slope positions within these mountain valleys, where groundwater flow from higher elevation may confer greater resistance to forest fires and disease in the face of the predicted climate changes.

Estimates of NRV for old forests in the upper Fraser were calculated by DeLong (2007) using a stochastic landscape model implemented in SELES (Spatially Explicit Landscape Event Simulator). This simulation was based on repeated runs using known fire return intervals calculated over a 1,000-year period. The natural range of variability in the cover of old forests in the Rocky Mountain Trench was estimated between 76 and 84% in wet-trench valley environments (ICH and SBS), and between 80 and 88% in wet-trench mountain environments (ESSF) (DeLong 2007). These NRV estimates represent an averaging of fire return intervals on different slope positions and hence would tend to underestimate NRV in wet site positions (i.e. toe-slope positions). However, even with this averaging they are still much narrower in range than those observed by Wimberly et al. (2000) in the Oregon Coast Range, where the calculated NRV for old forests ranged from 25 to 75%, or Agee (2003), whose NRV for old forests in the eastern Cascades fell between 38 and 63%. The relatively high NRV values obtained for the wet-trench landscape units of DeLong (2007) emphasize the significance of old-forest cover as a predominant seral stage in wet-trench landscapes of the upper Fraser River watershed.

In landscapes with a history of forest harvesting, old-forest cover values will typically fall below NRV targets. However, in wet-trench environments of the upper Fraser River watershed old-forest cover values of mountain forests (ESSF) fall close to NRV values, reflecting the inclusion of many non-commercial forests in the wet-trench mountain modelling area. In contrast, old-forest cover values for wet-trench valley forests, at 50% (Anonymous 2005), fall well below the historic NRV (76 to 84%) recommended by DeLong (2007), a disparity that may be even steeper when considering specific forest types, such as wet cedar-leading stands, which historically, rarely experienced stand level disturbances.

It is difficult to determine how great a disparity can be accommodated between NRV estimates of past old-forest cover and future (remnant) old-forest cover before the resilience of regional biota is placed at risk during future disturbance events (anthropogenic or natural). Little guidance is available in this regard. Most published studies, such as those of Agee (2003) or Wimberly et al. (2000), report on landscapes where present-day old-forest values fall well below NRV values. Several authors have speculated that risks to biodiversity are high if old-forest retention values within the landscape fall below 30% of the area determined by the minimum NRV estimates (Angelstam 1997). They make the important point that the longer old-forest patches remain as small refugia within the landscape, the greater the risk of extinction from stochastic events. Alternatively, risk to biodiversity will be lower when old-forest retention is nearer to 70% of the minimum range of NRV estimates (Angelstam 1997).

Given the pressing need for policy decisions in the upper Fraser watershed, DeLong (2002) recommended that the area of retained old forest be set between 41 and 61% of the area calculated to represent the natural range of variability for moist interior montane forests of the upper Fraser watershed. The implementation of this target resulted in a mandated requirement that 53% of ICH stands and 38% of SBS stands in the wet-trench valley be retained as old forest (that is forests greater than 140 years in age). It is important to note that this designation was aspatial in nature. In other words, these proportions have no fixed boundaries over time; rather it may be located in quite different parts of the landscape as harvesting proceeds regardless of tree species composition or site condition (i.e., wet versus dry soil moisture conditions). The strength of old-forest retention policies based on NRV estimates is that they force an examination of forestry practices at a regional level, where conservation biology planning has often been weak in the past. Their weakness, however, is that they cannot be used to dictate site level planning. The biological attributes of individual stands must also be considered in designated old-forest conservation biology priorities in regional landscapes.

## 3.2.   Conservation of High-Value Old-Forest Stands

An outstanding policy need in wet-trench forests is the development of mechanisms to recognize and protect old-forest sites with high ecological values. Recent research suggests that high levels of canopy lichen diversity can be found in productive valley bottom forests, especially in the so-called "antique" forest stands (>500 yrs since the last fire + uneven-aged stands), old and very old cedar-hemlock stands located on wet toe slope and bench topographic positions (figure 20.2) (Goward 1994). Many of these antique forest stands appear to have had no major stand-level disturbance for periods in excess of a thousand years or more. As a consequence, they support rich canopy lichen communities, including many species of lichens not found elsewhere in regional landscapes (Arsenault and Goward 2000; Goward and Spribille 2005). Unfortunately, the location of antique forest stands on mid-slope benches, in the same areas where rail and road corridors were developed, means that these stands were historically dispro- portionately targeted for harvest.

In recognition of the limitations of an approach for designating old-forest habitats based on "aspatial" regulations alone (a given percentage of that land- scape that was not fixed in any one location), a series of spatially designated old-growth management areas (OGMAs) were legislated in 2002. These OGMAs, which cover 19% of the overall forested land base in the wet trench (including both commercial and noncommercial forests) (Carson et al. 2002), were placed so as to protect spatially fixed features with high biodiversity value. Although the placement of OGMAs was initially limited to non-commercial forests, several were ultimately placed in commercial forests, in recognition that full representa- tion of habitats would not otherwise be obtained. In total, some 8% of valley stands in the wet trench were removed from the commercial forest harvesting land base through designation of OGMAs.

Given the longevity of cedars (often up to 1,000 yrs) as a dominant tree species in most stands designated as OGMAs, one might expect that the old forest attributes they support will be stable over long time periods. However, given the high edge-to-interior ratio of many of the designated OGMAs and the absence of any formal requirement for the establishment of buffers in adjacent stands, the long-term resilience of OGMAs remains open to question. The des- ignation of adjacent harvest blocks as partial-cut harvesting units would be one means of reducing these adjacency impacts. Current policy states that where adjacent forest harvesting or natural disturbance events are considered to have significantly impacted the value of an OGMA they can be considered for relocation

**Figure 20.2**

## Remnant western red cedar trees of exceptional stature and age can be found in old-growth antique forest stands located in wet toe-slope positions

These sites appear to have escaped disturbance for very long time periods. This tree, located on the "Ancient Forest Trail," falls within a stand that was designated for harvesting until the winter 2008, when its status was changed to that of an Old Growth Management Area.

(Carson et al. 2002). Options for future designation may be quite limited, however, particularly for high-value old-forest habitats, such as toe-slope antique forest stands.

One caveat on the implementation of old-forest retention targets in the upper Fraser River watershed has been the use of an age class threshold of 140 years for designating old forests. Campbell and Fredeen (2004) found that overall lichen diversity was much greater in old (250 years +) cedar-hemlock forest stands, compared to mature (140–250 years) or young forests (60–120 years) (figure 20.3). An even more striking difference between mature and old forest can be seen if one examines foliose cyanolichens. Radies and Coxson (2004) conducted paired comparisons on hemlocks of the same size and age growing in even-aged (120–140-year-old) and adjacent old-growth cedar-hemlock stands. Of nine cyanolichens species found growing on hemlocks in the old-growth stands, only one species was found in the younger even-aged stands. Clearly, as the accuracy of the forest inventory improves, designated thresholds for determining what qualifies as an old forest stand should be increased to recognize the greater biodiversity values in old (i.e., 250–500-year-old stands ) and antique (i.e. 500-year-old + multigenerational stands) cedar-hemlock stands.

### Figure 20.3
**Mean epiphytic macrolichen species richness (species per tree) for hemlock trees in three age classes in Interior Cedar-Hemlock stands**

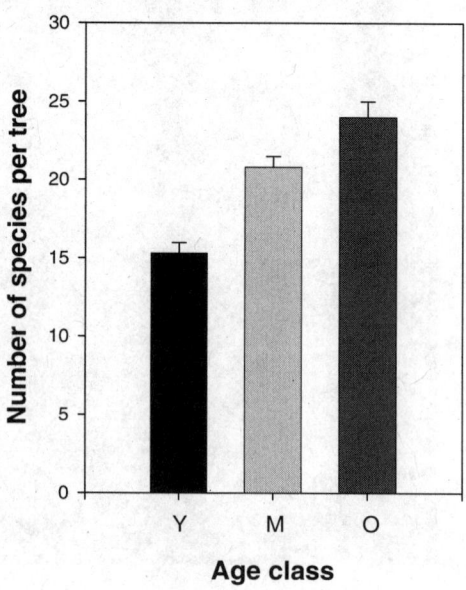

Legend: Young (Y): 60–120 years old. Mature (M): 140–250 years old. Old (O): >250 years. Global lichen diversity is higher in old stands than in mature and young stands. Adapted from Campbell and Fredeen (2000).

## 4. SPATIAL CONFIGURATION OF OLD FORESTS WITHIN REGIONAL LANDSCAPES

Spatial configuration of old residual forests is another major issue in managed landscapes. Small patches of old-growth forests maintained in highly fragmented landscapes can be more susceptible to edge effect and could also be affected by stochastic events.

DeLong (1998) found that the mean patch size for regenerating stands (after previous stand-destroying fire events) in wet cool montane forests was 74 ha, with a maximum patch size of 1,931 ha. Although DeLong's (1998) recommendations on the mean patch size of disturbances resulting from fire formed a part of the NRV calculations that ultimately led to the adoption of targets for overall retention of old forests within regional wet-trench landscapes (see section 3.1), his specific recommendations on patch size limits on harvest block size were not implemented. Given operational constraints faced by forest companies it was deemed infeasible to mandate patch size distribution of harvest blocks.

As an alternative to the direct management of patch size within the upper Fraser watershed, regulations have instead been adopted on the percentage of retained old forest in the landscape that must meet additional interior habitat◆ requirements. Current regulations state that 40% of the old forest designated for retention be maintained as interior habitat. The spatial designation of cut-blocks under these regulations can vary widely. DeLong (2007) compares two patterns resulting from the implementation of these regulations (figure 20.4). In the first

• Forest area that is not affected by edge effect. Following Burton (2001, 2002) recommendations, a forest located at 200 m or more from an edge is considered as interior forest.

### Figure 20.4
**Illustration of two areas that both meet the definition of interior old forest developed by the Landscape Objective Working Group for the Prince George Timber Supply Area**

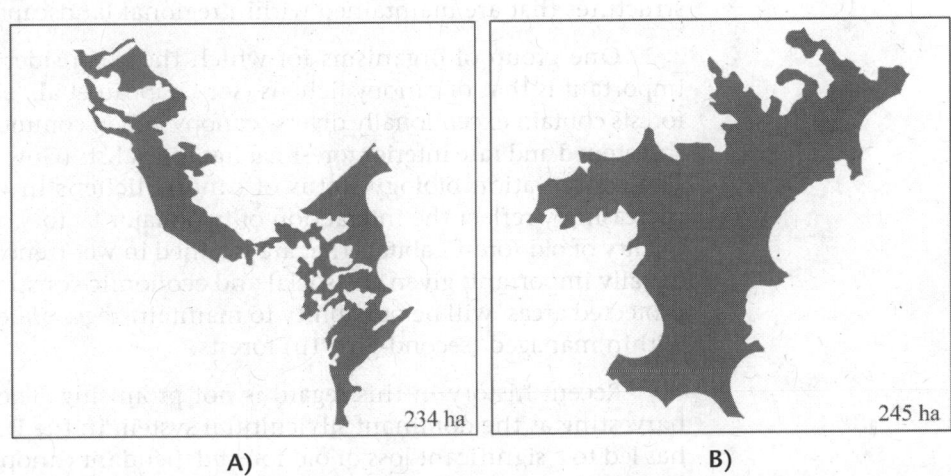

234 ha

A)

245 ha

B)

Figure A illustrates the disruption of continuity due to the pattern of dispersed block harvesting commonly practiced in B.C. Figure B illustrates a aggregated pattern of old-forest retention. In the two cases, the pattern respects current regulations in the Prince-George Timber Supply Area. From DeLong (2007).

example, the adoption of dispersed harvest blocks results in a landscape that is highly fragmented, notwithstanding the harvest blocks meeting regulations for the retention of interior forest (figure 20.4A). In the second example, the aggregation of retained forest patches resulted in much greater continuity of forest cover, albeit at the cost of greater harvesting impacts in adjacent areas (figure 20.4B).

Another major factor to consider is the quality of retained old-forest habitats. Many old-forest lichens in the ICH are sensitive to edge effects. When taken together, the two major components of regulations controlling old-forest retention (total old forest and proportion of interior old forest) will result in only 21% of wet-trench landscapes being retained as interior old-forest habitats. Although still a significant proportion of the landscape, the quality of these designated interior old-forest habitats must also be considered. Much of the designated old-forest habitat falls in sites that have little or no commercial forest harvesting values. In the wet ICH this often means hemlock-dominated stands growing on steep slopes with dry, nutrient-poor soils. Recent surveys suggest that these sites have low levels of canopy lichen biodiversity (D. Radies and D. Coxson, unpublished data).

## 5.  RESIDUAL HABITAT QUALITY AND IMPACT ON CANOPY LICHEN COMMUNITIES

The imposition of regulations that require relatively high levels of old-forest retention within wet-trench landscapes of the upper Fraser River watershed are unprecedented within a B.C. context and set standards that are much higher than are remnant levels of old-forest retention in much of the U.S. Pacific Northwest. However, these standards do not, in themselves, ensure that biological values will be maintained in the future within the upper Fraser River watershed. Consideration must also be given to the nature and type of internal stand structures that are maintained within regional landscapes.

One group of organisms for which these considerations are particularly important is that of canopy lichens (see Drapeau et al., chapter 14). Wet-trench forests contain exceptionally diverse canopy lichen communities, including many threatened and rare interior forest habitat specialists (Goward and Spribille 2005). The conservation biology status of canopy lichens in wet-trench forests will increasingly reflect the interaction of two major factors. The first of these is the quality of old-forest habitats that are retained in wet-trench landscapes. However, equally important, given the social and economic constraints on designation of protected areas, will be our ability to maintain/regenerate future lichen habitats within managed (second-growth) forests.

Recent history in this regard is not promising. The adoption of clear-cut harvesting as the dominant silvicultural system in the Rocky Mountain Trench has led to a significant loss of old forest dependant canopy lichen communities. Major lichen functional groups, such as Alectoriod lichens (Stevenson et al. 2001) and canopy cyanolichens (Radies and Coxson 2004), typically require in excess of 200 years in which to colonize wet-trench forests after disturbance events. As this time period considerably exceeds that anticipated for stand age at harvest (rotation age) within managed forests of the wet trench, we must consider the retention of canopy structural components within harvest blocks as a major element of any plan to maintain future canopy lichen communities.

Several partial-cut harvesting trials point to the value of retaining canopy structural components in harvest blocks. Coxson et al. (2002) found no significant decline of Alectoriod lichen communities (lichen loading on a per-tree basis) in group selection harvesting plots within the ESSF (figures 20.5 and 20.6). However, the same study also found that Alectoriod lichens showed a significant post-harvest decline in single-tree selection partial-cut harvest blocks. The nature and pattern of harvesting intensity is clearly important. Similarly in ICH stands, Coxson and Stevenson (2005) found that for three of four lichen functional groups (cyanolichen, foliose, and *Bryoria* group lichens) there were no significant treatment effects of group selection and group retention partial-cut harvesting on post-harvest lichen retention within marked trees (assessed two years after harvesting). Old-forest cyanolichens within wet-trench valley forests, may indeed, react positively to the creation of small canopy gaps, which emulate natural openings within old forest canopies, providing that the surrounding old-forest

##### Figure 20.5
## Oblique aerial view of single-tree selection (A) and group selection (B) partial-cut harvesting areas in Engelmann Spruce – Subalpine Fir forest at Pinkerton Mountain Silvicultural Systems Trial

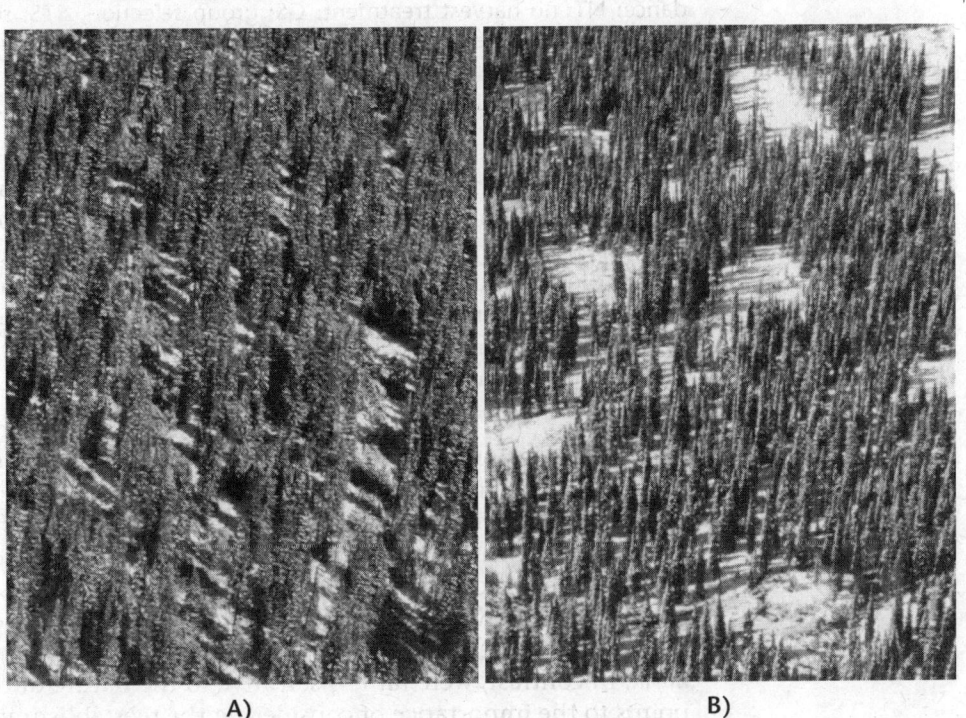

A)                                                                B)

From Coxson *et al.* (2002).

**Figure 20.6**
## Percent frequency distribution of changes in lichen abundance class by tree for three partial-cut treatments at Pinkerton Mountain

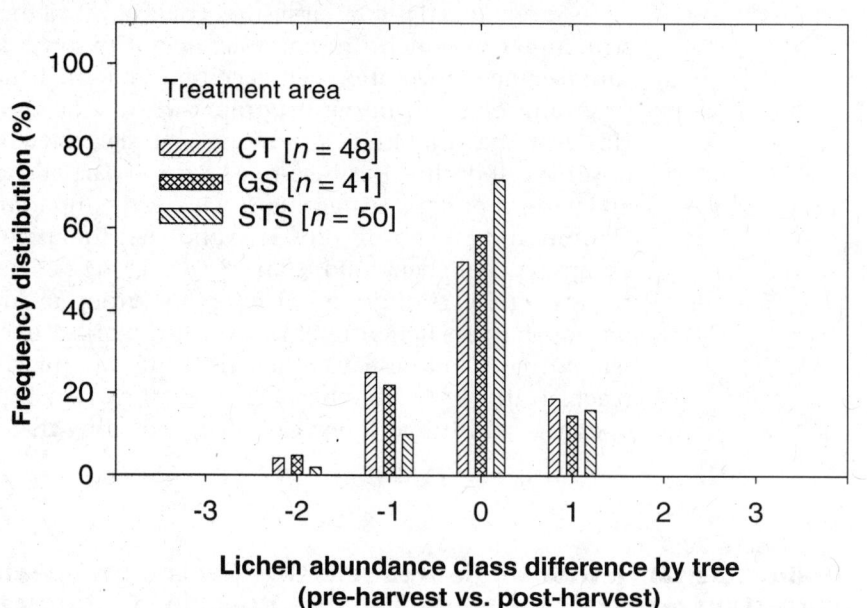

Lichen abundance class difference by tree
(pre-harvest vs. post-harvest)

Legend: 0: no difference in visual estimates; 5 the greatest differences in lichen abundance; NT: no harvest treatment; GS: group selection; STS: single-tree selection. The sample size (number of tress) for each group is indicated in brackets. From Coxson et al. (2002).

matrix is functionally intact and continues to support major relevant ecological values. Coxson and Stevenson (2007) found a strong relationship between growth rates in the canopy cyanolichen *Lobaria pulmonaria* and exposure to small gaps within old-forest canopies (figure 20.7). This attribute of lichen response is promising, to the extent that we can create or retain the structures of uneven-aged forest in second-growth stands (though contributing factors such as soil moisture, proximity to riparian corridors, and topographic position will clearly remain important site specific factors).

The adoption of forest harvesting practices that minimize edge effects may have an important influence on landscape level retention of canopy lichens. Coxson and Stevenson (2007b) found that growth of small thalli of *Lobaria pulmonaria* was much higher at stand edges when the adjacent harvest block was a variable-retention harvest block (creating a low-contrast or "soft" edge) compared to stands where the adjacent harvest block was a clearcut harvest block (creating a high-contrast or "hard" edge) (figure 20.8). Larger thalli of *L. pulmonaria*, in contrast, were far less sensitive to the nature of the adjacent edge. This points to the importance of considering the reestablishment phase when examining the impact of forest harvesting practices on canopy lichen communities.

**Figure 20.7**

## Cumulative percent lichen growth rates in Lunate and Viking stands as a function of canopy openness

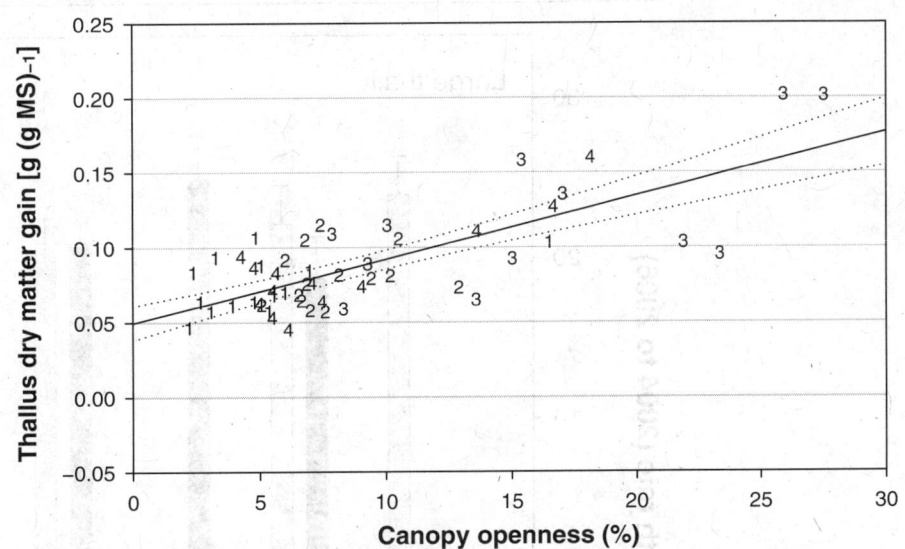

From June 2002 to June 2004, in even-aged stands (respectively plot symbols 1 and 2) and old-growth stands (respectively plot symbols 3 and 4). Linear regressions (solid lines) through scattergram, plus or minus 1 standard error (dotted line) are shown on plot. Adapted from Coxson and Stevenson (2007a).

To date there has been little adoption of partial-cut harvesting within commercial blocks in wet-trench forests. However, some forest licensees have adopted the practice of leaving significant amounts of advance regeneration (young and mature trees) on harvest blocks. In the short term this will help ameliorate edge effects within adjacent (remnant) old forest patches. It is less clear, however, to which degree these practices will contribute to the development of future old-forest characteristics. Much of this advanced retention comes from non-commercial species, mainly hemlock; species which tend to be more susceptible to natural disturbance events such as insect outbreaks and windthrow. Further, planned rotation age (time to the next harvest) in these harvest blocks would generally preclude the development of future old-forest structural characteristics.

## 6. IS NATURAL DISTURBANCE EMULATION POSSIBLE?

Trade-offs between the intensity of harvesting and retention of unfragmented old-forest patches are now at the heart of a major debate over the management of wet-trench landscapes. The challenge for forest managers is therefore to design silvicultural practices that both retain contiguous old-forest habitats and create harvest cut blocks that mimic natural disturbance processes.

**Figure 20.8**

**Percent cumulative growth rates of large and small *L. pulmonaria* thalli at defined transect positions along soft- and hard-edge transects for the period from fall 2004 to fall 2006**

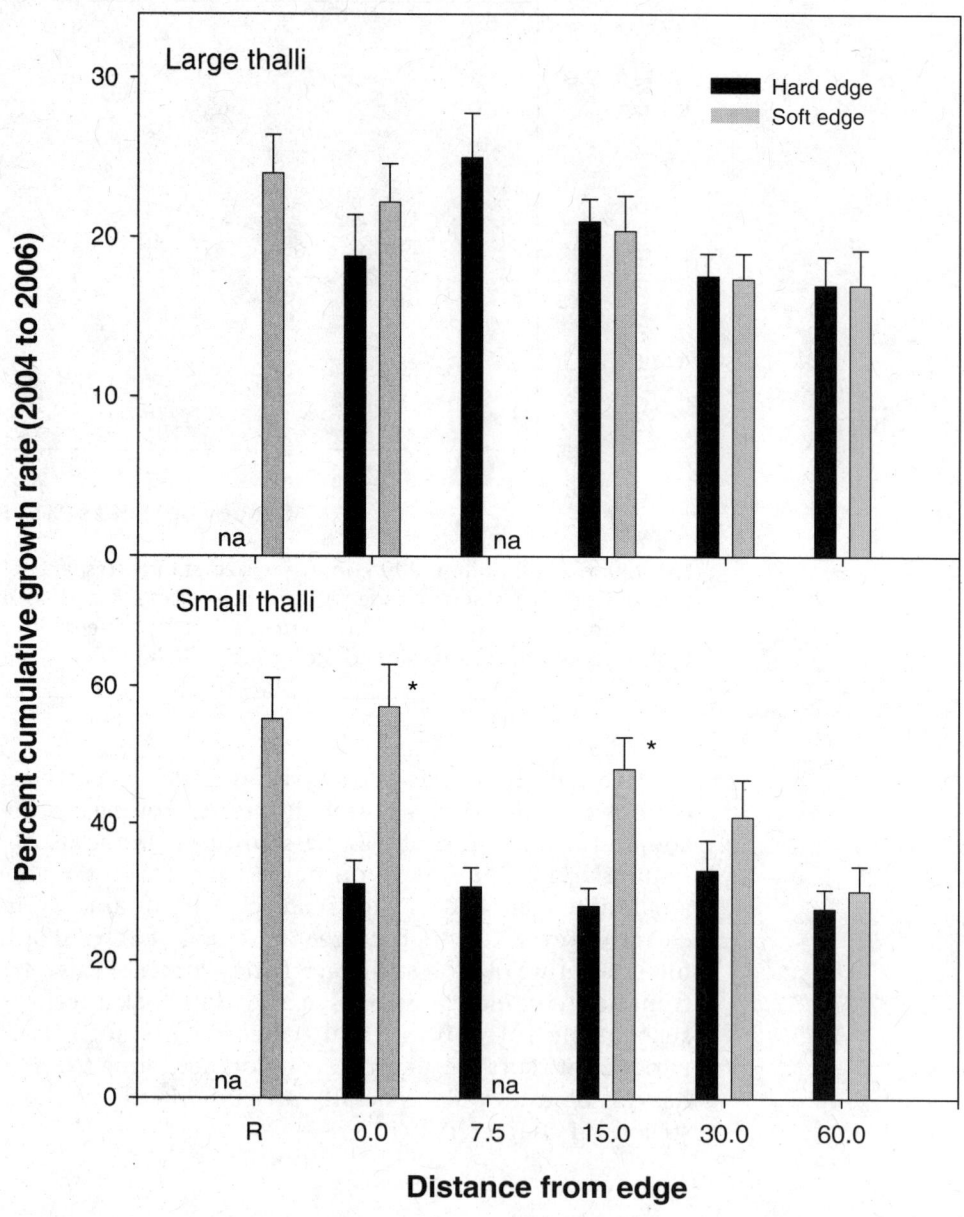

Mean and standard error of large (top) and small (bottom) *L. pulmonaria* thalli (n = 3 stands). R: residual trees located in cutblocks. ND: positions where comparisons between hard- and soft-edge transects were not available. Asterisks, where present, indicate where significant differences were found (Bonferonni t-tests, p <= 0.05) between growth rates of *L. pulmonaria* thalli at defined positions along hard- and soft-edge transects. From Coxson and Stevenson (2007b).

The different types of old-forest stands (e.g., wet north-facing toe-slope antique forest stands or dryer upslope and/or south-facing stands) in the upper Fraser River watershed have very likely been shaped by varying types and intensities of natural disturbance processes in the past. However, an important limitation at the present time on the implementation of harvesting practices based on natural disturbance emulation is the lack of baseline data on prevailing natural disturbance regimes within the upper Fraser River watershed. For example, Hoggett and Negrave (2001) examined impacts of a major western hemlock looper outbreak in the early 90s, but we have little information on the overall intensity of past outbreaks within regional landscapes.

Nonetheless, partial cut harvesting trials that have been conducted in wet-trench forests (Jull et al. 1999) provide a possible template upon which forest harvesting practices that emulate natural disturbance patterns could be based, if key elements of both forest canopy structure and tree species composition are maintained within harvest blocks. Coxson and Stevenson (2005), for instance, found that many of the structural characteristics retained within partial-cutting harvest blocks provided suitable habitat attributes for old-forest lichens (see also section 3.2). However, maintaining both of these objectives (natural disturbance emulation in harvest blocks and retention of significant unfragmented old forest patches) within landscapes of the upper Fraser River watershed would necessitate accepting reductions in the annual allowable cut (yearly wood supply), depending on the level of variable retention or partial-cut harvesting adopted.

If natural disturbance emulation is chosen as a desired form of ecosystem management it should be remembered that no one harvesting strategy can emulate natural disturbance patterns over the diverse wet-trench landscapes of the upper Fraser River watershed. Forestry prescriptions should instead be based on a consideration of site specific factors. Clear-cut harvesting may be appropriate in areas where stand-destroying fires do occur, such as hemlock-dominated forests on rocky soils; whereas in other areas, such as cedar-leading wet toe-slope positions, natural disturbance events occur mainly as the single-tree gap dynamics, and forestry practices should be scaled accordingly.

## 7.  CONCLUSION

The calculation of natural range of variability estimates for old-growth forest cover in wet-trench forests of the upper Fraser River valley has provided a valuable benchmark against which landscape level retention targets for old-growth forest management can be evaluated. However, these targets cannot substitute for the identification and protection of spatially designated high-value old-growth forest stands, particularly old-growth cedar stands located in wet toe-slope positions. The long-term retention of canopy lichen communities within regional landscapes will require both protection of these specific high-value areas and the greater adoption of partial-cut harvesting techniques in the surrounding forest harvesting land base. The adoption of forest harvesting practices which more closely emulate natural disturbance processes must be accompanied by a reduction in annual allowable cut, otherwise the net result will be one of landscapes that are highly fragmented and have few areas of viable interior forest habitat for lichen growth.

# REFERENCES

Agee, J.K. 2003. Historical range of variability in eastern Cascades forests, Washington, USA. Landsc. Ecol. **18**: 725–740.

Angelstam, P. 1997. Landscape analysis as a tool for the scientific management of biodiversity. Ecol. Bull. **46**: 140–170.

Anonymous. 2004. Recommended objectives for landscape level biodiversity conservation in the Prince George Timber Supply Area. Prince George Timber Supply Objective Working Group Report. March 2004. British Columbia Ministry of Sustainable Resource Management, Prince George, B.C.

Anonymous. 2005. Old forest retention results to March 31, 2005. Prince George Public Advisory Group, British Columbia Ministry of Forests, Prince George, B.C.

Arsenault, A. and Goward, T. 2000. Ecological characteristics of inland rainforests. Ecoforestry, **15**: 20–23.

Burton, P.J. 2001. Response of vascular vegetation to cut-block edges in the Sub-boreal Spruce zone of northwest-central British Columbia. *Presented at* Annual Meeting of the Canadian Botanical Association, June 25–27, 2001, Kelowna, B.C.

Burton, P.J. 2002. Effects of clearcut edges on trees in the Sub-boreal Spruce zone of northwest-central British Columbia. Silva Fenn. **36**: 329–352.

Campbell, J. and Fredeen, A.L. 2004. *Lobaria pulmonaria* abundance as an indicator of macrolichen diversity in interior cedar-hemlock forests of east-central British Columbia. Can. J. Bot. **82**: 970–982.

Carson, S., Brost, A., Nesbit, B., Spears, F., and Barry, S. 2002. Prince George Area Forest District Sustainable Resource Management Plan – Old Seral Chapter. British Columbia Ministry of Forests, Prince George, B.C.

Coxson, D.S. and Stevenson, S.K. 2005. Retention of canopy lichens after partial-cut harvesting in wet-belt interior cedar-hemlock forests, British Columbia, Canada. For. Ecol. Manag. **204**: 99–114.

Coxson, D.S. and Stevenson, S.K. 2007a. Growth rate responses of *Lobaria pulmonaria* to canopy structure in even-aged and old-growth cedar-hemlock forests. For. Ecol. Manag. **242**: 5–16.

Coxson, D.S. and Stevenson, S.K. 2007b. Influence of high-contrast and low-contrast forest edges on growth rates of *Lobaria pulmonaria* in the inland rainforest, British Columbia. For. Ecol. Manag. **253**: 103–111.

Coxson, D.S, Stevenson, S., and Campbell, J. 2002. Short-term impacts of partial cutting on lichen retention and canopy microclimate in an Engelmann spruce-subalpine fir forest in north-central British Columbia. Can. J. For. Res. **33**: 830–841.

DeLong, S.C. 1998. Natural disturbance rate and patch size distribution of forests in northern British Columbia: implications for forest management. Northwest Sci. **72**: 35–48.

DeLong, C. 2002. Natural disturbance units of the Prince George Forest Region: guidance for sustainable forest management. Unpublished report. British Columbia Ministry of Forests, Prince George, B.C.

DeLong, C. 2007. Implementation of natural disturbance-based management in northern British Columbia. For. Chron. **83**: 338–349.

Goward, T. 1994. Notes on old growth–dependent epiphytic macrolichens in the humid oldgrowth forests in inland British Columbia, Canada. Acta Bot. Fenn. **150**: 31–38.

Goward, T. and Spribille, T. 2005. Lichenological evidence for the recognition of inland rainforests in western North America. J. Biogeog. **32**: 1209–1219.

Hamann, A. and Wang, T. 2005. Models of climate normals for genecology and climate change studies in BC. Agric. For. Meteorol. **128**: 211–221.

Hoggett, A. and Negrave, R. 2001. Western hemlock looper and forest disturbance in the ICH wk3 of the Robson Valley – Stage 3: effects of western hemlock looper and disturbance classification – progress report and ecosystem management recommendations. [Online] <www.firthhollin.com/efmpp> (accessed November 7, 2007).

Jull, M., Stevenson, S., and Sagar, B. 1999. Group selection in old cedar hemlock forests: five-year results of the Fleet Creek partial-cutting trial. Prince George Forest Region Research Note # PG-20. Prince George, B.C.

Meidinger, D. and Pojar, J. (Eds), 1991. Ecosystems of British Columbia. Special Report Series 6. Research Branch, Ministry of Forests, Victoria, B.C.

Radies, D.N. and Coxson, D.S. 2004. Macrolichen colonization on 120–140-year-old *Tsuga heterophylla* in wet temperate rainforests of central-interior British Columbia: a comparison of lichen response to even-aged versus old-growth stand structures. Lichenologist, **36**: 235–247.

Reynolds, G. 1997. Climatic data summaries for the biogeoclimatic zones of British Columbia. British Columbia Ministry of Forests, Research Branch, Victoria, B.C.

Sloan, G. 1956. Forest resources of British Columbia. Report of the Commissioner. Victoria, B.C.

Stevenson, S.K., Armleder, H.M., Jull, M.J., King, D.G., McLellan, B.N, and Coxson, D.S. 2001. Mountain caribou in managed forests: recommendations for managers: second edition. British Columbia Ministry of Environment, Lands, and Parks. Wildlife Report No. R-26. Victoria, B.C.

Wimberly, M.C., Spies, T.A., Long, C.J., and Whitlock, C. 2000. Simulating historical variability in the amount of old forests in the Oregon Coast Range. Conserv. Biol. **14**: 167–180.

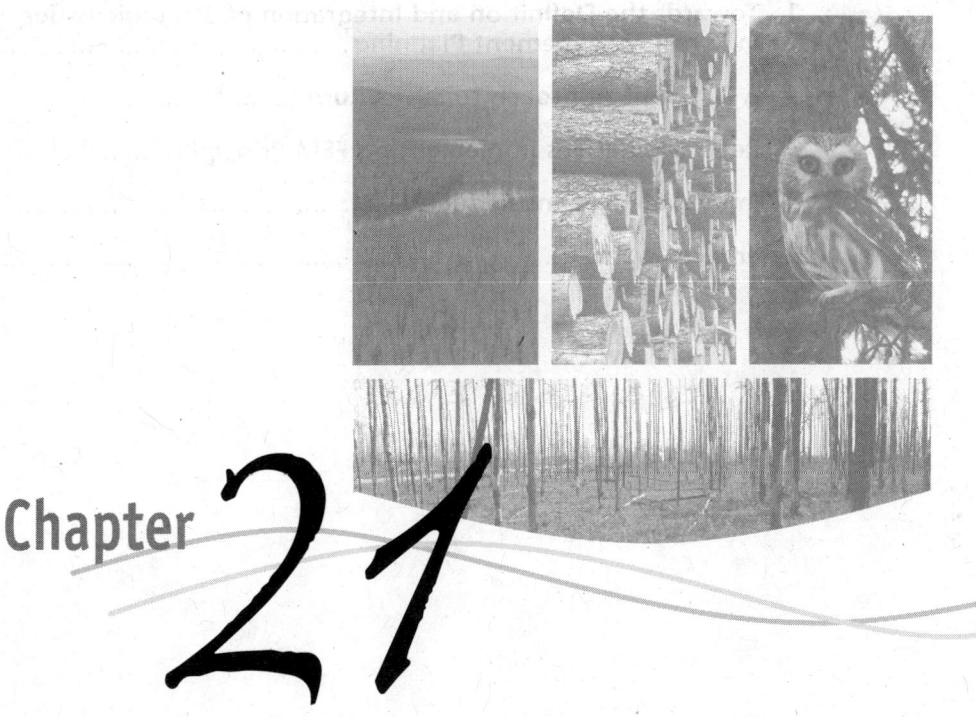

# Chapter 21

# Perspectives*

*Alain Leduc, Sylvie Gauthier, Marie-Andrée Vaillancourt,*
*Yves Bergeron, Louis De Grandpré, Pierre Drapeau,*
*Daniel Kneeshaw, Hubert Morin, and Dominic Cyr*

\* The authors are grateful to Gerardo Reyes for the translation of this chapter and to Pamela
Cheers and Isabelle Lamarre for the linguistic revision. The photos on this page were graciously
provided by Marie-Ève Sigouin, Virginie Arielle Angers, and Michel Robert (Canadian Wildlife
Service).

Throughout this book we have presented a number of examples that succinctly describe a forest ecosystem management approach. Far from being an exhaustive analysis, we consider this to be a reference point from which the development and application of FEM principles can be further refined and adapted for a regional forest management context. While adapting FEM poses considerable challenges for forest managers, it also offers several solutions to problems caused by current management practices. Above all, FEM marks a profound change in the manner in which we manage our forest resources.

## 1. TOWARDS THE DEFINITION AND INTEGRATION OF STRATEGIC ISSUES IN FOREST MANAGEMENT PLANNING

The FEM approach stems from the fact that current forestry practices have a substantial environmental impact. These practices, by their large spatial area and limited variety of silvicultural treatments applied at relatively short intervals (i.e., clearcutting), have considerably changed the age-class structure of the boreal forest cover, and greatly homogenized forest conditions. Studies reconstructing the historical natural disturbance regimes occurring in Québec and elsewhere in Canada show that the eastern Canadian boreal forest has rarely experienced periods wherein the age-class structure is as young as it is currently.

One of the major changes proposed by FEM is to emphasize the importance of evaluating the cumulative impacts of forest practices in space and time. Current spatio-temporal analyses show, for example, that with the agglomeration of harvested blocks over time, immense regeneration areas are created (Perron et al., chapter 6), resulting in the fragmentation and reduction of the amount of old-growth forest in managed areas (Gauthier et al., chapter 3).

Integrating environmental risk into forest management practices is a second major change proposed by FEM. Currently, the planning of forest operations, as well as calculating sustainable harvesting levels, are done without consideration for the various environmental risks. Yet natural disturbances such as fire (see Girardin et al., chapter 4) and insect outbreaks (see Morin et al., chapter 7; Sutton and Tardif, chapter 8) have always been an integral part of the eastern Canadian boreal forest dynamics, and despite all concerted efforts made to control them, they continue to occur in our managed forests. Le Goff et al. (chapter 5) propose a series of guidelines that make it easier to include environmental risks in forest management planning.

An FEM approach will therefore propose a forest management planning strategy that improves upon current management strategies by incorporating larger space and time horizon (see Bouchard, chapter 13). Moreover, FEM requires a clear definition of the management targets to be reached, as well as the insurance that forest conditions observed in managed landscapes do not deviate too much from a paired, reference natural landscape (see Harvey et al., chapter 18).

Thus, reconstruction of pre-industrial landscapes using dendrochronological and paleological techniques (see examples in Part 2; Belleau and Légaré, chapter 19) allows us to precisely define management targets; and in conjunction with analysis of the historical variability of natural disturbance regimes (Landres et al. 1999; box 21.1), we have information about the threshold levels of change that must not to be exceeded. The ecosystem approach can also be applied to forest landscapes that have been completely transformed by past management practices. In these forest landscapes, restoration measures need to be developed hand in hand with management practices (Kuuluvainen 2002; Boucher et al. 2006).

In short, under ecosystem management, forest management planning is not only concerned with the amount of wood to be harvested, but also considers the wood to be left in place at both the stand and landscape levels. Currently, the majority of the strategic issues related to forest management, i.e. those that involve spatio-temporal parameters that usually exceed the forest stand and the silvicultural scenario, are rarely included or are often overlooked in management plans and forestry models.

## Box 21.1
## Establishing management targets and maximum threshold levels of forest change as a function of historical variation in natural disturbance regimes

The intent of this box is to illustrate how to use studies on historical variability in natural disturbance regimes to determine management targets and the threshold levels of forest change not to be exceeded. The example uses fire frequency data; however, the same exercise can be undertaken using severity or size of disturbance data.

Information comes from the long-term variability of mean burn rates obtained from dendrochronological and paleoecological studies, which can subsequently be used to set management targets in terms of age of forest cover (see Cyr et al. 2009 for a concrete example). In our example, we have transposed the conservative and extended fire cycle ranges to determine the amount of old-growth forest in the landscape (figure 21.1). The conservative ranges were obtained with the aid of averages calculated for long periods of time, up to 1,000 years in landscapes where the fire cycle is generally long, whereas the extended ranges also include rarer situations that were nevertheless observed during brief periods of the post-glacial era in the landscape in question.

The proportion of old-growth forests that one would observe under each of these fire cycle estimates is modelled using the negative exponential distribution. The dark and lightly shaded areas correspond to conservative and extended fire cycle ranges observed, respectively (figure 21.1A). Scenarios for which current conditions are within the conservative range are acceptable (figure 21.1B). Scenarios wherein the representation of old-growth forests is outside the conservative range but within the extended range (figure 21.1C) are considered *of concern*, because even if these situations occurred within the landscape over the course of the post-glacial period, they were not very common nor did they last for long periods of time. They therefore represent extremes that are probably not appropriate as targets in a coarse filter approach. Furthermore, scenarios wherein the proportion of old-growth forests is outside the extended range of variability (figure 21.1D) are considered *alarming* given that these scenarios are not representative of the natural state of the landscape. They have to be considered as ecologically unacceptable management targets and should not be used. In the last two scenarios (figures 21.1C and D), management actions have to be planned in a manner in which the representation of old-growth forests lies within the natural range of variability. With caution in mind, we suggest favouring the conservative range of variability to set management targets in terms of age-class structure. The extended range of variability should be considered as a buffer or safeguard in the event of unexpected occurrences such as natural disturbances.

**Figure 21.1**
**Illustration of the conceptual approach (A) and the different possible scenarios (B, C, and D) that account for the historic proportion of older forests within the forest landscape**

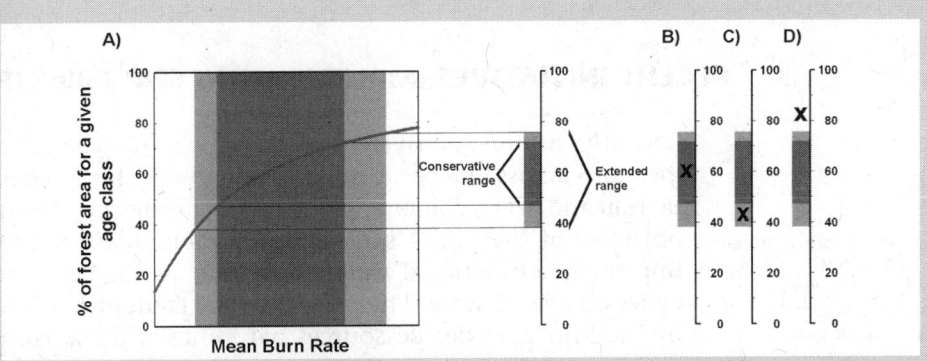

The curve represents the proportion of old forest under different fire cycles. The dark and light shaded areas correspond to conservative and expanded fire cycle ranges observed for a given territory, respectively. Crosses indicate the hypothetical present situation. From Cyr et al. (2009), figures 21.1B, C, and D inspired by Landres et al. (1999).

## 2. A DIFFERENT APPROACH TO SILVICULTURE

FEM proposes important changes to the manner in which we develop and plan silvicultural treatments. Under a FEM plan, management objectives will have repercussions on silvicultural treatment options. For example, to counter the domination of the forest landscape by younger age classes resulting from an even-aged harvesting system, FEM recommends the systematic use of a larger proportion of partial harvesting treatments (see Fenton et al., chapter 15; Bouchard, chapter 13). Harvest techniques that maintain a larger proportion of overstory forest cover help reduce the time needed to recover pre-harvest conditions, and also reduces the vulnerability associated with the open, young regenerating forest stage.

Greater knowledge of forest dynamics will also lead to the diversification of silvicultural treatments in ways that are better adapted to the various regional differences characteristic of the eastern Canadian boreal forest (see the regional examples in Part 2). For example, in areas sensitive to paludification, wherein a reduction in productivity is observed after harvest, FEM suggests using prescribed burns or scarification treatments to stimulate production (see Simard et al., chapter 11). Furthermore, in order to limit the proliferation of competitive species after harvest, partial harvesting treatments that allow for the maintenance of more closed forest cover should be prescribed for sites sensitive to competition. In all cases, FEM emphasizes diversified strategies for a productive forest ecosystem having different profiles of vulnerability to both environmental factors and forest management. The same type of forest stand can be prescribed different silvicultural treatments depending on the region or the ecological context under consideration. Accordingly, diversifying silvicultural techniques will limit homogenization of forest conditions within and among forest regions.

## 3.   RECENT INITIATIVES IMPLEMENTING FEM PRINCIPLES

Recently, a forest ecosystem management pilot study was initiated in three locations across Québec, Canada. Along with the Tembec project in Abitibi (see Belleau and Légaré, chapter 19), two other projects were established. The primary objective of the Triad◆ project in Mauricie, managed by AbitibiBowater, is to implement a functional zoning approach for the land and to develop FEM principles on a large scale. The project in the Laurentian Wildlife Reserve, managed by the Ministère des Ressources naturelles et de la Faune du Québec (MNRF), aims to develop an adaptive forest management approach that uses FEM as a reference point (Comité scientifique sur les enjeux de biodiversité 2007; Thiffault et al. 2007). In other locations across Canada, several forestry companies have begun to integrate FEM principles into harvest planning. In British Columbia, for example, several companies have adopted methods that follow the guidelines of the Clayoquot Sound Scientific Panel.◆ Since then, Alberta-Pacific Forest Industries Inc.◆ in Alberta, Mystik Management Ltd.◆ in Saskatchewan, and Tembec in Ontario◆ have adopted management strategies inspired by natural disturbances in their forest management planning. Without a doubt, the numerous advantages, including potential for forest certification and access to markets that demand forest products stemming from harvesting practices that respect the environment and are more socially acceptable, ensure that companies have much to gain by integrating a FEM approach into their planning and operations.

• For more information see: <www.projettriade.ca>; <www.iisaak.com>; <sciencepanel.html>; <www.alpac.ca>; <www.mistik.ca>; <tembec-frm-ontario.ca>.

## 4.   IMMEDIATE ACTIONS REQUIRED

Forest exploitation in the boreal forest zone of eastern Canada currently benefits from a vast capital of natural forests that are ready for harvest. In this context, management concerns are more the result of forest harvest planning than the restoration of the forest to its original condition. Over time, harvested areas have edged further and further into northerly locations that, in certain regions of eastern Canada, have reached the limit of contiguous boreal forest cover. This harvesting limit, which is continuing its northward progression, has left very few mature forests further south, which are often composed of residual stands divided into a multitude of smaller patches.

With the rapid progression of this harvesting limit, it is urgent that we develop measures to stop the fragmentation of mature forest and to preserve a number of large forested tracts. Action is more critical in more northern locations of the boreal forest zone as these forests appear to be less resilient to disturbance than in more southerly locales. Furthermore, the retreat of the boreal forest benefits the expansion of lichen woodlands (Girard et al. 2007), while opening of forest cover leads to the establishment of lichen woodlands further south, possibly because of the effects of more intense and frequent disturbances (Jasinski and Payette 2005). While hypotheses explaining these changes have been proposed, the mechanisms responsible for differences in resilience are more or less understood. Faced with these concerns, the ecosystem approach points out several courses of action to be undertaken and the fields of research need to be developed.

1. In response to the rapid and extensive changes to the forest matrix that have resulted in an age-class structure dominated by young forests, it is necessary to accelerate the process of implementing effective silvicultural techniques that favour the maintenance of a larger proportion of forest cover at the landscape level, and limit the impacts to changes in the age-class structure on species dependent on mature forest cover. Silvicultural techniques should strive to (i) leave more wood in place, (ii) maintain complex residual structures, and (iii) reduce the time needed for stand re-establishment after harvest. These techniques should also be based on a clear understanding of natural dynamics of forests in older stages of development (over-mature stands). Essentially, we need to propose methods of retaining live and dead wood material that reproduces natural forest dynamics observed in older forests.

2. The lack of a full understanding of forest dynamics should not limit the implementation of new forestry techniques, but rather should spur the initiation of monitoring programs (see Drapeau et al., chapter 14; Fenton et al., chapter 15). These programs would have several functions: (i) to evaluate the effects of harvest volumes and cutting cycles associated with partial harvesting, (ii) to verify if the structural attributes of over-mature stands have successfully been reproduced and maintained over time, and (iii) to verify if these new techniques ensure the maintenance of biodiversity *via* various groups of indicator organisms associated with mature and over-mature forests (see Drapeau et al., chapter 14). As more knowledge is obtained, silvicultural techniques will be tailored to better respond to the objectives for which they were developed. Note that this approach is in line with the adaptive management framework.

3. In light of the implementation of new silvicultural practices that involve a better understanding of natural forest dynamics, we must build and refine models that predict the development of forest stands beyond commercial harvesting age (see Saucier and Groot, chapter 16). Studying the development of stands undergoing silvicultural treatments that strive to generate or maintain an uneven-aged structure is of equal importance. In particular, it is necessary to refine models to include successional dynamics in the calculation of sustainable harvesting levels.

4. Given the incomplete knowledge of the resilience mechanisms of forests recovering after disturbance, it is important to proceed with cautious large-scale management approaches, particularly in the northern fringes of the boreal forest. One precaution would involve limiting, for the moment, the severity of forest interventions with respect to the percentage

of trees to be harvested until we obtain a clearer understanding of the mechanisms and risks involved in opening forest cover in more northern locales.

5. Lastly, to properly evaluate the efficiency of an ecosystem management approach, it is important to develop a network of protected areas wherein natural disturbances are left to run their course. These areas can then be used as ecological reference points for these new forest management systems and for monitoring the resilience of these forest ecosystems over time.

## REFERENCES

Boucher, Y., Arseneault, D., and Sirois, L. 2006. Logging-induced change (1930–2002) of a preindustrial landscape at the northern range limit of northern hardwoods, eastern Canada. Can. J. For. Res. **36**: 505–517.

Comité scientifique sur les enjeux de biodiversité. 2007. Enjeux de biodiversité de l'aménagement écosystémique dans la réserve faunique des Laurentides. Rapport préliminaire du comité scientifique. Ministère des Ressources naturelles et de la Faune, Québec, Que.

Cyr, D., Gauthier, S., Bergeron, Y., and Carcaillet, R. 2009. Forest management is driving the eastern North American boreal forest outside its natural range of variability. Front. Ecol. Environ. 7: doi:10.1890/080088

Girard, F., Payette, S., and Gagnon, R. 2007. Rapid expansion of lichen woodlands within the closed-crown boreal forest zone over the last 50 years caused by stand disturbances in eastern Canada. J. Biogeogr. doi:10.1111/j.1365-2699.2007.01816.x

Jasinski, J.P.P. and Payette, S. 2005. The creation of alternative stable states in the southern boreal forest, Québec, Canada. Ecol. Monogr. **75**: 561–583.

Kuuluvainen, T. 2002. Disturbance dynamics in boreal forests: defining the ecological basis of restoration and management of biodiversity. Silva Fenn. **36**: 5–12.

Landres, P.B., Morgan, P., and Swanson, F.J. 1999. Overview of the use of natural variability concepts in managing ecological systems. Ecol. Appl. **9**: 1179–1188.

Thiffault, N., Wyatt, S., Leblanc, M., and Jetté, J.-P. 2007. Adaptive forest management in Quebec: bits of the big and small pictures. Can. Silvic. May 2007. [Online] <www.canadiansilviculture.com/spr07/adaptive.html>.

# Authors' Contact Information

**Tuomas Aakala**, M. Sc., Ph.D. student
Department of Forest Ecology
University of Helsinki, P.O. Box 27, FIN-00014
University of Helsinki, Finland
E-mail: tuomas.aakala@helsinki.fi

**Claude Allain**, B. Sc., Ing. F.
Bureau du Forestier en chef
456, rue Arnaud, bureau 1.03
Sept-Îles (Québec) G4R 3B1
E-mail: claude.allain@fec.gouv.qc.ca

**Saida Amouch**, B. Sc., M. Sc. student
Université du Québec à Chicoutimi
555, boul. de l'Université
Chicoutimi (Québec) G7H 2B1
E-mail: samouch@uqac.ca

**Louis Bélanger**, Ph. D., professor
Faculté de foresterie et de géomatique
Département des sciences du bois et de la forêt
Pavillon Abitibi-Price
Université Laval
Québec (Québec) G1K 7P4
E-mail: Louis.Belanger@sbf.ulaval.ca

**Annie Belleau**, M. Sc., Ph.D. student
Chaire industrielle CRSNG-UQAT-UQAM en aménagement forestier durable
Université du Québec en Abitibi-Témiscamingue
445, boul. de l'Université
Rouyn-Noranda (Québec) J9X 5E4
E-mail: annie.belleau@uqat.ca

**Jonatan Belle-Isle**, M. Sc., biologist
Expertise immobilière
Hydro-Québec
800, de Maisonneuve Est, 20ᵉ étage
Montréal (Québec) H2L 4M8

**Yves Bergeron**, Ph. D., professor and director of the CRSNG-UQAT-UQAM industrial chair in Sustainable Forest Management and of the research chair in Ecology and Forest Management
Université du Québec en Abitibi-Témiscamingue
445, boul. de l'Université
Rouyn-Noranda (Québec) J9X 5E4
E-mail: yves.bergeron@uqat.ca

**Pierre Y. Bernier**, Ing. F., Ph. D., research scientist, ecophysiology and forest productivity
Natural Resources Canada
Canadian Forest Service – Laurentian Forestry Center
1055, rue du PEPS
C.P. 10380, succ. Sainte-Foy
Québec (Québec) G1V 4C7
E-mail: pbernier@cfl.scf.rncan.gc.ca

**Hervé Bescond**, M. Sc., Ph.D. student
Université du Québec en Abitibi-Témiscamingue
445, boul. de l'Université
Rouyn-Noranda (Québec) J9X 5E4
E-mail: herve.bescond@uqat.ca

**Claude-Michel Bouchard**, Ing. F., Forêt d'enseignement et de recherche du lac Duparquet
Université du Québec en Abitibi-Témiscamingue
445, boul. de l'Université
Rouyn-Noranda (Québec) J9X 5E4
E-mail: claude.bouchard@uqat.ca

**Mathieu Bouchard**, Ing. F., Ph. D., postdoctoral researcher
Direction de l'environnement forestier
Ministère des Ressources naturelles et de la Faune
880, chemin Sainte-Foy
Québec (Québec) G1S 4X4
E-mail: mathieu.bouchard@mrnf.gouv.qc.ca

**Dominique Boucher**, M. Sc., biologist
Natural Resources Canada
Canadian Forest Service – Laurentian Forestry Center
1055, rue du PEPS
C.P. 10380, succ. Sainte-Foy
Québec (Québec) G1V 4C7
E-mail: dobouch@nrcan-rncan.gc.ca

**Catherine Boudreault**, M. Sc., Ph.D. student
Département des sciences biologiques
Université du Québec à Montréal
Case postale 8888, Succursale Centre-Ville
Montréal (Québec) H3C 3P8
E-mail: boudreault.catherine@courrier.uqam.ca

**Suzanne Brais**, Ph. D., professor
Université du Québec en Abitibi-Témiscamingue
445, boul. de l'Université
Rouyn-Noranda (Québec) J9X 5E4
E-mail: suzanne.brais@uqat.ca

**Yves Claveau**, Ing. F., Ph. D., postdoctoral researcher
Centre for Forest Research
Département des sciences biologiques
Université du Québec à Montréal
Case postale 8888, Succursale Centre-Ville
Montréal (Québec) H3C 3P8
E-mail: y_claveau@sympatico.ca

**Darwyn Coxson**, Ph. D., professor
Ecosystem Science and Management Program
University of Northern British Columbia
Prince George (British Columbia) V2N 4Z9
E-mail: darwyn@unbc.ca

**Dominic Cyr**, M. Sc., Ph.D. student
Département de sciences biologiques
Université du Québec à Montréal
C.P. 8888, Succ. Centre-Ville
Montréal (Québec) H3C 3P8
E-mail: cyr.dominic@gmail.com

**Louis De Grandpré**, Ph. D., research scientist, ecology of plant communities
and forest dynamics
Canadian Forest Service – Laurentian Forestry Center
1055, rue du PEPS
C.P. 10380, succ. Sainte-Foy
Québec (Québec) G1V 4C7
E-mail: ldegrand@nrcan-rncan.gc.ca

**André de Römer**, M. Sc., biologist
Département de biologie, de chimie et de géographie
Université du Québec à Rimouski
300, allée des Ursulines, C.P. 3300
Rimouski (Québec) G5L 3A1
E-mail: a_de_romer@hotmail.com

**Pierre Drapeau**, Ph. D., professor
Département de sciences biologiques
Université du Québec à Montréal
C.P. 8888, Succ. Centre-Ville
Montréal (Québec) H3C 3P8
E-mail: drapeau.pierre@uqam.ca

**Margaret Donnelly**, B. Sc.
Donnelly Ecological Consulting Services Inc.
Box 146, Weymouth
NS B0W 3T0
E-mail: margdonn@ns.sympatico.ca

**Brock Epp**, M. Sc. (Botany)
Centre for Forest Interdisciplinary Research (C-FIR)
University of Winnipeg
515 Portage Avenue
Winnipeg (Manitoba) R3B 2E9
E-mail: brockepp@mts.net

**Nicole Fenton**, Ph. D., postdoctoral researcher
Université du Québec en Abitibi-Témiscamingue
445, boul. de l'Université
Rouyn-Noranda (Québec) J9X 5E4
E-mail: nicole.fenton@uqat.ca

**Mike D. Flannigan**, Ph. D., research scientist, fire researcher project leader
Natural Resources Canada
Canadian Forest Service – Great Lakes Forestry Centre
1219, Queen Street East
Sault Ste. Marie (Ontario) P6A 2E5
E-mail: mike.flannigan@nrcan.gc.ca

**Sylvie Gauthier**, Ph. D., research scientist, forest succession
Natural Resources Canada
Canadian Forest Service – Laurentian Forestry Center
1055, rue du PEPS
C.P. 10380, succ. Sainte-Foy
Québec (Québec) G1V 4C7
E-mail: sgauthier@cfl.forestry.ca

**Martin Girardin**, Ph. D., research scientist
Natural Resources Canada
Canadian Forest Service – Laurentian Forestry Center
1055, rue du PEPS,
C.P. 10380, succ. Sainte-Foy
Québec (Québec) G1V 4C7
E-mail: martin.girardin@rncan.gc.ca

**Art Groot**, Ph. D., research scientist, ecophysiology
Natural Resources Canada
Canadian Forest Service – Canadian Wood Fibre Centre
C.P. 490
1219, Queen Street East
Sault Ste. Marie (Ontario) P6A 2E5
E-mail: agroot@nrcan-rncan.gc.ca

**Brian Harvey**, Ing. F., Ph. D., professor and director of the Forêt d'enseignement
et de recherche du lac Duparquet
Université du Québec en Abitibi-Témiscamingue
445, boul. de l'Université
Rouyn-Noranda (Québec) J9X 5E4
E-mail: brian.harvey@uqat.ca

**Louis Imbeau**, Ph. D., professor
Université du Québec en Abitibi-Témiscamingue
445, boul. de l'Université
Rouyn-Noranda (Québec) J9X 5E4
E-mail: louis.imbeau@uqat.ca

**Jean-Pierre Jetté**, Ing. F.
Direction de l'environnement forestier
Ministère des Ressources naturelles et de la Faune
880, chemin Sainte-Foy
Québec (Québec) G1S 4X4
E-mail: jean-pierre.jette@mrnf.gouv.qc.ca

**Norm Kenkel**, Ph. D., professor
Department of Botany
University of Manitoba
Winnipeg (Manitoba) R3T 2N2
E-mail: kenkel@cc.umanitoba.ca

**Daniel Kneeshaw**, Ph. D., professor
Université du Québec à Montréal
C.P. 8888, Succ. Centre-Ville
Montréal (Québec) H3C 3P8
E-mail: kneeshaw.daniel@uqam.ca

**Timo Kuuluvainen**, Ph. D., professor
Department of Forest Ecology
P.O. Box 27, FIN-00014
University of Helsinki, Finland
E-mail: Timo.Kuuluvainen@helsinki.fi

**Danielle Laprise**, M. Sc., research professional
Université du Québec à Chicoutimi
555, boul. de l'Université
Chicoutimi (Québec) G7H 2B1
E-mail: dlaprise@uqac.ca

**Ève Lauzon**, M. Sc., biologist
Division St-Maurice, Projet TRIADE
AbitibiBowater
255, 1re Rue, C.P. 500
Grand-Mère (Québec) G9T 5L2
E-mail: eve.lauzon@gmail.com

**Paul Leblanc**, B. Sc. F., Ing. F.
District Forester
LP Canada Swan Valley
Forest Resources Division
Box 998
Swan River (Manitoba) R0L 1Z0
E-mail: Paul.Leblanc@LPCorp.com

**Nicolas Lecomte**, Ph. D., director, Valeur Nature
Valeur Nature
2709 boul. McDuff, local F
Valcanton (Québec) J0Z 1H0
E-mail: Nicolas.Lecomte@uqat.ca

**Alain Leduc**, Ph. D., researcher, Landscape ecology
Centre for Forest Research
Université du Québec à Montréal
C.P. 8888, Succ. Centre-Ville
Montréal (Québec) H3C 3P8
E-mail: alain.leduc@uqam.ca

**Sonia Légaré**, Ph. D., forest ecology specialist
Tembec Inc.
225, 9e Avenue, C.P. 2500
La Sarre (Québec) J9Z 2X6
E-mail: sonia.legare@tembec.com

**Héloïse Le Goff**, Ph.D.
Centre for Forest Research
Université du Québec à Montréal
C.P. 8888, Succ. Centre-Ville
Montréal (Québec) H3C 3P8
E-mail: heloise.legoff@nrcan.gc.ca

**Kim Logan**, B. Sc., analyst, forest fire and climate change
Canadian Forest Service – Great Lakes Forestry Centre
1219, Queen Street East
Sault Ste. Marie (Ontario) P6A 2E5

**Julie Messier**, M. Sc., biologist
Biology department
Stewart Building
McGill University
1205 Doctor Penfield Avenue
Montréal (Québec) H3A 1B1
E-mail: julie.messier@gmail.com

**Tom Moore**, M. Sc.
Spatial Planning Systems
Box 1389
56 Glendale Avenue
Deep River (Ontario) K0J 1P0
E-mail: tmoore@spatial.ca

**Hubert Morin**, Ph. D., professor
Université du Québec à Chicoutimi
555, boul. de l'Université
Chicoutimi (Québec) G7H 2B1
E-mail: Hubert_Morin@uqac.ca

**Jacques Morissette**, forest technician
Natural Resources Canada
Canadian Forest Service – Laurentian Forestry Center
1055, rue du PEPS
C.P. 10380, Succ. Sainte-Foy
Québec (Québec) G1V 4C7
E-mail: jacques.morissette@nrcan.gc.ca

**David Paré**, Ing. F., Ph. D., research scientist, forest soils, biochemistry
and ecosystem sustainability
Natural Resources Canada
Canadian Forest Service – Laurentian Forestry Center
1055, rue du PEPS
C.P. 10380, Succ. Sainte-Foy
Québec (Québec) G1V 4C7
E-mail: dpare@cfl.forestry.ca

**Sophie Périgon**, M. Sc., biologist
Centre for Forest Research (CEF)
Université du Québec à Montréal
C.P. 8888, Succ. Centre-Ville
Montréal (Québec) H3C 3P8

**Nathalie Perron**, Ph. D., biologist and coordonnator of the *Bilan d'aménagement
forestier durable*
Bureau du Forestier en chef
845, boulevard Saint-Joseph
Roberval (Québec) G8H 2L6
E-mail: Nathalie.Perron@fec.gouv.qc.ca

**Anh Thu Pham**, M. Sc., biologist
Centre for Forest Research
Université du Québec à Montréal
C.P. 8888, Succ. Centre-Ville
Montréal (Québec) H3C 3P8
E-mail: anhthu.pham@tpsgc.gc.ca

**David Radies**, M. Sc., Ph.D. student
Ecosystem Science and Management Program
University of Northern British Columbia
Prince George (British Columbia) V2N 4Z9
E-mail: darwyn@unbc.ca

**Robert S. Rempel**, Ph. D., research scientist
Centre for Northern Forest Ecosystem Research
Lakehead University
955, Oliver Road
Thunder Bay (Ontario) P7B 5E1
E-mail: rob.rempel@ontario.ca

**Gerardo Reyes**, Ph.D.
Centre for Forest Research
Université du Québec à Montréal
C.P. 8888, Succ. Centre-Ville
Montréal (Québec) H3C 3P8
E-mail: greyes@dal.ca

**Jean-Pierre Saucier**, Ing. F., Ph.D.
Ministère des Ressources naturelles et de la Faune
Division de l'analyse et de la diffusion des informations forestières et écologiques
Direction des inventaires forestiers
880, chemin Sainte-Foy, 5e étage
Québec (Québec) G1S 4X4
E-mail: Jean-Pierre.Saucier@mrnf.gouv.qc.ca

**Andrée-Anne Simard**, B. Sc., M. Sc. student
Université du Québec à Chicoutimi
555, boul. de l'Université
Chicoutimi (Québec) G7H 2B1
E-mail: AndreAnne_Simard@uqac.ca

**Martin Simard**, M. Sc., Ph.D. student
University of Wisconsin, Department of Zoology
Birge Hall, 430 Lincoln Dr.
Madison (Wisconsin)
USA, 53706
E-mail: simard@wisc.edu

**Alanna Sutton**, M. Sc., biologist
Centre for Forest Interdisciplinary Research (C-FIR)
University of Winnipeg
515 Portage Avenue
Winnipeg (Manitoba) R3B 2E9
E-mail: alannacolleen@hotmail.com

**Jacques C. Tardif**, Ph. D., professor and Canada Research Chair in Dendrochronology
Center for Forest Interdisciplinary Research (C-FIR), University of Winnipeg
515 Avenue Portage
Winnipeg (Manitoba) R3B 2E9
E-mail: j.tardif@uwinnipeg.ca

**Marie-Andrée Vaillancourt**, M. Sc., biologist
Natural Resources Canada
Canadian Forest Service – Laurentian Forestry Center
1055, du PEPS
C.P. 10380, Succ. Sainte-Foy
Québec (Québec) G1V 4C7
E-mail: mavaillancourt@gmail.com

**Laird Van Damme**, M. Sc. F.
KBM Forestry Consultants
349 Mooney Avenue
Thunder Bay (Ontario) P7B 5L5
E-mail: vandamme@kbm.on.ca

# Index